Design and Simulation
of Four-Stroke Engines

Also by Gordon P. Blair:

Design and Simulation of Two-Stroke Engines
(Order No. R-161)

For more information or to order this book, contact SAE at 400 Commonwealth Drive, Warrendale, PA 15096-0001; (724)776-4970; fax (724)776-0790; e-mail: publications @sae.org; web site: www.sae.org/BOOKSTORE.

Design and Simulation
of Four-Stroke Engines

Gordon P. Blair

SAE.
INTERNATIONAL

Society of Automotive Engineers, Inc.
Warrendale, Pa.

Library of Congress Cataloging-in-Publication Data

Blair, Gordon P.
    Design and simulation of four-stroke engines / Gordon P. Blair.
        p.   cm.
    Includes bibliographical references and index.
    ISBN 0-7680-0440-3
    1. Four-stroke cycle engines--Design and construction I. Title.
    TJ790.B577 1999
    621.43--dc21   99-27316
                                                                           CIP

Copyright © 1999   Society of Automotive Engineers, Inc.
                            400 Commonwealth Drive
                            Warrendale, PA 15096-0001 U.S.A.
                            Phone: (724)776-4841
                            Fax: (724)776-5760
                            E-mail: publications@sae.org
                            http://www.sae.org

ISBN 0-7680-0440-3

SAE Order No. R-186

The Last Mulled Toast

A Grand Prix race is very rough,
the going's fast, the pace is tough.
The four-stroke rules the world of cars,
in bikes it's two-strokes that are the stars.

Now, why is this you'd have to ask?
The rulemakers you can take to task.
For the intake air never needs to question,
"Is this the right bellmouth for my ingestion?"

The designer of both must surely know,
or else his engines will all be slow,
unsteady gas dynamic trapping
by right and left waves overlapping.

To model an engine is algebraic simple.
You sit on the gas like a veritable pimple,
solving the maths the waves to track
from valve to bellmouth in the intake stack.

At the inlet valve you scan induction,
count the air that's passed by suction
and just as the valve would shut the door,
you get a wave to ram home more.

In the exhaust it's furnace hot,
for the modeller 'tis a tropic spot.
Exhaust waves reflect but do the job
of sucking out the burned gas slob.

Some time ago I wrote two tomes
on two-strokes, including poems.
It seemed only fair to tell those with cars
that black-art tuning is best kept for bars.

This book informs the four-stroke tuner
what I wish I knew those decades sooner,
as Brian Steenson followed Agostini
with my exhaust on Mick Mooney's Seeley.

The pen's both strokes have now been told.
My writ is run, I'm pensioned old.
While I may be ancient and time is shrinking,
only *Dei voluntas* can stop me thinking.

Gordon Blair
1 November 1998

Foreword

Since 1990, I have written two books on the design and simulation of two-stroke engines. Not many in the four-stroke engine industry will read such books on the assumption that they are not relevant to them. I will not dwell on this issue as I have already dedicated a couple of stanzas to this very point, on the previous page. Hence, when I came to write this Foreword, and reread what I had set down in those previous books, I realized that much of what was written there for the two-stroke enthusiast was equally applicable to the reader of this book. So, if much of this reads like the Foreword in my previous books, I can only respond by saying that I know only one way of teaching this subject. So, if you have already absorbed that, then pass on.

This book is intended to be an information source for those who are, or wish to be, involved in the design of four-stroke engines. More particularly, the book is a design aid in the areas of gas dynamics, fluid mechanics, thermodynamics, and combustion. To stop you from instantly putting the book down in terror at this point, rest assured that the whole purpose of this book is to provide design assistance with the actual mechanical design of an engine in which the gas dynamics, fluid mechanics, thermodynamics, and combustion have been optimized so as to provide the required performance characteristics of power, or torque, or fuel consumption, or noise emission. Therefore, the book will attempt to explain the intricacies of, for example, intake ramming, and then provide you with empiricisms to assist you with the mechanical design to produce, to use the same example, better intake ramming in any engine design. Much of the engine simulation, with which I was involved at QUB over the last twenty-five years, and to which I have applied myself even more thoroughly in the three years since I formally retired from my alma mater, has become so complex, or requires such detailed input data, that the operator cannot see the design wood for the data trees. As a consequence, I wound this empiricism into visual software to guide me toward a more relevant input data set before applying it into an engine simulation computer model. Quite often, the simulation confirms that the empiricism, containing as it does the distilled experience of a working lifetime, was adequate in the first place. However, sometimes it does not and that becomes the starting point for a more thorough design and comprehension process by simulation. You will find many examples of that within this book. However, even that starting point is closer to a final, optimized answer than it would have been if mere guesses had been the initial gambit for the selection of input data to the engine simulation.

The opening of the book deals with the fundamentals of engine design and development, ranging from mechanical principles, to engine testing and the thermodynamics of engine cycles. To some it will read like the undergraduate text they once had; to undergraduates it will read like a tutorial by some pedantic professor, and to those who had no such formal education it will provide the thermodynamic backdrop they never had, but which they will need to follow the logic of the design and development of the four-stroke engine. It reminds you all, expert and novice alike, of the basics of the scene in which you wish to operate.

The acquisition of a fundamental understanding of unsteady gas dynamics is the first major step to becoming a competent engine designer. Hence, this book contains a major section dealing with that subject. It is little different than that within the more recent book on two-stroke engines*, although it is updated and extended with, it is hoped, all, typographical and theoretical errors removed. The fundamental theory of unsteady gas dynamics is the same for two-stroke and four-stroke engines, but I repeat myself as yet another stanza has already dealt with that. Nevertheless, without a basic understanding of unsteady gas dynamics, the mysteries of intake and exhaust tuning will remain just that.

The "flowing of cylinder heads" is a way of life for many developers of high-performance engines. As with all technologies, there is a right, and a wrong, way of going about it. I explain the only way to acquire the discharge coefficients of flow which will be meaningful if they are also required to be accurately applied with an engine simulation model.

The discussion of combustion follows a pragmatic approach, as distinct from one steeped in the chemistry of the subject. It provides data on the burn characteristics of a considerable range of actual engines, spark-ignition and compression-ignition, in a manner which gives real data input for those who wish to simulate a wide variety of power units with truly representative combustion characteristics.

The discussion on noise emission illustrates the point that actual silencers, intake, and exhaust, can be designed by simulation so that the trade-off in noise emission and performance characteristics can be thoroughly executed by the modeling of the entire engine together with its mufflers. It also makes the point that the traditional empiricism, which is based in acoustics, has a useful role to play in the design process as long as you do not believe implicitly in its predictions.

The majority of the book is devoted to the design of the spark-ignition engine, but there is also comprehensive treatment of the diesel or compression-ignition engine. The totality of the book is just as applicable to the design of the diesel as it is to the gasoline engine, for the only real difference between them is the methodology of the combustion process.

Much like this Foreword, the opening paragraphs of many of the chapters are very similar to those in the book on two-stroke engines. I suppose it is a simple statement, albeit a truism nevertheless, but having figured out a logical way to introduce you to the fundamentals of any given topic, and as those fundamentals do not change just because I am writing about the four-stroke engine, I decided that I would only sing the song badly if I attempted to change the lyrics.

I have had the inestimable privilege of being around at precisely that point in history when it became possible to unravel the technology of engine design from the unscientific black art which had surrounded it since the time of Otto, Diesel, and Clerk. That unraveling occurred because the digital computer permitted the programming of the fundamental unsteady gas-dynamic theory which has been in existence since the time of Rayleigh, Kelvin, Stokes, and

*See Chapter 1, Ref. [1.9]

Taylor. For me, that interest was stimulated by a fascination with high-performance engines in general, motorcycles in particular, and two-stroke engines even more particularly. It is a fascination that has never faded.

The marriage of these two interests, computers and racing engines, has produced this book and the material within it. For those in this world who are of a like mind, this book should prove to be useful.

Gordon P. Blair
23 October 1998

Acknowledgements

As I explained in the Foreword, this is the third book I have written, but the acknowledgements in those earlier books are still pertinent. The individuals who have influenced my life and work are still the very same people, so what else can I say.

The first acknowledgement is to those who enthused me during my school days on the subject of internal combustion engines in general, and motorcycles in particular. They set me on the road to a thoroughly satisfying research career which has never seen a hint of boredom. The two individuals were my father, who had enthusiastically owned many motorcycles in his youth, and Mr. Rupert Cameron, who had owned but one and had ridden it everywhere—a 1925 350 cc Rover. Of the two, Rupert Cameron was the greater influence, for he was a walking library of the Grand Prix races of the twenties and thirties and would talk of engine design, and engineering design, in the most knowledgeable manner. He was actually the senior naval architect at Harland and Wolff's shipyard in Belfast and was responsible for the design of some of the grandest liners ever to sail the oceans.

My father and Mr. Cameron talked frequently of two fellow Ulstermen, Joe Craig and Walter Rusk. They are both shown in the photograph in Plate 1.0, Joe Craig standing to the right and Walter Rusk astride the motorcycle. Walter Rusk is just about to start in the Ulster Grand Prix of 1935 on a works Norton; he crashed on the second lap while leading. The Ulster Grand Prix was the Grand Prix d'Europe that year, so it was an even bigger show than normal in a province where 10% of the entire population were known, and still are known, to turn up to watch a motorcycle race. Local media coverage ensures that racers, tuners, and engineers are household names. Joe Craig was the Chief Engineer of Norton Motorcycles from the early 1930s to the 1950s and he was responsible for the development of the single-cylinder 500 cc Manx Norton throughout that period. A cutaway drawing of the 1959 version of this engine is shown in Plate 1.2 and you can see its heredity etched in the lines of the 1935 engine in Plate 1.0. Joe Craig came from Ballymena in Co. Antrim. He designed and developed that same engine from 31 bhp in 1931 to 53 bhp in 1953, and as a schoolboy I listened to those tales and thought how grand it would be to know so much about tuning engines as to be able to do that. Walter Rusk came from nearby in Whitehead in Co. Antrim and went to my alma mater, Larne Grammar School. He was one of the top-bracket racers of the 1930s. He was killed while flying in the Royal Air Force and I looked up at his name, written in gold on the Roll of Honor for World War II, at school assembly every morning. These two people were my schoolboy heroes, but of the two it was Joe Craig, and his genius at engine tuning, who exerted the greater fascination.

I have to acknowledge that this book would not be written today but for the good fortune that brought Dr. Frank Wallace (Professor at Bath University since 1965 and now retired) to Belfast in the very year that I wished to do postgraduate research at The Queen's University of

Plate 1.0 Walter Rusk and Joe Craig at the 1935 Ulster Grand Prix.
(Courtesy of Norman Windrum)

Belfast (QUB). At that time, Frank Wallace was one of perhaps a dozen people in the world who comprehended unsteady gas dynamics, which was the subject area I already knew I had to understand if I was ever to be a competent engine designer. However, Frank Wallace taught me something else as well by example, and that is academic integrity. Others will judge how well I learned either lesson.

Professor Sir Bernard Crossland deserves a special mention, for he became the Head of the Department of Mechanical Engineering at QUB in the same year I started as a doctoral research student. His drive and initiative set the tone for the *engineering* research that has continued at QUB until the present day. I emphasize the word "engineering" because he instilled in me, and a complete generation, that real "know-how" comes from using the best theoretical science available, at the same time as conducting related experiments of a product design, manufacture, build, and test nature. That he became, in later years, a Fellow of the Royal Society, a Fellow of the Royal Academy of Engineering, and a President of the Institution of Mechanical Engineers, and was knighted, seems no more than justice.

I have been very fortunate in my early education to have had teachers of mathematics who taught me the subject not only with enthusiasm but, much more importantly, from the point of view of application. I refer particularly to Mr. T. H. Benson at Larne Grammar School and to Mr. Scott during my undergraduate studies at The Queen's University of Belfast. They gave me a lifelong interest in the application of mathematics to problem solving which has never faded.

The next acknowledgement is to those who conceived and produced the Macintosh computer. Without that machine, on which I have typed this entire manuscript, drawn every figure that is not from SAE archives, and developed all of the simulation software, there would be no book. In short, the entire book, and the theoretical base for much of it, is there because the Macintosh has such superbly integrated hardware and software allowing huge workloads to be tackled rapidly and efficiently.

The influence of Frank Wallace and Professor Bannister turned out to be even more profound than I had realized, for it was a reexamination of their approach to unsteady gas dynamics that lead me to produce the engine simulation techniques described herein. Professor Bannister was the external examiner for my PhD at QUB and came from the same University of Birmingham which educated Frank Wallace.

I wish to acknowledge the collaboration of all of my research students over the thirty-two years that I worked at QUB, commencing with the late Dr. John Goulburn and concluding with Dr. Dermot Mackey. The others will forgive me if I do not list them all—they are too numerous—but any glance at the References reveals their names. Without their intellect, support, enthusiasm, hard work, and, indeed, friendship, a great deal of that which is presented here would be missing material.

I am indebted to those who have provided many of the photographs and drawings that illustrate this book. Quite a few also provided experimental data, or theoretical predictions, which are found herein. I refer to, in no particular order of precedence:

Hans Hermann of Hans Hermann Engineering
Frank Honsowetz of Nissan Motorsports
Dr. Donald Campbell of Perkins Technology
Rowland White, Norman Windrum, and Bill McLeod
Mr. Rosenthal of Classic Bike
Paul Reinke of General Motors
Melvin Cahoon of Innovation Marine
Lennarth Zander of Volvo
Douglas Hahn of Volvo Penta
Ron Lewis of Ron Lewis Engineering
Fred Hauenstein of Mercury Marine
Dr. Barry Raghunathan of Adapco
Ing. Mario Mazuran of Seatek
Mr. Kometani and Mr. Motoyama of Yamaha Motor
Steve Wynne of Sports Motorcycles
Hau-Bing Lau (when an undergraduate), François Drouin (when a visiting student from the École Nationale Supérieure des Arts et Métiers), and Emerson Callender, Laz Foley and Graham Mawhinney (as doctoral students) at QUB

David Holland, a QUB engineering technician, requires a special mention for the expert production of many of the photographs that illustrate this book.

I cannot finish without recognizing those who helped me to establish QUB in motorcycle road racing, for without them our design skills would have been much less evident. I refer to the late Mick Mooney and the late Ronnie Conn of Irish Racing Motorcycles, the late Brian Steenson, and Colin Seeley and Ray McCullough. That QUB tradition continues to this very day.

Gordon P. Blair
25 October 1998

Contents

Nomenclature

Most parameters are expressed in strict SI units, but custom and practice often dictate the units below to be declared in metric but non-strict SI units. Where such units are used theoretically, unless in the simplest of equations where the units are declared locally, they must be employed as strict units such as m, s, N, kg, J, W, or K values.

NAME	SYMBOL	UNIT (SI)
Coefficients		
Coefficient of heat transfer, conduction	C_k	W/mK
Coefficient of heat transfer, convection	C_h	W/m^{2}K
Coefficient of heat transfer, radiation	C_r	W/m^{2}K^4
Coefficient of friction	C_f	
Coefficient of discharge	C_d	
Coefficient of discharge, actual	Cd_a	
Coefficient of discharge, ideal	Cd_i	
Coefficient of contraction	C_c	
Coefficient of velocity	C_s	
Coefficient of loss of pressure, etc.	C_L	
Squish area ratio	C_{sq}	
Coefficient of combustion equilibrium	K_p	
Area ratio of engine port to engine duct	k	
Manifold to port area ratio	C_m	
Local port area ratio	C_t	
Valve acceleration ratio	g_v	
Valve ramp lift ratio	C_r	
Valve lift ratio	L_r	
Modified valve lift-diameter ratio	LD	
Intake tuning ramming factor	C_{ir}	
Exhaust tuning factor for primary pipe length	C_{et}	
Exhaust tuning factor for collector tailpipe length	C_{tp}	
Exhaust collector pipe area ratio	C_{coll}	
Dimensions and physical quantities		
area	A	m^2
diameter	d	m
length	x	m

NAME	SYMBOL	UNIT (SI)
length of computation mesh	L	m
mass	m	kg
molecular weight	M	kg/kgmol
radius	r	m
time	t	s
volume	V	m3
force	F	N
pressure	p	Pa
pressure ratio	P	
pressure amplitude ratio	X	
mass flow rate	$\dot{m}$	kg/s
volume flow rate	$\dot{V}$	m^3/s
velocity of gas particle	c	m/s
velocity of pressure wave propagation	α	m/s
velocity of acoustic wave (sound)	a	m/s
Young's modulus	Y	N/m^2
wall shear stress	τ	N/m^2
gravitational acceleration	g	m/s^2

Dimensionless numbers

Froude number	**Fr**	
Grashof number	**Gr**	
Mach number	**M**	
Nusselt number	**Nu**	
Prandtl number	**Pr**	
Reynolds number	**Re**	

Energy, work, and heat related parameters

system energy	E	J
specific system energy	e	J/kg
internal energy	U	J
specific internal energy	u	J/kg
specific molal internal energy	$\bar{u}$	J/kgmol
potential energy	PE	J
specific potential energy	pe	J/kg
kinetic energy	KE	J
specific kinetic energy	ke	J/kg
heat	Q	J
specific heat	q	J/kg
enthalpy	H	J
specific enthalpy	h	J/kg

NAME	SYMBOL	UNIT (SI)
specific molal enthalpy	$\bar{h}$	J/kgmol
entropy	S	J/K
specific entropy	s	J/kgK
work	W	J
work, specific	w	J/kg
Engine, physical geometry		
number of cylinders	n	
cylinder bore	d_{bo}	mm
piston area	A_{bo}	m^2
cylinder stroke	L_{st}	mm
bore to stroke ratio	C_{bs}	
connecting rod length	L_{cr}	mm
crank throw	L_{ct}	mm
swept volume	V_{sv}	m^3
swept volume, trapped	V_{ts}	m^3
clearance volume	V_{cv}	m^3
compression ratio, geometric	CR	-
compression ratio, trapped	CR_t	-
speed of rotation	N	rev/min
speed of rotation	rpm	rev/min
speed of rotation	rps	rev/s
speed of rotation	ω	rad/s
mean piston speed	c_p	m/s
crankshaft position at top dead center	tdc	degrees
crankshaft position at bottom dead center	bdc	degrees
crankshaft angle before top dead center	°btdc	degrees
crankshaft angle after top dead center	°atdc	degrees
crankshaft angle before bottom dead center	°bbdc	degrees
crankshaft angle after bottom dead center	°abdc	degrees
crankshaft angle	θ	degrees
angle of obliquity of the connecting rod	ϕ	degrees
combustion period	b°	degrees
throttle area ratio	C_{thr}	-
exhaust blowdown time-area	ϕ_{bd}	s/m
exhaust pumping time-area	ϕ_{ep}	s/m
exhaust overlap time-area	ϕ_{eo}	s/m
intake ramming time-area	ϕ_{ir}	s/m
intake pumping time-area	ϕ_{ip}	s/m
intake overlap time-area	ϕ_{io}	s/m
intake valve opens	ivo	degrees

NAME	SYMBOL	UNIT (SI)
intake valve closes	ivc	degrees
exhaust valve opens	evo	degrees
exhaust valve closes	evc	degrees
Engine, performance related parameters		
mean effective pressure, brake	bmep	Pa
mean effective pressure, indicated	imep	Pa
mean effective pressure, friction	fmep	Pa
mean effective pressure, pumping	pmep	Pa
power output	$\dot{W}$	kW
power output, brake	$\dot{W}_b$	kW
power output, indicated	$\dot{W}_i$	kW
torque output	Z	Nm
torque output, brake	Z_b	Nm
torque output, indicated	Z_i	Nm
air-to-fuel ratio	AFR	
air-to-fuel ratio, stoichiometric	AFR_s	
air-to-fuel ratio, trapped	AFR_t	
equivalence ratio	λ	
equivalence ratio, molecular	λ_m	
specific emissions of hydrocarbons	bsHC	g/kWh
specific emissions of oxides of nitrogen	$bsNO_x$	g/kWh
specific emissions of carbon monoxide	bsCO	g/kWh
specific emissions of carbon dioxide	$bsCO_2$	g/kWh
specific fuel consumption, brake	bsfc	kg/kWh
specific fuel consumption, indicated	isfc	kg/kWh
air flow, scavenge ratio	SR	
air flow, delivery ratio	DR	
air flow, volumetric efficiency	η_v	
charging efficiency	CE	
trapping efficiency	TE	
scavenging efficiency	SE	
thermal efficiency	η_t	
thermal efficiency, brake	η_b	
thermal efficiency, indicated	η_i	
mechanical efficiency	η_m	
fuel calorific value (lower)	C_{fl}	MJ/kg
fuel calorific value (higher)	C_{fh}	MJ/kg
fuel latent heat of vaporization	h_{vap}	kJ/kg
mass fraction burned	B	
heat release rate	$\dot{Q}_{R_\theta}$	J/deg

NAME	SYMBOL	UNIT (SI)
combustion efficiency	η_c	
relative combustion efficiency wrt purity	η_{se}	
relative combustion efficiency wrt fueling	η_{af}	
index of compression	n_e	
index of expansion	n_c	
flame velocity	c_{fl}	
flame velocity, laminar	c_{lf}	
flame velocity, turbulent	c_{trb}	
squish velocity	c_s	
engine speed for intake ramming peaks	N_{rp}	rev/min
engine speed for intake ramming troughs	N_{rt}	rev/min
Gas properties		
gas constant	R	J/kgK
universal gas constant	$\bar{R}$	J/kgmolK
density	r	kg/m^3
specific volume	v	m^3/kg
specific heat at constant volume	C_v	J/kgK
specific heat at constant pressure	C_p	J/kgK
molal specific heat at constant volume	$\bar{C}_V$	J/kgmolK
molal specific heat at constant pressure	$\bar{C}_P$	J/kgmolK
ratio of specific heats	γ	
purity	Π	
temperature	T	K
viscosity	μ	kg/ms
kinematic viscosity	ν	m^2/s
volumetric ratio of a gas mixture	υ	
mass ratio of a gas mixture	ε	
Noise		
sound pressure level	β	dB
sound intensity	I	W/m^2
sound frequency	f	Hz
attenuation or transmission loss	β_{tr}	dB
wave length of sound	Λ	m
perforation opacity ratio	O	-
General		
vectors and coordinates	x, y, z	
differential prefixes,		
exact, inexact, partial and incremental	$d,\ \delta, f, \Delta$	

Chapter 1

Introduction to the Four-Stroke Engine

1.0 About This Book

It is generally accepted that the theoretical cycle on which the four-stroke engine is based was proposed by Beau de Rochas in 1876. The first practical demonstration of the engine was implemented by Otto in 1876. This book is not about the history of the internal-combustion engine, but realizing that some of you may wish to study it, it is recommended that you peruse the informed writings of Cummins, Obert, Taylor, Caunter, or Ricardo [1.1-1.5]. The book by Cummins [1.1] is quite an authoritative text in this historical context.

This book is also not about the detailed design of the mechanical components of an engine, such as crankshafts or connecting rods. For that, one reads elsewhere in the literature. Nor is it a comprehensive collection of design ideas for the cylinder head, valving, or ducting geometries of every configuration of four-stroke engine constructed in times past.

This book is about the design of the four-stroke engine so as to achieve its target performance characteristics for the application required, irrespective of whether that application is intended for Formula 1 car racing or a lawnmower. To do that, one must thoroughly understand the filling and emptying of the engine cylinders with air and exhaust gas and the combustion of the trapped charge within them. Hence, this book is about the unsteady gas dynamics and thermodynamics associated with the four-stroke engine. Nevertheless, to sensibly design for the performance characteristics, one must bring the real geometry of the engine, its cylinder head, combustion chamber, manifolding, and ducting into the gas dynamic and thermodynamic design process, otherwise the outcome is meaningless, not to mention useless. Therefore, very frequently, the real geometry and the measured test data from actual engines will be produced to illustrate a design point being made. To conduct such a design process, the only pragmatic approach is to simulate the unsteady gas dynamics and thermodynamics within the entire engine, basing the simulation on the physical geometry of that engine in the finest detail, from the aperture where air enters the engine initially to the aperture where the exhaust gas finally exits from the engine.

1.1 Fundamental Method of Operation of a Simple Four-Stroke Engine

1.1.1 The Four Strokes That Make Up the Cycle

The simple four-stroke engine is shown in Fig. 1.1, with the several phases of the filling and emptying of the cylinder illustrated as Figs. 1.1(A)–(D). The sketch shows the engine as a spark-ignited unit with access to the cylinder controlled by poppet valves actuated by cams,

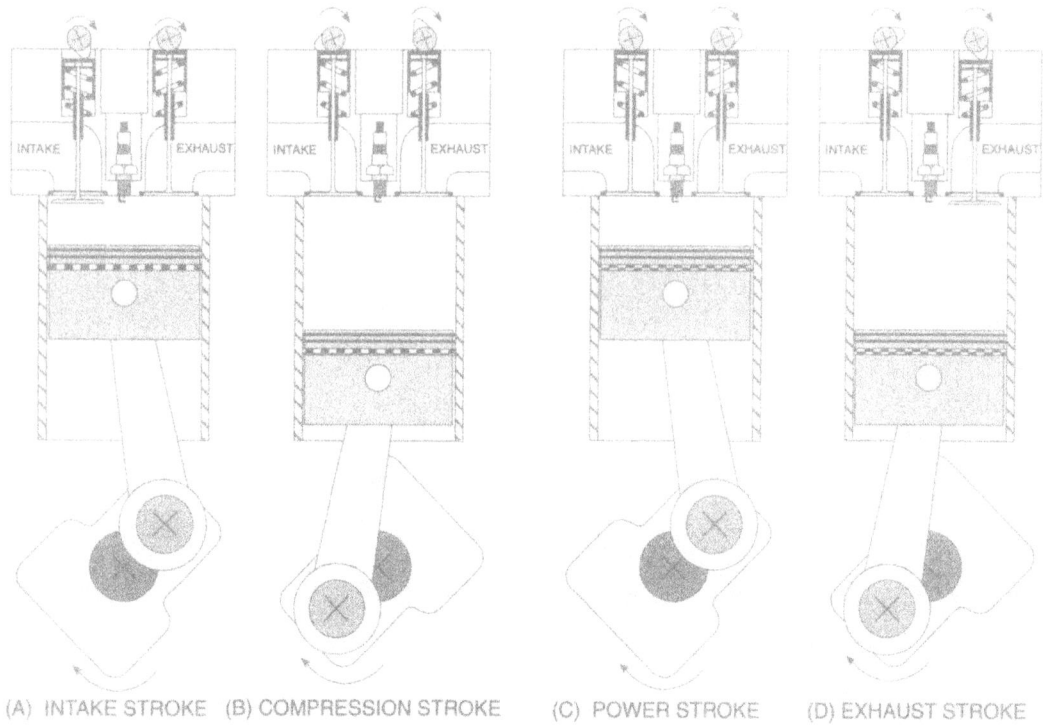

(A) INTAKE STROKE (B) COMPRESSION STROKE (C) POWER STROKE (D) EXHAUST STROKE

Fig. 1.1 The four strokes of the four-stroke engine.

tappets, and valve springs. The mechanical definition of some, if not all, of such words describing salient engine components is shown sketched on Fig. 1.2. The cylinder contains a piston with sealing effected by two piston rings and a third lower ring, which scrapes the excess lubricating oil off the cylinder walls back into the crankcase, which acts as an oil sump. The movement of the piston is controlled by a rotating crankshaft with the connecting rod linking the small-end bearing on the piston, through a gudgeon pin, to the big-end bearing on the crankpin. The word "gudgeon pin" is a British word; in the United States it is referred to as a "wrist pin." A flywheel is attached to the crankshaft to help smooth out the torque pulsations. In the sketch, Fig. 1.1, it is seen as a bob-weight flywheel.

Plate 1.1 shows a cross section through an engine much like that sketched in Fig. 1.1. All of the items described briefly above can be seen in this photograph, but it provides a realistic scaling of the many components. The intake tract is on the right of the picture and the exhaust duct is to the left.

It is clear from Fig. 1.1 that the maximum movement of the piston within the cylinder, i.e., the stroke of the piston, is, simplistically, twice the upward, or downward, movement of the length of the crank. The crank length is often referred to as the "crank throw." The maximum movement of the piston creates a volume swept by the piston, i.e., the swept volume, V_{sv}, and it is clear from the sketch that the piston stops short of the cylinder head face, thereby creating

2

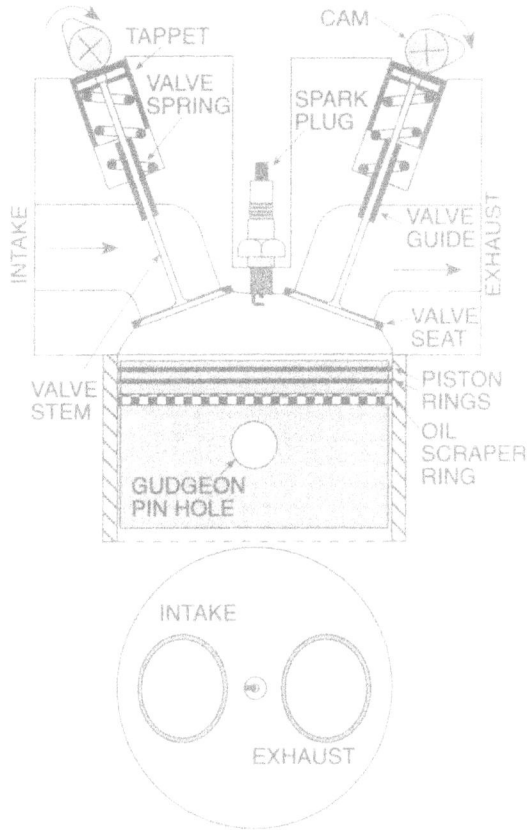

Fig. 1.2 An overhead valve engine with two valves per cylinder.

a minimum volume space in which the combustion of trapped air and fuel can take place. This minimum volume is defined as the clearance volume, V_{cv}. When the piston is at the top or minimum cylinder volume position, it is referred to as being at top dead center, tdc, and when it is at the bottom or maximum cylinder volume position it is described as being at bottom dead center, bdc. In certain books, mostly historical, tdc and bdc are sometimes called inner and outer dead center, idc and odc; they will not be referred to as such within this book. This volumetric behavior gives rise to the concept of a compression ratio for the engine. For the four-stroke engine this ratio is always employed as the geometric compression ratio, CR:

$$CR = \frac{\text{maximum volume}}{\text{minimum volume}} = \frac{V_{bdc}}{V_{tdc}} = \frac{V_{sv} + V_{cv}}{V_{cv}} \qquad (1.1.1)$$

The employment of a significant level of charge compression within an engine like this was the major inventive contribution of Otto [1.1].

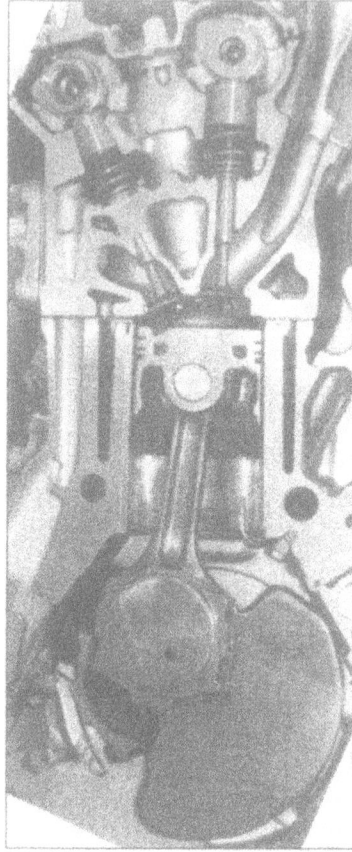

Plate 1.1 A cross section through a two-valve spark-ignition engine.

1.1.2 The Four Strokes

Air is induced into this engine by an intake stroke of the piston from tdc to bdc, thereby increasing the cylinder volume from its minimum to its maximum value. This is shown in Fig. 1.1(A). The intake poppet valve is actuated by the movement of the rotating cam depressing the tappet and spring. Ideally, it opens at tdc, lifts to a maximum value at about mid-stroke, and ideally closes at bdc. The effect is somewhat similar to when you open your mouth, sharply intaking air into your lungs, and then close your mouth again; such an intake process is caused by your lungs expanding and air entering because the atmosphere is then at a higher pressure than within your lungs. Thus, as the piston moves from tdc to bdc on the intake stroke, the cylinder fills—ideally with a mass of air equivalent to a swept volume at atmospheric pressure and temperature. Whether this air brings with it the requisite fuel for future combustion, either by a carburetor placed within the intake duct, or by a similarly placed fuel injector squirting fuel into the air as it passes by, or by a fuel injector that sprays fuel directly into the air now within the cylinder, is really a matter of implementation for any particular design configuration.

The air now trapped within the sealed cylinder experiences a compression stroke as the piston moves from bdc to tdc. The volume decreases from maximum to minimum, i.e., from V_{bdc} to V_{tdc}. The pressure and temperature of the air rises and the trapped fuel associated with this air vaporizes. This is illustrated in Fig. 1.1(B).

At the conclusion of the compression stroke, combustion takes place, initiated by an electric spark from the spark plug shown located between the valves in the cylinder head in Fig. 1.1. The pressure rises rapidly to many tens of atmospheres, and the temperature soars to several thousand degrees Celsius. Ideally, all of this takes place at tdc and instantaneously.

The piston now descends on the power stroke, pushed by the very high pressure difference across it from the upper side facing the cylinder to the lower side seeing the crankcase at atmospheric pressure. The cylinder volume goes from V_{tdc} to V_{bdc}. This force, i.e., a push, on the piston gives a torque on the crank, which in turn powers whatever the engine is driving. This is sketched in Fig. 1.1(C).

Because the cylinder now contains the products of combustion, i.e., an exhaust gas, which must be removed if ever the cylinder is to breathe air again and become a cyclic device, an exhaust stroke is initiated as shown in Fig. 1.1(D). The exhaust poppet valve is actuated by a well-timed turn of the rotating cam, which depresses the tappet and compresses the valve spring and lifts the valve. The exhaust valve ideally opens at bdc, lifts to a maximum value at about mid-stroke, and ideally shuts again at tdc. The exhaust gas is forcibly expelled by the upward sweeping movement of the piston through the aperture created by the annulus around the head of the lifted exhaust valve. The exhaust stroke occurs when the piston moves from bdc to tdc and the cylinder volume decreases from maximum to minimum, i.e., from V_{bdc} to V_{tdc}. Using the human body again as an analogy, the exhaust stroke is somewhat similar to you filling your lungs with air and opening your mouth, thereby letting your lungs decrease in volume with the exhalation; you then close your mouth again. That exhaust process is caused by your lungs deflating, and your personal exhaust gas exiting, because the atmosphere is at a lower pressure than within your lungs.

1.1.3 Analogy between the Human Body and an Engine

As I have often told my students, the analogy between the human body and the internal combustion engine is quite uncanny. We breathe in air, and breathe out exhaust gas, by varying the volume of our lungs. We consume fuel and some of us actually produce useful work! We are even the epitome of that thermodynamic ideal, expert practitioners of the isothermal heat transfer of the Carnot cycle [1.2]. We characterize the power output of the internal combustion engine by the horsepower, derived by a direct comparison with another animal engine. There are even exhaust emission analogies to be explored, but this text will not descend any further down this rich vein of humor, already infamously over-exploited during my academic teaching career to a ribald undergraduate audience. On a more serious note, and in this very context, only a century ago the Sanitation Officer of the City of New York, then populated by tens of thousands of horses, declared that he was looking forward to the day when the noxious odors, not to mention particulate emissions, producing health risks associated with these animals, would be eliminated by the arrival of the "horseless carriage and its clean engine." A hundred years later that debate has been turned on its head, with the internal combustion engine now relegated to the role of polluter.

1.2 Cylinder Head Geometry of Typical Spark-Ignition Engines
1.2.1 Two-Valve Overhead Valve Engine

The most common four-stroke engine produced is arguably the overhead valve (ohv) engine with two valves per cylinder. This type of engine is sketched in Fig. 1.2, with an elevation section through the valves at top; underneath is a plan view looking at the cylinder head face.

The sketch shows the actuation of the valves by two overhead camshafts, colloquially referred to as a double overhead camshaft engine, and often written simply as a dohc design. Plate 1.1 shows a cross section through just such an engine. Many famous racing engines were built just like this. The 350 or 500 cm³ Manx Norton racing motorcycles were outstanding examples in their own era of the high specific power output that could be extracted from this type of power unit; the engineering talents of the Chief Engineer of Norton, Joe Craig [1.35], have already been referred to in the Foreword. A drawing of the 1959 model of the 500cc Manx Norton engine is shown in Plate 1.2 with the overhead camshafts driven by a vertical shaft through bevel gear drives, and with the valve motion controlled by hairpin valve springs.

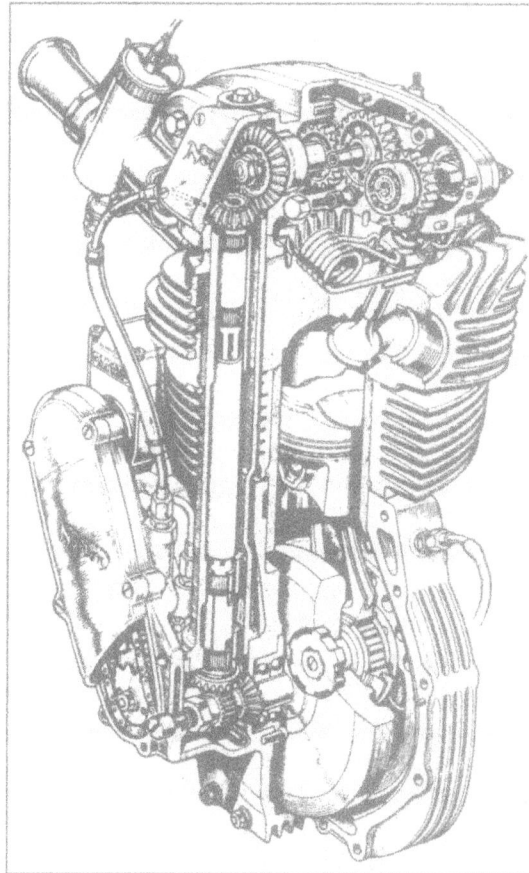

Plate 1.2 The 500cc Manx Norton racing motorcycle engine. (Courtesy of Classic Bike)

As already stated, although this book is not a compendium of the myriad ingenious methods that the engineer has designed to actuate the poppet valves in an engine, I propose to illustrate some relevant cylinder head geometry so that the discussion elsewhere, such as in Chapter 5, is more meaningful. Historically, the most popular method to actuate the two valves is the pushrod and rocker arm system, again often simply referred to as a dohv layout. The basic mechanism of its operation is shown in Fig. 1.3. The camshaft is driven from the engine crankshaft and pushes on a tappet, which lifts the pushrod, oscillates the rocker arm, and lifts and lowers the valve off, and back on to, the valve seat. An adjuster is employed to control the clearances of the elements of the mechanism. This is shown here as a simple mechanical device, but human ingenuity has yielded tappets that are sliding hollow cylinders filled with oil which can control the pushrod end clearances automatically; these are known as hydraulic tappets. Pushrod designs have ranged from solid steel rods to hollow aluminium tubes. Clearly, the higher is the location of the camshaft, the shorter is the pushrod and the less it will be inclined to bend or buckle when pushed by the tappet.

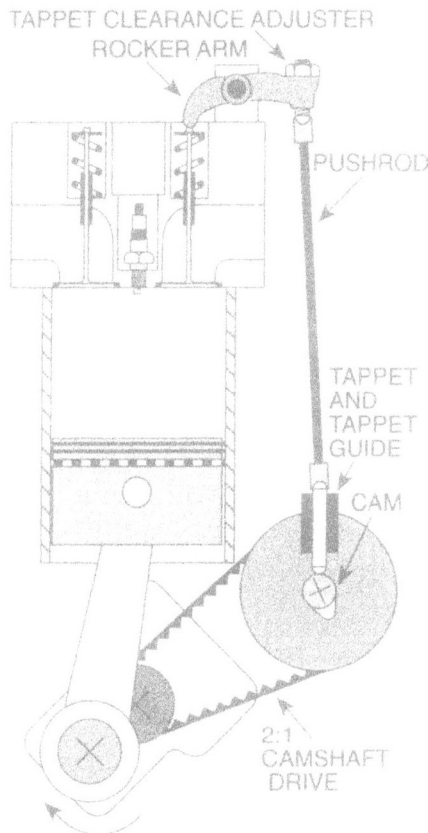

Fig. 1.3 Valve actuation by a pushrod and rocker arm.

A cutaway drawing of a two-valve ohv engine with pushrod actuation is shown in Plate 1.3. This is of the 500cc Gold Star BSA, which was used in both on-road and off-road racing throughout the world, and was particularly popular in the United States in the era when Daytona Beach racing was actually held on the beach! There was an on-road version using this engine, as a sports-touring machine, which must have been one of the most handsome motorcycles ever built. As a student at QUB in the 1950s, I virtually turned green with envy just looking at them through the showroom window of W.J. Chambers & Co. in Donegall Pass in Belfast.

Other options include a single overhead camshaft (sohc) moving rocker arms directly onto each valve that eliminates the need for pushrods (see Plate 5.6 in Chapter 5). References 1.1–1.8 include excellent drawings, sketches, and photographs of many engines developed throughout history and their methods of valve actuation.

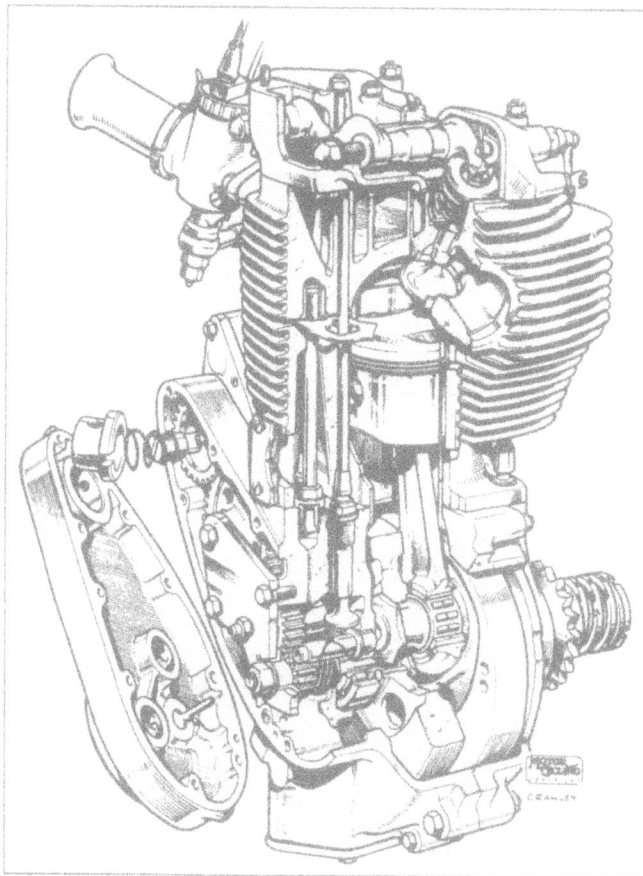

Plate 1.3 The 500cc Gold Star BSA racing motorcycle engine.
(Courtesy of Classic Bike)

Irrespective of how the valve is actuated, a typical valve-lift crank angle graph is as shown in Fig. 1.4, where the exhaust valve opens and closes at points marked as evo and evc, respectively. The intake valve opens and closes at points marked as ivo and ivc, respectively. The x-axis is crankshaft (or crank) angle. On this particular graph, this angle is zero at tdc on the power stroke, and lasts for 720°; the durations of the individual strokes are clearly marked.

Irrespective of the method of valve actuation, several fundamental points emerge from an examination of Figs. 1.2 and 1.4. The first point to be observed is that the valves do not, and cannot, lift instantaneously. They have mass, i.e., inertia, when accelerated, spring forces must be overcome, and friction resistance requires that force must be exerted to effect movement. All of this simply means that a real valve takes time to lift and requires work input to do it. The idealized set of strokes described in Sec. 1.1.2 assumes simplistically that the valves opened or closed instantaneously, at tdc or bdc, as the need arose. Fig. 1.4 makes clear the reality that an exhaust valve will typically open some 110 °atdc, i.e., 70° before bdc, and will close some 30 °atdc and not at tdc. All of this is necessary to ensure that the valve exposes sufficient aperture area to the cylinder so as to release the combustion products from it. It would be possible to drop the valve more rapidly back onto its seat at tdc. However, the ensuing bouncing behavior of what is, in effect, a spring-mass system would ensure that it would actually stay open for even longer and would incur such large impact forces on the cam and tappet as to seriously reduce its durability from excess wear and erosion. Nevertheless, the less the mass of this entire oscillating system, the higher is the rotational speed it can reliably run. Thus, in racing

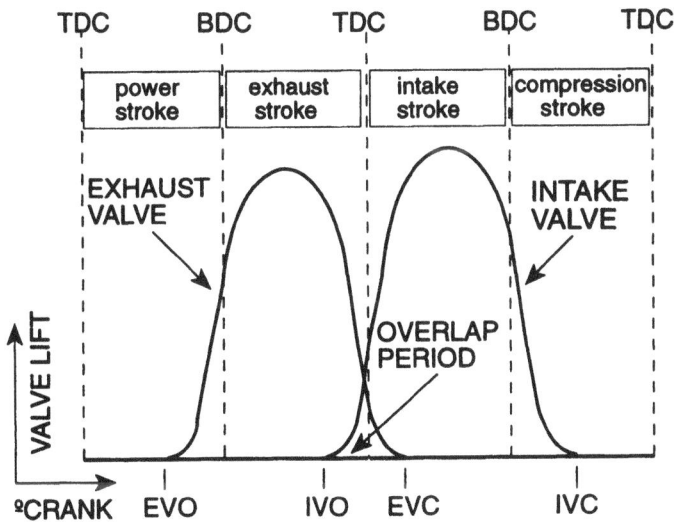

Fig. 1.4 Typical valve lift profiles with respect to crank angle.

engines it is not surprising to find double overhead camshaft systems employed. For lower-speed engines, it becomes possible to withstand the extra mass and inertia provided by rocker arms and pushrods, as shown in Fig. 1.3, without significant loss of valve lift or timing error at the maximum speed of operation.

The second point to be observed is that the thermodynamic cycle occupies four strokes of the piston whereas the total crankshaft angle it covers is 720°, i.e., two complete rotations of the crankshaft. However, because each valve is operated over the 360° period of the intake and exhaust strokes, and then must remain dormant during the 360° period of the two strokes that make up the compression and power strokes, it follows that the camshaft is driven at one-half the speed of rotation of the engine crankshaft. In short, as the crankshaft drives the camshaft, typically by either chain, toothed timing belt, or gears, the gearing ratio for this drive between crankshaft and camshaft is 0.5, i.e., the camshaft turns at half the engine speed.

The third point to be observed from Fig. 1.4 is that there is a period between the exhaust and intake strokes when the intake valve is opening and the exhaust valve is closing, i.e., both valves are open together. This is called the valve overlap period. The potential for mechanical or gas flow mayhem is obvious. Apart from the valves actually colliding, if the cylinder pressure is too high as the intake valve opens, then large quantities of exhaust gas can be shuttled into the intake tract. This gas is hot, maybe 1000°C, and can cause anything from backfire to coking-up fuel residues on the back of the intake valve. At the very least, every particle of exhaust gas sent up the intake tract will occupy space normally reserved for air and, when real cylinder inflow eventually occurs, only a reduced mass of air will be induced into the cylinder. As the air mass induced on each cycle ultimately equates to power attained, this can hardly be described as an optimum breathing procedure. On the other hand, if at the same juncture the exhaust pressure is lower than in either the intake tract or the cylinder, then the intake process can begin early on the intake stroke, so there exists the possibility of enhancing the mass of air inhaled. Overdoing this effect can actually send fuel-laced air into the exhaust pipe, which will ultimately appear in the atmosphere as an excess of unburned hydrocarbon emissions. Hence, based on these preliminary remarks, the design process for the engine during the valve overlap period is surely seen as a critical affair. But I get ahead of myself because this and many other similar subjects, such as the effectiveness of intake ramming behind a still-open intake valve on the compression stroke, are really what the rest of this book is all about!

The two-valve engine type, as sketched in Fig. 1.2, appears in many formats as can be seen in the literature [1.1-1.8]. Some have the inclined valves and a part-spherical combustion chamber, essentially as illustrated here. Such an engine is called a "hemi-head" engine, from the word "hemisphere," which rather describes the shape of the combustion area. Some have the valves vertical, as in the diesel, or compression-ignition, engines to be discussed later. Hence, most diesel engines have the combustion bowl within the piston crown.

1.2.2 Two-Valve Side Valve Engine

The simplest of all four-stroke engines is the side-valve engine, shown sketched in Fig. 1.5.

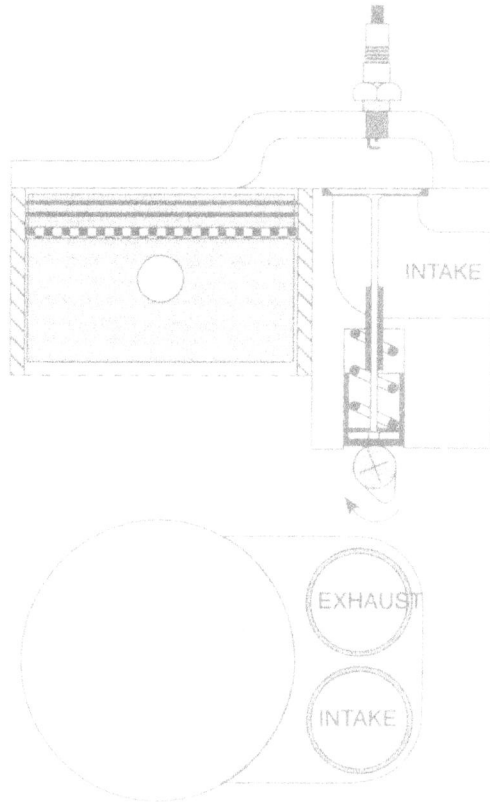

Fig. 1.5 A side-valve engine with two valves per cylinder.

Many a car in my youth used nothing else. Some were famous, such as the Ford V8 Pilot in the saloon car, and the 1172 cm³ Ford 100E four-cylinder engine that powered many a single-seat racing car in Britain and Ireland in the fifties and sixties, and is the forerunner of that used in today's very popular Formula Ford racing. The car in which I learned to drive, my father's Morris Series E, had a 1000 cm³ side-valve engine. Being possessed of rather indifferent suspension and steering, it was perhaps fortunate that its modest power output limited its top speed to a near-lethal 50 mph!

Fig. 1.5 shows that the valves are indeed at the "side" of the cylinders and, due to the physical limitations imposed by the bore dimension and normal inter-cylinder spacing, it is virtually impossible to employ other than one exhaust valve and one intake valve per cylinder. Moreover, because these limitations control the maximum size of the valves that may be employed, the breathing ability of the engine is compromised. The valve design is quite rigid, in that the camshaft is near the crankshaft. Indeed, in many ways it is as effective as an overhead camshaft engine design in this regard. The cylinder head is as simple as a loop-scavenged two-stroke engine and the total design is not only compact and light, but also very economical to manufacture. This is why tens of millions of these were made by Briggs and Stratton of Milwaukee for industrial, agricultural, and horticultural applications. The major design problem is that all gas flow into, or out of, the engine is very restricted by the throat between the valve chest area and the engine cylinder, causing pumping work losses in both directions. Upon the compression stroke, further pumping is required of the piston to get the fresh charge back over the valve chest area to be ignited by the spark plug. The combustion chamber shape is poor and prone to detonation, so a relatively low compression ratio must be used. Thus, heat transfer, combustion, and pumping work losses are high in this engine, which reduces its power output and fuel economy and raises its output of most of the legislated exhaust-gas emissions.

1.2.3 Four-Valve Pent-Roof Cylinder Head

This design for spark-ignition (si) engines has become very common in automobiles and motorcycles in the last decade of the twentieth century. Such an engine is shown sketched in Fig. 1.6 as a dohc layout and even that further sophistication is to be found in a high proportion of automobiles and motorcycles today. At one time, the four-valve dohc engine was reserved only for racing engines, but first the Japanese motorcycle industry, then the Japanese car industry, and then the rest of the world, employed it universally in general automotive production.

There have been many four-valve engines used throughout history. A Rudge motorcycle won the Ulster Grand Prix in 1928 and 1929, and the company subsequently produced a four-valve road-going equivalent of the very successful racer, and named it the Rudge "Ulster" model. My father owned one before the start of World War II and I can dimly remember him going to and fro in his work on this elegant black-and-gold machine. A cutaway drawing of the 1938 version of this engine is shown in Plate 1.4, where the valve actuation is by pushrods and rocker-arms. The intake valves are parallel, much as in a modern pent-roof head layout, but the exhaust valves are rather more radially disposed. The piston crown has a shallow dome.

A comparison of Figs. 1.2 and 1.6 shows that it is possible to get more valve head area in the four-valve design and so aid the engine to breathe more air during induction. The so-called pent-roof layout means that the valves are disposed on the sides of a wedge, or roof-like, combustion chamber with the added possibility of squish areas disposed around the cylinder. The spark plug can be located centrally, which gives equal flame travel paths to the remote corners of the combustion chamber. Although in the two-valve engine shown in Fig. 1.2 the spark plug are shown to be centrally located, it is more common to use larger valves than those sketched there, and to locate the plug toward one side of the axis of the two valves.

The 4v dohc design is now commonplace in current automobiles and the cutaway drawing of a 1998 2.4 liter I4 car engine by General Motors, shown in Plate 1.5, perfectly illustrates the genre. The camshaft drive is by a roller chain.

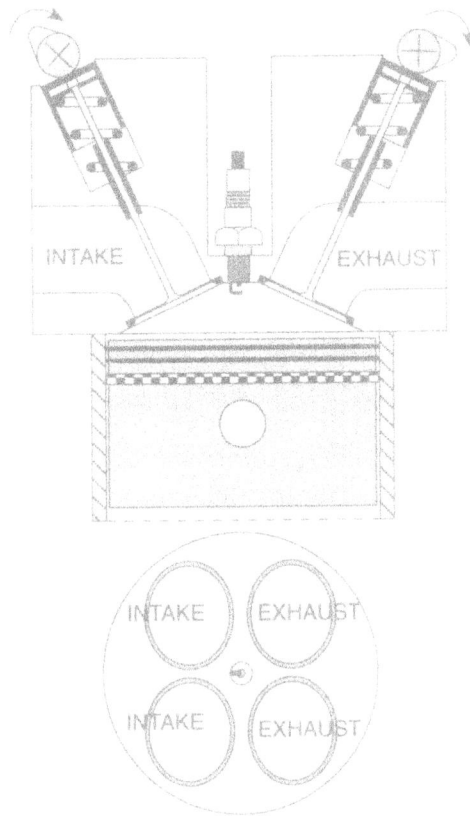

Fig. 1.6 A four-valve pent-roof cylinder head.

1.2.4 Porting the Cylinder Head by Other Means

There have been designs for porting the cylinder head by means other than the use of poppet valves. The first of real note is the use of sleeve valves which rotate—oscillate is a better word—around the engine cylinder and expose holes in the cylinder wall for the intake and exhaust processes, much as is seen for the porting of a two-stroke cycle engine. A cylinder wall port, almost by definition, opens more rapidly and exposes more aperture area to the cylinder than can a poppet valve [1.6, 1.9]. Hence, the engine has the possibility of breathing more air and expelling its exhaust gas more easily. Caunter [1.4] describes the Barr and Stroud and other, similar, sleeve valve engines at the turn of the twentieth century, and Ricardo [1.6] shows excellent scale drawings and photographs of many of these designs. A cutaway drawing of the 1922 350cc Barr and Stroud motorcycle engine is shown in Plate 1.6. The valve is known as the Burt-McCollom type and its motion is a combination of reciprocation and semi-

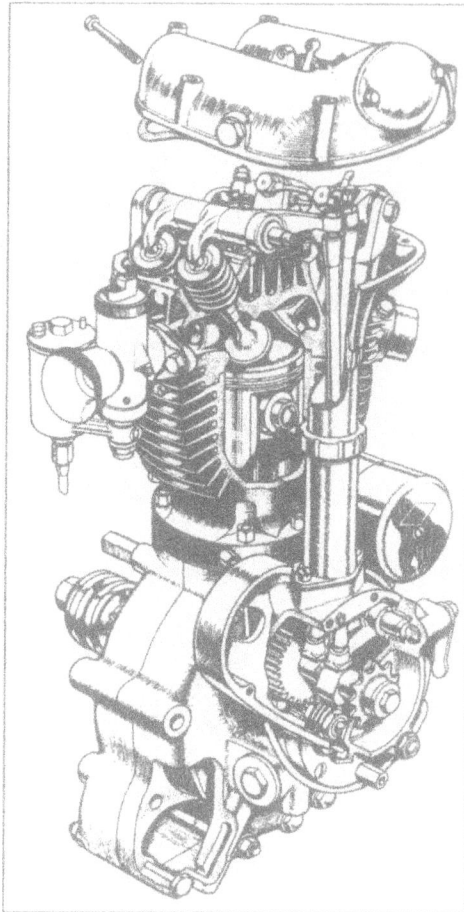

Plate 1.4 The 500cc Rudge Ulster motorcycle engine.
(Courtesy of Classic Bike)

rotation, i.e., as in a figure eight. Its mechanical reliability, because it was designed in the period before Sir Harry Ricardo established some logical design principles for sleeve-valve engines, was not outstanding.

Perhaps the most notable of these sleeve-valve designs, to operate on gasoline with spark-ignition, were the Bristol Hercules and Centaurus aircraft engines. They were far removed from simple designs, mechanically speaking, as the Hercules had a 14-cylinder, air-cooled, radial layout.

The second and different type of valving is the rotary valve installed in the cylinder head. Here, the poppet valves are replaced by a rotating valve which spins concentrically with the cylinder axis, or a barrel valve where the spin axis is at right angles to it. R.C. Cross of Bath was one of the originators of research into this type of engine [1.10], and I was greatly privileged to hear him deliver his Chairman's Address to the Automobile Division of the Institution of Mechanical Engineers, in Belfast in 1958, when I was a student. It was a brilliant lecture

14

Plate 1.5 A 2.4 liter in-line four-cylinder automobile engine. (Courtesy of General Motors)

but, as I listened to it quite enthralled, it certainly never occurred to me that I would follow him into his Chairman's role some thirty years later! The book by Hunter [1.11] provides an important reference for the geometries of virtually every rotary-valve engine ever built.

Due to the higher values of exposed breathing area, all of these engines, sleeve-valve or rotary-valve, in their day did deliver higher power outputs than the equivalent engines fitted with poppet valves. Their Achilles' heel was the exposure to the cylinder, and more importantly to its combustion chamber, of the rotating surfaces of the valves which are, however lightly, smeared with lubricating oil. The result was the partial burning of that oil, which raised the oil consumption rate of the engine, gave a somewhat smoky exhaust containing particulates, and tended to coke up the valve, thereby not only reducing its durability but also decreasing the breathing area of the aperture.

Wankel Engines

I do not intend to discuss the Wankel, or rotary-piston four-stroke engine within the pages of this book. Firstly, it is such a special case that it would distort every descriptive section of this book to the point where the debate on the conventional reciprocating engine would be lost in the caveats needed to cope with the Wankel engine. Secondly, as I write this book in 1998, I cannot think of a single Wankel engine that is in current production. So, there appears to be

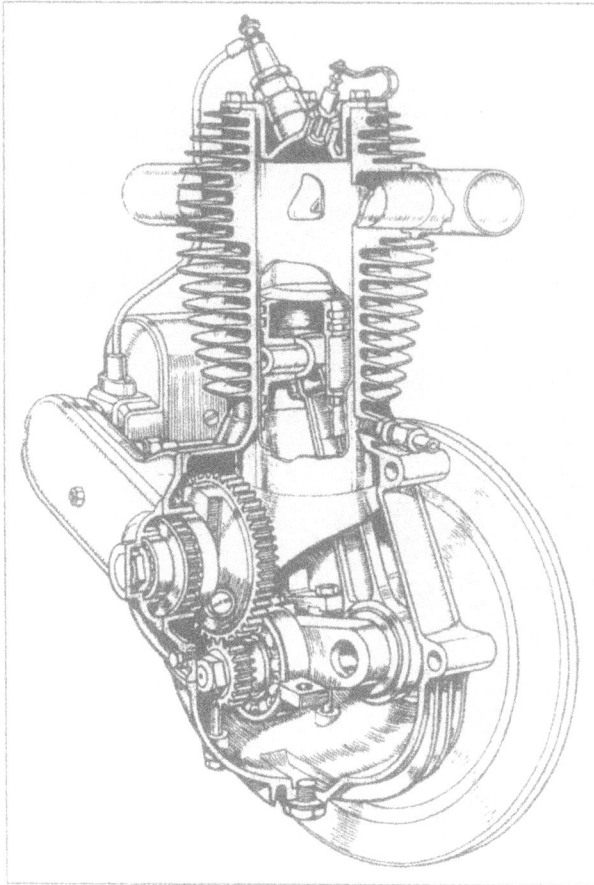

Plate 1.6 The 350cc Barr and Stroud sleeve-valve motorcycle engine.
(Courtesy of Classic Bike)

no current technical interest in the engine or its development. If there is such a Wankel engine in current production, then I apologize, but it has not come to my attention. Thirdly, I can well imagine why the Wankel is not in current production for automobiles and the automotive industry as a whole, but I can think of several areas where it ought to be seriously reconsidered for sound technical reasons. However, even to tell you about that would require a decent-sized book all by itself. Hence, the word "Wankel" will not reappear within these pages.

1.3 Cylinder Head Geometry of Typical Compression-Ignition Engines
1.3.1 General
This engine was invented by Rudolph Diesel in 1892 as a means of burning coal dust. He found out rather quickly that coal dust was inferior to liquid fuels for a compression-ignition combustion process. The engine carries his name to this day, being colloquially referred to as a "diesel." One argument holds that Ackroyd-Stuart in England produced an engine of this

16

type before Diesel. Be that as it may, today Rudolph Diesel conventionally gets the credit. Combustion in a compression-ignition (ci) engine occurs by using a compression ratio that is sufficiently high so as to produce an in-cylinder air temperature such that the vapor surrounding the injected fuel droplets is heated to its self-ignition temperature. Whereas the spark-ignition, using the more volatile gasoline fuel, has compression ratios between 8 and 11, the compression-ignition unit requires compression ratios that are typically between 18 and 21 to accomplish its combustion process. To prevent such a combustion process from being too vigorous, a fuel that contains heavier hydrocarbons, and is less volatile, than gasoline is employed. Thus, the spark-ignition engine uses a gasoline, which is typically based on octane of the paraffin family, with a chemical composition of C_8H_{18}, whereas the compression-ignition engine uses a fuel based on dodecane of the same paraffin family, with a chemical formula of $C_{12}H_{26}$. This subject is discussed in more detail, in Sec. 1.6.6 of this chapter, and in Chapter 4.

The basic process is one of heating air by compression, then using the hot air to vaporize a liquid fuel and heat the vapor to its self-ignition temperature. Such heating processes take time, so it is not surprising to find that diesel engines tend to run more slowly than spark-ignition engines. The basic mechanism that can be used to speed the process up is to raise the heat transfer rate between the hot air and the cold fuel droplets by increasing the relative velocity between these two fluids. This presents two options: (1) either have fuel droplets that move rapidly through the air, or (2) have slower-moving fuel be whirled by faster-moving air. The first option gives rise to the direct-injection diesel engine and the second to the indirect-injection type.

1.3.2 The Four-Valve Direct-Injection (DI) Engine

The necessity of having a high compression ratio is discussed briefly above. Because the needed compression ratio is typically about twice that of the gasoline engine, it follows that the clearance volume of the combustion chamber is about half that of the spark-ignition engine. It becomes almost impossible to achieve a such a small clearance volume and, at the same time, arrange the intake and exhaust valves any way other than vertical with respect to the cylinder axis. The layout of a four-valve engine is shown in Fig. 1.7 and the vertical disposition of the valves is clear. Because the piston runs very close to the head face at tdc, it becomes very difficult to achieve the small clearance volume except by having the minimum of valve overlap periods. With any other arrangement, the valves will contact the piston crown.

A high-pressure fuel injector is seen in the middle of the four valves and in the middle of the cylinder, squirting fuel in sprays of fast-moving fine droplets toward the edge of a combustion chamber, which is normally located as a bowl in the piston. To further enhance the process of heating the fuel, the air may be made to swirl around the vertical axis, albeit slowly by comparison with the indirect-injection engine discussed below in Sec. 1.3.3. The swirl ratio, i.e., the speed of rotation of this air vortex, is rarely more than five times higher than the engine speed. The in-cylinder air swirl is created during the induction process by suitably orienting the incoming air-direction past the intake valves; you may glance ahead at Plate 3.3. The spinning vortex is then somewhat accelerated by being compressed into the combustion chamber toward the end of compression. The fuel injection line pressure is typically in the range of 500 to 1200 atm, with the start of injection normally some 15 °btdc.

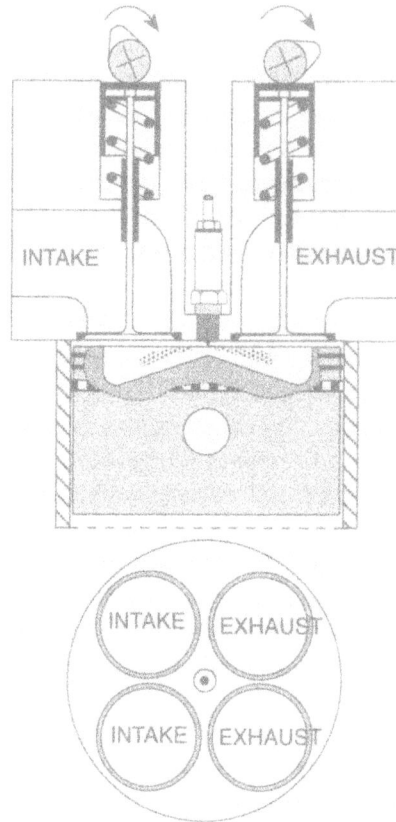

Fig. 1.7 A four-valve direct-injection (DI) diesel engine.

The particular shape of chamber sketched in Fig. 1.7 is known as, for obvious reasons, a "Mexican hat." Once again, engineering inventiveness has produced a plethora of designs for diesel combustion and you are referred to Lilly [1.12] for a compendium of their geometries. On a historical and personal note, this book is the successor to that founded by C.C. Pounder, who was the Chief Engineer of Harland and Wolff's shipyard in Belfast and a famous designer of marine diesel engines. Note, too, that fuel injection technology for both gasoline and diesel fuel will be discussed later in this book but, if you cannot wait, then Lilly may again be consulted [1.12].

The design shown in Fig. 1.7 has four valves per cylinder. Many designs throughout history had but two valves. The reason that four-valve designs have become more common is that diesel engines for automobiles have gained great popularity, particularly in Europe, because of good fuel economy and, in recent years by using turbo-chargers, good power output as well. The four-valve layout permits better air breathing and lower pumping losses, particularly considering the necessary minimum nature of the valve overlap period. Modern, turbo-charged DI engines for automobiles, with total capacities typically between 2 and 3 liters, now run to over 4000 rpm, and have specific outputs at around 55 kW per liter. For trucks, or marine pleasure craft, the cylinder size is usually about 1 liter per cylinder, but the engine speeds are lower and normally do not exceed 3000 rpm.

A cutaway drawing of a turbocharged truck diesel engine is shown in Plate 1.7. It is an I6 design, a "straight-six" in the jargon of the automotive world, with two valves per cylinder actuated by pushrods. The turbocharger can be seen at the left. The swept volume per cylinder of this engine is about 1 liter. The combustion bowl within the piston, for direct injection of the fuel, are clearly drawn as are the vertical disposition of the valves. A closer look at one of the combustion bowls in this engine is provided by the photograph in Plate 1.8. This design is known as a "squish lip" with the bowl being considerably reentrant. The compressed air is swirled and squished from the cylinder head area into the combustion bowl; evidence of these effects can be found in Sec. 4.5. Note also the large bearing areas at the small-end and the considerable length, bulk, and, inevitably, mass of this piston compared to that for the spark-ignition engine seen in Plate 1.1.

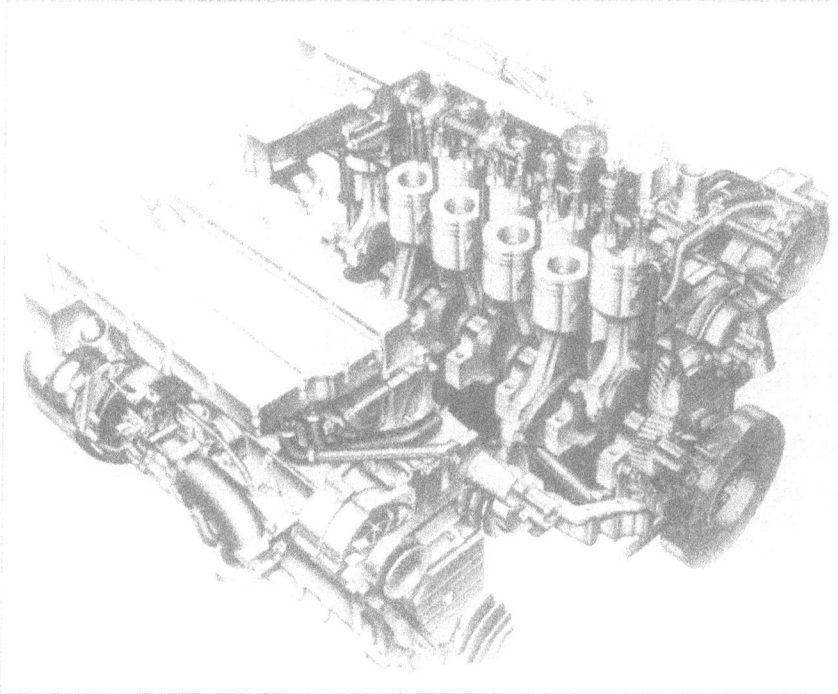

Plate 1.7 A cutaway drawing of a turbocharged direct-injection six-cylinder diesel engine. (Courtesy of Perkins Engines)

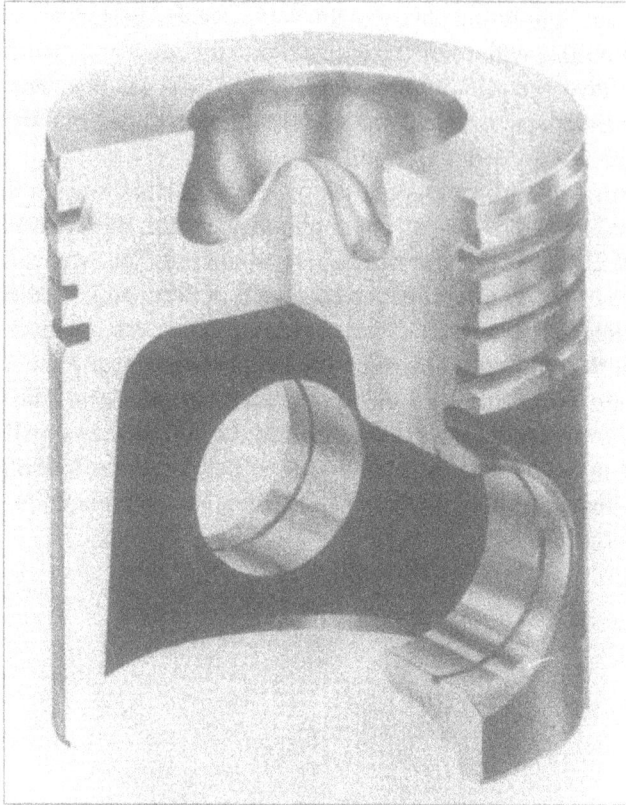

Plate 1.8 A cutaway drawing of the piston for a direct-injection diesel engine. (Courtesy of Perkins Engines)

1.3.3 The Three-Valve Indirect-Injection (IDI) Engine

This engine design typifies the other fuel-heating option of high-speed in-cylinder airflow and the slower moving fuel created by lower pressure fuel injection systems. Here, the fuel injection line pressure is normally in the range of 150 to 300 atm. The basic layout of the engine type is shown in Fig. 1.8.

Upon compression, the air is pumped into a side chamber and here the shape sketched mimics that developed by Ricardo [1.6] and is known as a Comet chamber. This is described as indirect injection, abbreviated as IDI, and is still widely employed in European diesel automobiles. The swirling air flow within the Comet chamber rotates at up to 25 times the engine speed. Because the engine can run to a higher speed than the DI design, i.e., up to 4500 rpm, this means that this air vortex spins at up to 90,000 rpm. The result is a "cleaner" combustion process than the DI engine, i.e., cleaner in terms of exhaust emissions.

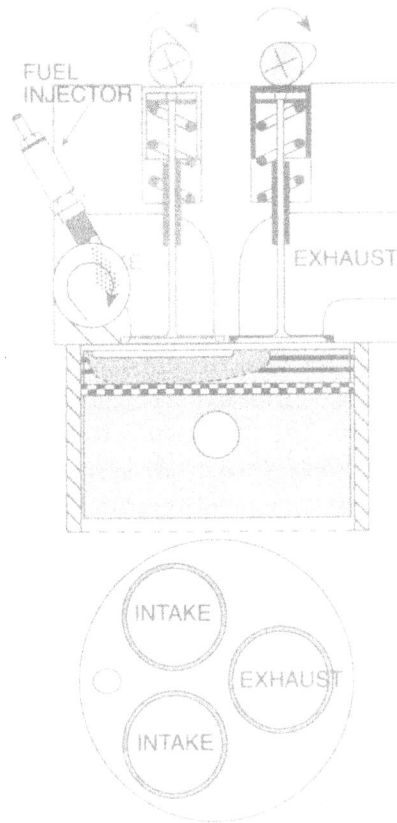

Fig. 1.8 A three-valve indirect-injection (IDI) diesel engine.

There is a price to pay for this smoother, and quieter-running, IDI diesel engine. The pumping process through the small orifice connecting the cylinder and the side chamber is paid for in pumping work put into it during compression. There are further losses on expansion as the combustion charge blows down through the same orifice to the main cylinder during the power stroke. The orifice-to-piston area ratio is quite severe, i.e., the area of the connecting orifice is only about 1% of the bore area. For example, if the cylinder bore is 90 mm, then the orifice diameter would be about 9 mm. The result of the energy consumed during these pumping processes is that there is about a 10% diminution of power and a similar diminution of the fuel efficiency of the engine.

As the air flows into the side chamber through the orifice upon compression, the air temperature falls in correspondence with the pressure drop. This air temperature decrease is considerable, which has implications for both normal running and cold start-up. The result is that the IDI engine requires a compression ratio about one ratio higher than the DI unit, and requires a glow plug to get the combustion going in a cold-start situation.

The design shown in Fig. 1.8 has three valves per cylinder, typically seen in recent PSA (Peugeot-Citroen) automobiles, and is probably the optimum cylinder head layout for this type of engine. Nevertheless, many cylinder head configurations have been proposed for the IDI engine type. Lilly [1.12] should be consulted for more diagrammatic design detail.

1.3.4 Exhaust Emissions

The diesel engine, like its spark-ignition counterpart, is under legislative pressure to conform to ever-tighter emissions standards. Even though it provides very low emissions of carbon monoxide and of hydrocarbons, the diesel engine does emit visible smoke in the form of carbon particulates and measurable levels of nitrogen oxides, the latter being considerably lower for the IDI than the DI engine. The level of emissions of both of these power units is under increasing environmental scrutiny and the diesel engine must conform to more stringent legislative standards by the year 2000. The combination of very low particulate and NO_x emission is a tough R&D proposition for the designer of diesel engines to be able to meet. As the combustion is lean of the stoichiometric mixture by some 50% at its richest setting in order to avoid excessive exhaust smoke, the exhaust gas is oxygen rich and so only a lean burn catalyst can be used on either a two-stroke or a four-stroke diesel engine. This does little, if anything at all, to reduce the nitrogen oxide emissions. This subject is covered more completely in Chapter 4.

1.4 Connecting Rod and Crankshaft Geometry

We have reached the point now where some mathematical treatment of design will begin. This will be conducted in a manner that can be followed by anyone with a mathematics education of university entrance level.

1.4.1 Units Used throughout this Book

Before embarking on this section, a word about units is essential. This book is written in SI units, and all mathematical equations are formulated in those units. Thus, all subsequent equations are intended to be used with the arithmetic values inserted for the symbols of the SI units. Units are listed in the Nomenclature section or within the text. If this practice is followed, then the value computed from any equation will appear also as the strict SI unit listed for that variable on the left-hand side of the equation. Should the user desire to change the unit of the ensuing arithmetic answer to one of the other units listed in the Nomenclature section, a simple arithmetic conversion process can be easily accomplished. One of the virtues of the SI system is that strict adherence to those units, in mathematical or computational procedures, greatly reduces the potential for arithmetic errors. I write this with some feeling as one who was educated with great difficulty, as one of my American friends once expressed it so succinctly, in the British "furlong, hundredweight, fortnight" system of units!

1.4.2 Swept Volume

If the cylinder of an engine has a bore, d_{bo}, and a stroke, L_{st}, as sketched in Fig. 1.9, then the swept volume, V_{sv}, of the cylinder is given by the product of cylinder bore area, A_{bo}, and stroke length:

$$A_{bo} = \frac{\pi}{4} d_{bo}^2 \qquad V_{sv} = A_{bo} \times L_{st} = \frac{\pi}{4} d_{bo}^2 L_{st} \qquad (1.4.1)$$

The total swept volume, V_{tsv}, often referred to simply as the "capacity," of an engine with a number of cylinders, n, is found from:

$$V_{tsv} = nV_{sv} = n\frac{\pi}{4} d_{bo}^2 L_{st} \qquad (1.4.2)$$

1.4.3 Compression Ratio

The geometrical compression ratio, CR, is already defined in Eq. 1.1.1, but this equation can be manipulated so that the clearance volume, V_{cv}, may be calculated from the swept volume of any cylinder and its compression ratio, thus:

$$V_{cv} = \frac{V_{sv}}{CR - 1} \qquad (1.4.3)$$

Theoretically, the actual compression process occurs after the intake valve closes, and the compression ratio after that point can be important in design terms. The volume swept by the piston from intake valve closure to tdc is referred to as the trapped swept volume, V_{ts}. This defines a trapped compression ratio, CR_t, which is then calculated from:

$$CR_t = \frac{V_{ts} + V_{cv}}{V_{cv}} \qquad (1.4.4)$$

1.4.4 Piston Position with Respect to Crankshaft Angle

The piston sweeps up and down the cylinder displacing volume and, in any thermodynamic simulation of an engine, it is vital to know the precise cylinder volume at any crankshaft angular position. The piston is connected to the crankshaft by a connecting rod of length L_{cr}, as seen in Fig. 1.9.

The throw of the crank is simplistically one half of the stroke under most normal circumstances, but is correctly designated as length L_{ct}. As with two-stroke engines, the ratios of connecting rod to crank throw are typically in the range between 3 and 4.

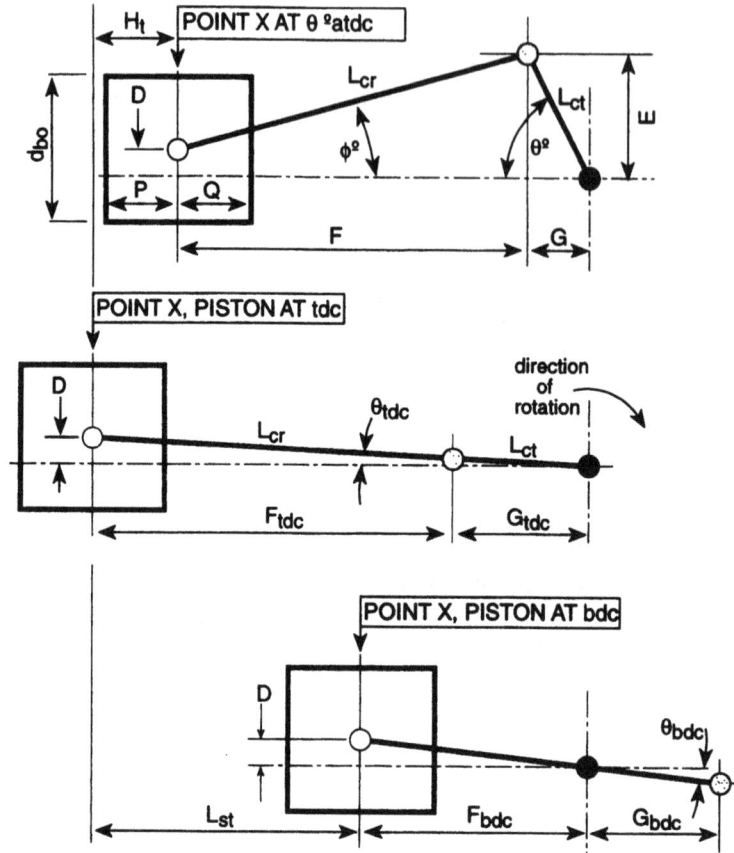

Fig. 1.9 The geometry of a connecting rod-crank mechanism.

At any given crankshaft angle, θ, after the tdc position of the crank, the connecting rod center line assumes an angle ϕ to the cylinder center line. This angle is often referred to in the literature as the "angle of obliquity" of the connecting rod. This is illustrated in Fig. 1.9. The bearing location where the connecting rod attaches to the piston is called the "small end" and, may be offset by an amount D to the cylinder center line. The same geometrical effect is given by the small-end bearing being located centrally on the piston, but the cylinder axis is offset by an amount D from the crankshaft center line. Clearly, it is possible to accommodate numerically any totality of pin and/or cylinder axis offset. Note that the offset value, D, is positive if that offset is toward the direction of crank rotation; this happens to be the situation shown in Fig. 1.9.

When the crank is located at its tdc angular position, i.e., when θ is zero, it is clear from the sketch that the piston may not be at its tdc position, i.e., its furthermost extension up the cylinder. If there is any pin or cylinder axis offset, i.e., if D is not zero, it is clear that this cannot be the case and the tdc and bdc piston positions will occur at θ_{tdc} and θ_{bdc}, respectively.

24

It is also clear that θ_{tdc} and θ_{bdc} are not equal. Using the nomenclature of the sketch it is possible to solve for the following unknown variables.

From the center panel of Fig. 1.9, with the piston at its maximum extension:

$$F_{tdc} + G_{tdc} = \sqrt{\left(L_{cr}^2 + L_{ct}^2 \right) - D^2} \tag{1.4.5}$$

hence

$$\theta_{tdc} = \tan^{-1}\left(\frac{D}{F_{tdc}} \right) \tag{1.4.6}$$

hence

$$G_{tdc} = L_{ct} \cos \theta_{tdc} \tag{1.4.7}$$

and

$$F_{tdc} = L_{ct} \cos \theta_{tdc} + \sqrt{\left(L_{cr}^2 + L_{ct}^2 \right) - D^2} \tag{1.4.8}$$

From the bottom panel of Fig. 1.9, with the piston at its minimum extension:

$$F_{bdc} = \sqrt{\left(L_{cr}^2 - L_{ct}^2 \right) - D^2} \tag{1.4.9}$$

hence

$$\theta_{bdc} = \tan^{-1}\left(\frac{D}{F_{bdc}} \right) \tag{1.4.10}$$

and

$$G_{bdc} = L_{ct} \cos \theta_{bdc} \tag{1.4.11}$$

The stroke of the piston is then given by:

$$L_{st} = F_{tdc} + G_{tdc} - F_{bdc} \tag{1.4.12}$$

If the gudgeon pin offset and/or cylinder axis offset is zero, i.e., if D is zero, then Eqs. 1.4.5-1.4.11, reduce to:

$$\theta_{tdc} = 0 \qquad \theta_{bdc} = 0$$

$$F_{tdc} = L_{cr} + L_{ct}$$

$$G_{tdc} = L_{ct}$$

$$F_{bdc} = L_{cr} - L_{ct}$$

and the length of the stroke becomes, from Eq. 1.4.12:

$$L_{st} = F_{tdc} + G_{tdc} - F_{bdc} = L_{cr} + L_{ct} - \left(L_{cr} - L_{ct}\right) = 2L_{ct} \qquad (1.4.13)$$

Consider the position within the cylinder of any point on a piston with respect to its motion from its tdc position to a point where the crank has turned through an angle θ from the tdc angular position of the crank. For convenience, a point, marked as X, is located at the small-end bearing center and its location down the cylinder from its tdc position is H_t. This is shown sketched on the upper part of Fig. 1.9.

The controlling trigonometric equations are, in solving for the length H_t:

$$H_t = \left(F_{tdc} + G_{tdc}\right) - \left(F + G\right) \qquad (1.4.14)$$

where
$$E = L_{ct} \sin \theta \qquad (1.4.15)$$

where
$$G = L_{ct} \cos \theta \qquad (1.4.16)$$

and
$$F = \sqrt{L_{cr}^2 - (E - D)^2} \qquad (1.4.17)$$

Consequently, using the information in Eq. 1.4.5 and Eqs. 1.4.14-1.4.17, the piston position is:

$$H_t = \sqrt{\left(L_{cr}^2 + L_{ct}^2\right) - D^2} - \sqrt{L_{cr}^2 - (L_{ct} \sin \theta - D)^2} - L_{ct} \cos \theta \qquad (1.4.18)$$

The angle of obliquity of the connecting rod, φ, is given by:

$$\phi = \tan^{-1}\left(\frac{E - D}{F}\right) = \tan^{-1}\left(\left(\frac{L_{cr}}{L_{ct} \sin \theta - D}\right)^2 - 1\right)^{-\frac{1}{2}} \qquad (1.4.19)$$

The angles employed during, or determined during, the analysis of the above equations are in radians. To convert any angle from radians to degrees, or vice-versa, it is useful to remember that 360°, i.e., a full circle, is equivalent to 2π radians.

A more complex trigonometrical question is often asked of the designer or modeller. If the distance from tdc on the piston is measured as H_t, then at what angle θ has the crank turned from its tdc angular position? From Eq. 1.4.14, where k becomes a known constant for any given data set:

$$F = (F_{tdc} + G_{tdc} - H_t) - G = k - G \qquad (1.4.20)$$

then
$$k = F_{tdc} + G_{tdc} - H_t$$

Squaring this equation, and inserting F and G from Eqs. 1.4.15-17 yields:

$$a \sin\theta + b \cos\theta + c = 0 \qquad (1.4.21)$$

where
$$a = -2DL_{ct} \qquad b = 2kL_{ct} \qquad c = k^2 - L_{ct}^2 + D^2 \qquad (1.4.22)$$

then let
$$t = \tan\alpha \qquad \text{where} \qquad \alpha = \frac{\theta}{2}$$

consequently
$$\sin\theta = \frac{2t}{1+t^2} \qquad \text{and} \qquad \cos\theta = \frac{1-t^2}{1+t^2}$$

Substitution of the values related to t into Eq. 1.4.21 reduces it to a quadratic equation, with t being the unknown variable:

$$(c - b)t^2 + (2a)t + (b + c) = 0 \qquad (1.4.23)$$

The solution for the crank angle, θ, becomes

$$\theta = 2 \tan^{-1}\left(\frac{-a \pm \sqrt{a^2 + b^2 - c^2}}{c - b}\right) \qquad (1.4.24)$$

Because the values of a, b, c, and k are all known, the solution for the crank angle, θ, at any piston position, H_t, can be determined directly.

1.4.5 Numerical Data on Piston Position With Respect to Crankshaft Angle
Sinusoidal Motion Compared to Connecting Rod-Crank Motion

Firstly, let us consider a simple numerical example of an engine with a bore dimension of 86 mm, a crank throw of 43 mm, and a connecting rod of length 165 mm. Let us also consider, in this first example, that the gudgeon pin offset, D, is zero. In that case, the piston position at its tdc position coincides with the crank at its tdc angular position. The above equations are solved and the piston position is plotted in Fig. 1.10, as a percentage of the maximum stroke which, by definition, is always a percentage of the swept volume. It can be seen, in Fig. 1.10, that the maximum stroke occurs at 180° crank angle. The stroke is 86 mm and the swept volume is 499.557 cm³. The connecting rod obliquity is calculated from Eq. 1.4.19 as 15.11° and occurs at a crank angle of 90°, and again at 270°, where it is -15.11°.

The motion of the piston is seen to be almost sinusoidal, which would be the solution for piston position in Eq. 1.4.18 if the connecting rod were infinitely long, in which case the value of H_t would be simply $L_{ct}\cos\theta$. This true sinusoidal motion is also plotted in Fig. 1.10 and it is quite clear that intolerable stroke and volumetric errors would occur if simple sinusoidal motion were to be assumed for the motion of the piston. The magnitude of that error is plotted in Fig. 1.11 and the maximum value of it is seen to occur at crank angle positions of 90° and 270° with a value close to 7%.

Effect of Gudgeon Pin Offset

Secondly, let us consider the same numerical example but with gudgeon pin offsets, D, of +2 mm and -2 mm.

If D is +2 mm, the stroke of the engine, L_{st}, becomes 86.007 mm as calculated using Eq. 1.4.12, and the swept volume is now 499.597 cm³ from Eq. 1.4.1. The piston tdc positions are no longer at 0° and 180° on the crank rotation, but at 0.551° and 180.939°, respectively. These angles correspond to those calculated as θ_{tdc} and θ_{bdc} in Eqs. 1.4.6 and 1.4.10. Maximum connecting rod obliquities still occur at 90° and 270° crank angle, but the values are now 14.39° and -15.83°, respectively.

If D is -2 mm, the stroke of the engine, L_{st}, remains at 86.007 mm as does the swept volume at 499.597 cm³. The piston tdc positions are also no longer at 0° and 180° on the crank rotation, but are now at 359.45° and 179.06°, respectively. Maximum connecting rod obliquities still occur at 90° and 270° crank angle, but the values are now changed to 15.83° and -14.39°, respectively.

Fig. 1.12 plots these differences in piston position with respect to a zero pin offset design, as a function of crank angle. The stroke error is shown as a percentage. It can be seen that stroke error is maximized at about 90° and 270° crank angle, with maximum values of ±0.6%.

During engine simulation, because cylinder volume can only be computed from piston position as a function of crank angle, if the gudgeon pin is offset, then the volumetric error, typified by the 0.6% calculated above, cannot be tolerated because this would give rise to computed pressure and temperature errors of at least the same magnitude. During any analysis of measured cylinder pressure diagrams, the error in the crank angle location of the piston

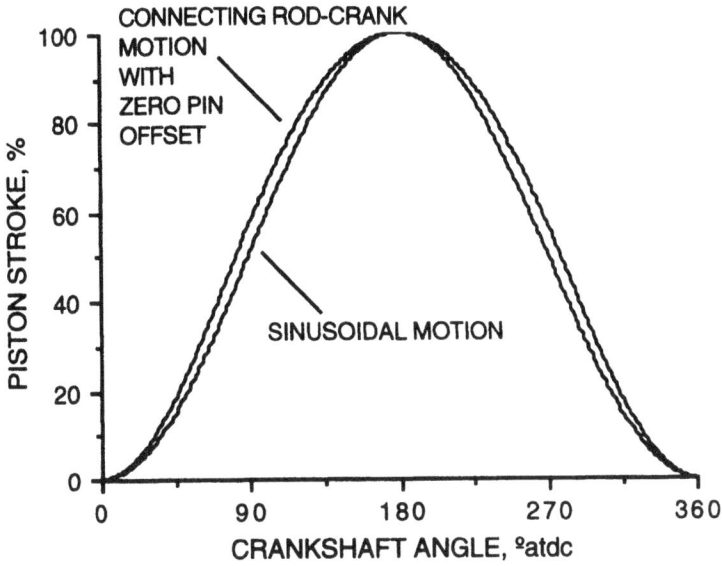

Fig. 1.10 Sinusoidal and connecting rod-crank motion.

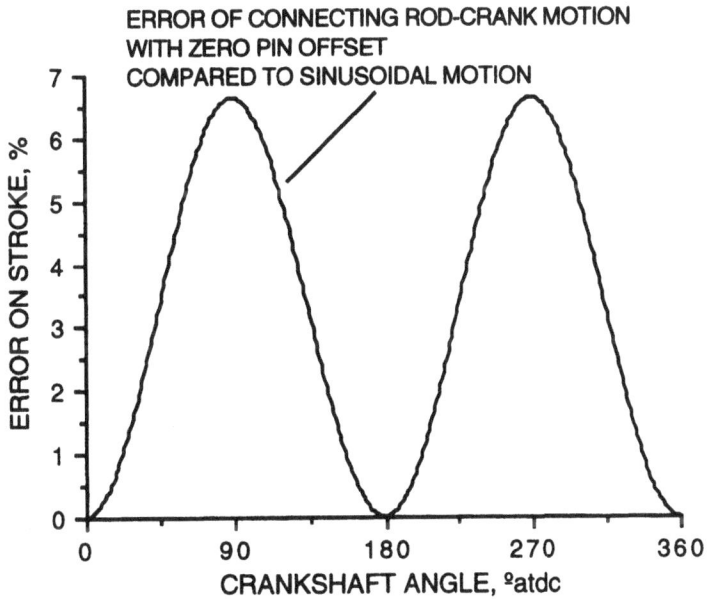

Fig. 1.11 The error by assuming sinusoidal motion for a connecting rod and crank.

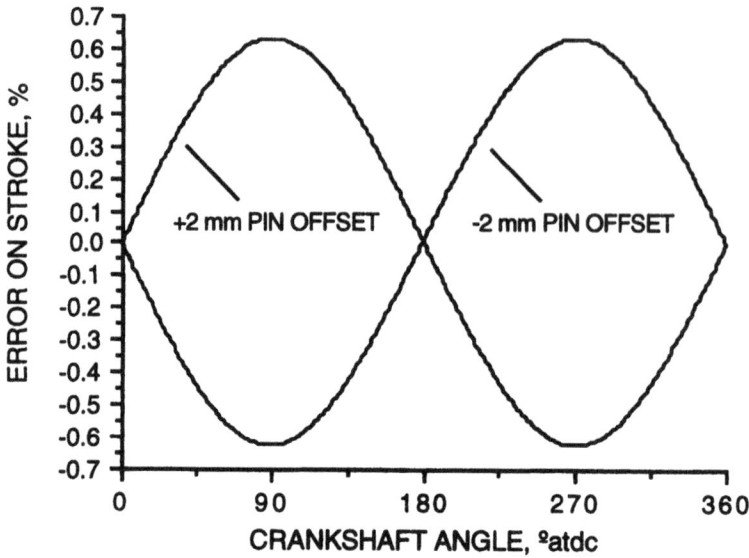

Fig. 1.12 The effect of gudgeon pin offset on piston motion.

position at tdc is known to be very significant [1.13] and will have a major bearing on the accuracy of calculation of engine indicated mean effective pressure and heat release. But I get ahead of myself again, for these matters are discussed later in Chapter 4.

It should be clear from the above that the engine modeller has no option other than to simulate accurately the geometry of the crankshaft and the piston motion using the theory set out in Eqs. 1.4.1-24. Anything less, or the use of assumptions or approximations, leads simplistically to a case of "gigo" syndrome, i.e., "garbage in is garbage out"!

1.5 The Fundamental Geometry of the Cylinder Head

The cylinder head of a four-stroke engine normally contains intake and exhaust poppet valves. As they lift, they expose area between the cylinder and the connecting duct, either an intake or an exhaust pipe, and gas will flow into, or out of, the cylinder depending on the prevailing pressure difference between them. During the simulation of an engine it is essential to be able to calculate these geometrical areas at any juncture in the rotation of the crankshaft. There are two aspects to this requirement. The first is the exposed valve area at any particular valve lift, and the second is the valve lift characteristic with respect to crank angle. This section is dedicated to solving these twin requirements for simulation purposes.

If the engine is a sleeve valve, or a rotary valve, engine you are referred to my book on two-stroke engines [1.9], where the debate on the aperture areas of ports at a cylinder is conducted in great detail, and the relevant calculation procedures are described.

1.5.1 Connecting the Valve Apertures with the Manifold

That four-stroke engines normally contain poppet valves within the cylinder head can be seen in Figs. 1.1-1.8. Parametric detail regarding the valve aperture and connecting duct areas is observed in Fig. 1.13.

In any given cylinder head, the number of valves for either the intake or the exhaust is designated as n_v and the valves are theoretically distinguished by the further subscript appellation of an 'i' or an 'e,' so the number of intake and exhaust valves is n_{iv} and n_{ev}, respectively. Similarly, as the general nomenclature for a port is A_p, the flow area at the port for each of these valves is A_{ip} and A_{ep}, respectively. If there is more than one intake, or more than one exhaust, valve then the total exposed port area is A_{ipt} and A_{ept}, respectively. Normally, the valve(s) connect to a single manifold outlet of area A_m, which gives rise to the nomenclature of A_{em} and A_{im} for the exhaust and intake manifold areas, respectively. This gives rise to the concept of pipe (manifold)-to-port area ratios, C_m, which are defined as follows for the exhaust and intake ducts in the cylinder head:

Intake:
$$C_{im} = \frac{A_{im}}{n_{iv} \times A_{ip}} = \frac{A_{im}}{A_{ipt}} \qquad (1.5.1)$$

Exhaust:
$$C_{em} = \frac{A_{em}}{n_{ev} \times A_{ep}} = \frac{A_{em}}{A_{ept}} \qquad (1.5.2)$$

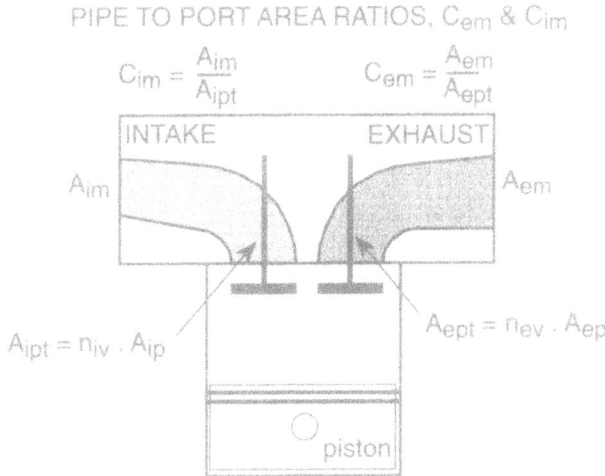

Fig. 1.13 The pipe-to-port area ratios for the four-stroke engine.

As will be shown in later chapters, these manifold-to-port ratios are critical for the performance of an engine for this area ratio, as either an expansion or a contraction, controls the amplitude of any pressure wave created within the ducting by the cylinder state conditions. Because the strength of any such unsteady gas flow tuning is a function of its pressure, the connection becomes obvious.

The bottom line of Eqs. 1.5.1 and 1.5.2 contains a term as yet not clearly defined, namely the throat or minimum port aperture area at any valving, A_{pt}. This is shown sketched on Fig. 1.14.

Inflow or outflow at any valve passes through the valve curtain areas which correspond to the side areas of a frustum of a cone. However, at the highest valve lifts, the minimum flow area is normally that at the inner port where the diameter is d_{ip}, which is further reduced by the presence of a valve stem of diameter d_{st}. These are the controlling aperture areas for the exhaust and intake ports at the valves, which are defined as follows:

$$A_{ept} = n_{ev}A_{ep} = n_{ev}\frac{\pi}{4}\left(d_{ip}^2 - d_{st}^2\right)_{exhaust} \qquad (1.5.3)$$

$$A_{ipt} = n_{iv}A_{ip} = n_{iv}\frac{\pi}{4}\left(d_{ip}^2 - d_{st}^2\right)_{inlet} \qquad (1.5.4)$$

Should any port not be reduced to d_{ip} from the inner seat diameter, d_{is}, then the value of d_{is} should replace d_{ip} in Eq. 1.5.3 or 1.5.4.

1.5.2 The Geometry of the Aperture Posed by a Poppet Valve

The physical geometry of the poppet valve and its location, as shown in Fig. 1.14, is characterized by a lift, L, above a seat at an angle ϕ, which has inner and outer diameters d_{is} and d_{os}, respectively. The valve curtain area, A_t, for this particular geometry is often simplistically, and quite incorrectly, expressed as the side surface area of a cylinder of diameter d_{is} and height L as:

$$A_t \approx \pi d_{is}L \qquad (1.5.5)$$

It is clear from the sketch that it is vital to calculate correctly the geometrical throat area of this restriction, A_t. In Fig. 1.14, the valve curtain area at the throat, when the valve lift is L, is that which is represented by the frustum of a cone defined by the side length dimension, x, the

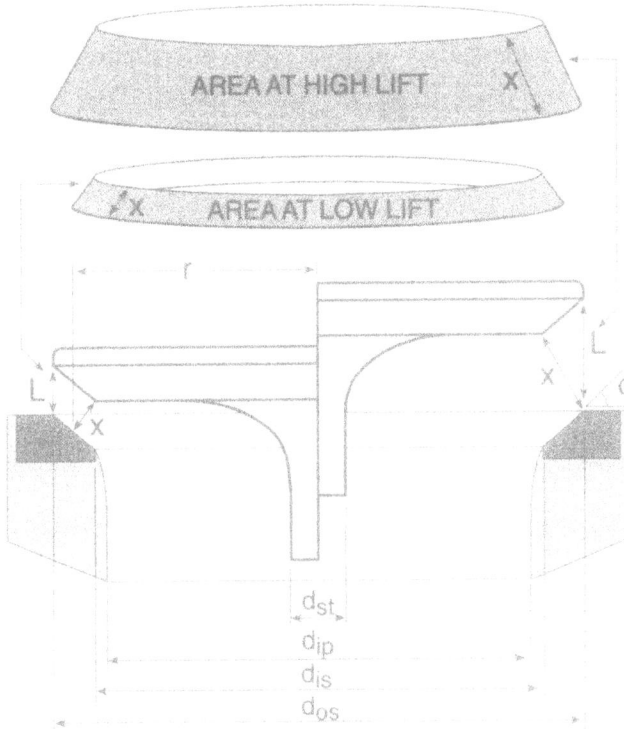

Fig. 1.14 Valve curtain areas at low and high valve lifts.

valve seat angle, ϕ, the inner or outer seat diameters, d_{is} and d_{os}, and the radius, r, all of which depend on the amount of valve lift, L. The side surface area of a frustum of a cone, A_s, is:

$$A_s = \pi \left(\frac{d_{major} + d_{minor}}{2} \right) x \qquad (1.5.6)$$

where x is the length of the sloping side and d_{minor} and d_{major} are its top and bottom diameters.

This is the maximum geometrical gas flow area through the seat of the valve for flow to, or from, the pipe beyond the port, where that minimum port area is A_{ipt} or A_{ept}, as defined above.

The dimension x through which the gas flows has two values, which are sketched in Fig. 1.14. On the left, the lift is sufficiently small that the value x is at right angles to the valve seat and, on the right, the valve has lifted beyond a lift limit, L_{lim}, where the value x is no longer normal to the valve seat at angle ϕ. By simple geometry, this limiting value of lift is given by:

$$L_{lim} = \frac{d_{os} - d_{is}}{2 \sin \phi \cos \phi} = \frac{d_{os} - d_{is}}{\sin 2\phi} \qquad (1.5.7)$$

For the first stage of poppet valve lift where:

$$L \leq L_{lim}$$

the valve curtain area, A_t, is given from the values of x and r as:

$$x = L \cos \phi \qquad (1.5.8)$$

$$r = \frac{d_{is}}{2} + x \sin \phi \qquad (1.5.9)$$

in which case $\qquad A_t = \pi L \cos \phi \left(d_{is} + L \sin \phi \cos \phi \right) \qquad (1.5.10)$

For the second stage of poppet valve lift where:

$$L > L_{lim}$$

the valve curtain area, A_t, is given from the higher value of x as:

$$x = \sqrt{\left(L - \frac{d_{os} - d_{is}}{2} \tan \phi \right)^2 + \left(\frac{d_{os} - d_{is}}{2} \right)^2} \qquad (1.5.11)$$

whence $\qquad A_t = \pi \left(\frac{d_{os} + d_{is}}{2} \right) \sqrt{\left(L - \frac{d_{os} - d_{is}}{2} \tan \phi \right)^2 + \left(\frac{d_{os} - d_{is}}{2} \right)^2} \qquad (1.5.12)$

If the seat angle is 45°, which is conventional, then tan(ϕ) is unity and Eq. 1.5.12 simplifies somewhat.

Thus, the total exposed annular flow areas of the intake and exhaust valves, A_{it} and A_{et}, are given at any particular valve lift by:

$$A_{et} = n_{ev}\left(A_t\right)_{exhaust} \tag{1.5.13}$$

$$A_{it} = n_{iv}\left(A_t\right)_{inlet} \tag{1.5.14}$$

A simple examination of these equations, and particularly the inaccurate approximation of Eq. 1.5.5, reveals that lifting any valve by more than about one quarter of its inner seat diameter would appear to be paradoxical, as the geometrical areas of the throat and port would become equal at that juncture. In practice, valves are lifted much higher to about 0.35 or 0.4 of the value of the inner seat diameter. The reason is that the effective area of flow through the valve seat is reduced by fluid mechanic flow losses, colloquially referred to as, and numerically encapsulated as, a discharge coefficient, Cd [1.16-1.18].

Considering the situation for a single valve layout, the flow "fills" the port or minimum area where that area is A_p. At a conventionally high valve lift, the valve throat area, A_t, normally exceeds that of the port area, A_p. If a discharge coefficient is defined as Cd, and will be seen in Chapter 3 to be measured for the totality of the port and valve throat restriction, then the effective area of that restriction becomes A_{te}, which is defined as follows:

$$\begin{aligned} A_t < A_p \qquad & A_{te} = Cd \times A_t \\ A_t \geq A_p \qquad & A_{te} = Cd \times A_p \end{aligned} \tag{1.5.15}$$

The topic of measurement and application of discharge coefficients is discussed at length in Chapter 3. It is not a trivial subject, but one of great importance, and has come under considerable reexamination in recent times [1.16-1.18].

To ensure accuracy regarding the prediction of mass flow rates and of the magnitude of pressure wave formation, the above analysis of the valve curtain area must be used in any simple empirical analysis and, equally importantly, within any competent engine simulation code, and in the derivation of the measured discharge coefficients [1.16-1.18].

1.5.3 The Lift Characteristics of a Poppet Valve

It can be seen in Fig. 1.4 that a poppet valve can neither be lifted nor dropped instantaneously. The several elements of a typical valve lift diagram are shown in Fig. 1.15.

The valve commences to lift at a crank angle vo, and upon closing returns to zero lift at crank angle vc. In this context, when examining manufacturers' specifications or technical papers [1.34] for valve lift, and the opening and closing crank angle locations, it is often found that these opening and closing points are quoted at some nominal lift value such as 0.5 or 1.0 mm. I have never heard a satisfactory technical explanation of why the valve timings and lifts

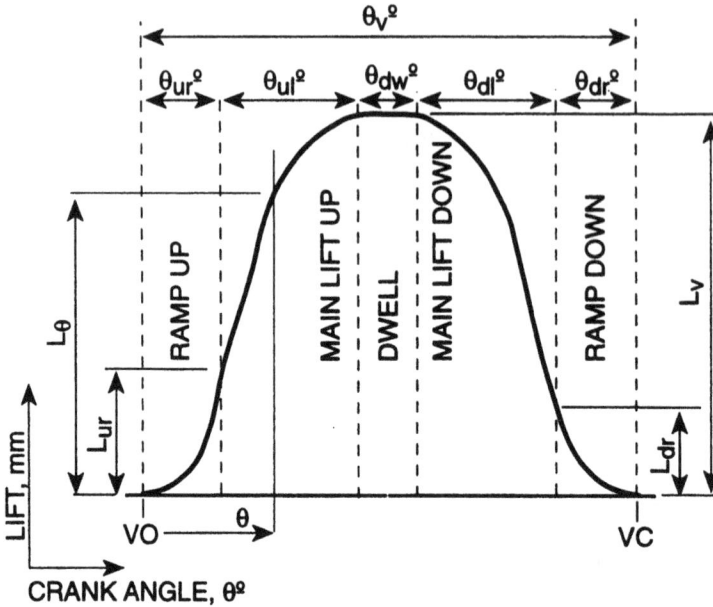

Fig. 1.15 Valve lift characteristics as a function of crank angle.

are quoted in this manner. Because this information is useless to the modeller, who must know the precise static timings of all valves so as to compute the valve aperture areas at any instant during crankshaft rotation, great care must be taken to ensure that the "gigo" syndrome has not developed.

A poppet valve lift is to be found in some five differing phases, as seen in Fig. 1.15. The maximum lift of the valve is L_v and the total duration is $\theta_v°$ crank angle.

The first phase is the opening ramp, designed to lift the valve gently off its seat; perhaps that should be restated as being as rapidly as one dares at the highest speed of engine operation. This phase lasts for a crank angle period of θ_{ur} during which the valve lifts from zero to a height of L_{ur}. In any discussion on valve lift, the conventional units for crank angle are in degrees and valve lift in mm dimensions. Note also that angular periods are quoted as crank rotation, and not cam rotation which is always one-half that of the crankshaft.

The second phase is the main lift from the end of the first ramp to the start of the dwell period around the maximum lift point. During this crank angle period of θ_{ur} degrees, the valve lifts from L_{ur} to L_v.

The third phase is a dwell around peak lift when the valve remains at L_v for period of θ_{dw} crank angle degrees.

The fourth phase is the valve drop from the end of dwell to the commencement of the final ramp. This lasts for θ_{dl} degrees and the valve falls from a lift of L_v to L_{dr}.

The fifth and final phase is the ramp to valve closing, which lasts for θ_{dr} degrees, with the valve falling from a lift of L_{dr} to zero.

In real engines, it is common practice to have the opening and closing ramps be identical in duration and to have similar values for L_{ur} and L_{dr}. The common practice in spark-ignition engines is to have values for ramp duration at 40° and, for such a duration, to have the valve lift by some 20% of the maximum lift value during that period. The common practice in (lower-speed) compression-ignition engines is for the ramp lift to be as high as 50% of the maximum lift. There is no common practice regarding the duration of the dwell period, but 5° is considered to be a fairly common value, although a dwell period is sometimes absent.

1.5.4 The Acceleration and Velocity Characteristics of a Poppet Valve

At any point during the lift, or drop, of a poppet valve, let us assume that the valve movement is dL during a time period dt when the engine speed is N rpm. The engine will rotate for a crank angle period of dθ during this time interval. Time and crank angle are related in the normal manner:

$$\frac{dt}{d\theta} = \frac{60/N}{360} = \frac{1}{6N} \qquad (1.5.16)$$

The velocity of the valve during this interval, c_v, is given by, if the lift is quoted in mm units:

$$c_v = \frac{1}{1000} \times \frac{dL}{dt} = \frac{1}{1000} \times \frac{dL}{d\theta} \times \frac{d\theta}{dt} = \frac{6N}{1000} \times \frac{dL}{d\theta} \quad m/s \qquad (1.5.17)$$

If the velocity of the valve changes by an amount dc_v during the time interval dt then the acceleration or deceleration of the valve, g_v, is given by:

$$g_v = \frac{dc_v}{dt} \times \frac{1}{g} = \frac{dc_v}{d\theta} \times \frac{d\theta}{dt} \times \frac{1}{g} = \frac{6N}{9.81} \times \frac{dc_v}{d\theta} \quad g \qquad (1.5.18)$$

where g is the acceleration due to gravity, which has a value of 9.81 m/s^2.

In practical terms, one has either measured, or has designed, a valve lift characteristic that reduces to a data file, which normally consists of valve lift, L_v, in mm, at crank angle intervals, dθ, which are typically in steps of about one degree. Consider the situation sketched in Fig. 1.16.

Fig. 1.16 consists of three points on a typical valve lift diagram where the lifts are L_1, L_2, and L_3 at crank angle positions θ_1, θ_2, and θ_3, respectively. Using the theory in Eq. 1.5.17 to determine the mean velocity in these two elements, c_{v1} and c_{v2}:

$$c_{v1} = \frac{6N}{1000} \times \frac{L_2 - L_1}{\theta_2 - \theta_1} \qquad c_{v2} = \frac{6N}{1000} \times \frac{L_3 - L_2}{\theta_3 - \theta_2} \quad m/s \qquad (1.5.19)$$

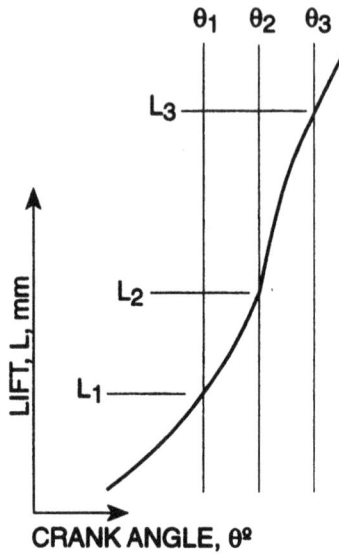

Fig. 1.16 Three adjacent points on a valve lift curve.

The mean acceleration, g_{v12}, for the lift process as motion from the median point of the first element, to the median point on the second element, is given by Eq. 1.5.18 as:

$$g_{v12} = \frac{6N}{9.81} \times \frac{c_{v2} - c_{v1}}{0.5\{(\theta_2 + \theta_3) - (\theta_1 + \theta_2)\}}\ g \qquad (1.5.20)$$

If it happens, as would be normal, that all crank angle intervals on the valve lift data file are identical, and have a value $\Delta\theta$, then Eqs. 1.5.19-1.5.20 reduce to:

$$c_{v1} = \frac{6N}{1000} \times \frac{L_2 - L_1}{\Delta\theta} \qquad c_{v2} = \frac{6N}{1000} \times \frac{L_3 - L_2}{\Delta\theta} \qquad g_{v12} = \frac{6N}{9.81} \times \frac{c_{v2} - c_{v1}}{\Delta\theta}$$

$$(1.5.21)$$

Thus, one can move from point to point on a known valve lift curve and find the velocity and acceleration characteristics at each point upon it, at any given speed of engine operation.

1.5.5 Designing the Valve Lift Characteristics of a Poppet Valve

During the modelling and simulation of an existing engine, the engineer will be certain to have available the measured lift characteristics of the valves in the actual engine. However, if the engine does not exist, the modeller has to be able to rapidly create a realistic valve lift file. Even the process of optimization of an existing engine will almost certainly include extending that optimization to its valving and valve events. So, here also arises the need for the creation of a different, yet realistic, set of valve lift files. The word "realistic" is used to describe a proposed valve lift file that would permit the design and manufacture of the rest of the valve train, i.e., cams, springs, tappets, pushrods, etc., without there being excessive accelerations or velocities on the several elements within it. The traditional route is to design a new cam based on, for example, polydyne principles, including the rest of the valve train in what is a very complex mathematical exercise in mechanics [1.18-1.21]. The modeller, being by instinct a gas dynamicist or a thermodynamicist, does not have the time, nor indeed the training, to carry out this exercise into a theoretical specialism all by itself. One could use a specific CAD design software package if it is readily available, but be assured that even the use of such software involves a very considerable amount of training.

I decided a long time ago that a simpler, yet effective and realistic, method for the rapid creation of a new valve lift file was essential [1.22]. An examination of the valve lift files from many engines showed that, if they were reduced to specific lift L_s, at specific angles θ_s, they exhibited very similar characteristics. Specific lift and specific angle, L_s and θ_s, are defined in terms of the nomenclature seen in Fig. 1.15, where the lift is L_θ at an angle θ from the commencement of lift, and where the maximum lift and valve opening duration are L_v and θ_v, respectively:

$$L_s = \frac{L_\theta}{L_v} \qquad\qquad (1.5.22)$$

$$\theta_s = \frac{\theta}{\theta_v} \qquad\qquad (1.5.23)$$

Further examination revealed that it would be extremely difficult, if not impossible, to formulate a single mathematical expression to accurately describe a specific valve lift curve. Furthermore, it would be very limiting in its coverage of all possible valve lift characteristics, even if it could be accomplished. It is essential to be able to design separately, in a mathematical sense, valve lifts that have differing up and down ramps, and varying amounts of dwell. Thus, the approach typified in Fig. 1.17 was adopted.

Fig. 1.17 shows that specific lift and specific angle relationships are described separately for a ramp period and a lift period. The same polynomial relationship is used for the "ramp down" as is used for the "ramp up," and similarly for the "main lift up" and the "main lift down." The relationship linking specific lift and angle is a third-order polynomial in each case, the coefficients of which are determined from an analysis of measured data. The functions are as follows, as shown in Fig. 1.17:

For ramp up or down: $\qquad L_s = k_{r0} + k_{r1}\theta_s + k_{r2}\theta_s^2 + k_{r3}\theta_s^3 \qquad\qquad (1.5.24)$

For main lift up or down: $L_s = k_{m0} + k_{m1}\theta_s + k_{m2}\theta_s^2 + k_{m3}\theta_s^3 \qquad\qquad (1.5.25)$

$$L_s = k_{m0} + k_{m1}\theta_s + k_{m2}\theta_s^2 + k_{m3}\theta_s^3 \qquad L_s = 1$$

MAIN LIFT $\qquad L_s = 0$

$\theta_s = 0$ $\qquad\qquad\qquad \theta_s = 1$

$L_s = 1$

$L_s = k_{r0} + k_{r1}\theta_s + k_{r2}\theta_s^2 + k_{r3}\theta_s^3$

RAMP $L_s = 0$

$\theta_s = 0$ $\qquad \theta_s = 1$

SPECIFIC LIFT, L_s

SPECIFIC ANGLE, θ_s

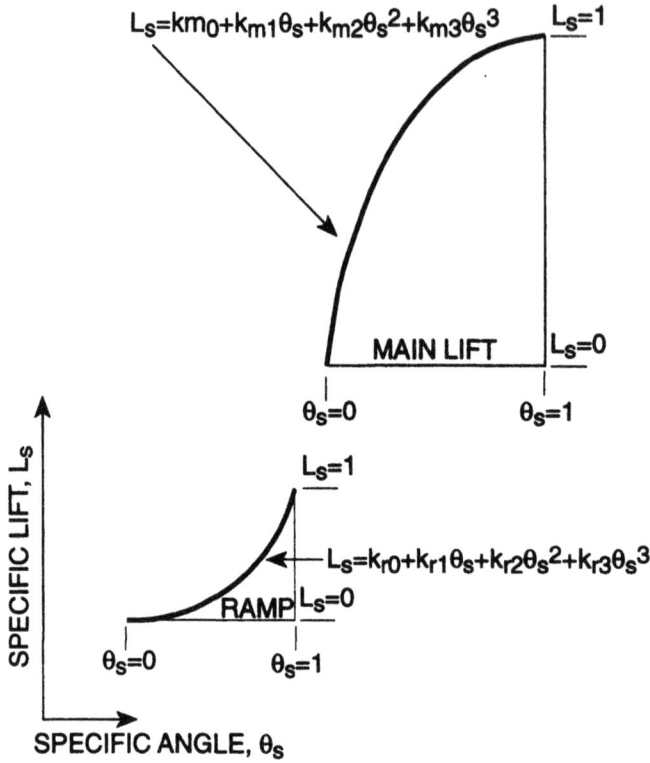

Fig. 1.17 Specific lift characteristics of a poppet valve.

The procedure for simulating any valve lift curve is to decide on the following basic parameters. The nomenclature here is as shown in Fig. 1.15. First, the opening and closing angles of the valve are "designed," i.e., numerical values are assigned to vo and vc, as crank angles atdc, where zero crank angle is tdc on the power stroke and extends for 720°. The valve opening duration, θ_v, is then given by:

$$\theta_v = vc - vo \qquad\qquad\qquad (1.5.26)$$

The duration of the dwell angle, θ_{dw}, is "designed" as a number of crank angle degrees, typically in the range of 0 to 10. The ramp-up and ramp-down periods, θ_{ur} and θ_{dr}, are "designed," i.e., each is assigned a number of crank angle degrees. Similarly, the main lift periods, up and down, θ_{ul} and θ_{dl}, are each assigned a number of crank angle degrees. The totality of these angular values must add up to θ_v, as seen in Fig. 1.15. Thus, the creation of the valve lift diagram, in total period or element duration terms, is completely flexible.

However, custom and practice, not to speak of simplicity of cam profile manufacture, dictates that the ramp periods, and the main lift periods, are normally identical. Hence, using this design process, I usually set the main lift periods to be identical and I normally use 40° for each ramp period. There are two reasons for this: (1) I have rarely found it necessary to do otherwise, and (2) all my polynomial coefficients listed below were reduced from measured data over a 40° crank angle ramp period. Hence:

$$\theta_{ul} = \theta_{dl} = \frac{\theta_v - \theta_{dw} - \theta_{ur} - \theta_{dr}}{2} \qquad (1.5.27)$$

The next "design" decision to be made is the amount of lift to be assigned to the up and down ramps, i.e., L_{ur} and L_{dr}. As stated earlier, as a proportion of the maximum lift, L_v, these numbers are rarely less than 20% for si engines or greater than 50% for ci units. Although there is no numerical problem in assigning different values to the "up" and "down" ramp lift ratios, symmetry of a valve lift profile is much more common in lift characteristics than asymmetry. Asymmetry, normally in the form of a higher lift ratio for the "ramp down" to the "ramp up," is sometimes seen in high-performance racing engines, particularly in Formula 1 engines. The lift ratios for the ramps, up and down, are defined as C_{ur} and C_{dr}, respectively.

Ramp lift ratios: $\qquad C_{ur} = \frac{L_{ur}}{L_v} \qquad C_{dr} = \frac{L_{dr}}{L_v}$ $\qquad (1.5.28)$

With these "design" decisions made, the valve lift curve can be computed by moving from element to element in sequence, as shown in Fig. 1.17, starting with valve opening and the opening "ramp up."

The Valve Lift Commences
At a crank angle θ of zero:

$$\theta = 0 \qquad L_\theta = 0 \qquad (1.5.29)$$

The Opening Ramp Up
At any angle θ, where:

$$0 < \theta \le \theta_{ur} \qquad (1.5.30)$$

Specific angle: $\qquad \theta_s = \frac{\theta}{\theta_{ur}}$ $\qquad (1.5.31)$

Insert θ_s into Eq. 1.5.24 and calculate L_s. Note that the opening point, where θ is zero, is not calculated here, but is positively declared as being zero above. The value of actual lift, L_θ, is found by translating specific lift into actual lift at the crank angle θ:

Actual valve lift: $$L_\theta = C_{ur}L_sL_v \qquad (1.5.32)$$

The Main Lift Up
    At any angle θ, where:

$$\theta_{ur} < \theta \le \theta_{ur} + \theta_{ul} \qquad (1.5.33)$$

Specific angle: $$\theta_s = \frac{\theta - \theta_{ur}}{\theta_{ul}} \qquad (1.5.34)$$

Insert θ_s into Eq. 1.5.25 and calculate L_s. The value of actual lift, L_θ, is found by translating specific lift into actual lift at the crank angle θ:

Actual valve lift: $$L_\theta = L_{ur} + L_s\left(L_v - L_{ur}\right) \qquad (1.5.35)$$

The Dwell Period
    At any angle θ, where:

$$\theta_{ur} + \theta_{ul} < \theta \le \theta_{ur} + \theta_{ul} + \theta_{dw} \qquad (1.5.36)$$

Specific angle: $$\theta_s = 1 \qquad (1.5.37)$$

The value of actual lift, L_θ, is simply the maximum lift:

Actual valve lift: $$L_\theta = L_v \qquad (1.5.38)$$

The Main Lift Down
    At any angle θ, where:

$$\theta_{ur} + \theta_{ul} + \theta_{dw} < \theta \le \theta_{ur} + \theta_{ul} + \theta_{dw} + \theta_{dl} \qquad (1.5.39)$$

Specific angle: $$\theta_s = \frac{\theta_{ur} + \theta_{ul} + \theta_{dw} + \theta_{dl} - \theta}{\theta_{dl}} \qquad (1.5.40)$$

It should be noted that the values of angle and lift are determined by their position from the start of the down ramp, i.e., the computation is operated in reverse for valve drop by comparison with valve lift. Insert θ_s into Eq. 1.5.25 and calculate L_s. The value of actual lift, L_θ, is found by translating specific lift into actual lift at the crank angle θ:

Actual valve lift:
$$L_\theta = L_{dr} + L_s\left(L_v - L_{dr}\right) \qquad (1.5.41)$$

The Ramp Down
At any angle θ, where:

$$\theta_{ur} + \theta_{ul} + \theta_{dw} + \theta_{dl} < \theta < \theta_v \qquad (1.5.42)$$

Specific angle:
$$\theta_s = \frac{\theta_v - \theta}{\theta_{dr}} \qquad (1.5.43)$$

It should be noted that the values of angle and lift are determined by their position from the start of the down ramp, i.e., the computation is operated in reverse for "ramp down" by comparison with "ramp up." Note also that the final point at θ_v is not calculated, but is reserved for a positive "shut" in the next segment below. Insert θ_s into Eq. 1.5.24 and calculate L_s. The value of actual lift, L_θ, is found by translating specific lift into actual lift at the crank angle θ:

Actual valve lift:
$$L_\theta = C_{dr}L_sL_v \qquad (1.5.44)$$

The Valve Shuts
At crank angle θ:

$$\theta = \theta_v \qquad L_\theta = 0 \qquad (1.5.45)$$

The use of positive zeroing of the valve lift curve, at opening and closing, takes care of the numeric problems caused by the polynomial coefficient k_{r0} in Eq. 1.5.24 not being an actual zero when the measured data for valve lift ramps are "best fitted" with a third-order polynomial function.

Smoothing the Computed Valve Lift Curve
Upon combining the several elements of a valve lift curve which are calculated by differing polynomial functions, it would be very surprising if they fitted together to form a smooth lift for the valve, particularly at the junctions between the elements. In other words, the "jagged edge" at the calculated boundary between, say, the opening ramp and the main lift, would provide unacceptably high levels of velocity and acceleration for the valve, if the cam were

ground to imitate precisely the calculated valve lift curve. This effect will be demonstrated numerically in the text and figures below. There is little point in using a "designed" valve lift curve within an engine simulation, so as to compute the engine performance characteristics, if that particular valve lift curve cannot be manufactured in practice. To correct this problem, a simple smoothing routine is easy to formulate and execute at the conclusion of the computation of a valve lift curve.

Consider an extension of Fig. 1.16, regarding a segment of a valve lift curve, to include the greater detail shown in Fig. 1.18.

On this segment of a valve lift diagram, there are three points where the lifts are L_1, L_2, and L_3 at crank angle positions θ_1, θ_2, and θ_3, respectively. Point L_2 may possibly be precisely located on one of those "jagged edges" at an inter-polynomial boundary, as mentioned above. To smooth point 2 so that it fits neatly into the valve lift progression from points 1 to 3, the lift of point 2 is easily adjusted from L_2 to L_{2a} as follows, using similar triangle theory:

$$L_{2a} = L_3 - \frac{\theta_3 - \theta_2}{\theta_3 - \theta_1}\left(L_3 - L_1\right) \qquad (1.5.46)$$

However, as would be conventional in any computation of valve lift, the angular interval between the three points is normally equal, in which case Eq. 1.5.46 simplifies to:

$$L_{2a} = \frac{L_1 + L_3}{2} \qquad (1.5.47)$$

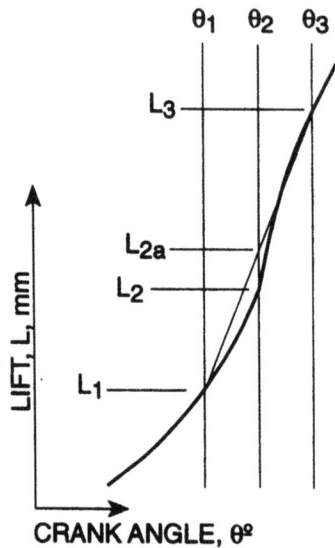

Fig. 1.18 Smoothing the computed valve lift profile.

To computationally smooth any given valve lift curve, the entire valve lift curve is indexed, every grouping of three points is sampled, and the lift of each middle point is adjusted to the mean of the first and last points, according to Eq. 1.5.47.

1.5.6 Numerical Examples of the Computation of Valve Lift and Area

To compute the valve lifts and areas, numbers must be assigned to all of the variables seen in the above equations. Some values refer generically to engine types, i.e., spark-ignition or compression ignition. Useful data for this purpose are presented below.

Values of the Ramp and Lift Coefficients for Spark-Ignition Engines

The numeric values of the coefficients are found from an analysis of measured data from si engines. The values of the coefficients for the ramps, i.e., k_{r0} to k_{r3}, are determined as 0.011172, -0.080336, 0.4686, and 0.6119, respectively. The values of the coefficients for the main lift, i.e., k_{m0} to k_{m3}, are determined as -0.007858, 1.7673, -0.45161, and -0.31158, respectively. Typical values of the lift ratios for the ramps, i.e., both C_{ur} and C_{dr} as defined in Eq. 1.5.28, are 0.2 and they are normally equal providing a symmetrical valve lift profile.

Values of the Ramp and Lift Coefficients for Compression-Ignition Engines

The numeric values of the coefficients are found from an analysis of measured data from ci engines. The values of the coefficients for the ramps, i.e., k_{r0} to k_{r3}, are found from such an analysis as -0.00018552, 0.045892, 1.9795, and -1.0256, respectively. The values of the coefficients for the main lift, i.e., k_{m0} to k_{m3}, are measured in the same manner as -0.0006083, 1.9236, -0.84122, and -0.082123, respectively. Typical values of the lift ratios for the ramps, i.e., both C_{ur} and C_{dr} as defined in Eq. 1.5.28, are almost double those for si engines, being in the range 0.4 to 0.5. They are usually equal providing a symmetrical valve lift profile. Diesel engines normally run slower than gasoline engines so the valves can be opened faster without incurring excessive velocity or acceleration values.

It will be observed that the zero coefficient, i.e., k_{r0} or k_{m0}, is not zero in either the si or the ci case, implying that when θ_s is zero there is some residual lift present. While this appears illogical, if one recalls Eq. 1.5.28 and 1.5.45, it will be noted that the valve lift is positively declared as zero at valve opening and closing, so the potential numeric problem does not arise. The non-zero value of these coefficients comes about by fitting third-order polynomial curves to measured valve lift data using mathematical regression techniques.

Comparison of Measured and Calculated Valve Lifts

Fig. 1.19 shows the intake valve lift from a four-valve DI diesel engine where the valve opens at 15° btdc and closes at 38° abdc, giving a total opening duration of 233° crank angle. It is a large engine with a 127 mm bore and 135 mm stroke dimensions. The measured intake valve lift is 11.1 mm. On Fig. 1.19, the measured data are plotted as a "scatter" graph, i.e., only the individual points are graphed. The calculated data using the theory from Eqs. 1.5.24-1.5.45 are computed at one-degree intervals and drawn as a line only. The lift ratio for the ramps, up and down, is 0.45, with this being a diesel engine. The computed valve lift data has been smoothed ten times sequentially, using the theory from Eqs. 1.5.46 and 1.5.47. It can be seen that the fit, from calculation to measurement, is very good.

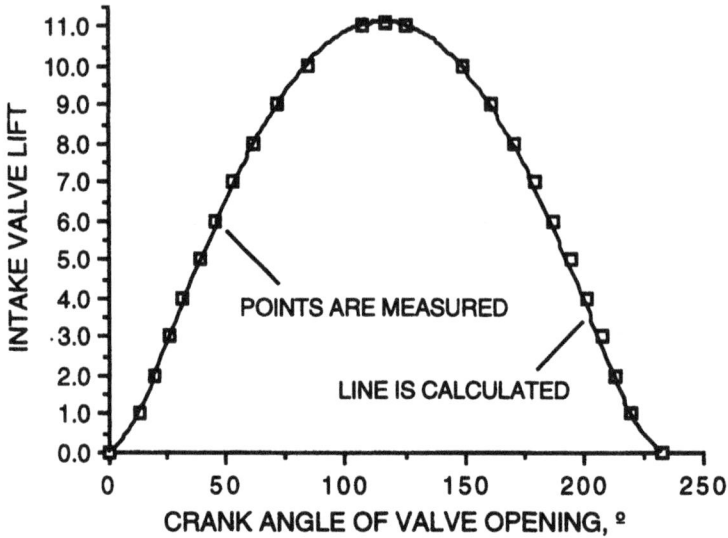

Fig. 1.19 Measured and calculated intake valve lift for a diesel engine.

The engine runs at 3100 rpm, which is quite a high maximum speed for an engine with such a long stroke. This engine speed is used in the calculations that also provide the velocity and acceleration diagrams after the smoothing exercise. They are shown in Fig. 1.20, having been computed using the theory given in Eqs. 1.5.16-1.5.18. The maximum velocities are seen to occur about halfway up the first, and halfway down the final, ramp with a value of 3 m/s. The maximum acceleration takes place even closer to the commencement of lift, and of valve closing. The maximum acceleration is just short of 300 g. Nevertheless, small acceleration ripples can still to be observed at the junctions of the ramp and the main lift periods. While this is possibly unacceptable as a production cam and valve system, any further smoothing needed for the cam would yield an altered profile involving valve lift changes of the order of microns (μm). From a simulation standpoint, this would have a quite negligible effect on the valve flow areas. More detail on this very subject is presented below.

Accuracy of Calculated Valve Flow Areas

Consider an engine that has a bore and stroke of 86 mm, a connecting rod length of 165 mm, runs at 6000 rpm, and has a two-valve hemi-head layout as seen in Fig. 1.2. The intake valve opens, ivo, at 35 °btdc and closes, ivc, at 75 °abdc. Using the notation of Fig. 1.15, it has inner and outer seat diameters, d_{is} and d_{os}, of 36 and 40 mm, and a valve seat angle, ϕ, of 45°. The up and down ramp lift ratios, C_{ur} and C_{dr}, are the same at 0.20 and a "ramp up" and a "ramp down" period of 40° crank angle is used. The dwell period at peak lift is 2° long. The maximum valve lift, L_v, is 13 mm and the duration, θ_v, from the valve timings given above, is 290°. Actually, engines with this cylinder capacity will be featured quite frequently throughout this entire text, with the data structures for them getting ever more detailed with the passing of the pages.

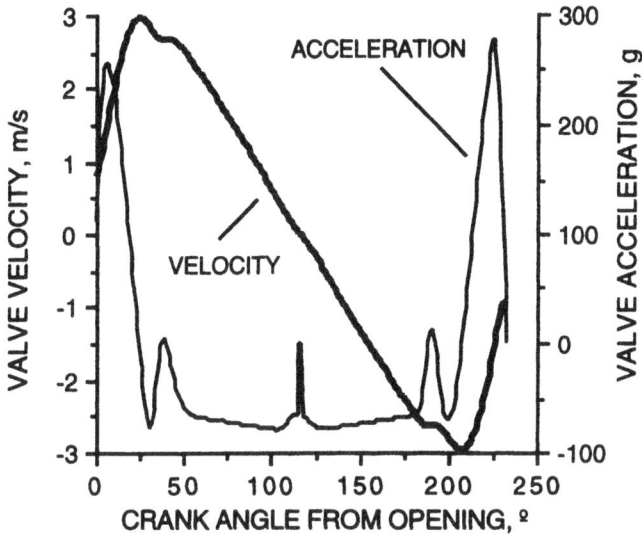

Fig. 1.20 Calculated valve velocity and acceleration diagrams for a diesel engine.

From a simulation viewpoint, the most important issue is to be able to predict accurately the valve flow areas. It is seen in Eq. 1.5.5 that there is a simple, albeit inaccurate, way to obtain the gas flow area presented by a valve as it lifts. An accurate method for this purpose is presented in Eqs. 1.5.10 and 1.5.12. The valve lift profile for this intake valve data is simulated by the methods presented above and subjected to ten smoothing sequences. The intake valve flow areas, A_{it}, are calculated by both the simplistic treatment and the more accurate method. The results are shown in Fig. 1.21.

It can be seen that using the simple method of Eq. 1.5.5 will tend to overpredict the valve flow area by about 10%, i.e., some 130 mm² on about 1300 mm² at the peak valve lift point. Such an error is bad enough. However, when the rest of the curve is examined for the error on flow area, at every point on the lift curve, the simple calculation method is revealed to be totally inadequate. The error at each point on the lift curve, as a percentage, is calculated by the following equation:

$$\text{error} = 100 \times \frac{A_{it_{simple}} - A_{it_{accurate}}}{A_{it_{accurate}}} \qquad (1.5.48)$$

This error function is plotted in Fig. 1.22. Here, it can be seen that the error is unacceptable at the lower values of valve lift where it rises to over 40%, whereas the 10% error at the peak is really its minimum value. Such inaccuracy, in calculating the valve flow area during a simulation, has serious repercussions for the accuracy of computation of the mass flow rate of gas through it, the filling or emptying of the cylinder contents behind it, and all of the thermodynamic state conditions within the cylinder and in the duct.

47

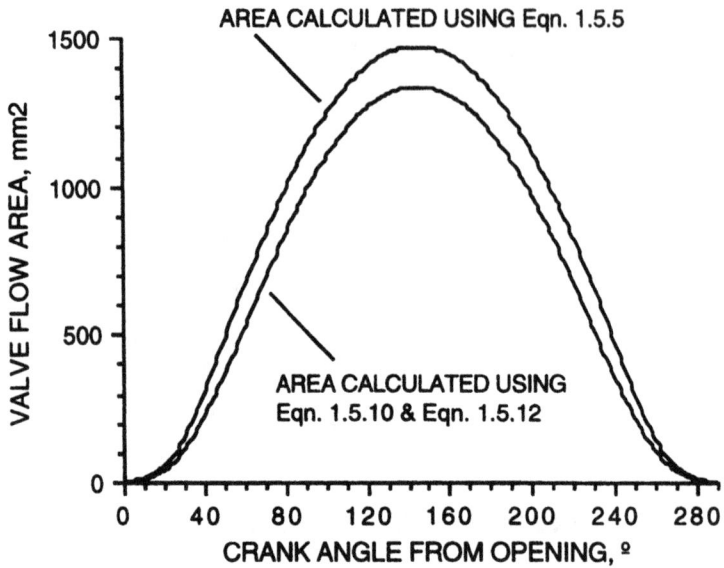

Fig. 1.21 Intake valve flow areas calculated by a simple and an accurate method.

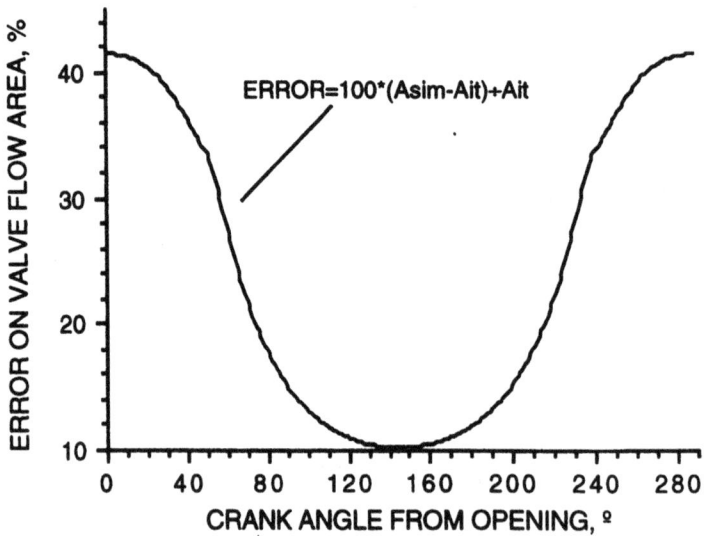

Fig. 1.22 The simple method of calculating valve flow area is inaccurate.

The Use of the Smoothing Technique on the Valve Lift Curve

In Fig. 1.18, the technique for smoothing the computed valve lift curve is sketched. Eqs. 1.5.46 and 1.5.48 describe the arithmetic method employed. Fig. 1.23 shows the result of the smoothing technique for the very intake valve lift data being discussed.

A segment of the computed curve is drawn around the end of the ramp-up period, from 36 to 44 degrees after the valve opens. Without any smoothing of the valve lift curve, the boundary, i.e., the "jagged edge" written about earlier, between the two different polynomials that detail the ramp up and the main lift up is clearly visible at 40° after the valve opening when the first ramp concludes. Because the maximum lift is 13 mm, and the ramp lift ratio is 0.2, the end of ramp lift should be located at 2.6 mm. That such is the case can be seen in Fig. 1.23. On the same figure is plotted the valve lift curve after the smoothing technique has been applied ten times. The "jagged edge" is gone. However, this jagged edge amounts to only 0.07 mm at the 41° point, and this difference in lift translates simplistically into a valve flow area correction of a mere 2.3%. It will be observed that the 2.6 mm lift at the ramp closing point has not been lost by the smoothing technique.

At the peak lift point, where the main lift up and main lift down segments meet the dwell period, the "jagged edge" comment seems even more apt. The unsmoothed valve lift curve is drawn in Fig. 1.24 between 135 and 155 degrees after valve opening.

The peak lift is at 145° crank angle. The jagged edge is very visible, albeit just 0.03 mm high. As a result of applying the smoothing technique ten times, plotted on this same figure, the protrusion has clearly been "filed off," at the expense of losing 0.03 mm off the peak of the valve lift curve, i.e., a valve flow area loss of about 0.23%.

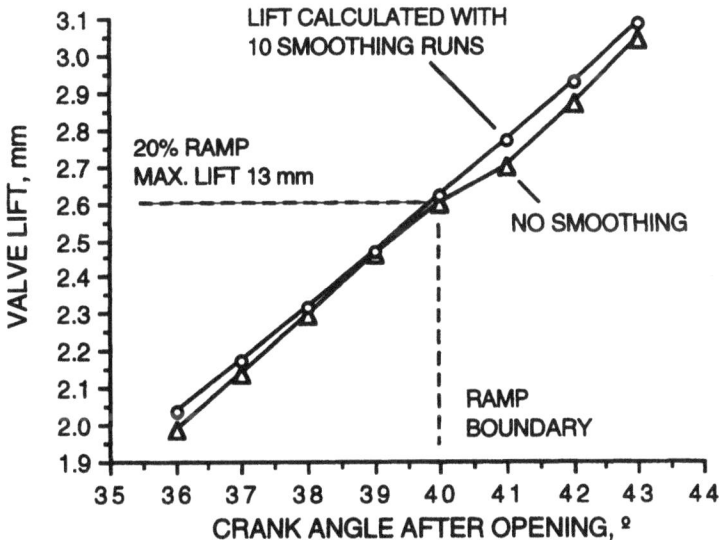

Fig. 1.23 The computational smoothing technique, as applied at a ramp boundary.

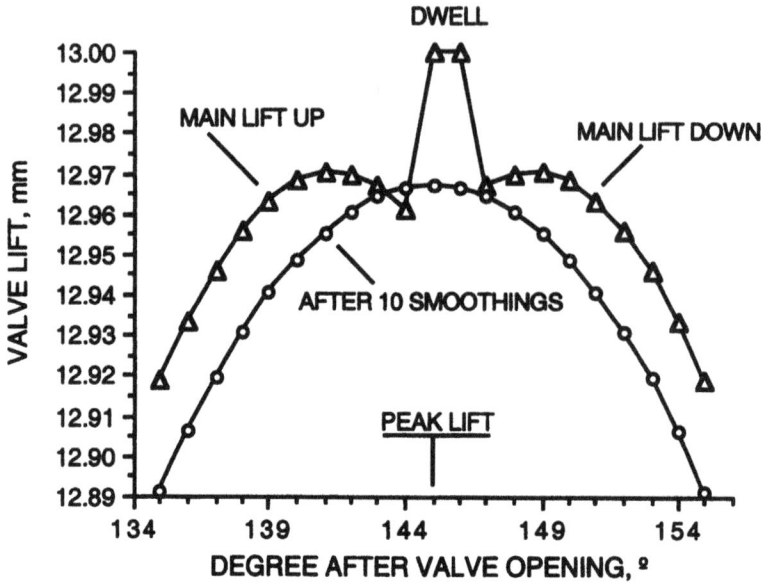

Fig. 1.24 The computational smoothing technique, as applied at peak lift.

Although the amount of alteration to the valve lift curve is dimensionally minimal, the effect on the velocity and acceleration curves is very significant. Figs. 1.25 and 1.26 show graphs of the velocity and acceleration curves for the intake valve operating at 6000 rpm; in both figures the effect of "no smoothing," and the use of ten applications of the smoothing technique, is illustrated.

In Fig. 1.25, the effect of the "jagged edges" at the ends of the ramps, and at peak lift, are clearly seen, as is the result of their removal by the smoothing technique. The velocity incursions that have been deleted are about 1 m/s on a maximum of about 6 m/s, which is about 17% at the several locations. The high-frequency ripple on the velocity curve is significant, as it will induce high acceleration rates locally. This is seen to be the case in Fig. 1.26, for the acceleration incursions run to unacceptably high values of ±3500 g, whereas when the valve lift curve is smoothed the peak accelerations are reduced to no more than ±700 g.

The smoothing technique is observed to be of real practical significance for the engine modeller. Using the simulation techniques described throughout this book for the optimization of the performance characteristics of an engine, the modeller can approach the cam and valve-train maker with a proposed set of valve lift characteristics that are quite practical as a manufacturing prospect. At one time, the engine modeller would have been derided by the mechanical designer for the naiveté of his thoughts on the design of valve train geometry. But not anymore; with these valve lift design techniques those days are history.

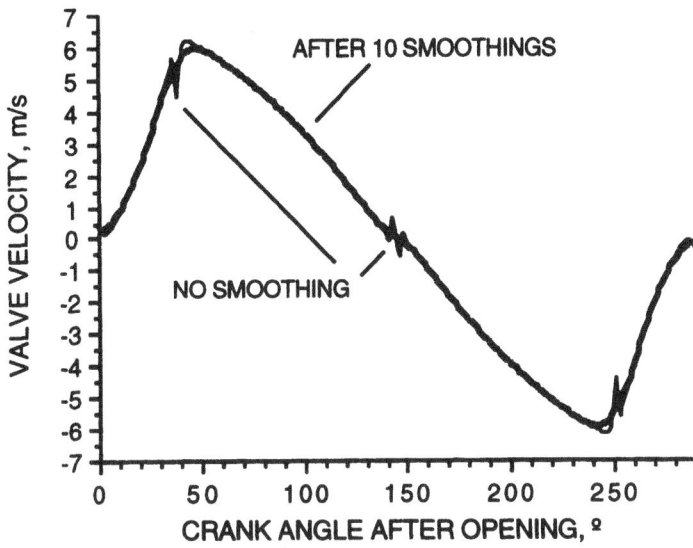

Fig. 1.25 The effect of the smoothing technique on valve velocity.

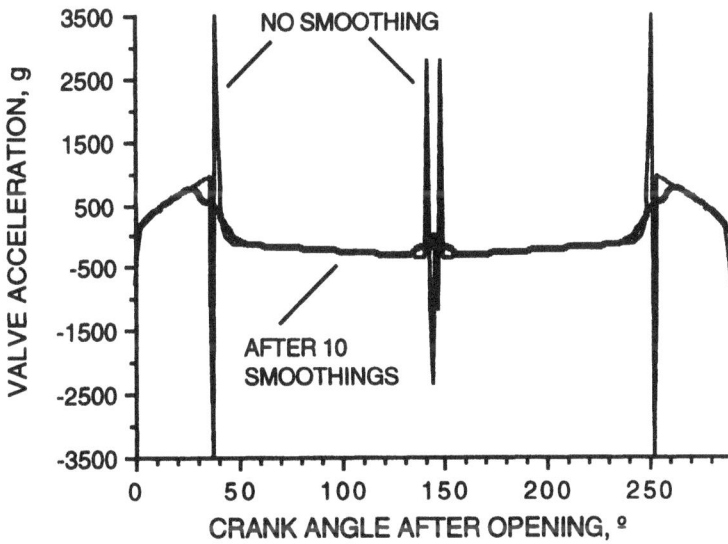

Fig. 1.26 The effect of the smoothing technique on valve acceleration.

The Use of Differing Lift Ratios for the Up and Down Ramps

Sec. 1.5.5 describes the possibility of varying the valve lift profile to be asymmetrical. This is done by setting the lift ratio for the "ramp up" period to be different from that for the "ramp down" event. To clarify this point, within the data for the intake valve of an si engine being used as the current arithmetic example, the present data has the ramp lift ratios for the two ramps, C_{ur} and C_{dr}, as identical at 0.20. These two data values are altered to 0.15 and 0.25, respectively, and the valve lift profile recalculated. The smoothing technique is executed for ten times sequentially and the results are plotted in Figs. 1.27-1.29 for valve lift, velocity, and acceleration.

In Fig. 1.27, the valve lifts for both the symmetrical and the asymmetrical case are plotted and compared. The lower lift during "ramp up," and the higher lift during the "ramp down," are clearly observed and conform to the decreed levels of the ramp lift ratios in both cases. The valve lift is now asymmetrical at 40° and 250° from valve opening, i.e., at the ramp end points, by a difference of 1.3 mm, yet still retains a visibly smooth profile.

You may well have made the not-illogical mental assumption that a lower value of lift ratio for the "ramp up" period, i.e., 0.15 instead of 0.20, would correspond to lower levels of maximum velocity and acceleration for the valve. From Figs. 1.28 and 1.29, this is seen not to be the case. In Figs. 1.28 and 1.29, although the initial values of velocity and acceleration are

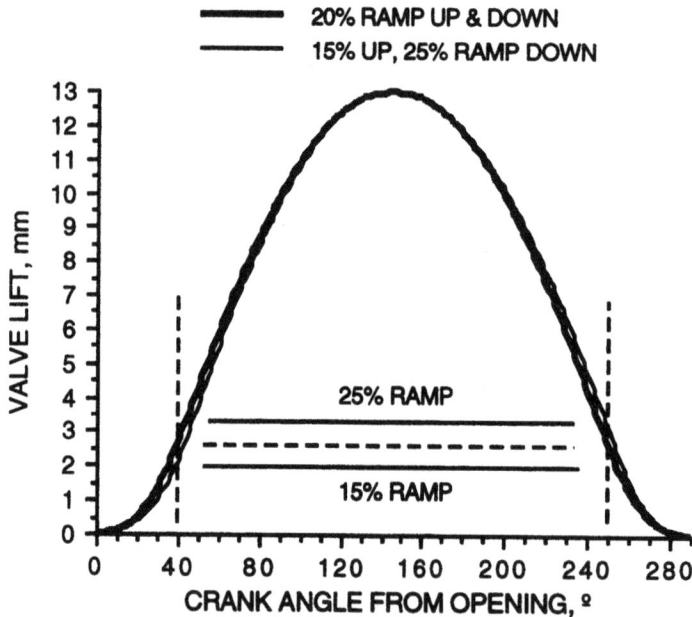

Fig. 1.27 Asymmetrical valve lift profiles.

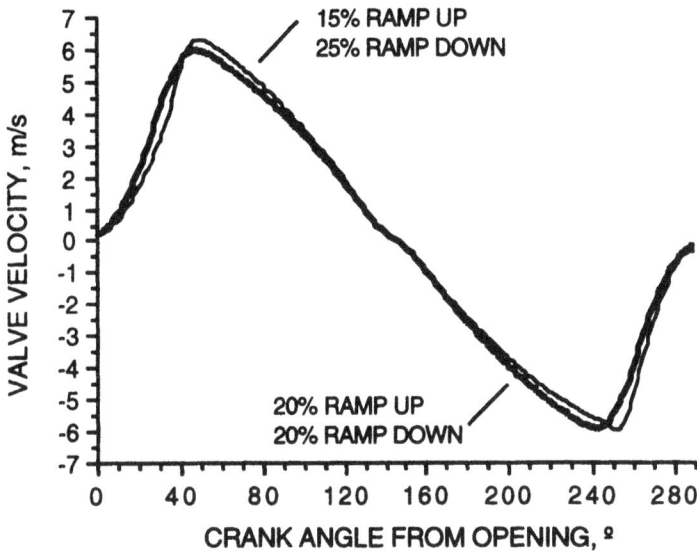

Fig. 1.28 Velocity diagrams for asymmetrical valve lift profiles.

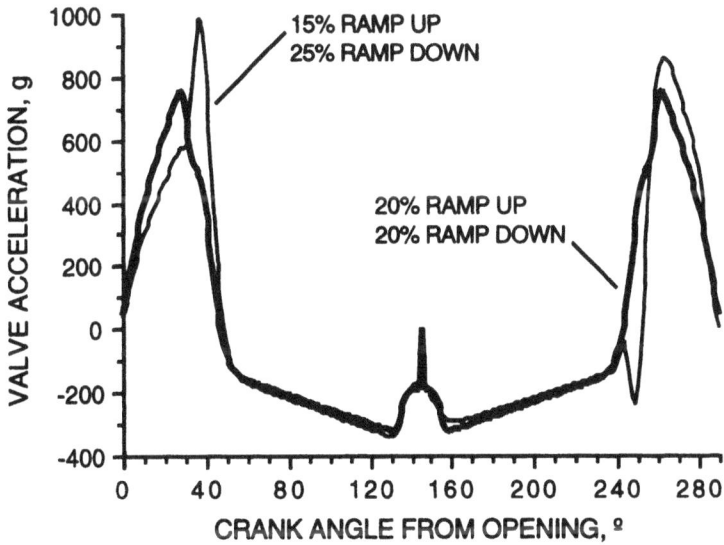

Fig. 1.29 Acceleration diagrams for asymmetrical valve lift profiles.

indeed lower during the first ramp up, because the valve still must lift to the same maximum lift of 13 mm, the velocity and acceleration at the beginning of the "main lift up" period now exceed those when the lift ratio was higher at 0.20. The maximum acceleration on the ramp up now approaches 1000 g, a rise of some 200 g above that for the symmetrical lift profile. This could be considered as excessive for a two-valve automobile engine at a peak operating speed of 6000 rpm. A similar, albeit more obviously intuitive, situation exists at the "ramp down" period, in the sense that a higher lift ratio ought to provide greater velocities and accelerations. In short, as within most engineering designs, compromise always must to be sought between the ideal and the practical.

1.6 Definitions of Thermodynamic Terms Used in Engine Design, Simulation, and Testing

Throughout this text, the units of any parameter to be inserted into thermodynamic equations are in strict SI units and all values of pressure and temperature are absolute values. To ensure a more complete understanding of the terms in the following discussion, you may find it helpful at this point to examine Appendix A1.1, where a precis of some fundamental thermodynamics is presented.

1.6.1 Volumetric Efficiency

In Fig. 1.1(A), the cylinder conducts an intake stroke, in which a mass of fresh air charge, m_{as}, is induced through the intake valve into the cylinder from the atmosphere. This is often known as the "mass of air supplied," hence the subscript notation. The local atmospheric conditions of pressure and temperature are conventionally referred to as ambient conditions. By measuring the local atmospheric pressure and temperature, p_{at} and T_{at}, the air density, ρ_{at}, is given by the thermodynamic equation of state, where R_a is the gas constant for air:

$$\rho_{at} = \frac{p_{at}}{R_a T_{at}} \qquad (1.6.1)$$

The volumetric efficiency, η_v, of the engine defines the mass of air supplied through the intake valve during the intake period by comparison with a reference mass, m_{vref}, which is that mass required to perfectly fill the swept volume under the prevailing atmospheric conditions, thus:

$$m_{vref} = \rho_{at} V_{sv} \qquad \eta_v = \frac{m_{as}}{m_{vref}} \qquad (1.6.2)$$

1.6.2 Delivery Ratio

In the same cylinder, conducting the same intake stroke, consider the same mass of fresh charge, m_{as}, being induced through the intake valve into the cylinder from the atmosphere. The standard reference atmospheric conditions for pressure and temperature, p_{dref} and T_{dref}, are defined as being 101325 Pa (1.01325 bar) and 20°C, respectively. The standard reference air density, ρ_{dref}, will be given by the thermodynamic equation of state, where R_a is the gas constant for air:

$$\rho_{dref} = \frac{p_{dref}}{R_a T_{dref}} \tag{1.6.3}$$

The delivery ratio, DR, of the engine defines the mass of air supplied through the intake valve during the intake period by comparison with a perfect inhalation into the swept volume of a standard reference mass of air, m_{dref}, at the standard reference density, ρ_{dref}. The standard reference mass and delivery ratio can then be expressed as:

$$m_{dref} = \rho_{dref} V_{sv} \qquad DR = \frac{m_{as}}{m_{dref}} \tag{1.6.4}$$

The scavenge ratio, SR, of a naturally aspirated engine defines the mass of air supplied during the intake period by comparison with filling the entire cylinder with a standard reference mass, m_{sref}, that mass which could fill the entire cylinder volume under standard reference atmospheric conditions:

$$m_{sref} = \rho_{dref}(V_{sv} + V_{cv}) \qquad SR = \frac{m_{as}}{m_{sref}} \tag{1.6.5}$$

Should the engine be supercharged or turbocharged, then the new reference mass, m_{bref}, for the estimation of a scavenge ratio for (what is often referred to as) a blown engine, SR_b, is calculated from the state conditions of pressure and temperature of the scavenge air supply, p_s and T_s. This is akin to describing the prevailing atmospheric conditions as being the supercharger (or turbocharger) line pressure and temperature, and the space being filled by it is the entire cylinder volume. These definitions devolve to the following equations:

$$\rho_s = \frac{p_s}{R_a T_s} \tag{1.6.6}$$

$$m_{bref} = \rho_s(V_{sv} + V_{cv}) \qquad SR_b = \frac{m_{as}}{m_{bref}} \tag{1.6.7}$$

Some General Comments Regarding Dimensionless Criteria for Air Flow

I prefer to use the definition of scavenge ratio for the naturally aspirated engine in all circumstances, as given in Eq. 1.6.5, on the grounds that it defines more clearly the amount of air being supplied. For example, if the SR value of a naturally aspirated engine is 1.0 and this engine is then blown to a pressure ratio of 2.0, the SR value, simplistically, would increase to about 2.0, but the SR_b value would remain at unity. I consider that to be both illogical and confusing, as the greater mass flow rate of air is not visible through the SR_b criterion, giving no expectation of a higher power output to be achieved through supercharging.

The same logic is applied to explain the separate functions required of the criteria of volumetric efficiency and delivery ratio. The former is used to denote changes in breathing ability volumetrically, and the latter provides dimensionless information about the mass of air ingested. It is important to note that it is air mass, and not volume, that ultimately devolves to power attained from any engine. Imagine an engine being tested at sea-level at standard atmospheric conditions and then being retested, at the very same engine speed, at an altitude where the local atmospheric density, ρ_{at}, is half that of the standard reference atmospheric value, ρ_{dref}. Let it be assumed that both the volumetric efficiency and the delivery ratio at the sea-level test are unity for both parameters. At the altitude test, if the engine breathed in precisely the same volume of air, which would not be abnormal, the volumetric efficiency would remain at unity, but the delivery ratio would be halved (see Eqs. 1.6.1-1.6.4). Thus, the volumetric efficiency parameter aligns itself with its description, namely it tells us that the engine breathing has been unimpaired with altitude. On the other hand, the delivery ratio criterion informs us that we can expect to achieve only half the sea-level power output of the engine. The reason for this is simple, and will be amplified thoroughly in later discussions. Briefly, because only half the mass of air is ingested in the test at altitude, we can only burn half the mass of fuel at the same air-fuel ratio. Hence, if we have available just half the heat energy in the fuel and burn it at the same combustion efficiency, the end result must approach a halving of the power output.

The SAE standard J604 [1.23] refers to, and defines, delivery ratio. However, the definition it provides is precisely that given above for volumetric efficiency. In the literature on four-stroke engines, delivery ratio and volumetric efficiency are commonly referred to and are sometimes defined in the fashion of Eqs. 1.6.1-1.6.4, sometimes as in SAE J604, and sometimes are not defined at all within this publication. Thus, one should always be skeptical of such numbers in a technical paper if the method of reducing the air flow to the dimensionless criteria of delivery ratio, volumetric efficiency, scavenge ratio and charging efficiency is not clearly defined. The definitions given for delivery ratio and volumetric efficiency through Eqs. 1.6.1-1.6.4 will be adhered to throughout this book.

All of the above theory has been discussed in terms of the ingested air flow compared to that perfectly inhaled into the cylinder swept volume, as if the engine were only a single-cylinder unit. However, if the engine is a multi-cylinder device, then it is the total air flow into the entire engine that is referenced to the total swept volume of the engine under consideration, i.e., that defined as V_{tsv}, in Eq. 1.4.2.

1.6.3 Trapping Efficiency

It may be that all of the air supplied to the cylinder, m_{as}, is not trapped within it at intake valve closure. For example, some may leave through the exhaust valve during the valve over-lap period. Definitions are found in the literature [1.9, 1.14] for trapping efficiency, TE. The mass of delivered air that has been trapped is defined as m_{tas}. Trapping efficiency is the capture ratio of the mass of delivered air that has been trapped by compared to that supplied:

$$TE = \frac{m_{tas}}{m_{as}} \qquad (1.6.8)$$

1.6.4 Scavenging Efficiency and Charge Purity
The Conditions within a Cylinder

The scavenging efficiency, SE, is defined as the mass of delivered air that has been trapped, m_{tas}, compared to the total mass of charge, m_{tr}, that is retained at intake valve closure. The trapped charge is composed of fresh charge trapped, m_{tas}, and the products of combustion from the previous cycle, m_{ex}. These products of combustion have the same composition as the exhaust gas that leaves the cylinder at release and enters the exhaust system. Semantic confusion tends to arise because in lean combustion not all of the air is consumed during a burn process and some air remains unburned from the previous cycle within the "products of combustion," i.e., the "exhaust gas" that departs the exhaust valve when it opens. By the trapping point when the intake valve closes, the mass of unburned air, m_{ar}, retained within the cylinder must be taken into consideration, as it will be available for, and affect, the next combustion process. In short, any combustion process takes place between all of the air in the cylinder and all of the fuel supplied to that cylinder. It is important to define the absolute purity of this trapped charge. The absolute trapped charge purity, Π_{abs}, is defined as the ratio of air trapped in the cylinder before combustion, m_{ta}, to the total mass of cylinder charge, m_{tr}, where:

$$\Pi_{abs} = \frac{m_{ta}}{m_{tr}} = \frac{m_{tas} + m_{ar}}{m_{tas} + m_{ex}} \qquad (1.6.9)$$

Because the absolute trapped charge purity, Π_{abs}, within the cylinder is traced during the course of a simulation by the combination of scavenging efficiency and the separate properties of the air and the combustion products, i.e., the exhaust gas, it is not featured any further within this book as a term to be discussed. However, the common terminology of "purity" appears very frequently and is required to trace the relative proportions of air and combustion products, i.e., the exhaust gas, within the cylinder and in the intake and exhaust ducting. You may well wonder why this is necessary, for the common perception is that the "exhaust gas" leaving the end of the exhaust tailpipe cannot contain any of the air that ever existed in the intake system, nor can any of the air that exists in the intake ducting ever be contaminated by "exhaust gas." By the time you have finished this book, such misconceptions will have been thoroughly dispelled.

Consequently, scavenging efficiency, SE, and the relative charge purity, Π, have common definitions for the in-cylinder conditions, both being defined as the relative mass proportions of the supplied air that has been trapped, m_{tas}, to the totality of the trapped charge:

$$SE = \Pi = \frac{m_{tas}}{m_{tr}} = \frac{m_{tas}}{m_{tas} + m_{ex}} \qquad (1.6.10)$$

In many technical papers and textbooks on two- and four-stroke engines, the word "charge purity" is often used carelessly by the authors, without the above caveats, which assumes prior knowledge on the part of the reader. Either they assume that all trapped charge is air, or that all charge supplied is trapped, or that the value of air retained post-combustion, m_{ar}, is zero. This latter assumption may be generally true for most spark-ignition engines and may well be correct when the combustion process is rich of stoichiometric, but it would not be true for diesel engines where the air is never totally consumed in the combustion process, and for similar reasons, it would not be true for a stratified combustion process in a gasoline-fuelled, direct-injection, spark-ignition (GDI) engine [1.15].

The Conditions within the Engine Ducting

During the valve overlap period, with both valves open, if the intake pressure exceeds the cylinder and exhaust duct pressures, then air can flow from the intake tract through the cylinder into the exhaust pipe. In addition, when the intake valve opens, if the cylinder pressure exceeds that in the intake system, combustion products, i.e., the "exhaust gas," can enter the intake duct and mix locally with its air contents. Hence, the totality of what is loosely-termed "exhaust gas" at some location within an exhaust duct, or "intake air" locally within an intake system, may well be composed of a mixture of "air" and "exhaust gas", i.e., the combustion products. During an engine simulation, all gas movements are tracked and the purity of the charge at any juncture in time and space is one of the properties that is recorded. Consider a specified zone within the ducting. At some instant in time, it contains a mass of air, m_{az}, and a mass of exhaust gas, m_{exz}. The charge purity, Π, is defined as the mass ratio of the air to the total mass present within the zone, and in a similar manner to that for scavenging efficiency and relative charge purity within an engine cylinder:

$$\Pi = \frac{\text{air mass in zone}}{\text{total mass in zone}} = \frac{m_{az}}{m_{az} + m_{exz}} \qquad (1.6.11)$$

Hence, if the zonal gas present is composed of air only it has a purity of unity, whereas if it is totally exhaust gas then its purity is zero. Being somewhat knowledgeable about two-stroke engines where the issue is of even greater importance, I am well aware of the semantic confusion that this topic may cause, so I trust that the thermodynamic debate in Sec. 2.18.9 further clarifies the situation.

1.6.5 Charging Efficiency

Charging efficiency, CE, expresses the ratio of the filling of the cylinder with air, compared to filling the cylinder swept volume perfectly with air at the onset of the compression stroke. For example, the design objective at the peak power point is to fill the cylinder with the maximum quantity of air in order to burn a maximum quantity of fuel with that same air. Hence, charging efficiency, CE, is given by:

$$CE = \frac{m_{tas}}{m_{dref}} \qquad (1.6.12)$$

where the reference mass, m_{dref}, is that defined in Eq. 1.6.4. This may be manipulated to show that charging efficiency is directly related to delivery ratio and trapping efficiency:

$$CE = \frac{m_{tas}}{m_{as}} \times \frac{m_{as}}{m_{dref}} = TE \times DR \qquad (1.6.13)$$

It should be made quite clear, again, that this is another definition that is not precisely as defined in SAE J604 [1.23]. In the nomenclature of that SAE standard, the reference mass is declared to be m_{vref} from Eq. 1.6.2, and not m_{dref} as used from Eq. 1.6.4. My defense of this further clash with the SAE standards is the same as that given above for the definition of delivery ratio, namely that Eq. 1.6.13 reflects a charging efficiency which is basically proportional to the torque produced by the engine, whereas the SAE definition does not necessarily do that.

1.6.6 Air-to-Fuel Ratio

This subject is discussed in much greater detail in Chapter 4, under the topic of Combustion, but some preliminary data is required to make more informative the Otto and Diesel cycle theory which is to be found in the later sections of this chapter.

The Gasoline, or Spark-Ignition, Engine

It is important to realize that there are narrow limits of acceptability for the spark-initiated combustion of air and a fuel such as gasoline. In the case of the gasoline used in the spark-ignition engine, the ideal fuel is octane, which is the eighth member of the family of paraffins whose general chemical formula is C_nH_{2n+2}. Consequently, octane is C_8H_{18}, which burns "perfectly" with air in a balanced chemical relationship, called the stoichiometric equation. Most students will recall that air is composed, volumetrically and molecularly, of 21 parts oxygen and 79 parts nitrogen, ignoring the (1%) argon and other trace gases commonly found in the atmosphere. Hence, the chemical equation for complete combustion becomes:

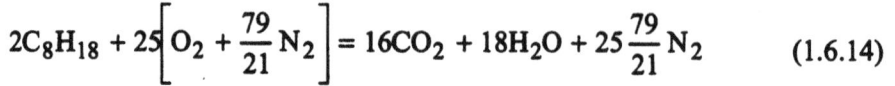

$$2C_8H_{18} + 25\left[O_2 + \frac{79}{21}N_2\right] = 16CO_2 + 18H_2O + 25\frac{79}{21}N_2 \qquad (1.6.14)$$

This produces the information the ideal stoichiometric air-to-fuel ratio, AFR, is such that for every two molecules of octane, twenty-five molecules of air are needed. This information is normally needed in mass terms. Because the molecular weights of O_2, H_2, and N_2 are simplistically 32, 2, and 28 respectively, and the atomic weight of carbon C is 12, the air-to-fuel ratio in mass terms becomes:

$$AFR = \frac{(25 \times 32) + \left(25 \times 28 \times \frac{79}{21}\right)}{2(8 \times 12 + 18 \times 1)} = 15.06 \qquad (1.6.15)$$

Because the equation is balanced, with the exact amount of oxygen being supplied to burn all of the carbon to carbon dioxide and all of the hydrogen to steam, such a burning process yields the minimum values of carbon monoxide, CO, emission and unburned hydrocarbons, HC. Mathematically speaking they are zero, and in practice they are also at a minimum level. Because this equation would also produce the maximum temperature at the conclusion of combustion, this gives the highest value of emissions of NO_x, the various oxides of nitrogen. Nitrogen and oxygen combine at high temperatures to give such gases as N_2O, NO, etc. Such statements, although based in theory, are almost exactly true in practice, as illustrated by the discussion in the Appendices to Chapter 4.

As far as combustion limits are concerned, although Chapter 4 will delve into this area more thoroughly, it may be helpful to point out at this stage that the rich misfire limit of a spark-initiated gasoline-air combustion probably occurs at an air-fuel ratio of about 9, peak power output at an air-fuel ratio of about 13, peak thermal efficiency (or minimum specific fuel consumption) at the stoichiometric air-fuel ratio of about 15, and the lean misfire limit at an air-fuel ratio of about 18. The air-fuel ratios quoted are those in the combustion chamber, at the onset of combustion of a homogeneous charge, and are referred to as the trapped air-fuel ratio, AFR_t. The air-fuel ratio derived in Eq. 1.6.15 is, more properly, the trapped air-fuel ratio, AFR_t, needed for stoichiometric combustion.

To briefly illustrate this point, in the engine shown in Fig. 1.1 it would be quite possible to scavenge the engine thoroughly with fresh air blown in under pressure and then supply the appropriate quantity of fuel by direct injection into the cylinder to provide an AFR_t of, say, 13. Due to a generous oversupply of scavenge air, the overall air-fuel ratio, AFR_o, could well be in excess of, say, 20.

The Diesel, or Compression-Ignition, Engine

In the case of the fuel used in the compression-ignition engine, the ideal diesel fuel is dodecane, the twelfth member of the family of paraffins, $C_{12}H_{26}$, which has the following stoichiometric equation:

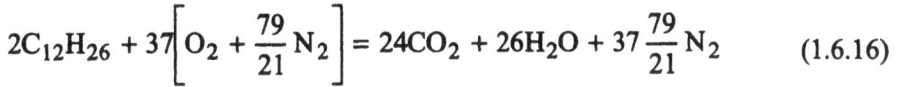

$$2C_{12}H_{26} + 37\left[O_2 + \frac{79}{21}N_2\right] = 24CO_2 + 26H_2O + 37\frac{79}{21}N_2 \qquad (1.6.16)$$

The ideal molecular stoichiometric air-to-fuel ratio for dodecane is such that thirty-seven molecules of air are needed for every two molecules of fuel. This translates to an AFR in mass terms of:

$$AFR = \frac{37 \times 32 + 37 \times 28 \times \dfrac{79}{21}}{2(12 \times 12 + 26 \times 1)} = 14.95 \qquad (1.6.17)$$

It will be observed that the stoichiometric AFR for gasoline and diesel fuel are virtually identical. Unlike gasoline, the rich limit for diesel combustion, in order to avoid an illegal excess of carbon particulates, i.e., a black smoke from the exhaust, is actually well lean of this stoichiometric AFR. It is at an air-fuel ratio of at least 20, and is probably closer to 23 in most modern engines designed to meet emissions legislation. Although an AFR of 20 may be the best available peak power point, the air-fuel ratio for peak thermal efficiency (or minimum specific fuel consumption) is probably between 23 and 25. The air-fuel ratio at idle (zero net power output) condition is at about 60. By definition, because the air-fuel ratio is always lean of stoichiometric, there is air left unused at the conclusion of any diesel combustion process.

Because the fuel is directly injected into the diesel cylinder, and because most diesel engines are "blown," the trapped air-fuel ratio, AFR_t, is nearly always different from the overall air-fuel ratio, AFR_o; indeed this latter ratio is of very little use in diesel engine design, development, or simulation.

Equivalence Ratio

In Sec. 1.9.1.5, is a preliminary discussion on the effect of using differing air-fuel ratios in spark-ignition engines, and in Sec. 1.9.2.2 is a similar debate regarding diesel engines. The concept of equivalence ratio is defined there, which numerically compares any given air-fuel mixture to its stoichiometric value.

1.6.7 Cylinder Trapping Conditions

The point of the foregoing discussion is to make you aware that the net effect of the cylinder induction process is to fill the cylinder with a mass of air, m_{tas}, within a total mass of charge, m_{tr}, at the trapping point where the intake valve closes. This total mass is highly dependent on the state conditions at trapping of pressure, p_{tr}, and temperature, T_{tr}, as the equation of state shows:

$$m_{tr} = \frac{p_{tr}V_{tr}}{R_{tr}T_{tr}} \qquad (1.6.18)$$

where:

$$V_{tr} = V_{ivc} + V_{cv} \qquad (1.6.19)$$

In any given situation, the trapping volume, V_{tr}, is a constant. This is also true of the gas constant, R_{tr}, for gas at the prevailing gas composition at the trapping point. The gas constant for exhaust gas, R_{ex}, is almost identical to the value for air, R_a. Because the cylinder gas composition is usually mostly air, the treatment of R_{tr} as being equal to R_a invokes little error. For any one trapping process, over a wide variety of breathing behavior, the value of trapping temperature, T_{tr}, would rarely change by 5%. Therefore, the value of trapping pressure, p_{tr}, is the significant variable. As stated earlier, the value of trapping pressure is directly controlled by the pressure wave dynamics of the intake and exhaust systems, be it a single-cylinder engine with, or without, tuned intake and exhaust systems, or a multi-cylinder power unit with a branched exhaust manifold. The methods of design and analysis for such complex systems are discussed in Chapters 2, 5, and 6. The value of the trapped fuel quantity, m_{tf}, can be determined from:

$$m_{tf} = \frac{m_{ta}}{AFR_t} \qquad (1.6.20)$$

1.6.8 Heat Released During the Burning Process

The total value of the heat that can be released from the combustion of this quantity of fuel is Q_R, and is controlled by combustion efficiency, η_c, and the (lower) calorific value, C_{fl}, of the fuel in question:

$$Q_R = \eta_c m_{tf} C_{fl} \qquad (1.6.21)$$

Sec. 1.9.1.6 presents a preliminary discussion of values that can be ascribed to combustion efficiency for spark-ignition engines, with respect to the air-fuel ratio found during any given burning process. It will come as no surprise that, in the real world, even the most perfect spark-initiated combustion process does not provide a combustion efficiency of unity! On the other hand, in Sec. 1.9.2 and in Eq. 1.9.9, it can be seen that the combustion efficiency, in the compression ignition of very lean mixtures, does approach unity.

1.6.9 The Thermodynamic Cycle for the Four-Stroke Spark-Ignition Engine

This is often referred to as the Otto cycle, and a full discussion can be found in many undergraduate textbooks on internal-combustion engines or thermodynamics, such as those by Obert [1.2], Taylor [1.3], or Heywood [1.25]. A precis of the thermodynamic theory for the ideal Otto cycle is given in Appendix A1.1, Sec. A1.1.8. The result of the calculation of a theoretical ideal Otto cycle can be observed in Figs. 1.30 and 1.31, by comparison with measured pressure-volume data from an engine with the same compression ratio. The correlation between measurement and calculation is not good, which you may well feel is an inauspicious start to the topic of simulation—in a book that has been declared as being dedicated to this topic!

In the theoretical case of the ideal Otto cycle—and this is clearly visible on the log(p)-log(V) plot in Fig. 1.31—the following assumptions are made regarding each stroke of the cycle:

(a) The compression stroke begins at bdc and is an isentropic, i.e., adiabatic, process.

(b) All heat release (combustion) takes place at constant volume at tdc.

(c) The expansion stroke begins at tdc and is an isentropic, i.e., adiabatic, process.

(d) A heat rejection process (exhaust) occurs at constant volume at bdc.

The compression and expansion processes occur under ideal, or isentropic and adiabatic, conditions with air as the working fluid, and so those processes are calculated, using the theory in Appendix A1.1, Sec. A1.1.3, as:

$$pV^\gamma = a \text{ constant} \tag{1.6.22}$$

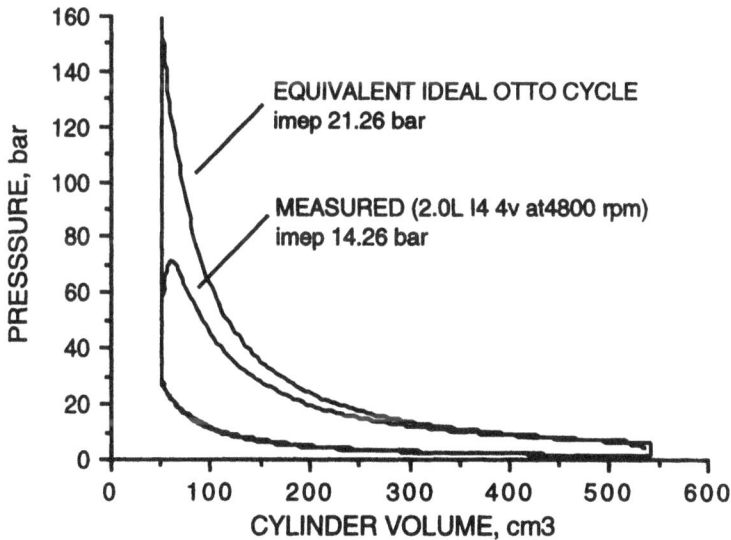

Fig. 1.30 Ideal Otto cycle comparison with experimental data.

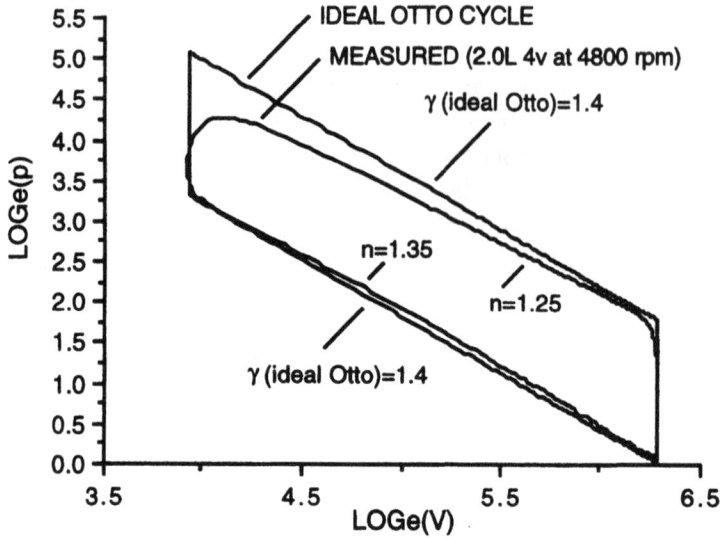

Fig. 1.31 Logarithmic plot of pressure and volume.

The exponent is the ratio of specific heats, γ, which has a value for air of 1.4. The theoretical analysis in Appendix A1.1, Sec. A1.1.8 shows that the thermal efficiency, η_t, of the cycle is given by:

$$\eta_t = 1 - \frac{1}{CR^{\gamma-1}}$$

(1.6.23)

where thermal efficiency is defined as:

$$\eta_t = \frac{\text{work produced per cycle}}{\text{heat available as input per cycle}}$$

(1.6.24)

In the measured case in Figs. 1.30 and 1.31, the cylinder pressure data is taken from a 2000-cm³ four-cylinder, four-valve, naturally aspirated, spark-ignition, four-stroke engine running at 4800 rev/min at wide-open throttle. Because the actual 2-liter engine has a compression value of 10, and from Eq. 1.6.23 the theoretical value of thermal efficiency, η_t, is readily calculated as 0.602, the considerable disparity between fundamental theory and experimentation becomes apparent, for the measured value is about one-half of that calculated, at around 30%.

Upon closer examination of Figs. 1.30 and 1.31, the theoretical and measured pressure traces look somewhat similar and the experimental facts do approach some of the presumptions of the ideal theory. However, the measured expansion and compression indices are at

1.25 and 1.35, respectively, which is rather different from the ideal value of 1.4 for an isentropic process in air. The data can be measured directly from the $\log_e(p)$-$\log_e(V)$ graph as the value of a polynomial, or an isentropic, index is the slope of the line on a log-log diagram. This is shown by manipulating Eq. 1.6.22:

$$\log_e p = -\log_e V^\gamma + \text{constant} = -\gamma \log_e V + \text{constant} \qquad (1.6.25)$$

From Appendix A1.1, Sec. A1.1.6, it is shown for the compression process, where the polytropic index is 1.35 and is less than the isentropic index of 1.4, that heat is being lost from the air in the cylinder to the walls. However, during the expansion process, because the polytropic index is 1.25 which is less than γ for air at 1.4, there is an implication from the same theory in Appendix A1.1, Sec. A1.1.6 that heat is being added to the cylinder gas from the walls. This is not the case. In the real engine, as will be shown in later chapters, the gas in the cylinder during expansion is not air but combustion products as in Eq. 1.6.14, and the γ value of that very hot burned gas is probably in the region of 1.2. Consequently, heat is still being lost to the walls during the expansion process.

The theoretical assumption of a constant-volume process for the combustion and exhaust processes is clearly in error compared to the data on the experimental pressure trace. The peak cycle pressures, calculated as 160 bar and measured as 70 bar, are demonstrably very different. Later in this book, but particularly in Chapters 4 and 5, more advanced theoretical analyses will be seen to approach the measurements ever more closely.

The work on the piston during the cycle is ultimately, and ideally, the work delivered to the crankshaft by the connecting rod. The word "ideal" in thermodynamic terms means that the friction or other losses, like leakage past the piston, are not taken into consideration. Therefore, the ideal work produced per cycle, as in Eq. 1.6.24, is the work carried out on the piston from the force F, created by the gas pressure p. Work, δW, is always the product of force and the distance moved by that force, dx. Hence, where A is the piston area:

$$\delta W = F dx = pA dx = p dV \qquad (1.6.26)$$

This is illustrated in Fig. 1.32, where the measured p-V diagram for the 2-liter car engine is repeated and further information is added.

The element of work, δW, shown above to be equal to the product, pdV, is seen to be an elemental area on the pressure-volume diagram. Therefore, the work produced during the cycle, which is for one crankshaft revolution only from tdc to tdc and includes only the compression and power strokes, is the cyclic integral of these elemental areas over the entirety of the pressure-volume diagram above the piston in the cylinder. The work, W_i, produced during the cycle is evaluated over the power mechanical cycle as:

$$W_i = \oint p dV = \int_{bdc}^{tdc} p dV + \int_{tdc}^{bdc} p dV \qquad (1.6.27)$$

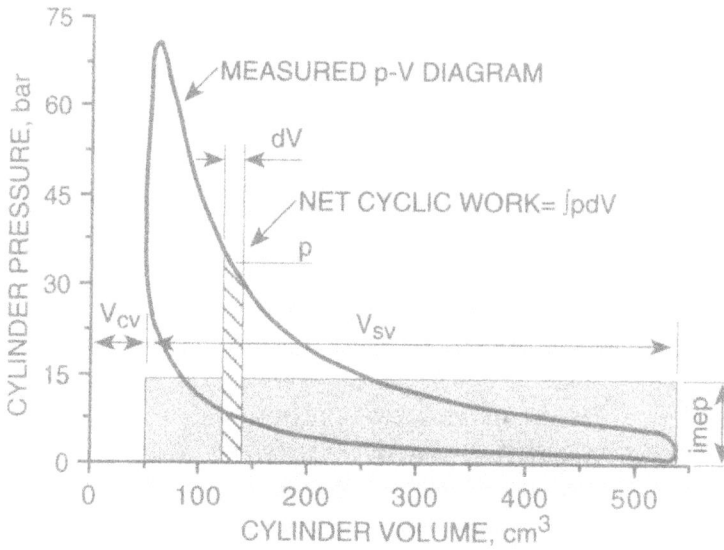

Fig. 1.32 Determination of imep from the cylinder pressure diagram.

Because the value of volume change, δV, during the compression stroke is negative—it is decreasing with every elemental step—the work put into the compression process is also negative. The compression work, W_c, is the first half of the cyclic integral given above:

$$W_c = \int_{bdc}^{tdc} p dV \qquad (1.6.28)$$

The value of volume change, dV, during the power stroke is positive—it is increasing with every elemental step—so the work produced during the power stroke, i.e., the expansion process, is positive. The expansion work on the power stroke, W_e, is the second half of the cyclic integral given above:

$$W_e = \int_{tdc}^{bdc} p dV \qquad (1.6.29)$$

The work produced from the ideal Otto cycle can now be seen as the enclosed area of the pressure-volume diagram, from the addition of a value of W_e, which is positive, and W_c, which is negative. The work produced during the ideal Otto cycle is evaluated only for the compression and power strokes, because no consideration is given to any work done during the exhaust or intake strokes, particularly because they are both assumed to have a zero work content in an ideal engine.

Pumping Work on the Exhaust and Intake Strokes

A real engine must inhale air and exhale exhaust gas. The work this takes to accomplish is called pumping work. The work required to remove the exhaust gas and to induce air must be taken into account. In Fig. 1.33 is graphed the pumping loop from a real engine, i.e., the cylinder pressure-volume diagram for the exhaust stroke and the intake stroke.

The pumping work required, W_p, is calculated over the pumping mechanical cycle as:

$$W_p = \oint pdV = W_{ex} + W_{in} = \int_{bdc}^{tdc} pdV + \int_{tdc}^{bdc} pdV \qquad (1.6.30)$$

The pumping cycle takes two strokes, the exhaust and the intake stroke, where the work required in each stroke, W_{ex} and W_{in}, respectively, makes up the two parts of the cyclic integral in Eq. 1.6.30. Because the volume change element, dV, is negative on the exhaust stroke and positive on the intake stroke, the net work required during the pumping cycle is the enclosed area on the cylinder pressure-volume diagram. When Eq. 1.6.30 is evaluated and W_{ex} is computed as being a negative number whose absolute value is higher than W_{in}, due to the pressure during the exhaust stroke being higher at all times than during the intake stroke, the pumping work, W_p, is also a negative number indicating that work is supplied to the engine.

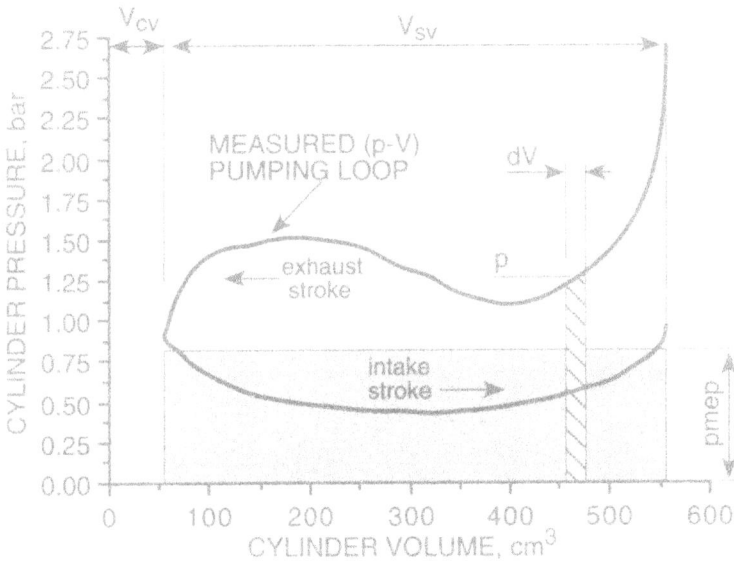

Fig. 1.33 Determination of pmep from the cylinder pressure diagram.

In a real engine, the deeper the suction during the intake stroke, or the greater the pressure bulge during the exhaust stroke, then the larger will be the enclosed area of the pumping loop and the more work will be required of it. To be blunt, the more the spark-ignition engine is throttled, the more work it takes to suck in a lesser mass of air! To be even more blunt, the creation of work costs fuel to be burned and we pay for the fuel.

Pumping work is assumed be zero in the ideal cycle because the cylinder pressures during the intake stroke and the exhaust stroke are assumed to be both constant and to be equal to each other. In short, although there is a pumping loop of work that must be done in reality, in the ideal case it becomes merely a line on the pressure-volume diagram with an enclosed area that is, by definition, zero.

1.6.10 The Concept of Mean Effective Pressure

As stated above, the enclosed pressure-volume diagram area is the work delivered to, or provided by, the piston in either the real or the ideal cycle. In Fig. 1.32 the rectangular shaded area is equal in area to the enclosed cylinder pressure-volume diagram. This rectangle is of height imep and of length V_{sv}, where imep is known as the indicated mean effective pressure and V_{sv} is the swept volume. The indicated mean effective pressure, imep, is evaluated during the power cycle by:

$$\text{imep} = \frac{W_i}{V_{sv}} = \frac{\left(\oint p\,dV\right)_{power}}{V_{sv}} \tag{1.6.31}$$

Fig. 1.32 shows measured data from an engine where the imep is found to be 14.26 bar. The shaded area with the height value as imep is drawn to scale on the figure and one can see the area correspondence with the enclosed power cycle loop.

Similarly, during the pumping work shown in Fig. 1.33, there is a rectangular shaded area that is equal in area to the enclosed cylinder pressure-volume diagram. This rectangle is of height pmep and of length V_{sv}, where pmep is known as the pumping mean effective pressure and V_{sv} is, again, the swept volume. The pumping mean effective pressure, pmep, is evaluated during the pumping cycle by:

$$\text{pmep} = \frac{W_p}{V_{sv}} = \frac{\left(\oint p\,dV\right)_{pumping}}{V_{sv}} \tag{1.6.32}$$

Fig. 1.33 shows measured data from an engine where the pmep is found to be 0.83 bar. On this figure, the shaded area of height pmep is drawn to scale and the correspondence between it and the enclosed loop of the pumping cycle can be observed.

In the text above, the word "indicated" is introduced as a definition and this may seem a strange choice of nomenclature. The word "indicated" stems from the historical fact that pressure transducers for engines used to be called "indicators" and the pressure-volume diagram of a steam engine, traditionally, was recorded on an "indicator card" [1.2]. In this cylindrical device with access to the engine cylinder, a pencil is connected to a spring-controlled piston which moves in correspondence with cylinder pressure. The pencil scribes on a cylindrical paper card which is made to oscillate by being connected through a cord to the crank as it rotates. The result is a pressure-volume diagram upon which one applied a planimeter to find the enclosed area of the work diagram and the imep. As a student, I used one on a low-speed steam engine quite a few times; the phrase, "an attention-getting experience," comes to mind.

The concept of mean effective pressure is extremely useful in relating one engine development to another for, while the units of imep are obviously those of pressure, the value is almost dimensionless. That remark is sufficiently illogical as to require careful explanation. The point is, any two engines of equal development or performance status will have identical values of mean effective pressure, even though they may be of totally dissimilar swept volume. In other words, if cylinder pressure diagrams are plotted as pressure-*volume ratio* plots, and the pressures happened to be identical at equal volume ratios, then the values of imep attained would also be identical, even for two engines of differing swept volume. You may glance ahead to Fig. 1.60 to get the point.

1.6.11 Indicated Values of Power and Torque and Fuel Consumption

It will be recalled that the indicated power cycle consists only of the compression and the power strokes. Power is defined as the rate of doing work. The engine rotation rate is recorded as revolutions per second, rps, or as revolutions per minute, rpm. Because the four-stroke engine has a power cycle every two crankshaft revolutions, the power delivered to the piston crown by the cylinder gas during the power cycle is called the indicated power output, $\dot{W}_i$, where, for an engine with n cylinders:

as: $\qquad\qquad \dot{W}_i = \text{net work per cycle} \times \text{work cycle rate}$

then:
$$\dot{W}_i = W_i \times \frac{rps}{2} = W_i \times \frac{rpm}{120}$$
$$= n \times imep \times V_{sv} \times \frac{rps}{2} = n \times imep \times V_{sv} \times \frac{rpm}{120} \qquad (1.6.33)$$

The indicated torque, Z_i, is the turning moment on the crankshaft and is related to indicated power output by the following equation:

$$\dot{W}_i = 2\pi Z_i rps = 2\pi Z_i \frac{rpm}{60} = \pi Z_i \frac{rpm}{30} \qquad (1.6.34)$$

As the engine consumes fuel of a calorific value, C_{fl}, at the measured (or at a theoretically calculated) mass flow rate of $\dot{m}_f$, the indicated thermal efficiency, η_i, of the engine is found from an extension of Eq. 1.6.24:

$$\eta_i = \frac{\text{power output}}{\text{rate of heat input}} = \frac{\dot{W}_i}{\dot{m}_f C_{fl}} \qquad (1.6.35)$$

Of great interest and in common usage in engineering practice is the concept of specific fuel consumption, i.e., the fuel consumption rate per unit power output. Hence, indicated specific fuel consumption, isfc, is given by:

$$\text{isfc} = \frac{\text{fuel consumption rate}}{\text{power output}} = \frac{\dot{m}_f}{\dot{W}_i} \qquad (1.6.36)$$

It will be observed from a comparison of Eqs. 1.6.35 and 1.6.36 that thermal efficiency and specific fuel consumption are inversely related to each other, but without the incorporation of the calorific value of the fuel. Because most petroleum-based fuels have virtually identical levels of calorific value, the use of specific fuel consumption as a comparator from one engine to another, rather than thermal efficiency, is quite logical and is more immediately helpful to the designer and the developer.

1.7 Laboratory Testing of Engines

1.7.1 Laboratory Testing for Power, Torque, Mean Effective Pressure, and Specific Fuel Consumption

Most of the testing of engines for their performance characteristics takes place under laboratory conditions. The engine is connected to a power-absorbing device, called a dynamometer, and the performance characteristics of power, torque, fuel consumption rate, and air consumption rate, at various engine speeds, are recorded. Many texts and papers describe this process and the Society of Automotive Engineers (SAE) provides a test code, J1349, for this very purpose [1.26]. There is an equivalent test code from the International Standards Organization in ISO 3046 [1.27].

There is little point in writing at length on the subject of engine testing and of the correction of the measured performance characteristics to standard reference pressure and temperature conditions, because these are covered in the standards and codes already referenced.

A laboratory engine testing facility is diagrammatically presented in Fig. 1.34.

The engine power output is absorbed in the dynamometer, for which the slang word is a "dyno" or a "brake." The latter word is particularly apt as the original dynamometers were, literally, friction brakes. The principle of any dynamometer operation is to allow the outer casing to swing freely. The reaction torque on the casing, which is exactly equal to the net engine torque, is measured on a lever of length L, from the center line of the dynamometer as a force F. This restrains the outside casing from revolving, or the torque and power would not be absorbed. Consequently, the reaction torque measured is the brake torque, Z_b, and is calculated by:

$$Z_b = F \times L \tag{1.7.1}$$

Therefore, the work output from the engine during an engine revolution is the distance "travelled" by the force F on a circle of radius L:

$$\text{work per revolution} = 2\pi F L = 2\pi Z_b \tag{1.7.2}$$

The measured power output, the brake power, $\dot{W}_b$, is the work rate at a rotational speed in units of revolutions per second, rps, or in units of revolutions per minute, rpm:

$$\dot{W}_b = (\text{Work per revolution}) \times (\text{revolutions / second})$$
$$= 2\pi Z_b \text{rps} \quad = \pi Z_b \frac{\text{rpm}}{30} \tag{1.7.3}$$

To some, this equation may clear up the apparent mystery of the use of the operator π in the similar theoretical equation, Eq. 1.6.34, when considering the indicated power output, $\dot{W}_i$.

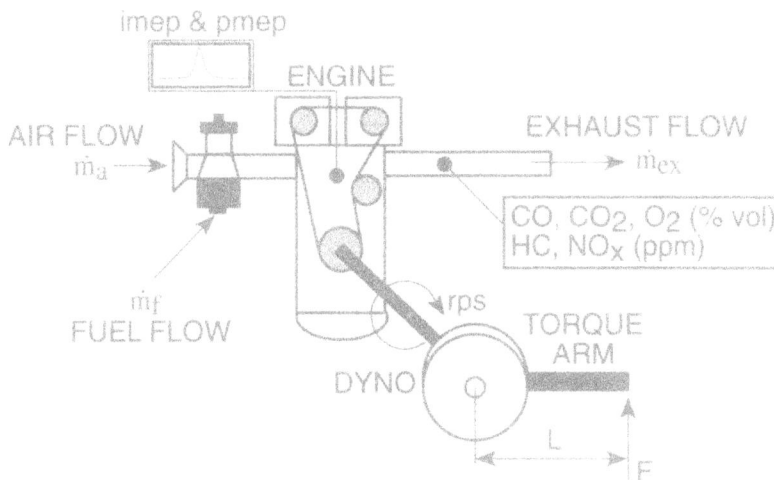

Fig. 1.34 Dynamometer test stand recording of performance parameters.

The brake thermal efficiency, η_b, is then given by the corresponding equation to Eq. 1.6.35:

$$\eta_b = \frac{\text{power output}}{\text{rate of heat input}} = \frac{\dot{W}_b}{\dot{m}_f C_{fl}} \qquad (1.7.4)$$

A similar situation holds for brake specific fuel consumption, bsfc, and the corresponding equation for the indicated values, Eq. 1.6.36:

$$bsfc = \frac{\text{fuel consumption rate}}{\text{power output}} = \frac{\dot{m}_f}{\dot{W}_b} \qquad (1.7.5)$$

However, it is also possible to compute a mean effective pressure corresponding to the measured power output of an engine with a number of cylinders, n. This is called the brake mean effective pressure, bmep, and is calculated from a manipulation of Eq. 1.6.33, but in terms of the measured values:

$$bmep = \frac{\dot{W}_b}{n \times V_{sv} \times \frac{rps}{2}} = \frac{\dot{W}_b}{n \times V_{sv} \times \frac{rpm}{120}} \qquad (1.7.6)$$

It is obvious that the brake power output and the brake mean effective pressure are the residue of the indicated power output and the indicated mean effective pressure, after the engine has lost power to internal friction and air pumping effects. These friction and pumping losses deteriorate the indicated performance characteristics by what is known as the mechanical efficiency of the engine, η_m. Friction and pumping losses are related simply by:

$$\text{net power} = \dot{W}_b = \dot{W}_i + \text{friction and pumping power} \qquad (1.7.7)$$

The mechanical efficiency, η_m, is defined as:

$$\eta_m = \frac{\dot{W}_b}{\dot{W}_i} = \frac{bmep}{imep} \qquad (1.7.8)$$

This raises the concept of the friction mean effective pressure, fmep, which is related from:

$$bmep = imep + pmep + fmep \qquad (1.7.9)$$

The discerning reader may query the veracity of Eq. 1.7.9. However, the pumping mean effective pressure, pmep, is evaluated from cylinder pressure data by Eqs. 1.6.30 and 1.6.32. The value of pmep is numerically negative. Historically, the engineer often mentally translates that into a positive number and attempts to apply it in Eq. 1.7.9 to equate imep, bmep, pmep, and fmep—and fails! Often, the only way to determine the friction content of the engine is to measure imep and pmep with cylinder-mounted pressure transducers, and the bmep on the dynamometer. These are all measurable parameters on a real engine, but the fmep is not nearly so amenable because much of its value will depend on the very magnitude of the cylinder pressure loading up the bearings of the connecting rod and the crankshaft.

It can be very difficult to segregate the separate contributions of friction and pumping in measurements taken in a laboratory, except by recording cylinder pressure diagrams in the manner described above. By measuring friction power using a motoring methodology that eliminates all pumping or cylinder pressure action at the same time, one negates some of the friction loss when it is actually pumping and firing. It is very easy to write this out in words, but it is much more difficult to accomplish this in practice. More information on this topic will be presented in Chapter 5. One method of measuring mechanical efficiency (reasonably) real-istically on a dynamometer test-bed for a multi-cylinder engine is to use the Willan's line method [1.6].

1.7.2 Laboratory Testing for Air Flow Rate and Exhaust Emissions
Air and Fuel Flow

SAE J1088 [1.28] deals with exhaust emission measurements for small utility engines, a category into which many four-stroke engines fall. Several interesting technical papers have been published in recent times questioning some of the correction factors used within such test codes for the prevailing atmospheric conditions. One of these by Sher [1.36] deserves further study.

The measurement of air flow rate into the engine is often best conducted using meters designed to conform to a standard, such as the British standard BS 1042 [1.29]. The total air flow rate into an engine of n cylinders, $\dot{m}_{as}$, can be reduced to delivery ratio, DR, or scavenge ratio, SR, values, using Eqs. 1.6.1 to 1.6.5:

$$DR = \frac{2\dot{m}_{as}}{n \times rps \times m_{dref}}$$ (1.7.10)

Upon measurement of the fuel flow rate, $\dot{m}_f$, the recording of the overall air-fuel ratio, AFR_o, is relatively straightforward as:

$$AFR_o = \frac{\dot{m}_{as}}{\dot{m}_f}$$ (1.7.11)

Continuing the discussion begun in Sec. 1.6.5, this overall air-fuel ratio, AFR_o, is also the trapped air-fuel ratio, AFR_t, if the engine is charged with a homogeneous supply of air and fuel that is not lean of the stoichiometric value, i.e., as in a normal carbureted design for a simple four-stroke engine. If the total fuel supply to the engine is, in any sense, stratified from the total air supply, AFR will not be equal to AFR_t; diesel engines are a classic example of where this is the case.

Exhaust Emissions

This area is so well covered in the literature that the inclusion of this topic in this book is merely as an introduction which will serve to assist you to comprehend a huge data base. Much of the test instrumentation available has been developed for, and specifically oriented toward, four-stroke cycle engine measurement and analysis.

The legislation on exhaust emissions details the limiting values of all pollutants on a mass basis at specified power output or load levels. Most exhaust gas analytical devices measure on a volumetric or molecular basis. It is necessary to convert such numbers from volumetric to mass values so as to permit the comparison of engines for their effectiveness in reducing exhaust emissions at equal power levels. Hence, it is necessary to describe here the fundamental theory required to derive measured, or brake specific, emission values for such pollutants as carbon monoxide, unburned hydrocarbons, oxides of nitrogen, and carbon dioxide. Of these pollutants, the first is toxic; the second and third are blamed for "smog" formation; the third is regarded as a major contributor to "acid rain"; and the last is saddled with the blame for a future "greenhouse effect."

As an example, consider any pollutant gas, G, with molecular weight, M_g, and a volumetric concentration in the exhaust gas of proportion V_{cg}. Often, such numbers are presented as parts per million (ppm), or as percentage by volume (%vol). The symbolisms for the unit in ppm, and the unit in %vol, are V_{ppmg} and $V_{\%g}$, respectively. Their arithmetic connection with the volumetric ratio is:

$$\text{volumetric ratio, } V_{cg} = V_{ppmg} + 10^6$$
$$= V_{\%g} + 100$$

The average molecular weight of the exhaust gas is M_{ex}. The measured power output is $\dot{W}_b$, and the fuel consumption rate is $\dot{m}_f$. The total mass flow rate of exhaust gas is $\dot{m}_{ex}$, which is connected by:

$$\dot{m}_{ex} = (1 + AFR_o)\dot{m}_f \quad \text{kg / s}$$
$$= \frac{(1 + AFR_o)\dot{m}_f}{M_{ex}} \quad \text{kgmol / s} \tag{1.7.12}$$

$$\text{pollutant gas flow rate} = \left(M_g V_{cg}\right) \frac{(1 + AFR_o)\dot{m}_f}{M_{ex}} \quad \text{kg/s} \qquad (1.7.13)$$

The brake specific pollutant gas flow rate, bsG, is then given by:

$$bsG = \frac{M_g V_{cg}}{\dot{W}_b} \times \frac{(1 + AFR_o)\dot{m}_f}{M_{ex}} \quad \text{kg/Ws} \qquad (1.7.14)$$

or,
$$bsG = \frac{(M_g V_{cg}) \times (1 + AFR_o) bsfc}{M_{ex}} \quad \text{kg/Ws} \qquad (1.7.15)$$

By quoting an actual example, this last equation is readily transferred into the usual numeric presentation units for the reference of any exhaust gas. If bsfc is employed in the conventional units of g/kWh and the pollutant measurement of, say, carbon monoxide, is in % by volume, the brake specific carbon monoxide emission rate, bsCO, in g/kWh units is given by:

$$bsCO = (1 + AFR_o)\, bsfc \left(\frac{V_{\%CO}}{100}\right)\left(\frac{28}{29}\right) \quad \text{g/kWh} \qquad (1.7.16)$$

where the average molecular weights of exhaust gas and carbon monoxide are assumed simplistically to be 29 and 28, respectively. To ensure total understanding, if in a practical example of a carbureted engine, the overall AFR_o is 13, the bsfc is 290 g/kWh, and the measured carbon monoxide volumetric concentration in the exhaust is 3%, then the brake specific carbon monoxide concentration is:

$$bsCO = \left(1 + 13\right) \times 290 \times \left(\frac{3}{100}\right) \times \left(\frac{28}{29}\right) = 117.6 \ \text{g/kWh}$$

The actual mass flow rate of carbon monoxide in the exhaust pipe is $\dot{m}_{CO}$ where the power output is $\dot{W}_b$ in kW units:

$$\dot{m}_{CO} = bsCO \times \dot{W}_b \quad \text{g/h} \qquad (1.7.17)$$

It should be pointed out that there are various ways of recording exhaust emissions as values "equivalent to a reference gas." In the measurement of hydrocarbons, either by an NDIR (non-dispersive infrared) device or by an FID (flame ionization detector), the readings are quoted as ppm hexane, or ppm methane, respectively. Therefore, the molecular weights for hexane, C_6H_{14}, or methane, CH_4, must be inserted in the appropriate equation if the pollutant gas G is to be regarded as equivalent hydrocarbons, HC, to either hexane or methane. Further, some FID meters use a $CH_{1.85}$ equivalent [1.28]; in that case, the molecular weight for the HC equivalent would have to be replaced by 13.85.

The same holds true for nitrogen oxide emissions. It is normally assumed that the measured value is to be compared to an NO equivalence and so the molecular weight for NO, which is 30, should be employed. Should the meter used in a particular laboratory be different, then, as for the HC example quoted above, the correct molecular weight of the reference gas must be inserted in any application of the above equations.

Trapping Efficiency from Exhaust Gas Analysis

In an engine where the combustion is sufficiently rich, or is balanced as in the stoichiometric equation, Eq. 1.6.15, it is a logical assumption that any oxygen in the exhaust gas must come from air that has been lost through the exhaust valve to the exhaust system during the overlap valve period. In practice, a stoichiometric air-fuel ratio will have some residual oxygen from combustion sources in the exhaust gas, so such an AFR should not be employed during measurements of trapping efficiency. In short, the experimental test for trapping efficiency should be conducted with a sufficiently rich mixture during the combustion process as to ensure that no free oxygen remains within the cylinder after the combustion period. If the combustion process is deliberately stratified, as in a diesel engine, this test technique cannot be used. However, on the assumption that the zero oxygen condition for burned gas can be met—and this is possible because most simple si four-stroke engines are homogeneously charged—the trapping efficiency, TE, can be calculated from the exhaust gas analysis as follows:

$$\text{exhaust gas mass flow rate} = \frac{(1 + AFR_o)\dot{m}_f}{M_{ex}} \quad \text{kg/mol·s} \qquad (1.7.18)$$

$$\text{exhaust } O_2 \text{ mass flow rate} = \frac{(1 + AFR_o)\dot{m}_f V_{\%O_2}}{100\,M_{ex}} \qquad (1.7.19)$$

$$\text{engine } O_2 \text{ mass inflow rate} = 0.2314\,\frac{\dot{m}_f\,AFR_o}{M_{O_2}} \quad \text{kg/mol·s} \qquad (1.7.20)$$

The numerical value of 0.2314 is the mass fraction of oxygen in air and 32 is the molecular weight of oxygen. The value noted as $V_{\%O2}$ is the percent by volume concentration of oxygen in the exhaust gas.

Trapping efficiency, TE, is defined as:

$$TE = \frac{\text{air trapped in cylinder}}{\text{air supplied}} \qquad (1.7.21)$$

hence

$$TE = 1 - \frac{\text{air lost to exhaust}}{\text{air supplied}} \qquad (1.7.22)$$

from Eqs. 1.7.19-1.7.20

$$TE = 1 - \frac{(1 + AFR_o)V_{\%O_2}M_{O_2}}{23.14 \times AFR_o M_{ex}} \qquad (1.7.23)$$

Assuming simplistically that the average molecular weight of exhaust gas is 29, that the molecular weight of oxygen is 32, and that atmospheric air contains 21% oxygen by volume, this equation becomes:

$$TE = 1 - \frac{(1 + AFR_o)V_{\%O_2}}{21 \times AFR_o} \qquad (1.7.24)$$

The methodology emanates from history in a paper by Watson [1.31] in 1908. Huber [1.32] basically uses the Watson approach, but provides an analytical solution for trapping efficiency, particularly for conditions where the combustion process yields some free oxygen. That it is still an important issue can be seen from more recent publications [1.33].

1.8 Potential Power Output of Four-Stroke Engines
At this stage of the book, it will be useful to be able to assess the potential power output of four-stroke engines. From Eq. 1.7.6, for an engine of n cylinders the power output $\dot{W}_b$ delivered at the crankshaft is seen to be:

$$\dot{W}_b = bmep \times n \times V_{sv} \times \frac{rps}{2} = bmep \times V_{tsv} \times \frac{rpm}{120} \qquad (1.8.1)$$

77

From experimental work on various types of four-stroke engines the potential levels of attainment of brake mean effective pressure are well known within quite narrow limits. The literature is full of experimental data for this parameter, and the succeeding chapters of this book provide further direct information on the subject, often predicted directly by engine modeling software. Some typical levels of bmep for a brief selection of engine types are given in Fig. 1.35.

The engines, listed as A-K, can be related to types that are familiar as production devices. Types A-G are spark-ignition engines running on gasoline; the nomenclature of na (naturally aspirated), sc (supercharged), and tc (turbo-charged), is quite conventional. The values of mean piston speed and bmep quoted are at the peak power point in the speed range.

For example, the type A engine could be a lawnmower or a small electric generator of less than 5 hp. The type B engine would appear in more sophisticated applications in the same field as engine A, as it is an overhead valve engine, which in the cost-conscious industrial engine market is a very significant factor. The type C and type D engines would be used for conventional on- or off-road cars or motorcycles. Nowadays, they are also found in outboard motors in an attempt to meet the oncoming exhaust emissions legislation for marine vehicles. The type E and type F engines are almost certain to be a car for the sports or luxury car market. The type G engine is a naturally aspirated racing engine used everywhere from sports car racing to Formula 1.

Types H-K are compression-ignition engines. The type H engine is a conventional IDI diesel car engine seen in the European market from about 1975-1995. In more recent times it has been replaced by turbo-charged IDI and DI engines such as types I and J. The modern truck engine, as type K, is normally DI and turbo-charged. Many of these engines are also used as inboard marine units for pleasure and sport fishing boats.

Engine Type	bmep bar	Piston Speed m/s	Bore/Stroke Ratio
Spark-Ignition			
A 2v na sv small industrial	8.0 - 9.0	10 - 12	1.0 - 1.3
B 2v na ohv small industrial	9.0 - 10.0	10 - 12	1.0 - 1.3
C 2v na ohv car/m'cycle	10.0 - 11.0	15 - 18	1.0 - 1.3
D 4v na ohc car/m'cycle	11.0 - 12.0	15 - 18	1.0 - 1.4
E 4v sc ohc car/m'cycle	14.0 - 16.0	15 - 17	1.0 - 1.3
F 4v tc ohc car/m'cycle	14.0 - 16.0	15 - 17	1.0 - 1.3
G 4v na ohc racing car/m'c	12.0 - 15.0	22 - 26	1.2 - 2.2
Compression Ignition			
H 2v na IDI car	8.0 - 9.0	11 - 13	0.85 - 1.0
I 3v tc IDI car	12.0 - 14.0	11 - 13	0.85 - 1.0
J 4v tc DI car	13.0 - 16.0	11 - 13	0.85 - 1.0
K 4v tc DI truck	14.0 - 17.0	9 - 11	0.80 - 1.0

Fig. 1.35 Potential performance criteria for some four-stroke engines.

Naturally, this table contains only the broadest of classifications and could be expanded into many subsets, each with a known band of attainment of brake mean effective pressure.

Therefore, it is possible to insert these data into Eq. 1.8.1, and for a given engine total swept volume V_{tsv}, at a rotation rate rps, determine the power output $\dot{W}_b$. It is quite clear that this might produce some optimistic predictions of engine performance, for example, by assuming a bmep of 14 bar for a single-cylinder spark-ignition engine of 500 cm³ capacity running at an improbable speed of 20,000 rpm. However, if that engine had ten cylinders, each of 50 cm³ capacity, it would be mechanically possible to safely rotate it at that speed—as Mr. Honda proved some thirty years ago in 50cc Grand Prix motorcycle racing! Thus, for any prediction of power output to be realistic, it becomes necessary to accurately assess the possible speed of rotation of an engine, based on relevant criteria related to its physical dimensions.

1.8.1 Influence of Piston Speed on the Engine Rate of Rotation

The maximum speed of rotation of an engine depends on several factors, but the principal one, as demonstrated by any statistical analysis of known engine behavior, is the mean piston speed, c_p. This is not surprising as a major limiting factor in the operation of any engine is the lubrication of the main cylinder components, the connecting rod, the piston, and the piston rings. In any given design, the oil film between those components and the cylinder liner will deteriorate at some particular rubbing velocity, and failure by piston seizure will result. The mean piston speed, c_p, is found from its conventional definition as:

$$c_p = 2 \times L_{st} \times rps \qquad (1.8.2)$$

Because one can vary the bore and stroke for any design within a number of cylinders, n, to produce a given total swept volume, the bore-stroke ratio, C_{bs}, is determined as follows:

$$C_{bs} = \frac{d_{bo}}{L_{st}} \qquad (1.8.3)$$

The total swept volume of the engine, originally formulated in Eq. 1.4.2, can be written as:

$$V_{tsv} = n\frac{\pi}{4}d_{bo}^2 L_{st} = n\frac{\pi}{4}C_{bs}^2 L_{st}^3 \qquad (1.8.4)$$

Substitution of Eqs. 1.8.2 and 1.8.4 into Eq. 1.8.1 reveals:

$$\dot{W}_b = \frac{bmep \times c_p}{4} \times (C_{bs} \times V_{tsv})^{0.666} \times \left(\frac{\pi n}{4}\right)^{0.333} \qquad (1.8.5)$$

This equation is strictly in SI units. Perhaps a more immediately useful equation in familiar working units, where the measured or brake power output, $\dot{W}$, is in kW, the bmep is in bar, and the total swept volume is in cm^3 units is:

$$\dot{W} = \frac{c_p \times bmep_{bar}}{400} \times \left(C_{bs} \times V_{tsv_{cm^3}} \right)^{0.666} \times \left(\frac{\pi n}{4} \right)^{0.333}$$

by removing the constants:

$$\dot{W} = \frac{c_p \times bmep_{bar}}{433.5} \times \left(C_{bs} \times V_{tsv_{cm^3}} \right)^{0.666} \times \left(n \right)^{0.333} \quad kW \qquad (1.8.6)$$

The values for bore-stroke ratio and piston speed, which are typical of the spark-ignition engines listed as types A-G, are shown in Fig. 1.35. It will be observed that the values of piston speed, which are quoted at peak horsepower, are normally in a common band from 15-17 m/s for most spark-ignition engines, and those with values above 22 m/s are for engines for racing or competition purposes which would have a relatively short lifespan. The values typical of diesel engines are slightly lower, reflecting not only the heavier cylinder components required to withstand the greater cylinder pressures, but also the reducing combustion efficiency of the Diesel cycle at higher engine speeds and the longer lifespan expected of this type of power unit. It will be observed that the bore-stroke ratios for petrol engines vary from "square" at 1.0 to "over-square" at 1.3, with some racing engines going above 2.0. The diesel engine, on the other hand, has bore-stroke ratios that range in the opposite direction to "under-square," reflecting the necessity for a suitable proportioning of the smaller combustion chamber in a much higher compression ratio power unit.

1.8.2 Influence of Engine Type on Power Output

With the theory developed in Eqs. 1.8.5 or 1.8.6, it becomes possible by the application of the bmep, bore-stroke ratio, and piston speed criteria to predict the potential power output of various types of engines. This type of calculation would be the opening gambit of theoretical consideration by a designer attempting to meet a required target. Naturally, the statistical information available would be of a more extensive nature than the broad bands indicated in Fig. 1.35, and would form what would be termed today as an "expert system." As an example of the use of such a calculation, four engines are examined by the application of the theory reflected in Eqs. 1.8.5 or 1.8.6. The results are shown in Fig. 1.36.

The engines are very diverse in character, including, for example, a small lawnmower engine, a racing motorcycle engine, a truck diesel powerplant, and a Formula 1 racing car engine. The input and output data for the calculation are declared in Fig. 1.36 and are culled from those applicable to the type of engine postulated in Fig. 1.35. The target power outputs in the data table are in kW (horsepower values in brackets). These values are 2.98 kW (4 bhp) for

Input Data	Rotary Lawnmower	Racing Motorcycle	Truck Turbo-Diesel	Formula 1 Racing Car
power, kW (bhp)	2.98 (4.0)	123 (165)	224 (300)	522 (700)
piston speed, m/s	6.0	25.2	10.0	25.1
bore/stroke ratio	1.0	1.5	0.95	2.1
bmep, bar	9.0	12.8	16.0	12.3
number of cylinders	1	2	6	10
Output Data				
bore, mm	53.0	98.4	108.7	92.7
stroke, mm	53.0	65.6	114.4	44.2
swept volume, cm^3	110.0	1000.0	6400	2990
engine speed, rpm	3300	11500	2600	17000

Fig. 1.36 Calculation output predicting potential engine performance.

the lawnmower; 123 kW (165 bhp) for the racing motorcycle engine; 224 kW (300 bhp) for the truck diesel engine; and 522 kW (7000 bhp) for the Formula 1 car engine. For those readers who are familiar with lawnmower engines, 1000cc v-twin Superbikes, truck turbo-diesels, or v10 Formula 1 racing engines, the numbers in the table in Fig. 1.36 have the ring of "bulls-eyes" scored! An expert system built up this way from measured data on real engines is the opening gambit in any design process. Designers take note: the more data gathered into an "expert system," the more logic will appear from its use as a preliminary design tool.

For the initial prediction of the potential power performance of an engine, the most useful part of this method is that considerable pragmatism is injected into the selection of the data for the physical dimensions and the speed of rotation of the engine.

1.9 The Beginnings of Simulation of the Four-Stroke Engine

To simulate an engine is to mathematically monitor all of the unsteady gas flow processes in, out, and through the entire engine and its ducting, and to track all of the heat transfer and related thermodynamic effects, such as combustion, in all segments of that ducting and its cylinders.

In this chapter, the concept of a simple thermodynamic cycle, the ideal Otto cycle, has already been introduced to trace the events in the closed cycle process constituting the power portion of the two mechanical cycles that together make up the four-stroke cycle engine. In Fig. 1.30, it is manifestly clear that the ideal Otto cycle is ineffective in simulating combustion in a spark-ignition engine, compared to data measured in a real engine. The problem is simply that an engine cannot conduct combustion at constant volume, i.e., instantaneously at tdc, because a real burning process takes time, the piston keeps moving, and the cylinder volume changes. If this latter problem could be remedied by keeping the piston stationary at tdc while combustion took place and then moving it down on the power stroke when all is burned, the imep and power would increase by some 50% (see Fig. 1.30). So also would the thermal efficiency improve as it would require no more fuel per cycle. Needless to add, many have

81

tried to induce stop-go piston motion characteristics into both two-stroke and four-stroke internal combustion engines. So far none has succeeded, using a plethora of mechanical contraptions and linkages, or at least none are known to be in mass production. The tone of these comments may imply that I think that they never will succeed, not so as I have accepted long ago that there is no limit to human inventiveness. To the contrary, I would actually encourage the world's inventors to keep on trying to accomplish this ic-engine equivalent of the "search for the Holy Grail."

On the assumption that engine simulation should primarily attempt to mimic the real world, it is essential that a time-related combustion process is modelled. When I was a student, in a pre-computer age, the university academics taught us about the ideal Otto cycle. Then, they very sensibly pointed out that further calculation work in this area for a reciprocating engine was well-nigh useless because the arithmetic required to model a phased combustion process simply could not be handled in any sensible time frame. Then, the arithmetic time frame was controlled by a slide rule, log tables, or graph paper. The academics of my day quickly moved on to the non-reciprocating gas turbine cycle where its simpler arithmetic could be accomplished rapidly and a slide-rule could be used to design the engine.

The question is, what does a real combustion process for a spark-ignition engine look like? Chapter 4 is devoted to this topic. There, it is shown how one can analyze measured cylinder pressure diagrams and deduce the rate at which heat is released into the cylinder, and the proportion of fuel burned, as the crank rotates. In Fig. 1.30, and again in Fig. 1.32, is a measured cylinder pressure diagram from a 2.0 liter automobile engine running at 4800 rpm at full throttle; the shorthand nomenclature (jargon) for full throttle is often written as wot, i.e., wide-open-throttle. Using the techniques described fully in Chapter 4, this measured cylinder diagram is analyzed to determine the rate at which fuel is consumed with time, which is called the mass fraction burned, B_θ. The result is shown in Fig. 1.37.

The time dependency of combustion is now evident. The spark plug ignited the mixture at 25 °btdc, but nothing transpired until 15 °btdc or 345 °atdc. The interval during which the flame gets going is called the "delay" period and here it lasts for 10° crank angle. The actual period of heat release takes place over a burn duration, b, of 50 °crank angle and the mass fraction burned curve clearly has an exponential profile while it is happening. On Fig. 1.37 are mentioned two Vibe coefficients to fit an exponential curve, a and m, which are assigned numbers. Although Chapter 4 could be read at this point, the relevant exponential curve is repeated here, giving the mass fraction burned, B_θ, at an angle θ, commencing at the onset of combustion, i.e., at the end of the delay period:

$$B_\theta = 1 - e^{-a\left(\frac{\theta}{b}\right)^{m+1}} \qquad (1.9.1)$$

This information can be programmed into a practical example of an engine and a comparison made of the effect it will have on the pressure-volume, and temperature-volume diagram, by comparison with those predicted by the ideal Otto cycle using the very same geometrical data. Such a comparison is made in Figs. 1.38 and 1.39.

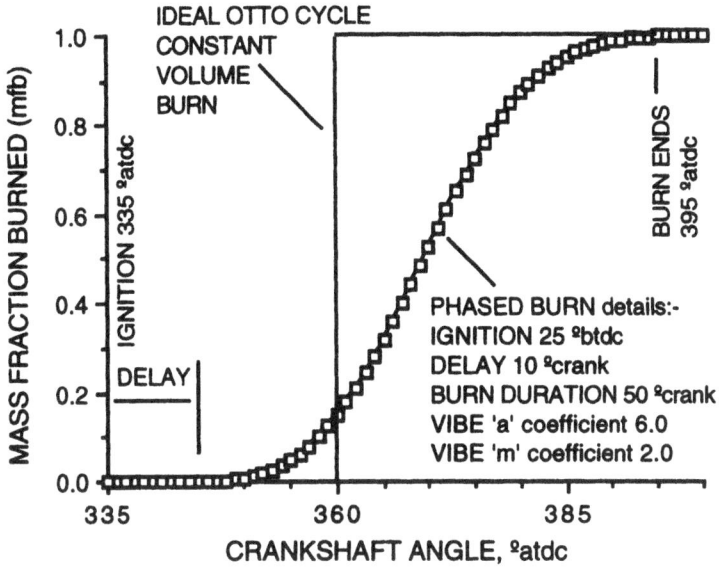

Fig. 1.37 Measured mass fraction burned curve for a spark-ignition engine.

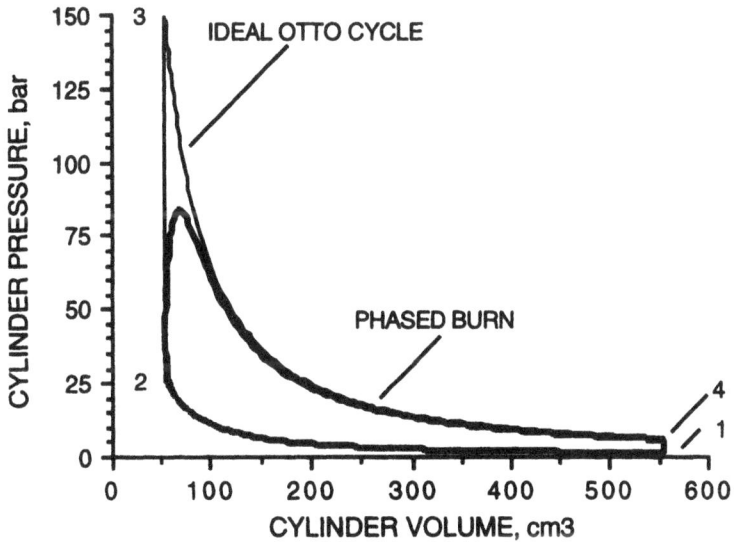

Fig. 1.38 P-V curves calculated for a practical and an ideal Otto cycle.

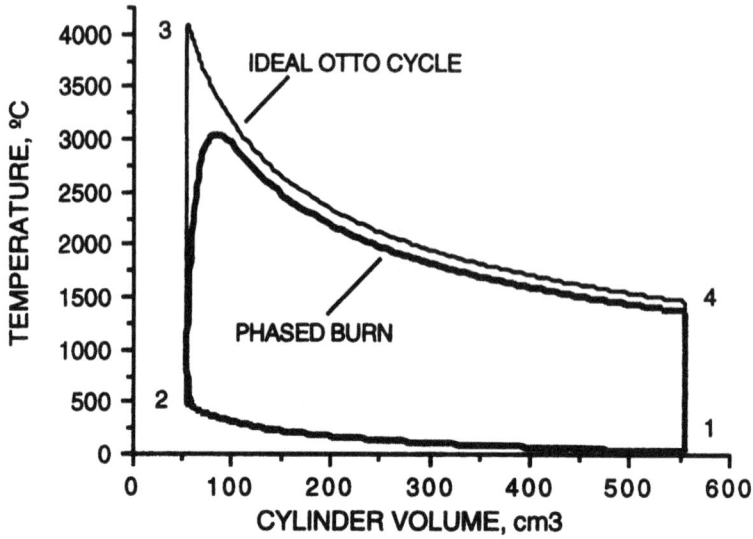

Fig. 1.39 T-V curves calculated for a practical and an ideal Otto cycle.

Clearly, the use of a "real" combustion curve has a profound influence on the similarity of the pressure-volume profile to that seen for the real engine in Figs. 1.30 and 1.32. The modeling process is obviously getting closer to reality and is now worth pursuing as a design aid. Why could this not have been done when I was a student? The answer is that it took a computer to analyze the measured cylinder pressure data to get the burn diagram, calculate the Otto cycle, and graph the results, all in a sufficiently rapid time frame. If I had tackled this same arithmetic when I was a student, it would have taken me a month. This practical approach to combustion simulation is applied to the Otto cycle, where it will be examined in more detail and its computations compared with those predicted by the ideal Otto cycle. At this early stage of simulation, we will not be concerned about how the engine gets filled with air, nor gets rid of the exhaust gas, nor of any scavenging during the valve overlap period. Rather, we will concentrate on the many fundamental lessons to be learned from basic thermodynamics applied to the closed system, i.e., the power cycle component of the overall four-stroke engine.

1.9.1 Power Cycle Analysis of the Otto Engine
The Test Engine Used as the Spark-Ignition Engine Example

Throughout the rest of this section on the Otto engine, a single cylinder unit of 500 cm^3 capacity, running at 4000 rpm, is used to illustrate all of the thermodynamic points to be made. It has a bore of 86 mm, a stroke of 86 mm, and a connecting rod with 150 mm centers. The gudgeon pin offset is zero. From the bore and stroke values, it is seen to be a "square" engine. The piston position at any point on the rotation of the crank is found using the theory of Sec. 1.4.4 and the cylinder volume by Eq. 1.4.1. The swept volume is 499.6 cm^3, but used in all thermodynamic equations as 499.6 × 10^{-6} m^3.

A compression ratio of 10 is employed. The clearance volume is obtained from Eq. 1.4.3. The clearance volume is 55.5 cm³, but used in all thermodynamic equations as 55.5×10^{-6} m³. The maximum and minimum cylinder volumes, i.e., at bdc and tdc, for this spark-ignition engine are 555.1 and 55.5 cm³, to be used thermodynamically as 555.1×10^{-6} and 55.5×10^{-6} m³.

The fuel is octane with a declared calorific value of 43.5 MJ/kg. From Eq. 1.6.15, the stoichiometric air-fuel ratio for octane is 14.95. In all spark-ignition analysis using a phased burn, the data presented in Fig. 1.37 will be used.

The working fluid within the engine cylinder is considered to be air with the properties given in Appendix A1.1, Secs. A1.1.1 and A1.1.2. However, my own eccentricity in conducting this type of thermodynamic analysis is to define that, at bdc, the swept volume is regarded as filled with "fresh" air. This is because the clearance volume is considered to be filled with exhaust, which is an inert gas, but which also has the properties of air. In short, only the swept volume at bdc is defined to contain a true "fresh" air, the significance of which theoretical chicanery will become clear in the section below!

1.9.1.1 Thermodynamic Navigation around the Ideal Otto Cycle

In Figs. 1.38 and 1.39 are seen the four numbered points that mark the events of the cycle. From Sec. 1.6.9, the four processes that make up the thermodynamic cycle are:

(a) Adiabatic and isentropic compression from points 1–2, where the index of compression is γ
(b) Constant volume heat addition (combustion) from points 2–3
(c) Adiabatic and isentropic expansion from points 3–4, where the index of expansion is γ
(d) Constant volume heat rejection (exhaust) from points 4–1

The cycle commences at bdc, at point 1. The mass of air in the cylinder is given by Eq. A1.5 from the declaration that the cylinder contains air at state conditions which are the same as standard reference atmospheric conditions, i.e., 1.01325 bar and 20°C. Any value of pressure and temperature can be employed, but for this ideal example the reference state conditions are to be used.

Initial Values for Total Mass, Air Mass, Density, and DR

From the state equation, Eq. A1.5, mass of gas in cylinder, m_1:

$$m_1 = \frac{p_1 V_1}{R T_1} = \frac{101325 \times 555.1 \times 10^{-6}}{287 \times 293} = 6.689 \times 10^{-4} \text{ kg}$$

The mass of air trapped in cylinder, m_{ta}:

$$m_{ta} = \frac{p_1 V_{sv}}{R T_1} = \frac{101325 \times 499.6 \times 10^{-6}}{287 \times 293} = 6.02 \times 10^{-4} \text{ kg}$$

From Eq. 1.6.1, the reference mass for DR is m_{dref} for which its density is required:

$$\rho_{at} = \frac{p_{at}}{RT_{at}} = \frac{101325}{287 \times 293} = 1.205 \text{ kg/m}^3$$

From Eqs. 1.6.3 and 1.6.4, delivery ratio is:

$$DR = \frac{m_{as}}{m_{dref}} = \frac{m_{ta}}{\rho_{at}V_{sv}} = \frac{6.02 \times 10^{-4}}{1.205 \times 499.6 \times 10^{-6}} = 1.0$$

The delivery ratio, DR, is precisely unity when the swept volume is filled with air at standard reference temperature and pressure and, by inference, density. This seems logical to me which explains the "chicanery" alluded to above. However, it is only fair to point out that many an ideal Otto cycle in many a good textbook does not have starting conditions defined this way.

Heat Available in Fuel to Be Released at tdc
From Eq. 1.6.20, the mass of fuel trapped, m_{tf}, is found as:

$$m_{tf} = \frac{m_{ta}}{AFR} = \frac{6.02 \times 10^{-4}}{15.06} = 4.0 \times 10^{-5} \text{ kg}$$

From Eq. 1.6.21, the heat transfer at tdc is equivalent to heat energy in fuel:

$$Q_2^3 = \eta_c m_{tf} C_{fl} = 1.0 \times 4.0 \times 10^{-5} \times 43.5 \times 10^6 = 1740 \text{ J}$$

Process 1-2, Adiabatic and Isentropic Compression
From Eqs. A1.26 and A1.50, pressure at end of compression, p_2:

$$p_2 = p_1 \left(\frac{V_2}{V_1} \right)^{-\gamma} = p_1 CR^{\gamma} = 101325 \times 10^{1.4} = 101325 \times 25.12$$

$$= 25.453 \times 10^5 \text{ Pa (i.e., 25.453 bar)}$$

From Eq. A1.55, temperature at end of compression, T_2:

$$\frac{T_2}{T_1} = \left(\frac{V_2}{V_1}\right)^{1-\gamma} = (CR)^{\gamma-1} \qquad T_2 = 293 \times 2.512 = 736 \text{ K}$$

From Eqs. A1.21 and A1.23, the work done during compression is negative:

$$W_1^2 = -m_1 C_v (T_2 - T_1) = -6.689 \times 10^{-4} \times 718 \times (736 - 293) = -212.8 \text{ J}$$

From Eq. A1.23, the change of internal energy done during compression is positive:

$$U_2 - U_1 = m_1 C_v (T_2 - T_1) = 6.689 \times 10^{-4} \times 718 \times (736 - 293) = 212.8 \text{ J}$$

Process 2-3, Constant Volume Combustion
The first law of thermodynamics is applied to the combustion process:

$$Q_2^3 = U_3 - U_2 + W_2^3 = m_1 C_v (T_3 - T_2) + 0$$

Then solving for T_3:

$$T_3 = T_2 + \frac{Q_2^3}{m_1 C_v} = 736 + \frac{1740}{6.689 \times 10^{-4} \times 718} = 4359 \text{ K}$$

Using the state equation, as in Eq. A1.44, solving for p_3:

$$p_3 = p_2 \times \frac{V_2}{V_3} \times \frac{T_3}{T_2} = p_2 \times \frac{T_3}{T_2} = 25.453 \times 10^5 \times \frac{4359}{736}$$
$$= 150.75 \times 10^5 \text{ Pa (i.e., 150.75 bar)}$$

Change of internal energy during the process from 2-3, using "first law" above:

$$U_3 - U_2 = Q_2^3 = 1740 \text{ J}$$

Process 3-4, Adiabatic and Isentropic Expansion
From Eq. A1.55, temperature at end of expansion, T_4, after process 3–4:

$$\frac{T_4}{T_3} = \left(\frac{V_4}{V_3}\right)^{1-\gamma} = (CR)^{1-\gamma} \qquad T_4 = \frac{4359}{2.512} = 1735\,K$$

From Eqs. A1.26 and A1.50, pressure at end of expansion, p_4:

$$p_4 = p_3\left(\frac{V_4}{V_3}\right)^{-\gamma} = p_3 CR^{-\gamma} = \frac{150.75 \times 10^5}{10^{1.4}} = 6.00 \times 10^5\,Pa\ \text{(which is 6.00 bar)}$$

From Eqs. A1.21 and A1.23, the work during expansion is positive:

$$W_3^4 = -m_1 C_v(T_4 - T_3) = -6.689 \times 10^{-4} \times 718 \times (1735 - 4359) = 1260\,J$$

From Eq. A1.23, the change of internal energy done during expansion is negative:

$$U_4 - U_3 = m_1 C_v(T_4 - T_3) = 6.689 \times 10^{-4} \times 718 \times (1735 - 4359) = -1260\,J$$

Process 4-1, to Complete the Cycle by Constant Volume Heat Rejection
From the first law of thermodynamics:

$$Q_4^1 = U_1 - U_4 + W_4^1 = m_1 C_v(T_1 - T_4) + 0$$

Change of internal energy during heat rejection from 4–1:

$$U_1 - U_4 = Q_4^1 = m_1 C_v(T_1 - T_4) = 6.689 \times 10^{-4} \times 718(293 - 1735) = -693\,J$$

The heat rejected is seen from the above as a negative number:

$$Q_4^1 = -693\,J$$

Obtaining Net Values for the Cycle

Net work output from the cycle, from Eq. A1.59:

$$\text{net work} = W_{net} = W_1^2 + W_3^4 = -213 + 1260 = 1047 \, J$$

Otto cycle thermal efficiency is given by Eq. A1.58:

$$\eta_t = \frac{\text{net work}}{\text{heat input}} = \frac{W_{net}}{Q_2^3} = \frac{1047}{1740} = 0.602$$

and is also given by Eq. A.1.66:

$$\eta_t = 1 - \frac{1}{CR^{\gamma-1}} = 1 - \frac{1}{2.512} = 0.602$$

and must also be the difference between heat added and rejected:

$$\eta_t = \frac{\text{heat input - heat rejected}}{\text{heat input}} = \frac{Q_2^3 - \left|Q_4^1\right|}{Q_2^3} = \frac{1740 - \left|-693\right|}{1740} = \frac{1047}{1740} = 0.602$$

The indicated mean effective pressure, imep, can be found using Eq. 1.6.31:

$$\text{imep} = \frac{W_{net}}{V_{sv}} = \frac{1047}{499.6 \times 10^{-6}} \approx 20.95 \times 10^5 \, \text{Pa (i.e., 20.95 bar)}$$

The power output of the engine can be found using Eq. 1.6.33:

$$\dot{W} = W_{net} \times \frac{\text{rpm}}{120} = 1047 \times \frac{4000}{120} = 34.9 \times 10^3 \, \text{W (i.e., 34.9 kW)}$$

The fuel consumption rate, $\dot{m}_f$, is found from:

$$\dot{m}_f = \text{mass per power cycle} \times \text{power cycles per second}$$
$$= m_{tf} \times \frac{\text{rpm}}{120} = 4.0 \times 10^{-5} \times \frac{4000}{120} = 1.333 \times 10^{-3} \, \text{kg / s}$$

From Eq. 1.6.36 the indicated specific fuel consumption can be calculated:

$$\text{isfc} = \frac{\dot{m}_f}{\dot{W}} = \frac{1.333 \times 10^{-3}}{34.9 \times 10^3} = 3.819 \times 10^{-8} \, \text{kg/Ws}$$

However, in more conventional units of kg/kWh, this becomes:

$$\text{isfc} = 3.819 \times 10^{-8} \, \frac{\text{kg}}{\text{Ws}} = 3.819 \times 10^{-8} \times 10^3 \times 3600 \, \frac{\text{kg}}{\text{Ws}} \frac{\text{W}}{\text{kW}} \frac{\text{s}}{\text{h}} = 0.138 \, \text{kg/kWh}$$

The Fundamentals Check Out Numerically
The assertion in Eq. A1.10 can be seen to be numerically accurate:

$$\oint \delta Q = \oint \delta W$$

as:

$$\oint \delta Q = \int_1^2 \delta Q + \int_2^3 \delta Q + \int_3^4 \delta Q + \int_4^1 \delta Q = 0 + 1740 + 0 - 693 = 1047 \, \text{J}$$

and:

$$\oint \delta W = \int_1^2 \delta W + \int_2^3 \delta W + \int_3^4 \delta W + \int_4^1 \delta W = -213 + 0 + 1260 + 0 = 1047 \, \text{J}$$

and from Eq. A1.11, the cyclic integral of the internal energy should also be zero:

$$\oint dU = \int_1^2 dU + \int_2^3 dU + \int_3^4 dU + \int_4^1 dU = 213 + 1740 - 1260 - 693 = 0 \, \text{J}$$

Figs. A1.3, 1.38, and 1.39 show graphs of pressures and temperatures as a result of a simulation of the ideal Otto cycle using the above theory and with the same numeric data for the engine and its initial state conditions at the commencement of the cycle. At the state points 1–4, the very same numeric values of pressure and temperature, as calculated above "by hand," are observed on this graph.

Carrying Out the Ideal Otto Cycle Simulation on a Computer
Figs. A1.3, 1.38, and 1.39 show the results of the above calculations for the ideal Otto cycle, but derived on a computer. The derivation procedure is identical to that formulated above, but the equations are coded in a calculation language such as Basic or Fortran. If the

language has graphics commands, which has been common in Basic on the Macintosh for near a decade, the graphs such as Figs. 1.38 and 1.39 can be shown directly on the computer screen with scaled axes, etc., and even made into a "movie" with the piston going to and fro as a function of either cylinder volume or crank angle. Such a movie is highly educational for students of engine design. Engineers always design as an extension of the picture in the brain, and the greater the extent of that picture library, the more inventive becomes the design process.

For the processes that involve changes of volume, i.e., processes 1–2 and 3–4, which are conducted above by a hand calculation in one pass from tdc to bdc, the computer will move the program in short steps of, say, 1° crank angle. At each step, a new piston position and a new cylinder volume will be found using the theory of Sec. 1.4.4. The altered values of pressure, temperature, change of work, change of internal energy, and change of heat transfer are computed over the volume increment involved, with the initial state conditions in each case being the final state conditions at the end of the previous step. At the end of any one step, the final state conditions calculated are swapped over to be the initial conditions for the ensuing step. The work, internal energy, and heat transfer calculated in any one step can each be added to their own computer store and printed out at the conclusion of the complete cycle. At that point all of the performance parameters of imep, power, or specific fuel consumption can be similarly computed. In short, the modern desktop computer can complete the entire simulation of the thermodynamic cycle, plot its graphics, and run its movie—as fast as I have typed this sentence.

Graphical Output from a Computer Simulation of the Ideal Otto Cycle

The graphs of p-V and T-V characteristics shown in Figs. A1.3, 1.38 and 1.39 are already referred to above. Fig. 1.40 shows the cumulative values of heat, Q, work, W, and internal energy, U, as the cycle progresses and with respect to crank angle. The convention in this chapter is that time starts on tdc at 0° crank angle at the beginning of the intake stroke, so the compression stroke starts at 180° and the power stroke commences at the next tdc at 360° crank angle.

The profile of cumulative work, W, shows it becoming progressively more negative toward tdc and is -213 J at that point. As the power stroke commences, the cumulative work becomes zero and then increasingly positive, finishing at +1047 J.

The profile of cumulative heat, Q, shows it remaining zero until tdc, when 1740 J is added at that point. During the power stroke this profile is flat at 1740 J. At bdc, the heat rejection process saps away 693 J, leaving the cumulative heat, Q, value at +1047 J.

The profile of cumulative internal energy, U, shows it becoming progressively positive toward tdc as the air gets hotter and is +213 J at that point. During the combustion at constant volume, the 1740 J added from the simulated combustion at constant volume raises it to 1953 J. During work output on the power stroke, the cylinder charge cools losing 1260 J of internal energy by bdc, leaving it at +693 J at that point. The final heat rejection process of -693 J reduces the cumulative internal energy to zero by the end of the cycle.

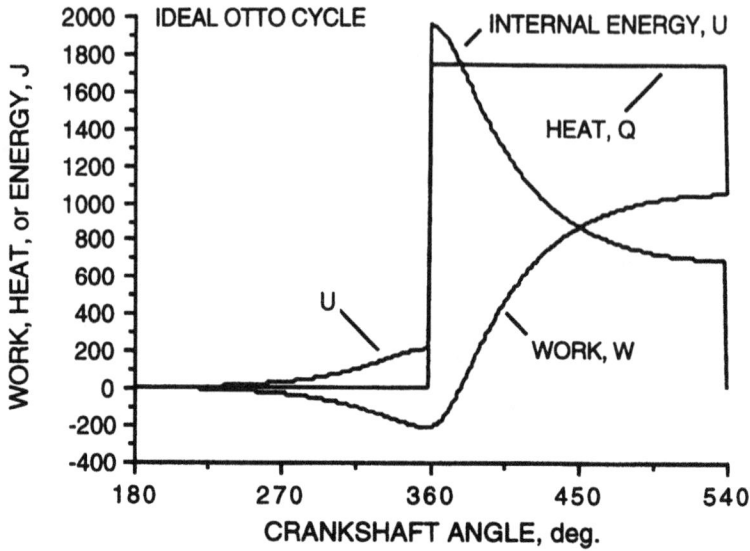

Fig. 1.40 Changes of heat, work, and internal energy in the ideal Otto cycle.

1.9.1.2 Thermodynamic Navigation Around an Otto Cycle with Phased Combustion

The computer simulation is repeated but with the heat addition to the cycle accomplished using the phased burn represented by the data in Fig. 1.37, otherwise all input data remain identical to those for the ideal Otto cycle. The only difference in the simulation, as far as the computer program is concerned, is that ideal adiabatic compression stops at 15 °btdc and another adiabatic process with internal heat transfer takes over for the next 50° crank angle. The internal heat transfer is the phased combustion process and it spans the end of compression and the beginning of expansion. The heat input function is defined by Eq. 1.9.1 and its input data are given in Fig. 1.37. Those data specify that ignition is at 25 °btdc, but a delay of 10° is also detailed, so heat input actually begins at 15 °btdc. In terms of the previous ideal Otto cycle diagram, point 2 is now brought back (advanced) to 15 °btdc and point 3 is delayed (retarded) to 25 °atdc. At all other locations in the cycle, the solution is exactly the same as for the ideal Otto cycle. During the combustion process between state points 2 and 3, as the simulation proceeds in steps of one degree crank angle, Eq. 1.9.1 is solved for the mass increment of fuel consumed and Eq. 1.6.21 revisited to obtain the heat input associated with it. The theory of Appendix A1.1, Sec. A1.1.4 is applied as a solution to what is assumed to be an adiabatic process in a closed system with both a change of volume and with internal heat transfer, so the ratio of specific heats remains at γ.

Due to the arithmetic complexity involved, I will not bore you, or myself, by setting out these calculations "by hand," but will simply show, on Figs. 1.38, 1.39, and 1.41, the computer output of the simulation.

Figs. 1.38 and 1.39 show the pressure-volume and temperature-volume graphs when the phased burn is incorporated into the solution. Here, too, are the same data for the ideal Otto cycle. At the beginning of this section, it is pointed out that the p-V diagram for the ideal Otto cycle bears little resemblance to the measured p-V diagram typified by Figs. 1.30 or 1.32, but the diagram for the phased burn in Fig. 1.38 certainly does. The resemblance does not end there, for the computed performance parameters for the phased burn Otto cycle are also very much closer to the reality of measured engine data. In the ideal Otto cycle, the indicated thermal efficiency is 0.602, and the imep and isfc are 20.94 bar and 0.138 kg/kWh, respectively. In the phased burn Otto cycle the indicated thermal efficiency, imep, and isfc are 0.512, 17.84 bar, and 162 kg/kWh, respectively. To put these number into a practical context, a conventional automobile engine with the same compression ratio and fuel will typically have an indicated thermal efficiency of 0.45, and imep and isfc values of about 15 bar and 0.190 kg/kWh, respectively. Although the use of a phased burn process has not immediately produced the same values as a conventional car engine, there has been a major shift toward realism from those provided by the ideal Otto cycle.

Fig. 1.41 shows the cumulative values of heat, Q, work, W, and internal energy, U, as the phased burn cycle progresses and with respect to crank angle. This figure may be compared with Fig. 1.40 where the equivalent data for the ideal Otto cycle are plotted.

All of the thermodynamic comments above regarding Fig. 1.40 are equally applicable to Fig. 1.41, except that the cumulative heat transfer, Q, and the cumulative internal energy change now have sloped profiles during combustion by comparison with the instantaneous, constant-volume process of the ideal Otto cycle. Although it is small, the increased negative work during the compression stroke just before tdc is just discernible.

1.9.1.3 Otto Cycle Data with Respect to Time Rather than Volume

When one employs a pressure transducer in the cylinder of an engine, or indeed anywhere in an engine, the signal recorded is of pressure with respect to time and not volume as has been the presentation format thus far in this chapter. Because much of the data in the rest of this book will compare measured and computed pressure diagrams, you must get used to seeing them in this fashion. Although the word "time" is mentioned above, rarely is time the x-axis for a graph; it is normally crank angle at the current rotational speed.

To make the point, the previous pressure and temperature data shown in Figs. 1.38 and 1.39 are redrawn against crank angle, θ, and are presented in Figs. 1.42 and 1.43.

The slope of pressure and temperature rise during combustion in the phased burn cycle is now very evident compared to the instantaneous process in the ideal Otto cycle. For the phased burn, the higher pressure before tdc is also very evident at the very end of the compression stroke, which is caused by heat release commencing before tdc itself. In Figs. 1.42 and 1.43, and indeed also in Figs. 1.38 and 1.39, the considerable drop in peak pressure and temperature due to phased combustion should be noted. The peak pressure is almost halved and the peak temperature has dropped by over 1000°C. No wonder the efficiency and power have deteriorated from the ideal; but that is life itself, it is never ideal. The inventor's trick is to attempt to move closer toward it.

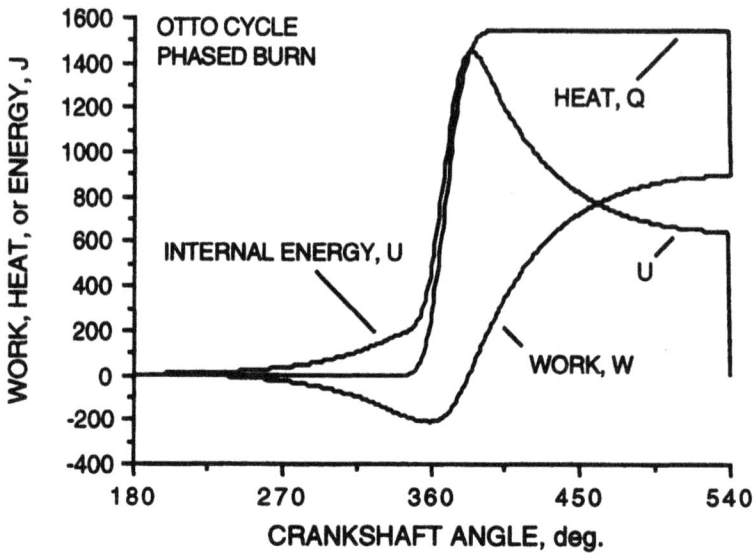

Fig. 1.41 Cumulative heat, work, and internal energy in a "phased burn" Otto cycle.

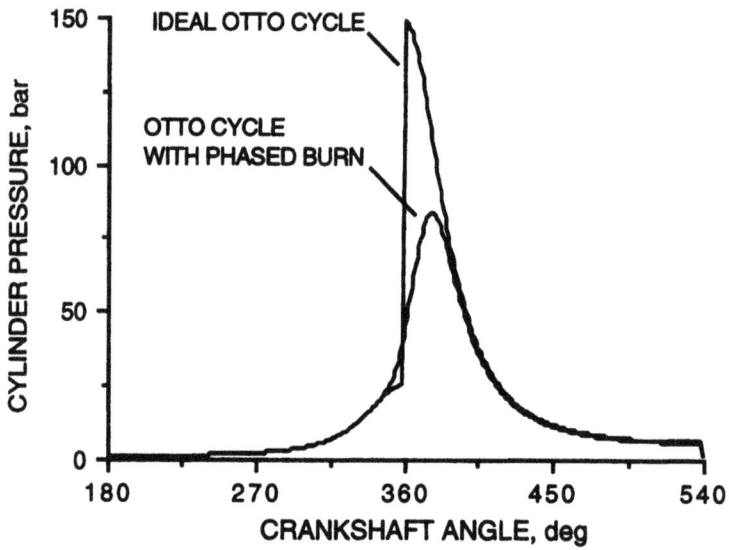

Fig. 1.42 P-θ curves calculated for a practical and an ideal Otto cycle.

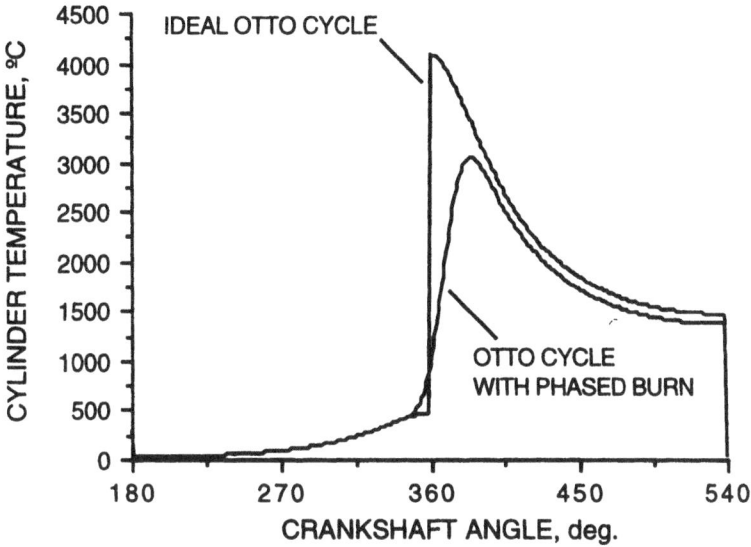

Fig. 1.43 T-θ curves calculated for a practical and an ideal Otto cycle.

One of the criteria of practical acceptability of any combustion process is the "rate of pressure rise" curve during a combustion process. The one for the phased burn is easily extracted from the computer simulation of the engine cycle. Rate of pressure rise, Δp_θ, is defined as follows and is normally obtained at one-degree crank angle intervals:

$$\Delta p_\theta = \frac{dp}{d\theta} = \left[\frac{\Delta p}{\Delta \theta}\right]_{\Delta\theta=1^\circ} \tag{1.9.2}$$

It can be seen that it is actually the slope of the p-θ graph at any point, but the incremental value is normally equally acceptable. The graph for this parameter is conventionally expressed in bar/deg or atm/deg units, and is shown in Fig. 1.44.

The maximum value is observed to occur at 5 °atdc at 3.55 bar/deg whereas peak pressure is 83.4 bar at 17 °atdc. On the same graph is the equivalent curve for a DI diesel engine. Further discussion on both profiles occurs later in Sec. 1.9.2.2.

1.9.1.4 The Behavior of the Otto Cycle Engine with Differing Compression Ratio
Thermal Efficiency

In Appendix A1.1, Sec. A1.8, in Eq. A1.66, it is shown that the higher the compression ratio, the higher is the cycle efficiency. This equation is plotted as a graph in Fig. 1.45, as it has already been shown above in the simulation of the ideal Otto cycle that it does yield an indicated thermal efficiency precisely according to this function. The fundamental lesson of this

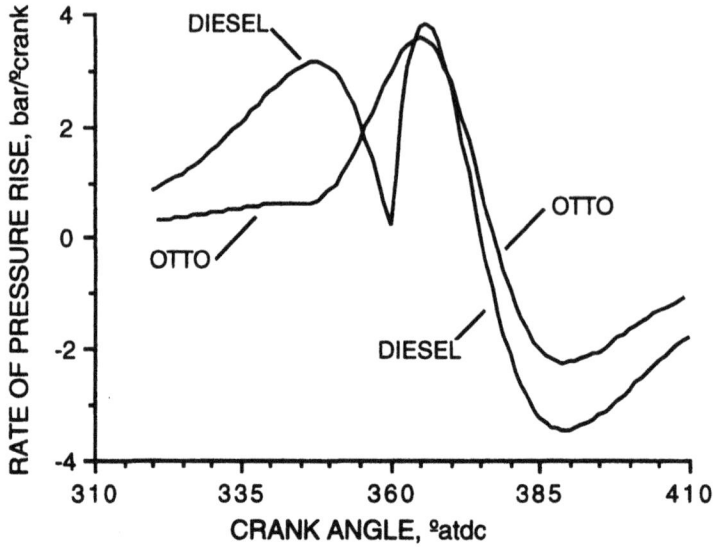

Fig. 1.44 Rate of pressure rise for a phased burn in the Otto cycle.

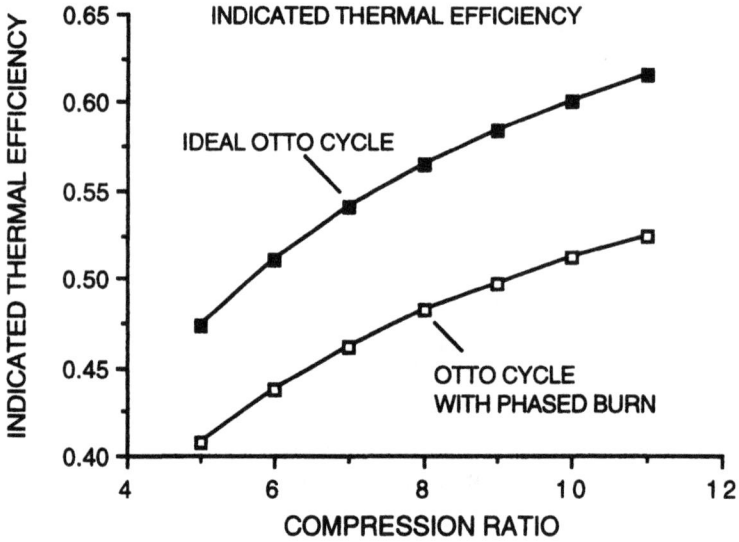

Fig. 1.45 Otto cycle thermal efficiency with respect to compression ratio.

figure for engine design has been known for many years. As a result, engine R&D, and the fuels research that accompanies it, has long been orientated to enable a spark-ignition engine to run at an ever-higher compression ratio. In the 1930s, a typical CR value for the si engine was 5 or 6. Today, the conventional automobile engine has a compression ratio of about 10.

However, the real engine does not burn according to the ideal Otto cycle. If the engine has a phased burn, does the indicated thermal efficiency also increase, and in the same proportion with change of CR as that for the ideal Otto cycle? With a computer-based simulation, we can easily begin to answer this question. The simulation model with the phased burn is run with CR data from 5 to 11 and the indicated thermal efficiency output is plotted in Fig. 1.45. It can be seen that the ideal Otto cycle is more optimistic regarding indicated thermal efficiency, not just at any one compression ratio, but with increasing compression ratio. It can be seen that η_t dropped from about 51% to 41% for a compression ratio change from 11 to 4. As seen below, this means that the power, the torque, and the imep changed by the same proportion.

Cylinder Pressure, Temperature, and Rate of Pressure Rise

Because each of the simulations commences with a DR of unity, the same mass of air is trapped and the same heat is released, either by a phased burn or in the ideal manner at constant volume. Thus, increasing compression ratio should be reflected in a greater work output, and higher cylinder pressures and temperatures. Fig. 1.46 shows the peak cylinder pressures for both the ideal Otto cycle and with a phased burn. It can be seen that, as in the case of the thermal efficiency, the disparity between them increases very markedly with increasing compression ratio. The profile for the phased burn is very much flatter than that for the ideal cycle, and at the highest compression ratio the peak pressure in the phased burn is actually less than half that for the ideal Otto cycle.

Fig. 1.47 shows the peak cycle temperature profiles with compression ratio for both the ideal Otto cycle and with the phased burn. If anything, the temperature change with compression ratio in the phased burn has an even flatter profile than that for pressure. With the phased burn, at a compression ration of 5, the peak cycle mean cylinder temperature is 3070°C, rising to 3326 °C at a CR of 11, i.e., a rise of 256 °C. In the ideal Otto cycle, the equivalent temperatures at a CR of 5 and 11 are 3776 and 4422 °C, respectively, i.e., a rise of 646°C. These temperatures are the mean gas temperatures within the cylinder, which is the average of that gas which is burning and that which is not yet burned. Later, in Chapter 4, the relationship between this mean temperature and those in the burned and unburned zones will be expanded much further.

Fig. 1.48 shows the p-V diagrams for all of the compression ratios and with the phased burn modeling the combustion process. The increasing cylinder pressure with greater compression ratio is easily seen, but the greater enclosed area of the p-V diagram is also very evident. The higher compression ratio means compression into a smaller volume by tdc, raising the end of compression pressure and temperature. The end of compression pressure rises from 9.5 bar when CR is 5; to 29 bar when the compression ratio is 11. That is, compression pressure virtually triples by doubling the compression ratio.

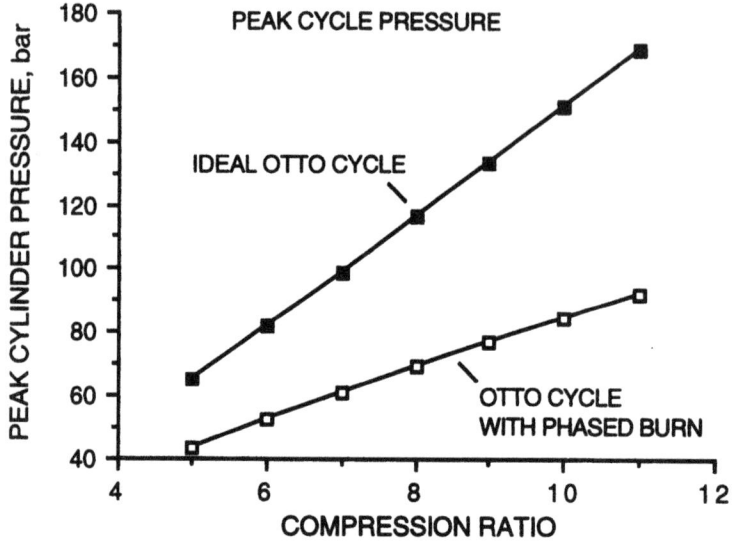

Fig. 1.46 Peak cycle cylinder pressure with respect to compression ratio.

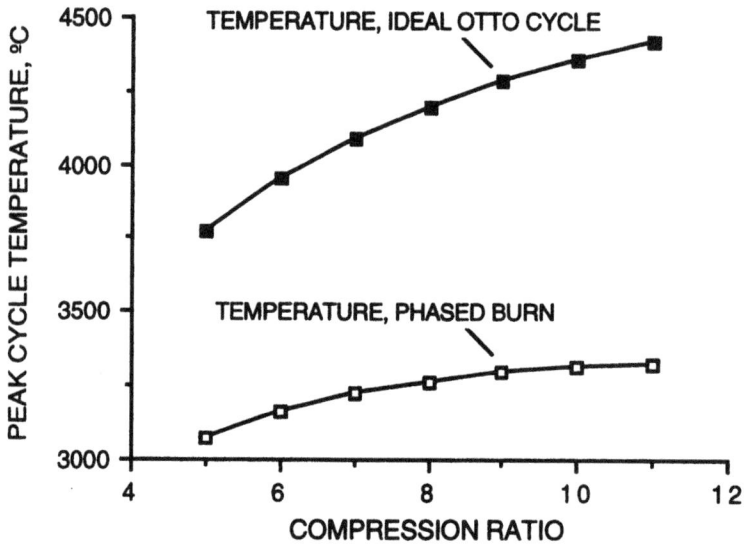

Fig. 1.47 Peak cycle cylinder temperature with respect to compression ratio.

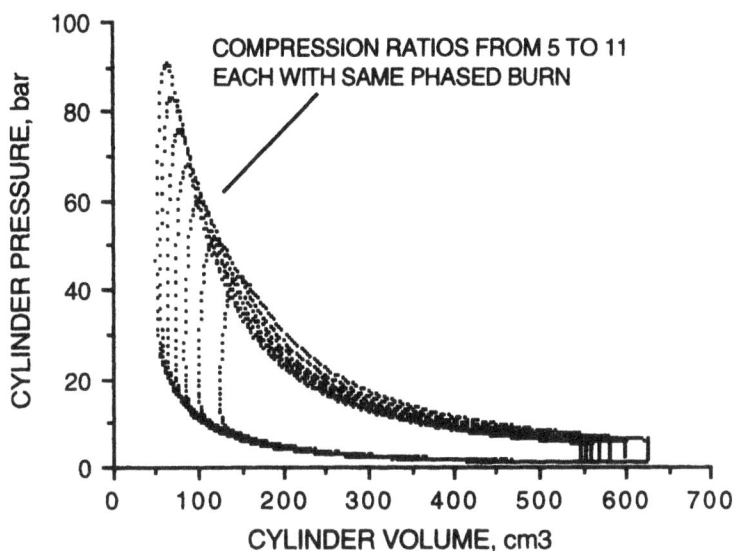

Fig. 1.48 p-V diagrams for a change of compression ratio.

The enclosed area on the p-V diagram is net work, thus Figs. 1.49 and 1.50 provide graphical evidence of the amount of extra work available from the higher CR values.

Fig. 1.49 shows cumulative work with respect to crank angle when a phased burn is used in the simulation for all CR values from 5 to 11. The extra compression work needed at the higher CR values can be observed. This work increases from -160 J when the CR is 5 to -225 J when the CR is 11. However, this pays off, for the net work by bdc at the end of the power cycle has risen to 912 J compared to 707 J, a net gain of 205 J as a result.

In Fig. 1.50, the imep is plotted with respect to compression ratio for both the ideal Otto cycle and the cycle with the phased burn. The imep in the ideal Otto cycle increases at a higher rate than that for the phased burn, a trend that complements that already seen for thermal efficiency in Fig. 1.45. The gain in imep, i.e., both torque and power at the rated speed, is considerable as they rise by 29% from a CR of 5 to a CR of 11.

Up to now, from the information presented, the designer will be tempted to use a very high compression ratio—perhaps too high. Thus far, everything points to the highest compression ratio giving the best answer for thermal efficiency or power output. However, Fig. 1.51 presents the warning note for those who would incorporate too high a compression ratio in a spark-ignition engine.

Fig. 1.51 shows the position of peak cylinder pressure, and the rate of pressure rise, for the phased burn simulations with the compression ratios from 5 to 11. It can be seen that the peak pressure location gets ever closer to tdc with increasing compression ratio. The rate of pressure rise, Δp_θ, as defined in Eq. 1.9.2, is seen to increase markedly and almost linearly from

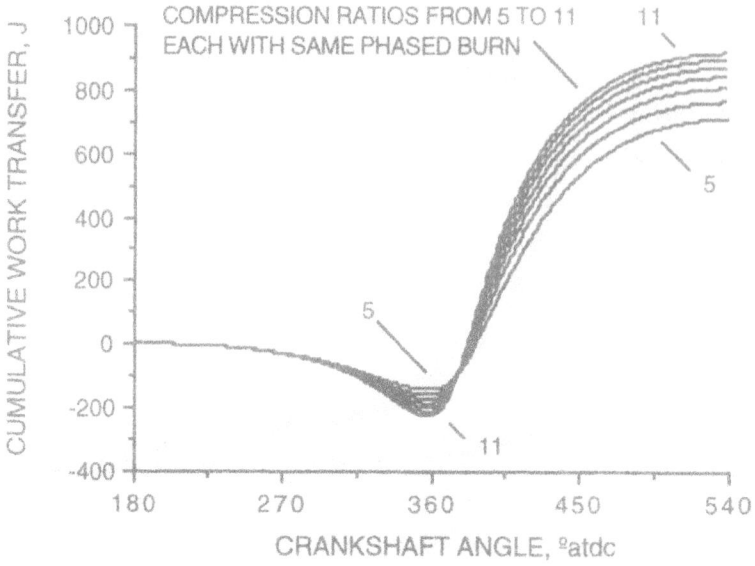

Fig. 1.49 Cumulative cycle work for a change of compression ratio.

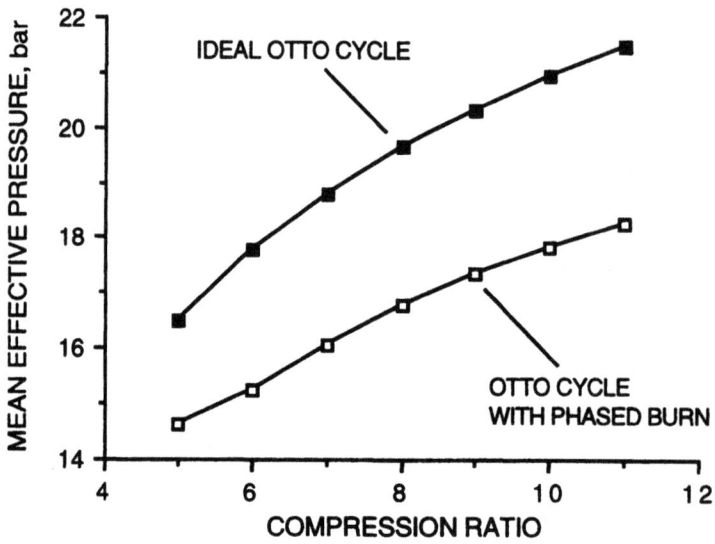

Fig. 1.50 The imep for the Otto cycle for a change of compression ratio.

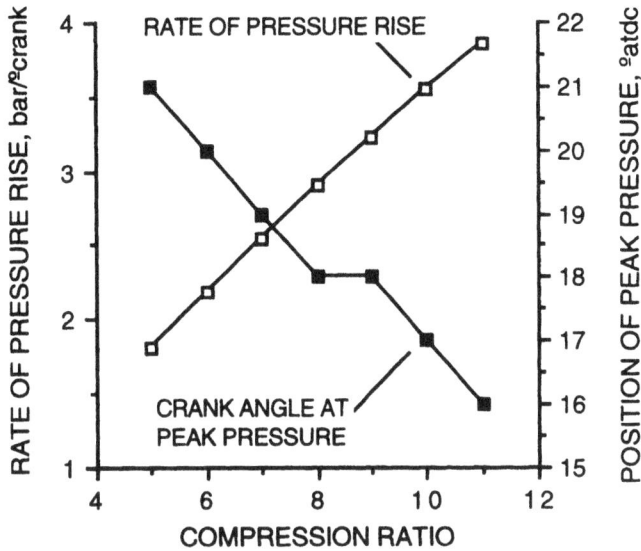

Fig. 1.51 Peak pressure location and rate of pressure rise.

1.7 to nearly 4.0 bar/deg over the same range of compression ratio. In Chapter 4, on combustion, the rate of pressure rise is cited as a marker for the onset of detonation in a combustion chamber, a phenomenon that will cause severe mechanical damage to the cylinder components. There is a limit to the amount that compression ratio can be increased in a si engine, and the rate of pressure rise in the cylinder is one of its principal tracers.

1.9.1.5 The Effect of Using Polytropic Indices of Expansion and Compression
Polytropic Index of Compression

In Appendix A1.1, Sec. A1.1.6, it is shown that if the index of compression is polytropic, i.e., it's n rather than γ, where n is less than the ratio of specific heats, γ, then heat is lost from the cylinder during the compression stroke. Naturally, if the index n is greater than γ, then the reverse would be the case, but I cannot think of a single example of such a situation in practice. Fig. 1.52 shows the effect of running the phased combustion Otto cycle simulation at the standard data value of compression ratio of 10, but where the index of compression is changed from its standard value of 1.4, which is the ratio of specific heats, γ, for air, to 1.3, then 1.2, and 1.1. Adiabatic expansion is retained so the index of expansion is kept at 1.4. Fig. 1.52 shows the effects on imep and indicated thermal efficiency. Loss of index of compression deteriorates both of these values quite significantly, i.e., about 6% over the range tested. It would be quite normal for a conventional spark-ignition engine to display values for the index of compression as low as 1.25.

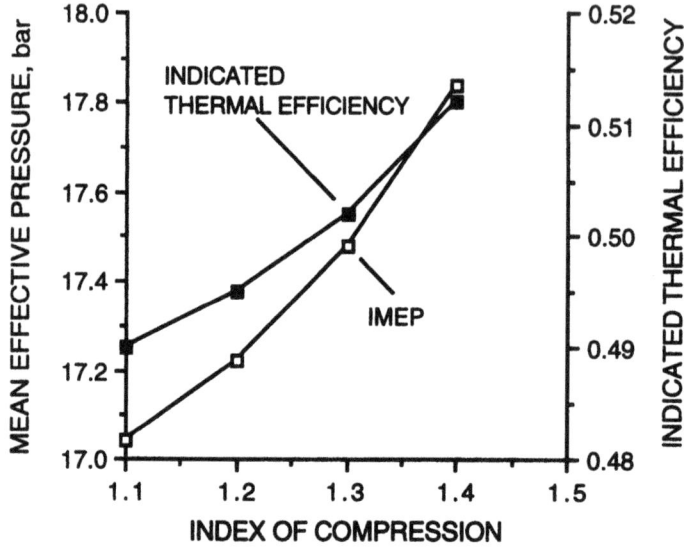

Fig. 1.52 Effect of index of compression on Otto cycle efficiency and work.

The manner in which this occurs is best seen in Figs. 1.53 and 1.54, which show cumulative heat and work transfer for the adiabatic case, where γ is 1.4, and one of the polytropic cases, where n is 1.1.

In Fig. 1.53, the loss of heat during the compression stroke is clear, and that carries over to the expansion stroke so that the cumulative heat is always diminished by that loss. In Fig. 1.54, the apparent curiosity is that it then takes less work in the polytropic case to compress the charge to tdc on the compression stroke. Even though the same heat is added during combustion, the cumulative work transfer still ends up short because a lesser peak pressure is attained during the burn period. The work diagram in Fig. 1.54 looks not unlike the scenario of running at one of the lower compression ratios, as seen in Fig. 1.49.

Polytropic Index of Expansion

Appendix A1.1, Sec. A1.1.6, shows that if the index of expansion is polytropic, i.e., it is n rather than γ, where n is greater than the ratio of specific heats, γ, then heat is lost from the cylinder during the expansion stroke. If the index n is less than γ, then the reverse would be the case. Fig. 1.55 shows the effect of running the phased combustion Otto cycle simulation at the standard data value of compression ratio of 10, but where the index of expansion is increased from its standard value of 1.4, which is the ratio of specific heats, γ, for air, to 1.425, then 1.45, 1.475, and 1.5. Adiabatic compression is retained so the index of compression is kept at 1.4.

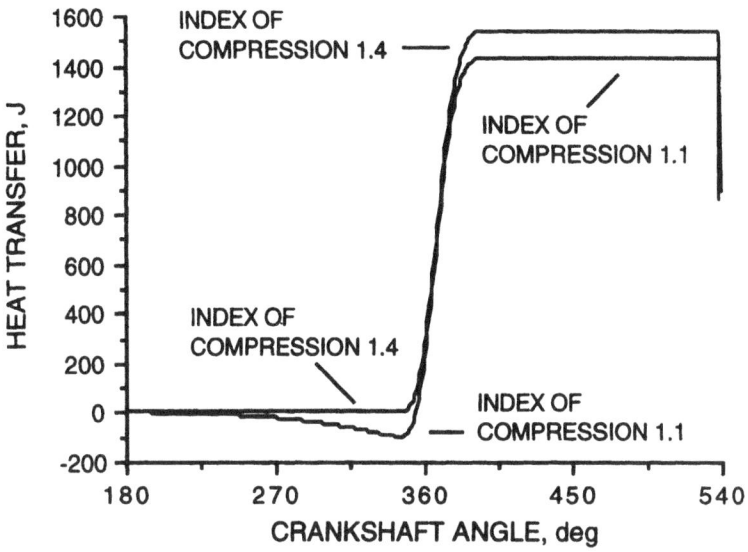

Fig. 1.53 Cumulative heat transfer for a change of index of compression.

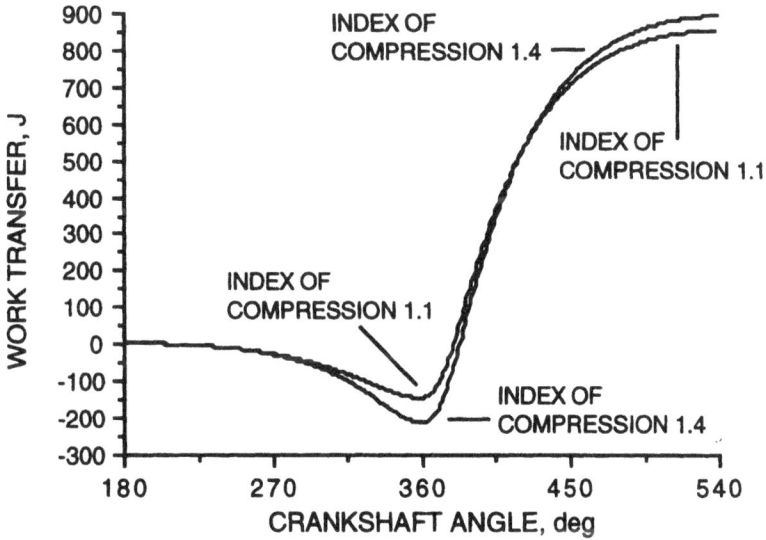

Fig. 1.54 Cumulative work transfer for a change of index of compression.

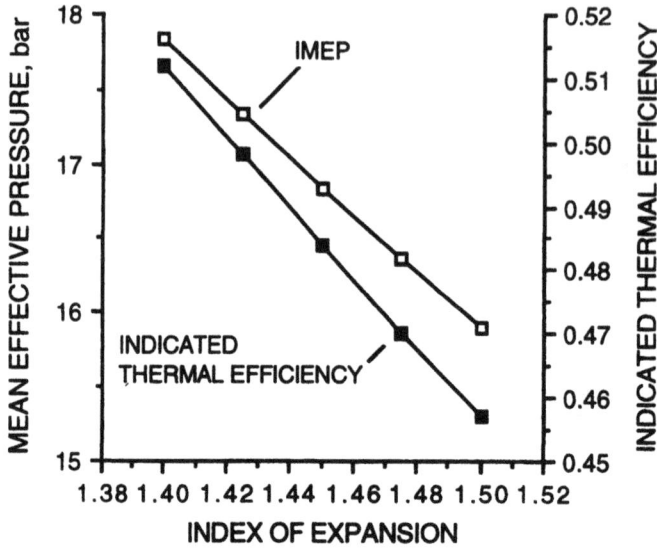

Fig. 1.55 Effect of index of expansion on Otto cycle efficiency and work.

Fig. 1.55 shows the effects on imep and indicated thermal efficiency. Increase of index of expansion deteriorates both of these values very significantly, i.e., about 15% over the range tested. The loss of thermal efficiency, or work output, is very much greater for a 0.1 shift upward for the index of expansion above the ratio of specific heats, compared to a 0.1 shift downward for the index of compression. In short, heat loss on the power stroke has a more significant effect on efficiency and power than on the compression stroke. In Fig. 1.56, which displays cumulative work transfer for a ratio of specific heats, γ, of 1.4 and a polytropic index of expansion of 1.5, the loss of net work output is seen to be 97 J. In Fig. 1.54, where the index of compression shift is 0.3, i.e., from 1.4 to 1.1, the net loss of cumulative work during the cycle is a mere 40 J by comparison.

Sec. 1.6.9 contains a discussion of the numeric value of the index of expansion. This value is obtained from the analysis of cylinder pressure diagrams and is normally found to be less than 1.4, which is the ratio of specific heats, γ, for air at reference atmospheric conditions. In that discussion, it is pointed out that (a) the gas in the cylinder of a real engine during expansion is not air, which is how the cylinder gas is being treated in these ideal simulations and (b) the air, or the post-combustion cylinder gas, is very hot and the ratio of specific heats of any real gas does decrease with rising temperature. Because the expanding gas in a real engine on the power stroke probably has a ratio of specific heats, γ, of about 1.2, then a measured polytropic index of 1.35 does indicate heat loss from the cylinder during expansion. This establishes that the fundamental theory of Appendix A1.1, Sec. 1.1.6 is correct in its presumptions and illustrates the magnitude of deterioration of thermal efficiency, or work output, coming from numeric differences between the ratio of specific heats and the polytropic indices of compression or expansion.

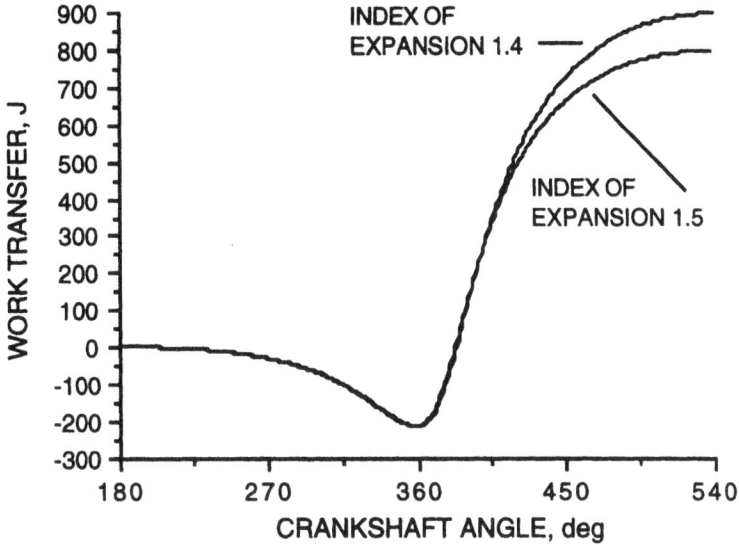

Fig. 1.56 Cumulative work transfer for a change of index of expansion.

1.9.1.6 The Effect of Using Differing Air-Fuel Ratios in the Simulation
Equivalence Ratio (lambda)

In Sec. 1.6.6, the concept of a stoichiometric air-fuel mixture is established to obtain the optimum oxidation of the carbon and hydrogen elements of a given fuel. If the air-fuel ratio is rich, then there is insufficient air for all of the carbon to be burned to carbon dioxide, i.e., some carbon monoxide is produced instead, and so less heat must be liberated per unit mass of fuel combusted. The amount of heat delivered by a combustion process, for whatever air-fuel ratio, is already defined in Eq. 1.6.21. This contains a combustion efficiency term which numerically controls the effect of burning rich, or lean, mixtures. The finer detail of this topic is the subject of Chapter 4. Here, an initial study of the effect on engine performance of changing the air-fuel ratio is conducted. To denote the proportion that any given air-fuel mixture is lean, or rich, of the stoichiometric value, the concept of an equivalence ratio is introduced, and is frequently referred to in ic engine jargon as "lambda," after the Greek letter that is conventionally employed to represent it.

Equivalence ratio: $$\lambda = \frac{AFR}{AFR_s}$$ (1.9.3)

The air-fuel ratio, AFR, is that found within the cylinder of any given engine at the onset of combustion, and AFR_s is the stoichiometric air-fuel ratio defined, for example, by Eq. 1.6.14 for the burning of octane with air.

105

The combustion efficiency, η_c, of a gasoline type fuel such as octane, the combustion efficiency of which is first mentioned in Eq. 1.6.21, can be expressed in terms of equivalence ratio from measured data as:

$$0.75 < \lambda < 1.2 \qquad \eta_c = \eta_{cmax}\left(-1.6082 + 4.6509\lambda - 2.0764\lambda^2\right) \qquad (1.9.4)$$

where the maximum possible value of the combustion efficiency, η_{cmax}, is typically about 0.9 in a spark-ignition engine using a gasoline fuel. The numeric output of Eq. 1.9.4 maximizes at about 12% lean of stoichiometric. The values calculated by use of Eqs. 1.9.3 and 1.9.4 can be inserted into Eq. 1.6.21 to determine the total amount of heat to be released at combustion under any given circumstance.

As with all polynomial equations, these equations are only truly representative over a limited range of values. In the case of Eq. 1.9.4, the range of effective λ values spans normal si combustion, i.e., from about 0.75 to about 1.20.

Changing Air-Fuel Ratio in the Spark-Ignition Simulation
In the phased burn simulations found thus far in this section, a stoichiometric air-fuel ratio is used in each case, so Eqs. 1.9.3 and 1.9.4 have also been used inherently within those several simulation experiments, albeit with the same answer for total heat input for combustion in each example. I apologize for not telling you this earlier, but my excuse is that I have always found the best form of thermodynamic education to be one of telling the student a series of white lies in sequence, as the thermodynamic black truth told all at once is far too large, and perhaps far too bitter, a pill for anyone to swallow in just one gulp!

Using Eqs. 1.9.3 and 1.9.4 above, and retaining the standard data with a phased burn, the simulations are repeated for equivalence ratios ranging from 0.75 to 1.1, in steps of 0.05. The variation of work output, graphed as indicated mean effective pressure, imep, is shown in Fig. 1.57. The effects on indicated thermal efficiency, η_t, are shown plotted in Fig. 1.58, and on its reciprocal relation, indicated specific fuel consumption, isfc, in Fig. 1.59.

Eq. 1.6.21 gives the heat available from combustion, Q_R. As the air-fuel ratio richens, the mass of fuel trapped, m_t, increases, but the combustion efficiency falls in line with Eq. 1.9.4. Thus, the heat input reaches a maximum at a particular air-fuel ratio. This occurs at the peak torque, or imep, point at an equivalence ratio, λ, of about 0.8, or some 20% rich of stoichiometric. This can be seen in Fig. 1.57 where the peak imep is 19.30 bar when λ is 0.8, compared to 17.84 bar when λ is unity. Power at rated speed, torque, or imep are seen to increase by some 8% over that obtained at the stoichiometric mixture. If the mixture is excessively rich, i.e., when λ is 0.75, some power and torque are lost from the peak torque level. At a 10% lean mixture, the imep is 16.63 bar, so some 7% torque or power is lost compared to that obtained at the stoichiometric mixture.

The bad news is that maximum power obtained by fuel enrichment is at the expense of thermal efficiency and specific fuel consumption. In Fig. 1.58, when λ is 0.8 at peak torque, the thermal efficiency is 0.444, compared to 0.512 at stoichiometric. This is a deterioration of 13% for a gain of 8% on imep, the same trend in fuel consumption as seen in Fig. 1.59. If one

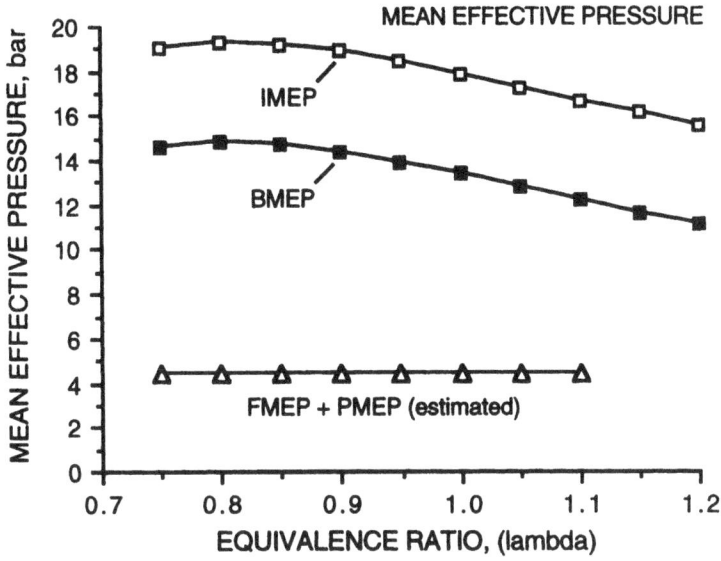

Fig. 1.57 Effect of air-fuel ratio on mean effective pressure.

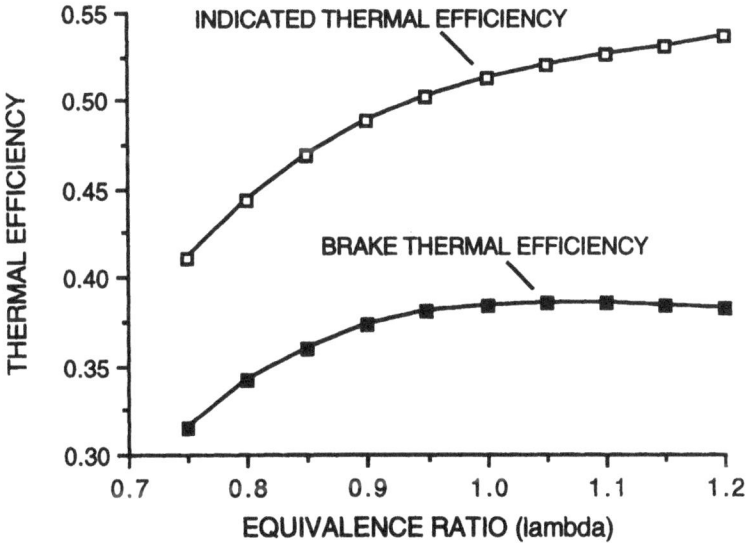

Fig. 1.58 Effect of air-fuel ratio on thermal efficiency.

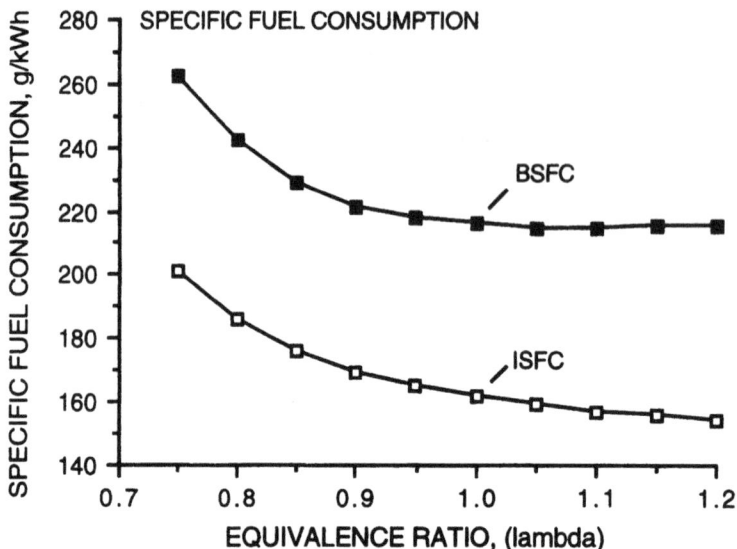

Fig. 1.59 Effect of air-fuel ratio on specific fuel consumption.

calculates the actual fuel consumption rates for the 4000 rpm test point at a stoichiometric AFR, the power is 34.9 kW and the actual fuel consumption rate is 5.65 kg/h. When λ is 0.8 and the AFR is 12.05, the power output is 37.78 kW, the isfc is 0.186 kg/kWh, and the fuel consumption rate is 7.03 kg/h. Hence, the fuel flow rate to the engine at the maximum power setting is 24% higher.

1.9.1.7 Including Friction and Pumping Losses in the Simulation

The discussion in Sec. 1.7.1 gives details of the relationship between the indicated work delivered to the piston compared to that measured on a dynamometer as the brake work. Eq. 1.7.9 shows that the brake work is the summation of the indicated work and the friction and pumping losses. Within the simulation described in this section, no knowledge can be obtained of the pumping loss as this simulation of the Otto cycle, be it with a phased burn or as the ideal Otto cycle, is conducted only over the power cycle. Hence, it generates absolutely no information for the other mechanical cycle of the engine where pumping takes place. In other words, the pumping loop diagram seen in Fig. 1.33 cannot be produced, nor the theory in Sec. 1.6.9, such as Eq. 1.6.32, indexed.

To demonstrate how brake-related data is found from a simulation, we will assume that friction and pumping losses are known for our standard engine and the combined total of these losses is 4.46 bar. In other words, and ensuring the numerical clarity required of the use of Eq. 1.7.9 by using the engineering sign convention for thermodynamics described in Appendix A1.1, Sec. A1.1.2, the brake mean effective pressure of the standard engine at the stoichiometric AFR, can be found from:

$$bmep = imep + pmep + fmep \qquad (1.9.5)$$

hence: $\text{bmep} = \text{imep} + \left(\text{pmep} + \text{fmep}\right) = 17.84 + (-4.46) = 13.38 \text{ bar}$ (1.9.6)

If Eq. 1.9.5 is applied in this fashion to all of the simulations conducted in Sec. 1.9.1.6, where the AFR is varied widely from rich to lean settings, then an equivalent set of simulation output becomes available which is related to brake, rather than indicated, data. So, on Figs. 1.57-1.59 are drawn all of the simulation output as brake data to be compared and contrasted with the indicated data. This output may also be compared with data typical of production si engines in order to assess how closely this simple simulation is approaching reality. The simulation output of brake-related values is for brake mean effective pressure in Fig. 1.57, for brake thermal efficiency in Fig. 1.58, and for brake specific fuel consumption in Fig. 1.59.

In Fig. 1.57, the friction and pumping loss at a combined 4.46 bar is seen drawn on the graph, giving a constant difference of 4.46 bar between the imep and bmep lines. When this is translated into brake thermal efficiency, η_b, it can be seen that the effect is no longer a linear shift from indicated thermal efficiency, η_t. The brake thermal efficiency profile is now very much flatter through the stoichiometric region, a trend that would be very typical of real si engines. This same trend is also seen in the bsfc graph in Fig. 1.59.

The peak bmep is now observed to be around 15 bar, a value that is still on the high side for a real si engine with a delivery ratio of 1.0, but that is not so hugely different from practice. The best bsfc point of about 220 g/kWh is about 14% too optimistic compared to one of today's production si automobile engines.

Even Closer to Reality

It will be recalled that these simulation data incorporate input data for the indices of compression and expansion at the ideal value of 1.4, which is equal to the ratio of specific heats for air. Recalling the discussion in Sec. 1.9.1.4 above, if these indices are each shifted by 0.1, to 1.3 and 1.5, respectively, and the present simulation rerun at a stoichiometric AFR, the bmep becomes 11.62 bar and the bsfc increases to 247 g/kWh, both of which numbers would be well in line with current practice.

1.9.2 Power Cycle Analysis of the Diesel Engine
Some Fundamental Differences between the Diesel and the Otto Cycles

Fig. 1.60 shows the measured cylinder traces from an Otto engine and a diesel engine. They are roughly at the same imep. The diesel engine is a 6.0 liter, six-cylinder, turbocharged, DI unit, and the Otto engine is a 2.0 liter, four-cylinder, naturally aspirated unit with electronic fuel injection (EFI). Because one has a cylinder swept volume that is twice the other, and the diesel engine has a CR of 17 while the Otto engine a CR of 10, the volume scale is drawn as volume ratio so that the engines may be readily compared. The higher CR of the diesel engine shows up during compression as a volume ratio that is about half that of the Otto engine.

The Otto engine would appear to have combustion at constant volume, although from the foregoing above we know that not to be the case. On the other hand, the diesel engine has a much higher end of compression pressure and, more importantly in design terms, temperature. The diesel burn also seems to be at constant volume, but these first impressions will be dispelled below.

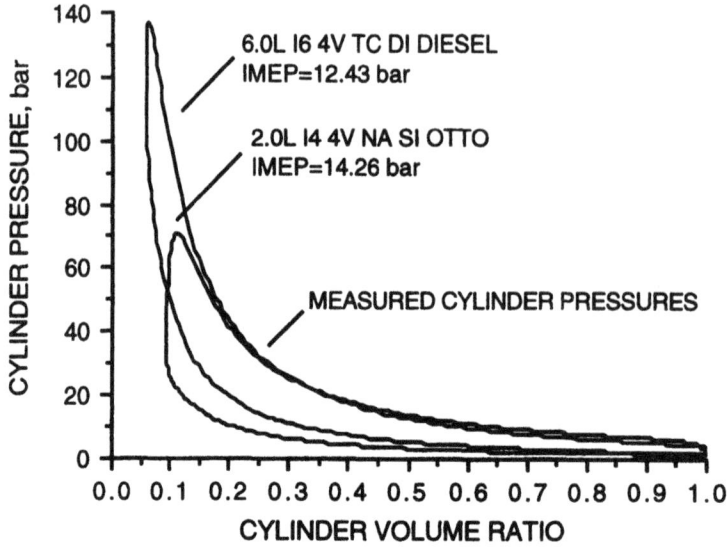

Fig. 1.60 Measured p-V diagrams for a diesel and an Otto engine.

There is no denying that the two p-V diagrams are not alike, either in shape or character. The Otto diagram is short and squat compared to the longer, leaner Diesel diagram.

The Definition of the Ideal Diesel Cycle

The ideal Diesel cycle is presented in some theoretical detail in Appendix A1.1, Sec. A1.1.10. It is illustrated there by Figs. A1.4 and A1.5, showing pressure-volume characteristics and temperature-volume characteristics, respectively. There is a common bond in fundamental thermodynamic thinking between the ideal Diesel and the ideal Otto cycle, the latter being described in Sec. 1.6.9. The elements of the ideal Diesel cycle are, referring to Figs. A1.4 and A1.5:

(a) Adiabatic and isentropic compression from state condition 1–2
(b) Constant volume combustion as heat transfer from 2–3
(c) Constant pressure combustion as heat transfer from 3–4
(d) Adiabatic and isentropic compression from state condition 4–5
(e) Constant volume heat rejection to represent exhaust process from 5–1

Thus, there is an extra heat addition process for diesel combustion compared to the ideal Otto cycle, otherwise the two cycles are identical. The original definition of the ideal Diesel cycle was that it had only constant pressure combustion whereas that described above is often called the Dual cycle [1.2]. I define either, or both, as an ideal Diesel cycle.

Appendix A1.1, Sec. A1.1.10 contains all of the thermodynamic theory required to resolve a given design in terms of its performance characteristics, or its state conditions at points 1–5 throughout the cycle. This theory will be used below to examine a test case for the ideal Diesel cycle.

The Test Engine Used as the Compression-Ignition Engine Example

Throughout this section on the diesel engine, a single cylinder unit of 997.5 cm³ capacity, running at 1800 rpm, is used to illustrate the thermodynamic points to be made. It is, in fact, one cylinder of a six-cylinder turbocharged DI diesel truck engine of 6 liters nominal capacity. It has a bore of 100 mm, a stroke of 127 mm, and a connecting rod with 217 mm centers. The gudgeon pin offset is zero. From the bore and stroke values, it is seen to be an "under-square" engine, the common feature of which has already been commented on in Sec. 1.8.1. The piston position at any point on the rotation of the crank is found using the theory of Sec. 1.4.4 and the cylinder volume by Eq. 1.4.1. The swept volume is 997.5 cm³, but is used in all thermodynamic equations as 997.5 x 10⁻⁶ m³.

A compression ratio of 17 is used, so the clearance volume is obtained from Eq. 1.4.3. This clearance volume is 62.3 cm³, but is used in all thermodynamic equations as 62.3×10^{-6} m³. The maximum and minimum cylinder volumes, i.e., at bdc and tdc, for this diesel engine are 1059.8 and 62.3 cm³, to be used in strict SI units as 1059.8×10^{-6} and 62.3×10^{-6} m³.

The fuel is dodecane with a declared calorific value of 43.5 MJ/kg. From Eq. 1.6.17, the calculated stoichiometric air-fuel ratio for dodecane is 14.95. The test engine is to be examined at a light load point, where the equivalence ratio, λ, is 2.95. Consequently, the air-fuel ratio in the chamber is 44.1. Let it be assumed that 10% of the fuel is to be burned at constant volume, leaving 90% for combustion at constant pressure.

The working fluid within the engine cylinder is considered to be air with the properties given in Appendix A1.1, Secs. A1.1.1 and A1.1.2. As stated before in Sec. 1.9.1, I define that at bdc the swept volume is regarded to be filled with "fresh" air because the clearance volume is considered to be filled with a gas that is inert, exhaust, but that also has the properties of air. In short, only the swept volume at bdc is defined to contain a true "fresh" air. Because many of the data calculated by the simple simulations below are to be compared with measured data from a firing engine, the state conditions selected at bdc at the onset of the compression stroke reflect the reality of turbocharging a diesel engine, compared to the standard reference pressure and temperature conditions selected for the naturally aspirated spark-ignition unit. The state conditions to be used at point 1 in Figs. A1.4 and A1.5 in Appendix A1.1 are 2.33 bar and 70°C. The pressure is known from the measured cylinder pressure data to be 2.33 bar, but the temperature I have had to estimate based on experience, due to the absence of real data.

1.9.2.1 Thermodynamic Navigation around the Ideal Diesel Cycle
Initial Values for Total Mass, Air Mass, Density, DR, and Heat from Fuel at State Point 1

The engine is turbocharged so the state condition 1 is at 2.33 bar and 70°C.
From the state equation, Eq. A1.5, mass of gas in cylinder, m_1:

$$m_1 = \frac{p_1 V_1}{RT_1} = \frac{2.33 \times 10^5 \times 1059.8 \times 10^{-6}}{287 \times 343} = 25.084 \times 10^{-4} \text{ kg}$$

The mass of air trapped in cylinder, m_{ta}:

$$m_{ta} = \frac{p_1 V_{sv}}{RT_1} = \frac{2.33 \times 10^5 \times 997.5 \times 10^{-6}}{287 \times 343} = 23.61 \times 10^{-4} \text{ kg}$$

From Eq. 1.6.1, reference mass, m_{dref}, for DR is:

$$\rho_{at} = \frac{p_{at}}{RT_{at}} = \frac{101325}{287 \times 293} = 1.205 \text{ kg/m}^3$$

From Eqs. 1.6.3 and 1.6.4, delivery ratio is:

$$DR = \frac{m_{as}}{m_{dref}} = \frac{m_{ta}}{\rho_{at} V_{sv}} = \frac{23.61 \times 10^{-4}}{1.205 \times 997.5 \times 10^{-6}} = 1.96$$

The delivery ratio, DR, is 1.96 and shows the considerable amount of air blown into the engine from the compressor section of the turbocharger.

From Eq. 1.6.20, the mass of fuel trapped, m_{tf}:

$$m_{tf} = \frac{m_{ta}}{AFR} = \frac{23.61 \times 10^{-4}}{44.1} = 5.354 \times 10^{-5} \text{ kg}$$

From Eq. 1.6.21, total heat transfer around tdc is equivalent to heat energy in fuel:

$$Q_R = \eta_c m_{tf} C_{fl} = 1.0 \times 5.354 \times 10^{-5} \times 43.5 \times 10^6 = 2329 \text{ J}$$

As the burn proportion at constant volume, k_{cv}, is decided at 0.1, then from Eq. A1.76:

$$Q_2^3 = k_{cv} Q_R = 0.1 \times 2329 = 232.9 \text{ J} \qquad Q_3^4 = (1 - k_{cv}) Q_R = 0.9 \times 2329 = 2096.1 \text{ J}$$

Process 1–2, Adiabatic and Isentropic Compression
From Eqs. A1.26 and A1.50, pressure at end of compression, p_2:

$$p_2 = p_1 \left(\frac{V_2}{V_1}\right)^{-\gamma} = p_1 CR^\gamma = 2.33 \times 10^5 \times 17^{1.4} = 2.33 \times 10^5 \times 52.8$$

$$= 123.02 \times 10^5 \text{ Pa (i.e., 123.02 bar)}$$

From Eq. A1.55, temperature at end of compression, T_2:

$$\frac{T_2}{T_1} = \left(\frac{V_2}{V_1}\right)^{1-\gamma} = (CR)^{\gamma-1} \qquad T_2 = 343 \times 3.106 = 1065.4 \quad K$$

From Eqs. A1.21 and A1.23, the work done during compression is negative:

$$W_1^2 = -m_1 C_v (T_2 - T_1) = -25.084 \times 10^{-4} \times 718 \times (1065.4 - 343) = -1301.1 \text{ J}$$

From Eq. A1.23, the change of internal energy done during compression is positive:

$$U_2 - U_1 = m_1 C_v (T_2 - T_1) = 25.084 \times 10^{-4} \times 718 \times (1065.4 - 343) = 1301.1 \text{ J}$$

Process 2–3, Constant Colume Combustion
The first law of thermodynamics is applied to the constant volume combustion process where the work output is zero.

$$Q_2^3 = U_3 - U_2 + W_2^3 = m_1 C_v (T_3 - T_2) + 0$$

then solving for T_3:

$$T_3 = T_2 + \frac{Q_2^3}{m_1 C_v} = 1065.4 + \frac{232.9}{25.084 \times 10^{-4} \times 718} = 1194.7 \quad K$$

Using the state equation, as in Eq. A1.44, solving for p_3:

$$p_3 = p_2 \times \frac{V_2}{V_3} \times \frac{T_3}{T_2} = p_2 \times \frac{T_3}{T_2} = 123.02 \times 10^5 \times \frac{1194.7}{1065.4} = 137.95 \times 10^5 \text{ Pa}$$
$$\text{(i.e., 137.95 bar)}$$

Change of internal energy during the process from 2–3:

$$U_3 - U_2 = Q_2^3 = 232.9 \quad J$$

Process 3–4 Constant Pressure Combustion

The first law of thermodynamics is applied to the constant pressure combustion process. From Eq. A1.77, solving for T_4:

$$T_4 = T_3 + \frac{Q_3^4}{m_1 C_p} = 1194.7 + \frac{2096.1}{25.084 \times 10^{-4} \times 1005} = 2026.1 \quad K$$

From Eq. A1.78, solving for p_4:

$$p_4 = p_3 \quad (\text{i.e., } 137.95 \text{ bar})$$

From Eq. A1.79:

$$V_4 = V_3 \frac{T_4}{T_3} = 62.3 \times 10^{-6} \times \frac{2026.1}{1194.7} = 105.66 \times 10^{-6} \quad m^3$$

The change of internal energy during the process from 3–4 is analyzed using the first law of thermodynamics for the closed system, from Eqs. A1.77 and A1.83:

$$U_4 - U_3 = m_1 C_v (T_4 - T_3) = 25.084 \times 10^{-4} \times 718 \times (2026.1 - 1194.7) = 1497.4 \quad J$$

By Eq. A1.84 the work output is given by:

$$W_3^4 = m_1 R (T_4 - T_3) = 25.084 \times 10^{-4} \times 287 \times (2026.1 - 1194.7) = 598.5 \text{ J}$$

Process 4–5, Adiabatic and Isentropic Expansion

From Eq. A1.80, temperature at end of expansion, T_5, after process 3–4:

$$\frac{T_5}{T_4} = \left(\frac{V_5}{V_4} \right)^{1-\gamma} \qquad T_5 = 2026.1 \times \left(\frac{1059.8 \times 10^{-6}}{105.66 \times 10^{-6}} \right)^{-0.4} = 805.6 \quad K$$

From Eq. A1.81, pressure at end of expansion, p_5:

$$p_5 = p_4 \left(\frac{V_4}{V_5} \right)^\gamma = 137.95 \times 10^5 \times \left(\frac{105.66 \times 10^{-6}}{1059.8 \times 10^{-6}} \right)^{1.4} = 5.47 \times 10^5 \text{ Pa (which is 5.47 bar)}$$

From Eqs. A1.84, the work done during expansion is positive:

$$W_4^5 = -m_1 C_v \left(T_5 - T_4 \right) = -25.084 \times 10^{-4} \times 718 \times \left(805.6 - 2026.1 \right) = 2198 \text{ J}$$

From Eq. A1.23, the change of internal energy done during expansion is negative:

$$U_5 - U_4 = m_1 C_v \left(T_5 - T_4 \right) = 25.084 \times 10^{-4} \times 718 \times \left(805.6 - 2026.1 \right) = -2198 \text{ J}$$

Process 5–1, to Complete the Cycle by Constant Volume Heat Rejection
From the first law of thermodynamics, as in Eq. A1.85:

$$Q_5^1 = m_1 C_v \left(T_1 - T_5 \right) = 25.084 \times 10^{-4} \times 718 \times \left(343 - 805.6 \right) = -833.1 \text{ J}$$

Change of internal energy during heat rejection from 4-1:

$$U_1 - U_5 = Q_5^1 = m_1 C_v \left(T_1 - T_5 \right) = -833.1 \text{ J}$$

Obtaining Net Values for the Cycle
Net work output from the cycle, from Eqs. A1.70 and A1.71:

$$W_{net} = W_1^2 + W_3^4 + W_4^5 = -1301.1 + 598.5 + 2198 = 1495.4 \text{ J}$$

Ideal Diesel cycle thermal efficiency is given by Eq. A1.71:

$$\eta_t = \frac{\text{net work}}{\text{heat input}} = \frac{W_{net}}{Q_R} = \frac{1495.4}{2329} = 0.642$$

The mean effective pressure, imep, can be found using Eq. 1.6.31 or Eq. A1.86:

$$imep = \frac{W_{nett}}{V_{sv}} = \frac{1495.4}{997.5 \times 10^{-6}} = 14.99 \times 10^5 \quad \text{Pa (i.e., 14.99 bar)}$$

The power output of one engine cylinder can be found using Eq. 1.6.33:

$$\dot{W} = W_{net} \times \frac{rpm}{120} = 1495.4 \times \frac{1800}{120} = 22.43 \times 10^3 \text{ W (i.e., 22.43 kW)}$$

The fuel consumption rate per cylinder, $\dot{m}_f$, is found from:

$$\dot{m}_f = \text{mass per power cycle} \times \text{power cycles per second}$$

$$= m_{tf} \times \frac{\text{rpm}}{120} = 5.354 \times 10^{-5} \times \frac{1800}{120} = 0.803 \times 10^{-3} \text{ kg / s}$$

From Eq. 1.6.36 the specific fuel consumption can be calculated:

$$\text{isfc} = \frac{\dot{m}_f}{\dot{W}} = \frac{0.803 \times 10^{-3}}{22.43 \times 10^3} = 3.58 \times 10^{-8} \text{ kg/Ws}$$

However, in more conventional units of kg/kWh, this becomes:

$$\text{isfc} = 3.58 \times 10^{-8} \frac{\text{kg}}{\text{Ws}} = 3.58 \times 10^{-8} \times 10^3 \times 3600 \frac{\text{kg}}{\text{Ws}} \frac{\text{W}}{\text{kW}} \frac{\text{s}}{\text{h}} = 0.129 \text{ kg/kWh}$$

Graphical Output from a Computer Simulation of the Ideal Diesel Cycle
Figs. A1.4 and A1.5, in Appendix A1.1, show the results of the above calculations for the ideal Otto cycle, but derived on a computer. The procedure and comments that are given in Sec. 1.9.1.1 pertaining to similar computations for the ideal Otto cycle are equally germane here, so they do not need to be repeated. Because Figs. A1.4 and A1.5 are drawn to scale, the numbers for pressure, temperature, and volume which are derived above can be seen at the relevant state positions numbered as 1-5.

Fig. 1.61 shows the cumulative work, heat transfer, and internal energy diagrams for this ideal Diesel cycle. The 10% jump in heat transfer, and accumulated internal energy, can be seen at the tdc point where the constant volume heat input takes place. The extra work required to compress the air in the Diesel cycle, compared to that for the Otto cycle in Fig. 1.40, is evident. It took just 213 J of work to compress the Otto cycle charge at a compression ratio of 10, but it took 1301 J in the diesel engine with the much higher compression ratio of 17. Although the diesel engine is double the capacity of the gasoline engine, that still translates to about three times the compression work for an equivalent diesel unit. This explains why a much more robust starter motor and battery are required to supply the necessary energy to get a car diesel engine going, as compared to its spark-ignition counterpart.

Ignition of the fuel in the diesel engine occurs by heating the injected liquid to vapor and then heating that vapor to its self-ignition temperature. The end of compression temperature, calculated above as T_2, is 1065.4 K or 792.4°C. A temperature at this level is more than sufficient to carry out the two heating processes described and to initiate ignition. However, under normal cold-start conditions the initial temperature at bdc would be more like 20°C and a recalculation of the end of compression temperature gives 910 K, or 637°C; starting is now more problematic. Under arctic air conditions of -30°C, the equivalent end of compression

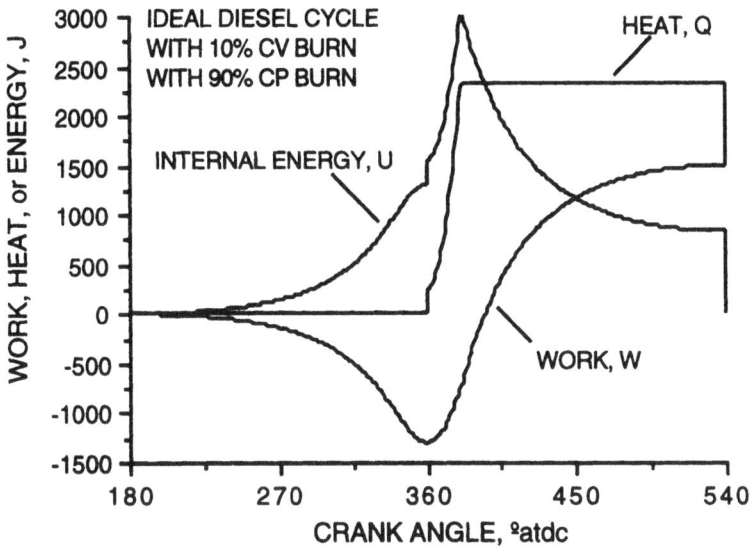

Fig. 1.61 Changes of heat, work, and internal energy in the ideal Diesel cycle.

temperature is 755 K, or 482°C; a conventional starting aid such as a glow plug may need to be seriously considered.

If the data for the ideal Diesel cycle are adjusted, it becomes possible to approach the shape of the p-V diagram and the imep output measured for the actual engine. By changing the input data for both the burn ratio, k_{cv}, and the equivalence ratio, λ, to 0.11 and 3.6, respectively, so that both the measured and calculated values of peak cycle pressure and imep coincide, the result is the p-V diagram plotted in Fig. 1.62.

The imep and peak cycle pressures are closely fitted and the calculated diagram is indeed roughly similar to that measured. However, the calculated values of indicated thermal efficiency and indicated specific fuel consumption, at 0.648 and 0.128 kg/kWh, respectively, are very wide of the reality mark. The real engine, at an equal load level, would have data for the same parameters of η_t and isfc at about 0.50 and 0.165 kg/kWh. An error of this order by a simulation, i.e., over-prediction by some 30%, makes it quite unusable as a design tool.

A second attempt is made by using the classic ideal Diesel cycle where all of the heat transfer to simulate combustion is applied in a constant pressure process only. The theory for this is given in Appendix A1.1, Sec. A1.1.10 using the alternative Eqs. A1.87-A1.89. The input data for both the burn ratio, k_{cv}, and the equivalence ratio, λ, are altered to zero and 3.5, respectively, so that both the measured and calculated value of imep coincides with the measurement. The result is the p-V diagram plotted in Fig. 1.63.

It can be seen that the peak cycle pressure is now poorly correlated, but that will come as no surprise. The fit of measured and calculated diagram is not as good as in Fig. 1.61, even though the comparison is being made at equal values of imep. The overall computed values of

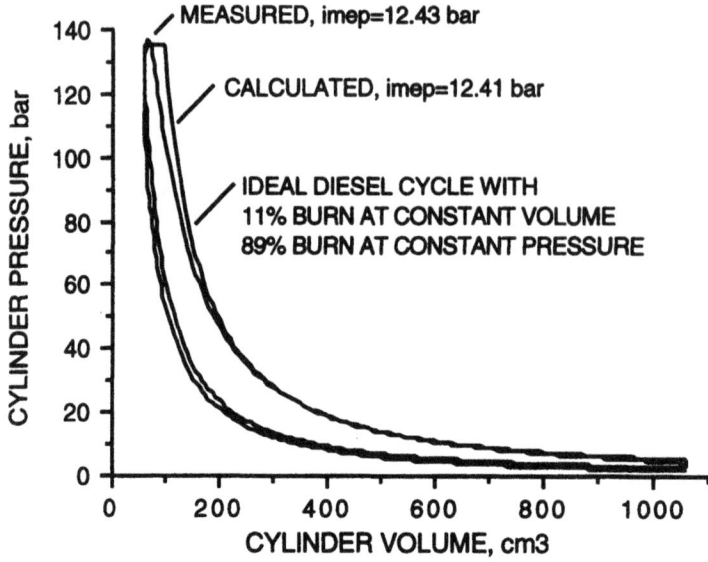

Fig. 1.62 Matching measured data with a 0.1 k_{cv} burn ratio.

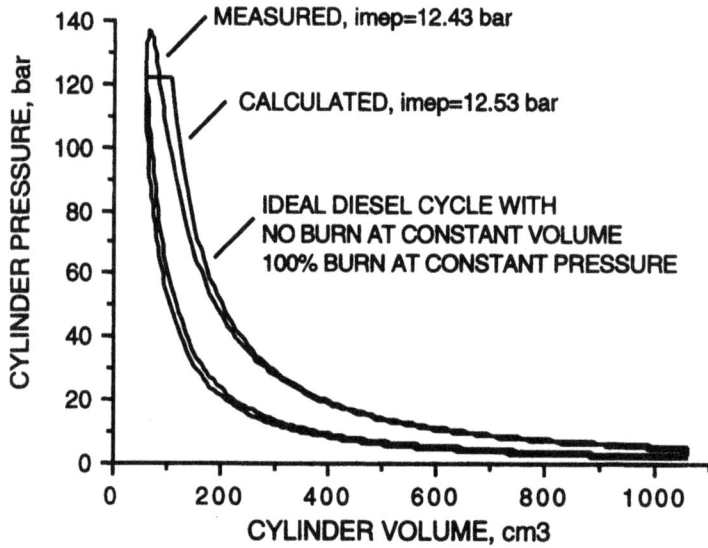

Fig. 1.63 Matching measured data with a constant volume burn only.

indicated thermal efficiency and indicated specific fuel consumption, are now 0.636 and 0.130 kg/kWh, respectively, but are still unacceptably optimistic compared to conventional measured data.

The main problem is that the heat input profile that gives the measured data in Figs. 1.62 and 1.63 is quite different from that envisaged for either version of the ideal Diesel cycle, i.e., as either a constant pressure process only, or with some of that total heat input also released in a constant volume process.

1.9.2.2 Thermodynamic Navigation around the Diesel Cycle with Phased Combustion

The theoretical approach typified by Eq. 1.9.1, which is explained in great detail in Chapter 4, provides the mass fraction burned diagram in Fig. 1.37 from measured cylinder pressure-volume data of a 2.0 liter, four-cylinder, spark-ignition Otto engine. The measured pressure-volume data, shown in Fig. 1.60, are taken from the 6.0 liter turbocharged diesel engine, and is analyzed using the same techniques. The mass fraction burned diagram obtained for the diesel engine is shown in Fig. 1.64. It will be recalled that the engine load level is quite modest, being at 12.43 bar imep, which is obtained at an equivalence ratio of about 3, whereas maximum usable power will be at a λ level of about 1.65. The word "usable," here means that the engine performs without having an illegal excess of black smoke, i.e., carbon particulates, in the exhaust gas.

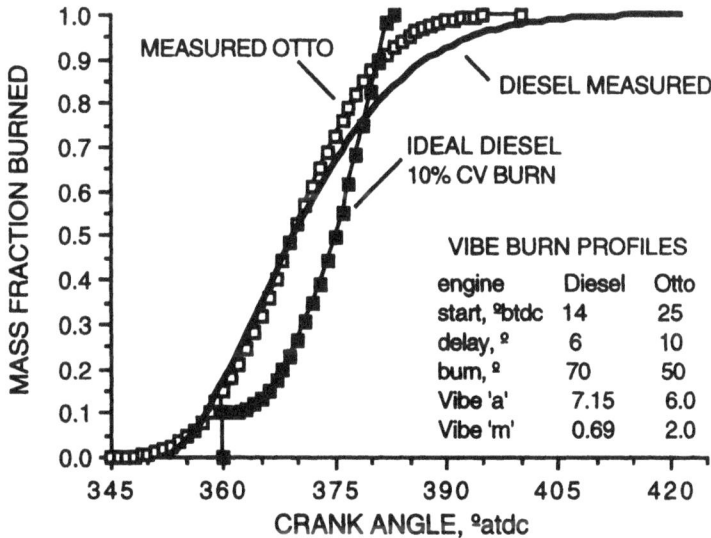

Fig. 1.64 Mass fraction burned for actual engines and for an ideal Diesel cycle.

To make comparisons more meaningful, on this same Fig. 1.64 is repeated the measured mass fraction burned curve for the spark-ignition engine as seen in Fig. 1.37. Also on Fig. 1.64 is drawn the mass fraction burned data which emanates from the very same ideal Diesel cycle analysis that produced Figs. A1.4 and A1.5, in Appendix A1.1, and Fig. 1.61. A summary of the numeric data for the Vibe functions of the measured Diesel and Otto burn curves is presented on Fig. 1.64; the symbolism for these numbers can be seen in Eq. 1.9.1 and from the discussion associated with this equation.

To further emphasize the point, the mass fraction burned curve is transformed into a heat release rate profile. Before you reach Sec. 4.2, which discusses the fundamental theory regarding the relationship between heat release rate, $\dot{Q}_{R_\theta}$, and mass fraction burned, B, the following simple equations will suffice to make the introduction. If the amount of heat added by combustion is δQ_{R_θ}, in any given crank angle interval $d\theta$, at a crank angle θ, from the beginning of the burn process, then the heat release rate is defined as:

$$\dot{Q}_{R_\theta} = \frac{\delta Q_{R_\theta}}{d\theta} \ \text{J/deg} \tag{1.9.7}$$

The mass fraction burned, B_θ, at any given angle θ, from initiation of combustion in a total burn period b is given by:

$$B_\theta = \frac{\sum\limits_{0}^{\theta} \delta Q_{R_\theta}}{\sum\limits_{0}^{b} \delta Q_{R_\theta}} = \frac{\sum\limits_{0}^{\theta} \delta Q_{R_\theta}}{Q_R} \tag{1.9.8}$$

So, mass fraction burned is the cumulative integral under the heat release rate curve to any angular period θ compared to that for the entire (crankshaft) angular period, b. For the total burn period it is the total amount released, Q_R, as defined by Eq. 1.6.21. Using this theory, the mass fraction burned curves, as measured for these Otto and diesel engines, are redrawn as the heat release rate diagrams graphed in Fig. 1.65; they are both rendered as specific diagrams to make comparison easier.

The first observation that can be made is that the measured mass fraction burned profiles for the diesel and Otto engine appear to be somewhat similar. However, an examination of the numeric data for their Vibe functions show that, while the shape may be somewhat similar, the only data value of any real similarity is the 50% value which occurs at the same location at some 9 °atdc. Otherwise, compared to that for the diesel engine, the Otto burn starts some 8° earlier, lasts for 20° longer, and has Vibe exponents that are significantly different. The more rapid input of heat around the tdc location for the diesel engine is more noticeable on the heat release diagram in Fig. 1.65, as is the long tail of slow end burning which is equally obvious in Fig. 1.64. What is very clear is that the mass fraction burned diagram for the ideal diesel cycle

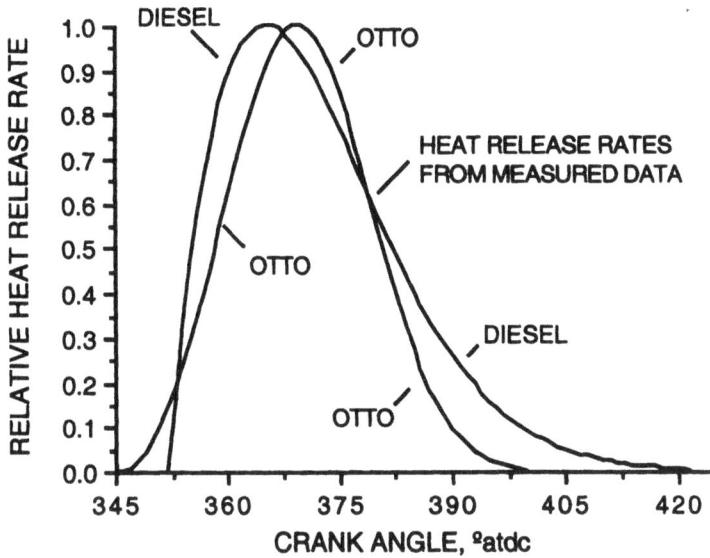

Fig. 1.65 Measured heat release rate curves from Otto and diesel engines.

has virtually nothing in common with that from measured data. The entire heat input rate is much too rapid, so it is perhaps not too surprising that its incorporation into a simulation gives only a modest correspondence with the measured p-V diagram and the overall performance characteristics.

The next step is to replay the measured mass fraction burned diagram into the simulation and find out, to what degree, it has improved the accuracy of this model of a diesel engine.

Calculation of the Diesel Cycle with Phased Combustion

The simulation is programmed to receive a phased burn as a Vibe function, and alternate changes of indices of compression and expansion, otherwise it is identical with the ideal cycle computation. The computer simulation is conceptually identical to that discussed above for the Otto cycle with a phased burn. The Vibe coefficient function data, as seen in Fig. 1.64, is used as input data. The index of compression is also an input data value and that found in the measured Diesel p-V diagram, i.e., 1.33, is inserted into the simulation. The index of expansion, as with the Otto cycle experiments earlier in Sec. 1.9.1.4, is raised above the ideal ratio of specific heats so that heat loss is simulated in the expansion process on the power stroke; a modest value of 1.45 is used as a data value. For completeness, the measured index of expansion is 1.35, but of course is not used as a data value in this simulation for all of the same reasons expounded in the discussions in Sec. 1.9.1.4 and Appendix A1.1, Sec. A1.1.6. Otherwise, all of the numeric data for the engine, given at the very beginning of this section (Sec. 1.9.2), are put into the input computer data files.

The results of the simulation, showing both measured and computed pressure-crank angle diagrams, are shown in Fig. 1.66.

121

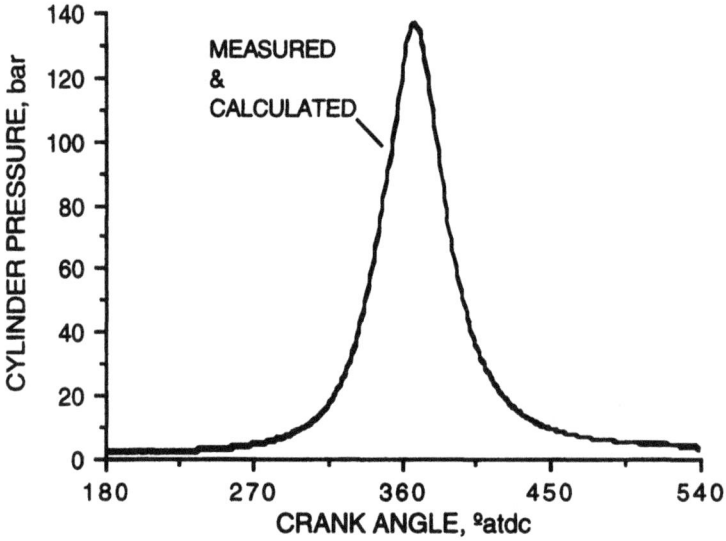

Fig. 1.66 Cylinder pressure—measured and calculated using a phased burn.

Because the simulation closely mimics the measured pressure-crank angle diagram, it becomes difficult to determine the level of accuracy attained. Hence, Fig. 1.67 is created, which shows the pressure difference between the two graphs in Fig. 1.66 In Fig. 1.67 it is observed that the majority of the error is around the tdc point at the end of compression and the beginning of combustion.

To illustrate further the conclusions of Sec. 1.9.1.4 regarding heat loss due to indices of compression and expansion not being identical to the ideal ratio of specific heats, the cumulative work, heat, and internal energy diagrams are drawn in Fig. 1.68.

The cumulative heat, Q, graphed in Fig. 1.68 illustrates the heat loss during compression. This goes negative until the heat input during the simulated combustion, then decreases again with further heat loss during expansion, until the cycle concludes with the final heat rejection "exhaust" process. The effect is considerable and can be seen by comparing Fig. 1.68 with Fig. 1.61, i.e., for the ideal Diesel cycle with the same heat input at tdc. The maximum cumulative heat input rises to just over 2000 J compared to 2329 J when the compression and expansion processes are adiabatic. The final maximum value of cumulative heat for the phased burn case is 1836 J, so 493 J, or 21%, is lost from the cylinder during the compression and expansion processes. Due to heat loss from the cylinder, the internal energy of the cylinder gas never attains the level it did through the adiabatic processes in the ideal cycle. The maximum internal energy attained in Fig. 1.61 is 3000 J and in Fig. 1.68 it is 2142 J. Such is the heat transfer realism that must be incorporated into any simulation which has the necessary pretensions of accuracy to become a useful design tool.

Considering the close correlation of measured and calculated pressure diagrams in Fig. 1.66, unsurprisingly the predicted value of indicated mean effective pressure, imep, and that measured are coincidental at 12.43 bar. The predicted value of indicated specific fuel consumption is 0.155 kg/kWh, i.e., some 20% worse than for the ideal Diesel cycle computations

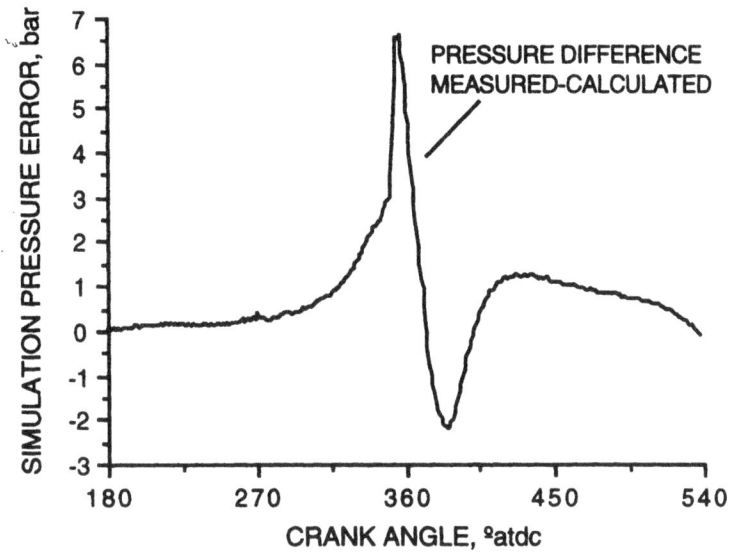

Fig. 1.67 Difference between measured and calculated cylinder pressure.

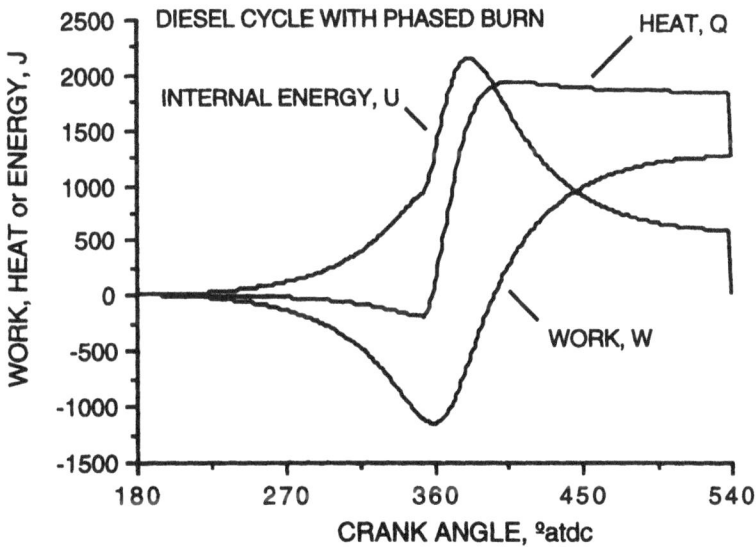

Fig. 1.68 Cumulative work, heat, and internal energy using a phased burn.

as seen from Figs. 1.62 and 1.63. "Worse" in this context means closer to reality, as the indicated thermal efficiency has dropped from an over-optimistic 0.648 to a more-realistic 0.533. Further comment on this point continues below.

Phased Combustion with Varying Fuel Injection Quantity

The diesel engine controls load solely by altering the injected fuel quantity between that which provides a zero work output, i.e., the idle condition, and a full load condition, i.e., that which is normally limited by the amount of black smoke permitted by law to appear in the exhaust gas. To fulfill these two criteria, the fuelling will range between an equivalence ratio, λ, of about 4.5 to about 1.5. The equivalence ratio is seen to be always lean of stoichiometric. Over this fueling range, the combustion efficiency, η_c, as defined in Eq. 1.6.21, for a diesel fuel such as dodecane, can be culled from measured data and expressed as a function in terms of equivalence ratio as:

$$
\begin{aligned}
1.33 < \lambda < 2.33 \quad & \eta_c = 0.35332 + 0.56797\lambda - 0.12472\lambda^2 \\
\lambda \geq 2.33 \quad & \eta_c = 1.0
\end{aligned}
\tag{1.9.9}
$$

This expression is not an absolute "law" and would have a somewhat, but not hugely, different characteristic from one particular DI diesel engine to another, and between DI and IDI engines in general.

If the fueling level is set so as to attain the maximum possible power at an equivalence ratio of about 1.4, the maximum cylinder pressure encountered may well exceed that value of cylinder pressure, and also temperature by inference, which is regarded as safe from the standpoint of durability of the cylinder components. In practice, the timing of the onset of fuel injection, and the rate at which it is then injected, are controlled so that a predetermined maximum cylinder pressure limit is not exceeded. Modern technological advances in electronics make this form of fueling control more accurate and precise. To illustrate these points, the equivalence ratio, λ, is changed in a series of computational experiments from 3.0, to 2.75, 2.5, 2.25, 2.0, 1.85, and 1.65. None of the other data values pertaining here is changed, with the exception that Eq. 1.9.9 is employed to find the combustion efficiency, η_c, at any given fueling level, and the timing point for the start of fuel injection is retarded to ensure that the same maximum cylinder pressure is reached at every fuelling condition. The start of fuel injection is set to 14, 13, 12, 11, 10, 9, and 8 °btdc, respectively, for each of the equivalence ratios shown. The shape and delay of the mass fraction burned diagram is retained, as is the amount of ignition delay, but the previous statement changes the point of fuel injection from 346 to 347, 348, 349, 350, 351, and 352 °atdc, respectively. Although this is not precisely what would happen in practice, the effect within this simple cycle simulation is sufficiently close to reality to illustrate the effect on an engine.

Fig. 1.69 shows that the result of controlling the injection timing is indeed a relatively constant maximum cylinder pressure with variable fueling. In this figure, all of the simulated pressure-crank angle diagrams are shown over the burn period. The effect of injection delay on the shape of the cylinder pressure profile with enrichment is quite clear. At the richest

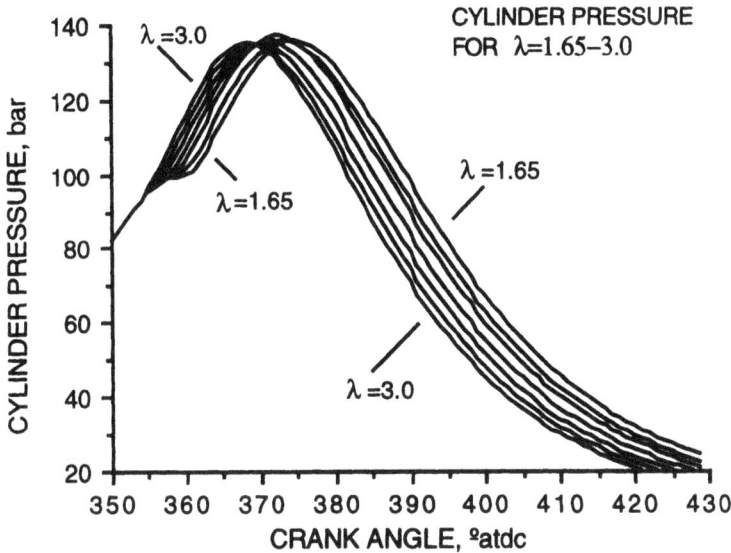

Fig. 1.69 Cylinder pressure diagrams from a phased burn with varying fueling.

setting, where λ is 1.65, the cylinder pressure only begins to rise at about the tdc point. The rate of pressure rise recorded is about 3.8 bar/deg, which makes it at least as high as the full load point for the Otto engine seen in Fig. 1.44 and, in effect, at least as noisy. That it is actually a noisier combustion process is best seen in Fig. 1.44, where the rates of pressure rise are plotted at full load for both the Otto and the diesel engine. For the diesel engine, the next derivative of this parameter (in bar/deg/deg units) has a value that is at least double that for the Otto engine. These rapid vibrations, transmitted through the metal of the cylinder head and block, provides the "rattle" that is so typical of Diesel combustion.

The mean cylinder gas temperature diagrams resulting from the simulation are plotted in Fig. 1.70. It is observed that, although the maximum cylinder pressure has been kept relatively constant, the maximum cycle temperature has not. The difference in maximum cycle temperature from the lowest to the highest loading is some 500°C.

The effect of fueling on the overall performance parameters for the indicated values of mean effective pressure, imep, and specific fuel consumption, isfc, and thermal efficiency, η_t, are shown in Figs. 1.71, 1.72, and 1.73, respectively. The imep profile in Fig. 1.71 shows a steady increase with fueling enrichment and the torque peak has not yet been reached by an equivalence ratio of 1.65; it would do so when λ approached 1.3, but the black smoke level in the exhaust would make the engine environmentally unacceptable. This trend is predicted by the graph of indicated thermal efficiency in Fig. 1.73 which shows that it begins to drop off as the rich limit approaches, due to the application of Eq. 1.9.9 within the simulation. The reciprocal picture of this comment is given in Fig. 1.72, where the specific fuel consumption starts to rise as the same rich limit approaches. At the leaner air-fuel ratio settings, i.e., when λ is greater than 2.0, the isfc and η_t curves are relatively flat.

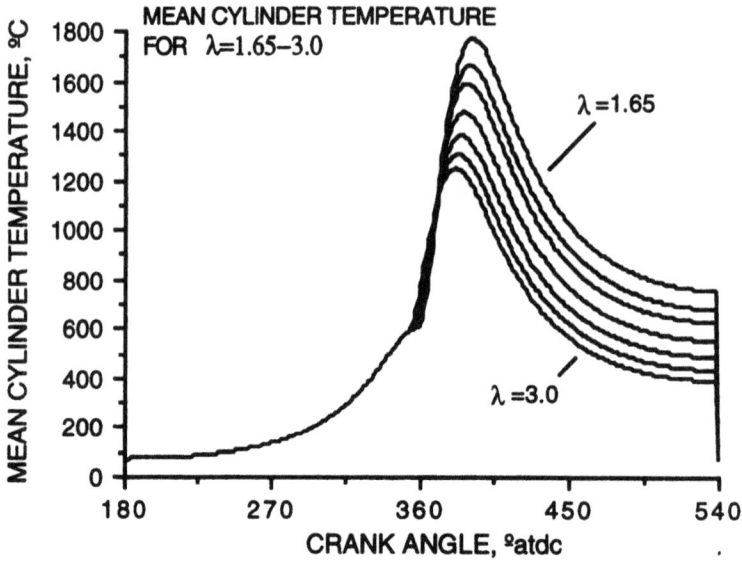

Fig. 1.70 Cylinder temperature diagrams from a phased burn with varying fueling.

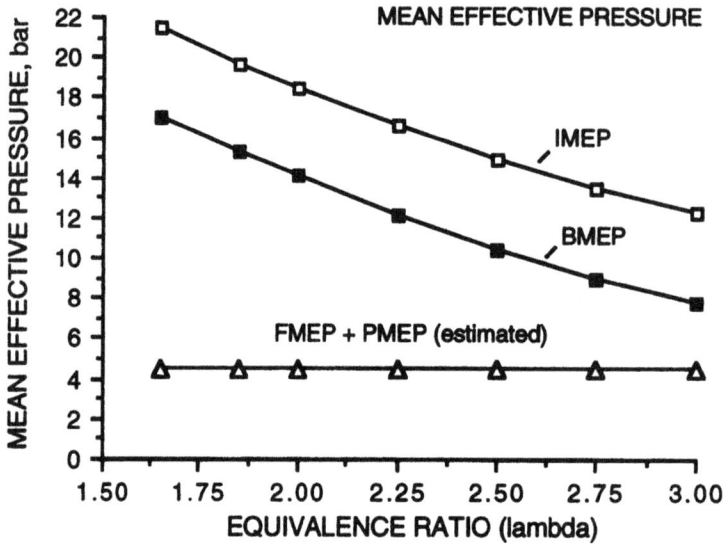

Fig. 1.71 The effects of fueling on mean effective pressure.

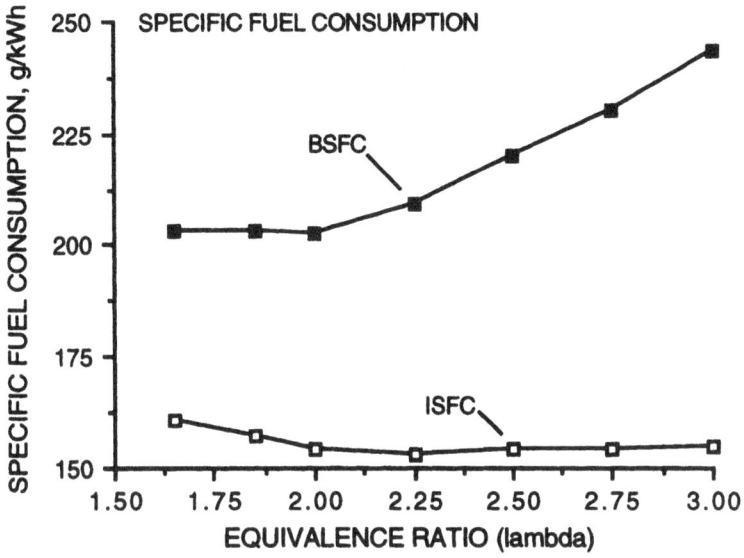

Fig. 1.72 The effects of fueling on specific fuel consumption.

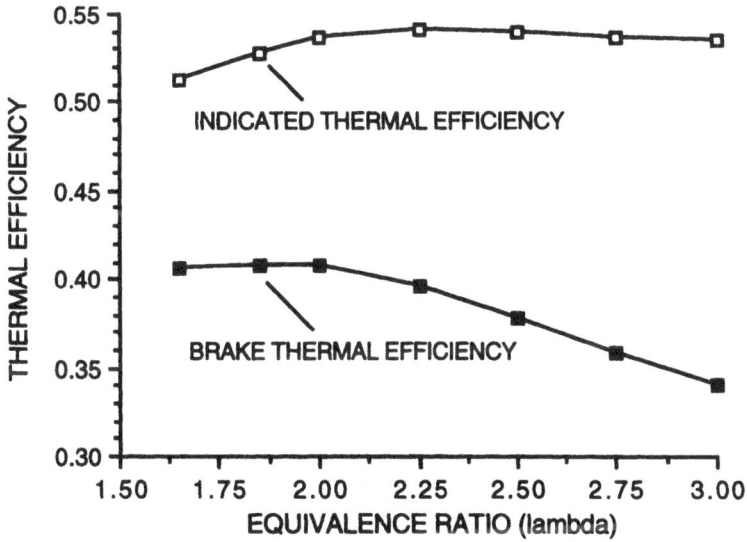

Fig. 1.73 The effects of fueling on thermal efficiency.

Including Friction and Pumping Losses in the Diesel Simulation

In a similar fashion to the simulation of the Otto engine described in Sec. 1.9.1.6, it is possible to assign to the above diesel engine a friction and pumping loss and so translate all simulation data from indicated parameters to brake-related parameters.

It will be assumed that friction and pumping losses are known for our standard diesel engine and the combined total of these losses is 4.46 bar. For the sake of a clear comparison, this number is retained to be exactly the same as that used for the Otto engine in Sec. 1.9.1.6, even though the numeric value for a diesel engine is generally somewhat higher than for an Otto engine. If the pumping and friction loss is applied in the manner of Eq. 1.9.6, as a combined fmep and pmep to all of the Diesel simulations above where the equivalence ratio is varied from rich to lean, then another set of simulation output appears which is related to brake, rather than indicated, data. So, on Figs. 1.71-1.73 is drawn the further simulation output of brake-related data. The graphs of brake-related values are for brake mean effective pressure in Fig. 1.71, for brake specific fuel consumption in Fig. 1.72, and for brake thermal efficiency in Fig. 1.73.

In Fig. 1.71, the consistent 4.46-bar difference appears on the plot, with the maximum usable bmep at 17.0 bar, a number that is quite consistent with current turbocharged DI practice. In Fig. 1.72, a different trend appears for bsfc, compared to the trend commented on above for isfc. Now, the best brake specific fuel consumption is closer to the highest usable load point and deteriorates by some 20% toward the lightest load point. In Fig. 1.73, the reciprocal comment applies, i.e., the best brake thermal efficiency is near the highest load point and drops as the load level decreases by reduced fuelling. The value of best point specific fuel consumption is 200 g/kWh, and its associated thermal efficiency is 41%, two values that are absolutely in line with those obtained by current, turbocharged, DI diesel engines used in trucks.

1.10 The End of the Beginning of Simulation of the Four-Stroke Engine

You have now been introduced to the concept of the thermodynamic simulation of an engine through the use of the ideal Otto and Diesel cycles. These ideal cycles are shown to be rather inaccurate for the purpose of design, although the fundamental lessons in thermodynamics are invaluable in pointing the way forward to enhancing that accuracy.

The first major step in this regard is the introduction of a combustion process that results from the analysis of measured cylinder pressure data and that relates the happenings in the combustion chamber to the burning of a real fuel with respect to time. When this alone is introduced into the simulation of the ideal cycle, be it Otto or Diesel, the cylinder pressure diagrams and the predicted overall performance parameters begin to more closely coincide with measured data.

The second major step forward is to introduce heat transfer loss into the simulation of the ideal cycle which, of course, makes it no longer an ideal cycle but moves it ever closer to reality. This heat transfer loss is effected by using indices of expansion and compression that differ from the ratio of specific heats and so negates the assumption that those processes are ideal, i.e., they are adiabatic and isentropic.

The third, and final, step forward is to assign friction and pumping losses to the engine cycle and transform the simulation output from indicated performance parameters to brake related data. That step gives engine performance characteristics for bmep and brake specific fuel consumption that are very similar to that measured for the Otto and diesel engines used as computation examples.

So If the Cycle Simulation Is Now That Good, What Is the Rest of this Book About?

Firstly, all simulation above is carried out by assuming the trapping state conditions of pressure and temperature at the beginning of compression. In short, the other half of the four-stroke cycle has been ignored, thus far. A simulation is required that will predict those very state conditions at the trapping point on the compression stroke, which is really at intake valve closure and not at bdc. The simulation must be extended to include all of the gas flow throughout all of the ducting of the engine, and in and through the cylinder on the intake and exhaust strokes, simply to ensure that those trapping conditions are predicted accurately. Heat transfer is taking place at every instant of time, to and from all of the gases throughout all of the engine, and failure to compute it correctly means just that. That is what Chapter 2 is all about.

Secondly, the flow through the valves, ports, and throttles, i.e., all ducting and cylinder elements, is far from an ideal process. The flow area geometry of poppet valves, as calculated in this chapter, will be shown to be an optimistic prediction of the actual flow area as the gas velocity changes or even the direction of that gas flow alters. So, measurements must be made of the losses accompanying real-world flow regimes, to accompany the gas flow theory of Chapter 2 to further enhance its simulation accuracy. That is what Chapter 3 is all about.

Thirdly, the gases that flow into an engine are mostly air and those that appear within it from combustion, and then flow out of it, are definitely not air. The properties of real gases differ considerably from air at any temperature, and all gases have properties that change with temperature. These facts alone deserve detailed study and incorporation within an accurate simulation of an engine. Combustion itself cannot be treated as a heat transfer process in the simplistic manner employed in this chapter. Properly carried out, one can predict the local and mean cylinder temperatures, and the time-varying gas composition, during the burn process so that one has more accurate information on the composition of the exhaust gas and of the cylinder state conditions throughout the power cycle. Heat transfer is taking place at every instant of time, to and from all of the gases present in the cylinder of the engine. Failure to analyze heat transfer in depth means that we must continue to supply the indices of expansion and compression to the simulation, rather than vice-versa. All of that theory needs to be incorporated into an accurate engine simulation. That is what Chapter 4 is all about.

In Chapter 5, we take stock of what has been learned in Chapters 2-4, and conduct some modelling of real engines to determine if simulation accuracy has actually improved, as a result of the ever-increasing sophistication of the thermodynamics and gas dynamics gained in the interim since Chapter 1.

In Chapter 6, we descend from the lofty heights of the best thermodynamics and gas dynamics and stoop to empiricism! Here, we search through the simulations to see if there are simplistic empirical relationships that can be culled, in order to save time in, and give direction to, the optimization of an engine. The problem with engine simulation is that there are at least ten times as many data values for the designer to assign as simulation input data as there are components in the engine. In the search for optimization, it is all too possible to miss the obvious data "tree" due to the presence of so many other data "trees" within the same engine "wood"! In Chapter 6, we carry out some very useful data tree-thinning. Much as in Sec. 1.8, we search for empirical simplicity to make more thorough a multi-faceted, complex, optimization procedure.

In Chapter 7, we discuss noise. All unsteady gas flow and combustion creates noise. The perfect silencer, intake or exhaust, produces inflow or outflow characteristics of constant gas velocity. The quietest engine never fires. The compromise, as always in engineering, must be sought between the ideal and reality. The unsteady gas flow theory for the ducting is extended to the propagation of that flow into the atmosphere beyond the ducting terminations so as to predict the noise it makes at some point in space. This theory is incorporated within the engine simulation to permit the design of silencers having high gas flow rates with a minimum of noise creation. The optimum design has a minimum impact on the trapping state conditions at the beginning of the compression stroke, thereby reducing the loss of power by silencing. That is what Chapter 7 is all about, which brings us back full circle to the start of Chapter 2.

In short, all of the above is what the rest of this book is all about.

References for Chapter 1

1.1 C.L. Cummins, *Internal Fire*, Society of Automotive Engineers, Warrendale, Pa., 1989.

1.2 E.F. Obert, *Internal Combustion Engines*, 10th Printing, International Textbook Company, Scranton, Pa., 1960.

1.3 C.F. Taylor and E.S. Taylor, *The Internal Combustion Engine*, International Textbook Company, Scranton, Pa., 1962.

1.4 C.F. Caunter, *Motor Cycles, a Technical History*, Science Museum, London, HMSO, 1970.

1.5 H.S. Ricardo, *The Pattern of My Life*, Constable, London, 1968.

1.6 H.S. Ricardo, *The High-Speed Internal-Combustion Engine*, 4th Edition, Blackie, London, 1953.

1.7 P.E. Irving, *Tuning for Speed*, 3rd Edition, Temple Press Books, London, 1956.

1.8 V. Willoughby, *Classic Motorcycles*, Hamlyn, London, 1975.

1.9 G.P. Blair, *Design and Simulation of Two-Stroke Engines*, R-161, Society of Automotive Engineers, Warrendale, Pa., 1996.

1.10 R.C. Cross, "Experiments with Internal Combustion Engines," Chairman's Address, Proc.I.Mech.E. (Automobile Division), p.1, 1957-58.

1.11 M.C.I. Hunter, *Rotary Valve Engines*, John Wiley, New York, 1946.

1.12 L.R.C. Lilly, *Diesel Engine Reference Book*, Butterworths, London, 1984.

1.13 W.L. Brown, "Methods for Evaluating Requirements and Errors in Cylinder Pressure Measurement," SAE International Congress, Detroit, SAE Paper No. 670008, Society of Automotive Engineers, Warrendale, Pa., 1967.

1.14 SAE J1349, Engine Power Test Code, Spark Ignition and Diesel, June 1985.

1.15 *Automotive Engineering*, Society of Automotive Engineers, Warrendale, Pa., p. 81, December 1997.

1.16 G.P. Blair, H.B. Lau, A. Cartwright, B.D. Raghunathan, and D.O. Mackey, "Coefficients of Discharge at the Apertures of Engines," SAE International Off-Highway Meeting, Milwaukee, Wisc., September 1995, SAE Paper No. 952138, pp. 71-85, Society of Automotive Engineers, Warrendale, Pa.

1.17 G.P. Blair, F.M. Drouin, "The Relationship between Discharge Coefficients and the Accuracy of Engine Simulation," SAE Motorsports Engineering Conference and Exposition, Dearborn Mich., December 8-10, 1996, SAE paper No. 962527, Society of Automotive Engineers, Warrendale, Pa.

1.18 G.P. Blair, D. McBurney, P. McDonald, P. McKernan, and R. Fleck, "Some Fundamental Aspects of the Discharge Coefficients of Cylinder Porting and Ducting Restrictions," Society of Automotive Engineers, International Congress, Detroit, Mich., February 1998, SAE Paper No. 980764, Society of Automotive Engineers, Warrendale, Pa.

1.19 F.Y. Chen, *Mechanics and Design of Cam Mechanisms*, Pergamon, Oxford, 1982.

1.20 *Cams and Cam Mechanisms* (Editor, J.R. Jones), I.Mech.E. Conference, Liverpool Polytechnic, 1974.

1.21 S. Molian, *The Design of Cam Mechanisms and Linkages*, Constable, London, 1968.

1.22 G.P. Blair, "Correlation of Measured and Calculated Performance Characteristics of Motorcycle Engines," Funfe Zweiradtagung, Technische Universität, Graz, Austria, 22-23 April 1993, pp. 5-16.

1.23 SAE J604, Engine Terminology and Nomenclature, June, 1995.

1.24 R.S. Benson, N.D. Whitehouse, *Internal Combustion Engines*, Volumes 1 and 2, Pergamon, Oxford, 1979.

1.25 J.B. Heywood, *Internal Combustion Engines Fundamentals*, McGraw-Hill, New York, 1988.

1.26 SAE J1349, Engine Power Test Code, Spark-Ignition and Diesel, June 1995.

1.27 ISO 3046, Reciprocating Internal Combustion Engines: Performance-Parts 1, 2, and 3, International Standards Organization, 1981.

1.28 SAE J1088, Test Procedure for the Measurement of Exhaust Emissions from Small Utility Engines, February 1993.

1.29 BS 1042, Fluid Flow in Closed Conduits, British Standards Institution, 1981.

1.30 G.J. Van Wylen and R.E. Sonntag, *Fundamentals of Classical Thermodynamics*, SI Version 2e, Wiley, New York, 1976.

1.31 W. Watson, "On the Thermal and Combustion Efficiency of a Four-Cylinder Petrol Motor," Proc. I. Auto. E., Vol. 2, p. 387, 1908-1909.

1.32 E.W. Huber, "Measuring the Trapping Efficiency of Internal Combustion Engines Through Continuous Exhaust Gas Analysis," SAE International Congress, Detroit, Mich., February, 1971, SAE Paper No. 710144, Society of Automotive Engineers, Warrendale, Pa.

1.33 D. Olsen, P. Puzinauskas, and O. Dautrebande, "Development and Evaluation of Tracer Gas Methods for Measuring Trapping Efficiency in 4-Stroke Engines," SAE Fuels and Lubricants Meeting, Dearborn, Mich., May 4-6, 1998, SAE Paper No. 981382, Society of Automotive Engineers, Warrendale, Pa.

1.34 F. J. Laimbock and R. Kirchberger, "Development of a 150cc, 4-Valve CVT Engine for Future Emission and Noise Limits," SAE International Off-Highway Meeting, Milwaukee, Wisc., September 1998, SAE Paper No. 982052, Society of Automotive Engineers, Warrendale, Pa.

1.35 N. Windrum, *The Ulster Grand Prix*, Blackstaff Press, Belfast, 1979.

1.36 E. Sher, "The Effect of Atmospheric Conditions on the Performance of an Air-Borne Two-Stroke Spark-Ignition Engine," SAE Paper No. 844962, 1984.

Appendix A1.1 Fundamental Thermodynamic Theory for the Closed Cycle

The thermodynamic statements that are set down here are but a precis of a full presentation to be found in formal undergraduate texts such as those by Van Wylen [1.30] or Heywood [1.25]. It is expected that the reader will either be fully aware of this subject matter and can safely ignore these jottings, or will find them useful as a memory jog from student days past, or will be unaware of basic thermodynamics in which case the reference text by Van Wylen [1.30] should be studied first. Nevertheless, as this and the following chapters are read and the theory studied, there will come moments when these brief statements will elucidate a stubborn theoretical line in a way that a thousand extra words at that juncture might not.

A1.1.1 The Equation of State

Gas properties are specified as pressure, p, and temperature, T, in absolute units of Pa and K, respectively. Pressure is normally measured by a gauge of some type and the units of such pressure are referred to as a gauge pressure, p_g, above the prevailing atmospheric pressure, p_a. The absolute pressure, p, is the addition of the gauge pressure and the atmospheric pressure:

$$p = p_g + p_a \qquad (A1.1)$$

The pressure may also be referred to as a dimensionless ratio, as a pressure ratio, P, defined as:

$$P = \frac{p}{p_0} \qquad (A1.2)$$

where the standard reference atmospheric pressure, p_0, is defined as 101325 Pa, or 1.01325 bar.

A measured temperature will almost certainly be recorded in units such as Celsius, i.e., as T_c °C. The absolute temperature, T, in Kelvin units is found from:

$$T = T_c + 273 \qquad (A1.3)$$

All gases have a gas constant, R, which is found from the universal gas constant, $\overline{R}$, which has a value of 8314.3 J/kg-molK. For air, which has a molecular weight, M, of 29 kg/kg-mol, the gas constant, R, is found by:

$$R = \frac{\overline{R}}{M} = \frac{8314.3}{29} = 287 \ \text{J/kgK} \qquad (A1.4)$$

In any volume of gas, V, which is known to be at a pressure, p, and temperature, T, the mass of gas contained therein, m, can be found from the equation of state. The units of volume are in strict SI units, which are m^3, and the units of mass are kg. The equation of thermodynamic state is:

$$pV = mRT \qquad (A1.5)$$

Hence, from Fig. A1.1 for the two geometries illustrated:

$$p_1V_1 = mRT_1 \qquad p_2V_2 = mRT_2$$

The density, ρ, and the specific volume, v, of a gas are reciprocally related to each other. The units of density and specific volume are kg/m^3 and m^3/kg, respectively. They are defined as, and related to each other, by:

$$v = \frac{V}{m} \qquad \rho = \frac{m}{V} \qquad \rho = \frac{1}{v} \qquad (A1.6)$$

The equation of state can be modified to incorporate density and specific volume as:

$$p = \rho RT \qquad pv = RT \qquad (A1.7)$$

Hence, in Fig. A1.1:

$$p_1 = \rho_1 RT_1 \qquad p_2v_2 = RT_2$$

A1.1.2 The First Law of Thermodynamics for a Closed System

The definition of a closed system is that the mass within the system is constant. The volume, the pressure, and/or the temperature may change, but the mass does not. As far as an engine is concerned, a closed system is typified by the gas within the cylinder when all valves or apertures to the cylinder are closed and the assumption made that the piston rings make a perfect gas seal with the cylinder walls. A typical situation is sketched in Fig. A1.1, where the piston in the side-valve engine is seen to move from position 1 to position 2, and the result of what is visibly a compression process is that the volume decreases from V_1 to V_2, and the pressure increases from p_1 to p_2. Work is expended by the piston to accomplish this compression process. These data on pressure and volume are graphed in Fig. A1.2. If the piston had been descending in the sketch, it would have been an expansion process instead.

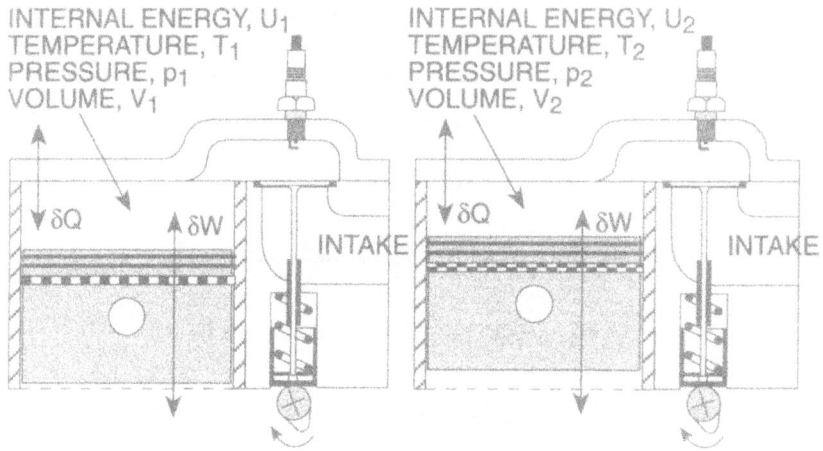

Fig. A1.1 A closed system process in a four-stroke engine.

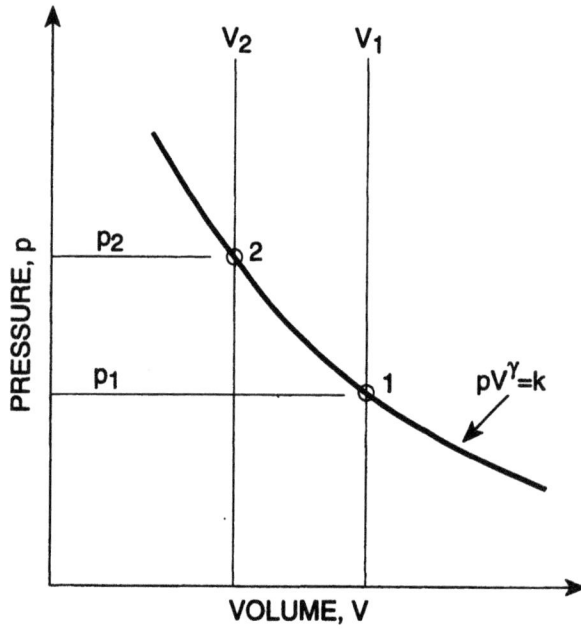

Fig. A1.2 The p-V changes for a closed system process.

If the gas within a closed system experiences a work process, δW, it is observed by a change of volume, dV, at pressure, p, and can be evaluated using the thinking behind Eq. 1.6.26 as:

$$\delta W = pdV \qquad (A1.8)$$

The first law of thermodynamics for a closed system states that any changes of heat transfer, δQ, and internal energy, dU, are related to any possible work change, δW, by:

$$\delta Q = dU + \delta W \qquad (A1.9)$$

Actually the above Eq. A1.9 is more completely stated as:

$$\delta Q = \left[dU + dKE + dPE + d\,? \right]_{system} + \delta W$$

where the term in the square brackets details the possible changes of system energy that could take place. However, because a closed system containing gas rarely has any significant content of kinetic energy (KE), or potential energy (PE), or some unknown such as magnetohydrodynamic energy, much less changes of them, i.e., dKE, etc., then the norm is that only the change of internal energy, dU, is of real significance.

Eq. A1.9 introduces a sign convention in engineering thermodynamics whereby heat into a system, and work out of a system, are defined as being numerically positive. The corollary is that the reverse direction, i.e., heat out of a system, and work into a system, are defined as being numerically negative.

The first law of thermodynamics for a closed system also states that cyclic processes can be evaluated directly, where the definition of a thermodynamic cycle is a series of processes that culminate in the initial thermodynamic state being restored:

$$\oint \delta Q = \oint \delta W \qquad (A1.10)$$

which means that:
$$\oint dU = 0 \qquad (A1.11)$$

The definition of a thermodynamic cycle should be noted carefully, for in the four-stroke cycle engine there are two mechanical cycles executed before the actual initial thermodynamic state is indexed and the totality of that thermodynamic cycle is completed.

The internal energy, dU, can be evaluated for any given process by:

$$dU = mC_v dT \qquad (A1.12)$$

where C_v is the specific heat at constant volume, in units of J/kgK. The implication is that for any process, such as that sketched in Fig. A1.1, where the temperature changes from T_1 to T_2, one can calculate the change of internal energy involved as follows:

$$U_2 - U_1 = \int_{T1}^{T2} mC_v dT = mC_v \int_{T1}^{T2} dT = mC_v(T_2 - T_1) \qquad (A1.13)$$

Moving the specific heat term to outside the integral sign means that specific heat has been designated a constant. Although real gases do not have specific heats that are constants with temperature, the temperature change would have to be many hundreds of degrees Celsius for it to become a numerical consideration of importance. Thus, for most calculations specific heat is considered to be a constant. However, when Chapter 4 is approached, and real combustion is considered theoretically, a different perspective will be required.

Further specific, i.e., per unit mass, definitions for heat transfer, work, and internal energy may be made:

Specific heat transfer, δq $\delta q = \dfrac{\delta Q}{m}$ (A1.14)

Specific internal energy, du $du = \dfrac{dU}{m}$ (A1.15)

Specific work transfer, δw $\delta w = \dfrac{\delta W}{m}$ (A1.16)

Thus, Eq. A1.9 can be redefined by dividing across by the mass, m, to give a specific formulation of the first law of thermodynamics:

$$\delta q = du + \delta w \qquad (A1.17)$$

You will notice that the differential term in front of heat and work is a δ and not a d, as in front of internal energy. This is because heat and work are path functions, i.e., the outcome of any process depends on the path, and internal energy is a point function, i.e., the outcome is dependent only on the initial and final states. Mathematically, this is of real significance:

$$\int_1^2 \delta Q = Q_1^2 \qquad \int_1^2 \delta W = W_1^2 = \int_1^2 p dV \qquad \int_1^2 dU = mC_v(T_2 - T_1)$$

In short, the internal energy can be found by knowing only the initial and final temperatures but, taking work as the example, unless we know the path function connecting p and V we have no possibility of evaluating the work integral. You can see that in Eq. A1.25, where the path is defined by the index n. If we do not know that number, we cannot evaluate the work content.

There are two specific heats to be defined for a gas: the specific heat at constant volume, C_v, and the specific heat at constant pressure, C_p. The formal definitions of each are:

$$C_v = \frac{du}{dT} \qquad C_p = \frac{d(u + pv)}{dT} = \frac{dh}{dt} \qquad (A1.18)$$

where the variable, h, is specific enthalpy, the definition of which is encapsulated in the above equation. It transpires that the ratio of these specific heats is a "constant" and is also a property of the gas. The fact is that specific heats for real gases do vary somewhat, but not hugely, with temperature. This will be discussed later in the text. The ratio of specific heats, γ, is defined as:

$$\gamma = \frac{C_p}{C_v} \qquad (A1.19)$$

It also transpires that these specific heats are related to the gas constant, R, from the formal definitions of enthalpy and internal energy, and the state equation as Eq. A1. 7:

$$dh = C_p dT = d(u + pv) = C_v dT + d(RT) = C_v dT + RdT$$

hence, $$C_p = C_v + R \qquad (A1.20)$$

Consequently, combining Eqs. A1.19 and 1.20:

$$C_v = \frac{R}{\gamma - 1} \qquad C_p = \frac{\gamma R}{\gamma - 1} \qquad (A1.21)$$

The value of specific heat at constant volume, C_v, for air is measured at 718 J/kgK (via Eq. A1.33; see below). From Eq. A1.20 the specific heat at constant pressure for air, C_p, is calculated to be 1005 J/kgK, as the gas constant for air, R, is a known quantity at 287 J/kgK. The ratio of specific heats, γ, for air is seen from Eq. A1.19 to be 1.4.

A1.1.3 An Adiabatic Work Process in a Closed System

By definition, an adiabatic process has zero heat transfer to, or from, the system across its boundary:

$$\delta Q = 0 \tag{A1.22}$$

If an adiabatic process occurs and it is as sketched in Fig. A1.1, where the pressure, temperature, and volume of the closed system change from p_1 to p_2, T_1 to T_2, and V_1 to V_2, respectively, the first law of thermodynamics will now state, from the integration of Eq. A1.9:

$$0 = \int_1^2 dU + \int_1^2 \delta W$$

With Eq. A1.21, this yields the work transfer, W_1^2:

$$W_1^2 = \int_1^2 \delta W = -\int_1^2 dU = -mC_v(T_2 - T_1) = \frac{p_2 V_2 - p_1 V_1}{1 - \gamma} \tag{A1.23}$$

Suppose this same adiabatic work process shown could be represented, as in Fig. A1.2, by a polynomial function connecting pressure and volume with an exponent n, and where the symbol k is a constant:

$$pV^n = k \tag{A1.24}$$

This work process could be evaluated by:

$$W_1^2 = \int_1^2 pdV = \int_1^2 kV^{1-n}dV = k\left|\frac{V^{1-n}}{1-n}\right|_1^2 = k\frac{V_2^{1-n} - V_1^{1-n}}{1-n} = \frac{p_2 V_2 - p_1 V_1}{1-n} \tag{A1.25}$$

which is found by remembering, from Eq. A1.24, that:

$$p_1 V_1^n = p_2 V_2^n = k$$

Comparison of Eqs. A1.25 and A1.23 shows that the polynomial indices, n, and γ, must be identical. In short, in an adiabatic process in a closed system the polynomial index n is the ratio of specific heats, γ. Hence, the work path function becomes:

$$pV^{\gamma} = k \qquad (A1.26)$$

A1.1.4 A Work Process in a Closed System with Heat Transfer
The sketch in Fig. A1.1 shows a work process that could well be taking place with heat transfer to, or from, the cylinder. Indeed, a simple combustion process could be simulated by assigning the burning of fuel as a heating process that occurs internally without any mass transfer, where the cylinder gas may be losing heat to the walls at the same time.

Heat Transfer To, or From, the System
First, let the simpler case of heat transfer to, or from, the cylinder walls be considered. Eq. A1.9 is integrated again, but this time the polynomial index is n, and n cannot be equal to γ, for the process is no longer adiabatic. Where the index, n, is not equal to γ, that index is known as a polytropic index to distinguish it from γ, which is labeled as the isentropic index.

$$\int_{1}^{2} \delta Q = \int_{1}^{2} dU + \int_{1}^{2} \delta W \qquad (A1.27)$$

$$Q_1^2 = mC_v\left(T_2 - T_1\right) + \frac{p_2 V_2 - p_1 V_1}{1 - n} \qquad (A1.28)$$

From the state equation:

$$p_1 V_1 = mRT_1 \qquad p_2 V_2 = mRT_2 \qquad (A1.29)$$

From the polytropic function:

$$p_1 V_1^{\,n} = p_2 V_2^{\,n} = k \qquad (A1.30)$$

In most numerical problems, enough data are known at the commencement of any step position, 1, to solve the three Eqs. A1.28-A1.30 to evaluate all unknown property data at step position 2.

For example, if the initial pressure, temperature, and volume, p_1, T_1, and V_1, the final volume, V_2, and the amount of heat transfer, Q_1^2, are all known, then the final pressure and temperature, p_2 and T_2, can be evaluated. In reverse, by measuring cylinder pressures and volumes, it becomes possible to evaluate the amount of heat transfer, Q_1^2. As shown later in Chapter 4, this is actually the theoretical basis for the evaluation of heat release from combustion.

An Adiabatic Process with Internal Heat Transfer

Second, let us use this solution to deal simply with a combustion process that is assumed to act as an internal heat transfer to an adiabatic system. The adiabatic definition refers to zero external heat transfer to, or from, the system. If the system is defined as adiabatic, then the above theory can be used, but the polytropic index, written above as n, reverts to the ideal, i.e., γ. The amount of internal heat transfer due to combustion becomes the numeric input value for Q_1^2.

A1.1.5 Heat Transfer Processes at Constant Volume in the Closed Cycle

Consider the situation in Fig. A1.1, but where the piston does not move. The volume will stay constant, i.e., V_2 equals V_1. In such a process all work transfer will be zero:

$$W_1^2 = \int_1^2 pdV = 0 \tag{A1.31}$$

However, heat transfer could take place, during which the temperature would rise or fall, depending on the direction of that heat transfer process. It will be evaluated, as usual, by the first law of thermodynamics for the closed system, as in the mass specific version of Eq. A1.17.

$$\delta q = du + \delta w = du + 0 = C_v dT \tag{A1.32}$$

Indeed, this is how the specific heat at constant volume can be measured for any gas by carrying out heat transfer experiments in a constant-volume, closed system. Integrating Eq. A1.32 over the process from state conditions 1 to 2 and including the system mass:

$$Q_1^2 = \int_1^2 \delta Q = mC_v \int_1^2 dT = mC_v \left(T_2 - T_1 \right) \tag{A1.33}$$

The system mass is constant and can be found from the state equation at the initial or final conditions at constant volume, where V_1 equals V_2:

$$m = \frac{p_1 V_1}{RT_1} = \frac{p_2 V_2}{RT_2} \tag{A1.34}$$

140

This process is precisely that specified in the discussion of the ideal Otto cycle in Sec. 1.6.8 as item (a) for the combustion process at tdc and as item (d) for the exhaust heat rejection process at bdc. These two processes are evaluated by the use of Eqs. A1.33 and A1.34. For example, the heat addition process at tdc is found by equating the heat transfer, Q_1^2, to the heat released by the fuel during the ideal constant volume process and solving for the unknown values of pressure and temperature at the end of the process, p_2 and T_2. More fundamental theory, which uses these thoughts regarding the ideal Otto cycle, is to be found in this appendix in Sec. A1.1.8.

A1.1.6 Direction of Heat Transfer in a Polytropic Process in a Closed Cycle

Consider the use of the first law for the analysis of a closed system where both polytropic work transfer and heat transfer are taking place. Using a modified version of the solution achieved in Eq. A1.28 for this type of process:

First Law in the specific format:

$$\delta q = du + \delta w$$

From Eq. A1.28:

$$\delta q = C_v dT + \frac{RdT}{1 - n} \tag{A1.35}$$

Rearranging:

$$\delta q = dT\left(C_v + \frac{R}{1 - n}\right) \tag{A1.36}$$

Inserting C_v from Eq. A1.21:

$$\delta q = dT\left(\frac{-R}{1 - \gamma} + \frac{R}{1 - n}\right) \tag{A1.37}$$

Rearranging:

$$\delta q = RdT\left(\frac{n - \gamma}{(1 - n)(1 - \gamma)}\right) \tag{A1.38}$$

A Compression Process

(i) In a compression process, the temperature, dT, will rise, i.e., dT is positive. Note that the values of the ratios of n and γ are always greater than unity.

If heat is added to the system, i.e., if δq is positive, then from Eq. A1.38:

$$n > \gamma \tag{A1.39}$$

If heat is lost from the system, i.e., if δq is negative, then from Eq. A1.38:

$$n < \gamma \tag{A1.40}$$

An Expansion Process
(ii) In an expansion process, the temperature, dT, will fall, i.e., dT is negative. If heat is added to the system, i.e., if δq is positive, then from Eq. A1.38:

$$n < \gamma \tag{A1.41}$$

If heat is lost from the system, i.e., if δq is negative, then from Eq. A1.38:

$$n > \gamma \tag{A1.42}$$

It will be observed, supporting the contentions of Eqs. A1.23-A1.25, that if n equals γ, then the heat transfer is always zero and the process must be adiabatic as a consequence.

A1.1.7 Gas Property Relationships in Adiabatic and Polytropic Processes

Consider a polytropic process proceeding from thermodynamic state 1 to state 2, as sketched in Fig. A1.1.
From the state equation:

$$p_1 V_1 = mRT_1 \qquad p_2 V_2 = mRT_2 \tag{A1.43}$$

Dividing these two equations:

$$\frac{p_1 V_1}{p_2 V_2} = \frac{mRT_1}{mRT_2} = \frac{T_1}{T_2} \tag{A1.44}$$

With specific volume instead:

$$\frac{p_1 v_1}{p_2 v_2} = \frac{mRT_1}{mRT_2} = \frac{T_1}{T_2} \tag{A1.45}$$

With density instead:

$$\frac{p_1}{p_2} = \frac{\rho_1 R T_1}{\rho_2 R T_2} = \frac{\rho_1 T_1}{\rho_2 T_2} \tag{A1.46}$$

From the polytropic equations:

$$pV^n = k \tag{A1.47}$$

which also can be expressed as: $\qquad pv^n = k \tag{A1.48}$

or as: $\qquad p = k\rho^n \tag{A1.49}$

Integrating between any two thermodynamic states 1 and 2, Eqs. A1.46-A1.48 become:

$$\frac{p_1}{p_2} = \left(\frac{V_1}{V_2}\right)^{-n} \tag{A1.50}$$

$$\frac{p_1}{p_2} = \left(\frac{v_1}{v_2}\right)^{-n} \tag{A1.51}$$

$$\frac{p_1}{p_2} = \left(\frac{\rho_1}{\rho_2}\right)^{n} \tag{A1.52}$$

Incorporating the temperature relationships in Eqs. A1.44-A1.46 will give modified versions of Eqs. A1.50-A1.52 as:

$$\frac{p_1 V_1}{p_2 V_2} = \frac{p_1}{p_2}\left(\frac{p_1}{p_2}\right)^{-\frac{1}{n}} = \left(\frac{p_1}{p_2}\right)^{\frac{n-1}{n}} = \frac{T_1}{T_2} \tag{A1.53}$$

or: $$\frac{p_1}{p_2} = \left(\frac{T_1}{T_2}\right)^{\frac{n}{n-1}} \tag{A1.54}$$

or:
$$\frac{p_1 V_1}{p_2 V_2} = \left(\frac{V_1}{V_2}\right)^{-n} \frac{V_1}{V_2} = \left(\frac{V_1}{V_2}\right)^{1-n} = \frac{T_1}{T_2} \qquad (A1.55)$$

or:
$$\frac{V_1}{V_2} = \frac{v_1}{v_2} = \left(\frac{T_1}{T_2}\right)^{\frac{1}{1-n}} \qquad (A1.56)$$

or:
$$\frac{\rho_1}{\rho_2} = \left(\frac{T_1}{T_2}\right)^{\frac{1}{n-1}} \qquad (A1.57)$$

It should be made clear, if the process is isentropic and adiabatic where the polynomial index is γ, rather than n as in a polytropic case, then the index γ will replace the index n in Eqs. A1.43-A1.57.

A1.1.8 The Thermal Efficiency of the Ideal Otto Cycle

The temperature-volume characteristics of this cycle are shown in Fig. A1.3. Sec. 1.6.9 describes how there are only two heat transfer processes in this cycle. The first is an internal heat addition process, i.e., to simulate combustion at tdc, taking place at constant volume as an adiabatic process from states labeled as 1 to 2 in Fig. A1.3.

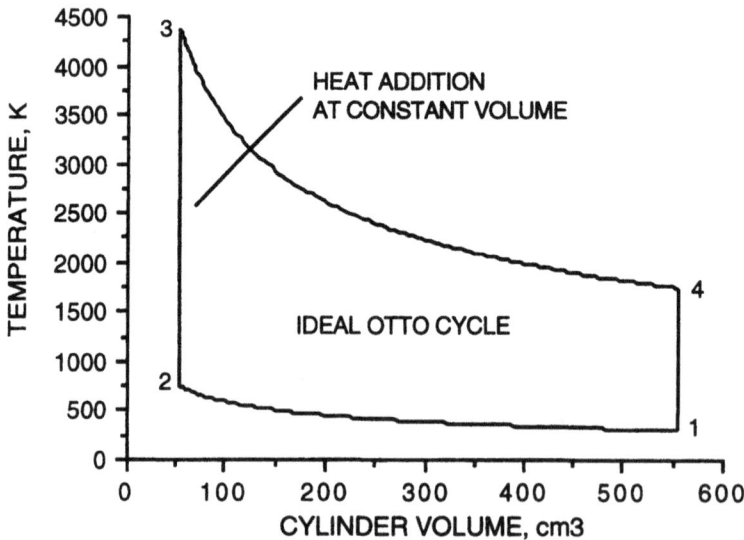

Fig. A1.3 The temperature-volume characteristics of the ideal Otto cycle.

The second is an internal heat rejection process, i.e., exhaust but without any mass flow at bdc, taking place at constant volume as an adiabatic process from states labeled as 4 to 1 in Fig. A1.3. The initial state 1 is so indexed and the thermodynamic cycle is completed. The other two processes labeled as 2–3 and 3–4 are adiabatic and isentropic, and are compression and expansion processes, respectively. By definition from Sec. A1.1.5, these processes have zero heat transfer characteristics. The thermal efficiency of this cycle is defined in Eq. 1.6.24 as

$$\eta_t = \frac{\text{net work output}}{\text{heat input}} \tag{A1.58}$$

There being only one heat input, and two work processes, reduces to:

$$\eta_t = \frac{W_1^2 + W_3^4}{Q_2^3} \tag{A1.59}$$

From Eqs. 1.23 and 1.33 for such work and heat processes, Eq. A1.59 becomes:

$$\eta_t = \frac{-mC_v\left(T_2 - T_1\right) - mC_v\left(T_4 - T_3\right)}{mC_v\left(T_3 - T_2\right)} \tag{A1.60}$$

Rearranging: $\qquad \eta_t = 1 - \dfrac{T_4 - T_1}{T_3 - T_2} = 1 - \dfrac{T_1}{T_2}\left(\dfrac{\dfrac{T_4}{T_1} - 1}{\dfrac{T_3}{T_2} - 1}\right) \tag{A1.61}$

However, from Eq. A1.55:

$$\left(\frac{V_1}{V_2}\right)^{1-\gamma} = \frac{T_1}{T_2} \qquad \left(\frac{V_4}{V_3}\right)^{1-\gamma} = \frac{T_4}{T_3} \tag{A1.62}$$

Compression ratio, CR, is defined as:

$$CR = \frac{V_1}{V_2} = \frac{V_4}{V_3} \tag{A1.63}$$

Hence, Eq. A1.62 shows:

$$\frac{T_1}{T_2} = \frac{T_4}{T_3} \qquad \text{(A1.64)}$$

Rearranging:

$$\frac{T_3}{T_2} = \frac{T_4}{T_1} \qquad \text{(A1.65)}$$

Using Eqs. A1.62, A1.63, and A1.65, Eq. A1.61 becomes:

$$\eta_t = 1 - \frac{T_1}{T_2} = 1 - CR^{1-\gamma} = 1 - \frac{1}{CR^{\gamma-1}} \qquad \text{(A1.66)}$$

A1.1.9 A Constant Pressure Process with Work and Heat Transfer in a Closed System

In the next section, a discussion will be presented on a thermodynamic process in a closed system, taking place at constant pressure, where there is both work and heat transfer. Let us examine the basic thermodynamics of it here. The first law of thermodynamics for a closed system can be applied to this case. From Eq. A1.17:

$$\delta q = du + \delta w$$

From Sec. A1.1.2:

$$\delta q = C_v dT + pdv \qquad \text{(A1.67)}$$

Because pressure is a constant, differentiating Eq. A1.7 yields:

$$d(pv) = d(RT)$$

The left side of the equation becomes:

$$d(pv) = pdv + vdp = pdv$$

The right side of the equation becomes:

$$d(RT) = RdT$$

Hence, the work term is:

$$\delta w = pdv = RdT \qquad \text{(A1.68)}$$

and heat transfer is:

$$\delta q = C_v dT + R dT = (C_v + R)dT = C_p dT \qquad (A1.69)$$

In short, the heat transfer in a constant pressure process in a closed system can be logically evaluated using the specific heat at constant pressure and the initial and final state conditions of temperature.

A1.1.10 The Thermal Efficiency of the Ideal Diesel Cycle

The pressure-volume and temperature-volume characteristics of the ideal version of this cycle are shown in Figs. A1.4 and A1.5.

It is seen that, compared to the Otto cycle in Fig. A1.3, there is an extra heat transfer process in this cycle. The total amount of heat to be added at the tdc location is carried out in two internal heat transfer processes: a constant-volume process followed by a constant-pressure process. In many texts [1.2, 1.3], the ideal Diesel cycle is defined as having all of the heat addition take place at constant pressure only, i.e., the constant-volume segment does not exist. The version of the ideal Diesel cycle that has both constant-volume and constant-pressure heat addition segments is often referred to as the Dual cycle [1.2]. This is semantics, as both are ideal versions of a cycle attempting to simulate the thermodynamics in a real diesel engine.

Apart from the mechanism of internal heat addition, the cycle is identical to the Otto cycle. The initial state 1 is indexed and the thermodynamic cycle commenced. Adiabatic isentropic compression takes place from state 1 to state 2. Constant volume heat addition takes place internally and adiabatically from state 2 to state 3. Constant pressure heat addition takes place internally and adiabatically from state 3 to state 4. Constant volume heat rejection in this closed system, to simulate the real exhaust process in an open system, takes place adiabatically as an internal heat transfer from state 4 to state 5.

The thermal efficiency of this, or any, thermodynamic cycle is defined in Eq. 1.6.24 as

$$\eta_t = \frac{\text{net work output}}{\text{heat input}} \qquad (A1.70)$$

There being two heat input, and three work processes, this reduces to:

$$\eta_t = \frac{W_1^2 + W_3^4 + W_4^5}{Q_2^3 + Q_3^4} \qquad (A1.71)$$

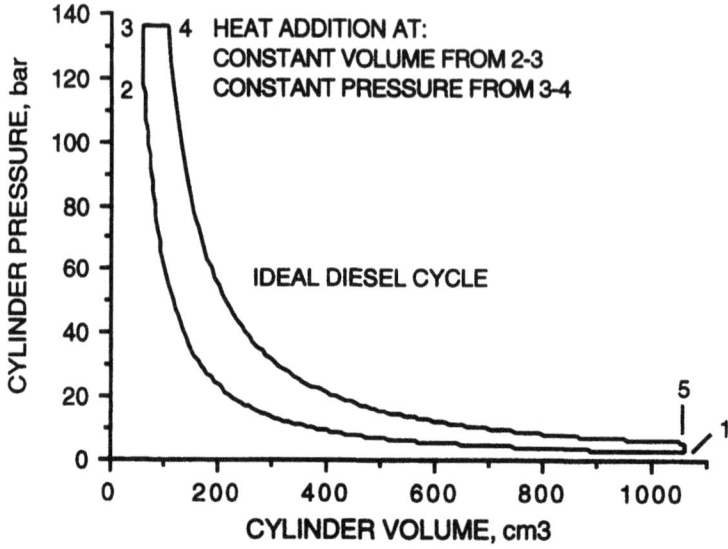

Fig. A1.4 The pressure-volume characteristics of the ideal Diesel cycle.

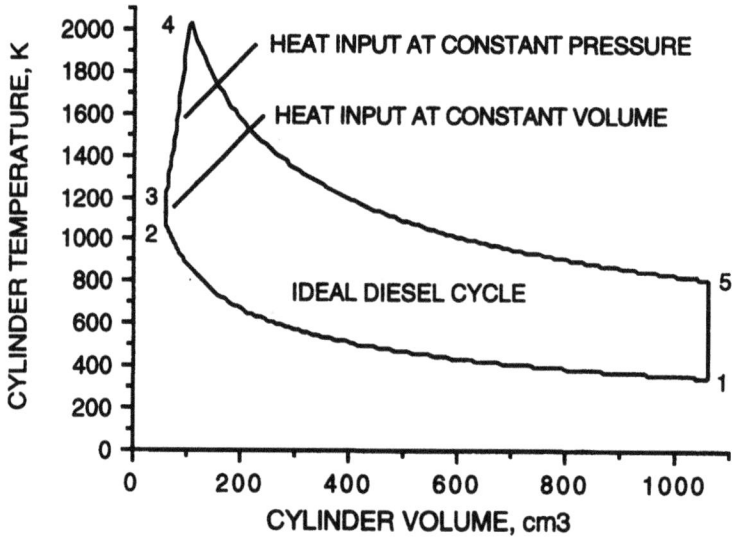

Fig. A1.5 The temperature-volume characteristics of the ideal Diesel cycle.

From Eqs. 1.23, 1.33, and A1.69 for such work and heat processes, Eq. A1.71 becomes:

$$\eta_t = \frac{-mC_v\left(T_2 - T_1\right) + mR\left(T_4 - T_3\right) - mC_v\left(T_5 - T_4\right)}{mC_v\left(T_3 - T_2\right) + mC_p\left(T_4 - T_3\right)} \qquad (A1.72)$$

Rearranging: $\qquad \eta_t = \dfrac{-\left(T_2 - T_1\right) + \left(\gamma - 1\right)\left(T_4 - T_3\right) - \left(T_5 - T_4\right)}{\left(T_3 - T_2\right) + \gamma\left(T_4 - T_3\right)} \qquad (A1.73)$

During any given analysis, the total heat, Q_R, to be added by heat transfer to simulate combustion will be a known numeric quantity. No numeric solution for thermal efficiency, or indeed any other cyclic parameter, is possible unless a decision is made regarding the burn proportion, k_{cv}, of the total heat transfer to be added during the constant volume segment at tdc. This burn proportion, k_{cv}, is defined as follows:

$$k_{cv} = \frac{\text{heat added at constant volume}}{\text{total heat added}} = \frac{Q_2^3}{Q_R} = \frac{Q_2^3}{Q_2^3 + Q_3^4} \qquad (A1.74)$$

If the value of the burn proportion, k_{cv}, is unity, then the solution reverts to that for the ideal Otto cycle in Sec. A1.1.8. If the value of k_{cv} is zero, then the result is an ideal Diesel cycle with heat added only at constant pressure, and falls under the classic definition of the ideal Diesel cycle. Any value of k_{cv} between 0 and 1 defines it, as Obert [1.2] would have done, as the Dual cycle. Because we are working with a Diesel cycle, it is assumed here that the value of k_{cv} must be other than unity.

Irrespective of the value of burn proportion used in any given circumstance, the determination of cyclic performance parameters requires, as usual, determination of the state conditions of pressure, temperature, and volume throughout the cycle. The initial state conditions must be defined, as must the total amount of heat to be internally added as simulated combustion. The swept volume and compression ratio must also be defined, which then provides the volumes at bdc and tdc, i.e., V_1 and V_2, and indeed V_3 as well by inference.

Because all processes in the engine are adiabatic, the first compression process from state 1 to 2, and the heat addition at constant volume can be analyzed as for the Otto cycle. Thus, to obtain T_3 the following equations are solved:

By definition:

$$Q_2^3 = k_{cv}Q_R \qquad Q_3^4 = \left(1 - k_{cv}\right)Q_R \qquad (A1.76)$$

From Eq. A1.33

$$Q_2^3 = m_1 C_v \left(T_3 - T_2 \right) \qquad \text{(A1.75)}$$

From Eq. A1.34:

$$p_3 = p_2 \frac{T_3}{T_2} \qquad \text{(A1.76)}$$

From Eq. A1.69

$$Q_3^4 = m_1 C_p \left(T_4 - T_3 \right) \qquad \text{(A1.77)}$$

By definition:

$$p_4 = p_3 \qquad \text{(A1.78)}$$

Hence, from Eq. A1.5

$$V_4 = V_3 \frac{T_4}{T_3} \qquad \text{(A1.79)}$$

From Eq. A1.55:

$$T_5 = T_4 \left(\frac{V_5}{V_4} \right)^{1-\gamma} \qquad \text{(A1.80)}$$

From Eq. A1.26:

$$p_5 = p_4 \left(\frac{V_4}{V_5} \right)^{\gamma} \qquad \text{(A1.81)}$$

From Eq. A1.23:

$$W_1^2 = -m_1 C_v \left(T_2 - T_1 \right) = \frac{p_2 V_2 - p_1 V_1}{1 - \gamma} \qquad \text{(A1.82)}$$

From Eq. A1.68:

$$W_3^4 = m_1 R(T_4 - T_3) = p_3(V_4 - V_3)$$ (A1.83)

From Eq. A1.23:

$$W_4^5 = -m_1 C_v(T_5 - T_4) = \frac{p_5 V_5 - p_4 V_4}{1 - \gamma}$$ (A1.84)

From Eq. A1.33:

$$Q_5^1 = m_1 C_v(T_1 - T_5)$$ (A1.85)

From Eq. 1.6.31:

$$imep = \frac{W_1^2 + W_3^4 + W_4^5}{V_{sv}}$$ (A1.86)

The thermal efficiency is found by using whichever of Eqs. A1.71-A1.73 is the most convenient, having evaluated all of the state conditions for points 1-5 and the work, heat transfer, and energy processes which connecting them.

The above analysis includes both constant-volume and constant-pressure segments for the heat transfer process which simulate combustion. In the event that a particular simulation does not call for any constant-volume combustion, then the cycle analysis is somewhat simplified.

Ideal Diesel Cycle with Constant-Pressure Heat Transfer Only
The following simplifications apply for insertion into Eqs. 1.70-A1.86

$$k_{cv} = 0 \qquad Q_2^3 = 0 \qquad Q_3^4 = Q_R$$ (A1.87)

$$p_3 = p_2 \qquad T_3 = T_2 \qquad U_3 = U_2$$ (A1.88)

In this case, it is very simple to show that the equation for thermal efficiency, Eq. A1.73, can be simplified as:

$$\eta_t = 1 - \left(\frac{1}{\gamma}\right)\left(\frac{T_5 - T_1}{T_4 - T_3}\right) = 1 - \left(\frac{1}{\gamma}\right)\left(\frac{T_5 - T_1}{T_4 - T_2}\right)$$ (A1.89)

Today, because numerical analysis by computer cares little for the algebraic simplifications that graced yesterday's textbooks [1.2], it hardly seems worth the effort to set it down here. Back when in my student days, and armed only with a slide rule to get it reduced to numbers, numerical analysis was a major issue. Trying to remember it for the (one and only) annual examination on the subject was even worse!

It is also possible to show [1.2] that this version of the ideal Diesel cycle has a thermal efficiency that is less than the ideal Otto cycle at equal compression ratios. Because any diesel engine operates at a compression ratio level that is about twice that of its Otto engine equivalent, this gratuitous information is of no real significance.

Chapter 2

Gas Flow through Four-Stroke Engines

2.0 Introduction

The gas flow processes into, through, and out of an engine are all unsteady. Unsteady gas flow is defined as that in which the pressure, temperature, and gas particle velocity in a duct are variable with time. In the case of exhaust flow, the unsteady gas flow behavior is produced because the cylinder pressure falls with the rapid opening of the exhaust valve or valves. This gives an exhaust pipe pressure that changes with time. In the case of induction flow into the cylinder through an intake valve whose area changes with time, the intake pipe pressure alters because the cylinder pressure is affected by the piston motion causing volumetric change within that space.

To illustrate the dramatic variations of pressure wave and particle motion caused by unsteady flow compared to that caused by steady flow, a series of photographs obtained by Sam Coates and me during research at QUB [7.2] is shown in Plates 2.1 to 2.4. These photographs were obtained using the Schlieren method [7.17], an optical means of observing the variation of the refractive index of a gas with its density. Each photograph was taken with an electronic flash duration of 1.5 μs and the view observed is around the termination of a 28-mm diameter exhaust pipe to the atmosphere. The exhaust pulsations occurred at a frequency of 1000 per minute. The first photograph, Plate 2.1, shows the front of an exhaust pulse about to enter the atmosphere. Note that this is a plane front, and its propagation within the pipe up to the pipe termination is clearly one-dimensional. The next photograph, Plate 2.2, shows the propagation of the pressure wave into the atmosphere in a three-dimensional fashion with a spherical front being formed. The beginning of rotational movement of the gas particles at the pipe edges is now evident. The third photograph, Plate 2.3, shows the spherical wave front fully formed and the particles being impelled into the atmosphere in the form of a toroidal vortex, or a spinning donut of gas particles or "smoke ring." This propagating pressure wave front arrives at the human eardrum, which deflects it, with the nervous system reporting it as "noise" to the brain. The final photograph of the series, Plate 2.4, shows that the propagation of the pressure wave front has now passed beyond the frame of the photograph, but the toroidal vortex of gas particles is proceeding downstream with considerable turbulence. Indeed, the flow through the eye of the vortex is so violent that a new acoustic pressure wave front is forming in front of that vortex. The noise that emanates from these pressure pulsations is composed of the basic pressure front propagation and also of the turbulence of the fluid

153

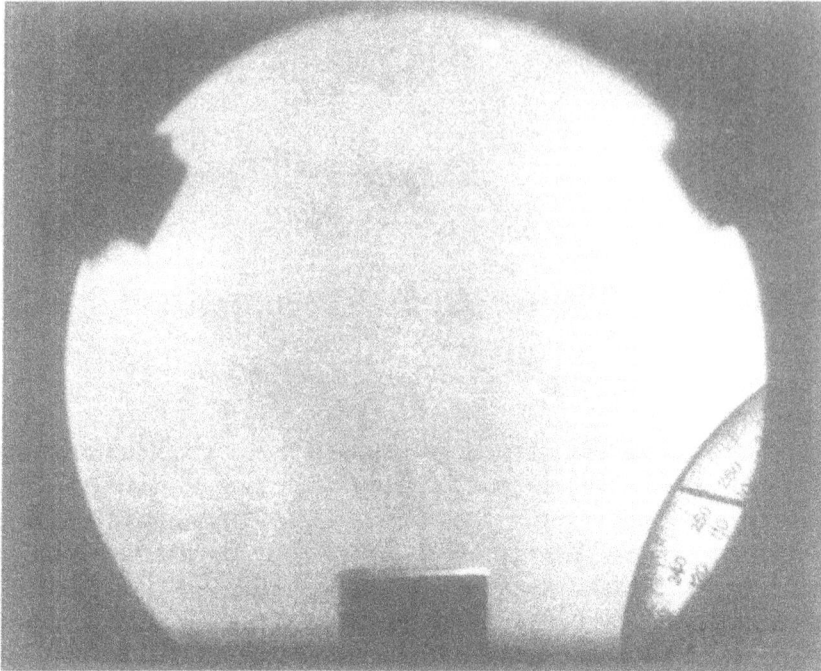

Plate 2.1 Schlieren picture of an exhaust pulse at the termination of a pipe.

motion in the vortex. Further discussion on the noise aspects of this flow is given in Chapter 7. I have always found this series of photographs to be particularly illuminating. When I was a schoolboy on a farm in Co. Antrim too many years ago, the milking machines were driven by a single-cylinder four-stroke Lister diesel engine with a long, straight exhaust pipe and, on a frosty winter's morning, it would blow a "smoke ring" from the exhaust on start-up. That schoolboy used to wonder how it was possible; I now know.

Because the performance characteristics of an engine are significantly controlled by this unsteady gas motion, it behooves the designer of engines to understand this flow mechanism thoroughly. This is true for all engines, whether they are destined to be a 2 hp lawnmower engine or a 1000 hp offshore boat racing engine. A simple example will suffice to illustrate the point. If one were to remove the tuned exhaust pipe from a single-cylinder racing engine while it was running at peak power output, the pipe being an "empty" piece of fabricated sheet metal, the engine power output would fall by at some 20% at that engine speed. The tuned exhaust pipe harnesses the pressure wave motion of the exhaust process to extract a greater mass of the exhaust gas from the cylinder during the exhaust stroke and initiate the induction process during the valve overlap period. Without it, the engine would only be able to inhale about 80% as much fresh air and fuel into the cylinder. To design such exhaust systems, and the engines that will take advantage of them, it is necessary to have a good understanding of the mechanism of unsteady gas flow. For the more serious student interested in a more

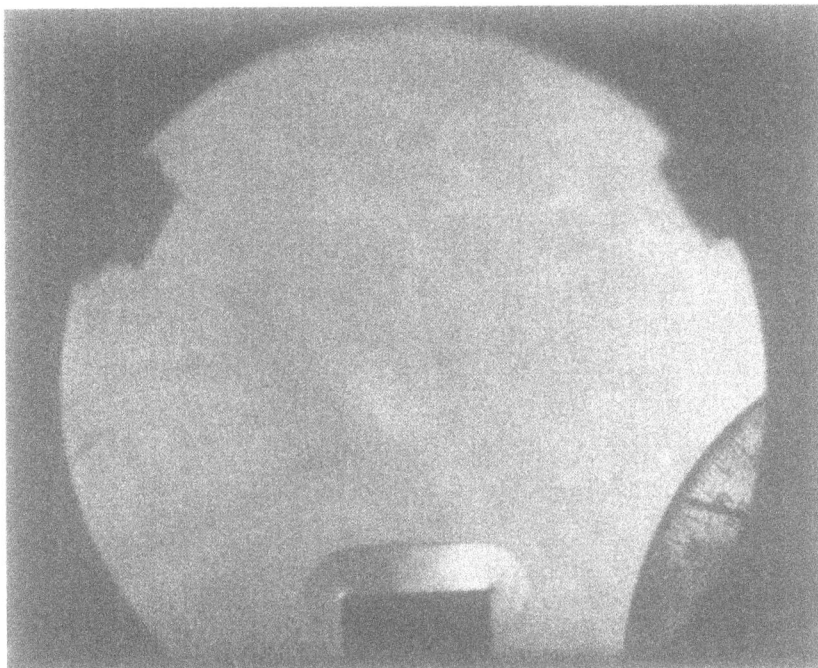

Plate 2.2   The exhaust pulse front propagates into the atmosphere.

in-depth treatment of the subject of unsteady gas dynamics, the series of lectures given by the late Prof. F. K. Bannister of the University of Birmingham [2.2] is an excellent introduction to the topic; so too is the book by Annand and Roe [3.12] and the books by Rudinger [2.3] and Benson [2.4]. The references cited in this chapter will give even greater depth to that study.

This chapter explains the fundamental characteristics of unsteady gas flow in the intake and exhaust ducts of reciprocating engines. Such fundamental theory is just as applicable to two-stroke engines as it is to four-stroke engines, although the bias of the discussion will naturally be toward the four-stroke engine. Throughout the chapter, the relevance of each unsteady gas flow topic is discussed in the context of the design of tuned exhaust and intake systems for four-stroke engines.

2.1 Motion of Pressure Waves in a Pipe
2.1.1 Nomenclature for Pressure Waves

The motion of pressure waves of small amplitude is already familiar to us through our experience with acoustic waves, or sound. Some of our experience with sound waves is helpful in understanding the fundamental nature of the flow of the much larger-amplitude waves to be found in engine ducts. As illustrated in Fig. 2.1, pressure waves, and sound waves, are of two types. They are either compression waves or expansion waves. In both Fig. 2.1(a) and (b), the undisturbed pressure and temperature in the pipe ahead of the pressure wave are p_0 and T_0, respectively.

155

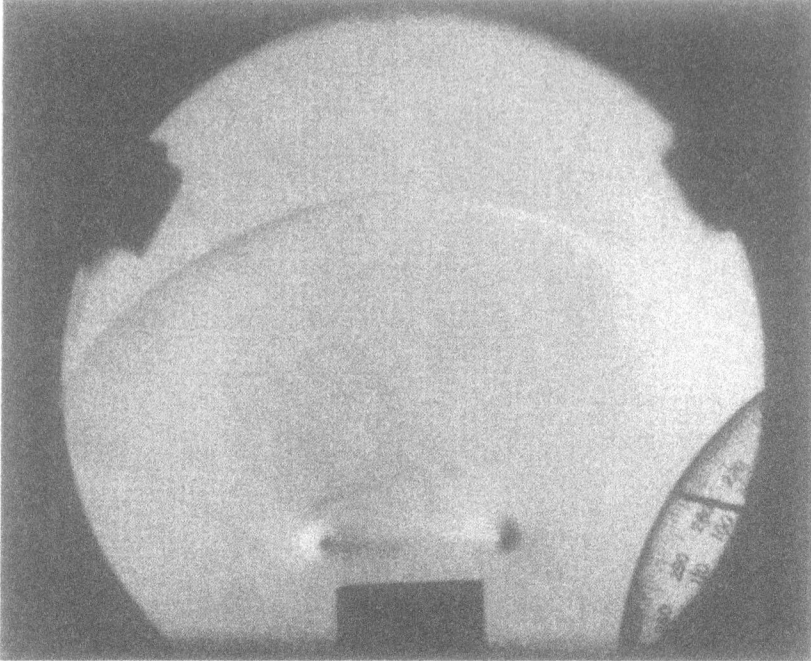

Plate 2.3 Further pulse propagation followed by the toroidal vortex of gas particles.

The compression wave in the pipe is shown in Fig. 2.1(a) and the expansion wave in Fig. 2.1(b). Both waves are propagating toward the right in the diagram. At a point on the compression wave, the pressure is p_e, where p_e is greater than p_0, and the wave is being propagated at a velocity α_e. It is also moving gas particles at a gas particle velocity c_e, in the same direction of propagation as the wave. At a point on the expansion wave, the pressure is p_i, where p_i is less than p_0, and the wave is being propagated at a velocity α_i. It is also moving gas particles at a gas particle velocity c_i, but in a direction opposite to the direction of propagation of the wave.

At this point, our experience with sound waves helps us to understand the physical nature of the statements made in the preceding paragraph. Imagine standing several meters away from another person, Fred. Fred produces a sharp exhalation of breath, for example, he says "boo" somewhat loudly. He does this by raising his lung pressure above the atmospheric pressure due to a muscular reduction of his lung volume. The compression pressure wave produced, albeit of small amplitude, leaves his mouth and is propagated at the local acoustic velocity, or speed of sound, to your ear. The speed of sound involved is on the order of 350 m/s. The gas particles comprising the "boo" leaving Fred's mouth have a much lower velocity, probably on the order of 1 m/s. However, the gas particle velocity is in the same direction as the propagation of the compression pressure wave, i.e., toward your ear. Contrast this simple experiment with a second test. Imagine that Fréd now produces a sharp inhalation of breath. This he accomplishes by expanding his lung volume so that his lung pressure falls

156

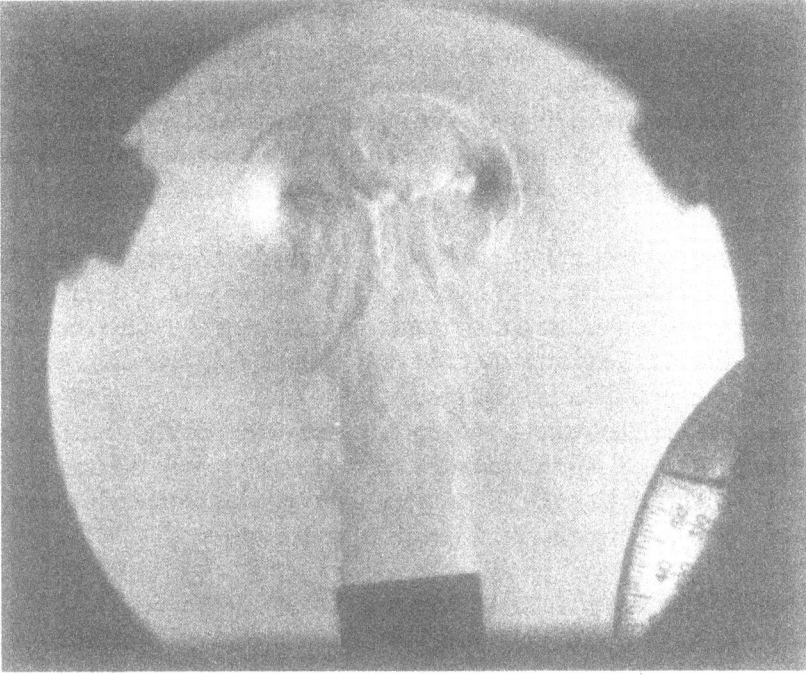

Plate 2.4 The toroidal vortex of gas particles proceeds into the atmosphere.

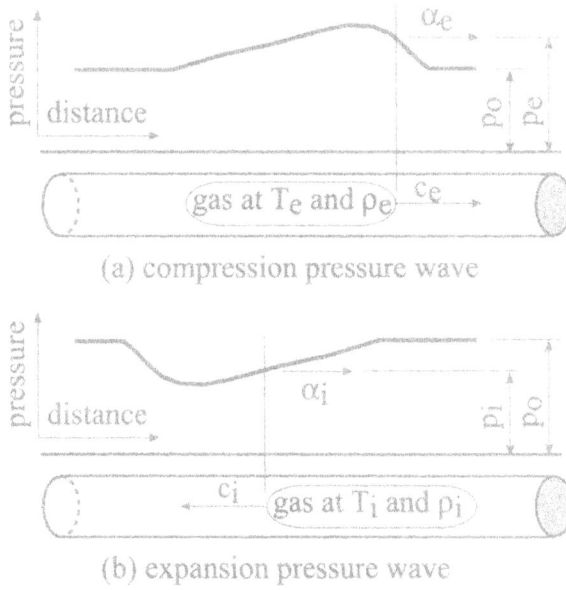

Fig. 2.1 Pressure wave nomenclature.

sharply below the atmospheric pressure. The resulting "u...uh" you hear is caused by the expansion pressure wave leaving Fred's mouth and propagating toward your ear at the local acoustic velocity. In short, the direction of propagation is the same as before with the compression wave "boo," and the propagation velocity is, for all intents and purposes, identical. However, because the gas particles manifestly entered Fred's mouth with the creation of this expansion wave, the gas particle velocity is clearly in the direction opposite to the expansion wave propagation.

It is obvious that exhaust pulses resulting from cylinder blowdown, when the exhaust valve opens, fall under the category of compression waves, whereas the waves generated by the rapidly falling cylinder pressure during induction with the intake valve open are expansion waves. However, as will be seen from the following sections, both expansion and compression waves appear in the inlet and the exhaust system.

NOTE: As in most technologies, other jargon is used in the literature to describe compression and expansion waves. Compression waves are variously called "exhaust pulses," "compression pulses," or "ramming waves." Expansion waves are often described as "suction pulses," "sub-atmospheric pulses," "rarefaction waves," or "intake pulses."

2.1.2 Propagation Velocities of Acoustic Pressure Waves

As already pointed out, acoustic pressure waves are pressure waves with small pressure amplitudes. Let dp be the pressure difference from atmospheric pressure, i.e., (p_e-p_0) or (p_0-p_i), for the compression or expansion wave, respectively. The value of dp for Fred's "boo" would be on the order of 0.2 Pa. The pressure ratio, P, for any pressure wave is defined as the pressure, p, at any point on the wave under consideration divided by the undisturbed pressure, p_0, more commonly called the reference pressure. This is normally the standard reference pressure, 101,325 Pa or 1.01325 bar. Here, the pressure ratio for Fred's "boo" would be:

$$P = \frac{p}{p_0} = \frac{101325.2}{101325} = 1.000002$$

For the loudest of acoustic sounds, say, a rifle shot at about 200 mm from the human ear, dp could be 2000 Pa and the pressure ratio would be 1.02. That such very loud sounds are still small in terms of pressure wave can be gauged from the fact that a typical exhaust pulse in an engine exhaust pipe has a pressure ratio of about 1.5.

According to Earnshaw [2.1], the velocity of a sound wave in air is given by a_0, where:

$$a_0 = \sqrt{\gamma R T_0} \qquad (2.1.1)$$

or

$$a_0 = \sqrt{\frac{\gamma p_0}{\rho_0}} \qquad (2.1.2)$$

The value denoted by γ is the ratio of specific heats for air. T_0 is the reference temperature and r_0 is the reference density, which are related to the reference pressure, p_0, by the state equation:

$$p_0 = \rho_0 R T_0 \qquad (2.1.3)$$

For sound waves in air, p_0, T_0, and r_0 are the values of the atmospheric pressure, temperature, and density, respectively, and R is the gas constant for the particular gas involved. Consult Appendix A1.1 for more information on these symbols and their definitions.

2.1.3 Propagation and Particle Velocities of Finite Amplitude Waves

Particle Velocity

Any pressure wave with a pressure ratio greater than that of an acoustic wave is called a wave of finite amplitude. Earnshaw [2.1] showed that the gas particle velocity associated with a wave of finite amplitude is given by c, where:

$$c = \frac{2}{\gamma - 1} a_0 \left[\left(\frac{p}{p_0} \right)^{\frac{\gamma-1}{2\gamma}} - 1 \right] \qquad (2.1.4)$$

Bannister's [2.2] derivation of this equation is explained with great clarity and is presented in Appendix A2.1. Within the equation, shorthand parameters can be employed which simplify the understanding of much of the further analysis. The symbol P is referred to as the pressure ratio of a point on a wave of absolute pressure, p. The notation of X is known as the pressure amplitude ratio, and G represents various functions of γ, which is the ratio of specific heats for the particular gas involved. These are set down as:

Pressure ratio

$$P = \frac{p}{p_0}$$

Pressure amplitude ratio

$$X = \left(\frac{p}{p_0} \right)^{\frac{\gamma-1}{2\gamma}} = P^{\frac{\gamma-1}{2\gamma}} \qquad (2.1.5)$$

Incorporation of the above shorthand notation into Eq. 2.1.4 gives:

$$c = \frac{2}{\gamma - 1} a_0 (X - 1) \qquad (2.1.6)$$

If the gas in which this pressure wave is propagating has the properties of air, then these properties are:

Gas constant $\qquad\qquad\qquad\qquad$ $R = 287$ J/kgK

Ratio of specific heats $\qquad\qquad\qquad$ $\gamma = 1.4$

Specific heat at constant pressure $\qquad$ $C_P = \dfrac{\gamma R}{\gamma - 1} = 1005$ J/kJK

Specific heat at constant volume $\qquad$ $C_V = \dfrac{R}{\gamma - 1} = 718$ J/kgK

Various functions of the ratio of specific heats, G_5, G_7, etc., which are useful as shorthand notation in many gas dynamic equations in this theoretical area, are given below. The logic of the notation for G is that the value of G_5 for air is 5, G_7 for air is 7, etc.

$$G_3 = \frac{4 - 2\gamma}{\gamma - 1} = 3 \text{ for air, where } \gamma = 1.4$$

$$G_4 = \frac{3 - \gamma}{\gamma - 1} = 4 \text{ for air, where } \gamma = 1.4$$

$$G_5 = \frac{2}{\gamma - 1} = 5 \text{ for air, where } \gamma = 1.4$$

$$G_6 = \frac{\gamma + 1}{\gamma - 1} = 6 \text{ for air, where } \gamma = 1.4$$

$$G_7 = \frac{2\gamma}{\gamma - 1} = 7 \text{ for air, where } \gamma = 1.4$$

$$G_{17} = \frac{\gamma - 1}{2\gamma} = \frac{1}{7} \text{ for air, where } \gamma = 1.4$$

$$G_{35} = \frac{\gamma}{\gamma - 1} = 3.5 \text{ for air, where } \gamma = 1.4$$

$$G_{67} = \frac{\gamma + 1}{2\gamma} = \frac{6}{7} \text{ for air, where } \gamma = 1.4$$

This useful notation simplifies analysis in gas dynamics, particularly because the equations are additive or subtractive with numbers, thus:

For example

$$G_4 = G_5 - 1 = \frac{2}{\gamma - 1} - 1 = \frac{2 - \gamma + 1}{\gamma - 1} = \frac{1 - \gamma}{\gamma - 1}$$

or $\qquad G_7 = G_5 + 2 \quad \text{or} \quad G_3 = G_5 - 2 \quad \text{or} \quad G_6 = G_3 + 3$

However, it should be noted in applications of such functions that they are generally neither additive nor operable, thus:

$$G_7 \neq G_4 + G_3 \quad \text{and} \quad G_3 \neq \frac{G_6}{2}$$

Gas mixtures are commonplace within engines. Air itself is a mixture—basically of oxygen and nitrogen. Exhaust gas is principally composed of carbon monoxide, carbon dioxide, steam, and nitrogen. Furthermore, the properties of gases are complex functions of temperature. A more detailed discussion of this topic is therefore necessary, and is given in Sec. 2.1.6.

If the gas properties are assumed to be as for air with the properties above, then Eq. 2.1.4 for the gas particle velocity reduces to the following:

$$c = \frac{2}{\gamma - 1} a_0(X - 1) = G_5 a_0(X - 1) = 5a_0(X - 1) \qquad (2.1.7)$$

where $\qquad X = \left(\frac{p}{p_0}\right)^{\frac{\gamma - 1}{2\gamma}} = \left(\frac{p}{p_0}\right)^{G_{17}} = \left(\frac{p}{p_0}\right)^{\frac{1}{7}} \qquad (2.1.8)$

Propagation Velocity

The propagation velocity at any point on a wave, where the pressure is p and the temperature is T, is like that of a small acoustic wave moving at the local acoustic velocity under those conditions, but on top of gas particles which are already moving. Therefore, the absolute propagation velocity of any point on a wave is the sum of the local acoustic velocity and the local gas particle velocity. The propagation velocity of any point on a finite amplitude wave is given by α, as:

$$\alpha = a + c \qquad (2.1.9)$$

where a is the local acoustic velocity at the elevated pressure and temperature of the wave point, p and T.

Acoustic velocity, a, is given by Earnshaw [2.1] from Eq. 2.1.1 as:

$$a = \sqrt{\gamma RT} \qquad (2.1.10)$$

Assuming the change in state conditions from p_0 and T_0 to p and T to be isentropic, and using the theory as given in Appendix A1.1, Sec. A1.1.7 we have:

$$\frac{T}{T_0} = \left(\frac{p}{p_0}\right)^{\frac{\gamma-1}{\gamma}} \qquad (2.1.11)$$

$$\frac{a}{a_0} = \sqrt{\frac{T}{T_0}} = \left(\frac{p}{p_0}\right)^{\frac{\gamma-1}{2\gamma}} = P^{G17} = X \qquad (2.1.12)$$

Hence, the absolute propagation velocity, α, defined by Eq. 2.1.9, is given by the expressions for a and c given by Eqs. 2.1.6 and 2.1.12:

$$\alpha = a_0 X + \frac{2}{\gamma-1}a_0(X-1) = a_0\left[\frac{\gamma+1}{\gamma-1}\left(\frac{p}{p_0}\right)^{\frac{\gamma-1}{2\gamma}} - \frac{2}{\gamma-1}\right] \qquad (2.1.13)$$

In terms of the G functions already defined,

$$\alpha = a_0\left[G_6 X - G_5\right] \qquad (2.1.14)$$

If the properties of air are assumed for the gas, then this reduces to:

$$\alpha = a_0\left[6X - 5\right] \qquad (2.1.15)$$

The density, ρ, at any point on a wave of pressure p is found from an extension of the isentropic relationships in Eqs. 2.1.11 and 2.1.14 as follows, using the isentropic theory of Appendix A1.1, Sec. A1.1.7:

$$\frac{\rho}{\rho_0} = \left(\frac{p}{p_0}\right)^{\frac{1}{\gamma}} = X^{\frac{2}{\gamma-1}} = X^{G5} \tag{2.1.16}$$

For air, where γ is 1.4, the density, ρ, at a pressure p, on the wave translates to:

$$\rho = \rho_0 X^5 \tag{2.1.17}$$

2.1.4 Propagation and Particle Velocities of Finite Amplitude Waves in Air

From Eqs. 2.1.4 and 2.1.15, the propagation velocities of finite-amplitude waves in air in a pipe are calculated by the following equations:

Propagation velocity $\qquad \alpha = a_0\left[6X - 5\right] \tag{2.1.18}$

Particle velocity $\qquad c = 5a_0(X - 1) \tag{2.1.19}$

Pressure amplitude ratio $\qquad X = \left(\frac{p}{p_0}\right)^{\frac{1}{7}} = P^{\frac{1}{7}} \tag{2.1.20}$

The reference conditions of acoustic velocity and density are found as follows:

Reference acoustic velocity $\qquad a_0 = \sqrt{1.4 \times 287 \times T_0} \ \ \text{m/s} \tag{2.1.21}$

Reference density $\qquad \rho_0 = \frac{p_0}{287 \times T_0} \ \ \text{kg/m}^3 \tag{2.1.22}$

It is interesting that these equations corroborate the experiment that we conducted with our imaginations regarding Fred's lung-generated compression and expansion waves.

Fig. 2.1 shows compression and expansion waves. Let us assume that the undisturbed pressure and temperature in both cases are at standard atmospheric conditions. In other words, p_0 and T_0 are 101,325 Pa and 20°C, or 293 K, respectively. The reference acoustic velocity, a_0, and reference density, ρ_0, are, from Eqs. 2.1.1 and 2.1.3 or Eqs. 2.1.21 and 2.1.22:

$$a_0 = \sqrt{1.4 \times 287 \times 293} = 343.1 \ \text{m/s}$$

$$\rho_0 = \frac{101,325}{287 \times 293} = 1.2049 \, \text{kg/m}^3$$

Let us assume that the pressure ratio, P_e, of a point on the compression wave is 1.2 and that of a point on the expansion wave, P_i, is 0.8. In other words, the compression wave has a pressure differential as much above the reference pressure as the expansion wave is below it. Let us also assume that the pipe has a diameter, d, of 25 mm.

The Compression Wave

First, consider the compression wave of pressure p_e:

$$p_e = P_e \times p_0 = 1.2 \times 101325 = 121590 \ \text{Pa}$$

The pressure amplitude ratio, X_e, is calculated as:

$$X_e = 1.2^{\frac{1}{7}} = 1.02639$$

Therefore, the propagation and particle velocities, α_e and c_e, are found from:

$$\alpha_e = 343.11 \times (6 \times 1.02639 - 5) = 397.44 \ \text{m/s}$$

$$c_e = 5 \times 343.11 \times (1.02639 - 1) = 45.27 \ \text{m/s}$$

From this it is clear that the propagation of the compression wave is faster than the reference acoustic velocity, a_0, and that the air particles move at a considerably slower rate. The compression wave is moving rightward along the pipe at 397.4 m/s and, as it passes from particle to particle, it propels each particle in turn in a rightward direction at 45.27 m/s. This is deduced from the fact that the signs of the numerical values of α_e and c_e are the same.

The local particle Mach number, M_e, is defined as the ratio of the particle velocity, c_e, to the local acoustic velocity, a_e, where:

$$M_e = \frac{c_e}{a_e} = \frac{G_5(X_e - 1)}{X_e} \qquad (2.1.23)$$

From Eq. 2.1.12:

$$a_e = a_0 X_e$$

hence, $\qquad a_e = 343.11 \times 1.02639 = 352.16 \quad \text{m/s}$

and the local particle Mach number, M_e,

$$M_e = \frac{45.27}{352.16} = 0.1285$$

The mass rate of gas flow, $\dot{m}_e$, caused by the passage of this point of the compression wave in a pipe of area A_e is calculated from the thermodynamic equation of continuity as the multiplication of density, area, and particle velocity:

$$\dot{m}_e = \rho_e A_e c_e$$

From Eq. 2.1.17:

$$\rho_e = \rho_e X_e^5 = 1.2049 \times 1.02639^5 = 1.2049 \times 1.1391 = 1.3725 \quad \text{kg/m}^3$$

The pipe area A_e is given by:

$$A_e = \frac{\pi d^2}{4} = \frac{3.14159 \times 0.025^2}{4} = 0.000491 \quad \text{m}^2$$

Therefore, because the arithmetic signs of the propagation and particle velocities are both positive, the mass rate of flow is in the same direction as the wave propagation as:

$$\dot{m}_e = \rho_e A_e c_e = 1.3725 \times 0.000491 \times 45.27 = 0.0305 \quad \text{kg/s}$$

The Expansion Wave
Second, consider the expansion wave of pressure p_i:

$$p_i = P_i \times p_0 = 0.8 \times 101325 = 81060 \quad \text{Pa}$$

The pressure amplitude ratio, X_i, is calculated as:

$$X_i = 0.8^{\frac{1}{7}} = 0.9686$$

Therefore, the propagation and particle velocities, α_i and c_i, are found from:

$$\alpha_i = 343.11 \times (6 \times 0.9686 - 5) = 278.47 \quad \text{m/s}$$

$$c_i = 5 \times 343.11 \times (0.9686 - 1) = -53.87 \quad \text{m/s}$$

From this it is clear that the propagation of the expansion wave is slower than the reference acoustic velocity, but the air particles move a little faster than the compression wave with the same dp value. The expansion wave is moving rightward along the pipe at 278.47 m/s and, as it passes from particle to particle, it propels each particle in turn in a *leftward* direction at 53.87 m/s. This is deduced from the fact that the numerical values of α_i and c_i are of *opposite* sign.

The local particle Mach number, M_i, is defined as the ratio of the particle velocity to the local acoustic velocity, a_i, where:

$$M_i = \frac{c_i}{a_i}$$

From Eq. 2.1.12,

$$a_i = a_0 X_i$$

hence

$$a_i = 343.11 \times 0.9686 = 332.3 \quad \text{m/s}$$

and local particle Mach number, M_i,

$$M_i = \frac{-53.87}{332.3} = -0.1621$$

The mass rate of gas flow, $\dot{m}_i$, caused by the passage of this point of the expansion wave in a pipe of area A_i is calculated by the continuity equation:

$$\dot{m}_i = \rho_i A_i c_i$$

From Eq. 2.1.17:

$$\rho_i = \rho_i X_i^5 = 1.2049 \times 0.9686^5 = 1.2049 \times 0.8525 = 1.0272 \quad \text{kg/m}^3$$

It is seen that the density is reduced in the more rarified expansion wave.

The pipe area is identical because the diameter is unchanged, i.e., A_i is 0.000491 m^2. The mass rate of flow is in the direction opposite to the wave propagation as:

$$\dot{m}_i = \rho_i A_i c_i = 1.0272 \times 0.000491 \times (-53.87) = -0.0272 \quad \text{kg/s}$$

You should note that, in these compression and expansion waves with the same dp value, the compression wave has the greater mass flow rate because its density is higher.

2.1.5 Distortion of the Wave Profile

It is clear from the foregoing that the value of propagation velocity is a function of the wave pressure and wave temperature at any point on that pressure wave. It should also be evident that, because all of the points on a wave are propagating at different velocities, the wave must change shape in its passage along any duct. To illustrate this, the calculations conducted in the previous section are displayed in Fig. 2.2. In Fig. 2.2(a) it can be seen that the front and tail of the wave both travel at the reference acoustic velocity, a_0, which is 53 m/s slower than the peak wave velocity. In their travel along the pipe, the front and the tail will keep station with each other in both time and distance. However, at the front of the wave, all of the pressure points between it and the peak are traveling faster and will inevitably catch up with it. Whether this will actually happen before some other event intrudes (for instance, the wave front could reach the end of the pipe) will depend on the length of the pipe and the time interval between the peak and the wave front. Nevertheless, there will always be the tendency for the wave peak to get nearer the wave front and farther away from the wave tail. This is known as "steep-fronting." The wave peak could, in theory, try to pass the wave front, which is what happens to a water wave in the ocean when it "crests." In gas flow, "cresting" is impossible and the reality is that a shock wave would be formed. This can be analyzed theoretically. Bannister [2.2] gives an excellent account of the mathematical solution for the particle velocity and the propagation velocity, α_{sh}, of a shock wave of pressure ratio P_{sh} propagating into an undisturbed gas medium at reference pressure p_0 and acoustic velocity a_0. The theoretically derived expressions for propagation velocity, α_{sh}, and particle velocity, c_{sh}, of a compression shock front are:

$$P_{sh} = \frac{P_e}{P_0}$$

$$\alpha_{sh} = a_0 \sqrt{\frac{\gamma+1}{2\gamma} P_{sh} + \frac{\gamma-1}{2\gamma}} = a_0 \sqrt{G_{67} P_{sh} + G_{17}} \qquad (2.1.24)$$

(a) distortion of compression pressure wave profile

(b) distortion of expansion pressure wave profile

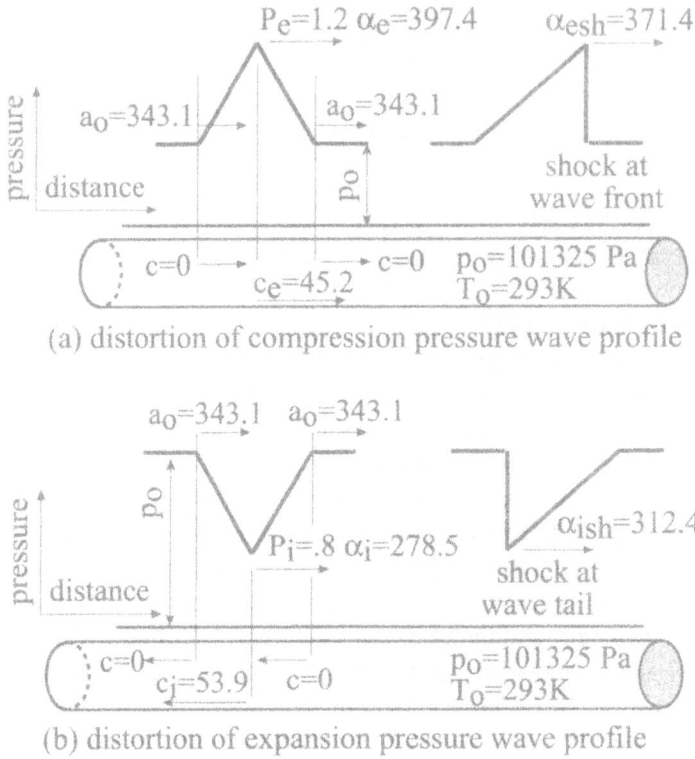

Fig. 2.2 Distortion of wave profile and possible shock formation.

$$c_{sh} = \frac{2}{\gamma+1}\left(\alpha_{sh} - \frac{a_0^2}{\alpha_{sh}}\right) = \frac{a_0(P_{sh}-1)}{\gamma\sqrt{G_{67}P_{sh}+G_{17}}} \qquad (2.1.25)$$

These equations are derived in Appendix A2.2.

The situation for the expansion wave, shown in Fig. 2.2(b), is the reverse, in that the peak is traveling 64.6 m/s slower than either the wave front or the wave tail. Thus, any shock formation taking place will be at the tail of an expansion wave, and any wave distortion will be where the tail of the wave attempts to overrun the peak.

In the case of shock at the tail of the expansion wave, the above equations also apply, but the "compression" shock is now at the tail of the wave and running into gas at acoustic state a_i and moving at particle velocity c_i. Thus, the propagation velocity and particle velocity of the

168

shock front at the tail of the expansion wave, which has an undisturbed state at p_0 and a_0 behind it, are given by:

$$P_{sh} = \frac{p_0}{p_i}$$

$$
\begin{aligned}
\alpha_{sh} &= \alpha_{sh \text{ relative to gas i}} + c_i \\
&= a_i \sqrt{G_{67} P_{sh} + G_{17}} + c_i \\
&= a_0 X_i \sqrt{G_{67} P_{sh} + G_{17}} + G_5 a_0 (X_i - 1)
\end{aligned}
\tag{2.1.26}
$$

$$
\begin{aligned}
c_{sh} &= c_{sh \text{ relative to gas i}} + c_i \\
&= \frac{a_i (P_{sh} - 1)}{\gamma \sqrt{G_{67} P_{sh} + G_{17}}} + c_i \\
&= \frac{a_0 X_i (P_{sh} - 1)}{\gamma \sqrt{G_{67} P_{sh} + G_{17}}} + G_5 a_0 (X_i - 1)
\end{aligned}
\tag{2.1.27}
$$

For the compression wave illustrated in Fig. 2.2, the use of Eqs. 2.1.24 and 2.1.25 yields finite amplitude propagation and particle velocities of 397.4 and 45.27 m/s, and for the shock wave of the same amplitude values of 371.4 and 45.28 m/s, respectively The difference in propagation velocity is some 7% less, but that for particle velocity is negligible.

For the expansion wave illustrated in Fig. 2.2, the use of Eqs. 2.1.26 and 2.1.26 yields finite amplitude propagation and particle velocities of 278.5 and -53.8 m/s, and for the shock wave of the same amplitude, values of 312.4 and 0.003 m/s, respectively The difference in propagation velocity is some 12% greater, but that for particle velocity is considerable in that the particle velocity at, or immediately behind, the shock is effectively zero.

2.1.6 The Properties of Gases

It will be observed that the propagation of pressure waves and the mass flow rate that they induce in gases is dependent on the gas properties, particularly on the gas constant, R, and the ratio of specific heats, γ. The value of the gas constant, R, is dependent on the composition of the gas. The value of the ratio of specific heats, γ, is dependent on both gas composition and temperature. It is essential to be able to index these properties at every stage of a simulation of gas flow in engines. Much of this information can be found in many standard texts on thermodynamics, but for reasons of clarity it is essential to repeat it here briefly, continuing the short introduction given in Appendix A1.1.

The gas constant, R, of any gas can be found from the relationship relating the universal gas constant $\overline{R}$ and the molecular weight, M, of the gas:

$$R = \frac{\overline{R}}{M} \tag{2.1.28}$$

The universal gas constant, $\overline{R}$, has a value of 8314.4 J/kgmolK. The specific heats at constant pressure and constant volume, C_P and C_V, are determined from their defined relationship with respect to enthalpy, h, and internal energy, u:

$$C_P = \frac{dh}{dT} \qquad C_V = \frac{du}{dT} \tag{2.1.29}$$

The ratio of specific heats, γ, is found simply as:

$$\gamma = \frac{C_P}{C_V} \tag{2.1.30}$$

It can be seen that if the gases have internal energies and enthalpies that are nonlinear functions of temperature, then neither C_P, C_V, nor γ, are constants. If the gas is a mixture of gases, then the properties of the individual gases must be assessed separately and then combined to describe the behavior of the mixture.

To illustrate the procedure to determine the properties of gas mixtures, let air be examined as a simple example of a gas mixture with an assumed volumetric composition, υ, of 21% oxygen and 79% nitrogen, while ignoring the small, and relatively unimportant, 1% trace concentration of argon. The molecular weights of oxygen and nitrogen are 31.999 and 28.013 respectively.

The average molecular weight of air, M_{air}, is then given by:

$$M_{air} = \Sigma(\upsilon_{gas}M_{gas}) = 0.21 \times 31.999 + 0.79 \times 28.013 = 28.85$$

The mass ratios, ε, of oxygen and nitrogen in air are given by:

$$\varepsilon_{O_2} = \frac{\upsilon_{O_2}M_{O_2}}{M_{air}} = \frac{0.21 \times 31.999}{28.85} = 0.233$$

$$\varepsilon_{N_2} = \frac{\upsilon_{N_2}M_{N_2}}{M_{air}} = \frac{0.79 \times 28.013}{28.85} = 0.767$$

The molal enthalpies, $\bar{h}$, for gases are given as functions of temperature with respect to molecular weight, where the κ values are constants:

$$\bar{h} = \kappa_0 + \kappa_1 T + \kappa_2 T^2 + \kappa_3 T^3 \quad \text{J/kgmol} \tag{2.1.31}$$

In this case, the molal internal energy of the gas is related thermodynamically to the enthalpy by:

$$\bar{u} = \bar{h} - \bar{R}T \tag{2.1.32}$$

Consequently, from Eq. 2.1.29, the molal specific heats are found by appropriate differentiation of Eqs. 2.1.31 and 32:

$$\bar{C}_P = \kappa_1 + 2\kappa_2 T + 3\kappa_3 T^2 \tag{2.1.33}$$

and
$$\bar{C}_V = \bar{C}_P - \bar{R} \tag{2.1.34}$$

The molecular weights and the constants, κ, for many common gases are listed in Table 2.1.1. These values are reasonably accurate for a temperature range of 300 to 3000 K. The values of the molal specific heats, internal energies, and enthalpies of the individual gases can be found at a particular temperature by using the values in Table 2.1.1.

Considering air as the example gas at a temperature of 20°C, or 293 K, the molal specific heats of oxygen and nitrogen are found using Eqs. 2.1.33 and 2.1.34 as:

Oxygen, O_2 $\bar{C}_P = 31192$ J/kgmolK $\bar{C}_V = 22877$ J/kgmolK
Nitrogen, N_2 $\bar{C}_P = 29043$ J/kgmolK $\bar{C}_V = 20729$ J/kgmolK

From a mass standpoint, these values are determined as follows:

$$C_P = \frac{\bar{C}_P}{M} \quad C_V = \frac{\bar{C}_V}{M} \tag{2.1.35}$$

Hence the mass related values are:

Oxygen, O_2 $C_P = 975$ J/kgK $C_V = 715$ J/kgK
Nitrogen, N_2 $C_P = 1037$ J/kgK $C_V = 740$ J/kgK

TABLE 2.1.1 PROPERTIES OF SOME COMMON GASES FOUND IN ENGINES.

GAS	M	κ_0	κ_1	κ_2	κ_3
O_2	31.999	-9.3039E6	2.9672E4	2.6865	-2.1194E-4
N_2	28.013	-8.503.3E6	2.7280E4	3.1543	-3.3052E-4
CO	28.011	-8.3141E6	2.7460E4	3.1722	-3.3416E-4
CO_2	44.01	-1.3624E7	4.1018E4	7.2782	-8.0848E-4
H_2O	18.015	-8.9503E6	2.0781E4	7.9577	-7.2719E-4
H_2	2.016	-7.8613E6	2.6210E4	2.3541	-1.2113E-4

For the mixture of oxygen and nitrogen known as air, the properties are given generally as:

$$R_{air} = \Sigma \left(\varepsilon_{gas} R_{gas} \right) \qquad C_{Pair} = \Sigma \left(\varepsilon_{gas} C_{Pgas} \right)$$

$$C_{Vair} = \Sigma \left(\varepsilon_{gas} C_{Vgas} \right) \qquad \gamma_{air} = \frac{C_{Pair}}{C_{Vair}}$$

$$(2.1.36)$$

Taking just one as a numeric example, the gas constant, R, which is not temperature dependent, is found by:

$$R_{air} = \Sigma(\varepsilon_{gas} R_{gas}) = 0.233 \left(\frac{8314.4}{31.999} \right) + 0.767 \left(\frac{8314.4}{28.011} \right) = 288 \text{ J/kgK}$$

and the specific heats for air at 293 K are found to be:

$$C_p = 1022 \text{ J/kgK} \quad C_V = 734 \text{ J/kgK} \quad \gamma = 1.393$$

It will be seen that the value of the ratio of specific heats, γ, is not precisely 1.4 at standard atmospheric conditions as stated earlier in Sec. 2.1.3. This is mostly because air contains argon, which is not included in the above analysis. Because argon has a γ value of 1.667, the value deduced above is weighted downward arithmetically.

The most important point to note here is that these properties of air are a function of temperature, so if the above analysis is repeated at 500 and 1000 K the following answers are found for air:

$$T = 500 \text{ K} \quad C_p = 1061 \text{ J/kgK} \quad C_V = 773 \text{ J/kgK} \quad \gamma = 1.373$$
$$T = 1000 \text{ K} \quad C_p = 1143 \text{ J/kgK} \quad C_V = 855 \text{ J/kgK} \quad \gamma = 1.337$$

Air can be found within an engine at these state conditions, thus is vital that any simulation takes these changes of property into account because they have a profound influence on the characteristics of unsteady gas flow.

Exhaust Gas

As a mixture of gases, exhaust gas clearly has quite a different composition compared to air. This matter is discussed in much greater detail in Chapter 4, and was already introduced in Chapter 1. However, consider now the simple and ideal case of stoichiometric combustion of octane with air. The chemical equation, which has a mass-based air-fuel ratio, AFR, of 15, is as follows:

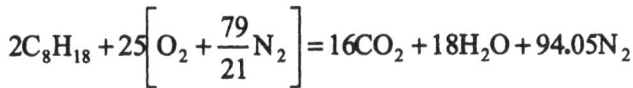

$$2C_8H_{18} + 25\left[O_2 + \frac{79}{21}N_2\right] = 16CO_2 + 18H_2O + 94.05N_2$$

If the total moles is taken to be 128.05, then the volumetric concentrations of the exhaust gas can be found by,

$$\upsilon_{CO_2} = \frac{16}{128.05} = 0.125 \qquad \upsilon_{H_2O} = \frac{18}{128.05} = 0.141 \qquad \upsilon_{N_2} = \frac{94.05}{128.05} = 0.734$$

This is precisely the same starting point as for the above analysis for air, so the procedure is the same for the determination of all of the properties of exhaust gas that ensue from an ideal stoichiometric combustion. A full discussion of the composition of exhaust gas as a function of air-fuel ratio is to be found in Chapter 4, Sec. 4.3.2, and an even more detailed debate in Appendices A4.1 and A4.2, on the changes to that composition, at any fueling level, as a function of temperature and pressure.

In reality, even at stoichiometric combustion there would be some carbon monoxide in existence and minor traces of oxygen and hydrogen. If the mixture were progressively richer than stoichiometric, then the exhaust gas would contain greater amounts of CO and a trace of H_2, but would show little free oxygen. If the mixture were progressively leaner than stoichiometric, the exhaust gas would contain lesser amounts of CO and no H_2, but would show higher concentrations of oxygen. The most important, perhaps obvious, issue is that the properties of exhaust gas depend not only on temperature, but also on the combustion process that created them. Tables 2.1.2 and 2.1.3 show the ratio of specific heats, γ, and gas constant, R, of exhaust gas at various temperatures emanating from the combustion of octane at various air-fuel ratios. An air-fuel ratio of 13 represents rich combustion; 15 is stoichiometric; and 17 is approaching the normal lean limit of gasoline burning. The composition of the exhaust gas is shown in the Table 2.1.2 at a low temperature of 293 K. The influence of exhaust gas composition on the value of gas constant and the ratio of specific heats is quite evident. Although the tabular values are quite typical of combustion products at these air-fuel ratios, naturally they are approximate as they are affected by more than the air-fuel ratio. For example, the local chemistry of the burning process and the chamber geometry, among many factors, will also have a profound influence on the final composition of any exhaust gas. Table 2.1.3 shows the composition of exhaust gas at higher temperatures. Compared with the data for exhaust gas at 293 K in Table 2.1.2, these same gaseous compositions show markedly different properties in Table 2.1.3, when analyzed by the same theoretical approach.

From this it is evident that the properties of exhaust gas are quite different from air, and while they are as temperature dependent as air, they are not influenced by air-fuel ratio, particularly with respect to the ratio of specific heats, as much as might be imagined. The gas constant for rich mixture combustion of gasoline is some 3% higher than that at stoichiometric and at lean mixture burning.

It is evident, however, that during any simulation of unsteady gas flow or of the thermodynamic processes within engines, it is imperative for the accuracy of the simulation to use the correct value of the gas properties at all locations within the engine.

2.2 Motion of Oppositely Moving Pressure Waves in a Pipe

In the previous section, you were asked to conduct an imaginary experiment with Fred, who produced compression and expansion waves by exhaling or inhaling sharply, producing a "boo" or an "u...uh," respectively. Once again, you are asked to conduct another experiment so as to draw on your experience with sound waves to illustrate a principle—in this case the behavior of oppositely moving pressure waves. In this second experiment, you and your friend Fred are going to say "boo" to each other from some distance apart, and at the same time. Each person's ears, being rather accurate pressure transducers, will record his own "boo" first, followed a fraction of time later by the "boo" from the other party. Obviously, the "boo" from each of you will have passed through the "boo" from the other and arrived at both Fred's ear

TABLE 2.1.2 PROPERTIES OF EXHAUST GAS AT LOW TEMPERATURE.

T=293 K AFR	% CO	%CO_2	% BY VOLUME %H_2O	%O_2	%N_2	R	γ
13	5.85	8.02	15.6	0.00	70.52	299.8	1.388
15	0.00	12.50	14.1	0.00	73.45	290.7	1.375
17	0.00	11.14	12.53	2.28	74.05	290.4	1.376

TABLE 2.1.3 PROPERTIES OF EXHAUST GAS AT ELEVATED TEMPERATURES.

AFR	T=500 K R	γ	AFR	T=1000 K R	γ
13	299.8	1.362	13	299.8	1.317
15	290.7	1.350	15	290.8	1.307
17	290.4	1.352	17	290.4	1.310

and your ear with no distortion caused by their passage through each other. If distortion does take place, then the sensitive human ear is able to detect it. The point of meeting, when the waves were passing through each other, is described as "superposition." This process occurs continuously within the ducts of engines. The theoretical treatment below is for air. This simplifies the presentation and will enhance your understanding of the basic theory; however, extension of the theory to the generality of gas properties is straightforward.

2.2.1 Superposition of Oppositely Moving Waves

Fig. 2.3 illustrates two oppositely moving pressure waves in air in a pipe. They are shown as compression waves, ABCD and EFGH, and are sketched as being square in profile, which is physically impossible, but it makes the task of mathematical explanation somewhat easier. In Fig. 2.3(a) the waves are about to meet. In Fig. 2.3(b) the process of superposition is taking place for the front EF on wave top BC, and for the front CD on wave top FG. The result is the creation of a superposition pressure, p_s, from the separate wave pressures, p_1 and p_2. Assume that the reference acoustic velocity is a_0. Assume also that the rightward direction is mathematically positive, and that the particle and the propagation velocity of any point on the wave top BC will be c_1 and α_1. From Eqs. 2.1.18 to 2.1.20:

$$c_1 = 5a_0(X_1 - 1) \qquad \alpha_1 = a_0(6X_1 - 5)$$

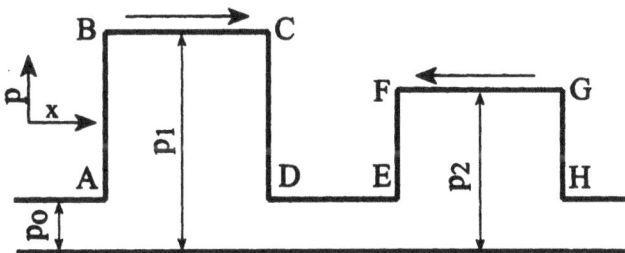

(a) two pressure waves approach each other in a duct

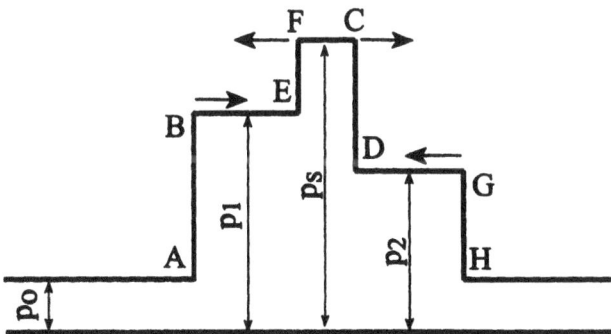

(b) two pressure waves partially superposed in a duct

Fig. 2.3 Superposition of pressure waves in a pipe.

Similarly, with rightward regarded as the positive direction, the values for the wave top FG will be:

$$c_2 = -5a_0(X_2 - 1) \qquad \alpha_2 = -a_0(6X_2 - 5)$$

From Eq. 2.1.14, the local acoustic velocities in the gas columns BE and DG during superposition will be:

$$a_1 = a_0 X_1 \quad a_2 = a_0 X_2$$

During superposition, the wave top F is now moving into a gas with a new reference pressure level at p_1. The particle velocity of F relative to the gas in BE will be:

$$c_{FrelBE} = -5a_1\left[\left(\frac{p_S}{p_1}\right)^{\frac{1}{7}} - 1\right] = 5a_0 X_1\left[\left(\frac{p_S}{p_1}\right)^{\frac{1}{7}} - 1\right] = -5a_0(X_S - X_1)$$

The absolute particle velocity of F, c_s, will be given by the sum of c_{FrelBE} and c_1, as follows:

$$c_S = c_{FrelBE} + c_1 = -5a_0(X_S - X_1) + 5a_0(X_1 - 1) = 5a_0(2X_1 - X_S - 1)$$

Applying the same logic to wave top C proceeding into wave top DG gives another expression for c_s, as F and C are at precisely the same state conditions:

$$c_S = c_{CrelDG} + c_2 = 5a_0(X_S - X_2) - 5a_0(X_2 - 1) = -5a_0(2X_2 - X_S - 1)$$

Equating the above two expressions for c_s for the same wave top FC gives two important equations as a conclusion, one for the pressure of superposition, p_s, and the other for the particle velocity of superposition, c_s:

$$X_S = X_1 + X_2 - 1 \qquad (2.2.1)$$

or

$$\left(\frac{p_S}{p_0}\right)^{\frac{1}{7}} = \left(\frac{p_1}{p_0}\right)^{\frac{1}{7}} + \left(\frac{p_2}{p_0}\right)^{\frac{1}{7}} - 1 \qquad (2.2.2)$$

then

$$c_S = 5a_0(X_1 - 1) - 5a_0(X_2 - 1) = 5a_0(X_1 - X_2) \qquad (2.2.3)$$

or
$$c_S = c_1 + c_2 \tag{2.2.4}$$

Note that the expressions for superposition particle velocity reserves the need for a sign convention, i.e., a declaration of a positive direction, whereas Eqs. 2.2.1 and 2.2.2 are independent of direction. The more general expression for a gas with properties other than air is easily seen from the above equations as:

$$X_S = X_1 + X_2 - 1 \tag{2.2.5}$$

then
$$\left(\frac{p_S}{p_0}\right)^{G17} = \left(\frac{p_1}{p_0}\right)^{G17} + \left(\frac{p_2}{p_0}\right)^{G17} - 1 \tag{2.2.6}$$

as
$$c_S = c_1 + c_2 \tag{2.2.7}$$

then
$$c_S = G_5 a_0 (X_1 - 1) - G_5 a_0 (X_2 - 1) = G_5 a_0 (X_1 - X_2) \tag{2.2.8}$$

At any location within the pipes of an engine, the superposition process is the norm as pressure waves continually pass to and fro. Further, if one places a pressure transducer in the wall of a pipe, it is the superposition pressure-time history that is recorded, not the individual pressures of the rightward and the leftward moving pressure waves. This makes it very difficult to interpret recorded pressure-time data in engine ducting. A simple example will make the point. In Sec. 2.1.4 and in Fig. 2.2, an example is presented of two pressure waves, p_e and p_i, with pressure ratio values of 1.2 and 0.8, respectively. Suppose that these pressure waves are in a pipe, but are the oppositely moving waves just discussed, and are in the position of precise superposition. There is a pressure transducer at the point where the wave peaks coincide and it records a superposition pressure, p_S. What will the value of p_S be and what is the value of the superposition particle velocity, c_S?

The values of X_e, X_i, a_0, c_e, and c_i were 1.0264, 0.9686, 343.1, 45.3, and -53.9, respectively, in terms of a positive direction for the transmission of each wave. In other words the properties of the two waves, p_e and p_i, are to be assigned to become those of waves 1 and 2, respectively, merely to reduce the arithmetic clutter both within the text and in your mind.

If wave p_1 is regarded as moving rightward and that is defined as the positive direction, then the second wave, p_2, is moving leftward in a negative direction. The application of Eqs. 2.2.1 and 2.2.2 shows:

$$X_S = X_1 + X_2 - 1 = 1.0264 - 0.9686 - 1 = 0.995$$

$$P_S = X_S{}^7 = 0.965 \qquad p_S = P_S p_0 = 0.965 \times 101325 = 97831 \ \text{Pa}$$

$$c_S = 45.3 + \left[-(-53.9)\right] = 99.2 \ \text{m/s}$$

177

Thus, the pressure transducer in the wall of the pipe would show little of this process, because the summation of the two waves reveals a trace that is virtually indistinguishable from the atmospheric line, and exhibits nothing of the virtual doubling of the particle velocity.

The opposite effect takes place when two waves of the same type undergo a superposition process. For example, if wave p_1 meets another similar wave, p_1, but going in the other direction and in the plane of the pressure transducer, then:

$$X_S = 1.0264 + 1.0264 - 1 = 1.0528 \quad \text{and} \quad P_S = X_S^{\,7} = 1.0528^7 = 1.434$$

$$c_S = 45.3 + \left[-(+45.3) \right] = 0 \text{ m/s}$$

The pressure transducer now shows a large compression wave with a pressure ratio of 1.434 and tells nothing of the zero particle velocity at the same spot and time.

This makes the interpretation of exhaust and intake pressure records within engine ducting a most difficult business if it is based on experimentation alone. Not unnaturally, the observing engineer will interpret a measured pressure trace exhibiting a large number of pressure oscillations as being evidence of lots of wave activity. This may well be so, but some of these fluctuations will almost certainly be periods approaching zero particle velocity, while yet other periods of the superposition trace exhibiting "calm" conditions could very well be operating at particle velocities approaching the sonic value! This is clearly a most important topic, and we will return to it in later sections of this chapter. It will also be a recurring theme throughout the text.

2.2.2 Wave Propagation during Superposition

The propagation velocity of the two waves during the superposition process must also take direction into account. The statement below is accurate for the superposition condition in which the rightward direction is considered positive.

$$a_S = a_0 X_S$$

The rightward superposition propagation velocity is given by the sum of the local acoustic and particle velocities, as in Eq. 2.1.9:

$$
\begin{aligned}
\alpha_{s\text{ rightward}} &= a_S + c_S \\
&= a_0 (X_1 + X_2 - 1) + G_5 a_0 (X_1 - X_2) \\
&= a_0 (G_6 X_1 - G_4 X_2 - 1)
\end{aligned}
\tag{2.2.9}
$$

$$\alpha_{s \text{ leftward}} = -a_s + c_s$$
$$= -a_0(X_1 + X_2 - 1) + G_5 a_0(X_1 - X_2) \qquad (2.2.10)$$
$$= -a_0(G_6 X_2 - G_4 X_1 - 1)$$

During superposition in air of the two waves, p_e and p_i, as presented above in Sec. 2.2.1, where the values of c_s, a_0, X_1, X_2, and X_s were 99.2, 343.1, 1.0264, 0.9686, and 0.995, respectively, this theory gives values of propagation velocity as:

$$a_s = a_0 X_s = 343.1 \times 0.995 = 341.38 \text{ m/s}$$

$$\alpha_{s \text{ rightward}} = a_s + c_s = 341.38 + 99.2 = 440.58 \text{ m/s}$$

$$\alpha_{s \text{ leftward}} = -a_s + c_s = -341.38 + 99.2 = -242.18 \text{ m/s}$$

This could have been determined more formally using Eqs. 2.2.9 and 2.2.10 as:

$$\alpha_{s \text{ rightward}} = a_0(G_6 X_1 - G_4 X_2 - 1) = 343.38(6 \times 1.0264 - 4 \times 0.9686 - 1) = 440.58 \text{ m/s}$$

$$\alpha_{s \text{ leftward}} = -a_0(G_6 X_2 - G_4 X_1 - 1) = -343.38(6 \times 0.9686 - 4 \times 1.0264 - 1) = -242.18 \text{ m/s}$$

The original propagation velocities of waves 1 and 2, when they were travelling "undisturbed" in the duct into a gas at the reference acoustic state, were 397.44 m/s and 278.47 m/s. It is thus clear from the above calculations that wave 1 has accelerated, and wave 2 has slowed down, by some 10% during this particular example of a superposition process. This effect is often referred to as "wave interference during superposition." In Sec. 6.4.5, you will be referred back to this section for a "refresher course" when it becomes apparent that this effect dominates the proceedings within a tuned exhaust pipe.

Consider some other basic examples, such as compression waves meeting each other, and a similar encounter with expansion waves.

(i) Rightward wave p_e meets leftward wave p_e going in the opposite direction.

$$\alpha_{s \text{ rightward}} = a_0(G_6 X_1 - G_4 X_2 - 1) = 343.38(6 \times 1.0264 - 4 \times 1.0264 - 1) = 361.5 \text{ m/s}$$

$$\alpha_{s \text{ leftward}} = -a_0(G_6 X_2 - G_4 X_1 - 1) = -343.38(6 \times 1.0264 - 4 \times 1.0264 - 1) = -361.5 \text{ m/s}$$

Hence, the two compression waves meeting in superposition slow each other's propagation down by some 10%, and the particle velocity during superposition is zero.

(ii) Rightward wave p_i meets leftward wave p_i going in the opposite direction.

$$\alpha_{s\,rightward} = a_0(G_6X_1 - G_4X_2 - 1) = 343.38(6 \times 0.9686 - 4 \times 0.9686 - 1) = 321.8 \text{ m/s}$$

$$\alpha_{s\,leftward} = -a_0(G_6X_2 - G_4X_1 - 1) = -343.38(6 \times 0.9686 - 4 \times 0.9686 - 1) = -321.8 \text{ m/s}$$

Hence, the two expansion waves meeting in superposition accelerate each other's propagation by some 15%, and the particle velocity during superposition is zero.

Because of the potential for arithmetic sign complexity, during computer calculations it is imperative to rely on formal equation statements, such as Eqs. 2.2.9 and 2.2.10, to provide computed values.

2.2.3 Mass Flow Rate during Wave Superposition

Because the superposition process accelerates some waves and decelerates others, the mass flow rate must also be affected. It is therefore necessary to be able to calculate mass flow rate at any position within a duct. The continuity equation provides the necessary information:

$$\text{mass flow rate} = \text{density} \times \text{area} \times \text{velocity} = \rho_s A c_s$$

where
$$\rho_s = \rho_0 X_s^{G5} \tag{2.2.11}$$

hence
$$\dot{m} = G_5 a_0 \rho_0 A (X_1 + X_2 - 1)^{G5}(X_1 - X_2) \tag{2.2.12}$$

In terms of the numerical example used in Sec. 2.2.2, the values of a_0, ρ_0, X_1, and X_2 were 343.1, 1.2049, 1.0264, and 0.9686 respectively. The pipe area is that of the 25 mm diameter duct, or 0.000491 m^2. The gas in the pipe is air.

We can solve for the mass flow rate by using the previously known superposition value of X_s, which was 0.995, or the value for particle velocity, c_s, which was 99.2 m/s, and determine the superposition density, ρ_s, thus:

$$\rho_s = \rho_0 X_s^{G5} = 1.2049 \times 0.995^5 = 1.1751 \quad \text{kg/m}^3$$

Hence, the mass flow rate, $\dot{m}$, during superposition is given by:

$$\dot{m}_{rightward} = 1.1751 \times 0.000491 \times 99.2 = 0.0572 \text{ kg/s}$$

The sign could have been found by knowing that the superposition particle movement was rightward and inserting c_s as +99.2 and not -99.2. Alternatively, the formal equation, Eq. 2.2.12, gives a numeric answer indicating the direction of mass or particle flow. This is obtained by solving Eq. 2.2.12 with the lead term in any bracket, i.e., X_1, as the value at which wave motion is considered to be in a positive direction.

Hence, mass flow rate *rightward*, with the direction of wave 1 being called positive, is:

$$\dot{m}_{rightward} = G_5 a_0 \rho_0 A (X_1 + X_2 - 1)^{G5}(X_1 - X_2)$$
$$= 5 \times 343.11 \times 1.2049 \times (1.0264 + 0.9686 - 1) \times (1.0264 - 0.9686)^5$$
$$= +0.0572 \quad kg/s$$

It will be observed, indeed it is imperative to satisfy the equation of continuity, that the superposition mass flow rate is the sum of the mass flow rate induced by the individual waves. The mass flow rates in the rightward direction of waves 1 and 2, computed earlier in Sec. 2.1.4, were 0.0305 and 0.0272 kg/s respectively.

2.2.4 Supersonic Particle Velocity during Wave Superposition

In typical engine configurations, it is rare for the magnitude of finite amplitude waves that occur to provide a particle velocity approaching the sonic value. The Mach number, **M**, is defined in Eq. 2.1.23 for a pressure amplitude ratio of X as:

$$\mathbf{M} = \frac{c}{a} = \frac{G_5 a_0 (X-1)}{a_0 X} = \frac{G_5(X-1)}{X} \tag{2.2.12}$$

For this to approach unity then:

$$X = \frac{G_5}{G_4} = \frac{2}{3-\gamma} \quad \text{and for air} \quad X = 1.25$$

In air, as seen above, this would require a compression wave of pressure ratio P, where:

$$P = \frac{p}{p_0} = X^{G7} \quad \text{and in air} \quad P = 1.25^7 = 4.768$$

Even in high-performance racing engines, a pressure ratio for an exhaust pulse of greater than 2.5 atmospheres is very unusual, thus sonic particle velocity emanating from such a source is not likely. A more realistic possibility is that a large exhaust pulse may encounter a strong oppositely moving expansion wave in the exhaust system and the superposition particle velocity may approach, or attempt to exceed, unity.

Unsteady gas flow does not permit supersonic particle velocity. It is self-evident that the gas particles cannot move faster than the pressure wave disturbance that is giving them the signal to move. Because this is not possible according to gas dynamics, the theoretical treatment supposes that a "weak shock" occurs and the particle velocity reverts to a subsonic value. The basic theory can be found in any standard text [2.4] and the resulting relationships are referred to as the Rankine-Hugoniot equations. The theoretical treatment is almost identical to that given here for moving shocks, where the particle velocity behind the moving shock is also subsonic (see Appendix A2.2).

Consider two oppositely moving pressure waves in a superposition situation. The individual pressure waves are p_1 and p_2, and the gas properties are γ and R, with a reference temperature and pressure denoted by p_0 and T_0. From Eqs. 2.2.5 to 2.2.8, the particle Mach number, M_s, is found from:

Pressure amplitude ratio
$$X_s = X_1 + X_2 - 1 \tag{2.2.14}$$

Acoustic velocity
$$a_s = a_0 X_s \tag{2.2.15}$$

Particle velocity
$$c_s = G_5 a_0 (X_1 - X_2) \tag{2.2.16}$$

Mach number (+ or ly)
$$M_s = \frac{c_s}{a_s} = \left| \frac{G_5 a_0 (X_1 - X_2)}{a_0 X_s} \right| \tag{2.2.17}$$

Note that the modulus of the Mach number is acquired. This eliminates directionality in any inquiry as to the magnitude of Mach number in absolute terms. If an inquiry reveals that M_s is greater than unity, then the individual waves p_1 and p_2 are modified by internal reflections to provide a shock to a gas flow at subsonic particle velocity. The superposition pressure after the shock transition to subsonic particle flow at Mach number M_{snew} is labelled as p_{snew}. The "new" pressure waves that travel onward after superposition is completed become labeled as p_{1new} and p_{2new}. The Rankine-Hugoniot equations describing this combined shock and reflection process are:

Pressure
$$\frac{p_{snew}}{p_s} = \frac{2\gamma}{\gamma+1} M_s^2 - \frac{\gamma-1}{\gamma+1} \tag{2.2.18}$$

Mach number
$$M_{snew}^2 = \frac{M_s^2 + \dfrac{2}{\gamma-1}}{\dfrac{2\gamma}{\gamma-1} M_s^2 - 1} \tag{2.2.19}$$

After the shock, the "new" pressure waves are related by:

Pressure
$$X_{snew} = X_{1new} + X_{2new} - 1 \qquad (2.2.20)$$

Pressure
$$p_{snew} = p_0 X_{snew}^{G7} \qquad (2.2.21)$$

Mach number
$$M_{snew} = \frac{c_{snew}}{a_{snew}} = \frac{G_5 a_0 (X_{1new} - X_{2new})}{a_0 X_{snew}} \qquad (2.2.22)$$

Pressure wave 1
$$p_{1new} = p_0 X_{1new}^{G7} \qquad (2.2.23)$$

Pressure wave 2
$$p_{2new} = p_0 X_{2new}^{G7} \qquad (2.2.24)$$

Using the knowledge that the Mach number in Eq. 2.2.17 has exceeded unity, Eqs. 2.2.18 and 2.2.19 of the Rankine-Hugoniot set provide the basis for the simultaneous equations needed to solve for the two unknown pressure waves p_{1new} and p_{2new} through the connecting information in Eqs. 2.2.20 to 2.2.22. For simplicity in presenting this theory, it is predicated that $p_1 > p_2$, i.e., that the sign of any particle velocity is positive. In any application of this theory, this assumption must be borne in mind and the direction of the analysis adjusted accordingly. The solution of the two simultaneous equations reveals, in terms of complex functions Γ_1 to Γ_4, which are composed of known pre-shock quantities:

$$\Gamma_1 = \frac{M_s^2 + \frac{2}{\gamma-1}}{\frac{2\gamma}{\gamma-1}M_s^2 - 1} \qquad \Gamma_2 = \frac{2\gamma}{\gamma+1}M_s^2 - \frac{\gamma-1}{\gamma+1} \qquad \Gamma_3 = \frac{\gamma-1}{2}\sqrt{\Gamma_1} \qquad \Gamma_4 = X_s \Gamma_2^{\frac{\gamma-1}{2\gamma}}$$

then
$$X_{1new} = \frac{1 + \Gamma_4 + \Gamma_3\Gamma_4}{2} \quad \text{and} \quad X_{2new} = \frac{1 + \Gamma_4 - \Gamma_3\Gamma_4}{2} \qquad (2.2.25)$$

The new values of particle velocity, Mach number, wave pressure, or other such parameters can be found by substitution into Eqs. 2.2.20 to 2.2.24.

Consider a simple numeric example of oppositely moving waves. The individual pressure waves are p_1 and p_2 with strong pressure ratios of 2.3 and 0.5, and the gas properties are those of air, where the specific heats ratio, γ, is 1.4 and the gas constant, R, is 287 J/kgK. The reference temperature and pressure are denoted by p_0 and T_0, and are 101325 Pa and 293 K, respectively. The conventional superposition computation as carried out previously in this section would show that the superposition pressure ratio, P_s, is 1.2474, the superposition temperature, T_s, is 39.1°C, and the particle velocity, c_s, is 378.51 m/s. This translates into a Mach number, M_s, during superposition of 1.0689, which is clearly just sonic. The application of the above theory reveals that the Mach number, M_{snew}, after the weak shock is 0.937, and that the ongoing pressure waves, p_{1new} and p_{2new}, have modified pressure ratios of 2.2998 and 0.5956 respectively.

From this example, it is obvious that it takes waves of uncommonly large amplitude to produce even a weak shock, and that the resulting modifications to the amplitude of the waves are quite small. Nevertheless, this analysis must be included within any computational modelling of unsteady gas flow that has pretensions of accuracy. Sec. 6.4.5, and the discussion related to Fig. 6.68, describe how this very effect delays the return of the expansion wave reflection at the end of a tuned exhaust pipe.

In this section I have implicitly introduced the concept that the amplitude of pressure waves can be modified by encountering some "opposition" to their perfect, i.e., isentropic, progress along a duct. This also implicitly introduces the concept of reflections of pressure waves, i.e., the taking of some of the energy away from a pressure wave and sending it in the opposite direction. This theme is one that will appear in almost every facet of the discussions that follow.

2.3 Friction Loss and Friction Heating during Pressure Wave Propagation

Particle flow in a pipe induces forces acting against the flow due to the viscous shear forces generated in the boundary layer close to the pipe wall. Virtually any text on fluid mechanics or gas dynamics will discuss the fundamental nature of this behavior in a comprehensive fashion [2.4]. The frictional effect produces a dual outcome: (1) the frictional force results in a pressure loss to the wave opposite to the direction of particle motion and, (2) the viscous shearing forces acting over the distance traveled by the particles with time means that the work expended appears as internal heating of the local gas particles. The typical situation is illustrated in Fig. 2.4, where two pressure waves, p_1 and p_2, meet in a superposition process.

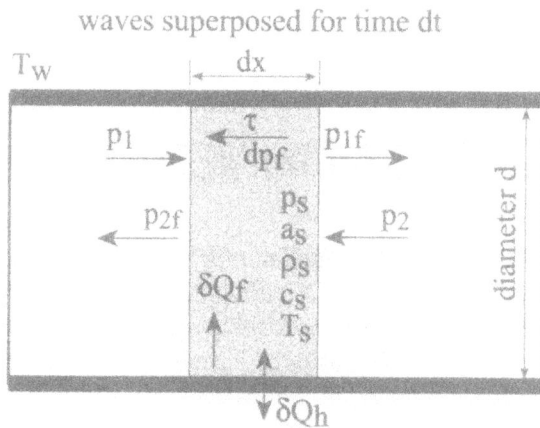

Fig. 2.4 Friction loss and heat transfer in a duct.

This makes the subsequent analysis more generally applicable. However, the following analysis applies equally well to a pressure wave, p_1, traveling into undisturbed conditions, as it remains only to nominate that the value of p_2 is the same as that of the undisturbed state, p_o.

In the general analysis, pressure waves, p_1 and p_2, meet in a superposition process and due to the distance dx traveled by the particles during a time dt engender a friction loss that gives rise to internal heating, dQ_f, and a pressure loss, dp_f. By definition, both of these effects constitute a gain of entropy, so the friction process is non-isentropic as far as the wave propagation is concerned.

The superposition process produces all of the velocity, density, temperature, and mass flow characteristics described in Sec. 2.2. However, the data regarding pressure loss and heat generated, and more importantly the altered amplitudes of pressure waves p_1 and p_2 after the friction process is completed, must be obtained from the theoretical analysis of friction pressure loss and heating.

The shear stress, τ, at the wall as a result of this process is given by:

Shear stress
$$\tau = C_f \frac{\rho_s c_s^2}{2}$$
(2.3.1)

The friction factor, C_f, is usually in the range 0.003 to 0.008, depending on factors such as fluid viscosity or pipe wall roughness. The direct assessment of the value of the friction factor is discussed later in this section.

The force F exerted at the wall on the pressure wave by the wall shear stress in a pipe of diameter d during the distance dx traveled by a gas particle during a time interval dt is expressed as:

Distance traveled
$$dx = c_s dt$$

Force
$$F = \pi d\tau dx = \pi d\tau c_s dt$$
(2.3.2)

This force acts over the entire pipe flow area, A, and provides a loss of pressure, d_{pf}, for the plane fronted wave that is inducing the particle motion. The pressure loss due to friction is found by incorporating Eq. 2.3.1 into Eq. 2.3.2:

Pressure loss
$$dp_f = \frac{F}{A} = \frac{\pi d\tau c_s dt}{\frac{\pi d^2}{4}} = \frac{4\tau c_s dt}{d} = \frac{2C_f \rho_s c_s^3 dt}{d}$$
(2.3.3)

It will be observed that this equation contains a cubed term for the velocity, and because there is a sign convention for direction, this results in a loss of pressure for compression waves and a pressure rise for expansion waves, i.e., a loss of wave strength and a reduction of particle velocity in either case.

Because this friction loss process is occurring during the superposition of waves of pressure p_1 and p_2 as in Fig. 2.4, values such as superposition pressure amplitude ratio, X_s, density, ρ_s, and particle velocity, c_s, can be deduced from the equations given in Sec. 2.2. They are repeated here:

$$X_s = X_1 + X_2 - 1 \qquad c_s = G_5 a_0 (X_1 - X_2) \qquad \rho_s = \rho_0 X_s^{G5}$$

The absolute superposition pressure, p_s, is given by:

$$p_s = p_0 X_s^{G7}$$

After the loss of friction pressure, the new superposition pressure, p_{sf}, and its associated pressure amplitude ratio, X_{sf}, will be, depending on whether it is a compression or expansion wave,

$$p_{sf} = p_s \pm dp_f \qquad X_{sf} = \left(\frac{p_{sf}}{p_0} \right)^{G17} \tag{2.3.4}$$

The solution for the transmitted pressure waves, p_{1f} and p_{2f}, after the friction loss is applied to both, is determined using the momentum and continuity equations for the flow regime before and after the event, thus:

Continuity $\qquad\qquad\qquad \dot{m}_s = \dot{m}_{sf}$

Momentum $\qquad\qquad\quad p_s A - p_{sf} A = \dot{m}_s c_s - \dot{m}_{sf} c_{sf}$

which becomes $\qquad\qquad A(p_s - p_{sf}) = \dot{m}_s c_s - \dot{m}_{sf} c_{sf}$

The mass flow is found from:

$$\dot{m}_s = \rho_s A c_s = G_5 a_s (X_1 - X_2) A \rho_0 X_s^{G5}$$

The transmitted pressure amplitude ratios, X_{1f} and X_{2f}, and superposition particle velocity, c_{sf}, are related by:

$$X_{sf} = (X_{1f} + X_{2f} - 1) \quad \text{and} \quad c_{sf} = G_5 a_0 (X_{1f} - X_{2f})$$

The momentum and continuity equations become two simultaneous equations for the two unknown quantities, X_{1f} and X_{2f}, which are found by determining the particle velocity, c_{sf}:

$$c_{sf} = c_s + \frac{p_{sf} - p_s}{\rho_s c_s} \tag{2.3.5}$$

where

$$X_{1f} = \frac{1}{2}\left(1 + X_{sf} + \frac{c_{sf}}{G_5 a_0}\right) \tag{2.3.6}$$

and

$$X_{2f} = 1 + X_{sf} - X_{1f} \tag{2.3.7}$$

Consequently the pressures of the ongoing pressure waves, p_{1f} and p_{2f}, after friction has been taken into account, are determined by:

$$p_{1f} = p_0 X_{1f}^{G7} \quad \text{and} \quad p_{2f} = p_0 X_{2f}^{G7} \tag{2.3.8}$$

Taking the data for the two pressure waves of amplitude 1.2 and 0.8 that have been used as examples in previous sections of this chapter, consider these waves to be superposed in a pipe of 25 mm diameter with the compression wave, p_1, moving rightward. The reference conditions are also as used before, that is, T_0 is 20°C or 293 K and p_0 is 101,325 Pa. However, pressure drop only occurs as a result of particle movement, so it is necessary to define a time interval for the superposition process to occur. Let it be considered that the waves are superposed at the pressure levels indicated for a period of 2° crankshaft in the duct of an engine running at 1000 rpm. This represents a time interval, dt, of:

$$dt = \frac{\theta}{360} \times \frac{60}{N} = \frac{\theta}{6N} \quad s \tag{2.3.9}$$

or

$$dt = \frac{2}{6 \times 1000} = 0.333 \times 10^{-3} \quad s$$

187

where the particle movement, dx, is given by:

$$dx = c_s dt = 99.1 \times 0.333 \times 10^{-3} = 33 \times 10^{-3} \text{ m}$$

From the above, the calculated numerical value for the superposition time element is 0.333 ms. Assuming a friction factor, C_f, of 0.004, the above equations in the section show that the loss of superposition pressure occurs over a distance, dx, of 33 mm within the duct and has a magnitude of 122 Pa. The pressure ratio of the rightward wave drops from 1.2 to 1.198, and that of the leftward wave rises from 0.8 to 0.8023. The rightward propagation velocity of the compression wave during superposition drops from 440.5 to 439.5 m/s, while that of the leftward expansion wave rises from 242.3 to 243.4 m/s. The superposition particle velocity drops from 99.1 m/s to 98.05 m/s.

It is evident that the pressure loss due to friction reduces the amplitude of compression waves and slows them down. The opposite effect applies to an expansion wave: the pressure loss due to friction raises its absolute pressure, i.e., weakens the wave, and thereby increases its propagation velocity from a subsonic value toward sonic velocity.

Although the likelihood of a single traverse of a pressure wave in a duct of an engine is remote, it should nevertheless be considered theoretically. By definition, friction opposes the motion of a pressure wave and does so continuously. This means that a train of pressure waves are sent off in the direction opposite to the propagation of the wave train and with a magnitude that can be calculated from the above equations. Using the data above, but with the exception that the single wave, p_1, traveling rightward, i.e., in the positive direction according to the sign convention, has a pressure ratio of 1.2. All other data remain the same and a fixed friction factor of 0.004 is used. All of the above equations can be used, inserting a value of p_2 equal to p_0, i.e., with a pressure ratio of 1.0. The results show that the ongoing wave pressure ratio, P_{1f}, is reduced to 1.1994 and the reflected wave, P_{2f}, is 1.0004. If the calculation is repeated to find the effect of friction on a single traverse of an expansion wave, i.e., by inserting a value of 0.8 for P_1 and a value of 1.0 for P_2, then P_{1f} and P_{1f} become 0.8006 and 0.9995, respectively.

In Sec. 2.19 the traverse of a single pressure wave in a duct is described as both theory and experiment.

2.3.1 Friction Factor during Pressure Wave Propagation

It is possible to predict the value of friction factor more closely by considering further information available in the literature on experimental and theoretical fluid mechanics. The thermal conductivity, C_k, and viscosity, μ, of air, or any gas, are functions of absolute temperature, T, and are required for the calculation of the shearing forces in air, from which friction factor can be assessed. The interconnection between friction factor and shear stress has been set out in Eq. 2.3.1. The thermal conductivity and viscosity of air can be found from data tables and curves fitted to provide sufficiently accurate data for values of T from 300 to 2000 K as:

$$C_k = 6.1944 \times 10^{-3} + 7.3814 \times 10^{-5} T - 1.2491 \times 10^{-8} T^2 \text{ W/mK} \qquad (2.3.10)$$

$$\mu = 7.457 \times 10^{-6} + 4.1547 \times 10^{-8}\,T - 7.4793 \times 10^{-12}\,T^2 \text{ kg/ms} \qquad (2.3.11)$$

Although the data above for air are not the same as those for exhaust gas, the differences are sufficiently small as to warrant describing exhaust gas by the same relationships used for air for the purposes of determining Reynolds, Nusselt, and other dimensionless parameters common in fluid mechanics.

The Reynolds number at any local point in space and time in a duct of diameter d at superposition temperature T_s, density ρ_s, and particle velocity c_s, is given by:

$$\mathbf{Re} = \frac{\rho_s d c_s}{\mu_{Ts}} \qquad (2.3.12)$$

Much experimental work in fluid mechanics has related friction factor to Reynolds number, and the expression ascribed to Blasius to describe fully turbulent flow is typical:

$$C_f = \frac{0.0791}{\mathbf{Re}^{0.25}} \quad \text{for} \quad \mathbf{Re} \geq 4000 \qquad (2.3.13)$$

Almost all unsteady gas flow in engine ducting is turbulent, but where it is not, and it is laminar or nearly laminar flow, it is simple and accurate to assign the friction factor as 0.01. This threshold number can be easily deduced from the Blasius formula by inserting a Reynolds number of 4000. ·

Using all of the data from the example cited in the previous section above, it was found that the calculated superposition particle velocity is 99.1 m/s. The superposition density is 1.1752 kg/m^3 and can be calculated using Eq. 2.2.11. The relationship for superposition temperature is given by Eq. 2.1.11, therefore:

$$\frac{T_s}{T_0} = \left(\frac{p_s}{p_0}\right)^{\frac{\gamma-1}{\gamma}} = X_s^2 \qquad (2.3.14)$$

Using the numerical data provided, the superposition temperature is 290.1 K or 17.1°C. The viscosity of air at 290.1 K is determined from Eq. 2.3.11 to be 1.888×10^{-5} kg/ms. Consequently, the Reynolds number, **Re**, in the 25-mm diameter pipe, from Eq. 2.3.12, is 154,210; it is clearly turbulent flow. Then, from Eq. 2.3.13, the friction factor, C_f, is calculated to be 0.00399, which is not far removed from the assumption used in the previous section and which appears frequently for air in the literature!

The use of this theoretical approach for the determination of friction factor, C_f, allied to the general theory regarding friction loss in the previous section, permits the accurate assessment of the friction loss in pressure waves in unsteady flow.

It is important to stress that the action of friction on a pressure wave passing through the pipe is a non-isentropic process. The manifestation of this is the heating of the local gas as friction occurs and the continual decay of the pressure wave due to its application. The heating effect can be calculated because the friction force, F, is available from the combination of Eqs. 2.3.1 and 2.3.2, and the work done by this force acting through distance dx appears as heat in the gas element involved.

$$F = \pi d \left[C_f \frac{\rho_s c_s^2}{2} \right] c_s dt \qquad (2.3.15)$$

Thus the work, δW_f, resulting in the heat generated, δQ_f, can be calculated by:

$$\delta W_f = Fdx = Fc_s dt = \frac{\pi d C_f \rho_s c_s^4 dt^2}{2} = \delta Q_f \qquad (2.3.16)$$

All of the relevant data for the numerical example used in this section are available to insert into Eq. 2.3.16, from which it is calculated that the internal heating due to friction is 1.974 mJ. Although this value may appear miniscule, it should be remembered that this is a continuous process occurring for a pressure wave during its excursion throughout a pipe, and that this heating effect of 1.974 mJ takes place in a time frame of 0.333 ms. This represents a heating rate of 5.93 W, putting the heating effect due to friction into a physical context that can be more readily comprehended.

One issue that must be taken into account by those concerned with the computation of wave motion is that all of the above equations use a length term within the calculation for friction force with respect to the work done or heat generated by opposition to it. This length term is quite correctly computed from the particle velocity c_s and the time period dt for the motion of those particles. However, should the computation method purport to represent a group of gas particles within a pipe by the behavior of those particles at the wave point under calculation, then the length term in the ensuing calculation for force F must be replaced by the length occupied by the said group of particles. The subsequent calculation to compute the work, δW_f, i.e., the heat quantity δQ_f generated by friction, is the force due to friction for all of the group of particles multiplied by the distance moved by any one of the particles in this group, with this distance remaining as the "$c_s dt$" term. This is discussed at greater length in Sec. 2.18.6.

2.3.2 Friction Loss during Pressure Wave Propagation in Bends in Pipes

This factor is seldom considered of pressing significance in the simulation of engine ducting because pressure waves travel around quite sharp kinks, bends, and radiused corners in ducting with pressure loss not much greater than that normally associated with friction, as has been discussed above. This is not a subject that has been researched to any great extent, as can be seen from the referenced literature of Blair et al. [2.20]. The basic mechanism of analysis is to compute the pressure loss, dp_b, in the segment of pipe length under analysis. The procedure is similar to that for friction:

$$\text{Pressure loss} \qquad dp_b = C_b \rho_s c_s^2 \qquad (2.3.17)$$

where the pressure loss coefficient, C_b, is principally a function of Reynolds number and the deflection angle per unit length around the bend:

$$C_b = f(\mathbf{Re}, \frac{d\theta}{dx}) \qquad (2.3.18)$$

The extra pressure loss as a result of deflection, by an angle appropriate to the pipe segment length, dx, can then be added to the friction loss term in Eq. 2.3.3 and the analysis can continue for the segment of pipe length, dx, under scrutiny. Sec. 2.14 contains a discussion of pressure losses at branches in pipes, and Eq. 2.14.1 gives an almost identical relationship for the pressure loss in deflecting flows around the corners of a pipe branch. All such pressure loss equations relate that loss to a gain of entropy through a decrease in the kinetic energy of the gas particles during superposition.

It is a subject deserving of painstaking measurement through research using the sophisticated experimental QUB SP apparatus described in Sec. 2.19. The aim should be to derive accurately the values of the pressure loss coefficient, C_b, and ultimately report them in the literature; I am not aware of such information having been published already.

2.4 Heat Transfer during Pressure Wave Propagation

Heat can be transferred to or from the wall of the duct by the gas as the unsteady flow process occurs. Although all three processes of conduction, convection, and radiation are potentially involved, it is much more likely that convection heat transfer will be the predominant phenomenon in most cases. This is certainly true of induction systems, but some of you will recall that exhaust manifolds glow red and ponder the potential errors of considering convection heat transfer as the sole mechanism involved. There is no doubt that in such circumstances radiation heat transfer should be seriously considered for inclusion in the theoretical treatment. But this is not an easy topic and the potential error of including radiation could actually be more serious than excluding it. As a consequence, only convection heat transfer will be discussed.

The information needed to calculate the normal and relevant parameters for convection is available from within the analysis of unsteady gas flow. The physical situation is illustrated in Fig. 2.4. A superposition process is underway. The gas is at temperature T_s, particle velocity c_s, and density ρ_s.

In Sec. 2.3.1, the computation of the friction factor, C_f, and the Reynolds number, **Re**, was described. From the Reynolds analogy of heat transfer with friction it is possible to calculate the Nusselt number, **Nu**, thus:

$$Nu = \frac{C_f \mathbf{Re}}{2} \tag{2.4.1}$$

The Nusselt number contains a direct relationship between the convection heat transfer coefficient, C_h, the thermal conductivity of the gas, C_k, and the effective duct diameter, d. The standard definitions for these parameters can be found in any conventional text on fluid mechanics or heat transfer [2.4]. The definition for the Nusselt number is:

$$Nu = \frac{C_h d}{C_k} \tag{2.4.2}$$

From Eqs. 2.4.1 and 2.4.2, the convection heat transfer coefficient can be determined:

$$C_h = \frac{C_k Nu}{d} = \frac{C_k C_f \mathbf{Re}}{2d} \tag{2.4.3}$$

The relationship for the thermal conductivity of air is given above in Eq. 2.3.10 as a function of the gas temperature, T. In any unsteady gas flow process, the time element, dt, and the distance of exposure of the gas element to the wall, dx, is available from the computation or from the input data. In such a case, knowing the pipe wall temperature, the heat transfer, δQ_h, from the gas to the pipe wall can be assessed. The direction of this heat transfer is clear from the ensuing theory:

$$\delta Q_h = \pi d C_h dx (T_w - T_s) dt \tag{2.4.4}$$

If we continue the numeric example with the same input data as given previously in Sec. 2.3, but with the added information that the pipe wall temperature is 100°C, then the output data using the theory shown in this section give the magnitude and direction of the heat transfer. Because the wall at 100°C is hotter than the gas at superposition temperature, T_s, at 17.1°C, the heat transfer direction is positive because it is added to the gas in the pipe. The value of heat transferred is 23.44 mJ, which is considerably greater than the 1.974 mJ attributable to friction. In this example, the Nusselt number, **Nu**, is calculated as 307.8 and the convection heat transfer coefficient, C_h, as 326.9 W/m²K.

The total heating or cooling of a gas element undergoing an unsteady flow process is a combination of the external heat transfer from convection and the internal heat transfer from friction. This total heat transfer is defined as δQ_{fh}, and is obtained thus:

$$\delta Q_{fh} = \delta Q_f + \delta Q_h \qquad (2.4.5)$$

For the same numeric data, the value of dQ_{fh} is the sum of +1.974 mJ and +23.44 mJ, an addition that would result in δQ_{fh} becoming +25.44 mJ.

2.5 Wave Reflections at Discontinuities in Gas Properties

As a pressure wave propagates within a duct, it is highly improbable that it will always encounter ahead of it gas that is at precisely the same state conditions or has the same gas properties as that through which it is currently traveling. In physical terms, many people have experienced an echo when they have shouted in foggy surroundings on a cold, damp morning, with the sun warming some sections of the local atmosphere more strongly than others. In acoustic terms, this is precisely the situation that commonly exists in engine ducting. It is particularly pronounced in the inlet ducting of a four-stroke engine that has experienced a strong back-flow of hot exhaust gas at the onset of the intake process. Two-stroke engines have an exhaust process that sends hot exhaust gas into the pipe during the blowdown phase and then short-circuits a large quantity of much colder air into this same duct during the scavenge process. Racing four-stroke engines with long overlap valve periods and tuned exhaust systems also induce some of the in-cylinder colder air into the hotter exhaust pipe. Pressure waves propagating through such pipes encounter gas at varying temperatures, and reflections ensue. Although this is referred to in the literature as a "temperature" discontinuity, this is quite misleading. It is actually a discontinuity of state, so pressure wave reflection will take place at such boundaries, even at constant temperature, if other gas properties such as the gas constant or the density are variable across it.

The physical and thermodynamic situation is sketched in Fig. 2.5, where a pressure wave p_1 is meeting pressure wave p_2. Clearly, a superposition process takes place. In Section 2.2.1, where the theory of superposition was set down, it was stated that once the superposition process is completed the pressure waves, p_1 and p_2, proceed onward with unaltered amplitudes. However, this superposition process is taking place with the arriving waves having traversed through gas at differing thermodynamic states. Wave p_1 is coming from side "a" where the gas has properties of γ_a and R_a, and reference conditions of density, ρ_{0a}, and temperature, T_{0a}. Wave p_2 is coming from side "b" where the gas has properties of γ_b and R_b, and reference conditions of density, ρ_{0b}, and temperature, T_{0b}. As with most reflection processes, the momentum equation is the equation that best describes a "bounce" behavior, and this proves to be so in this case. Because the superposition process occurs, and the reflection is taking place at the interface, the transmitted pressure waves assume amplitudes of p_{1d} and p_{2d}. The theory to describe this basically states that the laws of conservation of mass and momentum must be upheld.

Continuity $\qquad\qquad \dot{m}_{\text{side a}} = \dot{m}_{\text{side b}} \qquad (2.5.1)$

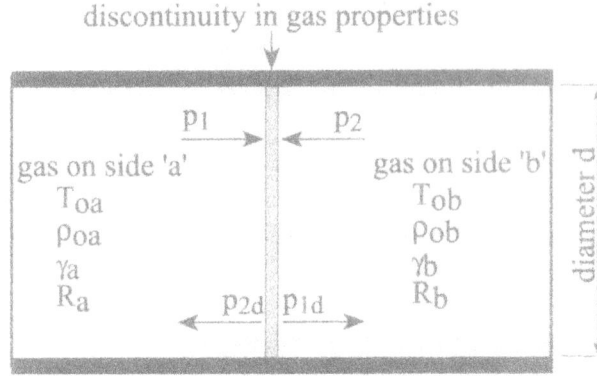

Fig. 2.5 Wave reflection at a temperature discontinuity.

Momentum $\quad A\left(p_{s \text{ side a}} - p_{s \text{ side b}}\right) = \dot{m}_{\text{side a}} c_{s \text{ side a}} - \dot{m}_{\text{side b}} c_{s \text{ side b}}$ $\qquad$ (2.5.2)

It can be seen that this produces a relatively simple solution whereby the superposition pressure and the superposition particle velocity are identical on either side of the thermodynamic discontinuity.

$$p_{s \text{ side a}} = p_{s \text{ side b}} \qquad (2.5.3)$$

$$c_{s \text{ side a}} = c_{s \text{ side b}} \qquad (2.5.4)$$

The solution divides into two different cases, one simple and the other more complex, depending on whether the gas composition is identical on both sides of the boundary.

(i) The Simple Case of Common Gas Composition
The following is the solution of the simple case where the gas is identical in composition on both sides of the boundary, i.e., where γ_a and R_a are identical to γ_b and R_b. Eq. 2.5.4 reduces to:

$$G_{5a} a_{oa}(X_1 - X_{2d}) = G_{5b} a_{ob}(X_{1d} - X_2) \qquad (2.5.5)$$

Because the gas composition is common, this reduces to:

$$\left(\frac{a_{0a} G_{5a}}{a_{0b} G_{5b}}\right)(X_1 - X_{2d}) = (X_{1d} - X_2) \qquad (2.5.6)$$

(NOTE: Although the G_5 terms are actually equal they are retained for completeness.)

Eq. 2.5.3 reduces to:

$$(X_1 + X_{2d} - 1)^{G7a} = (X_{1d} + X_2 - 1)^{G7b} \qquad (2.5.7)$$

In this simple case, the values of G_{7a} and G_{7b} are identical and are simply G_7:

$$X_1 + X_{2d} - 1 = X_{1d} + X_2 - 1 \qquad (2.5.8)$$

The solution becomes straightforward.

$$X_{2d} = \frac{2X_2 - X_1\left(1 - \dfrac{a_{0a}G_{5a}}{a_{0b}G_{5b}}\right)}{1 + \dfrac{a_{0a}G_{5a}}{a_{0b}G_{5b}}} \quad \text{hence} \quad p_{2d} = p_0 X_{2d}^{G7} \qquad (2.5.9)$$

$$X_{1d} = X_1 + X_{2d} - X_2 \quad \text{hence} \quad p_{1d} = p_0 X_{1d}^{G7} \qquad (2.5.10)$$

(ii) The More Complex Case of Variable Gas Composition

If the gas composition is different across the boundary, the simplicity of reducing Eq. 2.5.6 to Eq. 2.5.7, and also Eq. 2.5.7 to Eq. 2.5.8, is no longer possible because Eq. 2.5.7 remains as a polynomial function. The method of solution is to eliminate one of the unknowns from Eqs. 2.5.6 and 2.5.7, either X_{1d} or X_{2d}, and solve for the remaining unknown by the Newton-Raphson method. The final step is as in Eqs. 2.5.9 and 2.5.10, but with the gas composition inserted appropriately to the side of the discontinuity:

$$p_{2d} = p_0 X_{2d}^{G7a} \qquad (2.5.11)$$

$$p_{1d} = p_0 X_{1d}^{G7b} \qquad (2.5.12)$$

A numerical example will make the above theory easier to understand. Consider, as in Fig. 2.5, a pressure wave p_1 arriving at a boundary where the reference temperature, T_{01}, on side "a" is 200°C and T_{02} on side "b" is 100°C. The gas is colder on side "b" and is more dense. An echo should ensue. The gas is air on both sides, i.e., γ is 1.4 and R is 287 J/kgK. The situation is undisturbed on side "b," i.e., pressure p_2 is the same as the reference pressure p_0. The pressure wave on side "a" has a pressure ratio, P_1, of 1.3. Both the simple solution and the more complex solution will give precisely the same answer. The transmitted pressure wave into side "b" is stronger, with a pressure ratio, P_{1d}, of 1.32, whereas the reflected pressure wave, the echo, has a pressure ratio, P_{2d}, of 1.016.

If the above calculation is repeated with just one exception—the gas on side "b" is exhaust gas with the appropriate properties, i.e., γ is 1.36 and R is 300 J/kgK—then the simple solution can be shown to be inaccurate. In this new case, the simple solution will predict that the transmitted pressure ratio, P_{1d}, is 1.328, whereas the reflected pressure wave, the echo, has a pressure ratio, P_{2d}, of 1.00077. The accurate solution, solving the equations and taking into account the variable gas properties, will predict that the transmitted pressure ratio, P_{1d}, is 1.314, whereas the reflected pressure wave, the echo, has a pressure ratio, P_{2d}, of 1.011. The difference in these answers is considerable, making the unilateral employment of a simple solution, based on the assumption of a common gas composition throughout the ducting, inappropriate for use in engine calculations. In real engines, and therefore also in calculations with pretensions of accuracy, a variation of gas properties and composition throughout the ducting is the norm rather than the exception, and must be simulated as such.

2.6 Reflection of Pressure Waves

Oppositely moving pressure waves arise from many sources, but principally from reflections of pressure waves at the boundaries of the inlet or the exhaust duct. Such boundaries include junctions at the open or closed end of a pipe, and also any change in area, gradual or sudden, within a pipe. In the case of sound reflections, an echo is a classic example of a closed-ended reflection. All reflections are, by definition, a superposition process, as the reflection proceeds to move oppositely to the incident pressure wave causing it. Fig. 2.6 shows some of the possibilities for reflections in a three-cylinder engine. Whether the engine is a four-stroke or a two-stroke engine is of no consequence, but the air inflow and the exhaust outflow at the extremities are clearly marked. Expansion pressure waves are sent into the intake system by the induction process and exhaust pressure waves are shuttled into the exhaust system in the sequence of the engine firing order.

The potential for wave reflection in the ducting, and its location, is marked by a number. It is obvious that the location accompanies a change of section of the ducting or a change in its direction.

At point 1 is the intake bellmouth where the induction pressure wave is reflected at the atmosphere. At point 2 is a plenum chamber where pressure wave oscillations are damped out in amplitude. At point 3 is an air filter which provides a restriction to the flow and the possibility of minor "echos" taking place. At point 4 is a throttle in the duct which, depending on the throttle opening value, can pose either a major restriction or even none at all. At point 5 in the intake manifold is a four-way branch where the pressure wave must divide and send reflections back from the change of section.

At point 6 in the exhaust manifold are three-way branches where the pressure wave must also divide and send reflections back from the change of section. When a pressure wave arrives at a three-way junction, the values transmitted into each of the other branches will be a function of the areas of the pipes and all of the reflections present. When the induction pressure wave arrives at the open pipe end at 1, a reflection will take place which will be immediately superposed on the incident pressure pulse. In short, all reflection processes are superposition processes—the fundamental reason why the theoretical text up to this point has set down the wave motion within constant area ducting and the basics of the superposition process in some detail.

196

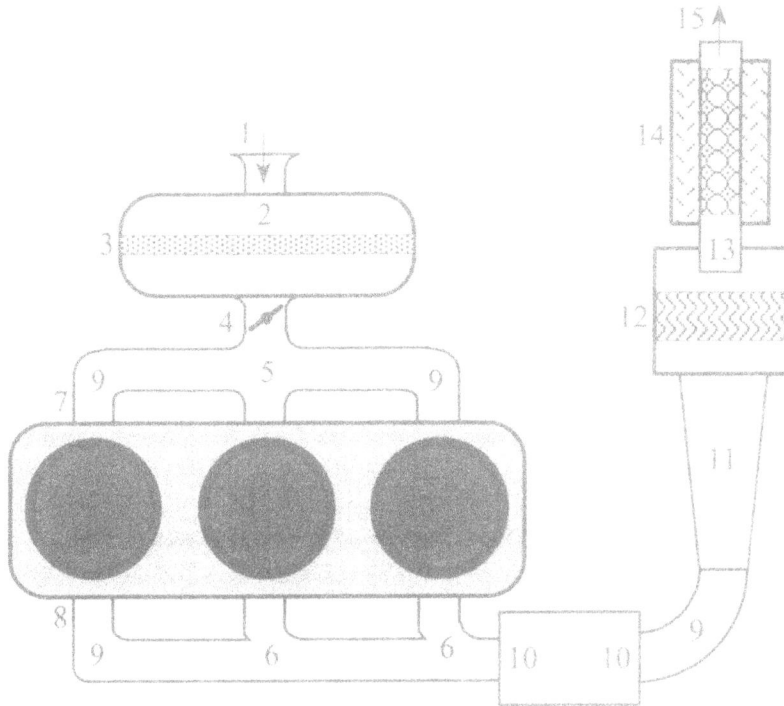

Fig. 2.6 Wave reflection possibilities in the manifolds of an engine.

At points 7 and 8 are the valves or ports into the cylinder which, depending on the valve or port opening schedule, behave as everything from a partially closed or open end to a cylinder at varying pressures, to a perfect "echo" location when the valves or ports are all closed. At point 9 are bends in the ducting where the pressure wave is reflected from the deflection process in major or minor part depending on the severity of the radius of the bend. At point 10 in the exhaust ducting are sudden expansion and contractions in the pipe. At point 11 is a tapered exhaust pipe which will act as a diffuser or a nozzle depending on the direction of the particle flow; wave reflections ensue in both cases. At point 12 is a restriction in the form of a catalyst, little different from an air filter except that chemical reactions are taking place at the same time—an explanation that is easily written but whose underlying theoretical calculations predicting the wave motion and the thermodynamics of the reaction are somewhat more complex. At point 13 is a reentrant pipe to a chamber, which is a very common element in any silencer design. At point 14 is an absorption silencer element, a length of perforated pipe surrounded by packing, which acts as both diffuser and the trimmer of sharp peaks on exhaust pulses. By definition, this element provides wave reflections. At point 15 is a plain-ended exhaust pipe entering the atmosphere. Pressure wave reflections also take place here.

The above "tour" of the pressure wave routes in and out of an engine is far from a complete description of the processes that take place. Nevertheless, it is meant to illustrate both the complexity of the events and to postulate that, without a complete understanding of every and all possibilities for wave reflection in and through an internal combustion engine, none can seriously claim to be a designer of engines.

The resulting pressure-time histories are very complex and beyond the memory-tracking capability of the human mind. A computer, however, is a methodical calculation tool that is ideal for this pedantic exercise. You will therefore be introduced to the use of computers for this purpose. Before that juncture, it is essential to comprehend the basic effect of each of these reflection mechanisms, as the mathematics of their behavior must be programmed in order to track the progress of all incident and reflected waves.

The sections that follow analyze virtually all of the above possibilities for wave reflection due to changes in pipe or duct geometry, and analyze the reflection and transmission process that takes place at each juncture.

2.6.1 Notation for Reflection and Transmission of Pressure Waves in Pipes

A wave arriving at a position where it can be reflected is called the incident wave. In the paragraphs that follow, all incident pressure waves, whether they be compression or expansion waves, will be designated by the subscript "i," i.e., pressure, p_i; pressure ratio, P_i; pressure amplitude ratio, X_i; particle velocity, c_i; density, ρ_i; acoustic velocity, a_i; and propagation velocity, α_i. All reflections will be designated by the subscript "r," i.e., pressure, p_r; pressure ratio, P_r; pressure amplitude ratio, X_r; particle velocity, c_r; density, ρ_r; acoustic velocity, a_r; and propagation velocity, α_r. All superposition characteristics will be designated by the subscript "s," i.e., pressure, p_s; pressure ratio, P_s; pressure amplitude ratio, X_s; particle velocity, c_s; density, ρ_s; acoustic velocity, a_s; and propagation velocity, α_s.

Where a gas particle flow regime is taking place, flow from gas in a regime is always subscripted with a "1" and flow to gas in a regime is subscripted with a "2." Thus the gas properties of specific heats ratio and gas constant in the upstream regime are γ_1 and R_1, while those in the downstream regime are γ_2 and R_2.

The reference condition of pressure is noted as p_0; temperature as T_0; acoustic velocity as a_0; and density as ρ_0.

2.7 Reflection of a Pressure Wave at a Closed End in a Pipe

When a pressure wave arrives at the plane of a closed end in a pipe, a reflection takes place. This is the classic echo situation, so it is no surprise to discover that the mathematics dictate that the reflected pressure wave is an exact image of the incident wave, but traveling in the opposite direction. The one certain fact available, physically speaking, is that the superposition particle velocity is zero in the plane of the closed end, as is shown in Fig. 2.7(a).

From Eq. 2.2.3:

$$c_s = c_i + c_r = 0 \tag{2.7.1}$$

or

$$c_r = -c_i \tag{2.7.2}$$

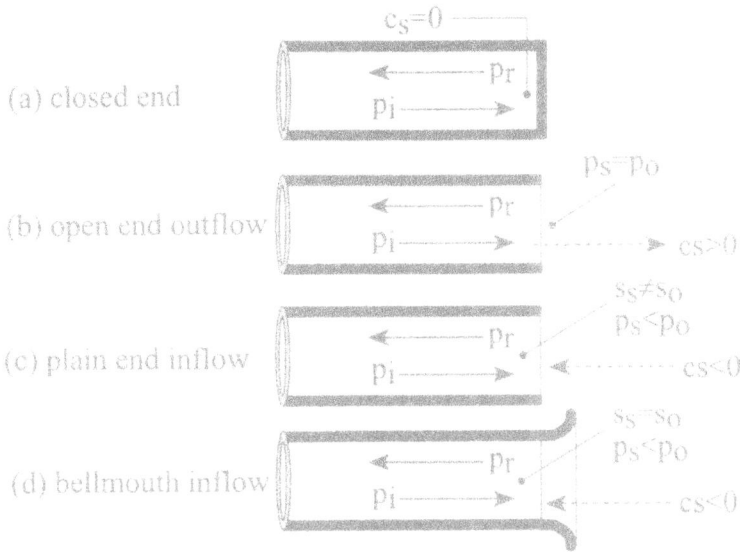

Fig. 2.7 Wave reflection criteria at some typical pipe.

From Eq. 2.2.2, because c_s is zero:

$$X_r = X_i \qquad\qquad (2.7.3)$$

hence
$$P_r = P_i \qquad\qquad (2.7.4)$$

From the combination of Eqs. 2.7.1 and 2.7.3:

$$X_s = 2X_i - 1 \qquad\qquad (2.7.5)$$

therefore
$$p_s = p_0(2X_i - 1)^{G7} \qquad\qquad (2.7.6)$$

2.8 Reflection of a Pressure Wave at an Open End in a Pipe

Here the situation is slightly more complex, in that the fluid flow behavior is different for compression and expansion waves. Compression waves are dealt with first. By definition, such waves must produce gas particle outflow from the pipe into the atmosphere.

2.8.1 Reflection of a Compression Wave at an Open End in a Pipe

In this case, as illustrated in Fig. 2.7(b), the first logical assumption that can be made is that the superposition pressure in the plane of the open end is the atmospheric pressure. The reference pressure, p_0, in such a case is the atmospheric pressure. From Eq. 2.2.1, this gives:

$$X_s = X_i + X_r - 1 = 1$$

hence
$$X_r = 2 - X_i \qquad (2.8.1)$$

Therefore, the amplitude of the reflected pressure wave, p_r, is:

$$p_r = p_0(2 - X_i)^{G7} \qquad (2.8.2)$$

For compression waves, because $X_i > 1$ and $X_r < 1$, the reflection is an expansion wave. From Eq. 2.2.6 for the more general case:

$$\begin{aligned}
c_s &= G_5 a_0(X_i - 1) - G_5 a_0(X_r - 1) \\
&= G_5 a_0(X_i - 1 - 2 + X_i + 1) \\
&= 2G_5 a_0(X_i - 1) \\
&= 2c_i
\end{aligned} \qquad (2.8.3)$$

This equation is applicable as long as the superposition particle velocity, c_s, does not reach the local sonic velocity, at which point the flow regime must be analyzed by further equations. The sonic regime at an open pipe end is an unlikely event in most engines, but not all. An analysis of this regime is described clearly by Bannister [2.2] and in Secs. 2.2.4 and 2.16 of this text.

It can also be shown that, because c_r equals c_i, and because the equation is sign dependent, the particle flow is in the same direction. This conclusion can also be reached from the earlier conclusion that the reflection is an expansion wave that impels particles opposite to its direction of propagation. This means that an exhaust pulse arriving at an open end sends suction reflections back toward the engine which will help to extract exhaust gas particles further down the pipe and away from the engine cylinder—or out of it if the valve is open. Clearly, this is a reflection that can be used by the designer to aid the scavenging, i.e., emptying of the cylinder of its exhaust gas. The numerical data below emphasise this point.

For a basic understanding of how this theory is used to calculate the reflection of compression waves at the atmospheric end of a pipe, consider an example using the compression pressure wave, p_e, previously used in Sec. 2.1.4.

You will recall that the wave, p_e, is a compression wave of pressure ratio 1.2. In the nomenclature of this section it becomes the incident pressure wave, p_i, at the open end. This pressure ratio is shown to have a pressure amplitude ratio, X_i, of 1.02639. Using Eq. 2.8.1, the reflected pressure amplitude ratio, X_r, is given by:

$$X_r = 2 - X_i = 2 - 1.02639 = 0.9736$$

or
$$P_r = X_r^{G7} = 0.9736^7 = 0.8293$$

That the reflection of a compression wave at the open end is a rarefaction wave is now evident numerically.

2.8.2 *Reflection of an Expansion Wave at a Bellmouth Open End in a Pipe*

This reflection process is connected with inflow, so it is necessary to consider the fluid mechanics of the flow into a pipe. Inflow of air in an intake system, which is the normal place to find expansion waves, is usually conducted through a bellmouth-ended pipe of the type illustrated in Fig. 2.7. This form of pipe end will be discussed in the first instance.

The analysis of gas flow to and from a thermodynamic system, which may also be experiencing heat transfer and work transfer processes, is analyzed by the first law of thermodynamics. The theoretical approach is to be found in any standard textbook on thermodynamics [1.30, 2.4]. In general, this is expressed as:

$$\Delta(\text{heat transfer}) + \Delta(\text{energy entering}) = \Delta(\text{system energy}) +$$
$$\Delta(\text{energy leaving}) + \Delta(\text{work transfer})$$

The First Law of Thermodynamics for an open system flow from the atmosphere to the superposition station at the full pipe area in Fig. 2.7(d) is as follows:

$$\delta Q_{\text{system}} + \Delta m_0 (h_0 + \frac{c_0^2}{2}) = dE_{\text{system}} + \Delta m_S (h_S + \frac{c_S^2}{2}) + \delta W_{\text{system}} \qquad (2.8.4)$$

If the flow at the instant in question can be presumed to be quasi-steady and steady-state flow without heat transfer, and also to be isentropic, then δQ, δW, and dE are all zero. The mass flow increments must be equal to satisfy the continuity equation. The difference in the enthalpy terms can be evaluated as:

$$h_s - h_0 = C_p(T_s - T_0) = \frac{\gamma R}{\gamma - 1}(T_s - T_0) = \frac{a_s^2 - a_0^2}{\gamma - 1}$$

Although the particle velocity in the atmosphere, c_0, is virtually zero, Eq. 2.8.4 reduces to:

$$c_0^2 + G_5 a_0^2 = c_s^2 + G_5 a_s^2 \qquad (2.8.5)$$

Because the flow is isentropic, from Eq. 2.1.12:

$$a_s = a_0 X_s \qquad (2.8.6)$$

Substituting Eq. 2.8.6 into Eq. 2.8.5 and now regarding c_0 as zero:

$$c_s^2 = G_5 a_0^2 (1 - X_s^2) \qquad (2.8.7)$$

From Eq. 2.2.8:

$$c_s = G_5 a_0 (X_i - X_r) \qquad (2.8.8)$$

From Eq. 2.2.1:

$$X_s = X_i + X_r - 1 \qquad (2.8.9)$$

Bringing these latter three equations together:

$$G_5 (X_i - X_r)^2 = 1 - (X_i + X_r - 1)^2$$

This becomes:

$$G_6 X_r^2 - \left(2G_4 X_i + 2\right) X_r + \left(G_6 X_i^2 - 2X_i\right) = 0 \qquad (2.8.10)$$

This is an equation that is quadratic in X_r. Ultimately neglecting the negative sign in the general solution to a quadratic equation, this yields:

$$X_r = \frac{\left(2G_4X_i + 2\right) \pm \sqrt{\left(2G_4X_i + 2\right)^2 - 4G_6\left(G_6X_i^2 - 2X_i\right)}}{2G_6}$$

$$= \frac{\left(1 + G_4X_i\right) \pm \sqrt{1 + X_i\left(2G_4 + 2G_6\right) + X_i^2\left(G_4^2 - G_6^2\right)}}{G_6} \tag{2.8.11}$$

For inflow at a bellmouth end when the incoming gas is air, where γ equals 1.4, this becomes,

$$X_r = \frac{1 + 4X_i + \sqrt{1 + 20X_i - 20X_i^2}}{6} \tag{2.8.12}$$

This equation shows that the reflection of an expansion wave at a bellmouth end of an intake pipe will be a compression wave. Take as an example the expansion pressure wave p_i of Sec. 2.1.4, which by coincidence retains its subscript nomenclature because it represents an incident wave, where that expansion wave has a pressure ratio, P_i, of 0.8 and a pressure amplitude ratio, X_i, of 0.9686. Substituting these numbers into Eq. 2.8.12 to determine the magnitude of the reflection of the wave at a bellmouth open end gives:

$$X_r = \frac{1 + 4 \times 0.9686 + \sqrt{1 + 20 \times 0.9686 - 20 \times 0.9686^2}}{6} = 1.0238$$

Thus,
$$P_r = X_r^7 = 1.0238^7 = 1.178$$

As predicted, the reflection is a compression wave and, if allowed by the designer to arrive at the intake valve while the intake process is still in progress, it will push further air charge into the cylinder. In the jargon of engine design, this effect is called "ramming." Ramming will be discussed later in this chapter and even more thoroughly in Chapters 5 and 6.

2.8.3 Reflection of an Expansion Wave at a Plain Open End in a Pipe

The reflection of expansion waves at the end of plain pipes can be dealt with in a similar theoretical manner. Bannister [2.2] describes the theory in some detail. It is presented here for completeness, although the generality of this boundary condition is laid out in Sec. 2.16. You can get a mental picture of the flow at this boundary by studying the results of a computational fluid dynamics (CFD) calculation at this very discontinuity, presented in Fig. 3.3.

The First Law of Thermodynamics still applies, as seen in Eq. 2.8.5, and c_0 is regarded as effectively zero.

$$G_5 a_0^2 = c_s^2 + G_5 a_s^2 \qquad (2.8.5)$$

However, as seen in Figs. 2.7(d) and 3.3, Eq. 2.8.6 cannot apply because the particle flow has a distinct vena contracta within the pipe end with an associated turbulent vortex ring, and cannot be regarded as an isentropic process. Thus:

$$a_s \neq a_0 X_s$$

but an entropy gain raises the value of a_0 to a_{00}, thus:

$$a_s = a_{00} X_s \qquad (2.8.13)$$

The most accurate statement is that the superposition acoustic velocity, using the format of Eq. 2.1.2, is given by:

$$a_s = \sqrt{\frac{\gamma p_s}{\rho_s}} \qquad (2.8.14)$$

Another gas dynamic relationship is required and, as is normal in a thermodynamic analysis of non-isentropic flow, the momentum equation is employed. Consider the flow from the atmosphere to the superposition plane of fully developed flow downstream of the vena contracta. Newton's Second Law of Motion can be expressed as:

Force = rate of change of momentum

where
$$p_0 A - p_s A = \dot{m}_s c_s - \dot{m}_0 c_0$$

or
$$p_0 - p_s = \rho_s c_s^2 \qquad (2.8.15)$$

Using the wave summation equations, and ignoring the entropy gain encapsulated in the a_{00} term so as to effect an approximate solution, the superposition pressure and particle velocity are related by:

$$X_s = X_i + X_r - 1$$

$$c_s = G_5 a_0 (X_i - X_r)$$

$$P_s = X_s^{G7}$$

Elimination of unwanted terms does not provide a simple solution, but one that requires an iterative approach by a Newton-Raphson method to arrive at the unknown value of reflected pressure X_r from a known value of the incident pressure X_i at the plain open end. The solution is expressed as a function of the unknown quantity X_r as follows:

$$f(X_r) = \frac{1}{2}\left[X_i + X_r - \sqrt{\frac{1}{G_5} \times \frac{1 - (X_i + X_r - 1)^{G7}}{1 + G_6(X_i + X_r - 1)^{G7}}} \right] - X_i = 0 \qquad (2.8.16)$$

This is solved by the Newton-Raphson method, i.e., by differentiation of the above equation with respect to the unknown quantity, X_r, and iterating to find the answer in the classic mathematical manner. Then the value of particle velocity, c_r, can be derived from substitution into the equation below once the unknown quantity, X_r, is determined.

$$c_s = c_i + c_r = G_5 a_0 (X_i - X_r) \qquad (2.8.17)$$

This is a far from satisfactory solution to an apparently simple problem describing what is, in reality, outflow from the atmosphere into a plain-ended pipe. That this is not simple at all can be seen in Sec. 2.16.

2.8.4 The Inadequacy of Simple Solutions

The above thermodynamics and gas dynamics, dealing with reflections of pressure waves at an open pipe end, probably appear sufficiently complex as to provide the definitive answer to our problem—at least for the case of particle inflow into a pipe with a bellmouth end. I have bad news, though: This is not correct. I have introduced these simple solutions in this section to provide you with correct mental and numeric images of what basically happens when pressure waves arrive at an open pipe end. In reality, the inflow process at a plain open-ended pipe, or a bellmouth-ended pipe for that matter, is a singular manifestation of what is generally called "cylinder to pipe outflow from an engine." In this case, the atmosphere is the very large "cylinder" flowing gas into a pipe! This complex problem is treated in great detail later in Sec. 2.16. The present boundary conditions for inflow at a pipe end can be calculated by the same theoretical approach as for pipe-to-cylinder flow, and the more complete thermodynamic and gas dynamic statements appear in those pages. At this point, you may well feel that I have wasted your time by asking you to study the simpler thermodynamics given above; not so, for the more complex theory to come is no more than a logical extension of that which you have been reading. Thus, when you get to Secs. 2.10, 2.12, 2.16, and 2.17, you will find comprehension that much easier. Even that is not the end of the matter, however, for you must wait until Chapter 3, which details the discharge coefficients of plain-ended and bellmouth-ended pipes, to learn how to solve these boundary conditions on a computer. However, I have good news, too. Within the pages of this book are all the equations and the information that you need to simulate the solution to these, and other, engine modeling problems. All you have to do is to write the code.

2.9 An Introduction to Reflection of Pressure Waves at a Sudden Area Change

It is quite common to find a sudden area change within a pipe or duct attached to an engine. In Fig. 2.6, sudden enlargements and contractions in pipe area are located at position 10. The basic difference—from a gas dynamics standpoint—between a sudden enlargement and contraction in pipe area, such as at position 10, and a plenum or volume, such as at position 2, is that the flow in the duct is considered to be one-dimensional, whereas in the plenum or volume it is considered to be three-dimensional. A subsidiary definition is that the particle velocity in a plenum or volume is considered to be so low as to be assigned as zero in any thermodynamic analysis.

For reflections at area changes within the duct, treating the flow as one-dimensional gives a change in amplitude of the transmitted pulse beyond the area change and also causes a wave reflection from it. Such sudden area changes are sketched in Fig. 2.8, where it can be seen that the pipe area contracts or expands at the junction. In each case, the incident wave at the sudden area change is depicted as propagating rightward, with the pipe nomenclature being "1" for the wave arrival pipe, with "i" signifying the incident pulse, "r" the reflected pulse, and "s" the superposition condition. The notation "p" is the conventional symbolism for absolute pressure. For example, at any instant, the incident pressure pulses at the junction are p_{i1} and p_{i2}. Depending on the areas A_1 and A_2, these incident pulses will give rise to reflected pulses, p_{r1} and p_{r2}.

In either expansion or contraction of the pipe area the particle flow is considered to be proceeding from the upstream superposition station 1 to the downstream superposition station 2. Therefore, the properties and composition of the gas particles, which are considered to be flowing, in any analysis based on quasi-steady flow are those of the gas at the upstream point. In all of the analyses presented here, this nomenclature is maintained. Therefore, the various functions of the gas properties are:

$$\gamma = \gamma_1 \quad R = R_1 \quad G_5 = G_{5_1} \quad G_7 = G_{7_1}, \text{ etc.}$$

It was Benson [2.4] who suggested a simple theoretical solution for such junctions. He assumed that the superposition pressure at the plane of the junction was the same in both pipes

206

at the instant of superposition. This assumption is inherently one of an isentropic process. Such a simple junction model will clearly have its limitations, but it is my experience that it is remarkably effective in practice, particularly if the area ratio changes, A_r, are in the band,

$$\frac{1}{6} < A_r < 6$$

The area ratio is defined as:

$$A_r = \frac{A_2}{A_1} \qquad (2.9.1)$$

From Benson,

$$p_{s1} = p_{s2} \qquad (2.9.2)$$

Consequently, from Eq. 2.2.1:

$$X_{i1} + X_{r1} - 1 = X_{i2} + X_{r2} - 1 \qquad (2.9.3)$$

From the continuity equation, equating the mass flow rate in an isentropic process on either side of the junction where,

$$\text{mass flow rate} = \text{density} \times \text{area} \times \text{particle velocity}$$

$$\rho_{s1} A_1 c_{s1} = \rho_{s2} A_2 c_{s2} \qquad (2.9.4)$$

Using the theory of Eqs. 2.1.17 and 2.2.2, where the reference conditions are p_0, T_0 and ρ_0, and with rightward decreed as positive particle flow, Eq. 2.9.4 becomes:

$$\rho_0 X_{s1}^{G5} A_1 G_5 a_0 (X_{i1} - X_{r1}) = -\rho_0 X_{s2}^{G5} A_2 G_5 a_0 (X_{i2} - X_{r2}) \qquad (2.9.5)$$

Because X_{s1} equals X_{s2}, this reduces to:

$$A_1 (X_{i1} - X_{r1}) = -A_2 (X_{i2} - X_{r2}) \qquad (2.9.6)$$

Combining Eqs. 2.9.1, 2.9.3, and 2.9.6, and eliminating each of the unknowns in turn, i.e., X_{r1} or X_{r2} :

$$X_{r1} = \frac{(1 - A_r)X_{i1} + 2X_{i2}A_r}{1 + A_r} \tag{2.9.7}$$

$$X_{r2} = \frac{2X_{i1} - X_{i2}(1 - A_r)}{(1 + A_r)} \tag{2.9.8}$$

To get a basic understanding of the results of employing Benson's simple "constant pressure" criterion for the calculation of reflections of compression and expansion waves at sudden enlargements and contractions in pipe area, consider an example using the two pressure waves, p_e and p_i, previously used in Sec. 2.1.4.

Wave p_e is a compression wave of pressure ratio 1.2 and wave p_i is an expansion wave of pressure ratio 0.8. Such pressure ratios have pressure amplitude ratios, X, of 1.02639 and 0.9686, respectively. Each of these waves in turn will be used as data for X_{i1} arriving in pipe 1 at a junction with pipe 2, where the area ratio will be either halved for a contraction or doubled for an enlargement to the pipe area. In each case, the incident pressure amplitude ratio in pipe 2, X_{i2}, will be taken as unity, which means that the incident pressure wave in pipe 1 is facing undisturbed conditions in pipe 2.

(a) An Enlargement, A_r = 2, for an Incident Compression Wave Where P_{i1} = 1.2 and X_{i1} = 1.02639
From Eqs. 2.9.7 and 2.9.8:

$$X_{r1} = 0.9912 \qquad X_{r2} = 1.01759$$

Hence, the pressure ratios, P_{r1} and P_{r2}, of the reflected waves are:

$$P_{r1} = 0.940 \qquad P_{r2} = 1.130$$

The sudden enlargement behaves like a slightly less effective "open end," as the completely open ended pipe in Sec. 2.8.1 gives a reflected wave with a pressure ratio of 0.8293 instead of 0.940. The onward transmitted pressure wave into pipe 2 is also one of compression, but it has a reduced pressure ratio of 1.13.

(b) An Enlargement, A_r = 2, for an Incident Expansion Wave Where P_{i1} = 0.8 and X_{i1} = 0.9686
From Eqs. 2.9.7 and 2.9.8:

$$X_{r1} = 1.0105 \qquad X_{r2} = 0.97908$$

Hence, the pressure ratios, P_{r1} and P_{r2}, of the reflected waves are:

$$P_{r1} = 1.076 \qquad P_{r2} = 0.862$$

As above, the sudden enlargement behaves as a slightly less effective "open end" because a "perfect" bellmouth open end to a pipe in Sec. 2.8.2 was shown to produce a stronger reflected wave with a pressure ratio of 1.178, instead of the weaker value of 1.076 found here. The onward transmitted pressure wave in pipe 2 is one of expansion, but it has a diminished pressure ratio of 0.862.

(c) A Contraction, A_r = 0.5, for an Incident Compression Wave where P_{i1} = 1.2 and X_{i1} = 1.02639
From Eqs. 2.9.7 and 2.9.8:

$$X_{r1} = 1.0088 \qquad X_{r2} = 1.0352$$

Hence, the pressure ratios, P_{r1} and P_{r2}, of the reflected waves are:

$$P_{r1} = 1.063 \qquad P_{r2} = 1.274$$

The sudden contraction behaves like a partially closed end, sending back a partial "echo" of the incident pulse. The onward transmitted pressure wave is also one of compression, but with an increased pressure ratio of 1.274.

(d) A Contraction, A_r = 0.5, for an Incident Expansion Wave where P_{i1} = 0.8 and X_{i1} = 0.9686
From Eqs. 2.9.7 and 2.9.8:

$$X_{r1} = 0.9895 \qquad X_{r2} = 0.9582$$

Hence, the pressure ratios, P_{r1} and P_{r2}, of the reflected waves are:

$$P_{r1} = 0.929 \qquad P_{r2} = 0.741$$

As in (c), the sudden contraction behaves like a partially closed end, sending back a partial "echo" of the incident pulse. The onward transmitted pressure wave is also one of expansion, but it should be noted that it is now a stronger rarefaction wave with a pressure ratio of 0.741.

The theoretical presentation here, following that of Benson [2.4], is clearly too simple to be completely accurate in all circumstances. The major objections to its use are that the assumption of "constant pressure" at the discontinuity in pipe area cannot possibly be tenable over all flow situations and that the underlying thermodynamic assumption is of isentropic flow in all circumstances. It is, however, a very good guide as to the magnitude of pressure wave reflection and transmission. A more complete theoretical approach is examined in detail in the following sections. A full discussion of the accuracy of Benson's simple assumption is illustrated by numeric examples in Sec. 2.12.2.

Please be assured that you have not wasted your time by studying Benson's simple "constant pressure" solution to a complex problem. Firstly, you now have a mental image of what happens when pressure waves arrive at expansions and contractions in ducts. Secondly, the more complex thermodynamic solution is one that is highly iterative on a computer. Because computer time is a function of the number of iterations required to gain the requisite accuracy for the solution, and completion of that goal depends on avoiding the omnipresent potential for the arithmetic "crashing" of such a calculation, the quality of the initial guess for the very first iteration is extremely important. Benson's simple solution gives the perfect choice for that initial guess.

2.10 Reflection of Pressure Waves at an Expansion in Pipe Area

This section contains the non-isentropic analysis of unsteady gas flow at an expansion in pipe area. The sketch in Fig. 2.8(a) details the nomenclature for the flow regime, in precisely the same manner as in Sec. 2.9. However, to analyze the flow completely, the further information contained in the sketches of Figs. 2.9(a) and 2.10(a) must also be considered.

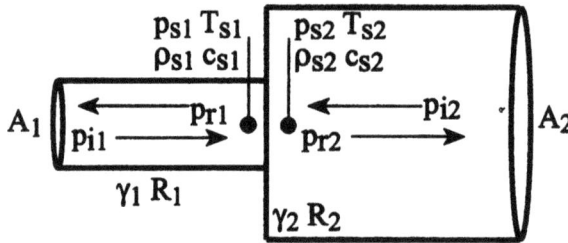

(a) sudden expansion in area in a pipe where $c_s > 0$

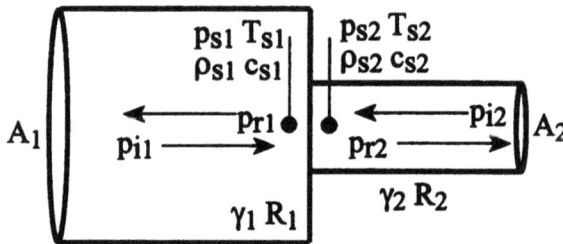

(b) sudden contraction in area in a pipe where $c_s > 0$

Fig. 2.8 Sudden contractions and expansions in area in a pipe.

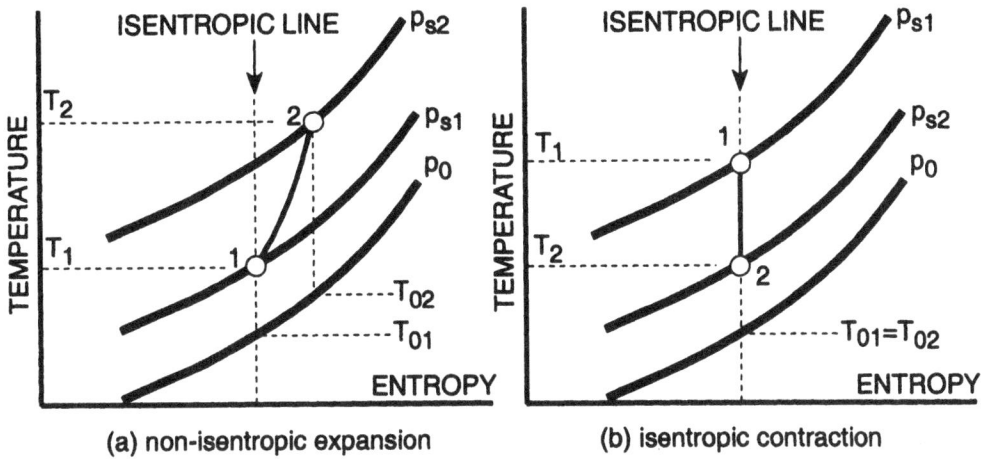

Fig. 2.9 Temperature entropy characteristics for simple expansions and contractions.

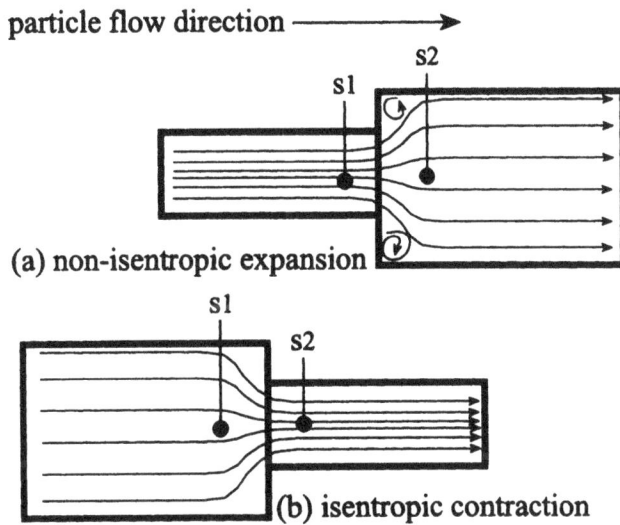

Fig. 2.10 Particle flow in simple expansions and contractions.

In Fig. 2.10(a) the expanding flow is seen to leave turbulent vortices in the corners of the larger section. That the streamlines of the flow give rise to particle flow separation implies a gain of entropy from area section 1 to area section 2. This is summarized on the temperature-entropy diagram in Fig. 2.9(a) where the gain of entropy for a diffusing flow rising from pressure p_{s1} to pressure p_{s2} is clearly visible.

As usual, the analysis of flow in this quasi-steady and non-isentropic context uses, where appropriate, the equations of continuity, the First Law of Thermodynamics, and the momentum equation. The properties and composition of the gas particles are those of the gas at the upstream point. Therefore, the various functions of the gas properties are:

$$\gamma = \gamma_1 \quad R = R_1 \quad G_5 = G_{51} \quad G_7 = G_{71}, \text{etc.}$$

The continuity equation for mass flow in Eq. 2.9.5 is still generally applicable and repeated here, although the entropy gain is reflected in the reference acoustic velocity and density at position 2:

$$\dot{m}_1 - \dot{m}_2 = 0 \qquad (2.10.1)$$

With rightward retained as the positive direction, this equation becomes:

$$\rho_{01} X_{s1}^{G5} A_1 G_5 a_{01}(X_{i1} - X_{r1}) + \rho_{02} X_{s2}^{G5} A_2 G_5 a_{02}(X_{i2} - X_{r2}) = 0 \qquad (2.10.2)$$

The First Law of Thermodynamics was introduced in Sec. 2.8 for such flow situations. The analysis required here follows similar lines of logic. The First Law of Thermodynamics for flow from superposition station 1 to superposition station 2 can be expressed as:

$$h_{s1} + \frac{c_{s1}^2}{2} = h_{s2} + \frac{c_{s2}^2}{2}$$

or

$$(c_{s1}^2 + G_5 a_{s1}^2) - (c_{s2}^2 + G_5 a_{s2}^2) = 0 \qquad (2.10.3)$$

The momentum equation for flow from superposition station 1 to superposition station 2 is expressed as:

$$A_1 p_{s1} + (A_2 - A_1) p_{s1} - A_2 p_{s2} + (\dot{m}_{s1} c_{s1} - \dot{m}_{s2} c_{s2}) = 0$$

The logic for the middle term in the above equation is that the pressure p_{s1} is conventionally presumed to act over the annulus area between the two ducts. The momentum equation, also taking into account the information from the continuity equation regarding mass flow equality, reduces to:

$$A_2(p_{s1} - p_{s2}) + \dot{m}_{s1}(c_{s1} - c_{s2}) = 0 \qquad (2.10.4)$$

As with the simplified "constant pressure" solution according to Benson, presented in Sec. 2.9, the unknown values will be the reflected pressure waves at the boundary, p_{r1} and p_{r2}, and also the reference temperature at position 2, namely T_{02}. There are three unknowns, necessitating three equations, namely Eqs. 2.10.2, 2.10.3, and 2.10.4. All other "unknown" quantities can be computed from these values and from the "known" values. The known values are the upstream and downstream pipe areas, A_1 and A_2, the reference state conditions at the upstream point, the gas properties at superposition stations 1 and 2, and the incident pressure waves, p_{i1} and p_{i2}.

It is recalled that:

$$X_{i1} = \left(\frac{p_{i1}}{p_0}\right)^{G17} \quad \text{and} \quad X_{i2} = \left(\frac{p_{i2}}{p_0}\right)^{G17}$$

The reference state conditions of density and acoustic velocity are:

Density
$$\rho_{01} = \frac{p_0}{RT_{01}} \quad \rho_{02} = \frac{p_0}{RT_{02}} \tag{2.10.5}$$

Acoustic velocity
$$a_{01} = \sqrt{\gamma RT_{01}} \quad a_{02} = \sqrt{\gamma RT_{02}} \tag{2.10.6}$$

The continuity equation, Eq. 2.10.2, reduces to:

$$\rho_{01}(X_{i1} + X_{r1} - 1)^{G5} A_1 G_5 a_{01}(X_{i1} - X_{r1})$$
$$+ \rho_{02}(X_{i2} + X_{r2} - 1)^{G5} A_2 G_5 a_{02}(X_{i2} - X_{r2}) = 0 \tag{2.10.7}$$

The First Law of Thermodynamics, Eq. 2.10.3, reduces to:

$$\left[\left(G_5 a_{01}(X_{i1} - X_{r1})\right)^2 + G_5 a_{01}^2 (X_{i1} + X_{r1} - 1)^2\right] -$$
$$\left[\left(G_5 a_{02}(X_{i2} - X_{r2})\right)^2 + G_5 a_{02}^2 (X_{i2} + X_{r2} - 1)^2\right] = 0 \tag{2.10.8}$$

The momentum equation, Eq. 2.10.4, reduces to:

$$p_0 A_2 \left[(X_{i1} + X_{r1} - 1)^{G7} - (X_{i2} + X_{r2} - 1)^{G7} \right]$$
$$+ \left[\rho_{01} (X_{i1} + X_{r1} - 1)^{G5} A_1 G_5 a_{01} (X_{i1} - X_{r1}) \right] \qquad (2.10.9)$$
$$\times \left[G_5 a_{01} (X_{i1} - X_{r1}) + G_5 a_{02} (X_{i2} - X_{r2}) \right] = 0$$

The three equations cannot be reduced any further, as they are polynomial functions of all three "unknown" variables. These functions can be solved by a standard iterative method for such problems. I have found that the Newton-Raphson method for the solution of multiple polynomial equations is stable, accurate, and rapid in execution. The arithmetic solution on a computer is conducted by a Gaussian Elimination method.

As with all numerical methods, the computer time required is heavily dependent on the number of iterations needed to acquire a solution of the requisite accuracy, in this case for an error no greater than 0.01% for the solution of any of the variables. As forecast in Sec. 2.9, the use of the Benson "constant pressure" criteria is invaluable in this regard because it considerably reduces the number of iterations required. This is because the initial guess from the Benson criteria is fairly close to the final answer. Numerical methods of this type are also arithmetically "frail," if the user makes ill-advised initial guesses for the value of any of the "unknowns." It is in this context that the use of the Benson "constant pressure" criteria is particularly indispensable. Numeric examples are given in Sec. 2.12.2.

Isentropic Analysis for Diffusing Flow
    If the duct between section 1 and 2 in Fig. 2.10(a) has a gradual increase in area, such as in a tapered pipe, rather than the sudden expansion illustrated, then it may be that the flow can be considered to occur isentropically. The analysis above is for non-isentropic flow, as evidenced by the theory showing that a_{02} and T_{02} are not equal to a_{01} and T_{01}. To solve for isentropic flow in this situation, these variables at section 2 now become constants:

Isentropic diffusion $\qquad a_{02} = a_{01} \qquad T_{02} = T_{01}$

In which case, as the number of variables in Eqs. 2.10.7 to 2.10.9 is reduced by one, so is the number of relevant equations. Hence, because the momentum equation encapsulates the non-isentropicity, it is abandoned, and the solution devolves to that for the continuity and energy equations, Eqs. 2.10.7 and 2.10.8. This solution can be used for smoothly tapered pipes in diffusing flow as long as the pipe taper is not too steep, and the particle velocity is not so high that the particle flow separates from its attachment to the walls of the diffuser. See Sec. 2.15 for a more thorough discussion on this point.

2.10.1 *Flow at Pipe Expansions Where Sonic Particle Velocity is Encountered*

In the above analysis of unsteady gas flow at expansions in pipe area, the particle velocity at section 1 will occasionally be found to reach, or even attempt to exceed, the local acoustic velocity. In thermodynamic or gas dynamic terms this is not possible because the particles in unsteady gas flow cannot move faster than the pressure wave signal that is impelling them. The highest particle velocity that is permissible is the local acoustic velocity at station 1, i.e., the flow is permitted to become choked. Therefore, during the mathematical solution of Eqs. 2.10.7, 2.01.8, and 2.10.9, the local Mach number at station 1 is monitored and retained at unity if it is found to exceed it.

$$M_{s1} = \frac{c_{s1}}{a_{s1}} = \frac{G_5 a_{01}(X_{i1} - X_{r1})}{a_{01}X_{s1}} = \frac{G_5(X_{i1} - X_{r1})}{X_{i1} + X_{r1} - 1} \qquad (2.10.10)$$

This immediately simplifies the entire procedure as it gives a direct solution for one of the unknowns:

$$\text{if } M_{s1} \geq 1 \quad M_{s1} = 1$$

hence
$$X_{r1} = \frac{M_{s1} + X_{i1}(G_5 - M_{s1})}{M_{s1} + G_5} = \frac{1 + G_4 X_{i1}}{G_6} \qquad (2.10.11)$$

The acquisition of all related data for pressure, density, particle velocity, and mass flow rate at both superposition stations follows directly from the solution of the three polynomials for X_{r1}, X_{r2}, and a_{02}, in the manner indicated in Sec. 2.9.

In many classic analyses of choked flow, a "critical pressure ratio" is determined for flow from the upstream point to the throat where sonic flow is occurring. This concept cannot be applied here, for it assumes zero particle velocity at the upstream point, which is clearly not the case here for unsteady gas flow with this geometry.

2.11 Reflection of Pressure Waves at a Contraction in Pipe Area

This section contains the isentropic analysis of unsteady gas flow at a contraction in pipe area. The sketch in Fig. 2.8(b) details the nomenclature for the flow regime, in precisely the same manner as in Sec. 2.9. However, to analyze the flow completely, the further information contained in the sketches of Figs. 2.9(b) and 2.10(b) must also be considered.

In Fig. 2.10(b) the contracting flow is seen to flow smoothly from the larger section to the smaller area section. The streamlines of the flow do not give rise to particle flow separation and so the flow is considered to be isentropic. This is in line with conventional nozzle theory as observed in many standard texts on thermodynamics. It is summarized in the temperature-entropy diagram in Fig. 2.9(b) where there is no entropy gain for the flow falling from pressure p_{s1} to pressure p_{s2}.

As usual, the analysis of quasi-steady flow in this context uses, where appropriate, the equations of continuity, the First Law of Thermodynamics, and the momentum equation. However, one less equation is required compared to the analysis for expanding or diffusing flow in Sec. 2.10. This is because the value of the reference state is known at superposition position 2 because the flow is isentropic:

$$T_{01} = T_{02} \quad \text{or} \quad a_{01} = a_{02} \tag{2.11.1}$$

Because there is no entropy gain, the equation normally reserved for the analysis of non-isentropic flow, the momentum equation, can be neglected in the ensuing analytic method.

The properties and composition of the gas particles are those of the gas at the upstream point. Therefore, the various functions of the gas properties are:

$$\gamma = \gamma_1 \quad R = R_1 \quad G_5 = G_{5_1} \quad G_7 = G_{7_1}, \text{etc.}$$

The continuity equation for mass flow in Eq. 2.9.5 is still generally applicable and is repeated here:

$$\dot{m}_1 - \dot{m}_2 = 0 \tag{2.11.2}$$

With rightward retained as the positive direction, this equation becomes:

$$\rho_{01} X_{s1}^{G5} A_1 G_5 a_{01} (X_{i1} - X_{r1}) + \rho_{02} X_{s2}^{G5} A_2 G_5 a_{02} (X_{i2} - X_{r2}) = 0 \tag{2.11.3}$$

or,
$$X_{s1}^{G5} A_1 (X_{i1} - X_{r1}) + X_{s2}^{G5} A_2 (X_{i2} - X_{r2}) = 0$$

The First Law of Thermodynamics was introduced in Sec. 2.8 for such flow situations. The analysis required here follows similar lines of logic. The First Law of Thermodynamics for flow from superposition station 1 to superposition station 2 can be expressed as:

$$h_{s1} + \frac{c_{s1}^2}{2} = h_{s2} + \frac{c_{s2}^2}{2}$$

or
$$(c_{s1}^2 + G_5 a_{s1}^2) - (c_{s2}^2 + G_5 a_{s2}^2) = 0 \tag{2.11.4}$$

As with the simplified "constant pressure" solution according to Benson, presented in Sec. 2.9, the unknown values will be the reflected pressure waves at the boundary, p_{r1} and p_{r2}. There are two unknowns, necessitating two equations, namely Eqs. 2.11.3, and 2.11.4. All other "unknown" quantities can be computed from these values and from the "known" values.

The known values are the upstream and downstream pipe areas A_1 and A_2, the reference state conditions at the upstream and downstream points, the gas properties at superposition stations 1 and 2, and the incident pressure waves, p_{i1} and p_{i2}.

You will recall that:

$$X_{i1} = \left(\frac{p_{i1}}{p_0}\right)^{G17} \quad \text{and} \quad X_{i2} = \left(\frac{p_{i2}}{p_0}\right)^{G17}$$

The reference state conditions are:

Density
$$\rho_{01} = \rho_{02} = \frac{p_0}{RT_{01}} \tag{2.11.5}$$

Acoustic velocity
$$a_{01} = a_{02} = \sqrt{\gamma RT_{01}} \tag{2.11.6}$$

The continuity equation, Eq. 2.11.3, reduces to:

$$\rho_{01}(X_{i1} + X_{r1} - 1)^{G5} A_1 G_5 a_{01}(X_{i1} - X_{r1})$$
$$+ \rho_{02}(X_{i2} + X_{r2} - 1)^{G5} A_2 G_5 a_{02}(X_{i2} - X_{r2}) = 0$$

or
$$(X_{i1} + X_{r1} - 1)^{G5} A_1 (X_{i1} - X_{r1}) + (X_{i2} + X_{r2} - 1)^{G5} A_2 (X_{i2} - X_{r2}) = 0 \tag{2.11.7}$$

The First Law of Thermodynamics, Eq. 2.11.4, reduces to:

$$\left[\left(G_5 a_{01}(X_{i1} - X_{r1})\right)^2 + G_5 a_{01}^2(X_{i1} + X_{r1} - 1)^2\right] -$$
$$\left[\left(G_5 a_{02}(X_{i2} - X_{r2})\right)^2 + G_5 a_{02}^2(X_{i2} + X_{r2} - 1)^2\right] = 0$$

or
$$\left[G_5(X_{i1} - X_{r1})^2 + (X_{i1} + X_{r1} - 1)^2\right] -$$
$$\left[G_5(X_{i2} - X_{r2})^2 + (X_{i2} + X_{r2} - 1)^2\right] = 0 \tag{2.11.8}$$

These two equations cannot be reduced any further as they are polynomial functions of the two variables. These functions can be solved by a standard iterative method for such problems. I have determined that the Newton-Raphson method for the solution of multiple polynomial equations is stable, accurate, and rapid in execution. The arithmetic solution on a computer is conducted by a Gaussian Elimination method. Actually, this is not strictly necessary, as a rather simpler solution can be effected as it devolves to two simultaneous equations for the two unknowns and the corrector values for each of the unknowns.

Sec. 2.9 contains some comments about how the use of the Benson "constant pressure" criteria, for the initial guesses for the unknowns to the solution, is indispensable. These comments are still very appropriate. Numeric examples are given in Sec. 2.12.2.

The acquisition of all related data for pressure, density, particle velocity, and mass flow rate at both superposition stations follows directly from the solution of the two polynomials for X_{r1} and X_{r2}.

2.11.1 Flow at Pipe Contractions where Sonic Particle Velocity is Encountered

In the above analysis of unsteady gas flow at contractions in pipe area, the particle velocity at section 2 can reach, and even attempt to exceed, the local acoustic velocity. In thermodynamic or gas dynamic terms, this is not possible. The highest particle velocity that is permissible is the local acoustic velocity at station 2, i.e., the flow is permitted to become choked. Therefore, during the mathematical solution of Eqs. 2.11.7 and 2.11.8, the local Mach number at station 2 is monitored and retained at unity if it is found to exceed it.

$$M_{s2} = \frac{c_{s2}}{a_{s2}} = \frac{G_5 a_{02}(X_{i2} - X_{r2})}{a_{02} X_{s2}} = \frac{G_5(X_{i2} - X_{r2})}{X_{i2} + X_{r2} - 1} \qquad (2.11.9)$$

This immediately simplifies the entire procedure for this gives a direct solution for one of the unknowns:

$$\text{if } M_{s1} \geq 1 \quad M_{s1} = 1$$

$$X_{r2} = \frac{M_{s2} + X_{i2}(G_5 - M_{s2})}{M_{s2} + G_5} = \frac{1 + G_4 X_{i2}}{G_6} \qquad (2.11.10)$$

In this instance of sonic particle flow at station 2, the entire solution can now be obtained directly by substituting the value of X_{r2} determined above into either Eq. 2.11.7 or 2.11.8 and solving it by the standard Newton-Raphson method for the one remaining unknown, X_{r1}.

2.12 Reflection of Waves at a Restriction between Differing Pipe Areas

This section contains the non-isentropic analysis of unsteady gas flow at restrictions between differing pipe areas. The sketch in Fig. 2.8 details much of the nomenclature for the flow regime, but essential subsidiary information is contained in a more detailed sketch of the geometry in Fig. 2.12. However, to analyze the flow completely, the further information contained in the sketches of Figs. 2.11 and 2.12 must be considered completely. The geometry is of two pipes of differing area, A_1 and A_2, which are butted together with an orifice of area A_t sandwiched between them. This geometry gives rise to my jargon referring to this discontinuity as a "butted joint." This geometry is very common in engine ducting; for example, it could be the throttle body of a carburetor with a venturi and a throttle plate. The details of this very geometry are analyzed in some depth in Chapter 3.

It could also be simply a sharp edged, sudden contraction in pipe diameter, where A_2 is less than A_1 and there is no actual orifice of area A_t at all. You may well feel that you have just read all about that in Sec. 2.11, but the reality is that in this case the flow forms a vena contracta with an effective area of throat A_t, which is less than A_2. This requires a more complete theoretical treatment. In short, the theoretical analysis to be given here is the more accurate and extended—and inherently more complex, version of that already presented in Sec. 2.11 for sudden expansions and contractions in pipe area.

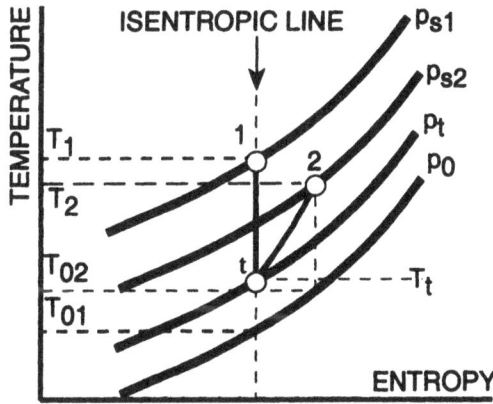

Fig. 2.11 Temperature-entropy characteristics for a restricted area change.

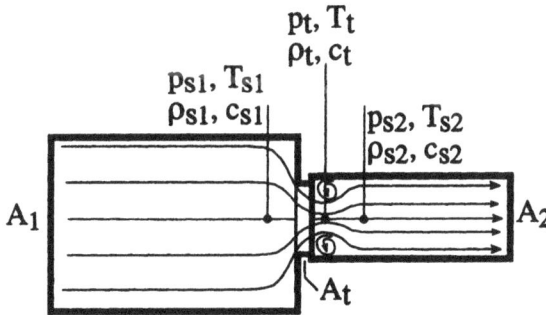

Fig. 2.12 Particle flow regimes at a restricted area change.

219

In Fig. 2.12, the expanding flow from the throat to the downstream superposition point 2 is seen to leave turbulent vortices in the corners of that section. That the streamlines of the flow give rise to particle flow separation implies a gain of entropy from the throat to area section 2. On the other hand, the flow from the superposition point 1 to the throat is contracting and can be considered to be isentropic in the same fashion as the contractions discussed in Sec. 2.11. This is summarized on the temperature-entropy diagram in Fig. 2.11, where the gain of entropy for the flow rising from pressure p_t to pressure p_{s2} is clearly visible. The isentropic nature of the flow from pressure p_{s1} to pressure p_t can also be observed as the vertical line on Fig. 2.11.

The properties and composition of the gas particles are those of the gas at the upstream point. Therefore, the various functions of the gas properties are:

$$\gamma = \gamma_1 \quad R = R_1 \quad G_5 = G_{51} \quad G_7 = G_{71}, \text{etc.}$$

As usual, the analysis of flow in this context uses, where appropriate, the equations of continuity, the First Law of Thermodynamics, and the momentum equation.

The reference state conditions are:

Density
$$\rho_{01} = \rho_{0t} = \frac{p_0}{RT_{01}} \quad \rho_{02} = \frac{p_0}{RT_{02}} \tag{2.12.1}$$

Acoustic velocity
$$a_{01} = a_{0t} = \sqrt{\gamma RT_{01}} \quad a_{02} = \sqrt{\gamma RT_{02}} \tag{2.12.2}$$

The continuity equation for mass flow in previous sections is still generally applicable and is repeated here, although the entropy gain is reflected in the reference acoustic velocity and density at position 2:

$$\dot{m}_1 - \dot{m}_2 = 0$$
$$\dot{m}_1 - \dot{m}_t = 0 \tag{2.12.3}$$

With rightward retained as the positive direction, these equations become:

$$\rho_{01}X_{s1}^{G5}A_1G_5a_{01}(X_{i1} - X_{r1}) + \rho_{02}X_{s2}^{G5}A_2G_5a_{02}(X_{i2} - X_{r2}) = 0$$
$$\rho_{01}X_{s1}^{G5}A_1G_5a_{01}(X_{i1} - X_{r1}) - \rho_t[C_cA_t][C_sc_t] = 0 \tag{2.12.4}$$

The above equation for the mass flow continuity for flow from the upstream station 1 to the throat contains the coefficient of contraction on the flow area, C_c, and the coefficient of velocity, C_s. These are conventionally connected in fluid mechanics theory to a coefficient of discharge, C_d, to give an effective throat area, A_{teff}, as follows:

$$C_d = C_c C_s \quad \text{and} \quad A_{teff} = C_d A_t$$

The equation of mass flow continuity, Eq. 2.12.4, now appears as:

$$\rho_{01} X_{s1}^{G5} A_1 G_5 a_{01}(X_{i1} - X_{r1}) - C_d \rho_t A_t c_t = 0$$

The First Law of Thermodynamics was introduced in Sec. 2.8 for such flow situations. The analysis required here follows similar lines of logic. The First Law of Thermodynamics for flow from superposition station 1 to superposition station 2 can be expressed as:

$$h_{s1} + \frac{c_{s1}^2}{2} = h_{s2} + \frac{c_{s2}^2}{2}$$

or

$$(c_{s1}^2 + G_5 a_{s1}^2) - (c_{s2}^2 + G_5 a_{s2}^2) = 0 \tag{2.12.5}$$

The First Law of Thermodynamics for flow from superposition station 1 to the throat can be expressed as:

$$h_{s1} + \frac{c_{s1}^2}{2} = h_t + \frac{c_t^2}{2}$$

or

$$C_p(T_{s1} - T_t) + \frac{c_{s1}^2 - c_t^2}{2} = 0 \tag{2.12.6}$$

The momentum equation for flow from the throat to superposition station 2 is expressed as:

$$A_2(p_t - p_{s2}) + \dot{m}_{s1}(c_t - c_{s2}) = 0 \tag{2.12.7}$$

The unknown values will be the reflected pressure waves at the boundaries, p_{r1} and p_{r2}, the reference temperature at position 2, namely T_{02}, and the pressure, p_t, and the velocity, c_t, at the throat. There are five unknowns, necessitating five equations, namely the two mass flow equations in Eq. 2.12.4, the two First Law equations, Eq. 2.12.5 and Eq. 2.12.6, and the momentum equation, Eq. 2.12.7. All other "unknown" quantities can be computed from these equations using the "known" values. The known values are the upstream and downstream pipe areas, A_1 and A_2, the throat area, A_t, the reference state conditions at the upstream point, the gas properties at superposition stations 1 and 2, and the incident pressure waves, p_{i1} and p_{i2}. A numeric value for the coefficient of discharge, C_d, is also required and further information on this difficult and often controversial subject is provided in Chapter 3. For the C_d values at this particular discontinuity of a "butted joint," consult Sec. 3.7.

You will recall that:

$$X_{i1} = \left(\frac{p_{i1}}{p_0}\right)^{G17} \quad \text{and} \quad X_{i2} = \left(\frac{p_{i2}}{p_0}\right)^{G17} \quad \text{and setting} \quad X_t = \left(\frac{p_t}{p_0}\right)^{G17}$$

Because there is isentropic flow from station 1 to the throat, the throat density and temperature are given by:

$$\rho_t = \rho_{01}X_t^{G5} \quad \text{and} \quad T_t = \frac{\left(a_{01}X_t\right)^2}{\gamma R}$$

The set of continuity equations in Eq. 2.12.4 reduce to:

$$\begin{aligned}&\rho_{01}(X_{i1} + X_{r1} - 1)^{G5}A_1 a_{01}(X_{i1} - X_{r1}) \\ &+ \rho_{02}(X_{i2} + X_{r2} - 1)^{G5}A_2 a_{02}(X_{i2} - X_{r2}) = 0\end{aligned} \tag{2.12.8}$$

$$(X_{i1} + X_{r1} - 1)^{G5}A_1 G_5 a_{01}(X_{i1} - X_{r1}) - X_t^{G5}C_d A_t c_t = 0 \tag{2.12.9}$$

The First Law of Thermodynamics in Eq. 2.12.5 reduces to:

$$\begin{aligned}&\left[\left(G_5 a_{01}(X_{i1} - X_{r1})\right)^2 + G_5 a_{01}^2(X_{i1} + X_{r1} - 1)^2\right] - \\ &\left[\left(G_5 a_{02}(X_{i2} - X_{r2})\right)^2 + G_5 a_{02}^2(X_{i2} + X_{r2} - 1)^2\right] = 0\end{aligned} \tag{2.12.10}$$

The First Law of Thermodynamics in Eq. 2.12.6 reduces to:

$$G_5\left\{\left(a_{01}(X_{i1} + X_{r1} - 1)\right)^2 - \left(a_{01}X_t\right)^2\right\} + \left(G_5 a_{01}(X_{i1} - X_{r1})\right)^2 - c_t^2 = 0 \tag{2.12.11}$$

The momentum equation in Eq. 2.12.7 reduces to:

$$p_0 A_2 \left[X_t^{G7} - (X_{i2} + X_{r2} - 1)^{G7} \right]$$

$$+ \left[\rho_{01} (X_{i1} + X_{r1} - 1)^{G5} A_1 G_5 a_{01} (X_{i1} - X_{r1}) \right] [c_t + G_5 a_{02} (X_{i2} - X_{r2})] = 0 \qquad (2.12.12)$$

The five equations, Eqs. 2.12.8 to 2.12.12, cannot be reduced any further because they are polynomial functions of all five "unknown" variables, X_{r1}, X_{r2}, X_t, a_{02}, and c_t. These functions can be solved by a standard iterative method for such problems. I have found that the Newton-Raphson method for the solution of multiple polynomial equations is stable, accurate, and rapid in execution. The arithmetic solution on a computer is conducted by the Gaussian Elimination method. Numeric examples are given in Sec. 2.12.2.

In this case, is not as easy to supply initial guesses that are sufficiently close to the final answers to ease the arithmetic of the iteration process on the computer, as it was in previous theoretical problems in this chapter. Although the Benson "constant pressure" assumption is of great assistance in this matter, it is not the universal answer. Even with the assistance of this assumption in arriving at a numerical solution, the theory must be programmed with great care to avoid arithmetic instability during its execution.

2.12.1 Flow at Pipe Restrictions Where Sonic Particle Velocity is Encountered

In the above analysis of unsteady gas flow at restrictions in pipe area, the particle velocity at the throat will quite commonly be found to reach, or even attempt to exceed, the local acoustic velocity. In thermodynamic or gas dynamic terms, this is not possible. The highest particle velocity that is permissible is the local acoustic velocity at the throat, i.e., the flow is permitted to become choked. Therefore, during the mathematical solution of Eqs. 2.12.8 to 2.12.12, the local Mach number at the throat is monitored and retained at unity if it is found to exceed it.

As, $$\mathbf{M}_t = \frac{c_t}{a_{01} X_t} = 1 \quad \text{then} \quad c_t = a_{01} X_t \qquad (2.12.13)$$

This simplifies the entire procedure somewhat, for this gives a direct relationship between two of the "unknowns," and so removes one of the previous equations from the solution. It is probably easier from an arithmetic standpoint to eliminate the momentum equation, but from a thermodynamic standpoint, it is certainly more accurate to retain it!

The acquisition of all related data for pressure, density, particle velocity, and mass flow rate at both superposition stations and at the throat follows directly from the solution of the remaining four polynomials for X_{r1}, X_{r2}, X_t, a_{02}, and c_t.

2.12.2 Examples of Flow at Pipe Expansions, Contractions and Restrictions

In Secs. 2.9 to 2.12 the theory of unsteady flow at pipe expansions, contractions, and restrictions has been presented. In Sec. 2.9 the simple theory of a "constant pressure" assumption by Benson was given whereas the more complete theory is in the subsequent sections. In these various sections the point was made repeatedly that the simple theory is reasonably accurate. Some numeric examples are given here which will illustrate that point and the extent of the inaccuracy. The input and output data all use the nomenclature of the theory sections and Figs. 2.8, 2.10 and 2.12. The input data for the diameters, d, are in mm units.

The input data include six different sets of numeric data. In Table 2.12.1 test data sets 1 and 2 illustrate an expansion of the flow, and test data sets 3 and 4 are for contractions. Test data set 5 is for a restricted pipe at a contraction, whereas test data set 6 is for a restricted pipe at an expansion in pipe area. The incident pulse for all tests has a pressure ratio of 1.2, the gas is air and is undisturbed on side 2, and the reference temperature at side 1 is always 20°C or 293 K. There are two output data sets: One in Table 2.12.2, where the Benson "constant pressure" theory is used, and one in Table 2.12.3, where the more complex theory of Secs. 2.10, 2.11, and 2.12 is used as appropriate. The more complex theory gives information on the pressure at the throat, P_t, and on the entropy gain exhibited by the reference temperature, T_{02}. The output shows the mass flow rate and the "error" when comparing the mass flow rate computed by the "constant pressure" theory to that calculated by the more complex theory. The column labeled "E or C" describes whether the flow encountered was at an expansion (E) or a contraction (C) in pipe area.

For simple expansions, as in test data sets 1 and 2, the "constant pressure" theory works remarkably well with an almost negligible error, i.e., less than 3% in terms of mass flow rate. Even if the expansion has a quite realistic coefficient of discharge of 0.85 applied to it, then the error on mass flow rate remains insignificant at 0.3% [3.18]. The magnitude of the reflected pressure waves compare quite favorably in most circumstances.

For the simple, sudden contraction in test data sets 3 and 4, with an unrealistic coefficient of discharge of 1.0, the mass flow rate error is 7.9% but changes little to a 9.1% error when a potentially logical C_d value of 0.7 for a sharp-edged joint is applied [3.18]. The magnitude of the reflected pressure waves resulting from the simple and the complex theories are not significantly different.

TABLE 2.12.1 INPUT DATA TO THE CALCULATIONS

No.	d_1	d_2	d_t	C_d	A_r	P_{i1}	P_{i2}
1	25	50	25	1.000	4.0	1.2	1.0
2	25	50	25	0.85	4.0	1.2	1.0
3	50	25	25	1.000	0.25	1.2	1.0
4	50	25	25	0.7	0.25	1.2	1.0
5	50	25	15	0.85	0.25	1.2	1.0
6	25	50	15	0.85	4.0	1.2	1.0

TABLE 2.12.2 OUTPUT DATA USING THE CONSTANT PRESSURE THEORY OF SEC. 2.9

No.	P_{r1}	P_{r2}	ṁ g/s	ṁ error %	E or C
1	0.8943	1.0763	45.15	-2.5	E
2	0.8943	1.0763	45.15	-0.3	E
3	1.1162	1.3357	52.68	7.90	C
4	1.1162	1.3357	52.68	9.10	C
5	1.1162	1.3357	52.68	42.3	C
6	0.8943	1.0763	45.15	45.7	E

TABLE 2.12.3 OUTPUT DATA FROM THE CALCULATIONS USING THE MORE COMPLEX THEORY

No.	P_{r1}	P_{r2}	ṁ g/s	P_t	Theory
1	0.8850	1.0780	46.34	1.065	SEC. 2.10
2.	0.8930	1.0770	45.31	1.061	SEC. 2.12
3	1.1230	1.3120	48.84	1.304	SEC. 2.11
4	1.1240	1.3080	48.25	1.279	SEC. 2.12
5	1.1460	1.2370	37.01	1.169	SEC. 2.12
6	0.9960	1.0540	30.98	1.030	SEC. 2.12

When any form of restriction is placed in the pipe, i.e., as in test data sets 5 and 6 for an expansion and a contraction respectively, the "constant pressure" theory is capable of providing some relevant information for the magnitude of the reflected waves but not for the mass flow rate. The error is similar for expansions than contractions, which is not the case for a simple junction when restrictions between the pipes are not present. There are significant entropy gains in tests 4 to 6.

Benson's "constant pressure theory" is seen to be reasonably accurate only for flow that encounters simple, sudden expansions in the ducting.

2.13 An Introduction to Reflections of Pressure Waves at Branches in a Pipe

A simple theoretical treatment for this situation was also suggested by Benson [2.4] and in precisely the same format as for the sudden area changes found in the previous section. A sketch of a typical branch is shown in Fig. 2.13. The sign convention for all of the branch theory presented here is that inward propagation of a pressure wave toward the branch will be regarded as "positive." Benson [2.4] postulates that the superposition pressure at the junction, at any instant of wave incidence and reflection, can be regarded as a constant. This is a straightforward extension of his thinking for the expansion and contractions given in Sec. 2.9.

The incident pressure waves are p_{i1}, p_{i2}, and p_{i3}, and the ensuing reflections are of pressures p_{r1}, p_{r2}, and p_{r3}. The superposition pressures are p_{s1}, p_{s2}, and p_{s3}.

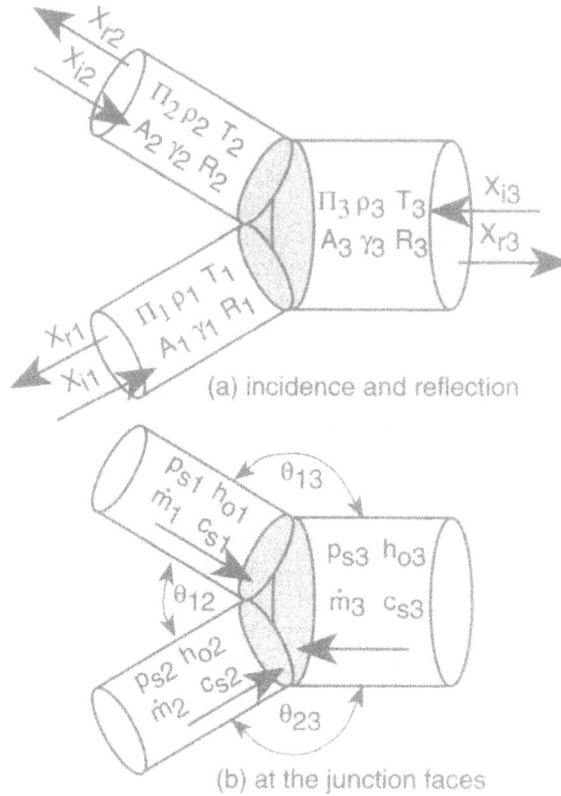

Fig. 2.13 Unsteady flow at a three-way branch.

Benson's "Constant Pressure" Solution at a Three-Way Branch

A three-way branch is very similar to the junction shown in Fig. 2.8, but with an extra pipe joining in. Therefore, the theoretical solution involves the extension of Eqs. 2.9.3 and 2.9.4 to deal with the superposition state and mass flow rate of the extra pipe 3 at the junction. Benson's criteria inherently assumes isentropic flow.

$$p_{s1} = p_{s2} = p_{s3}$$

or
$$X_{i1} + X_{r1} - 1 = X_{i2} + X_{r2} - 1 = X_{i3} + X_{r3} - 1 \qquad (2.13.1)$$

The net mass flow rate at the junction is zero:

$$A_1(X_{i1} - X_{r1}) + A_2(X_{i2} - X_{r2}) + A_3(X_{i3} - X_{r3}) = 0 \qquad (2.13.2)$$

There are three equations to solve for the three unknowns, X_{r1}, X_{r2} and X_{r3}. It is presumed that in the course of any computation, the values of all incident pressure waves from one calculation time step to another are known quantities.

Where the total area, A_t, is defined as:

$$A_t = A_1 + A_2 + A_3 \qquad (2.13.3)$$

The solution of the above simultaneous equations gives:

$$X_{r1} = \frac{2A_2X_{i2} + 2A_3X_{i3} + X_{i1}(A_1 - A_2 - A_3)}{A_t} \qquad (2.13.4)$$

$$X_{r2} = \frac{2A_1X_{i1} + 2A_3X_{i3} + X_{i2}(A_2 - A_3 - A_1)}{A_t} \qquad (2.13.5)$$

$$X_{r3} = \frac{2A_1X_{i1} + 2A_2X_{i2} + X_{i3}(A_3 - A_2 - A_1)}{A_t} \qquad (2.13.6)$$

Perhaps not surprisingly, the branched pipe can act as either a contraction or an enlargement of area to the gas flow. In short, two pipes may be supplying one pipe, or one pipe supplying the other two, respectively. Consider the two cases below, where the pressure waves employed as examples are the familiar pulses that have been used so frequently already in the text. For simplicity, all of the pipes are of equal area.

(a) Benson's Solution for a Compression Wave Arriving Down One of the Pipes

A compression wave comes down to the branch in pipe 1 through air, and all other conditions in the other branches are "undisturbed." The compression wave has an incident pressure ratio, P_{i1}, of 1.2, or the pressure amplitude ratio, X_{i1}, is 1.02639. The incident waves in the other two pipes have pressure ratios and pressure amplitude ratios of unity.

The input to the calculation is:

$$P_{i1}=1.2 \quad X_{i1}=1.02639 \quad X_{i2}=X_{i3}=1.0 \text{ as } P_{i2}=P_{i3}=1.0$$

The result of the calculation using Eq. 2.13.3 is:

$$X_{r1}=0.9911 \quad X_{r2}=X_{r3}=1.01759 \text{ hence } P_{r1}=0.940 \text{ and } P_{r2}=P_{r3}=1.13$$

As far as pipe 1 is concerned, the outcome is exactly the same as that for a 2:1 area ratio enlargement discussed in the previous section. At the branch, the incident wave divides evenly between the other two pipes, transmitting a compression wave onward and reflecting a rarefaction pulse into pipe 1. Because pipe 1 is supplying the other two pipes, the effect is an expansion.

(b) Benson's Solution for Two Compression Waves Arriving Down Two of the Pipes

Compression waves of pressure ratio 1.2 arrive as incident pulses in pipes 1 and 2 leading to the branch with pipe 3. Pipe 3 has undisturbed conditions, as its incident pressure ratio, P_{i3}, is 1.0. Now the branch behaves as a 2:1 contraction as far as pipes 1 and 2 are concerned.

The input to the calculation is:

$$\text{as } P_{i1} = P_{i2} = 1.2 \quad X_{i1} = X_{i2} = 1.02639 \quad X_{i3} = 1.0 \text{ as } P_{i3} = 1.0$$

The solution of Eq. 2.13.3 shows:

$$X_{r1} = X_{r2} = 1.0088 \quad X_{r3} = 1.0352 \text{ hence } P_{r1} = P_{r2} = 1.0632 \text{ and } P_{r3} = 1.274$$

Pipes 1 and 2 are supplying the third pipe, hence the unsteady gas dynamic effect is that of a contraction. The contraction effect is evidenced by the reflection of compression waves in pipes 1 and 2 and the transmission of a compression wave onward into pipe 3. These numbers are already familiar as computed data for pressure waves of identical amplitude at the 2:1 contraction discussed in the previous section.

When we have pipes with dissimilar areas and a mixture of compression and expansion waves incident upon a branch, the situation becomes much more difficult for the human mind to comprehend. At that point, the programming of the mathematics into a computer will leave the designer's mind free to concentrate more on the relevance of the information calculated and less on the arithmetic tedium of acquiring the data.

It is also obvious that the angles between the pipes at the branch must play some role in determining the amplitudes of the transmitted and reflected waves. This makes it even more obvious that a branch cannot transmit pressure waves without losses and entropy gains. This subject was studied in recent times by Bingham [2.19], Blair [2.20], and Fleck et al. [2.68] at QUB. Although the branch angles do have an influence on wave amplitudes, it is not as great in some circumstances as might be imagined. So, for those who wish to ensure accuracy for branched pipe calculations within an engine simulation, the theory in the following section is essential.

2.14 The Complete Solution of Reflections of Pressure Waves at Pipe Branches

Both historically and theoretically, the next step forward was to attempt to solve the momentum equation to cope with the non-isentropic realism that there are pressure losses for real flows changing direction when moving around the sharp corners posed by the pipes at the branch. Much has been written on this subject, and many of the references below provide a commentary on the subject sustained over many years. Suffice it to say that the practical approach adopted by Bingham [2.19], incorporating the use of a modified form of the momentum equation to account for the pressure losses around the branch, is the basis of the method used here. However, Bingham's solution was isentropic.

This same approach was also employed by McGinnity [2.39] in a non-isentropic analysis, but for a single composition fluid only. His solution was further complicated by using a non-homentropic Riemann variable method, which meant that, because gas properties were tied to path lines, they were not as clearly defined as in the method used here. It would appear from the literature [2.39] that the solution was reduced to the search for a single unknown quantity, whereas I deduce below that there are actually five such unknown quantities for a complete non-isentropic analysis of a three-way branch.

The merit of the Bingham method [2.19] is that it uses experimentally determined pressure loss coefficients at the branches, an approach that is capable of being enhanced further by information from data banks, such as that published by Miller [2.38]. Further data are also provided by Fleck et al. [2.68].

An alternative method, perhaps one that is more complete theoretically, is to resolve the momentum of the flow at any branch into its horizontal and vertical components and equate them both to zero. The disadvantage of this approach is that it does not include the fluid mechanic loss component which the Bingham method incorporates so pragmatically and realistically.

Although the discussion here is devoted exclusively to three-way branches, the theoretical process for the general case of multiple branches with n pipes, where n > 3, is almost identical to that reported below. It will be seen that the basic approach is to identify the pipes that are the suppliers, and the pipes that are supplied, at any junction. For an n-pipe junction, this is basically the only addition to the theory below, other than that the number of equations increases by the number of extra junctions. Using the Newton-Raphson method for solution of multiple polynomial equations, and handling the matrix arithmetic by the Gaussian Elimination method, the additional computational complexity is negligible.

The sign convention for particle flow is declared as "positive" toward the branch, and Fig. 2.13 inherently stipulates this convention. The pressure loss criteria for flow from one branch to another is set out by Bingham as,

Pressure loss $$\Delta p = C_L \rho_s c_s^2 \qquad (2.14.1)$$

where the loss coefficient, C_L, is given by the inter-branch angle, θ, in units of degrees:

$$\text{if } \theta < 167° \text{ then } C_L = 1.6 - \frac{1.6\theta}{167} \text{ else } C_L = 0 \qquad (2.14.2)$$

In any branch there are supplier pipes and supplied pipes. There are two possibilities in this regard and these lead to two assumptions for their solution. The more fundamental assumption in much of the theory is that the gas within the pipes is a mixture of two gases; in an engine context these are logically exhaust gas and air. Obviously, the theory is capable of being extended to mixtures of many gases, as indeed air and exhaust gas actually are. In addition, the theory is capable of being extended relatively easily to branches with any number of pipes at the junction. For greater ease of understanding, the theory set out below details a three-way branch for greater ease of understanding.

(a) One Supplier Pipe at a Three-Way Branch

Here there is one supply pipe and two pipes that are being supplied. In this case, the required solution is for the reflected wave amplitudes, X_r, in all three pipes, and for the reference acoustic velocities, a_0, for the gas going toward the two supplied pipes. (NOTE: The word "toward" is used here precisely.) The consequence of this is that the mathematical solution has five unknown values needing five equations. It is possible to reduce the number of unknowns by one, if we assume that the reference acoustic state toward the two supplied pipes is common. A negligible loss of accuracy accompanies this assumption. Using the notation of Fig. 2.13, it is implied that pipe 1 is supplying pipes 2 and 3; this notation will be used here only to "particularize" the solution, so as to aid understanding of the analysis. Because the pressure in the face of pipes 2 and 3 will normally be nearly the same, the difference between T_{02} and T_{03} should be small and, because the reference acoustic velocity is related to the square root of these numbers, the error is potentially even smaller. Irrespective of this assumption, it then follows absolutely that the basic properties of the gas entering the supplied pipes are those of the supplier pipe.

(b) Two Supplier Pipes at a Three-Way Branch

Here there is one supplied pipe and two pipes that are suppliers. In this case, the required solution is for the reflected wave amplitudes, X_r, in all three pipes, and for the reference acoustic velocity, a_0, in the supplied pipe. This means that there are four unknown values needing four equations. This is only possible by making the assumption that the superposition pressures at the faces of the two supplier pipes are equal; this is the same assumption used by Bingham and Blair [2.19] and McGinnity [2.39]. It then follows that the gas entering the supplied pipe has the properties of a mass flow related mixture of the gases in the supplier pipes. Using the notation of Fig. 2.13, it is implied that pipes 1 and 2 are supplying pipe 3, so the subscripts of the equations are juxtaposed if the supplied/supplier pipe scenario changes.

The subscript notation "e3" should be noted carefully, for this details the quantity and quality of the gas entering, i.e., going "toward," the mesh space beyond the pipe 3 entrance, whereas the resulting change of all of the gas properties within that mesh space is handled by the unsteady gas dynamic method of Sec. 2.18.9.

Continuity $\qquad\qquad\qquad\qquad \dot{m}_{e3} = \dot{m}_1 + \dot{m}_2 \qquad\qquad\qquad\qquad$ (2.14.3)

Purity $\qquad\qquad\qquad\qquad \Pi_{e3} = \dfrac{\dot{m}_1 \Pi_1 + \dot{m}_2 \Pi_2}{\dot{m}_{e3}} \qquad\qquad\qquad$ (2.14.4)

Gas constant
$$R_{e3} = \frac{\dot{m}_1 R_1 + \dot{m}_2 R_2}{\dot{m}_{e3}}$$
(2.14.5)

Specific heats ratio
$$\gamma_{e3} = \frac{C_{Pe3}}{C_{Ve3}} = \frac{\dot{m}_1 C_{P1} + \dot{m}_2 C_{P2}}{\dot{m}_1 C_{V1} + \dot{m}_2 C_{V2}}$$
(2.14.6)

The final analysis then relies on incorporating the equations resulting from all of these previous considerations regarding pressure losses, together with the continuity equation and the First Law of Thermodynamics. The notation of Fig. 2.13 applies, together with either Fig. 2.14(a) for two supplier pipes, or Fig. 2.14(b) for one supplier pipe.

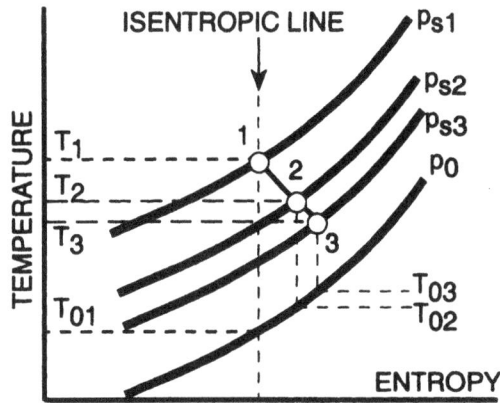

(a) one supplier pipe to the branch

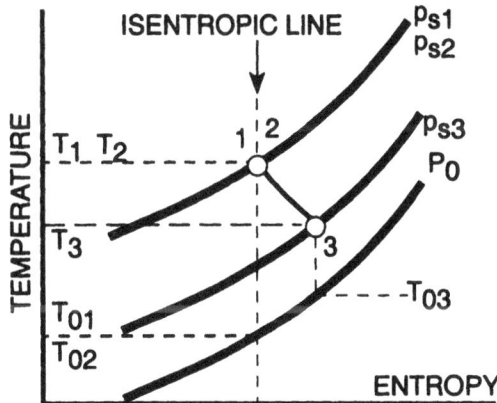

(a) two supplier pipes to the branch

Fig. 2.14 The temperature-entropy characteristics.

Density, Particle Velocity, and Mass Flow Rate at a Three-Way Branch
(a) For One Supplier Pipe
   For one supplier pipe, the following are the relationships for the density, particle velocity, and mass flow rate that apply to the supplier pipe and the two pipes that are being supplied:

$$\rho_{s1} = \rho_{01}\left(X_{i1} + X_{r1} - 1\right)^{G51} \quad c_{s1} = G_{51}a_{01}\left(X_{i1} - X_{r1}\right) \quad \dot{m}_1 = \rho_{s1}A_1c_{s1} \qquad (2.14.7)$$

$$\rho_{s2} = \rho_{0e2}\left(X_{i2} + X_{r2} - 1\right)^{G51} \quad c_{s2} = G_{51}a_{0e2}\left(X_{i2} - X_{r2}\right) \quad \dot{m}_2 = \rho_{s2}A_2c_{s2} \qquad (2.14.8)$$

$$\rho_{s3} = \rho_{0e3}\left(X_{i3} + X_{r3} - 1\right)^{G51} \quad c_{s3} = G_{51}a_{0e3}\left(X_{i3} - X_{r3}\right) \quad \dot{m}_3 = \rho_{s3}A_3c_{s3} \qquad (2.14.9)$$

(b) For Two Supplier Pipes
   For two supplier pipes, where the assumption is that the superposition pressure in the faces of pipes 1 and 2 are equal, the equations for pipes 2 and 3 become:

$$\rho_{s2} = \rho_{02}\left(X_{i2} + X_{r2} - 1\right)^{G52} \quad c_{s2} = G_{52}a_{02}\left(X_{i2} - X_{r2}\right) \quad \dot{m}_2 = \rho_{s2}A_2c_{s2} \qquad (2.14.10)$$

$$\rho_{s3} = \rho_{0e3}\left(X_{i3} + X_{r3} - 1\right)^{G51} \quad c_{s3} = G_{51}a_{0e3}\left(X_{i3} - X_{r3}\right) \quad \dot{m}_3 = \rho_{s3}A_3c_{s3} \qquad (2.14.11)$$

The First Law of Thermodynamics is expressed as,

$$\dot{m}_1h_{01} + \dot{m}_2h_{02} + \dot{m}_3h_{03} = 0 \qquad (2.14.12)$$

where the local work, heat transfer and system state changes are logically ignored, and the h_0 term is the stagnation specific enthalpy. This single line expression is the one to be used if the assumption is made that T_{02} equals T_{03} for a single supplier pipe situation; it is also strictly correct for a two supplier pipes model. To solve this equation without this assumption, it must be split into two and analyzed specifically for the separate flows from pipe 1 to 2 and from pipe 1 to 3. The reference densities in question result from this step, as shown below. The normal isentropic expressions for the reference densities are expressed as:

$$\rho_{01} = \frac{p_0}{R_1T_{01}} \quad \rho_{02} = \frac{p_0}{R_2T_{02}} \quad \rho_{03} = \frac{p_0}{R_3T_{03}} \qquad (2.14.13)$$

However, should gas be entering pipes 1 or 2 as a result of a non-isentropic process, then these expressions become, for pipes 2 and 3,

$$\rho_{0e2} = \frac{p_0}{R_{e2}T_{02}} \quad \rho_{0e3} = \frac{p_0}{R_{e3}T_{03}} \tag{2.14.14}$$

Stagnation Enthalpy
  Stagnation enthalpy is generally defined as the addition of enthalpy and kinetic energy as follows:

$$h_0 = h + \frac{c^2}{2} = C_P T + \frac{c^2}{2}$$

(a) Stagnation Enthalpy for One Supplier Pipe
  Where one pipe (in this example, pipe 1) is the supply pipe, the stagnation enthalpies are:

$$h_{01} = \frac{G_{51}a_{01}^2 X_{s1}^2 + c_{s1}^2}{2} \quad h_{02} = \frac{G_{5e2}a_{e2}^2 X_{s2}^2 + c_{s2}^2}{2} \quad h_{03} = \frac{G_{5e3}a_{e3}^2 X_{s3}^2 + c_{s3}^2}{2} \tag{2.14.15}$$

(b) Stagnation Enthalpy for Two Supplier Pipes
  As with continuity, when there are two supplier pipes, only the expression for stagnation enthalpy for pipe 2 is altered; the other expressions in Eq. 2.14.15 remain the same:

$$h_{02} = \frac{G_{52}a_{02}^2 X_{s2}^2 + c_{s2}^2}{2} \tag{2.14.16}$$

Pressure Losses at the Three-Way Branch
  The pressure loss equations first presented in Eq. 2.14.2 devolve to:

(a) For One Supplier Pipe
  For one supplier pipe, the flow is from pipe 1 to pipes 2 and 3,

$$C_{L12} = 1.6 - \frac{1.6\theta_{12}}{167} \quad C_{L13} = 1.6 - \frac{1.6\theta_{13}}{167} \tag{2.14.17}$$

$$p_0\left(X_{s1}^{G71} - X_{s2}^{G7e2}\right) = C_{L12}\rho_{s2}c_{s2}^2 \quad p_0\left(X_{s1}^{G71} - X_{s3}^{G7e3}\right) = C_{L13}\rho_{s3}c_{s3}^2 \tag{2.14.18}$$

(b) For Two Supplier Pipes

Here there are two supplier pipes, i.e., pipes 1 and 2 are supplying pipe 3,

$$C_{L12} = 0 \qquad C_{L13} = 1.6 - \frac{1.6\theta_{13}}{167} \qquad (2.14.19)$$

$$X_{s1}^{G71} - X_{s2}^{G72} = 0 \qquad p_0\left(X_{s1}^{G71} - X_{s3}^{G7e3}\right) = C_{L13}\rho_{s3}c_{s3}^2 \qquad (2.14.20)$$

In the analysis for the First Law of Thermodynamics, and the pressure loss terms, it should be noted that the superposition relationships for pressure amplitude ratio and particle velocity, X_s and c_s, for each pipe are written in full in the continuity equation to reveal the unknown variables, X_{r2}, X_{r2}, and X_{r3}.

These functions can be solved by a standard iterative method for such problems. I find the Newton-Raphson method for the solution of multiple polynomial equations to be stable, accurate, and rapid in execution. The arithmetic solution on the computer is conducted by a Gaussian Elimination method.

It should be noted that the "unknowns" are the three reflected wave pressures and T_{02} and/or T_{03}, depending on the thermodynamic assumptions discussed earlier. In practice, it has been found that Benson's "constant pressure" criterion provides excellent initial guesses for the unknown variables. This permits a final numerical solution to be obtained in a mere two or three iterations with a maximum error of only 0.05% of any one of the values.

This more sophisticated branched pipe boundary condition can be incorporated into an unsteady gas dynamic code for implementation on a digital compsuter. Papers describing such work have already been presented [2.41, 2.59, 2.68].

2.14.1 The Accuracy of Simple and More Complex Branched Pipe Theories

The assertion is made above that the "constant pressure" theory of Benson is reasonably accurate. The following example puts numbers to that statement. Consider a branch consisting of three pipes numbered 1 to 3, where the initial reference temperature of the gas is 20°C and the gas in each pipe is air. The angle θ_{12} between pipe 1 and pipe 2 is 30° and the angle θ_{13} between pipe 1 and pipe 3 is 180°, i.e., pipe 3 is lying straight through from pipe 1. The input data are shown in Table 2.14.1. The pipes are all of equal diameter, d, at 25 mm in tests 1, 3, and 4 in Table 2.14.1. In test 2, the diameter of pipe 3, d_3, is 35 mm. In tests 1 and 2 the incident pulse in pipe 1 has a pressure ratio, P_{i1}, of 1.4 and the other pipes have undisturbed wave conditions. In test 3 and 4, the incident pulse in pipe 1 has a pressure ratio, P_{i1}, of 1.4 and the incident pulse in pipe 3 has a pressure ratio, P_{i3}, of 0.8 and 1.1, respectively.

The output data for the calculations using the "constant pressure" theory are shown in Table 2.14.2. The usual symbols, P_{r1}, P_{r2}, and P_{r3}, are used for the pressure ratios of the three reflected pressure waves at the branch; and c_1, c_2, and c_3 are used for the superposition velocities. The output data obtained when the more complex theory is employed are shown in Table 2.14.3. The computed mass flow rates (in g/s units) $\dot{m}_1$, $\dot{m}_2$, and $\dot{m}_3$, are shown in Table 2.14.4. Table 2.14.5 shows the "errors" in the computed mass flow between the "constant pressure"

TABLE 2.14.1 INPUT DATA FOR THE CALCULATIONS

No.	P_{l1}	P_{l2}	P_{l3}	d_1	d_2	d_3	θ_{12}	θ_{13}	t_{01}
1	1.4	1.0	1.0	25	25	25	30	180	20
2	1.4	1.0	1.0	25	25	35	30	180	20
3	1.4	1.0	0.8	25	25	25	30	180	20
4	1.4	1.0	1.1	25	25	25	30	180	20

TABLE 2.14.2 OUTPUT DATA FROM THE CALCULATIONS USING "CONSTANT PRESSURE" THEORY

No.	P_{r1}	P_{r2}	P_{r3}	c_1	c_2	c_3
1	0.891	1.254	1.254	112.6	-56.3	-56.3
2	0.841	1.188	1.188	126.3	-42.7	-42.7
3	0.767	1.086	1.345	148.5	-20.4	-128.1
4	0.950	1.333	1.215	97.0	-72.0	-25.0

TABLE 2.14.3 OUTPUT DATA FROM THE CALCULATIONS USING COMPLEX THEORY

No.	P_{r1}	P_{r2}	P_{r3}	c_1	c_2	c_3
1	0.9073	1.2178	1.2760	108.2	-49.0	-60.8
2	0.8481	1.1646	1.1967	124.4	-37.7	-44.6
3	0.7680	1.0819	1.3482	148.0	-19.4	-128.6
4	0.9796	1.2734	1.2517	89.5	-60.3	-32.4

TABLE 2.14.4 OUTPUT ON MASS FLOW FROM CONSTANT PRESSURE AND COMPLEX THEORY

No.	CONSTANT PRESSURE THEORY			COMPLEX THEORY		
	$\dot{m}_1$	$\dot{m}_2$	$\dot{m}_3$	$\dot{m}_1$	$\dot{m}_2$	$\dot{m}_3$
1	78.3	-39.1	-39.1	76.0	-33.3	-42.8
2	84.5	-28.5	-55.9	83.6	-24.9	-58.7
3	93.2	-12.8	-80.4	93.0	-12.1	-80.9
4	70.4	-52.3	-18.1	66.4	-42.4	-24.0

TABLE 2.14.5 FURTHER OUTPUT DATA REGARDING ERRORS ON MASS FLOW

No.	$\dot{m}_1$ error %	$\dot{m}_2$ error %	$\dot{m}_3$ error %	ITERATIONS
1	3.0	17.4	-8.6	2
2	1.1	14.4	-4.8	2
3	0.2	5.7	-0.6	2
4	6.0	23.3	-24.6	2

theory and the more complex theory. The final column of Table 2.14.5 shows the number of iterations needed for the solution; the fact that only two iterations were required for a worst-case error of 0.05% on any variable reveals the true value of the Benson "constant pressure" theory in providing a very high-quality first guess for the "unknowns."

It can be seen in test 1 that the constant pressure theory takes no account of the branch angle, nor of the non-isentropic nature of the flow, and this induces mass flow differences of up to 17.4% compared to the more complex theory. The actual values of the reflected pressure waves are quite close for both theories, but the ensuing mass flow error is significant—and is a telling argument for the inclusion of the more complex theory in any engine simulation method requiring accuracy.

In test 2 the results for the two theories are closer, i.e., the mass flow errors are smaller, an effect that occurs by virtue of the fact that pipe 3 has a larger diameter of 35 mm, which reduces the particle velocity into pipe 2. The pressure loss around the intersection into pipe 2, which is angled back from pipe 1 at 30°, is seen from Eq. 2.14.1 to be a function of the square of the superposition velocity, c_{s2}^2. This decreases the pressure loss error within the computation and in reality.

This effect is exaggerated in test 3 where, even though the pipe diameters are equal, the suction wave incident at pipe 3 likewise reduces the gas particle velocity entering pipe 2. Here, the errors in mass flow are reduced to a maximum of only 5.7%.

The reverse effect is shown in test 4 where an opposing compression wave incident at the branch in pipe 3 forces more gas into pipe 2. The mass flow errors now rise to a maximum value of 24.6%.

In these tests comparing simple and complex theory, the amplitudes of the reflected pressure waves are quite similar. However, the compounding effect of any error on pressure settles on the density, and the non-isentropic nature of the flow alters the particle velocities, all of which gives rise to the more serious errors in the mass flow rate as computed by the "constant pressure" theory.

2.15 Reflection of Pressure Waves in Tapered Pipes

The presence of tapered pipes in the ducts of an engine is commonplace. The action of the tapered pipe in providing pressure wave reflections is often used as a tuning element to significantly enhance the performance of the engine. The fundamental reason for this effect is that the tapered pipe acts as either a nozzle or as a diffuser, in other words acts to produce a more gradual process of reflection of pressure waves experienced abruptly at the sudden expansions and contractions discussed in Secs. 2.10 and 2.11. Almost by definition the process is not only more gradual but is more efficient at reflecting wave energy because it is spread out in terms of time. As a consequence, any ensuing tuning effect on the engine is not only more pronounced but is effective over a wider speed range.

Because a tapered pipe acts to produce a gradual and continual process of reflection, irrespective of whether the pipe area is increasing or decreasing, it must be analyzed in a similar fashion. The ideal would be to conduct the analysis in very small distance steps over the tapered length, but this would be impractical because it would be a very time-consuming process—as indeed CFD is!

A practical method of analyzing the geometry of tapered pipes is shown in Fig. 2.15. For simplicity of treatment, I will assume that the pipes are circular in section, but the translation of the analytic principle to ducts of any cross-section profile will become obvious. The length L for the section or sections to be analyzed is usually selected to be compatible with the rest of the computation process for the ducts of the engine [2.31]. The tapered section of the pipe has a taper angle θ, which is the included angle of that taper. Having selected a length L over which the unsteady gas dynamic analysis is to be conducted, it is a matter of simple geometry to determine the diameters at the various locations on the tapered pipe. Consider sections 1 and 2 in Fig. 2.15. They are of equal length L. At the commencement of section 1, the diameter is d_a, and at its conclusion it is d_b. At the start of section 2, the diameter is d_b, and it is d_c at its conclusion.

Any reflection process for sections 1 and 2 is considered to take place at the interface as a "sudden" expansion or contraction, depending on whether the particle flow is acting in a diffusing manner as in Fig. 2.15(b) or in a nozzle fashion as in Fig. 2.15(c). In short, the flow proceeds in an unsteady gas dynamic process along section 1 in a parallel pipe of representative diameter, d_1, with area, A_1, and is then reflected at the interface to section 2, where the representative diameter is d_2, with area, A_2. This is the analytic case irrespective of whether the flow is acting in a diffusing manner as in Fig. 2.15(b) or in a nozzle fashion as in Fig. 2.15(c). The logical mean diameter, or mean area, for each of the sections is the mean area corresponding to the tapered volume within each section. Assuming a round pipe, the section volume, V, is the frustum of a cone. If the tapered section is irregular, or non-circular, then that geometry will have to be determined, but the principle is the same. The section volumes, V_1 and V_2, for tapered round pipes are:

$$V_1 = \frac{\pi L}{12}\left(d_a^2 + d_a d_b + d_b^2\right) \quad \text{and} \quad V_2 = \frac{\pi L}{12}\left(d_b^2 + d_b d_c + d_c^2\right) \qquad (2.15.1)$$

The mean diameters for each conical section are related to the mean areas and actual section volumes by:

$$A_1 = \frac{V_1}{L} \quad \text{hence} \quad d_1 = \sqrt{\frac{4A_1}{\pi}} \quad \text{and} \quad A_2 = \frac{V_2}{L} \quad \text{hence} \quad d_2 = \sqrt{\frac{4A_2}{\pi}} \qquad (2.15.2)$$

The analysis of the flow commences by determining the direction of the particle flow at the interface between section 1 and section 2 and the area change occurring at that position. If the flow is behaving as in a diffuser, then the ensuing unsteady gas dynamic analysis is conducted using the theory precisely as presented in Sec. 2.10 for sudden expansions. If the flow is behaving as in a nozzle, then the ensuing unsteady gas dynamic analysis is conducted using the theory precisely as presented in Sec. 2.11 for sudden contractions.

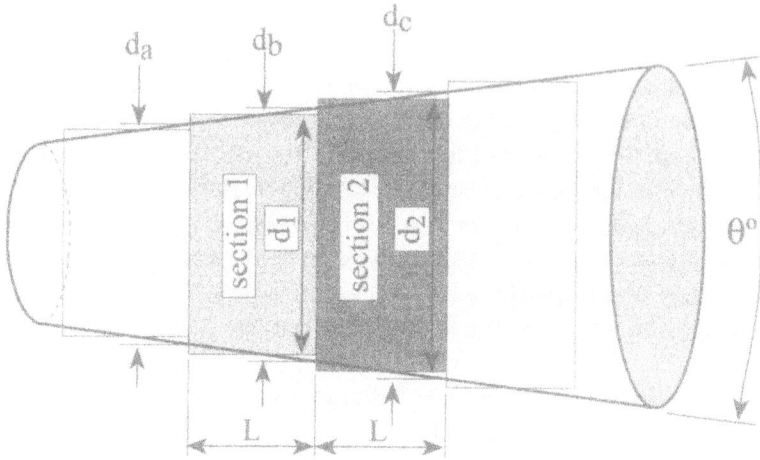

(a) the dimensioning of the tapered pipe

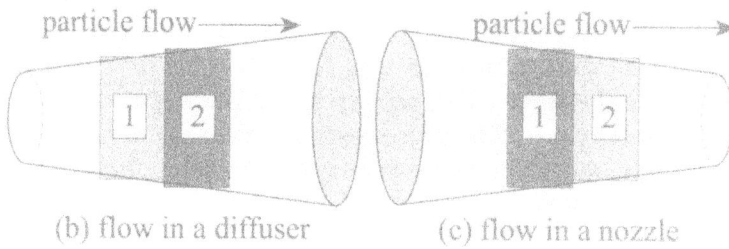

(b) flow in a diffuser   (c) flow in a nozzle

Fig. 2.15 Treatment of tapered pipes for unsteady gas-dynamic analysis.

2.15.1 Separation of the Flow from the Walls of a Diffuser

One of the issues that is always debated in the literature is flow separation from the walls of a diffuser, the physical situation being as in Fig. 2.15(b). Under such circumstances, the flow detaches from the walls into a central, highly turbulent core. As a consequence, the entropy gain is proportionately higher than in the thermodynamics shown in Fig. 2.9(a), for the pressure drop is not so great and the temperature drop is also lessened due to energy dissipation in turbulence. It is postulated that in such a flow separation the process becomes almost isobaric and can be represented as such in the thermodynamic analysis of Sec. 2.10. Therefore, if flow separation in a diffuser is assumed to occur, the analytic process set out in Sec. 2.10 should be amended to replace the equation that tracks the non-isentropic flow in the normal attached mode, namely the momentum equation, with another equation that better simulates the greater entropy gain of separated flow, namely a constant pressure equation.

Hence, in Sec. 2.10, Eq. 2.10.4 (or Eq. 2.10.9 in its final format) should be deleted from the set of equations to be analyzed, and it should be replaced with Eq. 2.15.3 (or its more detailed equivalent, Eq. 2.15.4) below.

The assumption is that the particle flow is moving, and diffusing, from section 1 to section 2 as in Fig. 2.15(b), and that wall separation has been detected. Constant superposition pressure at the interface between section 1 and 2 produces the following function, using the same variable nomenclature as in Sec. 2.10.

$$p_{s1} - p_{s2} = 0 \qquad (2.15.3)$$

This "constant pressure" equation is used to replace the final form of the momentum equation in Eq. 2.10.9. The "constant pressure" equation can be restated in the form below and used in the computational process:

$$X_{s1}^{G7} - X_{s2}^{G7} = 0 \qquad (2.15.4)$$

You may well inquire at what point in a computation should this change of analytical tack be implemented? In many gas dynamics texts, where *steady flow* is being described, either theoretically or experimentally, the conclusion reached is that flow separation will take place if the particle Mach number is greater than 0.2 or 0.3 and, more significantly, if the included angle, θ, of the tapered pipe is greater than a critical value, reported widely in the literature as lying between 5 and 7°. The work to date at QUB indicates that the angle, θ, is of very little significance, but that gas particle Mach number alone is the important factor controlling flow separation. Phrased mathematically, the current conclusion would be:

if $M_{s1} \geq 0.65$ then use the constant pressure equation, Eq. 2.15.4. (2.15.5)

if $M_{s1} < 0.65$ then use the momentum equation, Eq. 2.10.9 (2.15.6)

Future work in correlating theory with experiment, such as that seen in Sec. 2.19.7, may shed more light on this subject. Suffice it to say that there is enough evidence already to confirm that any computational method that universally employs the momentum equation for the solution of diffusing flow in steeply tapered pipes where the Mach number is high will inevitably produce a very inaccurate assessment of the unsteady gas flow behavior.

2.16 Reflection of Pressure Waves in Pipes for Outflow from a Cylinder

This situation is fundamental to all unsteady gas flow generated in the intake or exhaust ducts of a reciprocating ic engine. Fig. 2.16 shows an exhaust valve (or port) and pipe, or the throttled end of an exhaust pipe leading into a plenum such as the atmosphere or a silencer box. Anywhere in an unsteady flow regime where a pressure wave in a pipe is incident on a pressure-filled space, box, plenum, or cylinder, the following method is applicable to determine the magnitude of the mass outflow, of its thermodynamic state, and of the reflected

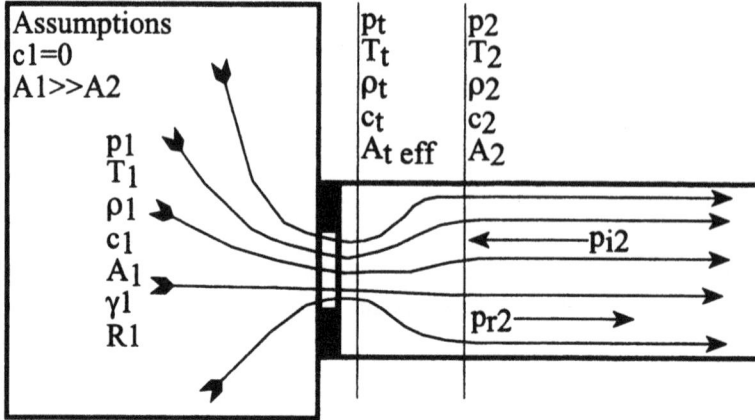

Fig. 2.16 Outflow from a cylinder or plenum to a pipe.

pressure wave. The theory to be generated is generally applicable to an intake valve (or port) and pipe for inflow into a cylinder, plenum, crankcase, or at the throttled end of an intake pipe from the atmosphere or a silencer box. However, the subtle differences for this analysis are given separately in Sec. 2.17.

You may well be tempted now to ask what then is the difference between this theoretical treatment and that given for the restricted pipe scenario in Sec. 2.12, for the drawings in Figs. 2.16 and 2.12 look remarkably similar. The answer is direct: In the theory presented here, the space from which the particles emanate is considered to be sufficiently large and the flow so three-dimensional as to give rise to the fundamental assumption that the particle velocity within the cylinder is considered to be zero, i.e., c_1 is zero.

The solution of the gas dynamics of the flow must include separate treatments for subsonic outflow and sonic outflow. The first presentation of the solution for this type of flow was by Wallace and Nassif [2.5], and their basic theory was used in a computer-oriented presentation by Blair and Cahoon [2.6]. Probably the earliest and most detailed exposition of the derivation of the equations involved is that by McConnell [2.7]. However, whereas all of these presentations declared that the flow was analyzed non-isentropically, a subtle error was introduced within the analysis which negated that assumption. Moreover, all of the earlier solutions, including that by Bingham [2.19], used fixed values of the cylinder properties throughout and solved the equations with either the properties of air ($\gamma = 1.4$ and $R = 287$ J/kgK) or exhaust gas ($\gamma = 1.35$ and $R = 300$ J/kgK). The arithmetic solution was stored in tabular form and indexed during the course of a computation. Today, that solution approach is inadequate, for the precise equations in fully non-isentropic form must be solved at each instant of a computation for the properties of the gas existing at that location, at that juncture.

Because a more complex solution, i.e., that for restricted pipes in Sec. 2.12, has already been presented, the complete solution for outflow from a cylinder or plenum in an unsteady gas dynamic regime will not pose any new theoretical difficulties.

The case of subsonic particle flow will be presented first and that for sonic flow is presented in Sec. 2.16.1.

Subsonic Outflow

In Fig. 2.16 the expanding flow from the throat to the downstream superposition point 2 is seen to leave turbulent vortices in the corners of that section. That the streamlines of the flow give rise to particle flow separation implies a gain of entropy from the throat to area section 2. On the other hand the flow from the cylinder to the throat is contracting and can be considered to be isentropic in the same fashion as the contractions discussed in Secs. 2.11 and 2.12. This is summarized on the temperature-entropy diagram in Fig. 2.17, where the gain of entropy for the flow rising from pressure p_t to pressure p_{s2} is clearly visible. The isentropic nature of the flow from p_1 to p_t—it is a vertical line on Fig. 2.17—can also be observed.

The properties and composition of the gas particles are those of the gas at the *exit* of the cylinder to the pipe. The word "exit" is used most precisely. For most cylinders and plenums, the process of flow within the cylinder is one of mixing. In such a case the properties of the gas at the exit from it for an outflow process is that of the mean of all of the contents. Not all internal cylinder flow is like that. Some engine cylinders have a stratified in-cylinder flow process. A two-stroke engine cylinder during scavenging would be a classic example of this situation. There, the properties of the gas exiting the cylinder would vary from combustion products only at the commencement of the exhaust outflow to a gas containing increasingly larger proportions of the air lost during the scavenge process; it would be mere coincidence if the exiting gas at any instant had the same properties as the average of all of the cylinder contents.

It is not inconceivable that the exhaust outflow process during the valve overlap period of the four-stroke engine is similar. This is illustrated in Fig. 2.25 where there are stratified zones labeled as CX surrounding the intake and exhaust apertures. The properties of the gas in those zones may well differ from the mean values for all of the cylinder, labeled in Fig. 2.25 as P_c, T_c, etc. and also the gas properties R_c and γ_c. In such a case a means of tracking the extent of the stratification must be employed and these variables determined as P_{cx}, T_{cx}, R_{cx}, γ_{cx}, etc., and employed for those properties that are subscripted with a 1 in the text below. Further debate on this issue is found in Sec. 2.18.10.

Although this singularity of stratification should always be borne in mind, and dealt with should it arise, the various gas properties for cylinder outflow are defined as:

$$\gamma = \gamma_1 \quad R = R_1 \quad G_5 = G_{51} \quad G_7 = G_{71}, \text{etc.}$$

As usual, the analysis of flow in this context uses, where appropriate, the equations of continuity, the First Law of Thermodynamics, and the momentum equation.

The reference state conditions are:

Density $$\rho_{01} = \rho_{0t} = \frac{p_0}{RT_{01}} \quad \rho_{02} = \frac{p_0}{RT_{02}} \tag{2.16.1}$$

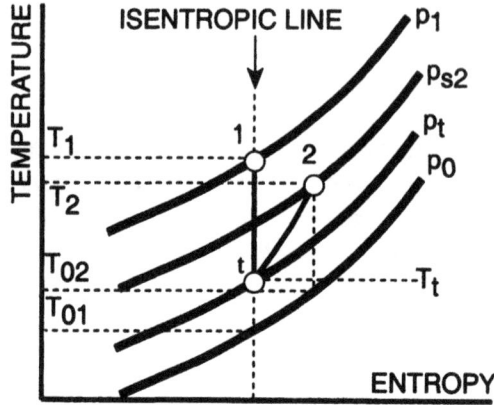

(a) temperature-entropy characteristics for subsonic outflow.

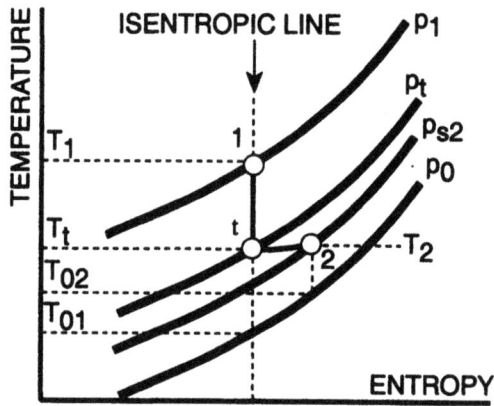

(b) temperature-entropy characteristics for sonic outflow.

Fig. 2.17 Temperature-entropy characteristics for cylinder or plenum outflow.

Acoustic velocity $\qquad a_{01} = a_{0t} = \sqrt{\gamma R T_{01}} \quad a_{02} = \sqrt{\gamma R T_{02}}$ $\qquad\qquad$ (2.16.2)

The continuity equation for mass flow in previous sections is still generally applicable and repeated here, although the entropy gain is reflected in the reference acoustic velocity and density at position 2:

$$\dot{m}_t - \dot{m}_2 = 0 \qquad\qquad (2.16.3)$$

The convention for particle flow direction is not significant, and this equation becomes:

$$\rho_t\left(C_cA_t\right)\left(C_sc_t\right)-\rho_{02}X_{s2}^{G5}A_2G_5a_{02}(X_{i2}-X_{r2})=0 \qquad (2.16.4)$$

The above equation, for the mass flow continuity for flow from the throat to the downstream station 2, contains the coefficient of contraction on the flow area, C_c, and the coefficient of velocity, C_s. These are conventionally connected in fluid mechanics theory to a coefficient of discharge, C_d, to give an effective throat area, A_{teff}, as follows:

$$C_d = C_cC_s \qquad \text{and} \qquad A_{teff} = C_dA_t$$

Chapter 3 is devoted to a full discussion on discharge coefficients, not only for poppet valves but for all types of ports and pipe end restrictions.

The equation of mass flow continuity, Eq. 2.16.4, becomes:

$$C_d\rho_tA_tc_t-\rho_{02}X_{s2}^{G5}A_2G_5a_{02}(X_{i2}-X_{r2})=0$$

The First Law of Thermodynamics was introduced in Sec. 2.8 for such flow situations. The analysis required here follows similar lines of logic. The First Law of Thermodynamics for flow from the cylinder to superposition station 2 can be expressed as:

$$h_1+\frac{c_1^2}{2}=h_{s2}+\frac{c_{s2}^2}{2}$$

or

$$G_5a_1^2-(G_5a_{s2}^2+c_{s2}^2)=0 \qquad (2.16.5)$$

The First Law of Thermodynamics for flow from the cylinder to the throat can be expressed as:

$$h_1+\frac{c_1^2}{2}=h_t+\frac{c_t^2}{2}$$

or

$$C_p(T_1-T_t)-\frac{c_t^2}{2}=0 \qquad (2.16.6)$$

The momentum equation for flow from the throat to superposition station 2 is expressed as:

$$A_2(p_t - p_{s2}) + \dot{m}_{s2}(c_t - c_{s2}) = 0 \qquad (2.16.7)$$

The unknown values will be the reflected pressure wave at the boundary, p_{r2}, the reference temperature at position 2, T_{02}, the throat pressure, p_t, and the velocity c_t, at the throat. This gives four unknowns, necessitating four equations, namely the mass flow equation in Eq. 2.16.4; the two First Law equations, Eq. 2.16.5 and Eq. 2.16.6; and the momentum equation, Eq. 2.16.7. All other "unknown" quantities can be computed from these values and from the "known" values. The known values are the downstream pipe area, A_2, the throat area, A_t, the gas properties exiting the cylinder, and the incident pressure wave, p_{i2}.

Recalling that,

$$X_1 = \left(\frac{p_1}{p_0}\right)^{G17} \quad \text{and} \quad X_{i2} = \left(\frac{p_{i2}}{p_0}\right)^{G17} \quad \text{and setting} \quad X_t = \left(\frac{p_t}{p_0}\right)^{G17}$$

then due to isentropic flow from the cylinder to the throat, the temperature reference conditions are given by:

$$a_1 = a_{01}X_1 \quad \text{or} \quad T_{01} = \frac{T_1}{X_1^2}$$

Because T_1 and X_1 are input parameters to any given problem, T_{01} is readily determined. In such a case, from Eqs. 2.16.1 and 2.16.2, the reference densities and acoustic velocities for the cylinder and throat conditions are also readily determined. As shown below, the density and temperature at the throat can also be related to the reference conditions.

$$\rho_t = \rho_{01}X_t^{G5} \quad \text{and} \quad T_t = \frac{\left(a_{01}X_t\right)^2}{\gamma R}$$

The continuity equation set in Eq. 2.16.4 reduces to:

$$\rho_{01}X_t^{G5}C_d A_t c_t - \rho_{02}(X_{i2} + X_{r2} - 1)^{G5} A_2 G_5 a_{02}(X_{i2} - X_{r2}) = 0 \qquad (2.16.8)$$

The First Law of Thermodynamics in Eq. 2.16.5 reduces to:

$$G_5(a_{01}X_1)^2 - \left[\left(G_5 a_{02}(X_{i2} - X_{r2}) \right)^2 + G_5 a_{02}^2 (X_{i2} + X_{r2} - 1)^2 \right] \qquad (2.16.9)$$

The First Law of Thermodynamics in Eq. 2.16.6 reduces to:

$$G_5 \left[\left(a_{01}X_1 \right)^2 - \left(a_{01}X_t \right)^2 \right] - c_t^2 = 0 \qquad (2.16.10)$$

The momentum equation Eq. 2.16.7 reduces to:

$$p_0 \left[X_t^{G7} - (X_{i2} + X_{r2} - 1)^{G7} \right]$$
$$+ \left[\rho_{02}(X_{i2} + X_{r2} - 1)^{G5} \times G_5 a_{02}(X_{i2} - X_{r2}) \right] \qquad (2.16.11)$$
$$\times [c_t - G_5 a_{02}(X_{i2} - X_{r2})] = 0$$

The four equations, Eqs. 2.16.8 to 2.16.11, cannot be reduced any further because they are polynomial functions of the four unknown variables, X_{r2}, X_t, a_{02}, and c_t. These functions can be solved by a standard iterative method for such problems. I use the Newton-Raphson method for the solution of multiple polynomial equations because it is stable, accurate, and rapid in execution. The arithmetic solution on a computer is conducted by a Gaussian Elimination method.

2.16.1 Outflow from a Cylinder Where Sonic Particle Velocity is Encountered

In the above analysis of unsteady gas outflow from a cylinder the particle velocity at the throat will quite commonly be found to reach, or even attempt to exceed, the local acoustic velocity. In thermodynamic or gas dynamic terms, this is not possible. The highest particle velocity that is permissible is the local acoustic velocity at the throat, i.e., the flow is permitted to become choked. Therefore, during the mathematical solution of Eqs. 2.16.8 to 2.16.11, the local Mach number at the throat is monitored and retained at unity if it is found to exceed it.

As,
$$M_t = \frac{c_t}{a_{01}X_t} = 1 \quad \text{then} \quad c_t = a_{01}X_t \qquad (2.16.12)$$

With c_t now fixed as a function of X_t, the First Law of Thermodynamics for outflow from the cylinder to the throat, in Eq. 2.16.10, now gives a direct solution for the pressure ratio from the cylinder to the throat. The combination of Eqs. 2.16.10 and 2.16.12 provides:

$$G_5\left\{\left(a_{01}X_1\right)^2 - \left(a_{01}X_t\right)^2\right\} - \left(a_{01}X_t\right)^2 = 0$$

Consequently,
$$\frac{X_t}{X_1} = \sqrt{\frac{G_5}{G_5+1}} \quad \text{or} \quad \frac{p_t}{p_1} = \left(\frac{2}{\gamma+1}\right)^{G35} \tag{2.16.13}$$

The pressure ratio from the cylinder to the throat where the flow at the throat is choked, i.e., where the Mach number at the throat is unity, is known as "the critical pressure ratio." The derivation of this ratio is also to be found in many standard texts on thermodynamics or gas dynamics. It is only applicable if the upstream particle velocity is considered to be zero. Consequently it is not a universal "law" and it must be applied only where the thermodynamic assumptions used in its creation are relevant. Significantly, it is not to be found in either Secs. 2.12.1 or 2.17.1.

This simplifies the sonic outflow case because it gives a direct solution for two of the unknowns and replaces two of the four equations in the subsonic solution. From an arithmetic standpoint, it is probably simpler to eliminate the momentum equation, and use only the continuity equation and the First Law of Eqs. 2.16.8 and 2.16.9, but from a thermodynamic standpoint, it is more accurate to retain it!

The acquisition of all related data for pressure, density, particle velocity and mass flow rate at both superposition stations and at the throat follows directly from the solution of the remaining two polynomial equations for X_{r2} and a_{02}.

2.16.2 Numerical Examples of Outflow from a Cylinder

The application of the above theory is illustrated by the calculation of outflow from a cylinder using the data given in Table 2.16.1. The nomenclature for these data is consistent with the theory and the associated sketch in Fig. 2.17. The units of the data, if inconsistent with strict SI units, are indicated in the several tables. The calculation output is shown in Tables 2.16.2 and 2.16.3.

The input data for tests numbered 1 and 2 pertain to a "blowdown" situation from gas at high temperature and pressure, with a small diameter port simulating a cylinder valve (or port) that has just commenced opening. The cylinder has a pressure ratio of 5.0 and a temperature of 1000°C. The exhaust pipe diameter, at 30 mm, is the same for all of the tests. In tests 1 and 2 the valve area is equivalent to a 3-mm diameter hole and has an assumed coefficient of discharge of 0.90. The gas in the cylinder and in the exhaust pipe in test 1 has a purity of zero, i.e., it is all exhaust gas.

TABLE 2.16.1 INPUT DATA TO CALCULATIONS OF OUTFLOW FROM A CYLINDER

No.	P_1	T_1 °C	Π_1	d_t mm	d_2 mm	C_d	$P_{|2}$	Π_2
1	5.0	1000	0.0	3	30	0.9	1.0	0.0
2	5.0	1000	1.0	3	30	0.9	1.0	1.0
3	1.8	500	0.0	25	30	0.75	1.0	0.0
4	1.8	500	0.0	25	30	0.75	1.1	0.0
5	1.8	500	0.0	25	30	0.75	0.9	0.0

TABLE 2.16.2 OUTPUT FROM CALCULATIONS OF OUTFLOW FROM A CYLINDER

No.	P_{r2}	P_{s2}	T_{s2} °C	P_t	T_t °C	$\dot{m}_{s2}$ g/s
1	1.0351	1.0351	999.9	2.676	805.8	3.54
2	1.036	1.036	999.9	2.641	787.8	3.66
3	1.554	1.554	486.4	1.319	440.0	85.7
4	1.528	1.672	492.5	1.546	469.5	68.1
5	1.538	1.392	479.9	1.025	392.9	94.3

TABLE 2.16.3 FURTHER OUTPUT FROM CALCULATIONS OF OUTFLOW FROM A CYLINDER

No.	c_t	M_t	c_{s2}	M_{s2}	a_{01} and a_{0t}	a_{02}
1	663.4	1.0	18.25	0.025	582.4	717.4
2	652.9	1.0	18.01	0.025	568.3	711.5
3	372.0	0.69	175.4	0.315	519.6	525.1
4	262.9	0.48	130.5	0.234	519.6	522.1
5	492.7	0.945	213.5	0.385	519.6	530.5

The purity, Π, defines the gas as being a mixture of air and exhaust gas, where the air is assumed to have a specific heats ratio, γ, of 1.4 and a gas constant, R, of 287 J/kgK. The exhaust gas is assumed to have a specific heats ratio, γ, of 1.36 and a gas constant, R, of 300 J/kgK. For further explanation see Eqs. 2.18.47 to 2.18.50.

In test 1, where the cylinder gas is assumed to be exhaust gas, the results of the computation are presented in Tables 2.16.2 and 2.16.3, and show that the flow at the throat is choked, i.e., that M_t is 1.0. It is shown in the table that a small pulse with a pressure ratio of just 1.0351 is sent into the exhaust pipe. The very considerable entropy gain is evident by the disparity between the reference acoustic velocities at the throat and at the pipe, a_{0t} and a_{02}, at 582.4 and 717.4 m/s, respectively. It is clear that any attempt to solve this flow regime as an isentropic process would be very inaccurate.

The presentation here of a non-isentropic analysis with variable gas properties is unique and its importance can be observed by comparing the results of tests 1 and 2. The cylinder gas in test data set 1 is exhaust gas. Test data set 2 is identical to set 1, except that the purity in the cylinder and in the pipe is assumed be unity, i.e., it is air. The mass flow rate calculated for data set 1 is 3.54 g/s and it is 3.66 g/s for data set 2; that is, there is an error of 3.4%. Mass flow errors in simulation translate ultimately into errors in the prediction of air mass trapped in a cylinder, the value of which is directly related to power output. This error of 3.4% is actually more significant than it appears because the effect is compounded throughout the entire engine cycle during the computer simulation.

Test data sets 3 to 5 illustrate the ability of pressure wave reflections to influence dramatically the "breathing" of an engine. The situation is one of exhaust flow from a cylinder with gas at high temperature and pressure. A large-area port simulates a valve that is at a well-open position. The cylinder gas has a pressure ratio of 1.8 and a temperature of 500°C. The exhaust pipe diameter, at 30 mm, is the same for all of the tests. The valve area is equivalent to a 25 mm diameter hole and has a typical coefficient of discharge of 0.75. The gas in the cylinder and in the exhaust pipe has a purity of zero, i.e., it is all exhaust gas. The only difference between data sets 3 to 5 is the amplitude of the pressure wave in the pipe incident on the exhaust port. In data set 3 it has a pressure ratio of 1.0, or undisturbed conditions exist; in data set 4 it has a pressure ratio of 1.1, providing a modest opposition to the outflow; in data set 5 it has a pressure ratio of 0.9, giving a modest suction effect on the cylinder. The results show very considerable variations in the ensuing mass flow rate exiting the cylinder, ranging from 85.7 g/s when the conditions are undisturbed in test 3, to 68.1 g/s when the incident pressure wave is of one compression, to 94.3 g/s when the incident pressure wave is one of expansion. These swings of mass flow rate represent variations of -20.5% to +10%. It will be observed that test 4 with the lowest mass flow rate, has the highest superposition pressure ratio, P_{s2}, at the pipe point, and that test 5 with the highest mass flow rate, has the lowest superposition pressure in the pipe. Because this pressure, P_{s2}, is that measured by a fast-response pressure transducer, you might be tempted now to conclude that test 3 is the one with the stronger wave action. Such is the folly of casually examining measured pressure traces in the exhaust ducts of engines—but this is nothing new to those who have studied Sec. 2.2.1!

This illustrates perfectly both the advantages of using pressure wave effects in the exhaust system of an engine to enhance the mass flow through it and the disadvantages of poorly designing the exhaust system. These simple numeric examples reinforce the opinions expressed earlier in Sec. 2.8.1 regarding the effective use of reflections of pressure waves in exhaust pipes. There is more of that ilk in Chapters 5 and 6.

2.17 Reflection of Pressure Waves in Pipes for Inflow to a Cylinder

This situation is fundamental to all unsteady gas flow generated in the intake or exhaust ducts of a reciprocating ic engine. Fig. 2.18 shows an inlet valve (or port) and pipe, or the throttled end of an intake pipe leading into a plenum such as the atmosphere or a silencer box. In an unsteady flow regime anywhere a pressure wave in a pipe is incident on a pressure filled space, box, plenum or cylinder, the following method is applicable to determine the magnitude of the mass inflow, of its thermodynamic state, and of the reflected pressure wave.

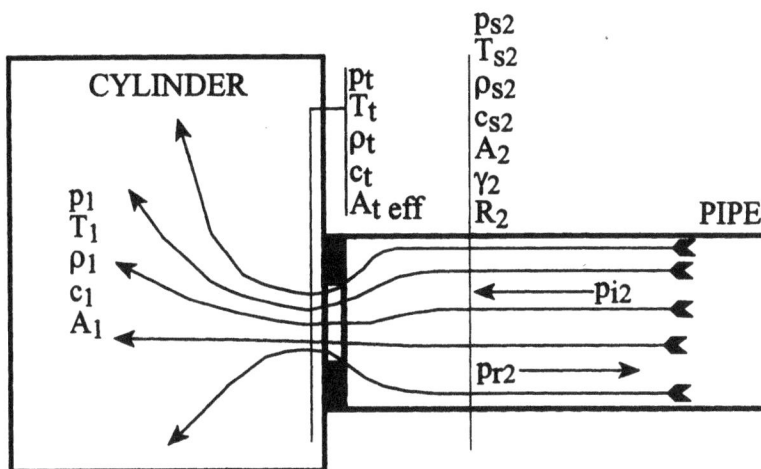

Fig. 2.18 Inflow from a pipe to a cylinder or a plenum.

In the theory presented here, the space into which the particles disperse is considered to be sufficiently large and also three-dimensional as to permit the fundamental assumption that the particle velocity within the cylinder is considered to be zero.

$$c_1 = 0 \qquad\qquad (2.17.1)$$

The case of subsonic particle flow will be presented first and that for sonic flow is presented in Sec. 2.17.1.

Subsonic Inflow

In Fig. 2.18, the expanding flow from the throat to the cylinder produces pronounced turbulence within the cylinder. The traditional assumption is that this dissipation of turbulence energy produces no pressure recovery from the throat of the port or valve to the cylinder. This assumption applies only where subsonic flow is maintained at the throat.

$$p_t = p_1 \qquad\qquad (2.17.2)$$

On the other hand, the flow from the pipe to the throat is contracting and can be considered to be isentropic in the same fashion as the other contractions discussed in Secs. 2.11 and 2.12. This is summarized on the temperature-entropy diagram in Fig. 2.19, where the gain of entropy for the flow rising from pressure p_t to cylinder pressure p_1 is clearly visible. The isentropic nature of the flow from p_{s2} to p_t—it is a vertical line on Fig. 2.19—can also be observed.

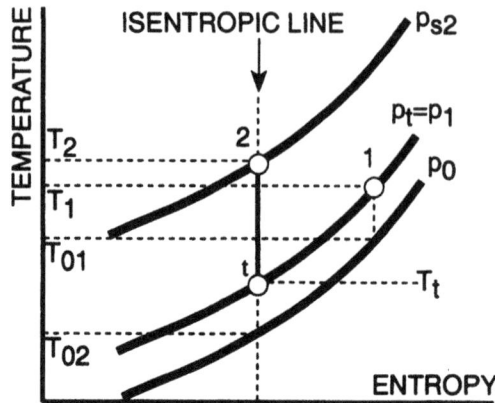

(a) temperature-entropy characteristics for subsonic inflow.

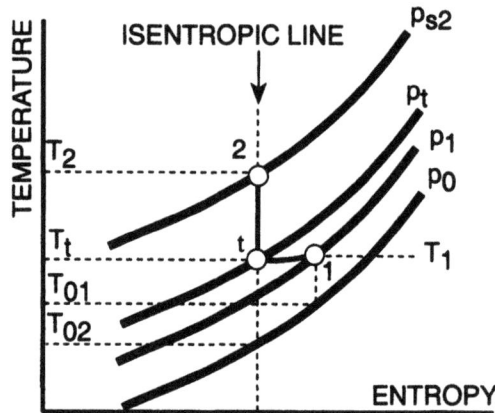

(b) temperature-entropy characteristics for sonic inflow.

Fig. 2.19 Temperature-entropy characteristics for cylinder or plenum inflow.

The properties and composition of the gas particles are those of the gas at the superposition point in the pipe. The various gas properties for cylinder inflow are defined as:

$$\gamma = \gamma_2 \quad R = R_2 \quad G_5 = G_{52} \quad G_7 = G_{72}, \text{etc.}$$

As usual, the analysis of flow in this context uses, where appropriate, the equations of continuity, the First Law of Thermodynamics, and the momentum equation. However, the momentum equation is not used in this particular analysis for subsonic inflow, because the constant pressure assumption used in Eq. 2.17.2 reflects an even higher gain of entropy, i.e., energy dissipation due to turbulence, than would be the case if the momentum equation were to be involved. The constant pressure assumption, in connection with a large entropy gain, has been used in this regard before within this text, notably in the section dealing with separation of diffusing flow in tapered pipes (Sec. 2.15.1).

The reference state conditions are:

Density
$$\rho_{02} = \rho_{0t} = \frac{p_0}{RT_{02}} \quad \rho_{01} = \frac{p_0}{RT_{01}} \tag{2.17.3}$$

Acoustic velocity
$$a_{02} = a_{0t} = \sqrt{\gamma RT_{02}} \quad a_{01} = \sqrt{\gamma RT_{01}} \tag{2.17.4}$$

The continuity equation for mass flow between the pipe and the throat is:

$$\dot{m}_t - \dot{m}_2 = 0 \tag{2.17.5}$$

This equation becomes:

$$\rho_t \left(C_c A_t\right)\left(C_s c_t\right) - \rho_{02} X_{s2}^{G5} A_2 G_5 a_{02}(X_{i2} - X_{r2}) = 0 \tag{2.17.6}$$

The above equation, for the mass flow continuity for flow from the upstream station 2 to the throat t, contains the coefficient of contraction on the flow area, C_c, and the coefficient of velocity, C_s. These are conventionally connected in fluid mechanics theory to a coefficient of discharge, C_d, to give an effective throat area, A_{teff}, as follows:

$$C_d = C_c C_s \quad \text{and} \quad A_{teff} = C_d A_t$$

Discharge coefficients are discussed at length in Chapter 3. The mass flow continuity equation becomes:

$$C_d \rho_t A_t c_t - \rho_{02} X_{s2}^{G5} A_2 G_5 a_{02}(X_{i2} - X_{r2}) = 0$$

The First Law of Thermodynamics for flow from the pipe to the throat can be expressed as:

$$h_2 + \frac{c_2^2}{2} = h_t + \frac{c_t^2}{2}$$

or
$$C_p(T_2 - T_t) + (\frac{c_2^2}{2} - \frac{c_t^2}{2}) = 0 \tag{2.17.7}$$

The unknown values will be the reflected pressure wave at the boundary, p_{r2}, and the velocity, c_t, at the throat. There are two unknowns, necessitating two equations, namely the mass flow equation in Eq. 2.17.6, and the First Law equation, Eq. 2.17.6. All other "unknown" quantities can be computed from these values and from the "known" values. The known values are the downstream pipe area, A_2, the throat area, A_t, the gas properties at superposition station 2 in the pipe, and the incident pressure wave, p_{i2}. You will recall that:

$$X_1 = \left(\frac{p_1}{p_0}\right)^{G17} \quad \text{and} \quad X_{i2} = \left(\frac{p_{i2}}{p_0}\right)^{G17} \quad \text{and setting} \quad X_t = \left(\frac{p_t}{p_0}\right)^{G17} = \left(\frac{p_1}{p_0}\right)^{G17}$$

The reference temperature T_{01} for the cylinder is given by:

$$a_1 = a_{01}X_1 \quad \text{or} \quad T_{01} = \frac{T_1}{X_1^2}$$

Because T_1 and X_1 are input parameters to any given problem, T_{01} is readily determined. In such a case, from Eqs. 2.17.1 and 2.17.2, the reference densities and acoustic velocities for the cylinder and throat conditions are also readily determined. The pipe point and the throat point are linked isentropically. Consequently, the density and temperature at the throat are related to their isentropic reference conditions and to the assumption regarding the throat pressure equality with cylinder pressure.

$$\rho_t = \rho_{02}X_t^{G5} = \rho_{02}X_1^{G5} \quad \text{and} \quad T_t = \frac{\left(a_{02}X_1\right)^2}{\gamma R}$$

Hence, the continuity equation set in Eq. 2.17.6 reduces to:

$$\rho_{02}X_1^{G5}C_dA_tc_t - \rho_{02}(X_{i2} + X_{r2} - 1)^{G5}A_2G_5a_{02}(X_{i2} - X_{r2}) = 0 \qquad (2.17.8)$$

and the First Law of Thermodynamics in Eq. 2.17.7 reduces to:

$$\begin{aligned} &\left[G_5(a_{02}(X_{i2} + X_{r2} - 1))^2 - G_5(a_{02}X_1)^2\right] \\ &+ \left[\left(G_5a_{02}(X_{i2} - X_{r2})\right)^2 - c_t^2\right] = 0 \end{aligned} \qquad (2.17.9)$$

Eqs. 2.17.8 and 2.17.9 cannot be reduced any further as they are polynomial functions of the two variables, X_{r1} and c_t. These functions can be solved by a standard iterative method for such problems. I have determined that the Newton-Raphson method for the solution of multiple polynomial equations is stable, accurate, and rapid in execution. The arithmetic solution on a computer is conducted by a Gaussian Elimination method.

2.17.1 Inflow to a Cylinder Where Sonic Particle Velocity is Encountered

In the above analysis of unsteady gas outflow from a cylinder, the particle velocity at the throat will quite commonly be found to reach, or even attempt to exceed, the local acoustic velocity. In thermodynamic or gas dynamic terms, this is not possible. The highest particle velocity that is permissible is the local acoustic velocity at the throat, i.e., the flow is permitted to become choked. Therefore, during the mathematical solution of Eqs. 2.17.8 and 2.17.9, the local Mach number at the throat is monitored and retained at unity if it is found to exceed it.

As, $$M_t = \frac{c_t}{a_{02}X_t} = 1 \quad \text{then} \quad c_t = a_{02}X_t \qquad (2.17.10)$$

The solution for a critical pressure ratio, as in Sec. 2.16.1, cannot be used here because the upstream particle velocity in the pipe is not zero.

Eq. 2.17.10 simplifies the entire procedure for this and gives a direct relationship between two of the unknowns. However, this does not reduce the solution complexity because it also eliminates a previous assumption for the subsonic flow regime that the throat pressure is equal to the cylinder pressure, i.e., the assumption made in Eq. 2.17.2. Therefore, Eqs. 2.17.8 and 2.17.9, must be revisited and that assumption removed.

After this revision, the continuity equation in Eq. 2.17.8 becomes:

$$a_{02}\rho_{02}X_t^{G6}C_dA_t - \rho_{02}(X_{i2} + X_{r2} - 1)^{G5}A_2G_5a_{02}(X_{i2} - X_{r2}) = 0 \quad (2.17.11)$$

After this revision, the First Law of Thermodynamics in Eq. 2.17.9 becomes:

$$\begin{aligned}
&\left[G_5(a_{02}(X_{i2} + X_{r2} - 1))^2 - G_5(a_{02}X_t)^2 \right] \\
&+ \left[(G_5a_{02}(X_{i2} - X_{r2}))^2 - (a_{02}X_t)^2 \right] = 0
\end{aligned} \qquad (2.17.12)$$

Eqs. 2.17.11 and 2.17.12, cannot be reduced any further because they are polynomial functions of the same two variables, X_{r1} and X_t. These functions can be solved by the same standard iterative method as used in the case of subsonic inflow.

The acquisition of all related data for pressure, density, particle velocity, and mass flow rate at the pipe superposition station and at the throat follows directly from the solution of the two polynomials for X_{r2} and X_t.

2.17.2 Numerical Examples of Inflow into a Cylinder

The application of the above theory is illustrated by the calculation of inflow into a cylinder using the data given in Table 2.17.1. The nomenclature for the data is consistent with the theory and the associated sketch in Fig. 2.18. The units of the data, if inconsistent with strict SI units, are indicated in the several tables. The calculation output is shown in Tables 2.17.2 and 2.17.3.

The input data for all of the tests are with reference to a normal intake situation from atmospheric air through a well-open valve, equivalent in area to a 20 mm diameter hole with a discharge coefficient of 0.75, simulating a cylinder valve or valve that is near to its maximum opening. The cylinder has a pressure ratio of 0.65 and a temperature of 70°C. The reference temperature at the pipe point is 70°C, an apparently high temperature that would not be abnormal for an intake system, the walls of which have been heated by conduction from the rest of the power unit. The intake pipe diameter, at 30 mm, is the same for all of the tests. The gas in the cylinder and in the induction pipe has a purity of 1.0, i.e., it is air.

Test data sets 1, 2, and 3 illustrate the ability of pressure wave reflections to influence the "breathing" of an engine. The difference between data sets 1 to 3 is the amplitude of the pressure wave in the pipe that is incident on the intake valve; in data set 1 it is at a pressure ratio of 1.0, or undisturbed conditions; in data set 2 it is at a pressure ratio of 0.9, indicating it has been throttled earlier upstream; in data set 3 it is at a pressure ratio of 1.2, giving a good ramming effect on the cylinder. The results show very considerable variations in the ensuing mass flow rate entering the cylinder, ranging from 36.0 g/s when the conditions are undisturbed in test 1, to 21.1 g/s when the incident "throttled" pressure wave is one of expansion, to

TABLE 2.17.1 INPUT DATA TO CALCULATIONS OF INFLOW INTO A CYLINDER

No.	P_1	T_{02} °C	Π_1	d_t mm	d_2 mm	C_d	P_{l2}	Π_2
1	0.65	70	1.0	20	30	0.75	1.0	1.0
2	0.65	70	1.0	20	30	0.75	0.9	1.0
3	0.65	70	1.0	20	30	0.75	1.2	1.0
4	0.65	40	1.0	20	30	0.75	1.2	1.0

TABLE 2.17.2 OUTPUT FROM CALCULATIONS OF INFLOW INTO A CYLINDER

No.	P_{r2}	P_{s2}	T_{s2} °C	P_t	T_t °C	$\dot{m}_{s2}$ g/s
1	0.801	0.801	48.9	0.65	30.3	36.0
2	0.779	0.698	36.5	0.65	30.3	21.1
3	0.910	1.095	78.9	0.65	30.3	57.3
4	0.910	1.095	48.2	0.65	3.75	60.0

**TABLE 2.17.3 FURTHER OUTPUT FROM CALCULATIONS OF
INFLOW INTO A CYLINDER**

No.	c_t	M_t	c_{s2}	M_{s2}	a_{01} and a_{0t}
1	201.9	0.578	58.0	0.161	371.2
2	118.2	0.338	37.4	0.106	371.2
3	321.4	0.921	73.8	0.196	371.2
4	307.0	0.921	70.5	0.196	354.6

57.3 g/s when the incident "ramming" pressure wave is one of compression. These variations in mass flow rate represent changes of -41.4% to +59.2%. The effect of "intake ramming," first discussed in Sec. 2.8.2, has a profound effect on the mass of air that can be induced by an engine. In addition, if the intake system is designed badly, and provides expansion wave reflections at the intake valve during that process, then the potential for the deterioration of induction of air is also self-evident. (Of course, if the design intention is throttling, then this criticism is eliminated!)

Test data set 4 is almost identical to test data set 2, except that the intake system has been cooled, or "intercooled" as in a modern turbo-diesel automobile engine, to give a reference air temperature at the pipe point of 40°C compared to test data set 2, which held hotter air at 70°C. The density of the intake air has been increased by 8.8%. This density increase does not translate linearly into the same increase of air mass flow rate. The computation shows that the air mass flow rate rises from 57.1 to 60.0 g/s, which represents a gain of 5.1%. However, because intake air flow and torque are almost directly related, the charge cooling represented by test data set 4 indicates a potentially useful gain of torque and power.

2.18 The Simulation of Engines by the Computation of Unsteady Gas Flow

Many computational methods have been suggested for the solution of this theoretical situation, such as Riemann variables [2.10], Lax-Wendroff [2.42] and other finite difference procedures [2.46], and yet others [2.12, 2.49, 2.69]. The basic approach adopted here is to reexamine the fundamental theory of pressure wave motion and adapt it to a mesh method interpolation procedure. At the same time, the boundary conditions for inflow and outflow, such as the filling and emptying of engine cylinders, are resolved for the generality of gas properties and in terms of the unsteady gas flow that controls those processes. The same generality of gas property and composition is traced throughout the pipe system. This change of gas property is very significant in two-stroke engines where the exhaust blowdown is followed by short-circuited scavenge air. For the four-stroke engine, especially high-performance spark-ignition engines or those with significant valve overlap periods, air can—indeed should—be drawn into the exhaust system to aid the induction process to attain a high volumetric efficiency. It is also very significant at varying load levels in diesel engines. Vitally important in this context is the solution for the continual transmission and reflection of pressure waves because they encounter both differing temperature gradients and gas properties, and both gradual and sudden changes of area throughout the engine ducting. Of equal importance is the ability of the calculation to predict the effect of internal heat generation within the duct or of external heat transfer with respect to it, and to be able to trace the effect of the ensuing gas temperature change on both the pressure wave system and the net gas flow.

The computation of unsteady gas flow through the cylinders of reciprocating internal combustion engines is a technology that is now nearly forty years old. The original paper by Benson et al. [2.10] formalized the use of Rieman variables as a technique for tracing the motion of pressure waves in the ducts of engines and the effect on the filling and emptying of the cylinders attached to them. Subsequently, other models have been introduced using finite difference techniques [2.29] and Lax-Wendroff methods [2.42, 2.46, 2.49]. Many technical papers and books have been published on this subject, including some by myself which are listed in the references.

For a more complete discussion of this subject, and in particular for the description of the previous computational procedures produced by authors from the University of Manchester (UMIST) and QUB, the textbooks by Benson [2.4] and Blair [2.25], and the papers by Chen et al. [2.47] and Kirkpatrick et al. [2.41] may be studied.

This introduction is somewhat brief, for a full discussion of the computational procedures developed by others would fill—indeed have filled—many textbooks, much less an introduction to this chapter. Consequently, many scores of worthy contributors to the literature have not been cited in the References and I hope that those who are not mentioned here will not feel slighted in any way.

I started working in the era when the graphical method of characteristics [2.8] was the only means of computing unsteady gas flow in pipes. While cumbersome in the arithmetic extreme before the advent of the digital computer, and hence virtually ineffective as a design tool, it had the very considerable merit of providing insights into unsteady gas flow in a manner that the Riemann variable, the finite difference, or the Lax-Wendroff methods, have never been able to. Doubtless, the programmers of such computational techniques have gained these insights, but this rarely, if ever, applied to the users of the software, i.e., the engineers and engine designers who had to employ it. As one who has taught such computational techniques to several generations of students, I can confirm that the insights gained by students always occurred most readily from lectures based on the papers produced by Wallace and Nassif [2.5] or the notes produced by Bannister [2.2]. These two publications were based on the fundamentals of pressure wave motion. Actually, one author, Jones [2.9], succeeded in solving the graphical method of characteristics by a computational procedure, and there are yet others who feel that this chimera is worth pursuing.

In 1990, I began reexamining many of these computational procedures to improve the quality of the design tool resulting from them, and found that, for one reason or another, all of the available methods fell short of engine design needs. Put simply, a design model firmly based on the motion of pressure waves is absolutely essential. At the same time, the computational model must be able to trace with great accuracy the rapidly changing gas properties in the cylinder and in the exhaust pipe, where the exhaust gas from blowdown may be rapidly followed by large amounts of fresh air, or fresh air and fuel, into the exhaust pipe and onward through an exhaust catalyst or silencer. In addition, for marine engines, where the exhaust is often internally water cooled, unless the model can readily incorporate the thermodynamic effects of that water "injection," it will be less than useful as a design tool. Diesel engines, be they two- or four-stroke cycle units also have gas properties in the exhaust system that vary very considerably as a function of the load level.

Like the four-stroke engine, the two-stroke engine is heavily dependent on pressure wave effects for the scavenging and charging processes and the model therefore must be based on pressure wave transmission, propagation, and reflections. The theoretical model described below must not be thought of as uniquely applicable to a two-stroke engine, for this is not the case. Indeed, the many recent papers on inlet manifold "tuning," employed as a means of modifying the volumetric efficiency-speed curve as a precursor to altering the torque curve, indicates the growing interest that designers of four-stroke engines for production automobile engines have in the prediction of unsteady gas motion. The designers of racing four-stroke engines already knew that the "black art" that accompanied their thinking, as often expressed to me, left much to be desired.

It should not be assumed that the lower-performance, four-stroke, spark-ignition engine is immune to disadvantageous pressure wave effects. The common occurrence of blowback into the inlet tract during the valve overlap period at the end of the exhaust stroke necessitates the tracking of variable gas properties within that duct. If a simulation model cannot do this, then a very inaccurate assessment of the ensuing air flow into the cylinder and of its trapped charge purity will result.

A simulation model that describes the complexities of pressure wave motion in an engine must meet one further requirement: Its principles must be understandable by the users, otherwise the process of optimizing a given design will not only be long-winded and tedious, but prone to decision-making on geometrical aspects of the engine that is illogically correct at one engine speed and load, yet disastrously wrong at another. Such dichotomies must be thoroughly checked out by a designer who understands the fundamental principles on which the model is based.

The basis of the GPB model employed here fulfills all of the criteria specified above.

2.18.1 *The Basis of the GPB Computation Model*

The unsteady gas flow process that is being modeled is illustrated in Fig. 2.20. This is the same process that has been discussed throughout this chapter. The computational procedure is somewhat similar to that found in other characteristics solutions. The pipe or pipes in the ducts of the engine being modeled are divided into meshes of a given length, L. The pressure waves propagating leftward and rightward are shown on Fig. 2.20, and at the instant that snapshot is taken for a particular mesh labeled as J, the pressure values at the left end are p_R and p_L and the pressure values at the right end of that mesh are p_{R1} and p_{L1}. The gas in the mesh space has properties of gas constant, R, and specific heats ratio, γ, that are assumed to be known at any instant in time. The reference density, ρ_0, and temperature, T_0, are also assumed known at any instant. Because the diameter, d, and the volume, V, of the mesh space are a matter of geometrical fact, the mass in the mesh space can be determined at that instant. The average

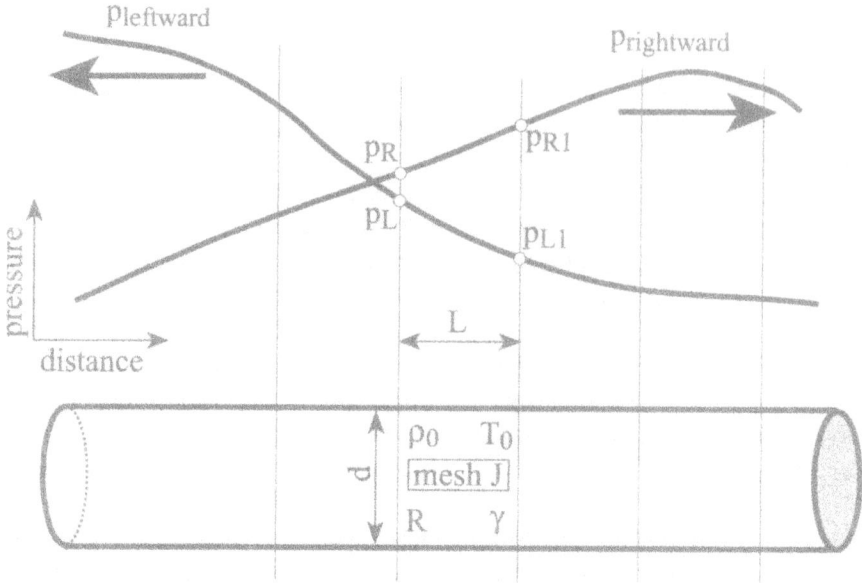

Fig. 2.20 Meshing of the duct for pressure wave propagation.

pressure throughout the mesh space can be considered to be the mean of the superposition pressures at each end of the mesh. The average superposition pressure amplitude ratio is given by X_J:

$$X_J = \frac{\left(X_R + X_L - 1\right) + \left(X_{R1} + X_{L1} - 1\right)}{2} \tag{2.18.1}$$

Consequently the average pressure, p_J, density, ρ_J, and temperature, T_J, are found from:

$$p_J = p_0 X_J^{G7} \tag{2.18.2}$$

$$\rho_J = \rho_0 X_J^{G5} \tag{2.18.3}$$

$$T_J = T_0 X_J^2 \tag{2.18.4}$$

The acoustic velocity, a_J, and mass, m_J, in the mesh are obtained by:

$$a_J = a_0 X_J \tag{2.18.5}$$

$$m_J = \rho_J V_J \quad \text{where} \quad V_J = \frac{\pi}{4} d^2 L \tag{2.18.6}$$

You will note that I have assumed any duct cross section to be "round" in all of the analysis, but any competent engineer will easily translate the theory for ducts of a differing cross-section profile.

The GPB computation model must determine:

(a) The effect of the motion of the pressure waves at either end of the mesh during some suitable time interval, dt, on the thermodynamics of the gas in the mesh space, J.

(b) The effect of the motion of the pressure waves at the right hand end of the mesh, J–1, or the left hand end of mesh, J+1, during some suitable time interval, dt, on the thermodynamics of the gas in the mesh space, J.

(c) The effect of the motion of the pressure waves propagating during time dt through mesh space J on their alteration in amplitude due to friction or area change.

(d) The effect on the pressure waves after arriving at the right hand end of the mesh, J, or at the left hand end of mesh, J, and encountering differing gas properties in adjacent mesh spaces, J+1 and J–1, respectively.

(e) The effect on the pressure waves after arriving at the right hand end of the mesh, J, or at the left hand end of mesh, J, and encountering a geometrical discontinuity such as a throttle, the valve or port of an engine cylinder, a branch in the pipe system, or an increase or decrease in duct area.

The first and most basic problem in any such computation model is to select the time interval at each step of the computation.

2.18.2 Selecting the Time Increment for Each Step of the Calculation

It is not essential that the mesh lengths in the inlet or exhaust pipes are equal, or are equal in any given section of an inlet or exhaust system. The reasoning behind these statements is to be found in the arithmetic nature of the ensuing iterative procedures, where interpolation of values is permissible, but extrapolation is arithmetically unstable [2.14]. This is best seen in Fig. 2.21. The calculation is to be advanced in discrete time steps. The value of the time step, dt, is obtained by sweeping each mesh space throughout the simulated pipe geometry and determining the "fastest" propagation velocity in the system. In the example illustrated in Fig. 2.21, this happens to be at some mesh other than mesh J. At either end of the mesh, the pressure waves, p_R and p_L, and p_{R1} and p_{L1}, induce left and right running superposition propagation velocities, α_{sR} and α_{sL}, and α_{sR1} and α_{sL1}, respectively. These superposition velocities can be determined from Eqs. 2.2.8 and 2.2.9.

It is assumed that there are linear variations within the mesh length, L, of pressure and propagation velocity.

Consequently, the time increment, dt, at mesh J is the least value of time taken to traverse the mesh J, where L is the mesh length peculiar to mesh J, by the fastest of any one of the four propagation modes at either end of mesh J:

$$dt = \frac{L}{\alpha_{sL}} \quad \text{or} \quad dt = \frac{L}{\alpha_{sL1}} \quad \text{or} \quad dt = \frac{L}{\alpha_{sR}} \quad \text{or} \quad dt = \frac{L}{\alpha_{sR1}}$$

Sweeping all meshes in the entire ducting to find the minimum value of dt:

$$dt_{minimum} = 0.99 \times \left| \frac{L}{\alpha_{s_{fastest\ in\ J}}} \right|_{J=1}^{J=total} \qquad (2.18.7)$$

This ensures that the subsequent iterative procedures for all mesh spaces are by interpolation, not extrapolation, thus satisfying the Courant, Friedrich, and Lewy "stability criterion" [2.14]. The addition of the 0.99 multiplier provides further arithmetic insurance that the internal mesh values, X_p and X_q, described in Sec. 2.18.3, also do not violate numerical stability.

2.18.3 The Wave Transmission during the Time Increment dt

This is illustrated in Fig. 2.21. Consider the mesh J, of length L. At the left end of the mesh the rightward moving pressure wave, p_R, is not propagating fast enough at α_{sR} to reach the right end of the mesh in time dt. Consequently, a value of superposition propagation velocity, α_p, from a wave point of pressure amplitude ratio, X_p, linearly related to its physical position, p, and the p_R and p_{R1} values, will just index the right end of the time-distance point at time dt, while it is being mutually superposed upon a leftward pressure wave, α_q, emanating from its physical position, q. This leftward propagating pressure wave point of amplitude Xq, will just indent the left-hand intersection of the time-distance mesh during the time increment dt. The values of Xq and α_q are also linearly related to their physical position and in terms of the leftward wave pressures, X_L and X_{L1}, at either end of mesh J.

In short, the calculation assumption is that, between any two meshes, there is a linear variation of wave pressure, wave superposition pressure, and superposition propagation velocity, both leftward and rightward, and that the values of Xp and Xq will—should no other effect befall them—become the new values of rightward and leftward pressure wave at either end of the mesh J at the conclusion of the time increment dt.

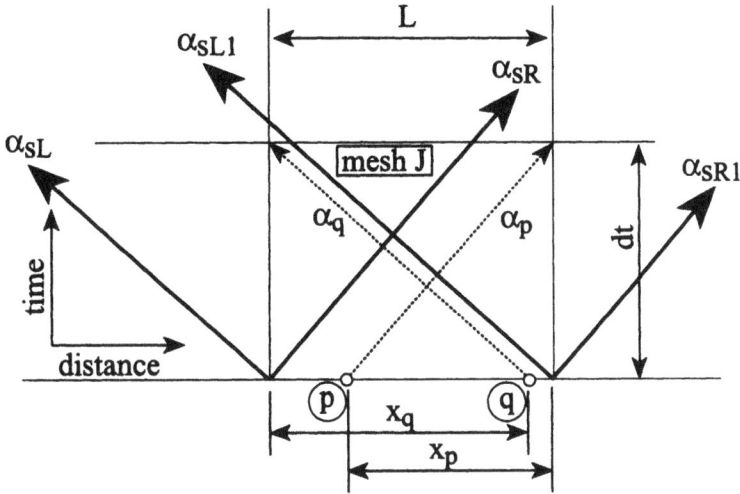

Fig. 2.21 The propagation of the pressure waves within mesh J.

2.18.4 The Interpolation Procedure for Wave Transmission through a Mesh

Having determined the time increment for a calculation step, and knowing the gas properties within any mesh volume for that transmission, the simulation must now determine the values of Xp and Xq, under the conditions outlined above. The situation is as sketched in Fig. 2.21.

The propagation of the rightward wave of amplitude Xp, through the leftward wave of amplitude Xq, is conducted at superposition velocities α_p and α_q, respectively. Retaining the sign convention that rightward motion is positive, from Eqs. 2.2.9 and 2.2.10, these values of propagation velocity are determined as,

$$\alpha_p = a_0\left(G_6 X_p - G_4 X_q - 1\right) \tag{2.18.8}$$

$$\alpha_q = -a_0\left(G_6 X_q - G_4 X_p - 1\right) \tag{2.18.9}$$

The time taken from their respective dimensional starting points, p and q, is the same, dt, where dt is equal to the minimum time step inferred from the application of the stability criterion in Sec. 2.18.2.

Therefore, and determining the arithmetic values of the lbcaths x_p and x_q,

$$x_p = \alpha_p dt \tag{2.18.10}$$

$$x_q = |\alpha_q| dt \tag{2.18.11}$$

The dimensional values x_p and x_q also relate to the numeric values of Xp and Xq as linear variations of the change of wave pressure between the two ends of the mesh J boundaries. These are found as:

$$X_p = X_R + \left(X_{R1} - X_R\right)\frac{L - x_p}{L} \tag{2.18.12}$$

$$X_q = X_{L1} + \left(X_L - X_{L1}\right)\frac{L - x_q}{L} \tag{2.18.13}$$

Eliminating x_p and x_q from Eqs. 2.8.10-2.18.13, produces two further equations,

$$\frac{X_{R1} - X_p}{X_{R1} - X_R} = \frac{a_0 dt}{L}\left(G_6 X_p - G_4 X_q - 1\right) \tag{2.18.14}$$

$$\frac{X_L - X_q}{X_L - X_{L1}} = \frac{a_0 dt}{L}\left(G_6 X_q - G_4 X_p - 1\right) \tag{2.18.15}$$

By defining the following groupings of variables as terms A to E, Eqs. 2.18.14 and 2.18.15 become Eqs. 2.18.16 and 2.18.17. The variables in the terms A to E are values that would be "known" quantities at the commencement of any mesh calculation.

$$A = E\left(X_{R1} - X_R\right)$$

$$B = E\left(X_L - X_{L1}\right)$$

$$C = \frac{X_{R1}}{A}$$

$$D = \frac{X_L}{B}$$

$$E = \frac{a_0 dt}{L}$$

Rearranging Eqs. 2.18.14 and 2.18.15 gives:

$$X_p\left(G_6 + \frac{1}{A}\right) - G_4 X_q - C - 1 = 0 \qquad (2.18.16)$$

$$X_q\left(G_6 + \frac{1}{B}\right) - G_4 X_p - D - 1 = 0 \qquad (2.18.17)$$

These simultaneous equations can be solved for the unknown quantities, X_p and X_q, after further collections of known terms within the parameters F_R and F_L are provided for simplification:

$$F_R = \frac{G_6 + \frac{1}{A}}{G_4}$$

$$F_L = \frac{G_6 + \frac{1}{B}}{G_4}$$

The final outcome is that:

$$X_p = \frac{1 + D + F_L + F_L C}{G_4(F_R F_L - 1)} \qquad (2.18.18)$$

$$X_q = \frac{1 + C + F_R + F_R D}{G_4(F_R F_L - 1)} \qquad (2.18.19)$$

Thus, assuming that the value of wave pressure is going to be modified by friction or area change during its travel during the time step, the new values of leftward and rightward wave pressures at the left and right hand ends of mesh J, X_p, and X_q, at the conclusion of the time step dt are given by:

$$X_{R1new} = X_p + \{\pm\text{friction effects}(\pm)\text{area change effects}\} \qquad (2.18.20)$$

$$X_{Lnew} = X_q + \{\pm\text{friction effects}(\pm)\text{area change effects}\} \qquad (2.18.21)$$

Each particle within the mesh space is assumed to experience this superposition process involving the pressure waves p_p and p_q. Consequently, the new values of the unsteady gas dynamic parameters attributed to the particles undergoing this superposition effect are:

Pressure amplitude ratio $\qquad X_s = X_p + X_q - 1 \qquad (2.18.22)$

Pressure $\qquad p_s = p_0 X_s^{G7} \qquad (2.18.23)$

Density $\qquad \rho_s = \rho_0 X_s^{G5} \qquad (2.18.24)$

Temperature $\qquad T_s = T_0 X_s^2 \qquad (2.18.25)$

Particle velocity $\qquad c_s = G_5 a_0 (X_p - X_q) \qquad (2.18.26)$

2.18.5 Singularities during the Interpolation Procedure

It is obvious that arithmetic problems could arise with the procedure given in Sec. 2.18.4 if the values of X_R and X_{R1}, or the values of X_L and X_{L1}, are equal. This would certainly be true in a model start-up situation, where the pipes would be "dead" and all of the array elements of pressure amplitude ratio, X_R, X_{R1}, X_L, and X_{L1}, would be unity. In this situation the values of A and B would be zero, C and D would be infinity and the calculation would collapse. Therefore, separate solutions are required for the wave pressures, X_p and X_q, in these unique situations, either at computation start-up or as arithmetic insurance during the course of a complete calculation. The solution is fairly straightforward using Eqs. 2.18.16 and 2.18.17, for the three possibilities (i) to (iii) involved.

In case (i), if X_R is equal to X_{R1}, the following is the solution:

if $\qquad X_p = X_{R1} \qquad (2.18.27)$

then
$$X_q = \frac{1 + D + G_4 X_p}{G_6 + \dfrac{1}{B}}$$
(2.18.28)

In case (ii), if X_L is equal to X_{L1}, the following is the solution:

if
$$X_q = X_{L1}$$
(2.18.29)

then
$$X_p = \frac{1 + C + G_4 X_q}{G_6 + \dfrac{1}{A}}$$
(2.18.30)

In case (iii), if X_R is equal to X_{R1} and X_L is equal to X_L, the following is the solution:

then
$$X_q = X_{L1}$$
(2.18.31)

and
$$X_p = X_{R1}$$
(2.18.32)

2.18.6 Changes Due to Friction and Heat Transfer during a Computation Step

The theoretical treatment for this is dealt with completely in Secs. 2.3 and 2.4. The values of the new wave pressure amplitude ratios at the mesh boundary, X_p and X_q, computed for mesh J are considered to be representative of the superposition process experienced by all of the particles in the mesh space during the time step dt. The pressure drop is computed as in Sec. 2.3, and the total heat transfer due to friction, δQ_f, can computed for all of the particles in mesh space J.

The wall surface area of mesh J is given by A_{sJ} where,

$$A_{sJ} = \pi dL$$
(2.18.33)

At this point, the warning given in the latter part of Sec. 2.3 regarding the computation of the heat generated in the mesh space due to friction should be heeded, and Eqs. 2.3.15 and 2.3.16 should be modified accordingly. The correct solution for the friction force, F, for all of the particles within the mesh space is:

$$F = C_f A_{sJ} \frac{\rho_s c_s^2}{2}$$
(2.18.34)

However, because each particle is assumed to move at superposition particle velocity c_s, during time interval dt, the work, δW_f, resulting in the heat generated, δQ_f, can be calculated by:

$$\delta W_f = Fdx = Fc_s dt = \left| \frac{C_f A_{sJ} \rho_s c_s^3 dt}{2} \right| = \delta Q_f \qquad (2.18.35)$$

The sign of this equation must always be positive because the heat due to friction is additive to the system comprising the gas particles within the mesh.

A similar process is adopted for the calculation of the heat transfer effect at the wall. This is discussed fundamentally in Sec. 2.4. It is assumed that the inner skin temperature of the wall, T_w, is a known quantity at the location of mesh J. The convection heat transfer coefficient, C_h, can be computed using the theory in Sec. 2.4. Consequently the heat transfer to or from the pipe walls at mesh J, with respect to the system comprising all of the particles within the boundaries of mesh J, is:

$$\delta Q_h = C_h A_J (T_w - T_s) dt \qquad (2.18.36)$$

2.18.7 Wave Reflections at the Inter-Mesh Boundaries After a Time Step

Sections 2.18.1-2.18.5 above, which discuss the simulation technique using the GPB method, have provided the theory for deducing the amplitudes of the leftward and rightward pressure waves, p_p and p_q, arriving at the left and right end boundaries of mesh J. This calculation method is employed similarly for all other meshes in all of the ducts of the engine. All meshes are assumed to have a constant diameter appropriate to that mesh. This applies even to tapered sections within any of the ducts, as a glance at Sec. 2.15 will confirm. However, the ability to ascertain the values of p_p and p_q is but half of the story. Consider the sketch of the model at this juncture as given in Fig. 2.22, which depicts parallel ducting and tapered ducting in panels (a) and (b), respectively. These situations are discussed in the sub-sections below.

(a) Parallel Ducting

Fig. 2.22(a) contains an illustration of two adjacent meshes, labeled as 1 and 2. After the application of the preceding theory in this section, the GPB model now has the information that pressure waves with pressure amplitude ratio, X_{p1} and X_{q1}, and X_{p2} and X_{q2}, have arrived at the left and right hand boundaries of meshes 1 and 2, respectively. The pressures are the "new" values, i.e., the values of X_p and X_q for each mesh having endured the loss of wave energy due to friction as described in Sec. 2.18.6. Consider the inter-mesh boundary between mesh 1 and 2 as being representative of all other inter-mesh boundaries within any one of the pipes in the entire ducting of the engine. At the commencement of the computation for the time step, the information for mesh 1 included the left and rightward pressure waves for that mesh at the left- and right-hand boundaries. As seen in Sec. 2.18.2 these pressure waves were labeled as p_R and p_L, and p_{R1} and p_{L1}; to denote that they are attached to these meshes, they will be relabeled here as $_1p_R$ and $_1p_L$, and $_1p_{R1}$ and $_1p_{L1}$. The equivalent values for mesh 2 at

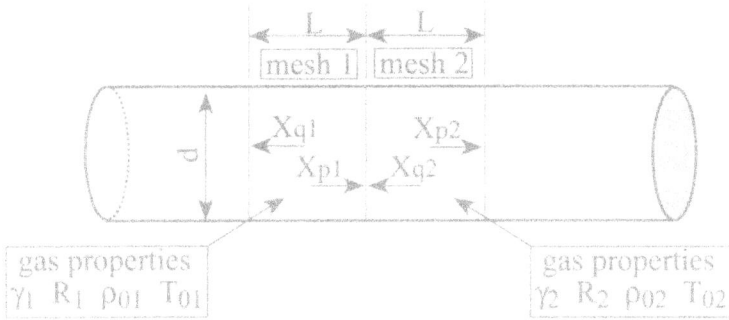

(a) two adjacent meshes in a parallel pipe

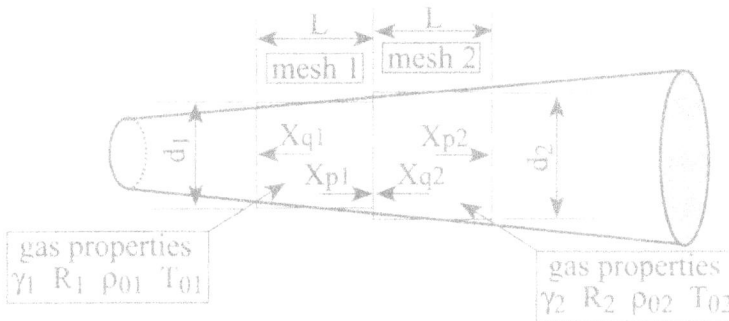

(b) two adjacent meshes in a tapered pipe

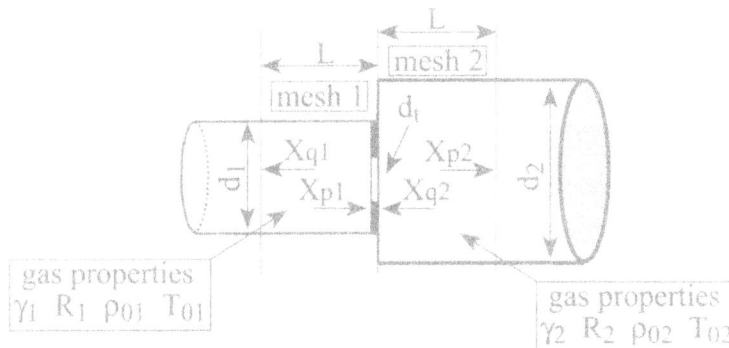

(c) two adjacent meshes in a restricted pipe

Fig. 2.22 Adjacent meshes in pipes of differing discontinuities.

the commencement of the computation time step would be $_2p_R$ and $_2p_L$, and $_2p_{R1}$ and $_2p_{L1}$. To begin a second time step, this set of information must be updated. The discussion elucidates the connection between the new values of X_{p1} and X_{q2}, i.e., two variables, for the inter-mesh boundary between mesh 1 and 2, and the updated values required for $_1p_R$ and $_1p_L$ and $_2p_{R1}$ and $_2p_{L1}$, i.e., four unknown values, at the same physical position.

If the gas properties and reference gas state are identical in meshes 1 and 2, then the solution is trivial, as follows:

$$_1X_{L1} = {}_2X_{q2} \text{ and } {}_2X_R = {}_1X_{p1} \qquad (2.18.37)$$

If, as is almost inevitable, the gas properties and reference gas state are not identical in meshes 1 and 2, then a temperature discontinuity exists at the inter-mesh boundary and the solution is conducted exactly as set out in Sec. 2.5. The similarity of the required solution and of the nomenclature is evident by comparison of Fig. 2.22(a) and Fig. 2.5.

(b) Tapered Pipes

The temperature discontinuity at the mesh boundary is solved and it remains to find the further reflections caused by change of area. A solution is required within the GPB model for two of the four unknowns, i.e., $_1p_{L1}$ and $_2p_R$, at the inter-mesh boundary in the tapered section of any pipe. In this situation, the known quantities are the values of $_1p_{R1}$ and $_2p_L$, as:

$$_1p_{R1} = p_0 X_{p1}^{G7} \qquad (2.18.38)$$

and

$$_2p_L = p_0 X_{q2}^{G7} \qquad (2.18.39)$$

The similarity of the sketches in Fig. 2.22(b) and Fig. 2.15 indicates that the fundamental theory for this solution is given in Sec. 2.15, with the base theory given before in Secs. 2.10 and 2.11. As related in Sec. 2.15, the basis of the method employed is to recognize whether the particle flow is diffusing or contracting. If it is contracting, then the flow is considered to be isentropic and the thermodynamics of the solution is given in Sec. 2.11 for subsonic particle flow, and in Sec. 2.11.1 for sonic flow. The Benson "constant pressure" criterion gives an excellent initial guess for the values of the unknown quantities and considerably reduces the number of iterations required for the application of the more complete gas dynamic theory in Sec. 2.11. If the particle flow is recognized to be diffusing, then the flow is considered to be non-isentropic and the solution is given in Sec. 2.10 for subsonic particle flow, and in Sec. 2.10.1 for sonic flow. As with contracting flow, the Benson "constant pressure" criterion gives an excellent initial guess for the values of the three remaining unknown quantities in diffusing flow. You should remember that, because the flow is non-isentropic, there is a third unknown quantity which is the reference temperature incorporating the entropy gain, on the downstream side of the expansion. For diffusing flow, special attention should be paid to the information on flow separation given in Sec. 2.15.1 where the pipe may be steeply tapered or the particle Mach number needs correction to the separation criterion.

2.18.8 Wave Reflections at the Ends of a Pipe after a Time Step

During computer modeling, the duct of an engine is meshed in distance terms and referred to within the computer program as a pipe which may have a combination of constant and gradually varying area sections between discontinuities. Thus, a tapered pipe is not a discontinuity in those terms, but a sudden change of section is treated as such. So too is the end of a pipe at an engine port or valve, or at a branch. This meshing nomenclature is shown in Fig. 2.22(c), with a restricted pipe employed as the example.

At first glance, the "joint" between mesh 1 and mesh 2 in Fig. 2.22(c), with the change of diameter from d_1 to d_2 from mesh 1 to mesh 2, appears to be no different from the previous case of tapered pipes discussed in Sec. 2.18.7. However, because there is an orifice of diameter d_t between the two mesh sections, and because the basic theory for a restricted pipe is presented in a separate section, Sec. 2.12, this situation cannot be discussed as just another inter-mesh boundary problem. It is in effect a pipe discontinuity. It is the "butted joint."

All engine configurations being modeled contain pipes, plenums, cylinders, and are fed air from, and feed exhaust gas to, the atmosphere. The atmosphere is nothing but another form of plenum. A cylinder during the open cycle is nothing but another form of plenum in which one of the walls move and holes, of varying size and shape, open and close as a function of time. It is the pipes that connect these discontinuities, i.e., cylinders, plenums, and the atmosphere together. However, pipes are connected to other pipes, either at branches or at the type of junction such as shown in Fig. 2.22(c). This latter case of the restricted pipe cannot be treated as an inter-mesh problem because it is more logical, from the standpoint of the overall organization of the structure of the computer software, to consider it as a junction at the ends of two separate pipes.

The point has already been made, in Sec. 2.18.7, that each mesh space is an "island" of information and that the results of the pipe analyses in Secs. 2.18.1 to 2.18.6 lead only to the determination of the new values of left and right moving pressure waves at the left- and right-hand boundaries of the mesh space, i.e., p_p and p_q. In Sec. 2.18.7, the analysis focused on an inter-mesh boundary and the acquisition of the remaining two unknown values for the pressure waves at each boundary of every mesh and at every mesh boundary except for that at the left-hand end of the extreme left-hand mesh of any pipe, and also for the right-hand end of the extreme right-hand mesh of any pipe.

For example, imagine in Fig. 2.22(a) that mesh 1 is at the extreme left hand end of the pipe. The value of X_{q1} automatically becomes the required value of X_L for mesh 1. To proceed with the next time step of the calculation, the value of X_R is required. What then is the value of X_R? The answer is that it will depend on what the left-hand end of the pipe is connected to, i.e., a cylinder, a plenum, etc. In addition, imagine in Fig. 2.22(a) that mesh 2 is at the extreme right-hand end of the pipe. The value of X_{p1} automatically becomes the required value of X_{R1} for mesh 1. To proceed with the next time step of the calculation, the value of X_{L1} is required. What then is the value of X_{L1}? The answer, as before, is that it will depend on what the right-hand end of the pipe is connected to, i.e., a branch, a restricted pipe, the atmosphere, etc.

The GPB modeling method stores information regarding the "hand" of the two end meshes in every pipe, and must be able to index that name of that "hand" at its connection to its own discontinuity, so that the numeric result of the computation of boundary conditions at the ends of each pipe is placed in the appropriate storage location within the computer.

Consider each boundary condition in turn: a restricted pipe, a cylinder, plenum or atmosphere, and a branch.

(a) A Restricted Pipe, as Sketched in Fig. 2.22(c)

Imagine, just as it is sketched, that mesh 1 is at the right-hand end of pipe 1 and that mesh 2 is at the left-hand end of pipe 2. Then:

$$_1X_{R1} = X_{p1} \text{ and } _2X_L = X_{q2} \tag{2.18.40}$$

The unknown quantities are the values of the reflected pressure waves, $_1p_{L1}$ and $_2p_R$, represented here by their pressure amplitude ratios, $_1X_{L1}$ and $_2X_R$. The analytic solution for these, and also for any entropy gain on the downstream side of whichever direction the particle flow takes, is to be found in Sec. 2.12. In terms of the notation in Fig. 2.8, which accompanies the text of Sec. 2.12, the value referred to as X_{p1} is clearly the incident wave, X_{i1}, and the value referred to as X_{q2} is obviously the incident pressure wave, X_{i2}.

(b) A Cylinder, Plenum, or the Atmosphere as Sketched in Figs. 2.16 and 2.18

A cylinder, plenum, or the atmosphere is considered to be a large "box" that is sufficiently large as to be able to consider the particle velocity within it to be effectively zero. For the rest of this section, a cylinder, plenum, or the atmosphere is often referred to as a "box." In the theory given in Secs. 2.16 and 2.17, the entire analysis is based on knowing the physical geometry at any instant. Depending on whether the flow is inflow or outflow, either the thermodynamic state conditions of the pipe or the "box" are known values respectively.

As discussed previously in this section, the "hand" of the incident wave at the mesh in the pipe section adjacent to the cylinder is an important element of the modelling process. Imagine that mesh 1, as shown in Fig. 22(a), is the mesh at the left-hand end of the pipe and attached to the cylinder exactly as it is sketched in either Fig. 2.16 or 2.18. In such a case at the end of the time step in computation, to implement the theory given in Secs. 2.16 and 2.17, the following is the nomenclature interconnection for that to occur:

$$_1X_L = X_{q1} \tag{2.18.41}$$

and

$$p_{i2} = p_{01}X_{q1}^{G7} \tag{2.18.42}$$

The properties of the gas subscripted as 2 in Fig. 2.16 or Fig. 2.18 are the properties of the gas in mesh 1 used here as the illustration.

At the conclusion of the computation for boundary conditions at the "box," using the theory of Secs. 2.16 and 2.17, the amplitude of the reflected pressure wave, p_{r2}, is provided. If it is outflow, the entropy gain appears in the form of the reference temperature, T_{02}. In nomenclature terms, this reflected pressure wave is the "missing" pressure wave value at the left-hand of mesh 1 to permit the computation to proceed to the next time step.

$$_1X_R = \left(\frac{p_{r2}}{p_0}\right)^{G17}$$

(2.18.43)

One important final point must be made regarding the rather remote possibility of encountering supersonic particle velocity in any of the pipe systems during the superposition process at any mesh point. The amplitudes of all of the left and right moving pressure waves have already been established at both ends of all meshes within the ducting of the engine. It is necessary now to search each of these mesh positions for the potential occurrence of supersonic particle velocity in the manner described completely in Sec. 2.2.4. If any occurrences are found, the corrective action of a weak shock to give the necessary subsonic particle velocity [2.15-2.17] must be applied. Unsteady gas flow does not permit supersonic particle velocity. It is self-evident that the gas particles cannot move faster than the disturbance that is giving them the signal to move.

2.18.9 Mass and Energy Transport along the Duct during a Time Step

In a real flow situation, energy and mass transport are conducted at the molecular level. Computational facilities are not large enough to accomplish this, so it is carried out in a mesh grid spacing of some 10 to 25 mm, a size commonly used in automotive engines. The size of mesh length is deduced by making the simple assumption that the calculation time step, dt, should translate to an advance of about 1° crankangle, and no more than 2°, for any engine design. The mesh length, in mm units, required is found as L_{mm} from this simple equation relating distance propagated by a wave at α (m/s) at an engine speed of N (rev/min) for a desired crankshaft interval of $d\theta$ (deg.):

$$L_{mm} = 1000 \times \alpha \times \frac{d\theta}{360} \times \frac{60}{N} = 1000\frac{\alpha \times d\theta}{6N}$$

This is best illustrated by a simple example. In the exhaust ducting, you might estimate under a given set of modeling circumstances that the reference temperature, T_0, is 450°C and that the largest exhaust pulse may be of 1.5 atm. The ratio of specific heats, γ, is probably 1.3, and the gas constant, R, is probably 300 J/kgK. The engine is running at 5000 rev/min, and the desired crankangle step, $d\theta$, is 1°. I will leave it to you to solve the above equation in conjunction with those in Sec. 2.1.2 for the propagation velocity, α, to show that the required exhaust mesh length is 25.3 mm.

In the same engine, but in the intake ducting, you might estimate under the same set of modeling circumstances that the reference temperature, T_0, is 30°C. The ratio of specific heats, γ, is almost certainly 1.4, and the gas constant, R, is as for air at 287 J/kgK. The engine is still running at 5000 rev/min, and the desired crankangle step, $d\theta$, remains at 1°. You will recall from Sec. 2.1.2 that the fastest moving parts of an expansion wave are the leading and trailing edges at 1.0 atm, where the propagation velocity is a_0, so I will leave it to you to solve for the required intake mesh length, which is 11.6 mm.

The necessary mesh length requirements for the hot exhaust system and the colder intake system, on the face of the above evidence, differ by a factor of about two. To compound the dilemma, there will be times during an engine cycle when the entire exhaust ducting may contain expansion waves from a tuned exhaust system, and there will be times when the entire intake system may contain compression waves for ramming at the intake valve, which changes some of the assumptions in the above calculations. Despite these comments, it is clear that the mesh length to be used for the intake system must be less than that in the exhaust system. If the intake system mesh length is made equal to that of the exhaust system, then a gross interpolation, and inherent inaccuracy, of the wave propagation process in the intake system would be the outcome. The most equitable solution, considering all of the above thinking, is to make the best possible estimate for the mesh length in the hottest possible exhaust duct, and then find the mesh length for all other ducts using the following ratio:

$$\frac{L_{\text{hottest pipe}}}{L_{\text{any other pipe}}} = \frac{a_{0\,\text{hottest pipe}}}{a_{0\,\text{any other pipe}}}$$

Within each mesh space, as shown in Fig. 2.20, the properties of the gas contained are assumed to be known at the start of a time step. During the subsequent time step, dt, as a result of the wave transmission, particles and energy are going to be transported from mesh space to mesh space, and heat transfer is going to occur by internal means, such as friction or a catalyst, or by external means through the walls of the duct.

To determine the effect of all of these mechanisms, the First Law of Thermodynamics is used. The physical situation is sketched in Figs. 2.23 and 2.24. Here, the energy transport across the boundaries for mesh space J will have unique values depending on the particle directions of this transport. This is illustrated in Fig. 2.23, where four different cases are shown to be possible.

All pressures, particle velocities, and densities are, by definition, for the superposition situation at the physical location at the appropriate end of mesh J. The left-hand end of any mesh is denoted as the "in" side and the right-hand end as the "out" side for flow of mass and energy, and as a sign convention when applying the First Law of Thermodynamics. Manifestly, as with case 1, there will be situations where the gas is flowing outward at what is nominally the "in" side of the mesh, but the arithmetic sign convention of the pressure wave analysis takes care of that problem automatically. It must be remembered that the GPB computer simulation must interrogate each mesh boundary and apply with precision the result of that interrogation for cases 1 to 4 (Fig. 2.23).

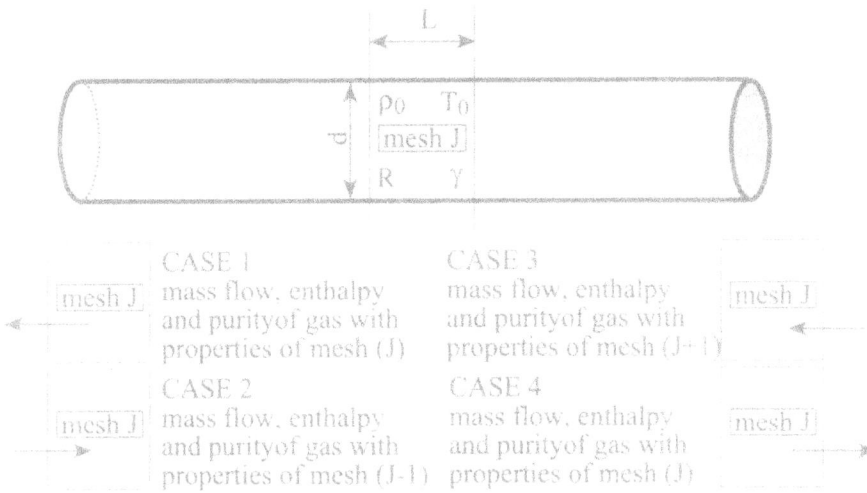

Fig. 2.23 Mass, energy, and gas species transport at mesh J during time interval dt.

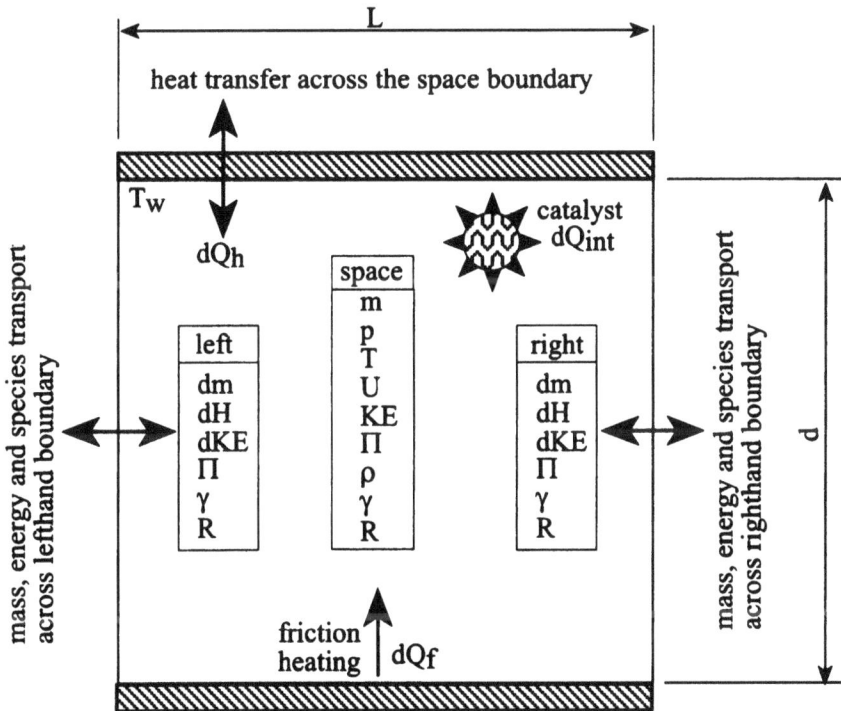

Fig. 2.24 The mesh J which encounters mass and energy transfer during time dt.

At the commencement of the time step, the system for mesh J has a known mass, pressure, temperature, and volume. The notation denotes the "before" and "after" situations during the time step, dt, for the mass, pressure, and temperature by a "b" and "a" prefix for the system properties. For example, this gives a symbolism for mesh J for the "before" conditions of mass, $_bm_J$; pressure ratio, $_bP_J$; and temperature, $_bT_J$.

Because each mesh has a constant flow area, A_J, volume, V_J, and an initial temperature, $_bT_J$, the four cases can be analyzed as follows:

(i) Case 1

Pressure $$_JX_{in} = {_J}X_R + {_J}X_L - 1$$

Particle velocity $$_Jc_{in} = {_J}G_5 \; _Ja_0({_J}X_R - {_J}X_L)$$

Density $$_J\rho_{in} = {_J}\rho_0 \; _JX_S^{{_J}G_5}$$

Specific enthalpy $$_J\,dh_{in} = {_J}C_P \; _bT_J + \frac{_Jc_{in}^2}{2}$$

Mass flow increment $$_J\,dm_{in} = {_J}\rho_{in} \; A_J \; _Jc_{in} \; dt$$

Enthalpy increment $$_J\,dH_{in} = {_J}\,dh_{in} \; _J\,dm_{in}$$

Air flow increment $$_J\,d\Pi_{in} = {_J}\Pi \; _J\,dm_{in}$$

(ii) Case 2

Pressure $$_JX_{in} = {_J}X_R + {_J}X_L - 1$$

Particle velocity $$_Jc_{in} = {_{J-1}}G_5 \; _{J-1}a_0({_J}X_R - {_J}X_L)$$

Density $$_J\rho_{in} = {_{J-1}}\rho_0 \; _JX_S^{{_{J-1}}G_5}$$

Specific enthalpy $$_J\,dh_{in} = {_{J-1}}C_P \; _bT_{J-1} + \frac{_Jc_{in}^2}{2}$$

Mass flow increment $$_J\,dm_{in} = {_J}\rho_{in} \; A_J \; _Jc_{in} \; dt$$

Enthalpy increment $\qquad$ $_J dH_{in} =_J dh_{in}\ _J dm_{in}$

Air flow increment $\qquad$ $_J d\Pi_{in} =_{J-1}\Pi\ _J dm_{in}$

(iii) Case 3

Pressure $\qquad$ $_J X_{out} =_J X_{R1} +_J X_{L1} - 1$

Particle velocity $\qquad$ $_J c_{out} =_{J+1} G_5\ _{J+1} a_0 (_J X_{R1} -_J X_{L1})$

Density $\qquad$ $_J \rho_{out} =_{J+1} \rho_0\ _J X_{out}^{J+1 G_5}$

Specific enthalpy $\qquad$ $_J dh_{out} =_{J+1} C_P\ _b T_{J+1} + \dfrac{_J c_{out}^2}{2}$

Mass flow increment $\qquad$ $_J dm_{out} =_J \rho_{out}\ A_J\ _J c_{out}\ dt$

Enthalpy increment $\qquad$ $_J dH_{out} =_J dh_{out}\ _J dm_{out}$

Air flow increment $\qquad$ $_J d\Pi_{out} =_{J+1}\Pi\ _J dm_{out}$

(iv) Case 4

Pressure $\qquad$ $_J X_{out} =_J X_{R1} +_J X_{L1} - 1$

Particle velocity $\qquad$ $_J c_{out} =_J G_5\ _J a_0 (_J X_{R1} -_J X_{L1})$

Density $\qquad$ $_J \rho_{out} =_J \rho_0\ _J X_{out}^{J G_5}$

Specific enthalpy $\qquad$ $_J dh_{out} =_J C_P\ _b T_J + \dfrac{_J c_{out}^2}{2}$

Mass flow increment $\qquad$ $_J dm_{out} =_J \rho_{out}\ A_J\ _J c_{out}\ dt$

Enthalpy increment $\qquad$ $_J dH_{out} =_J dh_{out}\ _J dm_{out}$

Air flow increment $$_Jd\Pi_{out} = _J\Pi \, _Jdm_{out}$$

For the end meshes, the required information for cases 1 to 4 is deduced from the boundary conditions of the flow which have been applied appropriately at the left or right hand edge of a mesh space. This means that the "hand" of the flow has to be taken into account when transferring the information from the solution resulting from the particular boundary condition for inflow or outflow with respect to the mesh space, as distinct from "inflow" or "outflow" from a cylinder, plenum, restricted pipe, or branch. However, the logic of the sign convention is quite straightforward in practice.

The acquisition of the numeric information regarding heating effects due to friction and heat transfer in any time step, δQ_f and δQ_h, have already been dealt with in Sec. 2.18.6. Should a catalyst or some similar internal heating device be present within the mesh space and in operation, during the time increment dt, then it is assumed that the quantity of heat emanating from it, δQ_{int}, can be determined and used as further numeric input to the analysis below. It should be pointed out that catalysts are commonplace within exhaust systems at this point in history, as a means of reducing exhaust emissions. However, a cooling device may be employed instead to give an internal heat transfer effect; and water injection would be one such example. In both examples, the chemical composition of the gas will also change and this requires a further extension to any analysis given below.

During the time step, the new system mass, $_am_J$, is derived from the continuity equation:

$$_a m_J = _b m_J + _Jdm_{in} - _Jdm_{out} \tag{2.18.44}$$

The First Law of Thermodynamics for the system consisting of the mesh space is:

heat transfer + energy in = change of system state + energy out + work done

$$\left(\delta Q_{int} + \delta Q_f + \delta Q_h\right)_J + _JdH_{in} = dU_J + _JdH_{out} + P_JdV_J \tag{2.18.45}$$

The work term is clearly zero. All of the terms except that for the change of system state are already known through the theory given above in this section. Expansion of this unknown term reveals:

$$dU_J = _a\left[_am_J\left(_au_J + \frac{_ac_J^2}{2}\right)\right]_J - _b\left[_bm_J\left(_bu_J + \frac{_bc_J^2}{2}\right)\right]_J \tag{2.18.46}$$

Although the velocities at either end of a mesh are almost identical, and the difference between the kinetic energy terms is small, it is unwise to exclude the system kinetic energy from the analysis. The system particle velocity is assumed to be the mean of that at either end of the mesh:

$$_a c_J = \frac{1}{2}\left(\frac{_J c_{in}^2}{2} + \frac{_J c_{out}^2}{2} \right)$$

Eq. 2.18.45 becomes:

$$dU_J = {_J}C_V(_a m_J \, _a T_J - _b m_J \, _b T_J) + \frac{1}{2}\left(_a m_J \, _a c_J^2 - _b m_J \, _b c_J^2 \right) \tag{2.18.47}$$

This can be solved directly for the system temperature, $_a T_J,$ after the time step.

The gas properties in the mesh space will have changed due to the mass transport across its boundaries. As with the case of mass and energy transport, direction is a vital consideration and the four cases presented in Fig. 2.23 are applicable to the discussion. For all engine calculations, the gases to be found within the ducting are either (i) products of combustion, or (ii) air, or (iii) a mixture of these two gases. However, only two will be formally tracked by the simulation, namely air and the combustion products, i.e., "exhaust gas." Unfortunately, we have the bad semantic habit of labeling the contents of an exhaust system as "exhaust gas." The semantic confusion arises because the four-stroke engineer subconsciously tends to consider the contents of the exhaust system as having precisely the same properties as the products of combustion, whereas the two-stroke designer is all-too-conscious that large amounts of air enter an exhaust system during a scavenge process [2.67]. Irrespective of the perspective, the more general case where a multiplicity of gases is present throughout the system must be handled with simplicity. After all, air and combustion products are each composed of a multiplicity of gases, and combustion products, particularly after a diesel combustion process, may well contain a significant amount of unburned air!

Throughout the ducting, any mixture of air and combustion products has a purity, Π, which is defined in Eq. 1.6.11 and is repeated here as:

$$\Pi = \frac{\text{mass of air}}{\text{total mass}} = \frac{\text{mass of air}}{\text{mass of air } + \text{ mass of exhaust gas}} \tag{2.18.48}$$

Hence, the reasoning for including the air flow increment in the four cases of mass and energy transport presented above and in Fig. 2.23.

The new purity, $_a\Pi_J$, in mesh space J is found by the mass average as follows:

$$_a\Pi_J = \frac{_bm_J\ _b\Pi_J + _Jd\Pi_{in} - _Jd\Pi_{out}}{_am_J} \tag{2.18.49}$$

Remembering that the phrase "exhaust gas" in the following statements really means "combustion products," and that the gas properties of air and "exhaust gas" are denoted by their respective gas constants and specific heat ratios as R_{air} and R_{exh} and γ_{air} and γ_{exh}, the new properties of the gas in the mesh space after the time step are:

Gas constant
$$_aR_J = _a\Pi_J R_{air} + (1-_a\Pi_J)R_{exh} \tag{2.18.50}$$

Specific heats ratio
$$_a\gamma_J = \frac{_aC_{P_J}}{_aC_{v_J}} = \frac{_a\Pi_J C_{P_{air}} + (1-_a\Pi_J)C_{P_{exh}}}{_a\Pi_J C_{V_{air}} + (1-_a\Pi_J)C_{V_{exh}}} \tag{2.18.51}$$

It should be noted that all of the gas properties employed in the theory must be indexed for their numerical values based on the gas composition and temperature at every step in time and at every location, using the theoretical approach given in Sec. 2.1.6.

From this point, it is possible, using the theory of Sec. 2.18.3, to establish the remaining properties in space J, in particular the reference acoustic velocity, density, and temperature. The average superposition pressure amplitude ratio, $_aX_J$, in the duct is derived using Eq. 2.18.1 with the updated values of the pressure amplitude ratios at either end of the mesh space. The connection for the new reference temperature, $_aT_0$, is given by Eq. 2.18.4 as:

$$_aT_0 = \frac{_aT_J}{_aX_J^2} \tag{2.18.52}$$

Consequently the other reference conditions are:

Reference acoustic velocity, $_aa_0$:

$$_aa_0 = \sqrt{_a\gamma_J\ _aR_J\ _aT_0} \tag{2.18.53}$$

Reference density, $_a\rho_0$:

$$_a\rho_0 = \frac{p_0}{_aR_J \, _aT_0} \tag{2.18.54}$$

All of the properties of the gas at the conclusion of a time step dt have now been established at all of the mesh boundaries and in all of the mesh volumes.

2.18.10 The Thermodynamics of Cylinders and Plenums during a Time Step

During a time step in calculation, the pipes of the engine being simulated are connected to cylinders and plenums. For simplicity, as in Sec. 2.18.8, an engine cylinder or a plenum will be referred to occasionally and collectively as a "box." In calculation terms, a plenum is but a cylinder of constant volume. During the time step, due to mass flow entering or leaving the box, the state conditions of the box will change. For example, during exhaust outflow from an engine cylinder, the pressure and the temperature fall and its mass is reduced. In a GPB simulation method proceeding in small time steps of 1° or 2° of crankshaft angle at some rotational speed, the situation is treated as quasi-steady flow for that period of time. To proceed to the next time step of the simulation, the new state conditions in all cylinders and plenums must be determined. You may recall from Chapter 1 that I stated that the whole point of the simulation process is to predict the mass and state conditions of the gas in the cylinder at the conclusion of the open cycle, as influenced by the pressure wave action in the ducting, so that a closed cycle computation may provide the requisite data of power, torque, fuel consumption, or emissions.

The application of the boundary conditions given in Secs. 2.16 and 2.17 provides all of the information on mass, energy, and air flow at the mesh boundaries adjacent to the cylinder or plenum. Fig. 2.25 shows the cylinder with state conditions of pressure, P_C, temperature, T_C, volume, V_C, mass, m_C, etc., at the commencement of a time step. The gas properties are defined by the purity, Π_C, a data value that leads directly to the gas constant, R_C, and specific heats ratio, γ_C, in the manner shown by Eqs. 2.18.50 and 2.18.51, and by applying the theory in Sec. 2.1.6 regarding gas properties.

It will be noted that the box has "inflow" and "outflow" apertures, which implies that there are only two apertures. The values of mass flow increment during the time step, shown as either dm_I and dm_E, are in fact the combined total of all of those ports or valves designated as being "inflow" or "outflow." The same reasoning applies to the energy and airflow terms, dH_I and $d\Pi_I$, or dH_E and $d\Pi_E$; they are the combined totals of the energy and airflow terms of the (perhaps) several intake or exhaust ducts at a cylinder. All of these terms are the direct equivalents—indeed those at the boundary edges of a pipe mesh adjacent to a box are identical to—those quoted as cases 1 to 4 in Sec. 2.18.9.

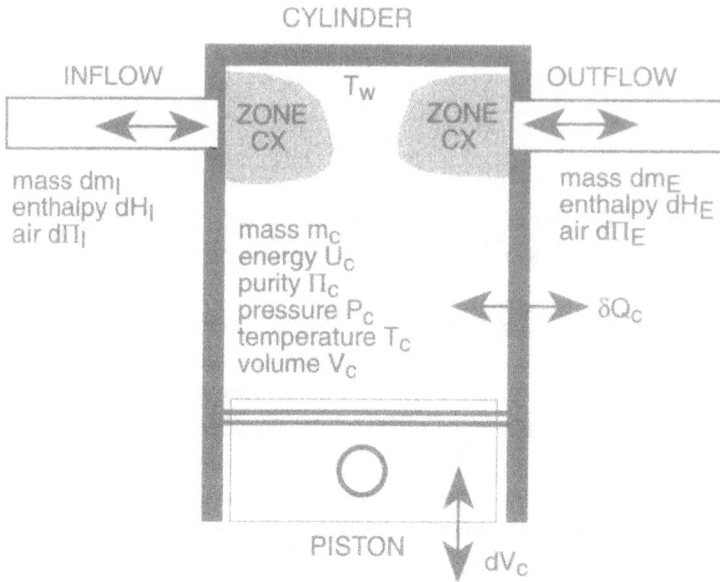

Fig. 2.25 The thermodynamics of open cycle flow through a cylinder.

It will be noted that the arrow at either "inflow" or "outflow" in Fig. 2.25 is bidirectional. In short, simply because a port is designated as inflow does not mean that the flow is always toward the cylinder. All ports and valves experience backflow at some point during the open cycle. The words "inflow" and "outflow" are, in the case of an engine cylinder, a convenient method of denoting ports and valves whose nominal job is to supply air into, or exhaust gas out of, the cylinder of an engine. When dealing with an intake plenum, or an exhaust silencer box, "inflow" and "outflow" become a directionality denoted at the whim of the modeler—but having made a decision on the matter, the ensuing sign convention must be adhered to rigidly. The sign convention, common in engineering thermodynamics, is that inflow is "positive" and that "outflow" is also "positive." This sign convention is employed not only in the theory below, but throughout this text. Backflow, by definition, is flow opposite to that which is decreed as positive and is then a negative quantity. The mesh computation must then reorient in sign terms the numeric values for the "right" and "left" hand ends of pipes during inflow and outflow boundary calculations as appropriate to those junctions defined as "inflow" or "outflow" at the cylinders or plenums. Although the computer software logic for this is trivial, care must be taken not to confuse the thermodynamic needs of the mesh spaces defined in Sec. 2.18.9 with those of the cylinder or plenum being examined here.

Heat transfer is defined as positive for heat added to a system. Work out is also defined as positive. Using this convention, the First Law of Thermodynamics is:

$$\text{heat transfer} + \text{energy in} = \text{change of system state} + \text{energy out} + \text{work done}$$

The entire computation at this stage makes the assumption that the previous application of the boundary conditions, using the theory of Secs. 2.16 and 2.17, has produced the correct values of the terms dm_I, dH_I, $d\Pi_I$, dm_E, dH_E, and $d\Pi_E$. To reinforce an important point that has been made before, during the application of the boundary conditions, in the case of a cylinder where the outflow is stratified, the local properties of zone CX, which are pressure, P_{cx}, temperature, T_{cx}, purity, Π_{cx}, etc., are those that replace the mean values of pressure, P_c, temperature, T_c, purity, Π_c, etc. In such a case, the solution of the continuity equation and the First Law of Thermodynamics is as accurate and as straightforward as it was in Sec. 2.18.9 in Eqs. 2.18.44 to 2.18.49. Further discussion on this topic is given elsewhere [2.32, 2.67]. I am not convinced that flow into, through, and out of the cylinder during the long valve overlap periods of the high-performance engine can be considered as a "perfect mixing process," using terminology that would be better understood by a two-stroke engine designer. I have this strong suspicion that a modicum of short-circuiting is involved in such circumstances, but it has never been proven. To comprehend these points, Chapter 3 in Ref. [2.67] should prove useful, for therein is described an experimental method that could be used to assess the extent, if any, of the short-circuiting of the intake flow through the open exhaust valve(s) and a theoretical method to adapt those experimental results into this theory.

The subscript notation for the properties and state conditions within the box after the time step dt is "C1," i.e., the new values are pressure, P_{C1}, temperature, T_{C1}, purity, Π_{C1}, etc.

The heat transfer, δQ_c, to or from the box in the time step, dt, is given by the local convection heat transfer coefficient, C_h, the total surface area, A_C, and the average wall temperature of the box, T_w.

$$\delta Q_C = C_h A_C (T_W - T_C) dt \qquad (2.18.55)$$

The value of the heat transfer coefficient, C_h, to be employed during the open cycle is the subject of much research, of which the work by Annand [2.58] is noteworthy. The approach by Annand is recommended for the acquisition of heat transfer coefficients for both the open and the closed cycle within the engine cylinder. For all engines, it should be noted that at some period during the closed cycle, an allowance must be made for the cooling of the cylinder charge due to the vaporization of the fuel. For spark ignition engines, it is normal to permit this to happen linearly from the trapping point to the onset of ignition. For compression ignition units it is conventional to consider that this occurs simultaneously with each packet of fuel being burned during a computational time step. The work of Woschni [2.60] has also provided significant contributions to this thermodynamic field. Expansion of Eq. 2.18.55 to provide separate heat transfer contributions by differing surfaces within the box, i.e., a piston crown or a cylinder head, is straightforward.

The continuity equation for the process during the time step dt is given by:

$$m_{C1} = m_C + dm_I - dm_E \qquad (2.18.56)$$

The First Law of Thermodynamics for the cylinder or plenum system becomes:

$$\delta Q_C + dH_I = dU_C + dH_E + \frac{P_C + P_{C1}}{2} dV_C \qquad (2.18.57)$$

The work term is clearly zero for a plenum of constant volume. All of the terms except that for the change of system state, dU_C, and the cylinder pressure, P_{C1}, are already known through the theory given above in this section. Expansion of one unknown term reveals:

$$dU_C = m_{C1} u_{C1} - m_C u_C \qquad (2.18.58)$$

This reduces to:

$$dU_C = m_{C1} C_{V_{C1}} T_{C1} - m_C C_{V_C} T_C \qquad (2.18.59)$$

where the value of the specific heat at constant volume is that appropriate to the properties of the gas within the cylinder at the beginning and end of the time step:

$$C_{V_C} = \frac{R_C}{\gamma_C - 1} \quad \text{and} \quad C_{V_{C1}} = \frac{R_{C1}}{\gamma_{C1} - 1}$$

Normally, as with the debate on the gas constant, R, below, the values of C_V should be those at the beginning and end of the time step. However, if the computation is conducted in just one time step, and the values of the enthalpies and kinetic energies entering and exiting the mesh system are computed using the "initial" values, then it is incorrect to update the specific heat at constant volume, or the gas constant, to find the new system temperature through Eq. 2.18.59. In single-step calculation mode, this becomes:

$$dU_C = C_{V_C} \left(m_{C1} T_{C1} - m_C T_C \right)$$

If the computation is repeated to update all of the "after" energy flows, then the original form of Eq. 2.18.59 is used; my own view is that I have never found this to improve the modeling accuracy, so I conduct the computation process in a single iteration. An example of a further exception is during a combustion process where the temperature changes during a given time step are so extreme that the gas properties must be indexed correctly using the theory of Sec. 2.1.6 and, if necessary, an iteration undertaken several times to acquire sufficient accuracy. In this case, the valves are closed because it is a closed cycle computation, so there are no entering or exiting energy flows to complicate an already complex issue. This area needs greater study, and because the closed cycle case in the cylinder holds a combustion process, this subject is debated throughout Chapter 4.

At the conclusion of the time step, the cylinder volume is V_{C1} caused by the piston movement, and this is:

$$V_{C1} = V_C + dV_C \qquad (2.18.60)$$

The mass of the cylinder, m_{C1}, given by Eq. 2.18.56, and the new cylinder pressure and temperature are related by the state equation:

$$P_{C1}V_{C1} = m_{C1}R_C T_{C1} \qquad (2.18.61)$$

Eq. 2.18.56 to 2.18.61 may be combined to produce a direct solution for T_{C1} for a cylinder or plenum as:

$$T_{C1} = \frac{\delta Q_C + dH_I - dH_E + m_C C_{V_C} T_C - \dfrac{P_C dV_C}{2}}{\left(m_C + dm_I - dm_E\right)\left(C_{V_C} + \dfrac{R_C dV_C}{2V_C}\right)} \qquad (2.18.62)$$

This can be solved directly for the system temperature, T_{C1}, after the time step, and with dV_C as zero in the event that any plenum or cylinder has no volume change. The cylinder pressure is found from Eq. 2.18.61.

The gas properties in the box will have changed due to the mass transport across its boundaries. For almost all engine calculations, the gases within the box are either exhaust gas or air. This situation will be discussed here because it is the norm, but the more general case where a multiplicity of gases is present throughout the system can be handled with equal simplicity. After all, air and exhaust gas are composed of a multiplicity of gases. This argument, with the same words, is precisely that mounted in the previous section Sec. 2.18.9 for flow through the mesh spaces.

The new purity, Π_{C1}, in the box is found simply as follows:

$$\Pi_{C1} = \frac{m_C \Pi_C + d\Pi_I - d\Pi_E}{m_{C1}} \qquad (2.18.63)$$

The new properties of the gas in the box after the time step are:

Gas constant $\qquad\qquad R_{C1} = \Pi_{C1}R_{air} + \left(1 - \Pi_{C1}\right)R_{exh} \qquad (2.18.64)$

Specific heats ratio $\quad \gamma_{Cl} = \dfrac{C_{P_{Cl}}}{C_{V_{Cl}}} = \dfrac{\Pi_{Cl}C_{P_{air}} + (1-\Pi_{Cl})C_{P_{exh}}}{\Pi_{Cl}C_{V_{air}} + (1-\Pi_{Cl})C_{V_{exh}}}$ (2.18.65)

From this point, it is possible, using the theory given in Secs. 2.18.3 and 2.1.6, to establish the remaining properties in the box, in particular the reference acoustic velocity, density, and temperature. The connection for the new reference temperature, T_0, is given by the isentropic relationship between pressure and temperature as:

$$\frac{T_{Cl}}{T_0} = \left(\frac{p_{Cl}}{p_0}\right)^{\frac{\gamma_{Cl}}{\gamma_{Cl}-1}}$$ (2.18.66)

Consequently the other reference conditions to be employed at the commencement of the next time step are:

Reference acoustic velocity $\quad a_0 = \sqrt{\gamma_{Cl}R_{Cl}T_0}$ (2.18.67)

Reference density $\quad \rho_0 = \dfrac{p_0}{R_{Cl}T_0}$ (2.18.68)

All of the properties of the gas at the conclusion of a time step dt have now been established at all of the mesh boundaries, in all of the mesh volumes, and in all cylinders and plenums of the engine being modeled. It becomes possible to proceed to the next time step and continue with the GPB simulation method, replacing all of the "old" values with the "new" ones acquired during the progress described in the entirety of this section. A new time step may now commence with all data stores refreshed with the updated numeric information.

Information on the outcome of the modeling process of the open cycle at the cylinder and in all of the ducts may need to be collected. This is discussed in the next section.

2.18.11 Air Flow, Work, and Heat Transfer during the Modeling Process

The modeling of an engine, or of any device, that inhales and exhales in an unsteady fashion, is oriented toward the determination of the effect of that unsteady process on many facets of the operation. For example, the engine designer will wish to know the totality of the airflow into each cylinder of the engine as well as the quantification of it with respect to time or crankshaft angle, i.e., to determine the delivery ratio of the engine as well as the extent, or lack, of backflow at certain periods of crankshaft rotation which may deteriorate the overall value. Thus, during a calculation, summations of quantities will be made by the modeler to aid the design process. To illustrate this, the example of delivery ratio will be used, in the first instance.

Air Flow into an Engine

The air flow into an engine is the summation of all of the increments of air flow at each time step at any point in the intake tract. Any mesh point can be selected for this purpose, because the net mass of air flow should be identical at every mesh point over a long time period, such as a complete engine cycle. Consider the general case first. A parameter, B, is required to be assessed for its mean value, $\overline{B}$, over a period of time t, at a particular mesh location, J. The time period starts at t_1 and ends at t_2. In addition, for an engine running at engine speed N, this can be carried out over, and is more informative during, specific periods of crankshaft angle, θ, ranging from θ_1 to θ_2. This is effected by:

$$\overline{B} = \frac{\sum\limits_{t=t1}^{t=t2} B\,dt}{\sum\limits_{t=t1}^{t=t2} dt} = \frac{\sum\limits_{\theta=\theta1}^{\theta=\theta2} B\,d\theta}{\sum\limits_{\theta=\theta1}^{\theta=\theta2} d\theta} \qquad (2.18.69)$$

For an engine, the relationship between crankshaft angle, in degree units, and time is:

$$\theta = \frac{60}{360N} = \frac{1}{6N} \qquad (2.18.70)$$

For the specific case of air flow into the engine, the modeler will select a mesh, J, for the assessment point, and the shrewd modeler will select mesh J as being the value beside the intake valves or ports of the engine. The total mass of air flow, m_a, passing this point will then be:

$$m_a = \frac{\sum\limits_{\theta=0}^{\theta=720} \dot{m}_J \Pi_J\,d\theta}{\sum\limits_{\theta=0}^{\theta=720} d\theta} \qquad (2.18.71)$$

The crankshaft period selected will be noted as 720 degrees. This is correct for a four-stroke engine, for that is its total cyclic period. On the other hand, for a two-stroke engine the cyclic period is 360 degrees and the modeler should perform the summation appropriately. If the engine is a multi-cylinder unit of n cylinders, then the accumulation process will be repeated at all of the intake ports of the unit. As in Sec. 1.6.2, to transfer that data value to a delivery ratio, DR, value, the conventional theory is used where ρ_{ref} is the reference density for the particular industry standard employed, and V_{sv} is the swept volume of one of the cylinders of the engine:

$$\text{Delivery ratio} \qquad DR = \frac{\displaystyle\sum_{cylinder=1}^{cylinder=n} m_a}{n\rho_{ref}V_{sv}} \qquad (2.18.72)$$

The cyclic air flow rate for a two-stroke engine is called the scavenge ratio, SR, where the reference volume term is the entire cylinder volume. This is given by:

$$\text{Scavenge ratio} \qquad SR = \frac{\displaystyle\sum_{cylinder=1}^{cylinder=n} m_a}{n\rho_{ref}\left(V_{cv} + V_{sv}\right)} \qquad (2.18.73)$$

The above example is only for one of the many terms that are required for design assessment during modeling. However, the procedure is identical for the others, many of which are defined in Sec. 1.6. Further examples of such terms will be given here.

Work Done during an Engine Cycle

The work output of an engine is that caused by the cylinder pressure, p_C, acting on the piston(s) of the power unit. The in-cylinder work is known as the "indicated work" and it is often reduced to the pseudo-dimensionless value called the "indicated mean effective pressure." Let us assume that there are n cylinders, each with an identical bore area, A_{bo}, with swept volume, V_{sv}. The piston movement in any one cylinder at each time step is a variable, dx, as is the time step, dt, and the volume change, dV.

The work done, δW, at each time step in each cylinder is,

$$\delta W = p_C A_{bo} dx = p_C dV \qquad (2.18.74)$$

The total work done, W_C, for that cylinder over a cycle is given by the summation of all such terms for that period. It should be remembered that the thermodynamic cycle period, θ_C, is 720 degrees for a four-stroke engine, and 360 degrees for a two-stroke engine. Thus, the accumulated work done over the cycle is:

$$W_C = \frac{\displaystyle\sum_{\theta=0}^{\theta=\theta_C} \delta W d\theta}{\displaystyle\sum_{\theta=0}^{\theta=\theta_C} d\theta} \qquad (2.18.75)$$

The total work done by the engine, W_E, over a cycle is given by,

$$W_E = \sum_{cylinder=1}^{cylinder=n} W_C \qquad (2.18.76)$$

The indicated power output of the engine calculated by the simulation is the rate of execution of this work at N rpm, and is:

Four-stroke engine $\qquad \dot{W}_I = W_E \dfrac{N}{120} \qquad (2.18.77)$

Two-stroke engine $\qquad \dot{W}_I = W_E \dfrac{N}{60} \qquad (2.18.78)$

The indicated mean effective pressure, imep, for the engine is the pressure that would act on the pistons throughout the thermodynamic cycle and produce the same work output, thus:

$$imep = \frac{W_E}{nV_{sv}} \qquad (2.18.79)$$

It is quite clear that the imep values in each cylinder can be assessed separately.

The brake values, i.e., those that would be measured on a "brake" or dynamometer, are any or all of the above values of indicated performance multiplied by the mechanical efficiency, η_m.

Other work and heat transfer related terms, such as pumping mean effective pressure during the exhaust and intake strokes of a four-stroke engine, or pumping mean effective pressure for the induction into the crankcase of a two-stroke engine, or heat loss during the open cycle of the engine can be determined throughout the GPB modeling process, either cycle by cycle or cumulatively over many cycles, in exactly the same fashion as for imep. Similarly, the designer may wish to know, and relate to measured terms, the mean pressures and temperatures at significant locations throughout the engine; all such mean values are found using a methodology similar to that given above.

2.18.12 The Modeling of Engines using the GPB Finite System Method

Correlations of experiment and this theory have been reported in the literature of several international conferences and meetings. The references list these publications [2.30, 2.31, 2.32, 2.33, 2.34, 2.41, 2.42, 2.59 and 3.21]. The closed cycle modeling is discussed in much greater detail in Chapter 4, as is the acquisition of discharge coefficients in Chapter 3. The use of the GPB simulation software as a design tool is contained within Chapters 5 to 7 of this book.

2.19 The Correlation of the GPB Finite System Simulation with Experiments

In design engineering, a theoretical simulation process that has not been checked for accuracy against relevant experiments is, depending on its intended purpose, at best potentially misleading and at worst inapplicable as a design tool. In the technology allied to the simulation of unsteady gas flow, many experiments have been carried out by the researchers involved. Virtually every reference in the literature cited below carries evidence, relevant or irrelevant, of experimentation designed to test the validity of the theories presented by their author(s).

I was closely associated with a series of experiments, reported by Kirkpatrick et al. [2.41, 2.65, 2.66], designed specifically to test the validity of the theories of unsteady gas flow, in particular those presented here, and to compare and contrast the GPB finite system simulation with both experiments and other simulation methods such as Riemann characteristics, Lax-Wendroff, Harten-Lax-Leer, etc. The experimental apparatus is quite unique and is detailed fully by Kirkpatrick [2.41]. It will be described here sufficiently well so that the presentation of the experimental test results may be fully understood. Although the main purpose is to determine the extent of the accuracy of the GPB simulation method, the test method and the experimental results illustrate many of the contentions in the theory presented above.

2.19.1 The QUB SP (Single Pulse) Unsteady Gas Flow Experimental Apparatus

Most experimenters and modelers in unsteady gas dynamics have correlated measured pressure-time diagrams in the ducts of engines, firing or motored, against their theoretical contentions. Because all unsteady gas flow within the ducts of engines is in a state of superposition, this makes the process of correlation very difficult indeed. It is almost impossible to tell which pressure wave is traveling in which direction. Although fast-response pressure transducers are still the best experimental tool that the theoretician possesses, the simple truth is that they are totally directionally insensitive. That much is manifestly clear in just about every numeric example quoted up to this point in the text. Worse, the correlation of mass flow is in an even more parlous state when working with engines, either motored or firing. Although the cylinder pressure may be recorded accurately, a density record in the same place is not possible because a temperature, or purity, or density transducer with a sufficiently fast response has yet to be invented. Thus, while the experimenter may infer the mass of trapped charge, or as a matter of even greater necessity the mass of trapped air charge, in the cylinder from the overall engine air consumption and the cylinder pressure transducer record, the blunt truth is that he is "whistling in the wind." To those few readers who are my contemporaries in this technology, be assured that I readily admit to being as guilty of perfidy in my technical publications as you are. Perhaps it was this "guilt complex" that led to the design of the QUB SP (single pulse) apparatus.

The criteria for the design of the apparatus were straightforward:

(i)   The device should be based on the assumption that a fast response pressure transducer is the only accurate experimental tool readily available.

(ii)   The pipe(s) attached to the device must be sufficiently long as to permit visibility of a pressure wave traveling left or right without undergoing superposition in the plane of the transducer while recording some particular phenomenon of interest.

(iii)   The cylinder of the device must be capable of having the mass and purity of its contents recorded with absolute accuracy.

(iv)   The cylinder of the device must be capable of containing any gas desired, and at a wide variety of state conditions, prior to the commencement of the experiment, which could be the simulation of either an exhaust or an induction process.

(v)   The pipes attached to the cylinder must be capable of containing any gas desired, over a wide variety of state conditions, prior to the commencement of the experiment, which could be the simulation of either an exhaust or an induction process.

(vi)   The pipes attached to the device must be capable of holding any of the discontinuities known in engine technology, i.e., diffusers; cones; bends; branches; sudden expansions, contractions and restrictions; throttles, carburetors, catalysts, silencers, air filters, poppet valves, etc.

These design criteria translated into the single pulse device shown in Fig. 2.26. The cylinder is a rigid, cast iron container of 912 cm^3 volume, and can contain gas up to pressure of 10 bar and a temperature of 500°C. The gas can be heated electrically (H). The cylinder has valves, V, that permit the charging of gas into the cylinder at sub-atmospheric or supra-atmospheric pressure conditions. The port, P, at the cylinder has a hole of 25 mm diameter which mates precisely with an aluminum exhaust pipe of 25 mm diameter. The valve mechanism, S, is a flat, polished nickel-steel plate with a hole of 25 mm which mates perfectly with the port and pipe at maximum opening. It is actuated by a pneumatic impact cylinder and its movement is recorded by an attached comb with a 2 mm pitch that is sensed by an infrared source and integral photo-detector. Upon impact, the valve slider opens the port from a "perfect" sealing of the cylinder, gas flows from (or into, in an induction process) the cylinder and

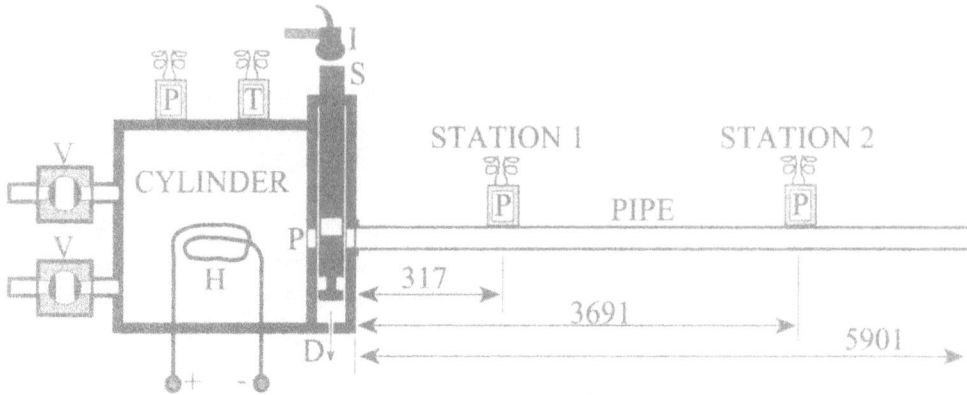

Fig. 2.26 QUB SP apparatus with a straight pipe attached.

seals it again upon the conclusion of its passing. A damper, D, decelerates the valve to rest after the port has already been closed. An exhaust pulse generated in this manner is typical in time and amplitude of, say, an engine at 3000 rpm. A typical port opening lasts for 0.008 seconds.

The cylinder gas properties of purity, pressure, and temperature are known at commencement and upon conclusion of an event; in the case of purity, by chemical analysis through a valve, V, if necessary. The pressure is recorded by a fast-response pressure transducer. The temperature is known at commencement and at conclusion, without concern for the time response of that transducer, so that the absolute determination of cylinder mass can be conducted accurately.

The coefficients of discharge, C_d, of the cylinder port, under wide variations of cylinder-to-pipe pressure ratio giving rise to inflow or outflow and valve opening as port-pipe area ratio, are determined under steady flow conditions with air. They are presented in Figs. 2.27(a) and 2.27(b), for the outflow and inflow directions, and are a classic picture of the variation of C_d with respect to such parameters. Their measurement is necessary for the accurate correlation of experiment with theory, as would be the case for any engine simulation to be accurate, as inspection of Eq. 2.16.4 will indicate. In this context, the discussion in Chapter 3 requires careful study, for the traditional methods employed for the reduction of the measured data for the coefficients of discharge, C_d, have been found to be inadequate for use in conjunction with a theoretical engine simulation.

It is acknowledged that the initial design and development work on the QUB SP rig was carried out by R.K. McMullan and finalized by S.J. Kirkpatrick. All of the pressure diagram data presented in Figs. 2.28-2.53 were recorded by S.J. Kirkpatrick, with the exception of those in Figs. 2.55 to 2.57, which were measured by D.O. Mackey, and the discharge coefficients in Fig. 2.27, which were taken by L. Foley on an updated version of the SP apparatus. The above-named individuals carried out this work as part of their research work for a PhD at The Queen's University of Belfast.

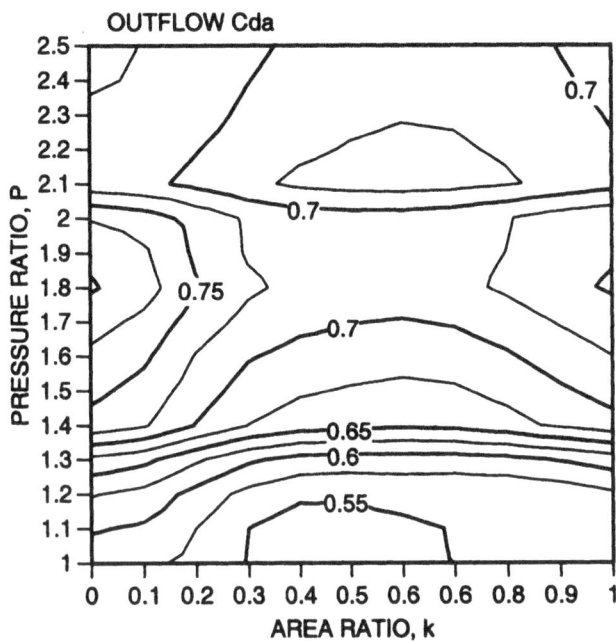

Fig. 2.27(a) Coefficients of discharge for the QUB SP apparatus for outflow.

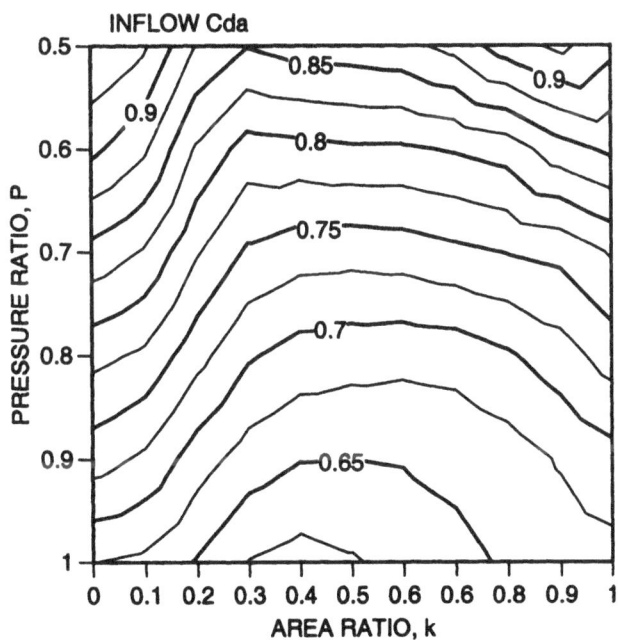

Fig. 2.27(b) Coefficients of discharge for the QUB SP apparatus for inflow.

2.19.2 A Straight Parallel Pipe Attached to the QUB SP Apparatus

The experiments simulate both exhaust and intake processes with a straight pipe attached to the cylinder. Fig. 2.26 shows a straight aluminium pipe, with a length of 5.9 m and an internal diameter of 25 mm, attached to the port and ending at the atmosphere. The pipe has a relatively smooth bore typical of the quality of ducting common in an engine design. There are pressure transducers attached to the pipe at stations 1 and 2 at the length locations indicated.

(i) The Outflow Process Producing Compression Pressure Waves in the Exhaust Pipe

The cylinder was filled with air and the initial cylinder pressure and temperature were 1.5 bar and 293 K. Figs. 2.28 to 2.30 show the pressure-time records in the cylinder and at stations 1 and 2. The result of computations using the GPB modeling method are shown on the same figures. The results are so close that it is sometimes difficult to distinguish between theory and experiment, but where differences can be discerned they are indicated on the figure.

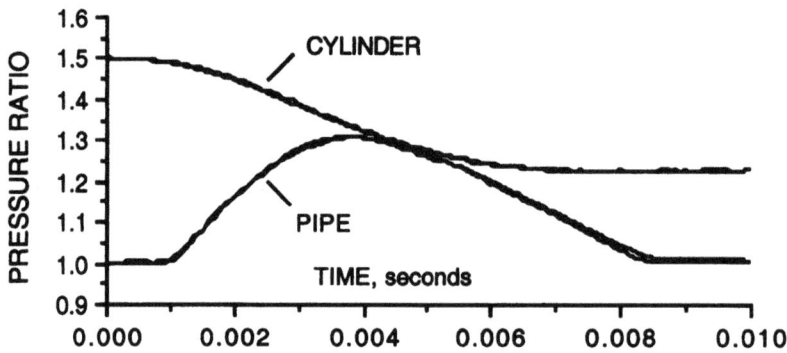

Fig. 2.28 Measured and calculated cylinder and pipe pressures.

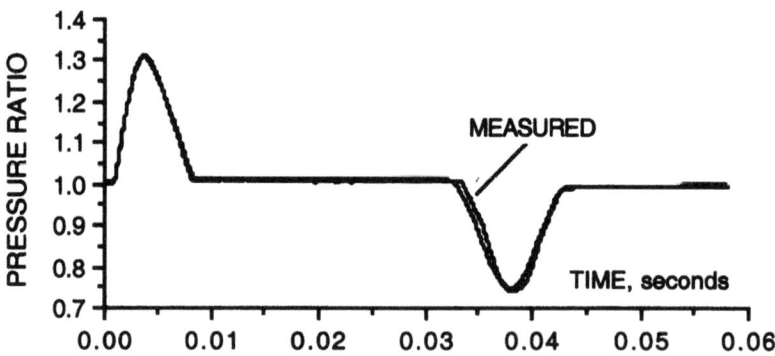

Fig. 2.29 Measured and calculated pressures at station 1.

292

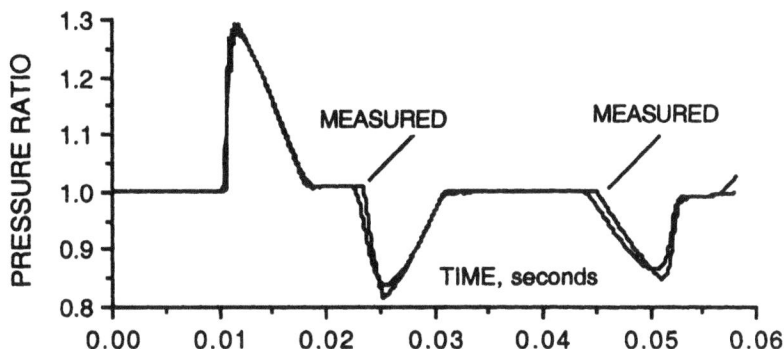

Fig. 2.30 Measured and calculated pressures at station 2.

The correlation of measured and calculated mass is virtually exact. The criteria used is the ratio of the final to the initial cylinder mass, and in this case both the calculated and the measured value was 0.866. As can be seen from these figures, no wave action impinged on the port to influence the cylinder emptying process, so this is more of a tribute to the accuracy of the cylinder-to-pipe boundary conditions of Sec. 2.16, rather than any comment on the GPB finite system modeling of the wave action in the pipe system!

The main event is the creation of the exhaust pulse as a wave of compression, the peak of which is passing station 1 at 0.00384 seconds. It reflects at the open end as an expansion wave, and the peak returns to station 1 at 0.0382 seconds. The basic theory of the motion of finite amplitude waves is discussed in Sec. 2.1.4, and that of reflections of compression waves at a plain open end, in Sec. 2.8.1. The reference acoustic velocity in air at 20°C is 343 m/s. This is estimated to be the average propagation velocity, on the simplistic grounds that the expansion wave travels as much below sonic velocity as the compression wave is supersonic above it. Thus, the expected return point in time, by a very simple computation is:

$$\text{time, t} = 0.00384 + \frac{2(5901 - 317)}{1000 \times 343} = 0.0364 \text{ seconds}$$

This approximate calculation can be seen to give an answer that is close to reality. In Sec. 6.4.5, however, this same simplistic approach is found to be totally inadequate because of the very same wave superposition effects that the SP apparatus is designed to avoid.

Figs. 2.28 to 2.30 the individual waves can be seen clearly. The creation of the exhaust pulse is accurate as can be seen in Figs. 2.28 and 2.29. Later it passes station 2 after 0.01 seconds. Steep-fronting of the compression wave has occurred as discussed in Sec. 2.1.5. In Fig. 2.30, the steep-fronted compression wave reflects at the plain open end as an expansion wave. The profile has changed little by the time it passes station 2 going leftward, but upon returning to station 2, it has steepened at the rear of the expansion wave, again as discussed in Sec. 2.1.5. It will be seen that a shock did not develop on the waves at any time, even though the long pipe runs provided the waves with the space and time to do so.

The size and scale of the diagrams make it somewhat difficult to observe, but it still can be seen that the pressure at 0.01 seconds in Fig. 2.28 and at 0.02 seconds in Fig. 2.29 is slightly above atmospheric. It can be seen that the pressure at 0.04 seconds in Fig. 2.30 is slightly below atmospheric. These effects arise due to the main compression wave, and the expansion wave reflection of it, sending their continual reflections due to friction in the direction opposite to their propagation. Because this is an important topic, a separate graph is presented in Fig. 2.58(a) which expands the time scale for the time period between 0.007 and 0.01 seconds. The close correlation between the measured and calculated pressures, and the residual waves attributable to the friction reflections after 0.0085 seconds, is now observable by zooming in on this time period. As for friction causing deterioration of the peak of the pressure wave, as it passed station 1 it had a measured peak pressure ratio of 1.308 (calculated at 1.307), and when it passed station 2 this had reduced to 1.290 (calculated at 1.273). Sec. 2.3 presents a theoretical discussion of this subject.

It can be seen that the GPB modeling method provides a very accurate pressure-time history of the recorded events for the motion of compression waves and of the exhaust process from a cylinder.

(ii) The Inflow Process Producing Expansion Pressure Waves in the Intake Pipe

The cylinder was filled with air and the initial cylinder pressure and temperature were 0.8 bar and 293 K. Figs. 2.31 to 2.33 show the pressure-time records in the cylinder and at stations 1 and 2. The result of computations using the GPB modeling method are shown on the same figures. The results are sufficiently close as to warrant the description of being good, but where differences can be discerned they are indicated on the figure.

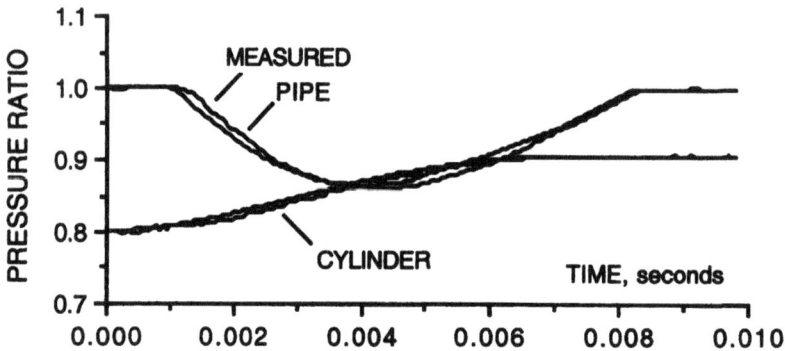

Fig. 2.31 Measured and calculated cylinder and pipe pressures.

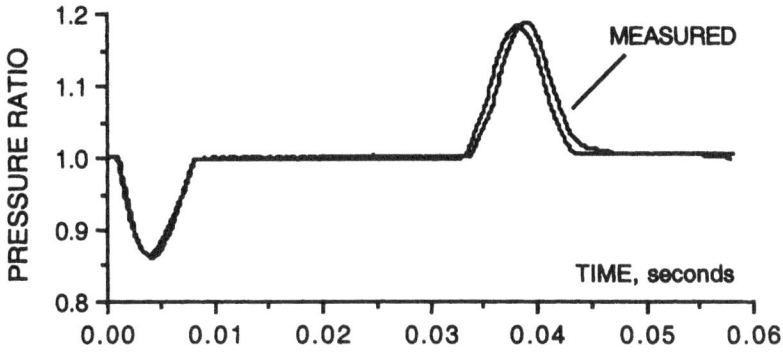

Fig. 2.32 Measured and calculated pressures at station 1.

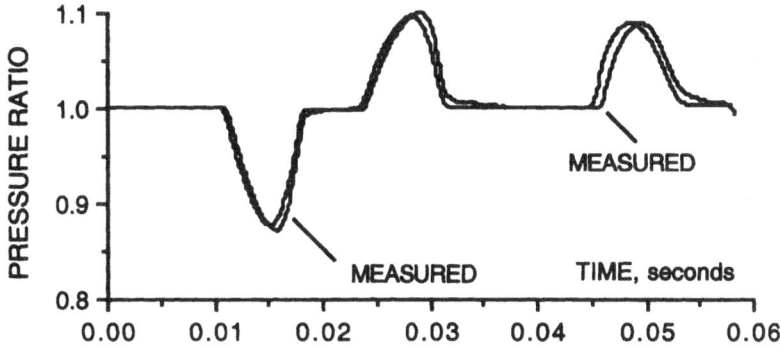

Fig. 2.33 Measured and calculated pressures at station 2.

The correlation of measured and calculated mass is also good. The criteria used is the ratio of the final to the initial cylinder mass. The calculated and the measured values are 1.308 and 1.311 respectively, which represents an error of 0.2%. As with the exhaust process in (i) above, no wave action other than friction reflections impinged on the port to influence the cylinder emptying process.

The main event is the creation of the intake pulse as a wave of expansion, the peak of which is passing station 1 at 0.004 seconds. It reflects at the open end as a compression wave and the peak returns to station 1 at 0.0383 seconds. The basic theory of the motion of finite amplitude waves is discussed in Sec. 2.1.4 and that of reflections of expansion waves at a plain open end, in Sec. 2.8.3. The reference acoustic velocity in air at 20°C is 343 m/s. This is estimated to be the average propagation velocity on the simplistic grounds that the expansion wave travels as much below sonic velocity as the compression wave is supersonic above it. Thus, the expected return point in time, by a very simple computation is:

$$\text{time, t} = 0.004 + \frac{2(5901-317)}{1000 \times 343} = 0.0366 \text{ seconds}$$

This approximate calculation can be seen to give an answer that is similar to the experimental value. You can see that it is the same as for the exhaust process in the same pipe length.

In Figs. 2.31 to 2.33, the individual waves can be seen clearly. The creation of the induction pulse is accurate as can be seen in Figs. 2.31 and 2.32. Later it passes station 2 after 0.01 seconds. Steepening of the tail of the expansion wave has occurred as discussed in Sec. 2.1.5. It will be seen that a shock did not develop on the wave at any time.

The effect of friction is presented in Fig. 2.58(b) which expands the time scale for the time period between 0.007 and 0.01 seconds. The close correlation between the measured and calculated pressures, and the residual waves attributable to the friction reflections after 0.0085 seconds causing the sub-atmospheric pressure on the traces, are now observable by zooming in on this time period. As for friction causing deterioration of the peak of the pressure wave, as it passed station 1 it had a measured peak pressure ratio of 0.864 (calculated at 0.859), and when it passed station 2 this had increased to 0.871 (calculated at 0.876). Sec. 2.3 contains a theoretical discussion of this subject.

It can be seen that the GPB modeling method provides a very accurate pressure-time history of the recorded events for the motion of expansion waves and of the induction process into a cylinder.

2.19.3 A Sudden Expansion Attached to the QUB SP Apparatus

The experiment simulates an exhaust process with a straight pipe incorporating a sudden expansion attached to the cylinder. Fig. 2.34 shows a straight aluminium pipe that is 3.394 m long, with an internal diameter of 25 mm, attached to the port, followed by a 2.655 m length of pipe with an internal diameter of 80 mm ending at the atmosphere. There are pressure transducers attached to the pipe at stations 1, 2, and 3 at the length locations indicated. The basic theory of pressure wave reflections at sudden expansions and contractions is given in Secs. 2.10 and 2.12.

The cylinder was filled with air and the initial cylinder pressure and temperature were 1.5 bar and 293 K. Figs. 2.35 to 2.37 show the measured pressure-time records in the cylinder and at stations 1, 2, and 3. The result of the computations using the GPB modeling method are shown on the same figures, and the correlation is very good. Where differences can be discerned between measurement and computation they are indicated on that figure.

In Fig. 2.35, the basic action of reflection at a sudden expansion is observed. The exhaust pulse passes the pressure transducer at station 1 at 0.004 seconds and the sudden expansion in pipe area sends an expansion wave reflection back to arrive at 0.024 seconds. It is echoed off

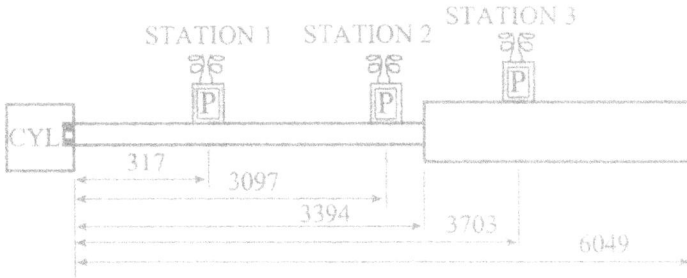

Fig. 2.34 QUB SP rig with a sudden expansion attached.

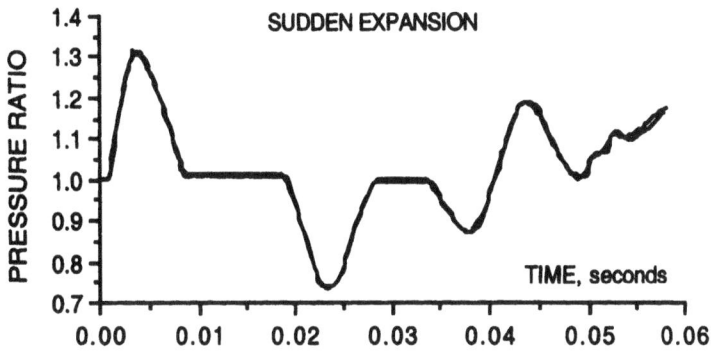

Fig. 2.35 Measured and calculated pressures at station 1.

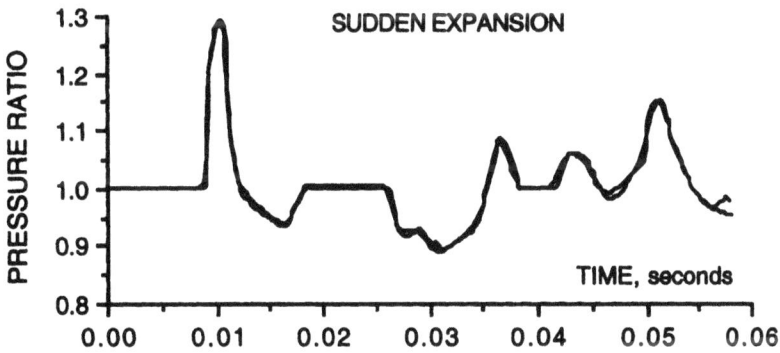

Fig. 2.36 Measured and calculated pressures at station 2.

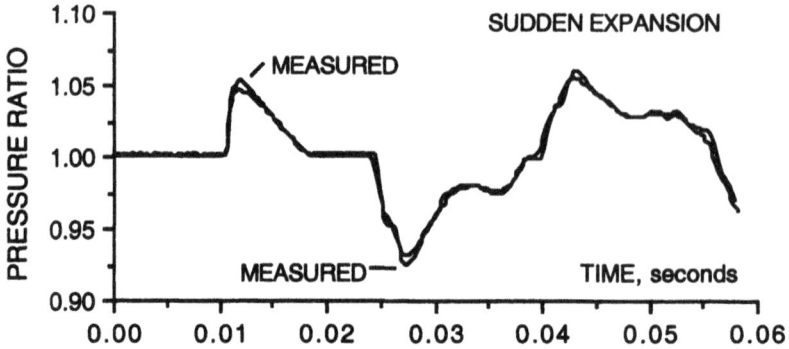

Fig. 2.37 Measured and calculated pressures at station 3.

the closed end, is reflected again at the sudden expansion, but this time as a compression wave which returns to station 1 at 0.044 seconds preceded by an expansion wave at 0.038 seconds. Where does this expansion wave come from?

In Fig. 2.37, the pressure record at station 3 at 0.012 seconds shows the onward transmission of the residue of the original exhaust pulse having traversed the sudden expansion in pipe area. In short, the original exhaust pulse commenced its journey with a pressure ratio of 1.3, which is reduced to 1.05 in the 80 mm pipe when passing station 2. This onward transmitted wave reflects off the open end as an expansion wave, returns to the sudden area change, which it now sees as a sudden contraction. The onward transmission leftward of that reflection process is seen in Fig. 2.36 to proceed past the transducer at station 2 at 0.03 seconds and arrive at station 1 at 0.038 seconds, thereby answering the question posed in the previous paragraph.

In Fig. 2.36 the pressure transducer is sufficiently near the sudden change of pipe area that all pressure records are of a superposition process. This is evident between 0.01 and 0.02 seconds, where the rightward propagating exhaust pulse is partially superposed on the leftward travel of its own reflection from the sudden expansion in pipe area. The measured superposition processes are followed closely in amplitude and phase by the theory.

It can be observed that the GPB finite system modeling by computer provides a very accurate pressure-time history of the recorded events for the reflection of compression pressure waves at sudden increases in pipe area. There is also evidence, because the second-order reflections are a mixture of expansion and compression waves, which see the sudden area change as both an expansion and a contraction in pipe area, that the theory is generally applicable to all such boundary conditions.

2.19.4 A Sudden Contraction Attached to the QUB SP Apparatus

The experiment simulates an exhaust process with a straight pipe incorporating a sudden contraction attached to the cylinder. Fig. 2.38 shows a short, straight aluminium pipe that is 108 mm long, with an internal diameter of 25 mm, attached to the port, followed by a 2.346 m length of pipe with an internal diameter of 80 mm; the final length to the open end to the atmosphere is a 2.21 m length of pipe with an internal diameter of 25 mm. This type of pipe is to be found as a diffusing silencer element as discussed in Chapter 7, or as the exhaust mid-section of a sports automobile engine to assist the attainment of the required higher-performance specification. There are pressure transducers attached to the pipe at stations 1, 2, and 3 at the length locations indicated. The basic theory of pressure wave reflections at sudden expansions and contractions is given in Secs. 2.10 and 2.12.

The cylinder was filled with air and the initial cylinder pressure and temperature were 1.5 bar and 293 K. Figs. 2.39 to 2.41 show the measured pressure-time records in the cylinder and at stations 1, 2, and 3. The result of computations using the GPB modeling method are shown on the same figures, and the correlation is very good. Where differences can be observed between measurement and computation they are indicated on that figure.

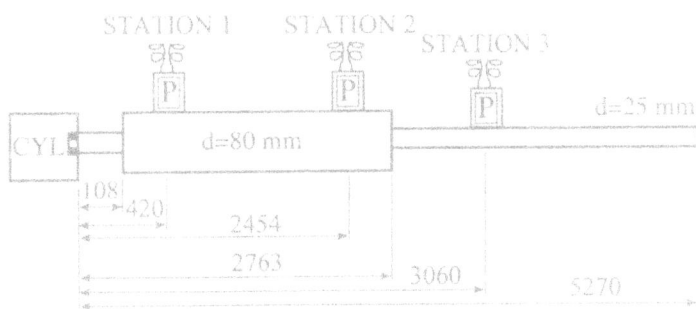

Fig. 2.38 QUB SP rig with a sudden contraction attached.

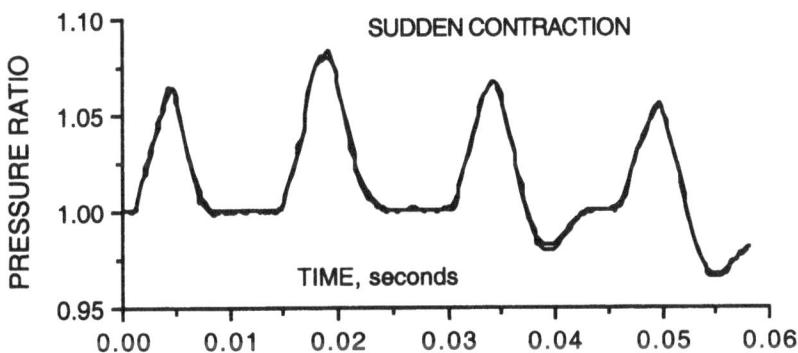

Fig. 2.39 Measured and calculated pressures at station 1.

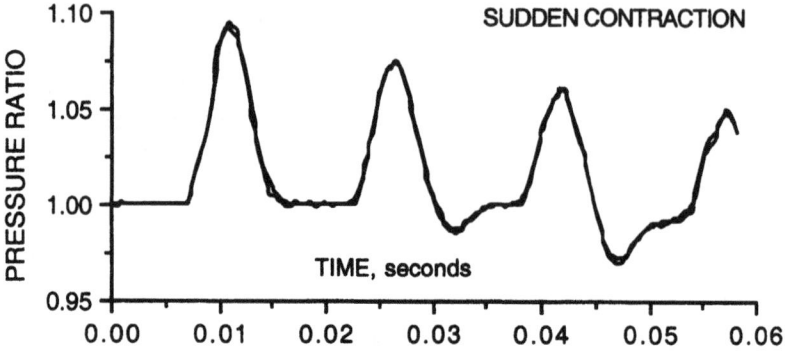

Fig. 2.40 *Measured and calculated pressures at station 2.*

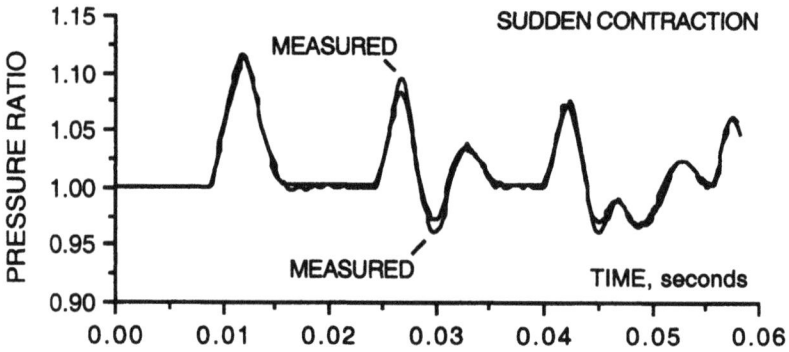

Fig. 2.41 *Measured and calculated pressures at station 3.*

The short 108 mm length means that the pressure wave encounters a sudden expansion before proceeding to the point of conduction of this particular test, i.e., the sudden contraction posed to the exhaust pressure wave. In the previous section, this was shown to be accurate, and so it is no surprise to see in Fig. 2.39 that the measured and computed pressure waves at station 1 are in close correlation at 0.05 seconds. The amplitude of the exhaust pressure wave passing station 1 has a pressure ratio of 1.065.

The sudden contraction sends a compression wave reflection back from it and can be seen in a superposition condition at station 1 at 0.018 seconds, and at station 2 earlier at 0.011 seconds. The clean and non-superposed onward transmission of the exhaust pulse can be found passing station 3 at 0.012 seconds with a pressure ratio of 1.12. This is a higher amplitude than the original exhaust pulse, of 1.06 atm, but the wave proceeding to the outlet has been contracted from a 80 mm diameter pipe into a 25 mm tail pipe.

It can be observed that the GPB finite system modeling gives a very accurate representation of the measured events for the reflection of compression pressure waves at sudden decreases in pipe area.

2.19.5 A Divergent Tapered Pipe Attached to the QUB SP Apparatus

The experiment simulates an exhaust process with a straight pipe incorporating a divergent taper attached to the cylinder. Fig. 2.42 shows a straight aluminium pipe that is 3.406 m long, with an internal diameter of 25 mm, attached to the port, followed by a 195 mm length of steeply tapered pipe at a 12.8° included angle to a diameter of 68 mm; the final length to the open end to the atmosphere is a 2.667 m length of pipe with an internal diameter of 68 mm. This form of steep taper is very commonly found within the ducting of ic engines, particularly in the exhaust diffuser sections of highly tuned engines. There are pressure transducers attached to the pipe at stations 1, 2, and 3 at the length locations indicated. The basic theory of pressure wave reflections in tapered pipes is given in Sec. 2.15.

The cylinder was filled with air and the initial cylinder pressure and temperature were 1.5 bar and 293 K. Figs. 2.43 to 2.45 show the measured pressure-time records in the cylinder and at stations 1, 2, and 3. The result of computations using the GPB modeling method are shown on the same figures, and the correlation is very good. Where differences can be observed between measurement and computation they are indicated on that figure.

The tapered pipe acts as an expansion to the area of the pipe system and sends an expansion wave reflection back from it. It is observed arriving back at station 1 at 0.024 seconds in Fig. 2.43, albeit in a superposition condition from the nearby closed end.

Actually, because it is a short, steeply tapered pipe, there is a close resemblance between the behavior of this pipe and the sudden contraction discussed in Sec. 2.19.3 and illustrated in Figs. 2.34 to 2.37. The cylinder release conditions were identical in both cases, so the exhaust pulses that arrived at the discontinuity in areas were the same. Consequently there are great similarities between Fig. 2.35 and 2.43 for the pressure transducer at station 1, and between

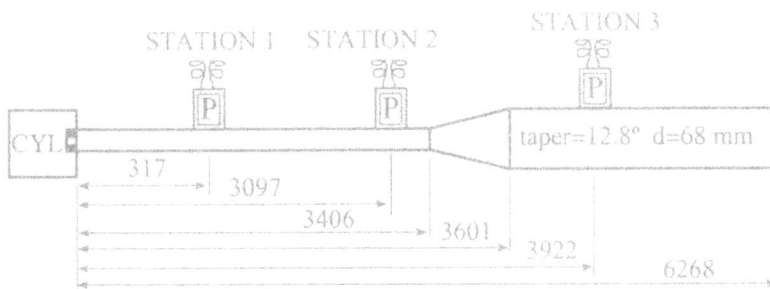

Fig. 2.42 QUB SP rig with a divergent taper attached.

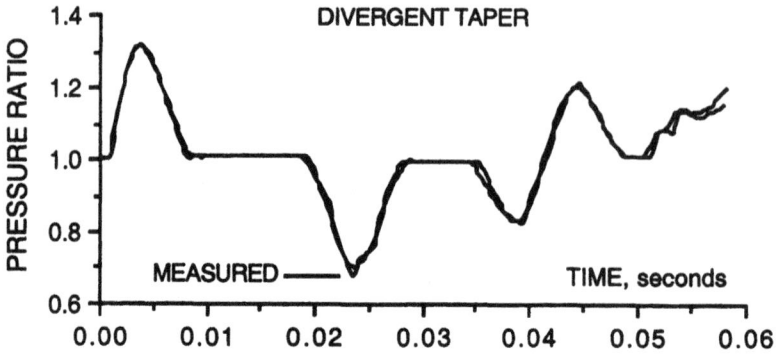

Fig. 2.43 Measured and calculated pressures at station 1.

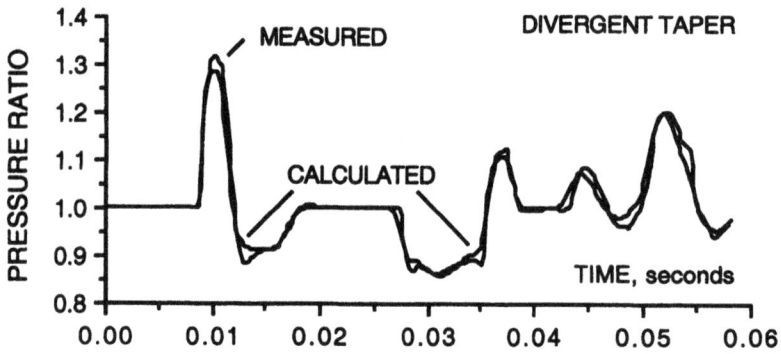

Fig. 2.44 Measured and calculated pressures at station 2.

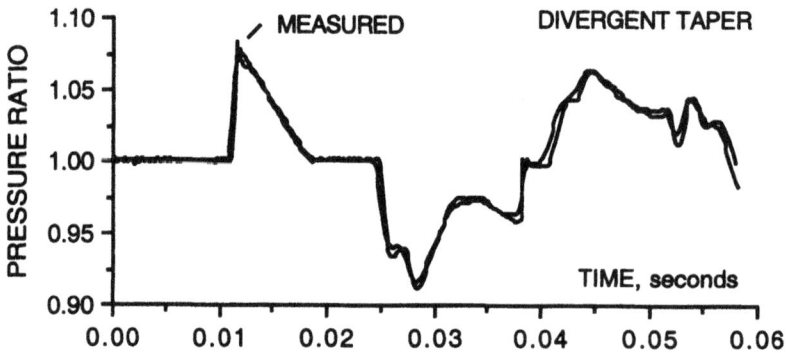

Fig. 2.45 Measured and calculated pressures at station 3.

Figs. 2.36 and 2.44 for the pressure transducer at station 2. The records for the tapered pipe look slightly more "peaky" in certain places. However, that effect is seen more strongly in Fig. 2.45, where the onward transmission of the residual of the exhaust pulse is observed to have developed a shock front as it passes the pressure transducer at station 3. In other words, the tapered pipe has contributed to, and exaggerated, the normal distortion process of a compression wave profile. It is also interesting to compare the amplitudes of the transmitted and reflected waves, as the tapered pipe should be more efficient at this than a sudden expansion. The peak of the suction wave from the tapered pipe undergoing superposition at 0.024 seconds at station 1 is 0.68 atm; for the sudden expansion it is 0.74 atm or with less strength as an expansion wave reflection. The peak of the transmitted wave from the tapered pipe passing station 3 undisturbed at 0.011 seconds is 1.08 atm; for the sudden expansion it is 1.05 atm, which means that an exhaust pulse of reduced strength is delivered further down the pipe system.

It can be observed that the GPB finite system modeling gives a very accurate representation of the measured events for the reflection of compression pressure waves at a steeply tapered pipe segment within ducting.

2.19.6 A Convergent Tapered Pipe Attached to the QUB SP Apparatus

The experiment simulates an exhaust process with a straight pipe incorporating a convergent taper attached to the cylinder. Fig. 2.46 shows a straight aluminium pipe that is 108 mm long, with an internal diameter of 25 mm, attached to the port, followed by a 2.667 m length of 68 mm parallel section pipe, then a steeply tapered pipe at 12.8° included angle convergent to 25 mm diameter over a 195 mm length; the final length of pipe to the open end to the atmosphere is 2.511 m long with an internal diameter of 25 mm. This form of steep taper is very commonly found within the ducting of ic engines, particularly for the rear cone sections of expansion chambers of highly tuned two-stroke engines. There are pressure transducers attached to the pipe at stations 1, 2, and 3 at the length locations indicated. The basic theory of pressure wave reflections in tapered pipes is given in Sec. 2.15.

The cylinder was filled with air and the initial cylinder pressure and temperature were 1.5 bar and 293 K. In Figs. 2.47 to 2.49 show the measured pressure-time records in the cylinder and at stations 1, 2, and 3. The result of computations using the GPB modeling method are shown on the same figures, and the correlation is very good. Where differences can be observed between measurement and computation they are indicated on that figure.

The converging tapered pipe acts as a contraction to the area of the pipe system and sends a compression wave reflection back from an incident compression wave; or an expansion wave reflection back from an incident expansion wave, if that happened to be the case in point. It is observed arriving back at station 1 at 0.02 seconds in Fig. 2.47, albeit in a superposition condition from the nearby contraction.

Actually, because it is a short, steeply tapered pipe, there is a close resemblance between the behavior of this pipe and the sudden contraction discussed in Sec. 2.19.4 and illustrated in Figs. 2.38 to 2.41. The cylinder release conditions were identical in both cases, so the exhaust

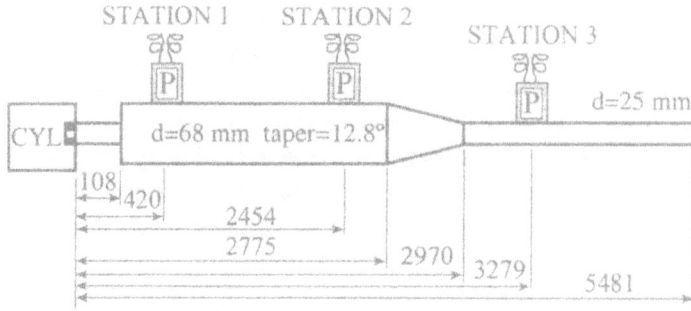

Fig. 2.46 *QUB SP rig with a convergent cone attached.*

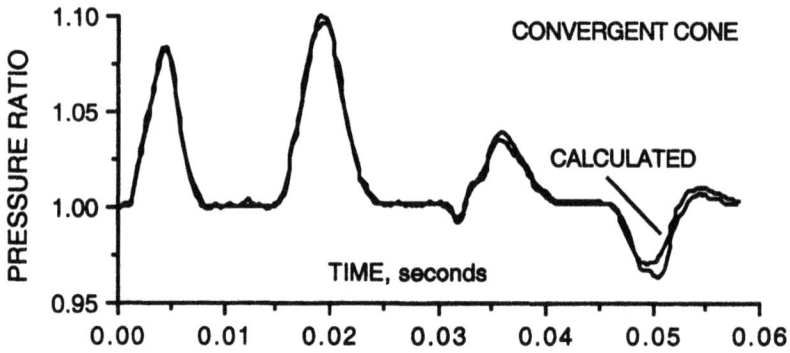

Fig. 2.47 *Measured and calculated pressures at station 1.*

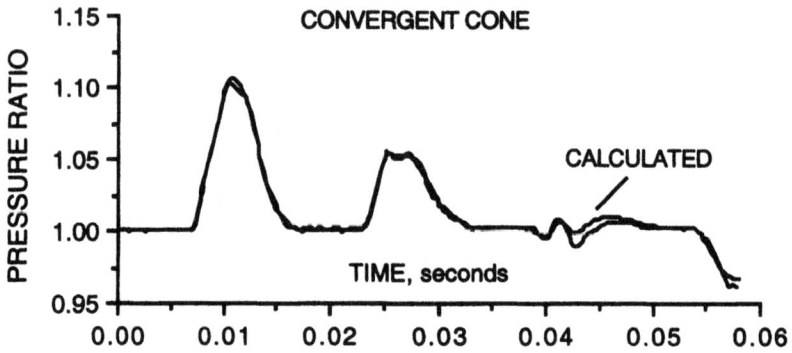

Fig. 2.48 *Measured and calculated pressures at station 2.*

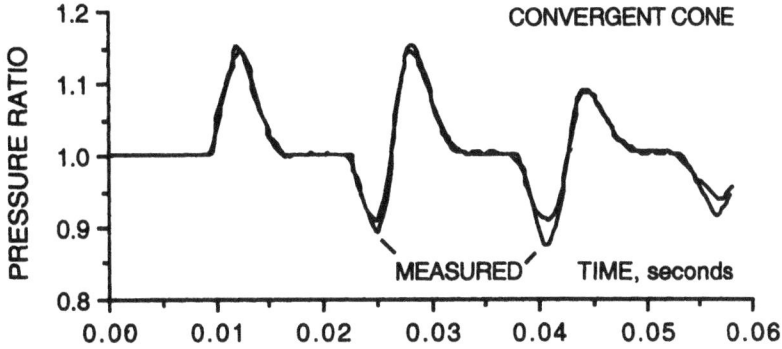

Fig. 2.49 Measured and calculated pressures at station 3.

pulses that arrived at the discontinuity in areas were the same. Consequently there are great similarities between Fig. 2.39 and 2.47 for the pressure transducer at station 1, and between Figs. 2.40 and 2.48 for the pressure transducer at station 2. This close comparison also extends to the pressure transducer at station 3, in the tapered cone version in Fig. 2.49, and in Fig. 2.41 for the sudden expansion. This was not the case in the previous section, when it was observed that the tapered diffuser pipe had distorted the profile of the exhaust pulse into a shock front as it reached the station 3 pressure transducer. It is germane to this discussion, and it is fully covered in the theoretical analysis in Secs. 2.10, 2.12 and 2.15, that a process in a diffuser is non-isentropic, whereas that in a nozzle is isentropic. In Fig. 2.45, shock formation is clearly a manifestation of non-isentropic behavior, whereas the undistorted profile emanating from the nozzle in Fig. 2.49 is obviously more efficient. Even the profile in Fig. 2.37 of the transmitted exhaust pulse onward from the sudden expansion exhibits elements of steep front formation.

It is also interesting to compare the amplitudes of the transmitted and reflected waves, as the nozzle should be more efficient at this than a sudden contraction. The peak of the compression wave from the tapered pipe undergoing superposition at 0.02 seconds at station 1 is 1.105 atm; for the sudden contraction it is weaker at 1.08 atm. The peak of the transmitted wave from the converging tapered pipe passing station 3 undisturbed at 0.012 seconds is 1.155 atm; for the sudden contraction it is 1.12 atm. Hence, an exhaust pulse of reduced strength is delivered onward down the pipe system by the tapered nozzle.

One final word on the above sections, which illustrates many of the theoretical considerations from Secs. 2.10, 2.12 and 2.15, is that loose labels such as "cones," "nozzles," "diffusers," etc. applied to area changes are only relevant when taken in the context of the direction of the particle velocity at that instant of the discussion. In short, what is a "nozzle" for a particle traveling rightward at one instant in time is a "diffuser" for another traveling leftward in another time frame.

It can be observed that the GPB finite system modeling gives a very accurate representation of the measured events for the reflection of compression pressure waves at a steeply tapered pipe segment within ducting. Taking heed of the warning in the previous sentence, it should be noted that the word "nozzle" is not mentioned in this last statement.

2.19.7 A Longer Divergent Tapered Pipe Attached to the QUB SP Apparatus

The experiment simulates an exhaust process with a straight pipe attached to the cylinder and incorporating a long divergent taper attached at the end of that straight pipe. Fig. 2.50 shows a straight aluminium pipe that is 3.417 m long, with an internal diameter of 25 mm, attached to the port, followed by a 600 mm length of tapered pipe at 8° included angle to 109 mm diameter at the open end to the atmosphere. This form of steep taper is very commonly found within the ducting of ic engines, particularly for the diffuser outlet sections of highly tuned four-stroke engines. These are often referred to as "megaphones," for obvious reasons. There is a pressure transducer attached to the pipe at the length location indicated. The basic theory of pressure wave reflections in tapered pipes is presented in Sec. 2.15.

The cylinder was filled with air and the initial cylinder pressure and temperature was 2.0 bar and 293 K. Figs. 2.51 to 2.53 show the measured pressure-time records at the pressure transducer location; this is the same on each graph. The megaphone acts as an expansion to the area of the pipe system and sends a strong expansion wave reflection back from it. This is chased back along the pipe system by the reflection at the atmosphere of the residue of the exhaust pulse which makes it to the open end. It returns as an expansion wave, but now strengthening as it nozzles its way back to the smaller area of the parallel pipe leading back to the cylinder. It is observed arriving back at the pressure transducer at 0.021 seconds in Figs. 2.5 to 2.53, in a virtually undisturbed condition, apart from the residual friction waves. The original exhaust pulse has a pressure ratio of 1.55 and the suction reflection has a peak of 0.63 atm. The suction reflection has steepened at its tail, a feature already observed in regard to the action of tapered pipes in Sec. 2.19.5. It will be seen from the discussion below that the superposition of oppositely moving expansion and compression waves gives very high particle velocities within the pipe and diffuser system, in excess of a Mach number of 0.7.

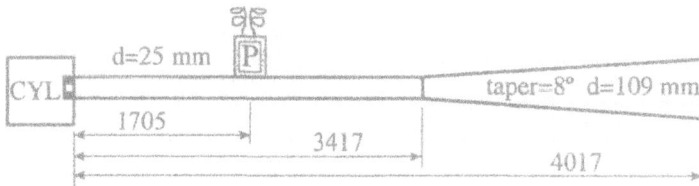

Fig. 2.50 QUB SP rig with a megaphone exhaust attached.

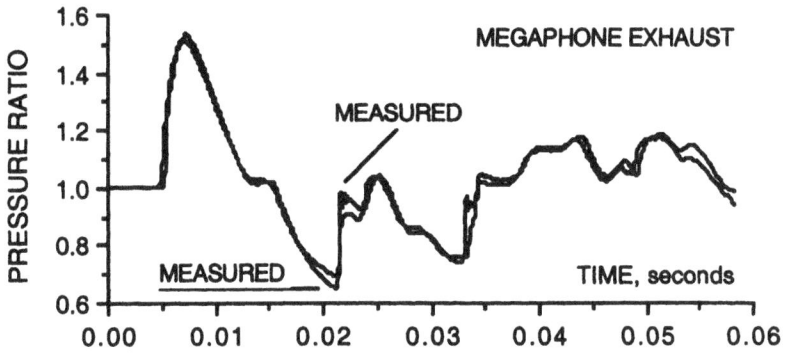

Fig. 2.51 Measured and calculated pressures with Mach 0.5 criterion.

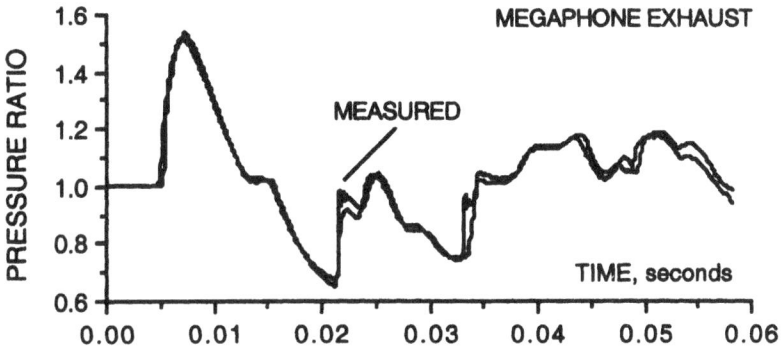

Fig. 2.52 Measured and calculated pressures with Mach 0.6 criterion.

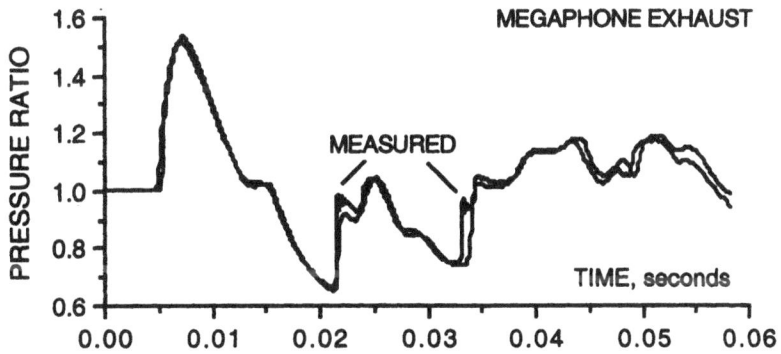

Fig. 2.53 Measured and calculated pressures with Mach 0.7 criterion.

Consider the strong suction wave evident in the several figures at 0.02 seconds. If it arrives at the exhaust valves or ports of an engine, particularly a four-stroke engine during the long valve overlap period of a high-performance unit, it can induce a considerable flow of air through the combustion chamber, removing the residue of the combustion products and initiating a high volumetric efficiency. The earlier discussion in Sec. 2.8.1 alludes to this potential for the enhanced cyclic charging of the engine with air.

The result of the computations using the GPB modeling method are shown on the same figures with the theoretical criterion for flow separation from the walls of a diffuser, as discussed in Sec. 2.15.1, differing in each of those three figures. It will be recalled from the discussion in that section, and by examining the criteria declared in Eq. 2.15.5, that gas particle flow separation from the walls of a diffuser will induce deterioration of the amplitude of the reflection of a compression wave as it traverses a diffuser. The taper of 8° included angle employed here would be considered in steady gas flow to be sufficiently steep as to give particle flow separation from the walls. The theory used to produce the computations in Figs. 2.51 to 2.53, was programmed to record the gas particle velocity at every mesh within the diffuser section. The statement made in Eq. 2.15.3 was implemented at every mesh at every time step, except that the computational switch was set at a Mach number of 0.5 when computing the theory presented in Fig. 2.51; at a Mach number of 0.6 for the theory presented in Fig. 2.52; and at a Mach number of 0.7 for the theory presented in Fig. 2.53. It will be seen that the criterion presented in Eq. 2.15.5 provides the accuracy required. It is also clear that any computational method that cannot accommodate such a fluid mechanic modification of its thermodynamics will inevitably provide considerable inaccuracy. Total reliance on the momentum equation alone for this calculation gives a reflected wave amplitude at the pressure transducer of 0.5 atm. It is also clear that flow separation from the walls occurs only at very high Mach numbers. Because the criterion of Eq. 2.15.5 is employed for the computation of the theory in Figs. 2.43 to 2.45, where the taper angle is at 12.8° included, it is a reasonable assumption that wall taper angle in unsteady gas flow is not the most critical factor.

The differences between measurement and computation are indicated on each figure. It can be observed that the GPB finite system modeling gives a very accurate representation of the measured events for the reflection of compression pressure waves at a tapered pipe ending at the atmosphere.

2.19.8 A Pipe with a Gas Discontinuity Attached to the QUB SP Apparatus

The experiment simulates an exhaust process with a straight pipe attached to the cylinder. Fig. 2.54 shows a straight aluminium pipe that is 5.913 m long, with an internal diameter of 25 mm diameter attached to the port and ending at a closed end with no exit to the atmosphere. There are pressure transducers attached to the pipe at stations 1, 2, and 3 at the length locations indicated. However, at 3.401 m from the cylinder port, there is a sliding valve, S, with a 25 mm circular aperture which, when inserted, seals off both segments of the exhaust pipe. Prior to the commencement of the experiment, the cylinder was filled with air and the initial cylinder pressure and temperature were 1.5 bar and 293 K. The segment of the pipe between the cylinder and valve S is filled with air at a pressure and temperature of 1.0 bar and 293 K, and the segment between valve S and the closed end is filled with carbon dioxide at the same state conditions. The experiment is conducted by retracting valve S to the fully open position and impacting the valve at the cylinder port open at the same instant, i.e., pneumatic impact cylinder I in Fig. 2.26 opens cylinder port P, and then closes it in the manner described in Sec. 2.19.1. An exhaust pulse is propagated into the air in the first segment of the pipe and encounters the carbon dioxide contained in the second segment, before echoing off the closed end. In the quiescent conditions for the instant of time between retracting valve S and sending a pressure wave to arrive at that position some 0.015 seconds later, it is not anticipated that either the carbon dioxide or the air will have migrated far from their initial positions, if at all. This is a classic experiment examining the boundary conditions for pressure wave reflections at discontinuities in gas properties in engine ducting, as described in Sec. 2.5. The gas properties of carbon dioxide are significantly different from air as the ratio of specific heats, γ, and the gas constant, R, are 1.28 and 189 J/kgK, respectively. The reference densities in the two segments of the pipe this experiment, and employed in the GPB modeling process, are then given by:

Air reference density $\qquad _{air}\rho_0 = \dfrac{p_0}{R_{air}T_0} = \dfrac{101325}{287 \times 293} = 1.205 \text{ kg/m}^3$

CO_2 reference density $\qquad _{CO_2}\rho_0 = \dfrac{p_0}{R_{CO_2}T_0} = \dfrac{101325}{189 \times 293} = 1.830 \text{ kg/m}^3$

It will be observed that the reference density of carbon dioxide is significantly higher than air by some 52%. The reference acoustic velocities, a_0, which profoundly affect the propagation and the particle velocities are also very different as:

a_0 in air $\qquad _{air}a_0 = \sqrt{\gamma_{air}R_{air}T_0} = \sqrt{1.4 \times 287 \times 293} = 343.1 \text{ m/s}$

a_0 in CO_2 $\qquad _{CO_2}a_0 = \sqrt{\gamma_{CO_2}R_{CO_2}T_0} = \sqrt{1.28 \times 189 \times 293} = 266.2 \text{ m/s}$

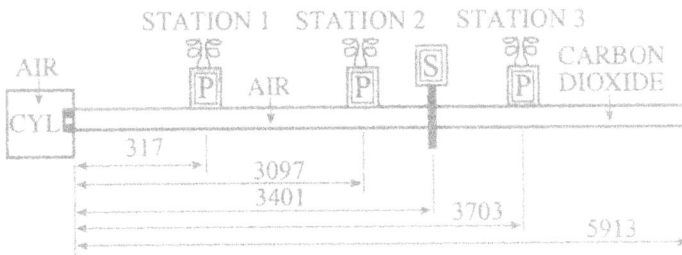

Fig. 2.54 QUB SP rig with pipe containing a gas discontinuity.

The reference acoustic velocity for air is some 28.9 % higher than that in carbon dioxide. At equal values of pressure wave amplitude this provides significant alterations to the motion of the pressure wave in each segment of the gas in the pipe. Consider the theory of Sec. 2.1.4 for a compression wave with a pressure ratio of 1.3 at the above state conditions in a 25 mm diameter pipe, i.e., a pipe area, A, of 0.000491 m^2.

Air Properties

For air, the results for pressure amplitude ratio, X, particle velocity, c, propagation velocity, α, density, ρ, and mass flow rate, $\dot{m}$, would be:

X
$$X = P^{G17} = 1.3^{\frac{1}{7}} = 1.0382$$

c
$$c = G_5 a_0 (X - 1) = 5 \times 343.1 \times 0.0382 = 65.5 \text{ m/s}$$

α
$$\alpha = a_0 \left(G_6 X - G_5 \right) = 343.1 \times (6 \times 1.0382 - 5) = 421.7 \text{ m/s}$$

ρ
$$\rho = \rho_0 X^{G5} = 1.205 \times 1.0382^5 = 1.453 \text{ kg/m}^3$$

$\dot{m}$
$$\dot{m} = \rho A c = 1.453 \times 0.000491 \times 65.5 = 0.0467 \text{ kg/s}$$

Carbon Dioxide Properties

For carbon dioxide, the results for pressure amplitude ratio, X, particle velocity, c, propagation velocity, α, density, ρ, and mass flow rate, $\dot{m}$, would be:

X
$$X = P^{G17} = 1.3^{0.1094} = 1.0291$$

c
$$c = G_5 a_0 (X - 1) = 7.143 \times 266.2 \times 0.0291 = 55.3 \text{ m/s}$$

α
$$\alpha = a_0 \left(G_6 X - G_5 \right) = 266.2 \times (8.143 \times 1.0291 - 7.143) = 329.3 \text{ m/s}$$

ρ
$$\rho = \rho_0 X^{G5} = 1.830 \times 1.0291^{7.143} = 2.246 \text{ kg/m}^3$$

$\dot{m}$
$$\dot{m} = \rho A c = 2.246 \times 0.000491 \times 55.3 = 0.061 \text{ kg/s}$$

The GPB finite system simulation method has no difficulty in modeling a pipe system to include these considerable disparities in gas properties and the ensuing behavior in terms of wave propagation. The computation is designed to include the mixing and smearing of the

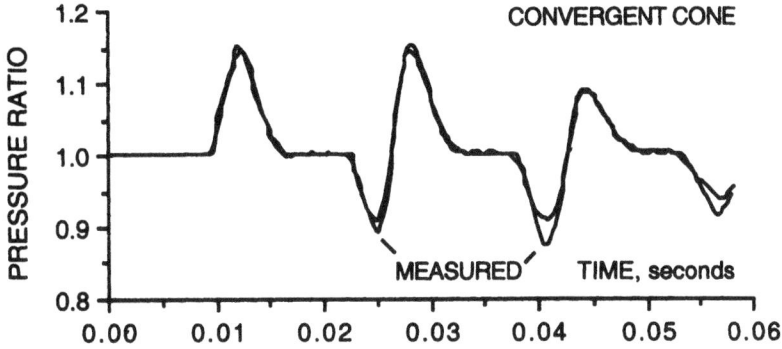

Fig. 2.49 Measured and calculated pressures at station 3.

pulses that arrived at the discontinuity in areas were the same. Consequently there are great similarities between Fig. 2.39 and 2.47 for the pressure transducer at station 1, and between Figs. 2.40 and 2.48 for the pressure transducer at station 2. This close comparison also extends to the pressure transducer at station 3, in the tapered cone version in Fig. 2.49, and in Fig. 2.41 for the sudden expansion. This was not the case in the previous section, when it was observed that the tapered diffuser pipe had distorted the profile of the exhaust pulse into a shock front as it reached the station 3 pressure transducer. It is germane to this discussion, and it is fully covered in the theoretical analysis in Secs. 2.10, 2.12 and 2.15, that a process in a diffuser is non-isentropic, whereas that in a nozzle is isentropic. In Fig. 2.45, shock formation is clearly a manifestation of non-isentropic behavior, whereas the undistorted profile emanating from the nozzle in Fig. 2.49 is obviously more efficient. Even the profile in Fig. 2.37 of the transmitted exhaust pulse onward from the sudden expansion exhibits elements of steep front formation.

It is also interesting to compare the amplitudes of the transmitted and reflected waves, as the nozzle should be more efficient at this than a sudden contraction. The peak of the compression wave from the tapered pipe undergoing superposition at 0.02 seconds at station 1 is 1.105 atm; for the sudden contraction it is weaker at 1.08 atm. The peak of the transmitted wave from the converging tapered pipe passing station 3 undisturbed at 0.012 seconds is 1.155 atm; for the sudden contraction it is 1.12 atm. Hence, an exhaust pulse of reduced strength is delivered onward down the pipe system by the tapered nozzle.

One final word on the above sections, which illustrates many of the theoretical considerations from Secs. 2.10, 2.12 and 2.15, is that loose labels such as "cones," "nozzles," "diffusers," etc. applied to area changes are only relevant when taken in the context of the direction of the particle velocity at that instant of the discussion. In short, what is a "nozzle" for a particle traveling rightward at one instant in time is a "diffuser" for another traveling leftward in another time frame.

It can be observed that the GPB finite system modeling gives a very accurate representation of the measured events for the reflection of compression pressure waves at a steeply tapered pipe segment within ducting. Taking heed of the warning in the previous sentence, it should be noted that the word "nozzle" is not mentioned in this last statement.

2.19.7 A Longer Divergent Tapered Pipe Attached to the QUB SP Apparatus

The experiment simulates an exhaust process with a straight pipe attached to the cylinder and incorporating a long divergent taper attached at the end of that straight pipe. Fig. 2.50 shows a straight aluminium pipe that is 3.417 m long, with an internal diameter of 25 mm, attached to the port, followed by a 600 mm length of tapered pipe at 8° included angle to 109 mm diameter at the open end to the atmosphere. This form of steep taper is very commonly found within the ducting of ic engines, particularly for the diffuser outlet sections of highly tuned four-stroke engines. These are often referred to as "megaphones," for obvious reasons. There is a pressure transducer attached to the pipe at the length location indicated. The basic theory of pressure wave reflections in tapered pipes is presented in Sec. 2.15.

The cylinder was filled with air and the initial cylinder pressure and temperature was 2.0 bar and 293 K. Figs. 2.51 to 2.53 show the measured pressure-time records at the pressure transducer location; this is the same on each graph. The megaphone acts as an expansion to the area of the pipe system and sends a strong expansion wave reflection back from it. This is chased back along the pipe system by the reflection at the atmosphere of the residue of the exhaust pulse which makes it to the open end. It returns as an expansion wave, but now strengthening as it nozzles its way back to the smaller area of the parallel pipe leading back to the cylinder. It is observed arriving back at the pressure transducer at 0.021 seconds in Figs. 2.5 to 2.53, in a virtually undisturbed condition, apart from the residual friction waves. The original exhaust pulse has a pressure ratio of 1.55 and the suction reflection has a peak of 0.63 atm. The suction reflection has steepened at its tail, a feature already observed in regard to the action of tapered pipes in Sec. 2.19.5. It will be seen from the discussion below that the superposition of oppositely moving expansion and compression waves gives very high particle velocities within the pipe and diffuser system, in excess of a Mach number of 0.7.

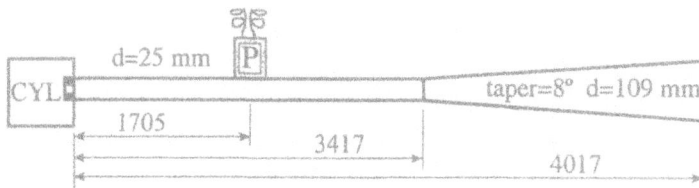

Fig. 2.50 QUB SP rig with a megaphone exhaust attached.

gases at the interface between mesh systems which have different properties at every mesh within the computation, and the reflections of the pressure waves at the inter-mesh boundaries as a function of those differing gas properties. The theory for this is presented in Sec. 2.18.

Figs. 2.55 to 2.57 show the measured pressure-time records in the cylinder and at stations 1, 2, and 3. The result of the computations using the GPB modeling method are shown on the same figures. The pressure traces are so close together that it is difficult to distinguish between theory and experiment, but where differences can be discerned they are indicated on that figure.

The first point of interest is on Fig. 2.55, where there is a "bump" of pressure at 0.023 seconds. This is the echo of the initial exhaust pulse, having propagated through the air in the first pipe segment, off the more dense carbon dioxide, which commences at position S in the second pipe segment, returning to station 1. At that point it is observed as a superposition process bouncing off the closed end at the port.

The second point of interest is the overall accuracy of the computation. The phasing of the pressure waves is very accurate, considering the disparity of the propagation velocities of waves of equal pressure ratio of 1.3, illustrated above, i.e., some 28% faster in air. Thus, the phasing error if the computation had been carried out with air only would have been very considerable. In Fig. 2.55, the return time, peak-to-peak of the exhaust pulse, passing station 1 to return to station 1, is 0.03 seconds. Very approximately and ignoring interference due to superposition as in Eqs. 2.2.8 and 2.2.9, in a pipe filled with air only, that return time would have been:

Return time, t, for air only $\quad t = \dfrac{2\left(5.913 - 0.317\right)}{421.7} = 0.0265 \text{ s}$

Because the peak of the exhaust pulse passed station 1 at 0.004 seconds originally, this means that it would have returned there at 0.0265 + 0.004, or at 0.0305 seconds, and not at the observed value of 0.035 seconds. The phase error resulting from a simulation process detailing a completely air-filled pipe should be highly visible in Fig. 2.55.

In fact, on Figs. 2.55 to 2.57 there is a third pressure trace where the computation has been conducted with the pipe filled only with air. The traces are marked as "air only." The phase error deduced very simply above is seen to be remarkably accurate, for on Fig. 2.55 the "air only" returning reflection does arrive at station 1 at 0.031 seconds. On the other graphs, the "air only" computation reveals the serious error that can occur in an unsteady gas dynamics simulation when the correct properties of the actual gases involved are not included. Needless to add, in Fig. 2.55, for an "air only" computation, there is no sign of the pressure wave "bump" at 0.025 seconds, for there is no CO_2 in that computation!

From the measured and computed pressure traces it can be seen that the waves have steep fronted and that the computation follows that procedure very accurately. Some minor phase error is seen by 0.047 seconds at station 3, but this is after some 15.5 meters of travel by the initial pressure wave.

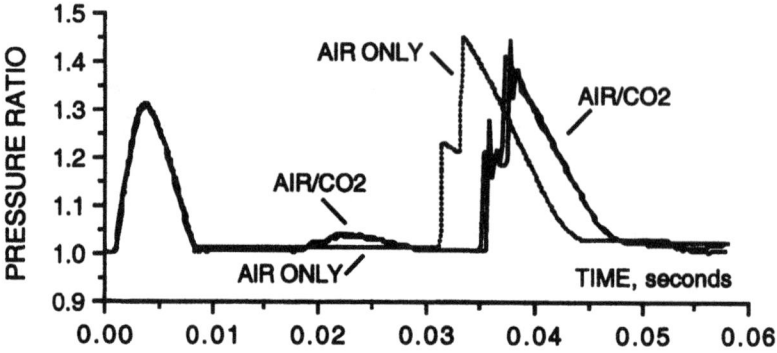

Fig. 2.55 Measured and calculated pressures at station 1.

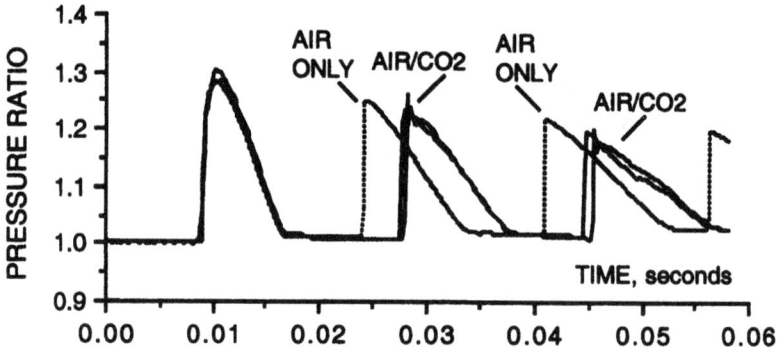

Fig. 2.56 Measured and calculated pressures at station 2.

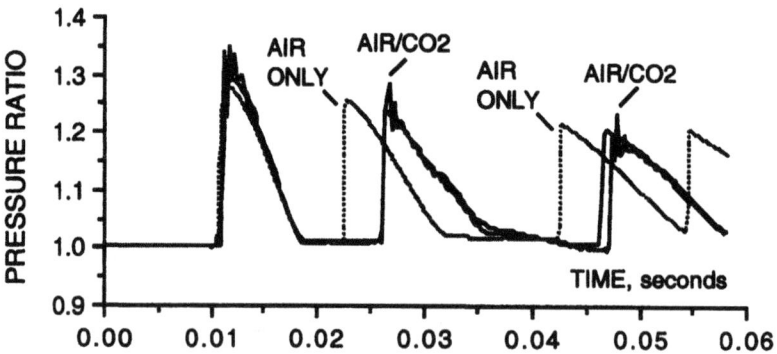

Fig. 2.57 Measured and calculated pressures at station 3.

It can be observed that the GPB finite system modeling gives a very accurate representation of the measured events for the reflection of compression pressure waves at gas property variations within ducting.

2.20 Computation Time

One of the important issues for any computer code is the speed of its operation. Kirkpatrick et al. [2.41] conclude that the GPB finite system simulation method and the Lax-Wendroff (+Flux Corrected Transport; +FCT) are equal in computational speed and both are several times faster than the non-homentropic method of characteristics. Of great importance is the inherent ability of the GPB code to automatically take into account the presence of variable gas properties and variable gas species throughout a duct. This is not the case with the Lax-Wendroff (+FCT) code, and the inclusion of variable gas properties and variable gas species within a duct has been reported to slow that computer code down by anything from a factor of 1.6 for a minimum acceptable level of accuracy to a factor of 5 to attain complete accuracy [2.62, 2.63].

2.21 Concluding Remarks

The GPB finite system simulation technique can be used with some confidence regarding its accuracy, for the thermodynamics and unsteady gas dynamics of flow into, through, and from engine cylinders, and throughout the ducting attached to an internal-combustion engine.

It is an interesting commentary on the activities of those who fund research in engineering, in this case the Science and Engineering Research Council of the UK, that when they were approached some years ago for the funding of the theoretical and experimental work described in this chapter, a "committee of peers" rejected the application for such funding on the grounds that it had "all been done before." While the chagrin and the pique has somewhat faded during the intervening decade, I still reckon that the last line of the "Second Mulled Toast" should be item number one on any agenda for meetings of organizations that fund research in engineering. However, as you can see from this chapter, lack of funds never stopped anybody at QUB from thinking. On the other hand, thinking costs nothing.

References for Chapter 2

2.1 S. Earnshaw, "On the Mathematical Theory of Sound," *Phil.Trans.Roy.Soc.*, Vol. 150, p. 133, 1860.

2.2 F.K. Bannister, "Pressure Waves in Gases in Pipes," *Ackroyd Stuart Memorial Lectures*, University of Nottingham, 1958.

2.3 G. Rudinger, *Wave Diagrams for Non-Steady Flow in Ducts*, Van Nostrand, New York, 1955.

2.4 R.S. Benson, *The Thermodynamics and Gas Dynamics of Internal Combustion Engines*, (ed. J.H. Horlock, D.E. Winterbone), Vols. 1 and 2, Clarendon Press, Oxford, 1982 and 1986, respectively.

2.5 F.J. Wallace and M.H. Nassif, "Air Flow in a Naturally Aspirated Two-Stroke Engine," *Proc.I.Mech.E.*, Vol. 1B, p. 343, 1953.

2.6 G.P. Blair and W.L. Cahoon, "A More Complete Analysis of Unsteady Gas Flow through a High Specific Output Two-Cycle Engine," SAE Automotive Congress, Detroit, Mich., January, 1972, SAE Paper No. 720156.

2.7 J.H. McConnell, "Unsteady Gas Flow Through Naturally Aspirated Four-Stroke Cycle Internal Combustion Engines," Doctoral Thesis, The Queen's University of Belfast, March, 1974.

2.8 P. De Haller, "The Application of a Graphic Method to Some Dynamic Problems in Gases," *Sulzer Tech. Review*, Vol. 1, p. 6, 1945.

2.9 A.D. Jones, "Noise Characteristics and Exhaust Process Gas Dynamics," Ph.D. Thesis, University of Adelaide, Australia, p. 256, 1978.

2.10 R.S. Benson, R.D. Garg, and D. Woollatt, "A Numerical Solution of Unsteady Flow Problems," *Int.J.Mech.Sci.*, Vol. 6, pp. 117-144, 1964.

2.11 D.R. Hartree, "Some Practical Methods of Using Characteristics in the Calculation of Non-Steady Compressible Flow," US Atomic Energy Commission Report, AECU-2713, 1953.

2.12 M. Chapman, J.M. Novak, and R.A. Stein, "A Non-Linear Acoustic Model of Inlet and Exhaust Flow in Multi-Cylinder Internal Combustion Engines," Winter Annual Meeting, ASME, Boston, Mass., 83-WA/DSC-14, November, 1983.

2.13 P.J. Roache, *Computational Fluid Dynamics*, Hermosa Publishers, USA, 1982.

2.14 R. Courant, K. Friedrichs, and H. Lewy, Translation Report NYO-7689, *Math. Ann*, Vol. 100, p. 32, 1928.

2.15 G.P. Blair and J.R. Goulburn, "The Pressure-Time History in the Exhaust System of a High-Speed Reciprocating Internal Combustion Engine," SAE Mid-Year Meeting, Chicago, Ill., May 15-19, 1967, SAE Paper No. 670477.

2.16 G.P. Blair and M.B. Johnston, "Unsteady Flow Effects in the Exhaust Systems of Naturally Aspirated, Crankcase Compression, Two-Stroke Cycle Engines," SAE Farm, Construction and Industrial Machinery Meeting, Milwaukee, Wisc., September 11-14, 1968, SAE Paper No. 680594.

2.17 G.P. Blair and J.R. Goulburn, "An Unsteady Flow Analysis of Exhaust Systems for Multi-Cylinder Automobile Engines," SAE Mid-Year Meeting, Chicago, Ill., May 19-23, 1969, SAE Paper No. 690469.

2.18 G.P. Blair and M.B. Johnston, "Simplified Design Criteria for Expansion Chambers for Two-Cycle Gasoline Engines," SAE Automotive Engineering Congress, Detroit, Mich., January 12-16, 1970, SAE Paper No. 700123.

2.19 J.F. Bingham and G.P. Blair, "An Improved Branched Pipe Model for Multi-Cylinder Automotive Engine Calculations," *Proc.I.Mech.E.*, Vol. 199, No. D1, 1985.

2.20 A.J. Blair and G.P. Blair, "Gas Flow Modelling of Valves and Manifolds in Car Engines," *Proc.I.Mech.E.*, International Conference on Computers in Engine Technology, University of Cambridge, April, 1987, C11/87.

2.21 J.S. Richardson, "Investigation of the Pulsejet Engine Cycle," Doctoral Thesis, The Queen's University of Belfast, March, 1981.

2.22 M. Kadenacy, British Patents 431856, 431857, ...,484465, 511366.

2.23 E. Giffen, "Rapid Discharge of Gas from a Vessel into the Atmosphere," *Engineering*, August 16, 1940, p. 134.

2.24 G.P. Blair, "Unsteady Flow Characteristics of Inward Radial Flow Turbines," Doctoral Thesis, The Queen's University of Belfast, May, 1962.

2.25 G.P. Blair, "The Basic Design of Two-Stroke Engines," R-104, Society of Automotive Engineers, Warrendale, Pa., p. 672.

2.26 R. Fleck, R.A.R. Houston, and G.P. Blair, "Predicting the Performance Characteristics of Twin Cylinder Two-Stroke Engines for Outboard Motor Applications," Society of Automotive Engineers International Off-Highway and Powerplant Congress, Milwaukee, Wisc., September 12-15, 1988, SAE Paper No. 881266, p. 25.

2.27 F.A. McGinnity, R. Douglas, and G.P. Blair, "Application of an Entropy Analysis to Four-Cycle Engine Simulation," SAE International Congress and Exposition, Detroit, Mich., February 1990, SAE Paper No. 900681, p. 15.

2.28 R.Douglas, F.A. McGinnity, and G.P. Blair, "A Study of Gas Temperature Effects on the Prediction of Unsteady Flow," Conference on Internal Combustion Engine Research in Universities and Polytechnics, Proc.I.MechE, C433/036, 30-31, London, pp. 47-55, January 1991.

2.29 M. Chapman, J.M. Novak, and R.A. Stein, "Numerical Modelling of Inlet and Exhaust Flows in Multi-Cylinder Internal Combustion Engines", Proc. ASME Winter Meeting, Phoenix, Ariz., November 1982.

2.30 D.E. Winterbone, "The Application of Gas Dynamics for the Design of Engine Manifolds," IMechE Paper No. CMT8701, 1987.

2.31 G.P. Blair, "An Alternative Method for the Prediction of Unsteady Gas Flow through the Reciprocating Internal Combustion Engine," SAE International Off-Highway and Powerplant Congress, Milwaukee, Wisc., September 9-12, 1991, SAE Paper No. 911850.

2.32 G.P. Blair, "Correlation of an Alternative Method for the Prediction of Engine Performance Characteristics with Measured Data," SAE, International Congress, Detroit, Mich., March 1993, SAE Paper No. 930501, p. 20.

2.33 G.P. Blair, "Correlation of Measured and Calculated Performance Characteristics of Motorcycle Engines," Funfe Zweiradtagung, Technische Universität, Graz, Austria, 22-23 April 1993, pp. 5-16.

2.34 G.P. Blair, R.J. Kee, R.G. Kenny, and C.E. Carson, "Design and Development Techniques Applied to a Two-Stroke Cycle Automobile Engine," International Conference on Comparisons of Automobile Engines, Verein Deutscher Ingenieuer (VDI-GFT), Dresden, 3-4 June 1993, pp. 77-103.

2.35 G.P. Blair and S.J. Magee, "Non-Isentropic Analysis of Varying Area Flow in Engine Ducting," SAE International Off-Highway and Powerplant Congress, Milwaukee, Wisc., September 13-16, 1993, SAE Paper No. 932399, p. 22.

2.36 R.S. Benson,, D. Woollatt, and W.A. Woods, "Unsteady Flow in Simple Branch Systems," Proc.I.Mech.E., Vol. 178, Pt. 3I(iii), No.10, 1963-64, p. 21.

2.37 B.E.L. Deckker and D.H. Male, "Unsteady Flow in a Branched Duct," Proc.I.Mech.E., Vol. 182, Pt. 3H, No.10, 1967-68, p. 104.

2.38 D.S. Miller, "Internal Flow Systems," BHRA, Fluid Engineering Series, No. 5, 1978, p. 290.

2.39 F.A. McGinnity, "The Effect of Temperature on Engine Gas Dynamics," PhD thesis, The Queen's University of Belfast, December, 1989.

2.40 G.P. Blair, "Non-Isentropic Analysis of Branched Flow in Engine Ducting," SAE, International Congress, Detroit, Mich., March 1994, SAE Paper No. 940395, pp. 55-72.

2.41 S.J. Kirkpatrick, G.P. Blair, R. Fleck, and R.K. McMullan, "Experimental Evaluation of 1-D Computer Codes for the Simulation of Unsteady Gas Flow Through Engines—A First Phase," SAE International Off-Highway and Powerplant Congress, Milwaukee, Wisc., September 14-16, 1994, SAE Paper No. 941685, pp. 77-96.

2.42 P.D. Lax and B. Wendroff, "Systems of Conservation Laws," Communs. Pure Appl. Math., Vol. 15, 1960, pp. 217-237.

2.43 R.D. Richtmyer and K.W. Morton, "Difference Methods for Initial Value Problems," 2nd Ed., John Wiley & Sons, Inc., New York, 1967.

2.44 H. Niessner and T. Bulaty, "A Family of Flux-Correction Methods to Avoid Overshoot Occurring With Solutions of Unsteady Flow Problems," Proceedings of GAMM 4th Conf., 1981, pp. 241-250.

2.45 D.L. Book, J.P. Boris, and K. Hain, "Flux-Corrected Transport II: Generalization of the Method", J. Comp. Phys., Vol. 18, 1975, pp. 248-283.

2.46 A. Harten, P.D. Lax, and B. Van Leer, "On Upstream Differencing and Gudunov Type Schemes for Hyperbolic Conservation Laws," SIAM Rev., Vol. 25, 1983, pp. 35-61.

2.47 C. Chen, A. Veshagh, and F.J. Wallace, "A Comparison Between Alternative Methods for Gas Flow and Performance Prediction of Internal Combustion Engines," SAE Paper No. 921734, 1992.

2.48 T. Ikeda and T. Nakagawa, "On the SHASTA FCT Algorithm for the Equation," Maths. Comput., Vol. 33, No. 148, October 1979, pp. 1157-1169.

2.49 S.K. Gudunov, "A Difference Scheme for Numerical Computation of Discontinuous Solutions of Equations of Fluid Dynamics," Mat. Sb., Vol. 47, 1959, pp. 271-290.

2.50 S.C. Low and P.C. Baruah, "A Generalised Computer Aided Design Package for an I.C. Engine Manifold System," SAE Paper No. 810498, 1981.

2.51 G.P. Blair and R. Fleck, "The Unsteady Gas Flow Behaviour in a Charge Cooled Rotary Piston Engine," SAE Paper No. 770763, 1977.

2.52 F.K. Bannister and G.F. Mucklow, "Wave Action Following the Release of Compressed Gas from a Cylinder," Proc.I.Mech.E., Vol. 159, 1948, p. 269.

2.53 R.S. Benson, "Influence of Exhaust Belt Design on the Discharge Process in Two-Stroke Engines," Proc.I.Mech.E., Vol. 174, No. 24, 1960, p. 713.

2.54 W.A. Woods, "On the Formulation of a Blowdown Pulse in the Exhaust System of a Two-Stroke Cycle Engine," Int. J. Mech. Sci., Vol. 4, 1962, p. 259.

2.55 W.L. Cahoon, "Unsteady Gas Flow Through a Naturally Aspirated Two-Stroke Internal Combustion Engine," Ph.D. Thesis, Dept. Mech. Eng., The Queen's University of Belfast, November, 1971.

2.56 F.J. Wallace and R.W. Stuart-Mitchell, "Wave Action Following the Sudden Release of Air Through an Engine Port System," Proc. I. Mech. E., Vol. 1B, 1953, p. 343.

2.57 F.J. Wallace and M.H. Nassif, "Air Flow in a Naturally Aspirated Two-Stroke Engine," Proc. I. Mech. E., Vol. 168, 1954, pp. 515.

2.58 W.J.D. Annand, "Heat Transfer in the Cylinders of Reciprocating Internal Combustion Engines," Proc.I.Mech.E, Vol. 177, p. 973, 1963.

2.59 G.P. Blair, W.L. Cahoon, and C.T. Yohpe, "Design of Exhaust Systems for V-Twin Motorcycle Engines," SAE Motorsports Engineering Conference and Exposition, Dearborn, Mich., December 1994, SAE Paper No. 942514, pp. 119-130.

2.60 G. Woschni, "A Universally Applicable Equation for Instantaneous Heat Transfer in the Internal Combustion Engine", SAE Congress, Detroit, Mich., 1967, SAE Paper No. 670931.

2.61 D.D. Agnew, "What is Limiting Engine Air Flow. Using Normalised Steady Air Flow Bench Data", SAE Motorsports Engineering Conference and Exposition, Dearborn, Mich., December 1994, SAE Paper No. 942477, pp. 31-39.

2.62 D.E. Winterbone and R.J. Pearson, "A Solution of the Wave Equation Using Real Gases," Int.J.Mech.Sci., Vol. 34, No. 12, pp. 917-932, 1992.

2.63 R.J. Pearson and D.E. Winterbone, "Calculating the Effects of Variations in Composition on Wave Propagation in Gases," Int.J.Mech.Sci., Vol. 35, No. 6, pp. 517-537, 1993.

2.64 J.F. Bingham, "Unsteady Gas Flow in the Manifolds of Multicylinder Automotive Engines," Ph.D. Thesis, Dept. Mech. Eng., The Queen's University of Belfast, October, 1983.

2.65 G.P. Blair, S.J. Kirkpatrick, and R. Fleck, "Experimental Evaluation of a 1D Modelling Code for a Pipe Containing Gas of Varying Properties," SAE International Congress, Detroit, Mich., February 28-March 3, 1995, SAE Paper No. 950275, p. 14.

2.66 G.P. Blair, S.J. Kirkpatrick, D.O. Mackey, and R. Fleck, "Experimental Evaluation of a 1D Modelling Code for a Pipe System Containing Area Discontinuities," SAE International Congress, Detroit, Mich., February 28-March 3, 1995, SAE Paper No. 950276, p. 16.

2.67 G.P. Blair, *Design and Simulation of Two-Stroke Engines*, R-161, Society of Automotive Engineers, Warrendale, Pa., 1996, p. 623.

2.68 R. Fleck, D. Thornhill, P. Long, and G.P. Blair, "Validation of a New Non-Isentropic Pressure Loss, Branched Pipe Junction Model," SAE International Off-Highway and Powerplant Congress, Milwaukee, Wisc., September 14-16, 1998, SAE Paper No. 982055.

2.69 H. Seifert, "20 Jahre Erfolgreiche Entwicklung des Programmsystemes PROMO," MTZ 51, 1990.

Appendix A2.1 The Derivation of the Particle Velocity for Unsteady Gas Flow

This section owes much to the text by Bannister [2.2], which was a vital component of my education during the period 1959–1962. His lecture notes have remained a model of clarity and a fine example of the manner in which matters theoretical should be written by those who wish to elucidate others.

The exact differential equations employed by Earnshaw [2.1] in his solution of the propagation of a wave of finite amplitude are those established using the notation of Lagrange.

Fig. A2.1(a) shows a frictionless pipe of unit cross section, containing gas at reference conditions of density and pressure, r_0 and p_0. Element AB is of length dx, and at distance x from an origin of time and distance. Fig. A2.1(b) shows the changes that have occurred in the same element AB by time t, due to the influence of a pressure wave of finite amplitude. The element face, A, has now been displaced to a position L further on from the initial position. Thus, at time t, the distances of A and B from the origin are no longer separated by dx, but by a dimension that is a function of that very displacement; this is shown on Fig. A2.1(b). The length of the element is now

$$\left(1 + \frac{\partial L}{\partial x}\right)dx$$

The density, ρ, in this element at this instant is related by the fact that the mass in the element is unchanged from its initial existence at the reference conditions and that the pipe area, A, is unity;

$$\rho_0 A dx = \rho A\left(1 + \frac{\partial L}{\partial x}\right)dx \tag{A2.1.0}$$

Hence,
$$\frac{\rho_0}{\rho} = 1 + \frac{\partial L}{\partial x} \tag{A2.1.1}$$

Since the process is regarded as isentropic,

$$\frac{p}{p_0} = \left(\frac{\rho}{\rho_0}\right)^{\gamma}$$

Then,
$$\frac{p}{p_0} = \left(1 + \frac{\partial L}{\partial x}\right)^{-\gamma} \tag{A2.1.2}$$

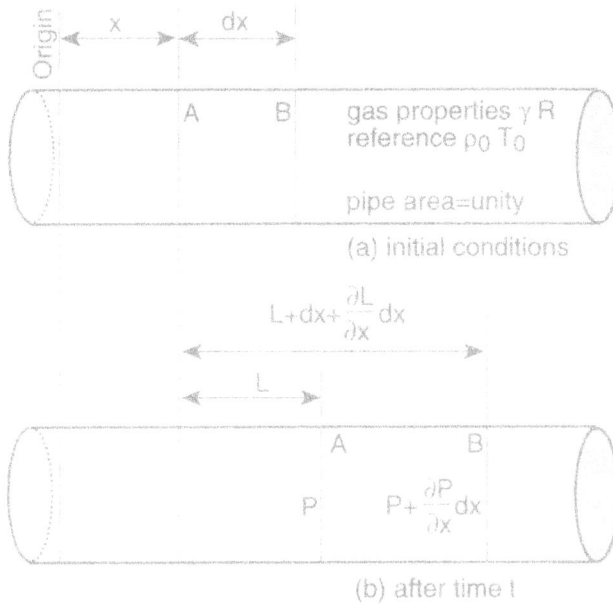

Fig. A2.1 Lagrangian notation for a pressure wave.

Partial differentiation of this latter expression with respect to distance gives:

$$\frac{\partial p}{\partial x} = -\gamma p_0 \left(1 + \frac{\partial L}{\partial x}\right)^{-\gamma - 1} \frac{\partial^2 L}{\partial x^2} \qquad (A2.1.3)$$

The accelerating force applied to the gas element in the duct of unit area A at time t is:

$$A\left(p - \left(p + \frac{\partial p}{\partial x} dx\right)\right) \quad \text{or} \quad -\frac{\partial p}{\partial x} A dx$$

The mass in the element is $\rho_0 A dx$, from Eq. A2.1.1 and from Newton's Laws where "force equals mass times acceleration":

$$-\frac{\partial p}{\partial x} A dx = \rho_0 A dx \frac{\partial^2 L}{\partial t^2} \qquad (A2.1.4)$$

319

Eliminating the area, A, and substituting from Eq. A2.1.4 gives:

$$\frac{\gamma p_0}{\rho_0}\left(1+\frac{\partial L}{\partial x}\right)^{-\gamma-1}\frac{\partial^2 L}{\partial x^2}=\frac{\partial^2 L}{\partial t^2} \qquad (A2.1.5)$$

The reference acoustic velocity, a_0, can be stated as:

$$a_0=\sqrt{\gamma R T_0}=\sqrt{\frac{\gamma p_0}{\rho_0}}$$

and replacing the distance from the origin with y, i.e., replacing (x+L), Eq. A2.1.5 becomes:

$$a_0^2\left(\frac{\partial y}{\partial x}\right)^{-\gamma-1}\frac{\partial^2 y}{\partial x^2}=\frac{\partial^2 y}{\partial t^2} \qquad (A2.1.6)$$

This is the fundamental thermodynamic equation and the remainder of the solution is merely mathematical "juggling." This is carried out quite normally by making logical substitutions until a solution emerges. The substitution that works is:

Let
$$\frac{\partial y}{\partial t}=f\left(\frac{\partial y}{\partial x}\right)$$

in which case the following are the results of this substitution:

$$\frac{\partial^2 y}{\partial t^2}=f'\left(\frac{\partial y}{\partial x}\right)\frac{\partial}{\partial t}\left(\frac{\partial y}{\partial x}\right) \quad\text{where}\quad f'\left(\frac{\partial y}{\partial x}\right)=\frac{d\left(f\left(\frac{\partial y}{\partial x}\right)\right)}{d\left(\frac{\partial y}{\partial x}\right)}$$

Thus, by transposition:

$$\frac{\partial^2 y}{\partial t^2}=f'\left(\frac{\partial y}{\partial x}\right)\frac{\partial}{\partial x}\left(\frac{\partial y}{\partial t}\right)=f'\left(\frac{\partial y}{\partial x}\right)\frac{\partial}{\partial x}\left(f\left(\frac{\partial y}{\partial x}\right)\right)$$

Hence
$$\frac{\partial^2 y}{\partial t^2} = f'\left(\frac{\partial y}{\partial x}\right) f'\left(\frac{\partial y}{\partial x}\right) \frac{\partial^2 y}{\partial x^2} = \left(f'\left(\frac{\partial y}{\partial x}\right)\right)^2 \frac{\partial^2 y}{\partial x^2}$$

This relationship is substituted into Eq. A2.1.6, which produces:

$$a_0^2 \left(\frac{\partial y}{\partial x}\right)^{-\gamma-1} \frac{\partial^2 y}{\partial x^2} = \left(f'\left(\frac{\partial y}{\partial x}\right)\right)^2 \frac{\partial^2 y}{\partial x^2}$$

or
$$\left(\frac{\partial y}{\partial x}\right)^{\frac{-\gamma-1}{2}} = \pm \frac{1}{a_0} f'\left(\frac{\partial y}{\partial x}\right)$$

Integrating this expression introduces an integration constant, k:

$$\left(\frac{\partial y}{\partial x}\right)^{\frac{1-\gamma}{2}} = \pm \frac{\gamma-1}{2a_0} f\left(\frac{\partial y}{\partial x}\right) + k$$

As
$$y = x + L$$

then
$$\frac{\partial y}{\partial x} = 1 + \frac{\partial L}{\partial x} = \frac{\rho_0}{\rho} = \left(\frac{p}{p_0}\right)^{\frac{-1}{\gamma}}$$

and from the original substitution, the fact that x is a constant, and as the gas particle velocity is c, which is the rate of change of the length dimension, L, with time:

$$f\left(\frac{\partial y}{\partial x}\right) = \frac{\partial y}{\partial t} = \frac{\partial(x + L)}{\partial t} = \frac{\partial L}{\partial t} = c$$

Therefore
$$\left(\frac{p}{p_0}\right)^{\frac{\gamma-1}{2\gamma}} = \pm \frac{(\gamma-1)c}{2a_0} + k$$

If, at the wave head the following facts are correct, the integration constant, k is found as:

$$p = p_0 \quad c = 0 \quad \text{then} \quad k = 1$$

where the equation for the particle velocity, c, in unsteady gas flow is deduced:

$$c = \pm \frac{2a_0}{\gamma - 1} \left[\left(\frac{p}{p_0} \right)^{\frac{\gamma - 1}{2\gamma}} - 1 \right]$$

The positive sign is the correct one to adopt because if $p > p_0$, then c must be a positive value for a compression wave.

Appendix A2.2 Moving Shock Waves in Unsteady Gas Flow

The text in this section owes much to the oft-quoted lecture notes by Bannister [2.2]. The steepening of finite amplitude waves is discussed in Sec. 2.1.5 resulting in a moving "shock" wave. Consider the case of the moving shock wave, AB, illustrated in Fig. A2.2. The propagation velocity is α, and it is moving into stationary gas at reference conditions ρ_0 and p_0. The pressure and density behind the shock front are p and ρ, while the associated gas particle velocity is c. Imagine imposing a mean gas particle velocity of α on the entire system illustrated in Fig. A2.2(a) so that the regime in Fig. A2.2(b) becomes "reality." This would give a stationary shock, AB, i.e., the moving front would be brought to rest and the problem is now reduced to one of steady flow. Consider that the duct area is A and is unity.

The continuity equation shows across the now stationary shock front:

$$(\alpha - c)\rho A = \alpha \rho_0 A \qquad (A2.2.1)$$

The momentum equation gives, where force is equal to the rate of change of momentum,

$$(\alpha - (\alpha - c))\alpha \rho_0 A = (p - p_0)A$$

or
$$c\alpha \rho_0 = p - p_0 \qquad (A2.2.2)$$

This can be rearranged as:

$$\frac{p}{\rho} = \frac{p_0}{\rho} + \frac{c\alpha \rho_0}{\rho} \qquad (A2.2.3)$$

The First Law of Thermodynamics across the now stationary shock front shows:

$$C_P T_A + \frac{(\alpha - c)^2}{2} = C_P T_B + \frac{c^2}{2}$$

or
$$\frac{\gamma R}{\gamma - 1} T_A + \frac{(\alpha - c)^2}{2} = \frac{\gamma R}{\gamma - 1} T_B + \frac{c^2}{2}$$

or
$$\frac{1}{\gamma - 1}\frac{p}{\rho} + \frac{p}{\rho} + \frac{(\alpha - c)^2}{2} = \frac{1}{\gamma - 1}\frac{p_0}{\rho_0} + \frac{p_0}{\rho_0} + \frac{c^2}{2}$$

(a) shock wave AB moving rightward

(b) shock wave imagined to be stationary

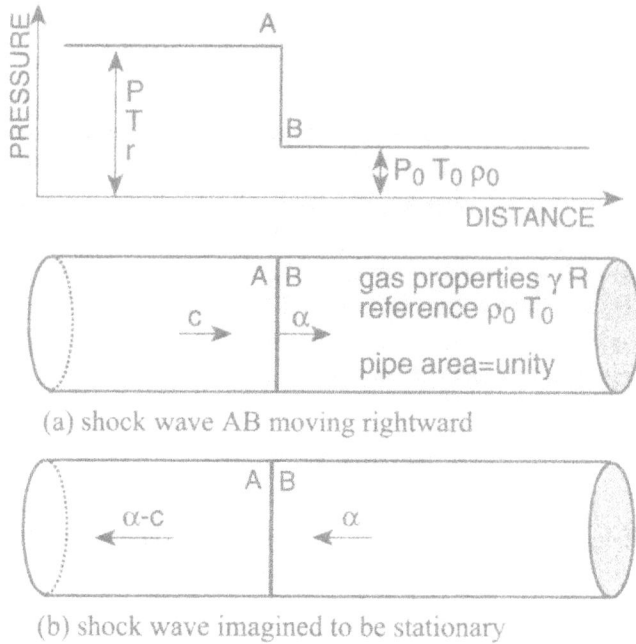

Fig. A2.2 The moving shock wave.

or

$$-\alpha c + \frac{c^2}{2} = \frac{1}{\gamma-1}\frac{p}{\rho} + \frac{p}{\rho} - \frac{1}{\gamma-1}\frac{p_0}{\rho_0} - \frac{p_0}{\rho_0}$$

Substituting from Eq. A2.2.3 for $\frac{\rho_0}{\rho}$ and replacing the a_0^2 term with $\frac{\gamma p_0}{\rho_0}$ produces:

$$\alpha^2 - \frac{\gamma+1}{2}\alpha c - a_0^2 = 0 \qquad\qquad (A2.2.4)$$

Therefore

$$c = \frac{2a_0}{\gamma+1}\left(\frac{\alpha}{a_0} - \frac{a_0}{\alpha}\right) \qquad\qquad (A2.2.5)$$

Combining Eqs. A2.2.2 and A2.2.4:

$$\alpha^2 - a_0^2 = \frac{\gamma+1}{2}\left(\frac{p-p_0}{\rho_0}\right)$$

Dividing throughout by a_0^2 and substituting $\frac{\gamma p_0}{\rho_0}$ for it provides the relationship for the propagation velocity of a moving compression shock wave as:

$$\alpha = a_0 \sqrt{\frac{\gamma + 1}{2\gamma} \frac{p}{p_0} + \frac{\gamma - 1}{2\gamma}} \qquad (A2.2.6)$$

Substituting this latter expression for the shock propagation velocity, α, into Eq. A2.2.5 gives a direct relationship for the gas particle velocity, c:

$$c = \frac{\dfrac{a_0}{\gamma}\left(\dfrac{p}{p_0} - 1\right)}{\sqrt{\dfrac{\gamma + 1}{2\gamma} \dfrac{p}{p_0} + \dfrac{\gamma - 1}{2\gamma}}} \qquad (A2.2.7)$$

The temperature and density relationships behind the shock are determined as follows, firstly from the equation of state:

$$\frac{p}{p_0} = \frac{\rho R T}{\rho_0 R T_0} = \frac{\rho T}{\rho_0 T_0} \qquad (A2.2.8)$$

which when combined with Eq. A2.2.1 gives:

$$\frac{T}{T_0} = \frac{p}{p_0}\left(\frac{\alpha - c}{\alpha}\right)$$

which when combined with Eq. A2.2.5 gives:

$$\frac{T}{T_0} = \frac{p}{p_0}\left(\frac{\gamma - 1}{\gamma + 1} + \frac{2}{\gamma + 1}\frac{a_0^2}{\alpha^2}\right)$$

and in further combination with Eq. A2.2.6 reveals the temperature relationship:

$$\frac{T}{T_0} = \frac{p}{p_0}\left(\frac{\frac{\gamma-1}{\gamma+1}\frac{p}{p_0}+1}{\frac{p}{p_0}+\frac{\gamma-1}{\gamma+1}}\right) \tag{A2.2.9}$$

and in further combination with Eq. A2.2.8 reveals the density relationship:

$$\frac{\rho}{\rho_0} = \frac{\frac{p}{p_0}+\frac{\gamma-1}{\gamma+1}}{\frac{\gamma-1}{\gamma+1}\frac{p}{p_0}+1} \tag{A2.2.10}$$

The functions in Eqs. 2.2.9 and 2.2.10 reveal the non-isentropic nature of the flow. For example, an *isentropic* compression would give the following relation between pressure and density:

$$\frac{p}{p_0} = \left(\frac{\rho}{\rho_0}\right)^{\gamma}$$

This isentropic expression is clearly quite different from that deduced non-isentropically for the moving shock wave in Eq. A2.2.10. The non-isentropic functions relating pressure, temperature, and density for a moving shock wave are often described in the literature as the Rankine-Hugoniot equations. They arise again in the discussion in Sec. 2.2.4.

Chapter 3

Discharge Coefficients of Flow within
Four-Stroke Engines

3.0 Introduction to Discharge Coefficients
3.0.1 The Need for Measured Discharge Coefficients
The inclusion of discharge coefficients within gas dynamic theory is described in Chapter 2, in Secs. 2.16 and 2.17 in connection with the flow into and out of the cylinders of an engine from the attached intake and exhaust ducting. The inclusion of discharge coefficients can also been found in the theory for the throttled pipe section, the so-called "butted joint," in Sec. 2.12. Because the flow at the end of a duct to the atmosphere, be it outflow from the pipe as in an exhaust process or inflow as in an intake process, is a flow in which the atmosphere is simply considered realistically as an infinitely large air-filled cylinder, the theory of Secs. 2.16 and 2.17 will still apply. Consequently, because the theoretical solution for the unsteady gas flow at the ends of all ducts contains a term for a discharge coefficient, the numerical value of this coefficient must be known with some precision, otherwise this theory cannot be accurately applied in practice. During the execution of an engine simulation model, where this boundary condition theory is applied at all duct ends at every time step, data on discharge coefficients must be stored and be available to be indexed at each step. The accurate computation of either the pressure wave trains, or the instantaneous gas mass flow rates throughout the engine, is not possible if this information is not available. Hence, maps of discharge coefficients are required for flow in both directions at all duct boundaries where valving, porting, plenums, cylinders, the atmosphere, catalysts, air filters, throttles, etc., are located. In short, maps of discharge coefficients are needed everywhere that a pressure wave can reflect at a discontinuity in the ducting. In order to acquire the discharge coefficients, the only solution is to experimentally measure them as they are not readily amenable to theoretical deduction.

3.0.2 The Published Literature on Discharge Coefficients
The literature contains many technical papers on the discharge coefficients of valves, ports, and most of the conceivable geometrical boundaries at the ends of ducts. A mere sample of that collection of papers is given in the References [3.1-3.14]. If you now make the obvious assumption that the entirety of this published information may be used as a source of discharge coefficients within an engine simulation model, you are automatically forgiven. I have already made that mistake myself, so it would not be right for me to chastise you for thinking likewise.

The above remarks imply that all of these researchers who studied this before [3.1-3.14] got it wrong. Because there are some "household" names in that list, such as the Taylors, Kastner, Williams, Annand, Watanabe and Woods, I am obviously treading on the toes of some famous reputations, not to mention skating on thin experimental or theoretical ice. The fact is that there is nothing inaccurate about any of these papers [3.1-3.14]—indeed they are a very valuable data record of the methodology of the measurement techniques for discharge coefficients and the relative merits of one geometrical arrangement vis-à-vis another. The problem is that the number they deduce as a discharge coefficient for any particular valving or porting geometry is, to varying degrees, inaccurate when replayed into an engine simulation model. This arises because of the traditional method of reducing the experimental data to obtain the numerical value of a discharge coefficient, while still using the conventional experimental measurement technique. In short, there is nothing wrong with the experimental method, only with the analysis of the data which it provides.

3.1 The Traditional Method for the Measurement of Discharge Coefficients

The apparatus consists of a means of blowing, or sucking, air through an aperture at the end of a duct, where the aperture is the one whose discharge coefficient is being tested. A typical example of this type of apparatus, and one that is employed at QUB, is shown in Fig. 3.1. In this example, the apparatus is being used to find the discharge coefficients of plain pipe ends, bellmouths, and orifices at the ends of round pipes.

The suction flow comes from a large 150 m^3 tank that can be evacuated to -700 mm Hg, a substantial vacuum. The smaller settling tank shown in Fig. 3.1 is closer to the measurement section. It is about 1 m^3 in volume and is connected to the vacuum tank by a flow measurement section and a pressure reducing valve. The flow measurement section is designed with d and 0.5d tappings to conform to the British Standard 1042 [3.15]. A plethora of air flow rate measurement techniques can be used and are normally designed to correspond to the dictates of a "standards" body, such as BS (British Standards Institution), ISO (International Standards Organization), or ASME (American Society of Mechanical Engineers). In this case, the mass flow rate of air through the test section, $\dot{m}_{ex}$, is measured at the BS 1042 measuring section. The duct and aperture section that are to be tested are on top of the settling tank. A simple example of flow to an open, sharp-edged, round steel pipe from the atmosphere will be used to illustrate the measurement technique and the following section, Sec. 3.2, shows that it is possible to obtain three different values of discharge coefficient by analyzing the same measured data. This straightforward example serves to illustrate not only the simplicity of measurement of a discharge coefficient but also the complexity of its numerical analysis. Although this may seem to be a very simple example, it is very necessary to have stored within an engine simulation such data on discharge coefficients, because the particle flow will reverse at the ends of all exhaust pipes from time to time throughout an engine cycle as compression and expansion wave trains arrive there.

In this example of flow to the pipe from the atmosphere, which is in reality an example of plenum outflow in the unsteady gas dynamic sense, the pipe is of round section with a 22.82 mm internal diameter, d_2, and a 25.4 mm external diameter, $\varnothing_{od}$, and the ends are machined square and are sharp-edged. The internal finish of the thin steel pipe is smooth, but not polished. Pressure and temperature measurements are taken some four diameters, in this case 100 mm,

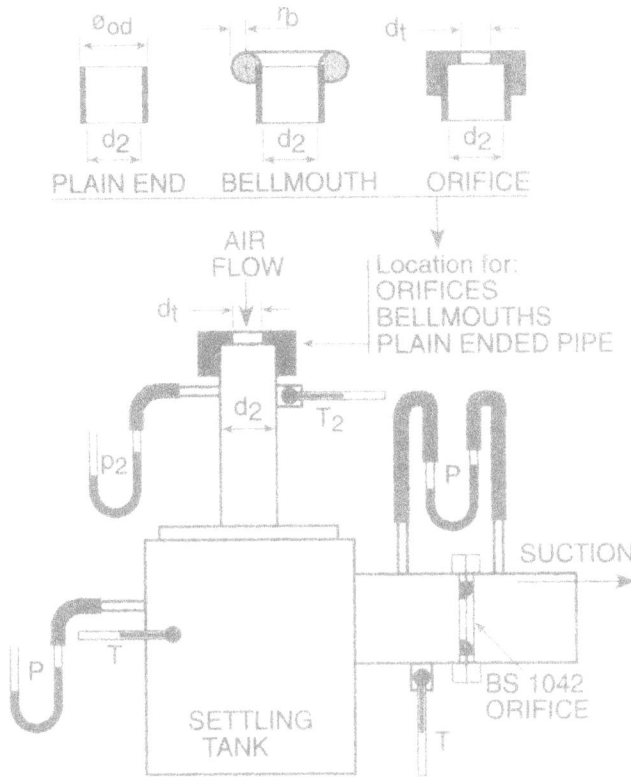

Fig. 3.1 Measuring the discharge coefficients of orifices, bellmouths and plain pipe ends.

from the open pipe end, both of which are static measurements at the inside of the pipe wall. The type of pressure measurement device is non-significant as long as it is sufficiently accurate. Manometers filled with water or mercury are quite adequate for the purpose of pressure measurement, as are thermometers for temperature measurement, but fully automatic sensors for data logging via a computer are also equally acceptable as long as they are accurately calibrated. Although they are more time-consuming to read, manometers have the inestimable advantage of being self-calibrating.

The discharge coefficient at a given pressure ratio across it for this outflow process must be known. To ensure that this process is fully understood, consult again the relevant theory for an outflow process, which is given in Sec. 2.16. To provide further understanding in this regard, Fig. 2.16 is suitably modified to represent this particular flow regime, and is shown here as Fig. 3.2.

The flow is proceeding from the atmosphere at pressure, p_1, to the pipe at pressure, p_2. The pressure ratio, P, across this aperture from "cylinder" to pipe, is given by:

$$P = \frac{\text{cylinder pressure}}{\text{pipe pressure}} = \frac{p_1}{p_2} \qquad (3.1.1)$$

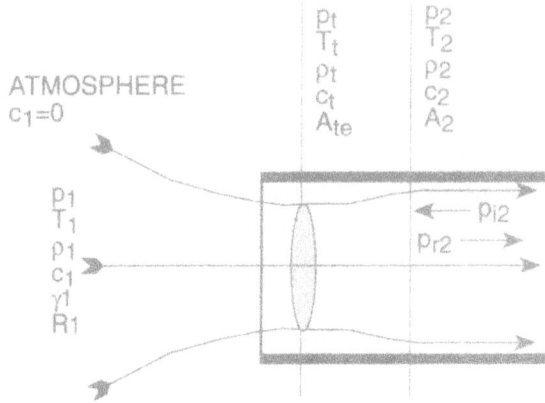

Fig. 3.2 The outflow process from a reservoir to a pipe.

The port-to-pipe aperture ratio, k, for this plain pipe end is given by the conventional equation using the symbolism of Fig. 3.2. For such a plain pipe end, where there is no geometrical throat area, it is manifestly unity:

$$k = \frac{\text{throat area}}{\text{pipe area}} = \frac{A_t}{A_2} = 1.0 \qquad (3.1.2)$$

At any given pressure ratio, P, across the plain pipe end, the mass flow rate of air, $\dot{m}_{ex}$, is measured at the BS 1042 measuring section. To obtain a discharge coefficient in the traditional manner, this measured mass flow rate is compared with that which would have theoretically flowed in an ideal thermodynamic process at the same pressure ratio through the same pipe end. The geometrical nozzle area in this case is the pipe area, because there is no apparent physical restriction to air entering the full pipe area.

This same flow process, in the same physical geometry at a pressure ratio of 1.083 from the atmosphere to the pipe, is analyzed by computational fluid dynamics (CFD) using a commercially available CFD code, StarCD™. The results for the air velocity profiles are shown in Fig. 3.3. The similarity between the sketch of Fig. 3.2 and the computation shown in Fig. 3.3 is quite clear.

The air velocities range from -31.5 m/s in a recirculation zone centered on the outer pipe edge, to +185 m/s in the very center of the pipe in what is commonly termed the "vena contracta" [3.16]. The pressure ratio employed is 1.083 from an atmosphere at a pressure of 760 mm Hg and a temperature of 20°C. It is clear that the entering air flow does not fill the full pipe diameter, and that there is a pronounced throat area through which air flows with a velocity profile that is maximized at the pipe center line. The computation is carried out with a relatively coarse mesh of some 60,000 cells.

The CFD computation provides numeric information on particle velocity, temperature, pressure, etc. at every cell location. Fig. 3.4 shows the result of the computation for the particle

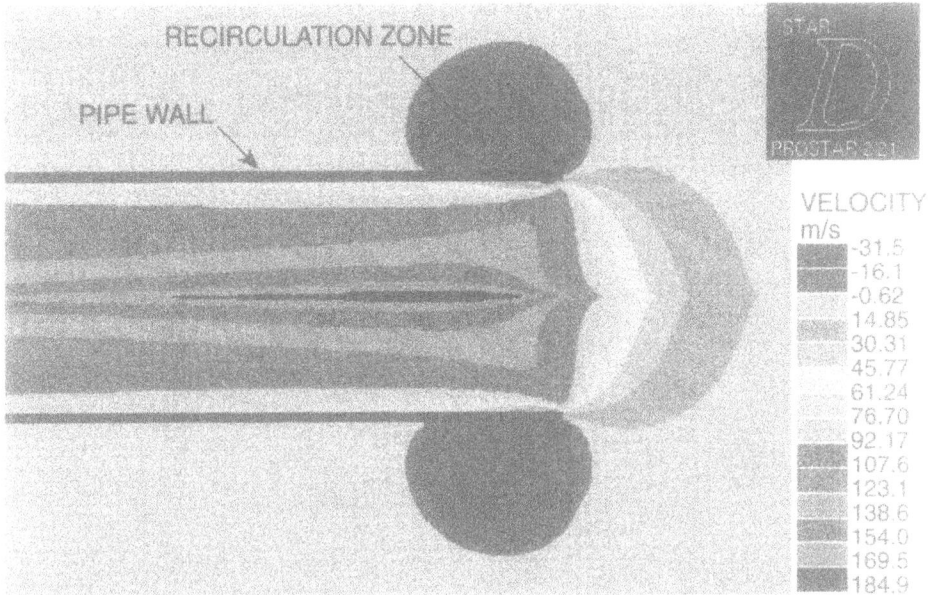

Fig. 3.3 The outflow process from the atmosphere to a pipe calculated by CFD.

velocity profile across the pipe at one and four diameters, 22.8 and 91.2 mm, respectively, from the pipe end.

The graph is in two halves, illustrating the output at two pressure ratios, one of which is at 1.083 (that used in Fig. 3.3), and another at a higher value of 1.305. It is seen that the velocity profiles conform to the conventional image of turbulent flow with increasing pressure ratio. However, by comparing Figs. 3.2 and 3.4, it is obvious that, in the real fluid-mechanics world that a CFD computation theoretically approaches, there is no single effective throat area, A_{te}, and there is no single particle velocity, c_t, that can be assigned to it as presumed by the one-dimensional analysis of Sec. 2.16. On the other hand, the one-dimensional analysis takes micro-seconds to complete on a desktop computer, whereas the CFD process requires several weeks of geometrical meshing followed by some days of computation on a high-performance computer work station.

3.2 The Reduction of Measured Data to Determine a Discharge Coefficient
3.2.1 The Several Analyses for the Discharge Coefficient for Outflow at a Plain End on a Pipe

There are three methods available to deduce the discharge coefficient for this simple example of outflow from the atmosphere into a plain open-ended pipe. The experimentally measured data is not extensive as it includes only five values, namely the atmospheric pressure and temperature, the pipe pressure and temperature, and the air mass flow rate. Sec. 3.2.2 presents the traditional analytic method as found in the literature [3.1-3.14] the result of which is labeled as an isentropic discharge coefficient, Cd_{is}. Sec. 3.2.3 presents the complete thermodynamic solution for the outflow process from the atmosphere to the pipe. In this case,

CFD PREDICTION OF FLOW INTO PLAIN ENDED PIPE

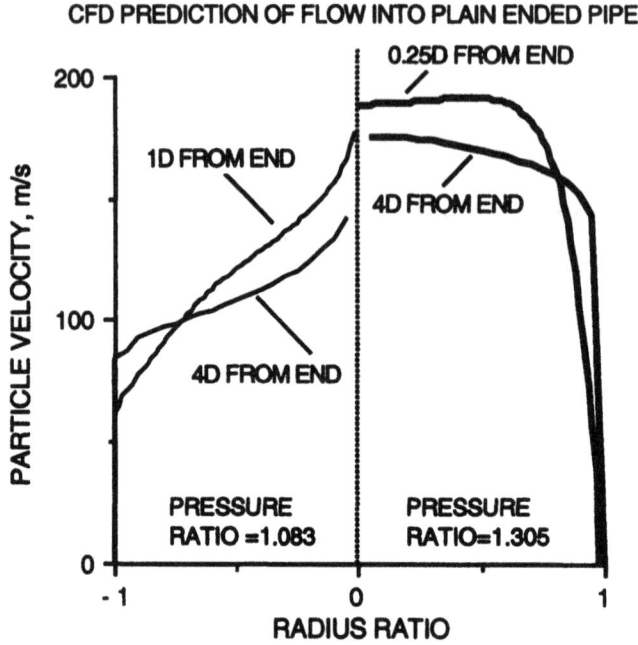

Fig. 3.4 The particle velocities within the pipe calculated by CFD.

because the outcome is termed the "ideal mass flow rate," the discharge coefficient so determined is labeled as an ideal discharge coefficient, Cd_{id}. Sec. 3.2.4 presents the analytic technique that is used to compute what is called the "actual discharge coefficient," Cd_a. By definition, when the actual discharge coefficient is incorporated into the same complete thermodynamic theory over the same pressure ratio, the computational result will be the measured mass flow rate.

3.2.2 The Analysis of the Isentropic Discharge Coefficient for Outflow at a Plain End on a Pipe

The discharge coefficient, which in this case is the isentropic discharge coefficient, Cd_{is}, is defined by:

$$Cd_{is} = \frac{\text{measured mass flow rate}}{\text{isentropic nozzle mass flow rate}} = \frac{\dot{m}_{ex}}{\dot{m}_{is}} \qquad (3.2.1)$$

In many of the papers listed in the references [3.1-3.14], the isentropic nozzle mass flow, $\dot{m}_{is}$, is usually given by an analysis of the First Law of Thermodynamics, which assumes isentropic flow from the cylinder to the throat and on to the pipe.

$$h_1 = h_t + \frac{c_t^2}{2} = h_2 + \frac{c_2^2}{2} \qquad (3.2.2)$$

332

This is combined with the continuity equation from throat to pipe, which yields:

$$\dot{m}_{is} = \rho_t A_t c_t = \rho_2 A_2 c_2 \qquad (3.2.3)$$

The ensuing algebraic solution is not simple. From a collection of papers [3.1-3.7] written in a pre-computer era, it is therefore not surprising that assumptions are made to yield arithmetic that could be dealt with on a slide rule. The popular solution approach is given by Annand and Roe [3.12], and its use continues into more modern times, for example, it is quoted again by Agnew [3.14]. The solution is expressed in terms of the effective area of the throat, A_{te}, as:

$$A_{te} = \cfrac{\dot{m}_{ex}\sqrt{\gamma R T_1}}{\gamma p_1 \left(\dfrac{p_2}{p_1}\right)^{\frac{1}{\gamma}} \sqrt{\dfrac{2}{\gamma - 1}\left(1 - \left(\dfrac{p_2}{p_1}\right)^{\frac{\gamma-1}{\gamma}}\right)}} \qquad (3.2.4)$$

Consequently, the isentropic discharge coefficient becomes:

$$Cd_{is} = \frac{\text{effective throat area}}{\text{geometrical throat area}} = \frac{A_{te}}{A_t} = \frac{A_{te}}{A_2} \qquad (3.2.5)$$

3.2.3 The Analysis of the Ideal Discharge Coefficient for Outflow at a Plain End on a Pipe

The ideal discharge coefficient, Cd_{id}, follows the definition of Eq. 3.2.1 more strictly, and is defined by:

$$Cd_{id} = \frac{\text{measured mass flow rate}}{\text{ideal mass flow rate}} = \frac{\dot{m}_{ex}}{\dot{m}_{id}} \qquad (3.2.6)$$

Using, and solving, the theory of Sec. 2.16 on a computer presents no real problem with the technology available today. The upstream condition is that of a cylinder, a plenum or the atmosphere where the particle velocity is assumed to be zero. Within the numerical solution, the cylinder pressure and temperature, p_1 and T_1, are held at the measured values and the values of the incident pressure amplitude ratio, X_{i2}, are varied until the calculated superposition pressure corresponds precisely to the measured pipe pressure, p_2. The value of the throat area is retained at the geometrical value, A_t, which, of course, in this instance is the pipe area, A_2. The complete solution, which includes non-isentropic flow from the throat to the pipe point, yields the "ideal" mass flow rate, $\dot{m}_{id}$, as if there were no losses at the pipe entry point. Eq. 3.2.6 is then applied to find the ideal discharge coefficient, Cd_{id}.

3.2.4 The Analysis of the Actual Discharge Coefficient for Outflow at a Plain End on a Pipe

The theory of Sec. 2.16 is used, and solved, on a computer using the numerical techniques detailed in that section. Within the numerical solution, the cylinder pressure and temperature, p_1 and T_1, are held at the measured values, and the values of the incident pressure amplitude ratio, X_{i2}, are varied until the calculated superposition pressure corresponds precisely to the measured pipe pressure, p_2. As the incident pressure amplitude ratio, X_{i2} is varied, the numeric value of the throat area is also changed until a value of effective area, A_{te}, is found which gives a theoretical mass flow rate that corresponds precisely with the measured mass flow rate, $\dot{m}_{ex}$. The double iterative process involved is not simple and, as with all such numerical procedures, requires internal arithmetic guidance to avoid either an excessive number of iterations or computational failure. The actual discharge coefficient, Cd_a, is now defined by:

$$Cd_a = \frac{\text{effective throat area}}{\text{geometrical throat area}} = \frac{A_{te}}{A_t} = \frac{A_{te}}{A_2} \tag{3.2.7}$$

3.2.5 Which Analysis of the Discharge Coefficient is Correct?

Consider the flow at a pressure ratio of 1.305 from the atmosphere to the 22.82-mm internal diameter round steel pipe with a 1.29-mm thick wall. The actual measured test conditions in the experiment are as follows: The atmospheric pressure and temperature are 753 mm Hg and 24°C. The pipe pressure is -175.8 mm Hg, and the temperature is 9°C. The mass rate of the air flow, $\dot{m}_{ex}$, is measured at 67.1 g/s. This gives an atmospheric pressure of 1.0039 bar and a pipe pressure of 0.769 bar, which translates to an overall pressure ratio, P, of 1.3047, as defined by Eq. 3.1.1. The area ratio, k, for the flow is 1.0, as defined by Eq. 3.1.2.

Using the theory given in Secs. 3.2.2 to 3.2.4, the several variants of the discharge coefficient are computed. The isentropic discharge coefficient, Cd_{is}, is calculated as 0.8102. The ideal discharge coefficient, Cd_{id}, is calculated as 0.8182, which is very similar to the isentropic Cd. However, the actual discharge coefficient, Cd_a, is found to be very different at 0.7007. There is a variation of some 14% between the actual discharge coefficient and those determined by the more traditional approaches.

The pipe entry wall thickness is 1.29 mm. If this sharp pipe edge at entry is rounded very slightly, a small improvement in the air mass flow rate is to be expected, and such is the case. With the same atmospheric conditions and with a pipe pressure of -176.1 mm Hg, i.e., a very similar overall pressure ratio of 1.3054, the mass rate of the air flow, $\dot{m}_{ex}$, is measured as having increased to 70.2 g/s. It is expected that the discharge coefficients should be higher and such is the case; the isentropic discharge coefficient, Cd_{is}, is now 0.8471, the ideal discharge coefficient, Cd_{id}, is now 0.8556, and the actual discharge coefficient, Cd_a, is now 0.7335. Thus, the two traditional Cd values reflect correctly the improvement in the air flow characteristics and are totally adequate as a means of providing comparative data on the relative merits of one duct-end aperture geometry vis-à-vis another. What they do not give, and what only the analysis of the actual discharge coefficient can provide, is a numerical value for a discharge coefficient which, when replayed into an engine simulation for a particular aperture geometry, is a computed mass flow that is identical to the measured mass flow rate at the same overall

334

pressure ratio. This criterion is essential if the measurement of discharge coefficients for all pipe end discontinuities is to yield data on discharge coefficients that will provide accuracy when used within an engine simulation model. Only the analysis that deduces the actual discharge coefficient, Cd_a, satisfies this criterion. The other two, the isentropic and the ideal, Cd_{is} and Cd_{id}, are shown to be considerably in error in this regard.

3.2.6 The Discharge Coefficient for Outflow at a Plain End on a Pipe as a Function of Pressure

The experiments concerning flow into the sharp-edged plain pipe end from the atmosphere are conducted over a range of pressure ratios, and the data are analyzed for both the ideal and the actual discharge coefficients [3.16]. The results are plotted in Fig. 3.5.

It is observed that the actual discharge coefficient, Cd_a, is a considerable function of pressure ratio, P. The mathematical form of that function is given here:

$$1 < P < 1.4 \qquad Cd_a = -23.543 + 60.686P - 51.04P^2 + 14.387P^3 \qquad (3.2.8)$$

$$P \geq 1.4 \qquad Cd_a = 0.838 \qquad (3.2.9)$$

Stated more simply, the actual discharge coefficient varies with pressure ratio in a near-linear fashion from about 0.5 at a no-flow condition, where P is 1.0, to about 0.7 where the pressure ratio is 1.3. By contrast, the ideal discharge coefficient, and by inference its close relative the isentropic, hardly varies at all with pressure ratio. Previous researchers [3.1-3.14]

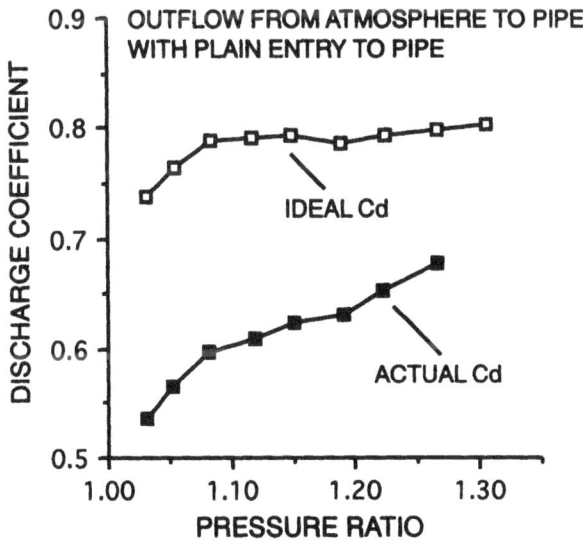

Fig. 3.5 The discharge coefficients for an outflow process from the atmosphere to a pipe.

who used the traditional approach to data reduction and determined either Cd_{is} or Cd_{id} data only, probably observed the apparent lack of relevance of pressure ratio on their discharge coefficients and, not surprisingly, upgraded it to the status of a general conclusion. This probably explains why much of the data published previously were obtained at but a single pressure ratio, with a pressure ratio of 1.2 being a very popular choice for the experimental value of that parameter.

Consider further information from Fig. 3.5, based on experimental data of flow into the sharp-edged pipe at a pressure ratio of 1.2. The ideal discharge coefficient, Cd_{id}, is seen to be 0.78 and, when employed at this pressure ratio into the theory of Sec. 2.16, the computation yields a mass flow rate of air of 69 g/s. The actual discharge coefficient, Cd_a, at this pressure ratio is 0.65, and a similar employment of this value in Sec. 2.16 theory gives, as it must by definition, the measured mass flow rate of 59.5 g/s. As the computed mass flow rates are different by some 16% this demonstrates why it is quite intolerable to use a traditionally determined discharge coefficient in an engine simulation model.

To obtain data on discharge coefficients that are effective and accurate within an engine simulation, it is clearly essential to deduce the actual discharge coefficient, Cd_a, over as wide a range of pressure ratio as is practical within the constraints imposed by the mass flow rate potential of the actual test apparatus.

3.2.7 Comparison of 1D and 3D Computation for Outflow at a Plain End on a Pipe

The three-dimensional computation of this flow process by CFD has already been introduced in Sec. 3.1 and Figs. 3.3 and 3.4 illustrate the outcome. The computations are carried out at two pressure ratios, namely 1.083 and 1.305, from an atmosphere at a pressure of 760 mm Hg and a temperature of 20°C. Using the CFD, the air mass flow rates at these two pressure ratios are computed as 46.4 and 64.84 g/s, respectively. The measured values, at the same two pressure ratios in the experiments to determine the discharge coefficients, are 42.3 and 67 g/s, respectively. This shows the accuracy of the CFD to be very reasonable, considering the relatively coarse mesh structure of 60,000 cells that is employed.

As already pointed out above with reference to Figs. 3.2 and 3.4, the CFD shows that there is no single effective throat area, A_{te}, and there is no single particle velocity, c_t, that can be assigned to it. Yet these are the unavoidable assumptions which must be made in a one-dimensional analysis, even in one as complete as the non-isentropic version of Sec. 2.16.

During the analysis of the experimental data to deduce the actual discharge coefficient, Cd_a, using the theory of Sec. 2.16, the complete sets of thermodynamic values at the three stations in Fig. 3.2, are calculated. In all cases, by the definitions in Sec. 3.2.4, the measured and calculated mass rates of flow of air correspond precisely.

Data from the 1D Analysis at a Pressure Ratio of 1.083

Consider the analysis of the experimental data at the pressure ratio of 1.083. The discharge coefficient is determined to be 0.596, as seen in Fig. 3.5. The effective throat area, A_{te}, is calculated to be 241.9 mm², or a diameter of 17.55 mm. This corresponds to a radius ratio of 0.77, which can be seen to correlate quite well with the cross-over point of the CFD-predicted velocity profiles at the one- and four-diameter locations on the left half of Fig. 3.4. The 1D code predicts that the throat velocity, c_t, is 162.8 m/s, which is about the average of the CFD predictions at the pipe center line at the one- and four-diameter locations. The 1D code further predicts that the pipe velocity, c_2, is 91.6 m/s, which is near the average of the CFD predictions for the pipe location at four-diameters from the end.

Data from the 1D analysis at a Pressure Ratio of 1.305

Consider the analysis of the experimental data at the pressure ratio of 1.305. The discharge coefficient is determined to be 0.701, as seen in Fig. 3.5. The effective throat area, A_{te}, is calculated to be 285.2 mm^2, or a diameter of 19.05 mm. This corresponds to a radius ratio of 0.84, which can be seen to correlate very well with the CFD-predicted velocity profiles at the quarter-diameter pipe location on the right half of the Fig. 3.4. In this higher pressure case, the CFD result shows more of a pronounced throat with a relatively constant velocity of 190 m/s across it, compared to the lower pressure ratio flow on the left half of the same diagram. However, the 1D code predicts that the throat velocity, c_t, is 308 m/s, which is at least 30% higher than the CFD prediction at the pipe center line at the quarter-diameter location. This is not too surprising, as the 1D code, with a Cd_a of 0.70, basically assumes that there is no flow at all over 30% of the cross-section area of the pipe at the throat! The 1D code further predicts that the pipe velocity over the entire cross-section area, c_2, is 170 m/s, which is very close to the average of the CFD predictions at the four-diameter pipe location. This, too, is not so surprising, as both the 1D and the 3D code are computing over the same total pipe area.

3.3 The Discharge Coefficients of Bellmouths at an Open End to a Pipe

The Intake Process into an Engine

The use of a bellmouth at the end of an intake pipe is the conventional method employed to improve the mass rate of flow of air into the intake pipe from the atmosphere. If the engine has an intake silencer box, in order to provide either noise reduction and/or filtering of the intake air, to maximize the air flow into the engine, all pipes leaving this plenum will have bellmouth entries. From the foregoing discussion on plain-ended pipes, within an engine simulation it is clearly essential to have available the data on the actual discharge coefficients, Cd_a, of bellmouths over a significantly wide range of pressure ratios.

Measured Discharge Coefficients for Bellmouth Entry to Pipes

The apparatus used to determine the discharge coefficients of bellmouthed pipe entries is shown in Fig. 3.1. The plain-ended pipe is fitted with a series of "wrap-around" bellmouths, which are circular in cross-section. The internal pipe diameter, d_2, is 22.82 mm and the cross-section radii, r_b, of the three bellmouths tested are 4, 5, and 6 mm, respectively. The tests are conducted over a pressure ratio range of 1.0 to about 1.3. The results are plotted in Fig. 3.6 for both the ideal discharge coefficient, Cd_{id}, and the actual discharge coefficient, Cd_a [3.16].

As with the results for the plain-ended pipe seen in Fig. 3.5, the ideal discharge coefficient, Cd_{id}, is higher than the actual discharge coefficient and is virtually unaffected by differing pressure ratios. It can also be seen that the larger the radius of the bellmouth, the higher the

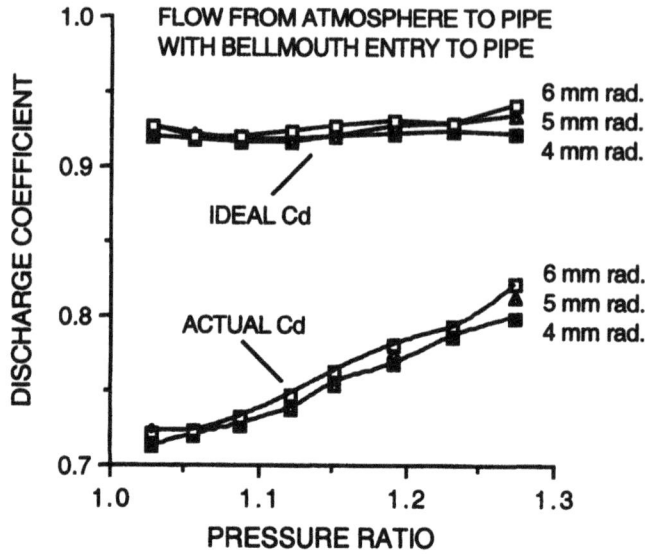

Fig. 3.6 The discharge coefficients for outflow from the atmosphere to a bellmouthed pipe.

discharge coefficient, although there is only a 1% or 2% advantage for one vis-à-vis the other. The actual discharge coefficients for the several bellmouths can be fitted with polynomial curves and used within an engine simulation model:

For $r_b = 4.0$ mm: $1 < P < 1.4$ $Cd_a = 0.64133 - 0.17437P + 0.23533P^2$ (3.3.1)

For $r_b = 5.0$ mm: $1 < P < 1.4$ $Cd_a = 1.5106 - 1.7002P + 0.90592P^2$ (3.3.2)

For $r_b = 6.0$ mm: $1 < P < 1.4$ $Cd_a = 1.1887 - 1.1604P + 0.68420P^2$ (3.3.3)

For all radii: $P \geq 1.4$ $Cd_a = 0.905$ (3.3.4)

It is reasonably clear that a bellmouth improves the discharge coefficient of the flow. This can be seen for the ideal and the actual discharge coefficients in Figs. 3.5 and 3.6. In Fig. 3.7, using Cd_a values only, the bellmouth clearly improves the flow at the pipe end.

It can be seen in Fig. 3.7 that the pressure-related profiles of the actual discharge coefficients of both pipe ends are very similar, but those for the bellmouth end are consistently 0.13 higher than for the plain pipe end. This means that, for the same pressure ratio, the air mass flow rate into the pipe is typically 20% higher with a bellmouth end. The adoption of a bellmouth at the end of an intake pipe is clearly the best design option and one that is well worth optimizing

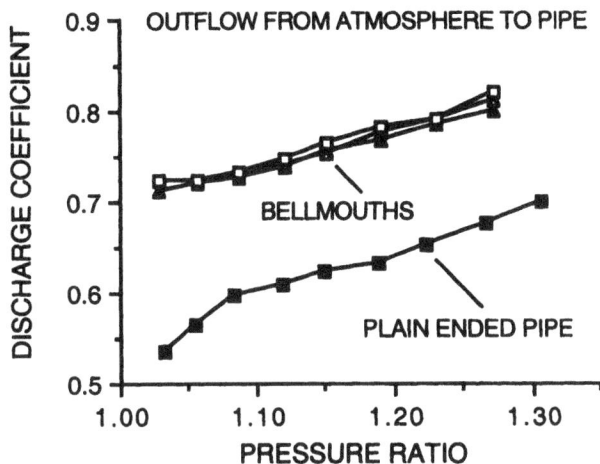

*Fig. 3.7 The Cd_a values for outflow from the atmosphere
to plain and bellmouthed pipes.*

on an experimental apparatus to improve its discharge coefficient. It is interesting to note that the technical literature is not exactly bulging with information on this topic which is manifestly important to the designer of engines, particularly of high-performance racing engines.

3.4 The Discharge Coefficients of a Throttled End to a Pipe
The Restriction at a Pipe End Facing a Large Plenum

A restriction at the end of a pipe is found very commonly within the geometry of an engine. The poppet valve poses a restriction to flow at the end of intake or exhaust ducts. In a two-stroke engine, the piston-controlled port provides a time-varying restriction to the inflow, or outflow, of gas at a cylinder or a crankcase. It is common practice to crimp pipe ends within an exhaust silencer (muffler) to enhance the silencing characteristics by reducing the mass flow rate and raising the transmitted noise frequency of the pipe element. The turbocharger nozzle ring poses a restriction to the exhaust pressure wave motion at the end of the exhaust manifold so as to raise the particle velocity of the gas at the turbine rotor entry. Naturally, to provide the necessary data for accurate engine simulation, all of these elements must be tested, and the actual discharge coefficients found, over a suitable range of pressure ratios for the particle flow in both directions.

With any technology, it is always instructive know the behavior of a fundamental version of a complex device. Hence, it is useful to initially test a fundamental pipe end restriction and then compare poppet valves, two-stroke ports, throttled pipe ends, or turbocharger nozzle rings for their characteristic behavior with respect to it. The most fundamental restriction is a sharp-edged orifice. Fig. 3.1, shows a thin, sharp-edged orifice placed at the open end of the pipe facing the atmosphere [3.16]. As shown in Fig. 3.1, this is a gas dynamic case of outflow from an infinitely large plenum, i.e., the atmosphere, to a pipe, thus the theory of Sec. 2.16 will apply. If the pipe is inverted into the settling tank, with the orifice retained in position at the pipe end, this becomes a case of inflow from a pipe through the orifice to a large plenum, i.e., the settling tank; in this case, the theory of Sec. 2.17 now applies.

The Orifices Tested at the End of the Pipe
  The internal pipe diameter, d_2, is 22.82 mm, and is the same pipe as described above. There are five orifices to be placed on the pipe end, all of which are centrally located. These orifices are of varying diameter to give port-to-pipe area ratios, k, of 0.1, 0.3, 0.5, 0.7, and 0.9. The orifice wall is 1.5 mm thick and all edges at the orifice are sharp-edged. The tests are conducted for flow in both directions, i.e., for outflow as sketched in Fig. 3.1, and for inflow with the pipe inverted. The pressure ratios employed experimentally range up to 2.0 for both flow directions, although for the larger-diameter orifices the apparatus could not sustain constant velocity flow at the very highest pressure ratios. The definition of pressure ratio is different for inflow, compared to outflow, so that the quoted pressure ratio, P, always ranges upward from 1.0 at a no-flow condition.

For outflow: $$P = \frac{\text{cylinder pressure}}{\text{pipe pressure}} = \frac{p_1}{p_2} \qquad (3.4.1)$$

For inflow: $$P = \frac{\text{pipe pressure}}{\text{cylinder pressure}} = \frac{p_2}{p_1} \qquad (3.4.2)$$

  Naturally, the "cylinder" pressure in the inflow tests is that in the settling tank, whereas in outflow tests, it is the atmospheric pressure. The discharge coefficients, both actual and ideal, are extracted from the experimental data in the manner already described in Secs. 3.2.3 and 3.2.4, and are plotted using software that I wrote. Having already made the point that the ideal discharge coefficients have no application within an accurate engine simulation, only the actual discharge coefficients, Cd_a, will be plotted and discussed in the rest of this section. If you wish to examine the difference between the ideal and actual discharge coefficients for this configuration, information on this can be found in the literature [3.16].

Mapping the Discharge Coefficients for the Orifices Tested at the End of the Pipe
  The experimental data are analyzed and the actual discharge coefficients, Cd_a, together with the associated pressure ratio, P, and the area ratio, k, are taken into my software mapping program. The results of the mapping process for outflow are presented in Fig. 3.8.
  This figure shows what the user sees on the computer screen, which is an isometric projection of a three-dimensional map with discharge coefficient as the z-axis, pressure ratio as the y-axis, and area ratio as the x-axis. The only difference is that the screen map is in color! Actually, this is rather important because the experimental points—the short vertical lines seen poking through the "planes"—are colored either red or blue to indicate if the fit of the line to the "plane" is above or below it. The lines have been deliberately left somewhat longer than normal, i.e., the fit of any experimental point to the "plane" has been deteriorated so that you may observe it clearly! An experimental line colored red indicates that the Cd value is at the

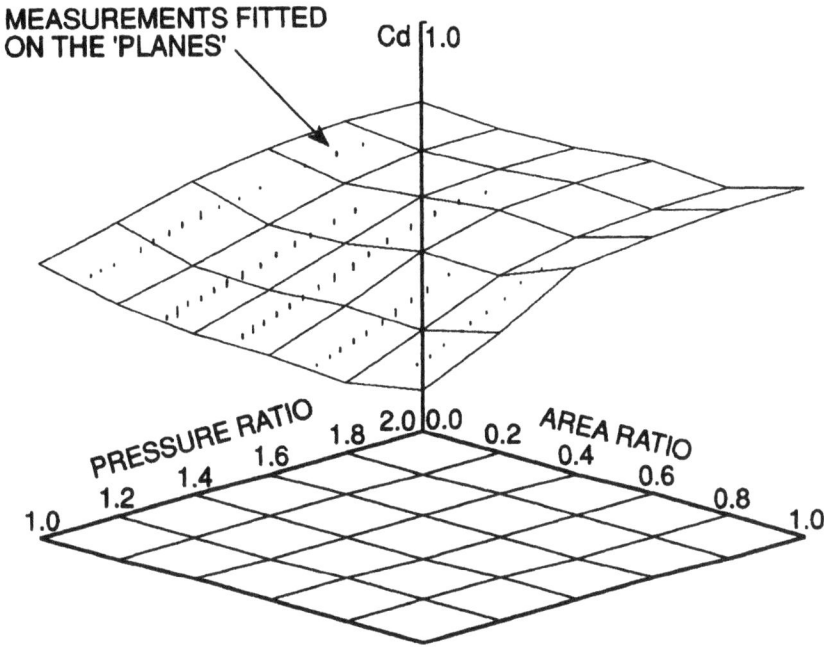

Fig. 3.8 The mapping of Cd_a values for outflow from the atmosphere to a throttled pipe.

top of the line and its fit to the "plane" is at the bottom, whereas a line colored blue indicates that the Cd value is at the bottom of the line and its fit to the "plane" is at the top. Naturally, in the black and white rendition of Fig. 3.8, this cannot be observed.

The fit of the experimental points to the twenty-five "planes" can be seen. Most of you will be aware that it is not normally possible to mathematically fit a flat plane between any four points in space; this is only normally possible between three points in space. Thus, the fit of the experimental points to the twenty-five "planes" you are observing is actually the fit of those points to fifty triangles in space, which explains why the word "plane" thus far in this discussion has been enclosed in quotation marks. Later on in this chapter, in Sec. 3.8 the entirety of this mapping exercise will be explained in great detail, for its application in reverse is how the engine simulation model decodes the data on any Cd map and reemploys it at the relevant duct discontinuity.

There are several advantages of mapping the discharge coefficients in this manner. Not every piece of experimental data is "kosher." Manometers may be misread, temperatures may not be recorded accurately, a joint may be leaking air at some test point, or a pressure transducer attached to its computerized data logging system may malfunction. The result is a Cd value giving a line length off a "plane" on the map that is visibly at odds with its near neighbors and can be rejected from the set and the mapping process. Frankly, if any set of experimental data is analyzed blindly without recourse to a means of rejecting bad data points, the outcome is a Cd map that is in error. In this computer-oriented era there's a natural tendency to trust implicitly data acquisition systems and the information supplied to them by the transducer(s), often with scant regard for transducer calibration drift due to part-faulty connections and sealing, aging, temperature, vibration, humidity, etc., etc. While thousands of data points can be so measured and analyzed in very short time frames, if a few transducer signals among those thousands are in part-error, and no procedure is used to find and eliminate them, the unwitting outcome is useless.

Fig. 3.8 shows a series of experimentally fitted points at various pressure ratios running up a line where the value of the area ratio, k, is consistently either 0.1, 0.3, 0.5, 0.7, or 0.9, i.e., the k values declared above. The highest pressure ratio, P, that could be employed by the experimenter, H.B. Lau [3.16], can be seen to be 1.75 at a k value of 0.1; 1.5 at a k value of 0.3; 1.7 at a k value of 0.5; 1.38 at a k value of 0.7; and 1.4 at a k value of 0.9. These were the highest pressure ratios that allowed the apparatus to run at steady state. Beyond these pressure levels on the map, the Cd values at the plane corners are set purely by the eye of the experimenter and constitute extrapolation. You are quite entitled to treat such extrapolation with the traditional contempt it deserves. Within my computer software which draws this map to the screen, each of the plane corners has an individual "button" which permits the user to raise or lower the Cd value until the fit, either of interpolation among the points or of extrapolation beyond them, is regarded as satisfactory. When the mapping process is complete and the best estimate of the surface profile over the total range of pressure ratio, P, from 1 to 2 and area ratio, k, from 0 to 1 is obtained, the Cd data at the corners of the planes can be extracted for use within the engine simulation model. They may also be plotted on an xyz contour chart, in the manner presented in Fig. 3.9, allowing you to acquire the data directly, to a sufficiently high accuracy by imposing a simple grid over the figure and interpolating for its Cd values. It can be confidently used within any engine simulation model you create, particularly if the boundary conditions are computed with the theory given in Secs. 2.16 and 2.17. The mapped data for the throttled pipe end for outflow from the atmosphere through a central, sharp-edged orifice is shown in Fig. 3.9.

On Fig. 3.9, it can be seen that, in this fundamental example of throttled end outflow, the Cd value is comparatively insensitive to area ratio, but is very much a function of pressure ratio. Thus, because the bulk of the traditional approach tends to quote testing at just one pressure ratio [3.1-3.14], and typically with a pressure ratio of 1.2, the inadequacy of that approach to acquire simulation-relevant data is, yet again, self-evident.

3.4.1 The Discharge Coefficients for Inflow for the Orifices Tested at the End of the Pipe

When the pipe, with the orifice at its end, is inverted into the tank in Fig. 3.1, in the manner described above, it becomes possible to repeat the experimental series to measure the discharge coefficients for inflow, but this time the flow is from the pipe through the orifice into the very large plenum that is the settling tank [3.16]. The theory of Sec. 2.17 now applies and is used in the extraction of the discharge coefficient from the experimental data. The map profiling process described above is carried out on the ensuing Cd, P, and k values, and ultimately the xyz contour map can be plotted, which is shown in Fig. 3.10.

On Fig. 3.10 where the actual discharge coefficients, Cd_a, are plotted, it is clear that the inflow map is quite different from that for outflow shown in Fig. 3.9. Not only are the numbers different as the Cd values are generally higher, but the profile is also dissimilar. At high and low area ratio values, i.e., k values above 0.8 and below 0.2, the discharge coefficient is not particularly sensitive to pressure ratio. In the middle zone, where the area ratio ranges from 0.2 to 0.8, the Cd value is a modest function of pressure ratio.

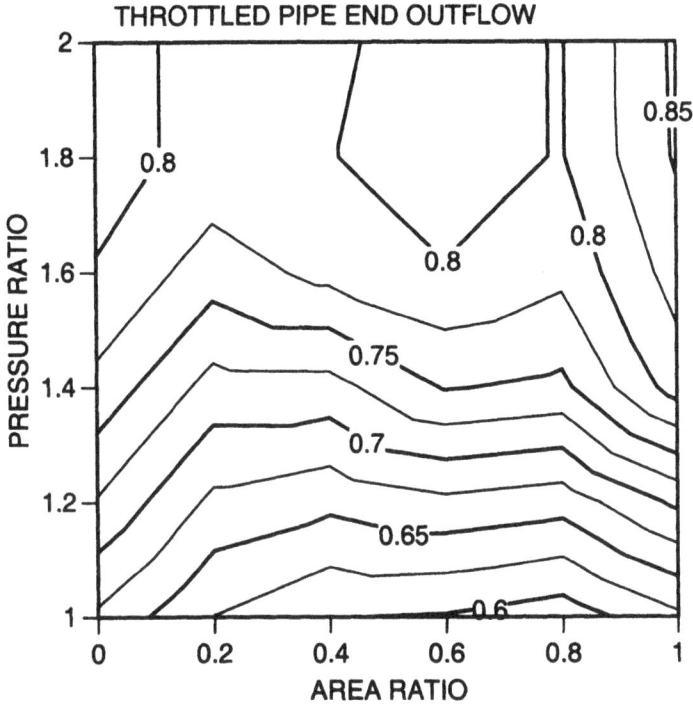

Fig. 3.9 The Cd_a values for outflow from the atmosphere to a throttled pipe.

As with outflow, testing at a constant pressure ratio is clearly inadequate as a means of acquiring Cd data for engine simulation purposes. The need to measure Cd characteristics at a pipe end discontinuity for flow in both directions is again highlighted. You can also draw a simple grid over the contour map in Fig. 3.10 and extract the Cd data for simulation purposes.

3.4.2 The Discharge Coefficients for a Fundamental Restriction at the End of a Pipe

At the beginning of this section it is pointed out that inflow and outflow at these centrally located pipe end orifices constitutes a fundamental experiment in flow at restricted pipe ends. This provides typical Cd data against which all other pipe end restrictions may be compared. In the publication by Blair et al. [3.16], the test data for other fundamental restrictions are published, such as for round, sharp-edged orifices not located co-axially with the pipe center line, and rectangular orifices with radiused corners.

3.5 The Discharge Coefficients of a Port in the Cylinder Wall of a Two-Stroke Engine

3.5.1 The Restriction at a Pipe End Facing the Cylinder of a Two-Stroke Engine

At this point, some of you, having paid good money for this book in order to learn about the design and simulation of four-stroke engines, may well wish to be immediately reimbursed upon reading the title of this section. However, there is a fundamental rationale for including this section, and I am sure that, when you have finished reading it, you will agree with my

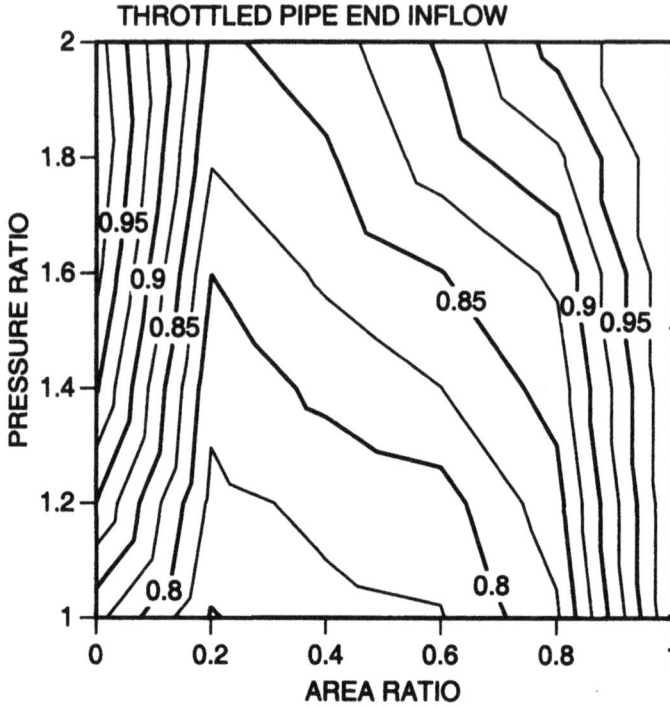

Fig. 3.10 The Cd_a values for inflow to the settling tank from the throttled pipe.

viewpoint! In the interim, could I remind you that the apertures of the sleeve valve, or a rotary valve, for a four-stroke engine have the same geometry as the cylinder wall port of a two-stroke engine.

Most two-stroke engines breathe air in, and eject exhaust gas through, ports cut in the cylinder wall [3.17]. These ports are normally rectangular is shape with radiused corners. The basic geometry of such a port in a cylinder wall is shown in Fig. 3.11.

The top edge of the piston controls the timing of the opening and closing of the port and, by definition, the area of the port, A_t, exposed at any instant during the piston movement. At the extreme movement of the piston, the port is normally fully uncovered giving an exposed port area, A_p, which at that juncture is also the port throat area, A_t. The gas flows between the port and the pipe where the manifold diameter is d_m with an area A_m which, in the nomenclature of either Fig. 2.16 for outflow, or Fig. 2.18 for inflow, are also d_2 and A_2, respectively. In Secs. 3.2 to 3.4 the discharge coefficient is found with respect to pressure ratio, P, and to the port-to-pipe area ratio, k. Remember from Eq. 3.1.2, that the port-to-pipe area ratio, k, is defined as:

$$k = \frac{\text{throat area}}{\text{pipe area}} = \frac{A_t}{A_m} = \frac{A_t}{A_2} \qquad (3.5.1)$$

344

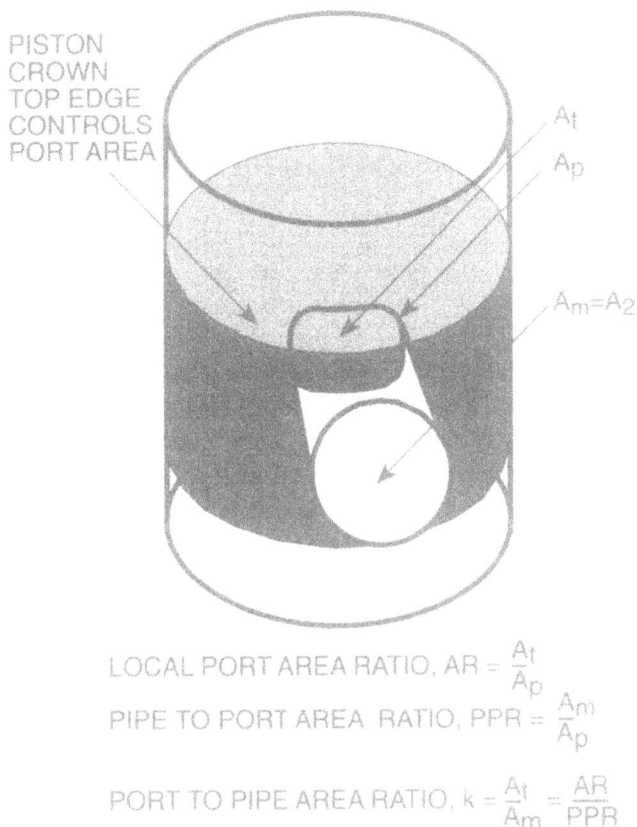

Fig. 3.11 The exposed throat, port, and pipe areas in a two-stroke engine cylinder.

Up to now in this chapter, the area of the pipe at the port, A_p, has always equaled the downstream pipe area, but now, because A_p rarely equals the pipe area, A_2, in a two-stroke engine, something of a dilemma is faced. The measurement technique, together with the appropriate analytic technique to deduce the discharge coefficient, is unaffected by this new geometry, but the plotting and mapping of the data is more problematic. The reason is that there are now three possible area ratios to be considered with relevance to the plotting process, two of which are variables with the piston movement, and one of which is a constant. These area ratio relationships are summarized in Fig. 3.11. One is shown above in Eq. 3.5.1, and the other two, referred to as the local port area ratio, C_t, and the pipe-to-port area ratio, C_m, are repeated here.

Local port area ratio: $$C_t = \frac{A_t}{A_p} \qquad (3.5.2)$$

Pipe-to-port area ratio: $$C_m = \frac{A_2}{A_p}$$ (3.5.3)

The C_m value is manifestly a constant and connects the other two area ratios:

Port-to-pipe area ratio: $$k = \frac{A_t}{A_2} = \frac{C_t}{C_m}$$ (3.5.4)

Thus, when a map of discharge coefficients is to be produced, which varying area ratio, i.e., k or C_t, should be used for the x-axis in the manner previously seen in Figs. 3.8 to 3.10? In other words, does the pipe-to-port area ratio, C_m, have a major influence on the discharge coefficient or is the discharge coefficient influenced solely by the constriction characteristics at the throat of the port? I have already reported the fundamental research on this topic [3.18], the conclusion of which is that the pipe-to-port area ratio, C_m, of the duct leading from the port to the pipe has no influence on the discharge coefficient *for any given geometry of port*. This comes as a relief to the engine modeler, otherwise every time an exhaust or intake *pipe* diameter would be changed within the data for an engine design, a different Cd map would need to be acquired and indexed as further data for the simulation. Hence, when the Cd map for a port in a cylinder wall, such as the two-stroke engine under discussion, is prepared, the x-axis of the map uses the local port area ratio, C_t, and not the pipe-to-port area ratio, k.

3.5.2 The Discharge Coefficients for Ports in the Cylinder Wall of a Two-Stroke Engine

The experiments reported by Blair *et al.* [3.18] examine a simple rectangular-shaped port with corner radii in a cylinder with a bore and stroke of 50 mm. The port is 31.93 mm wide, and 21.98 mm high, and has 5 mm corner radii. The port is profiled out to the pipe, where the pressure and temperature, p_2 and T_2, are measured. The experimental apparatus is as shown in Fig. 3.12, with the engine cylinder placed so as to record the outflow discharge coefficients.

The profile of the top edge of the port and of the duct leading to the pipe is "horizontal," i.e., it is at right angles to the cylinder axis, exactly as sketched in Fig. 3.12. In the Blair et al. paper [3.18], the duct connecting the port to the pipe is varied to give port to pipe area ratios of 1.0, 1.125, 1.25, 1.5, and 2.0. The discharge coefficients for both inflow and outflow are recorded for these five duct configurations with an identical port in each case so that the several Cd maps may be compared to investigate the influence of the port-to-pipe area ratio, C_m. No significant influence on Cd is found to be caused by the expansion in area from the port to the pipe. The Cd maps with a port-to-pipe area ratio, C_m, of 1.125 are shown plotted in Figs. 3.13 and 3.14 for outflow and inflow, respectively. These are the actual discharge coefficients, Cd_a.

In line with the above discussion, it can be seen that the x-axis of the xyz contour plots in Figs. 3.13 and 3.14 is drawn using the local port area ratio, C_t.

In Sec. 3.4, the fundamental nature of discharge coefficients of pipe end orifices facing the atmosphere is investigated. It is instructive, therefore, to correlate the outflow characteristics of a two-stroke engine port with this fundamental restriction, by comparing Figs. 3.9 and 3.13. For outflow, the basic dependence on pressure ratio of the sharp-edged orifice is seen in Fig. 3.9.

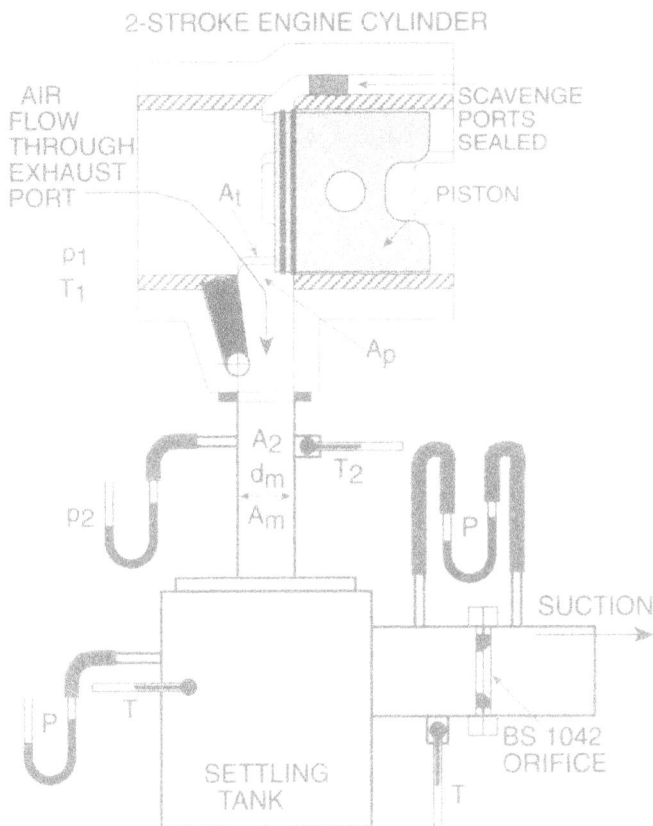

Fig. 3.12 Apparatus to measure the discharge coefficients of two-stroke engine ports.

This dependence is also found in the Cd map for the two-stroke port, but only at area ratios above 0.2. In Fig. 3.13, below the C_t value of 0.2, there is almost no influence from pressure ratio on the Cd value. It would appear that the close proximity of the cylinder wall to the upper edge of the port in the engine geometry, compared to the infinitely open space of the atmosphere for the central sharp-edged round orifice, has an influence on the Cd value of a two-stroke engine port at small port openings when the width-to-height ratio of the port is very large, i.e., the port shape is akin to a thin slit.

Comparing inflow from Figs. 3.10 and 3.14—although the numbers are not exactly the same—it can be seen that the basic shape of the Cd map for the fundamental restriction is very like that of the port in the two-stroke engine.

The similarity between the Cd maps in Figs. 3.13 and 3.14 for a port in the cylinder wall of a two-stroke engine with those shown in Fig. 2.27 for the sliding circular port of the QUB SP rig, is to be expected because they are both examples of ports in the wall of a chamber. The totality of information from these figures, and from the fundamental restriction of the throttled pipe end in Figs. 3.9 and 3.10, should give you a mental picture of the Cd maps that could be expected from sleeve valves or rotary valves and, possibly anticipate, experiments on poppet valves.

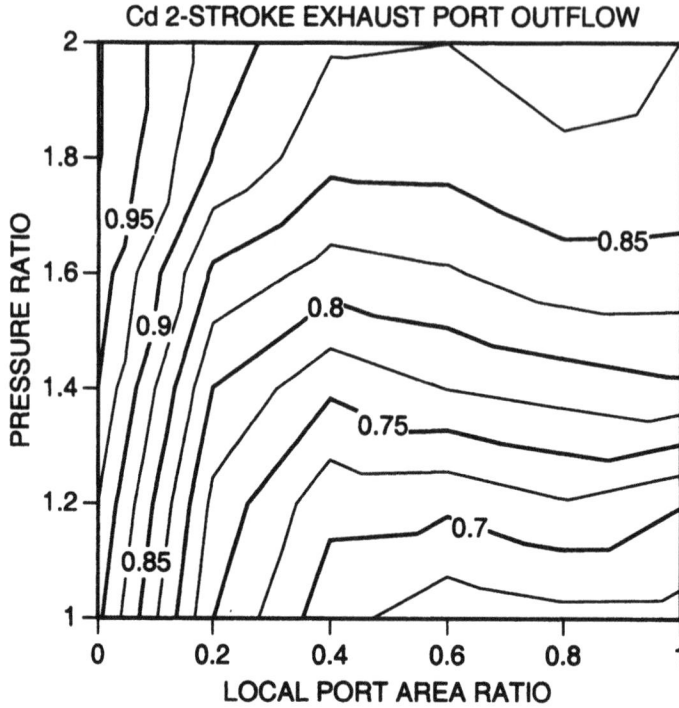

Cd 2-STROKE EXHAUST PORT OUTFLOW

Fig. 3.13 The discharge coefficients for outflow of a two-stroke engine port.

3.5.3 The Effect of the Duct on the Discharge Coefficients for Exhaust Outflow

Although it is attested above [3.18] that the area ratio, C_m, of the duct connecting any given port to the pipe has no influence on the discharge coefficients for inflow or outflow, a test result will now be given that apparently contradicts that statement. In an SAE paper, Jante [3.19] stated that an exhaust port in a two-stroke engine that is inclined "down" by 10° will assist the exhaust outflow process from it. The word "down" means that the opening timing edge of the duct, and the center line of the duct itself, is directed at an angle, in this case 10°, away from the cylinder head and toward the piston; this is an anti-clockwise direction on the sketch in Fig. 3.12. The logic is that the duct profile conforms more closely to the direction the exiting gas flow is attempting to follow.

In the tests reported by Blair et al. [3.18], another cylinder with a port-to-pipe area ratio, C_m, of 1.125, is manufactured with a "downward" inclination of the duct of 10°, and tested for its outflow discharge coefficients. Apart from the duct inclination angle which extended from the pipe to the port, the cylinders are identical. Because all of the cylinders in this test series [3.18] were made using stereo-lithography, dimensional similarity from one cylinder to another can be assured.

The outflow discharge coefficients of this modified cylinder are acquired and plotted and are compared with the outflow discharge coefficients of the cylinder with the same C_m of 1.125, but with a standard port declination of zero. To do so, the Cd values at the "plane"

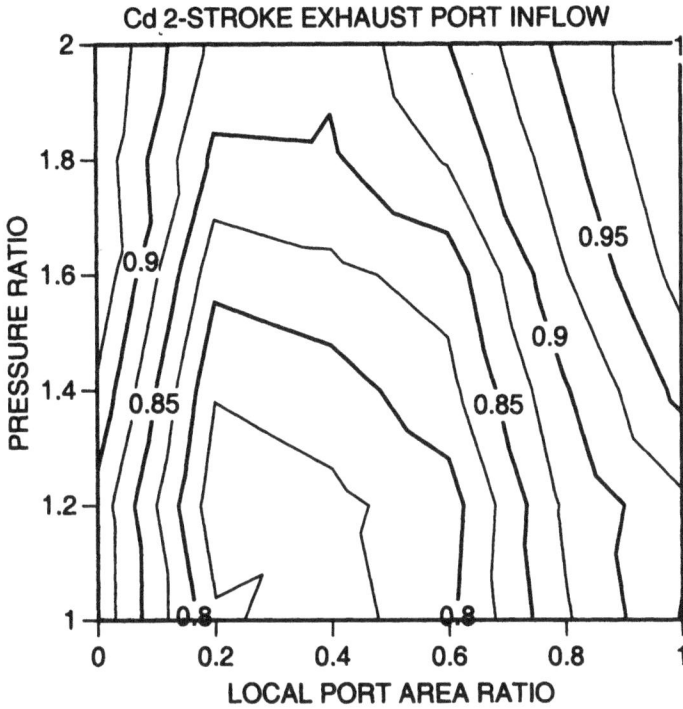

Fig. 3.14 The discharge coefficients for inflow of a two-stroke engine port.

corners of the two maps are subtracted, and the differences reduced to a percentage of those from the standard cylinder.

$$\text{At each map point:}\quad \text{Cd}_\% = 100 \times \left(\frac{\text{Cd}_{10° \text{ 'down'}} - \text{Cd}_{\text{standard}}}{\text{Cd}_{\text{standard}}} \right) \qquad (3.5.5)$$

The result of this data reduction is shown in Fig. 3.15, where positive values indicate a higher value of Cd for the cylinder with the 10° port declination.

This improvement, as indicated by Jante [3.19], is as high as 15%, particularly at local port area ratios, C_t, in the middle band of 0.3 to 0.6, when the port is half-open. Actually, Jante stated that the discharge coefficient is improved during the blowdown period of exhaust opening, i.e., at C_t values from 0 to about 0.3, but the maximum improvement is seen to be at somewhat greater port openings.

That the discharge coefficient has been changed is unarguable. The basic question is now fundamental: Has the discharge coefficient been altered by a modification to the duct profile or by a modification to the port at its exit point from the cylinder? Although this question must remain unanswered in a fluid mechanics sense, as far as the engine modeler is concerned, the availability of both data maps will provide the precise outcome regarding the simulated engine

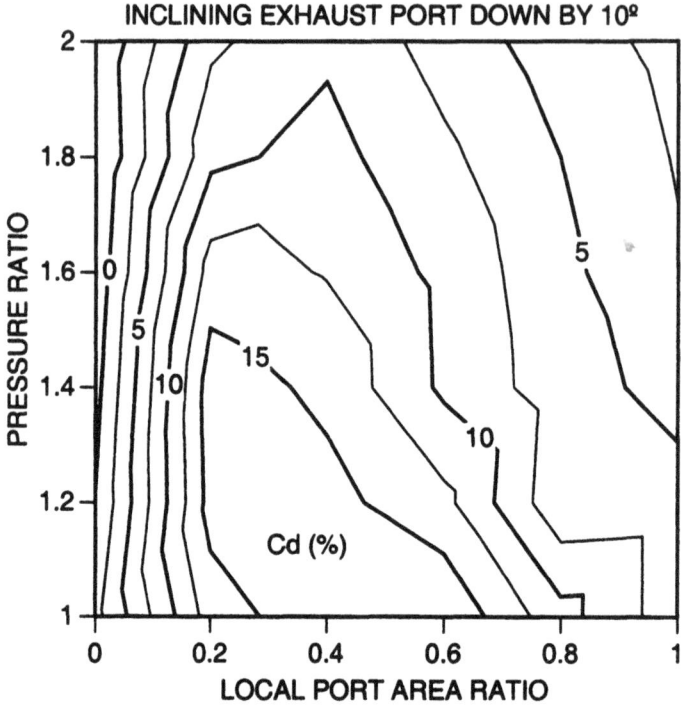

Fig. 3.15 Improving the outflow discharge coefficients of a port by duct declination.

performance characteristics. Although there is no complete experimental proof of this statement, the implications thus far are that, if another series of engine cylinders with varying C_m values were manufactured and each had a $10°$ port declination angle, any Cd experiments on outflow would show them to be indistinguishable one from another. In short, for outflow, the fluid mechanics of the flow through the throat at the port dictates the Cd value to the greatest extent, but the duct shape from the port to the pipe also has an influence, albeit lesser. For inflow, not enough experimental evidence is available to come to any, or similar, conclusions.

The fundamental significance of these results regarding ports and orifices must be borne in mind when reading the following sections, which are mainly devoted to the discharge coefficients of poppet valves.

3.6 The Discharge Coefficients of Poppet Valves in a Four-Stroke Engine
3.6.1 The Restriction at a Pipe End Facing the Cylinder of a Four-Stroke Engine
In Sec. 1.5, the geometrical flow area, A_t, created by the lift of any poppet valve is presented in great detail. At the port, the geometrical flow area, A_p, the combination of that posed by the valve seat and port and the valve stem, is also described in equally fine detail. From the

port the gas flows to, or from, the duct at a point where the cross-section area, A_2, normally becomes constant. This geometry is summarized in the sketch in Fig. 3.16. Put simply, the valve, valve stem, and port pose a restriction at the end of an intake or an exhaust pipe much like the orifices or two-stroke ports discussed above.

It is possible to define area ratios for this poppet valve geometry, just as it was for the throttled pipe end and the two-stroke engine port. The following are the conventional descriptions for these several area ratios, which were found to be helpful in the mapping of discharge coefficients for the other pipe end restrictions. However, as will be demonstrated shortly, these descriptions are inadequate for the same purpose for poppet valves.

Local port area ratio: $$C_t = \frac{A_t}{A_p} \qquad (3.6.1)$$

Pipe (manifold)-to-port area ratio: $$C_m = \frac{A_2}{A_p} \qquad (3.6.2)$$

As heretofore, the C_m value remains as a constant.

Port-to-pipe area ratio: $$k = \frac{A_t}{A_2} = \frac{C_t}{C_m} \qquad (3.6.3)$$

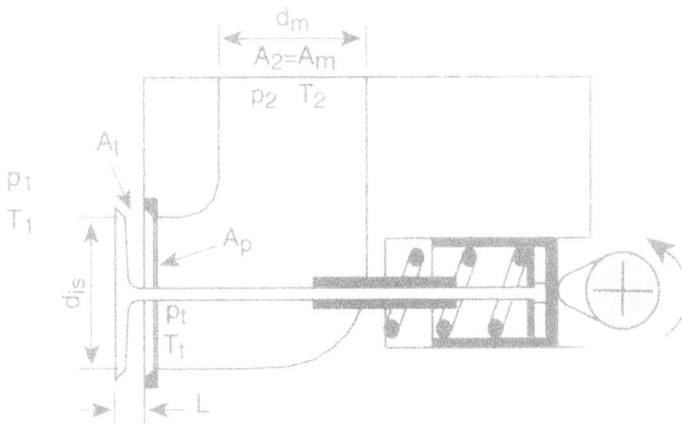

Fig. 3.16 The geometry at a poppet valve in the cylinder head.

However, for a poppet valve geometry as described in Sec. 1.5.1, at high levels of valve lift, L, particularly for those approaching conventional levels of maximum valve lift, L_v, the flow area at the valve throat normally exceeds the flow area at the port, i.e., at a conventional high valve lift:

$$A_t > A_p$$

Hence, the local port area ratio, C_t, is now greater than unity. In short, at high valve lifts, it is the port that becomes the major geometrical flow restriction and not the valve throat area. Due to this duality for the flow area at a poppet valve, in order to assess a discharge coefficient for the flow past it, it becomes necessary to specify the geometrical flow area of the restriction, A_r, much more precisely than for throttled pipe ends or two-stroke engine ports:

$$A_t < A_p \qquad A_r = A_t \tag{3.6.4}$$

$$A_t \geq A_p \qquad A_r = A_p \tag{3.6.5}$$

The pipe area, A_2, has either the same area as the port, A_p, as is conventional for a well-designed intake duct, or is larger, as is normal in an exhaust duct in the cylinder head. This topic is discussed more thoroughly in Chapter 6. However, it is possible—indeed is not uncommon in intake ducts within the cylinder head—for the port area, A_p, to be larger than the pipe area, A_2. In this case, as the theory in Secs. 2.16 and 2.17 shows for particle flow between a pipe and a port, it is the pipe that apparently becomes the major restriction at high valve lifts:

$$A_2 > A_p \quad A_p = A_2 \quad A_r = A_2 \tag{3.6.6}$$

Taking into account the above arguments by assigning the correct restriction area at the port, A_r, it becomes possible to accurately formulate the various area ratios for a poppet valve, firstly in the context of gas flow between a single valve and a single manifold inlet or outlet:

Local port area ratio: $$C_t = \frac{A_r}{A_p} \tag{3.6.7}$$

Pipe-to-port area ratio: $$C_m = \frac{A_2}{A_p} \tag{3.6.8}$$

Port-to-pipe area ratio:
$$k = \frac{A_r}{A_2} = \frac{C_t}{C_m} \qquad (3.6.9)$$

In a cylinder head design where two or more valves, n_v, are connected to a single intake or exhaust manifold connection, these equations become:

Local port area ratio:
$$C_t = \frac{n_v A_r}{n_v A_p} = \frac{A_r}{A_p} \qquad (3.6.10)$$

Pipe-to-port area ratio:
$$C_m = \frac{A_2}{n_v A_p} \qquad (3.6.11)$$

Port-to-pipe area ratio:
$$k = \frac{n_v A_r}{A_2} = \frac{A_r}{A_p} \times \frac{n_v A_p}{A_2} = \frac{C_t}{C_m} \qquad (3.6.12)$$

It can be seen that the C_m value is a constant and that the maximum value of any of the other area ratios is unity. In the case of single or multiple valve arrangements, the local port area ratio, C_t, cannot exceed unity due to the precise definition of all of the possible restrictions at a pipe end where a poppet valve is located. Furthermore, it is quite conventional for C_t to have a value of unity over a range of valve lifts in the region of maximum valve lift.

3.6.2 The Local Port Area Ratio Is No Longer Adequate as a Cd Plotting Parameter

In the case of throttled pipe ends and two-stroke engine ports, the maximum value of the local port area ratio, C_t, is unity. For a port that is opened by the piston of a two-stroke engine, the C_t value varies continuously and only reaches unity when the port is fully uncovered and no further geometrical area change takes place at that port.

Where there is a poppet valve at the end of the pipe, it is clear that the maximum geometrical restriction area, A_r, is reached at some valve lift, L, that is less than the maximum valve lift, L_v, so the local port area ratio, C_t, is maximized at unity. Any further valve lift will change the flow characteristics through the port and past the valve so that the discharge coefficient will almost certainly be altered. Hence, while the discharge coefficient for a poppet valve is obviously still connected directly to the valve lift, the local port area ratio can no longer be used logically as an x-axis plotting parameter of an xyz contour plot of the discharge coefficients. What plotting parameter should then be used for the x-axis?

The first obvious plotting parameter for the x-axis is valve lift, L. However, this would make every Cd contour plot unique to the valve lift characteristics of a particular valve. This leaves the engine modeler in the highly unsatisfactory position of being unable to specify the Cd values for valve/duct combinations of similar geometry and possibly even minor variations of the same valve geometry. Each x-axis would have different values and would not run from 0 to 1.0.

The second obvious plotting parameter for the x-axis is the valve lift-diameter ratio, LD_v, where the diameter in question is the inner seat diameter, d_{is}, and is defined as follows:

$$LD_v = \frac{L}{d_{is}} \qquad (3.6.13)$$

This has the basic advantage of being in common usage in the literature [3.12, 3.14], but has the distinct disadvantage of producing a value of LD that has no fixed range. Although LD rarely exceeds 0.4, it is still a linear function of the actual valve lift, L. This is inconsistent with all other Cd contour plots for pipe end restrictions, including the plotting procedures and data reduction systems within an engine simulation, which employ a zero-unity range as an x-axis plotting parameter. Area ratio, or a local port area ratio, are obvious examples. Because such area ratios can provide numbers that always lie between zero and unity, it is vital to deduce a plotting parameter for poppet valve Cd with a similar numerical range. There are two possible options: (1) use the valve lift ratio, L_r, or (2) define a modified lift-diameter ratio, LD, so that it lies in a range that can span from zero to unity. These options are defined as follows.

Valve lift ratio: $\qquad L_r = \dfrac{\text{valve lift}}{\text{maximum valve lift}} = \dfrac{L}{L_v} \qquad (3.6.14)$

Modified lift-diameter ratio: $LD = \dfrac{2.5 \times L}{d_{is}} = \dfrac{10}{4} \times \dfrac{L}{d_{is}} \qquad (3.6.15)$

Both of these ratios numerically range from 0 to 1, with that in Eq. 3.6.14 doing this precisely for all valve configurations, and the latter "ten-L-over-four-D" value not quite reaching unity for all but the most extreme racing valve geometries. It little matters which of these two criteria is employed. However, all of the Cd data for poppet valves are presented here with the x-axis as the valve lift ratio, L_r. It can be logically argued that the lift-diameter ratio, LD, is more dimensionally dimensionless for use within a simulation. In other words, it may be a better criterion as a base from which to assess the Cd of an unknown poppet valve from a measured Cd plot of a similar geometry. For those who disagree with this argument, for all of the contour plots presented below I have also quoted the maximum valve lift and the valve inner seat diameter so that you may redraw them in this fashion, i.e., with LD rather than L_r on the x-axis.

3.6.3 Measuring the Discharge Coefficients of Poppet Valves in a Cylinder Head

The experimental arrangements used to measure the discharge coefficients for outflow and inflow are shown in Figs. 3.17 and 3.18, respectively.

As with the throttled pipe ends, or the two-stroke engine ports, the main measurement parameters are the air mass flow rate, $\dot{m}_{ex}$, and the cylinder and pipe pressures and temperatures, p_1 and p_2, and T_1 and T_2, respectively. The pipe and cylinder locations correspond to those shown on the basic sketch, Fig. 3.16. The valve is lifted at discrete intervals, and the above data are recorded at various pressure ratios ranging above 1.0 and as close to 2.0 as permitted by the mass flow capability of the apparatus. The results are analyzed, using the theoretical approach given in Sec. 3.2.4, for the actual discharge coefficient, Cd_a, at each test point. If, for comparison purposes only and not for design use within a simulation, the ideal discharge coefficient, Cd_{id}, is required, the same test data are re-analyzed using the theoretical techniques of Sec. 3.2.3. When an entire data set has been analyzed for the Cd_a values, the mapping technique, as illustrated in Fig. 3.8 and discussed in Sec. 3.4, is employed to find the discharge coefficients at the "plane" corners. This exercise is repeated, if needed, for the data

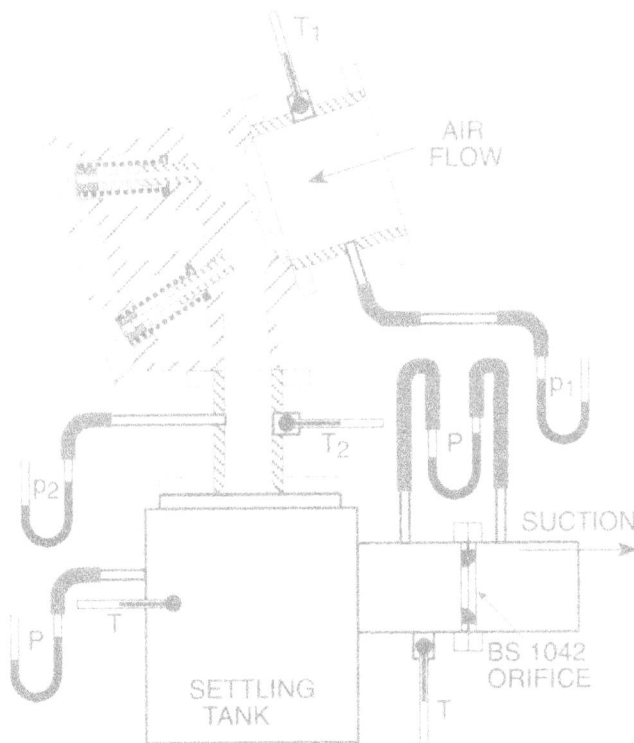

Fig. 3.17 Apparatus used to measure the discharge coefficients for outflow of poppet valves.

355

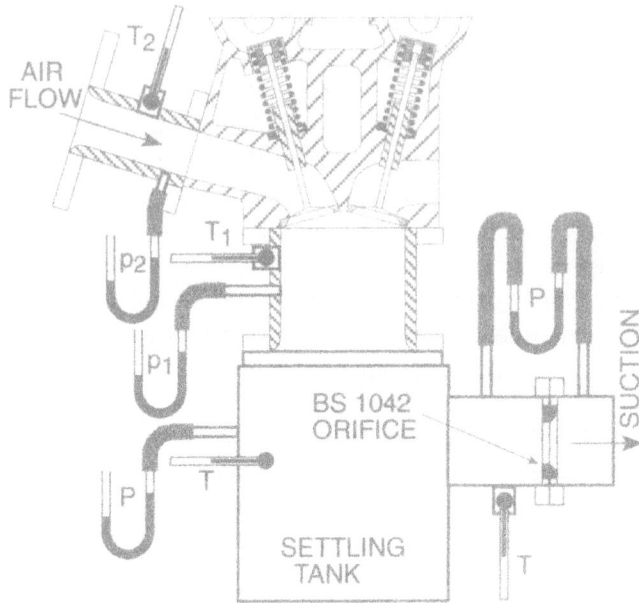

Fig. 3.18 Apparatus used to measure the discharge coefficients for inflow of poppet valves.

set of ideal discharge coefficients. Finally, the Cd values are plotted as xyz contours for presentation within this text. The main difference between the Cd map for a poppet valve, and those shown previously for ports and throttled pipes, is that the x-axis is the valve lift ratio, L_r, and not the area ratios, such as C_t or k, in line with the above discussion.

3.6.4 The Discharge Coefficients of Poppet Valves in a Two-Valve Honda Moped Engine

The first results to be presented are those from a small engine used by Honda in a production moped which has a nominal 50 cm³ capacity. I am indebted to Emerson Callender, a doctoral student at QUB, for providing all of the test data reported here for moped engines. The engine bore and stroke are 39 mm and 41 mm, respectively, giving a swept volume of 49 cm³. The sketch in Fig. 1.14 illustrates this valve geometry type. The intake valve has inner and outer seat diameters, d_{is} and d_{os}, of 20.6 mm and 22.8 mm, respectively. The exhaust valve has inner and outer seat diameters of 18.0 mm and 19.8 mm, respectively. The inner port diameters, d_{ip}, are 16.4 for the exhaust and the intake ports. The manifold diameters, d_2, are 20 mm and 17.5 mm for the exhaust and intake ducts at the cylinder head, giving pipe-to-port ratios, C_{em} and C_{im}, of 1.67 and 0.96, respectively. Both valve stem diameters are 5.4 mm and the maximum lift of each valve is the same at 5.0 mm. This gives modified lift-diameter ratios, LD, as defined in Eq. 3.6.15, of 0.694 and 0.607 for the exhaust and intake valves, respectively.

The actual discharge coefficients for the exhaust valve for outflow and inflow are shown in Figs. 3.19 and 3.20, respectively. The equivalent Cd_a maps for the intake valve for outflow and inflow are given in Figs. 3.21 and 3.22, respectively.

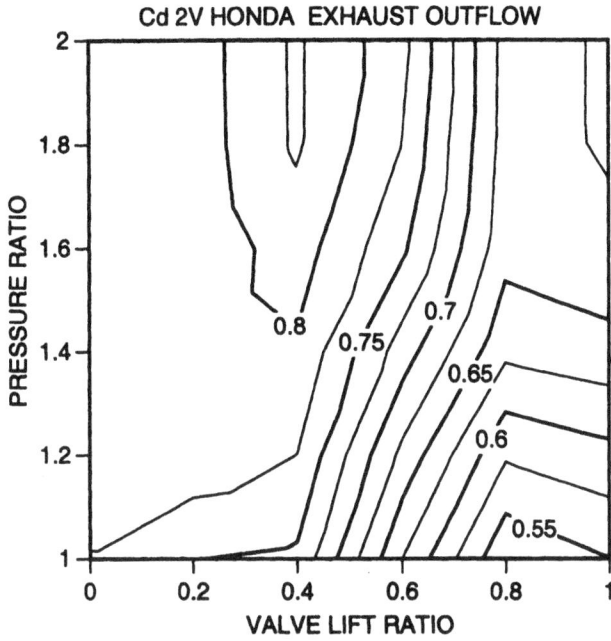

Fig. 3.19 The discharge coefficients for exhaust outflow of a 2v Honda 50 cm³ engine.

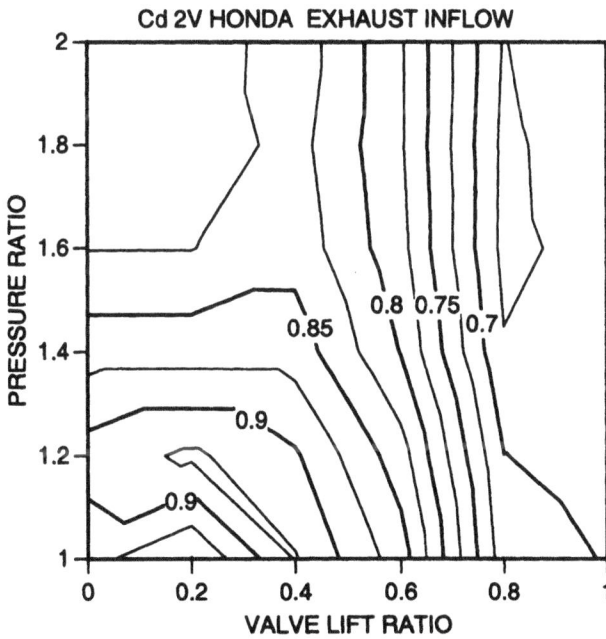

Fig. 3.20 The discharge coefficients for exhaust inflow of a 2v Honda 50 cm³ engine.

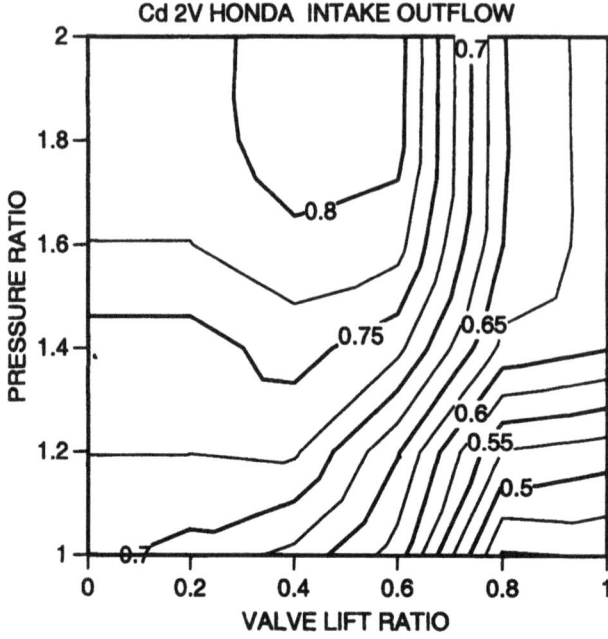

Fig. 3.21 The discharge coefficients for intake outflow of a 2v Honda 50 cm³ engine.

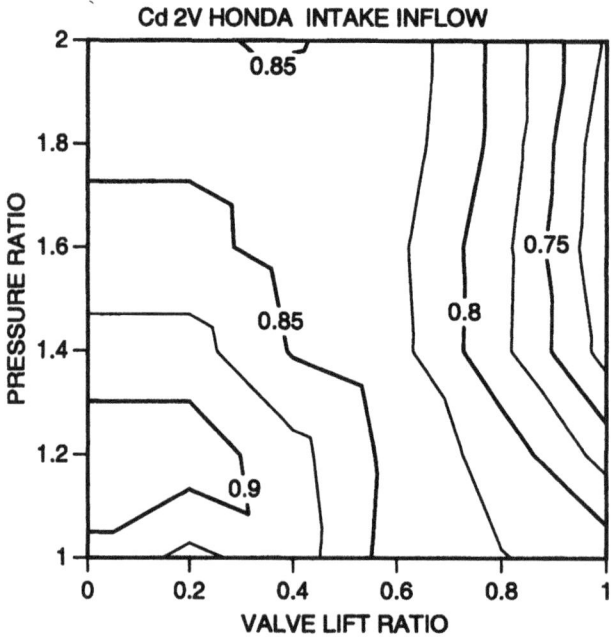

Fig. 3.22 The discharge coefficients for intake inflow of a 2v Honda 50 cm³ engine.

The Cd_a Values for Inflow in the 2v Honda Moped Engine

Perhaps the first Cd map of obvious interest is that for intake inflow, in Fig. 3.22, and it is certainly the one that receives the majority of attention in the historical literature [3.1-3.14]. The values range from 0.9 at low valve lift and low pressure ratio to 0.675 at maximum lift and high pressure ratio. There is some dependence on pressure ratio, P, at low valve lifts, and also at high valve lifts at pressure ratios up to 1.3, a range that is not untypical of that encountered within an engine during the induction process. What is clear, however, is that this Cd map bears no resemblance to that for inflow at throttled pipe ends in Fig. 3.10, or for two-stroke engine ports in Fig. 3.14. Poppet valve inflow is manifestly different. The inflow picture for the 2v Honda exhaust valve, i.e., reverse flow as far as the exhaust valve is concerned, is shown in Fig. 3.20. This bears a close resemblance to that for intake inflow. The Honda engineers probably paid some attention to the quality of the intake flow at the intake valve, which is evidenced by intake valve discharge coefficients at high lift that are some 10% better than those for the same flow regime at the exhaust valve; in backflow at the exhaust valve, it is probable that they had little experimental interest. Remember that this is the very engine with which Mr. Honda started a "legend" [3.20].

The Cd_a Values for Outflow in the 2v Honda Moped Engine

The exhaust outflow characteristics are of obvious importance. The Cd map in Fig. 3.19 shows that there is very little dependence of the discharge coefficient on pressure ratio, except at the highest valve lifts. One aspect to be observed, which will be repeated in other figures of poppet valve flow, is the rise of Cd as the valve is lifted to its maximum from a lift ratio, L_r, of about 0.8. In other words, there is a very good reason to lift a poppet valve beyond the supposed maximum value of lift-diameter ratio, LD_r, of 0.25; the simplistic maximum value is discussed in Sec. 1.5.2. The discharge coefficient actually improves beyond the point where the throat area of the valve, A_t, is greater than the port area, A_p. This is the point at which the area of the geometrical restriction, A_r, is now A_p and not A_t, as seen in Eqs. 3.6.4 and 3.6.5. Hence, the effective area of the combined valve and port aperture, A_{te}, as seen by manipulating Eq. 3.2.7, has increased because the discharge coefficient is higher:

$$A_{te} = C_{da} \times A_r \qquad (3.6.16)$$

If you glance back at Eq. 2.16.8 in Sec. 2.16, you will see that the "C_dA_t" term in that equation is precisely the same as the "$C_{da}A_r$" term in Eq. 3.16.6 above. Because the pipe area, A_2, is common terminology, it follows that more mass flow will take place for common state conditions in the pipe and the cylinder.

To put some numbers to this discussion, the geometrical port area, A_{ep}, at the Honda exhaust valve, as defined by Eq. 1.5.3, is 188.3 mm^2. When the exhaust valve lift is 4.1 mm, which is at a valve lift ratio, L_r, of 0.802, the valve curtain area as calculated by Eq. 1.5.12 is also precisely 188.3 mm^2. Thus, the cross-over point for the combined restriction of the valve throat and its port, as defined by Eqs. 3.6.4 and 3.6.5, has occurred in the valve lift profile. It will be observed this valve lift ratio of 0.802 corresponds precisely to the "bend" in the Cd contours on Fig. 3.19 at pressure ratios up to 1.6, when the flow becomes choked, i.e., at a particle Mach number of unity at the inner port. When the exhaust valve lift is a maximum at 5.0 mm, when the valve lift ratio is unity, the valve curtain area is calculated to be 249.2 mm^2, so the restriction area, A_r, remains at 188.3 mm^2. Selecting a representative common pressure ratio of 1.25 in each case, the Cd values are seen from Fig. 3.19 to be 0.59 and 0.61 at lift ratios of 0.802 and 1.0, respectively. Consequently, the effective area of the restriction, A_{te}, as defined by Eq. 3.6.16, is improved by the numerical ratio of 61 to 59, which is 1.034. This would equate to some 3% increase of gas mass flow rate at that juncture. Doesn't sound like much? Designers must always remember the fundamental lessons of Chapters 1, 5, and 6, that accumulating a series of 3% increases in values on exhaust gas removed, air induced, and air mass trapped will soon compound to become 10 hp increases in 100 hp engines.

The two maps shown in Figs. 3.19 and 3.21 for exhaust outflow and intake outflow, i.e., reverse flow at the intake valve, exhibit many of the same features of, but clearly have nothing in common with, the equivalent maps for outflow at throttled pipe ends or two-stroke ports shown in Figs. 3.9 and 3.13, respectively. Outflow from ports and throttled pipe ends has a distinct dependence on pressure ratio; indeed the locale of small port openings where the Cd for the two-stroke port showed itself to be reasonably independent of pressure ratio is the very same region of low poppet valve lift that is influenced by pressure ratio. As with inflow, poppet valve outflow has manifestly different Cd characteristics from the more fundamental restrictions posed by ports or, more particularly, pipe-end orifices.

The Prediction by CFD of Outflow and Inflow at a Poppet Valve

In obtaining a clear mental picture of the fluid mechanics of the flow at a poppet valve and in the duct attached to the porting, the use of computational fluid dynamics (CFD) provides great illumination. The profile of the duct leading to, or from, a poppet valve has an influence on the discharge coefficient for inflow or outflow through it. In computational fluid dynamics, the duct and the valve are finely meshed and the three-dimensional calculation proceeds in steady or unsteady flow with a stationary, or moving, valves and piston, respectively. Clearly, the unsteady flow computation is much more complex and requires guidance from a 1D engine simulation for the time-varying state properties of the gas within the cylinder at the piston surface and at the extremity of the mesh spaces within the ducting. The meshing of the entire space involved is also a complex and time-consuming affair, although automatic meshing is becoming more commonplace.

Plate 3.1 shows the meshing of the cylinder head and part of the intake duct of a two-valve spark-ignition engine. The outline of the exhaust valve seat can also be observed. The steeply-inclined intake duct is profiled in a manner very similar to that seen in Plate 1.1.

Plate 3.2 shows the meshing of the intake duct and cylinder head of a four-valve, pent-roof design for a gasoline engine, the internals of which are similar to the FZR Yamaha head shown in Plate 3.4. The two exhaust valve seats can be seen in outline.

Plate 3.3 shows the intake duct and cylinder head of a two-valve direct-injection compression-ignition engine. The cylinder head is flat, and at tdc on the power stroke it mates closely with a flat-topped piston in which is contained a combustion bowl similar to that shown in Plate 1.8. You can see that the intake tract is in the form of a helix which swirls the air into the

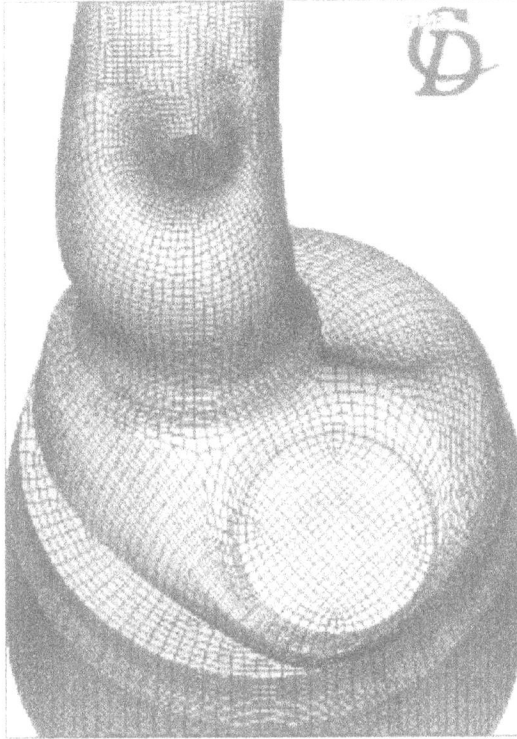

Plate 3.1 The CFD mesh structure of a two-valve spark-ignition engine.
(Courtesy of Adapco)

cylinder and, during compression, swirls it even faster into that bowl. These squish-enhanced swirl events are depicted in Figs. 4.39 (a) and (b). In this chapter, the coefficient of discharge is being discussed and it is fairly self-evident that the swirling action at the flow annulus of the poppet valve must deteriorate the coefficient of discharge [3.25].

Figs. 3.23 and 3.24 show the predictions by computational fluid dynamics (CFD) of the gas particle velocities during exhaust outflow and intake inflow, respectively. They are calculated for a two-valve spark-ignition engine in a meshed cylinder head and duct, much as shown in Plate 3.1. I am indebted to Dr. Barry Raghunathan of Adapco for these diagrams and for Plates 3.1-3.3.

Figs. 3.23(a) and 3.24(a) show the velocity contours of the flow, with the velocity value shown in greyscale. Figs. 3.23(b) and 3.24(b) show the velocity vectors of the flow in a computation cell, with the velocity magnitude represented by the length and size of the arrow.

The exhaust outflow, in Fig. 3.23, is clearly at a low valve lift. The high-velocity region at the valve throat fills it with flow so, not surprisingly, the Cd values will be high, even approaching unity, as already seen in Fig. 3.19 and, later in the text, in Fig. 3.31. The high-velocity plume is attached to the back of the valve and surrounds the valve stem and, as it is headed for

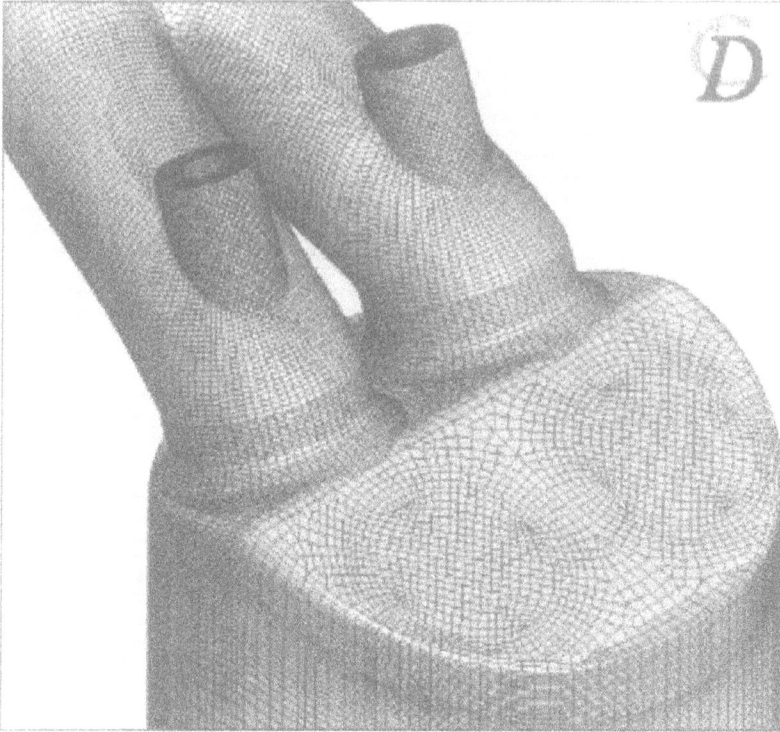

Plate 3.2 The CFD mesh structure of a four-valve spark-ignition engine.
(Courtesy of Adapco)

a collision with the valve guide and the outer wall of the exhaust duct, some further vortex activity and turbulence generation can be expected. This will reduce its Cd value before it reaches the exhaust manifold. The well-ordered nature of the flow can be seen from the velocity vectors in Fig. 3.23(b), with any recirculation zones being confined to within the cylinder at the edges of the valve.

The intake inflow, in Fig. 3.24, is at a high valve lift position. The high-velocity regions of the flow do not fill the valve throat, indeed they are not even symmetrical at the two edges of the valve. At the bottom edge, high-velocity flow is seen to separate from the duct wall and attach to the valve seat, but it does not fill the valve throat. At the upper edge, the high-velocity plume is observed in the center of the valve throat and also fails to fill it. Hence, the effective area of the valve throat is reduced considerably at this high valve lift ratio, giving rise to the Cd values of about 0.75 already seen in Fig. 3.22 and, later on, in Figs. 3.29 and 3.34. The less-ordered nature of the flow is observed from the velocity vectors in Fig. 3.24(b), as there are recirculation zones at the inner port on the left-hand side and within the cylinder at both edges of the valve [3.24].

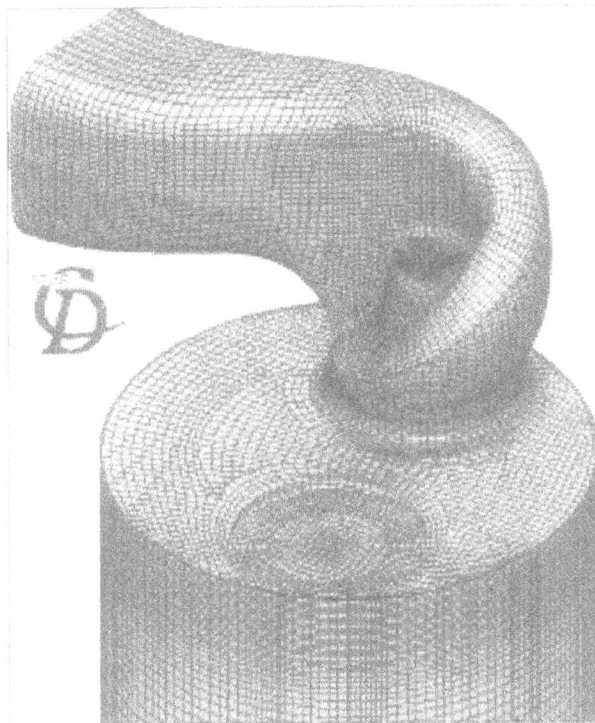

Plate 3.3 The CFD mesh structure of a two-valve direct-ignition diesel engine.
(Courtesy of Adapco)

The Reverse Flow Cd Maps

From the engine developer's standpoint, the reverse flow Cd maps present an interesting technical challenge. These are the maps for inflow at the exhaust valve and outflow at the intake valve, i.e., as shown in Figs. 3.20 and 3.21, respectively. The optimum answer for the numbers that appear on these maps would be zero, as this would eliminate the backflow which can be so deleterious to engine performance. Perhaps the worst example of such backflow is that of exhaust gas into the intake tract upon intake valve opening during the latter stages of the exhaust stroke—although sucking exhaust gas back into the cylinder on the intake stroke through a late-closing exhaust valve is not much better as an outcome! The current situation, as far as this Honda moped engine is concerned, is that the Cd maps for inflow or outflow at the exhaust and the intake valves, are very similar. There is as at least as much potential gain in actual engine performance by experimentally improving the reverse flow Cd maps, i.e., making the numbers on them lower, as there is in experimentally raising the numerical values of Cd for inflow at the intake valve or outflow at the exhaust valve. Naturally, to pick one example, any experimental R&D, or even theoretical work by CFD, on the intake valve and its duct must ensure that a reduction of the Cd for outflow is not also accompanied by a reduction of the Cd for inflow! I am unaware of any guidance for you in this area in the literature.

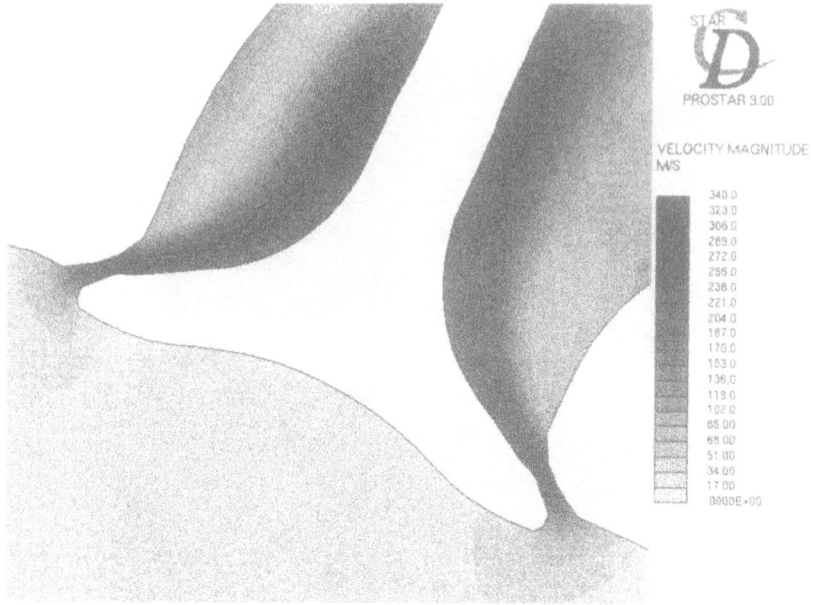

Fig. 3.23(a) Velocity contours for outflow at an exhaust poppet valve. (Courtesy of Adapco)

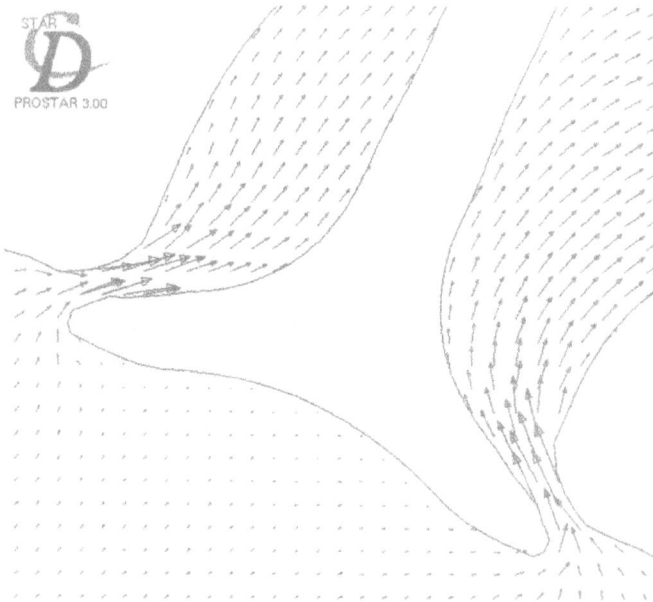

Fig. 3.23(b) Velocity vectors for outflow at an exhaust poppet valve. (Courtesy of Adapco)

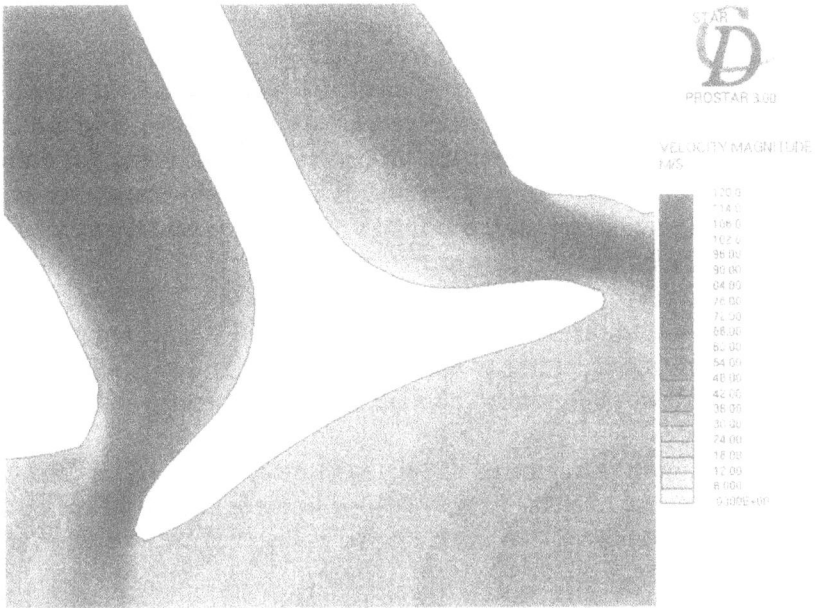

Fig. 3.24(a) Velocity contours for inflow at an intake poppet valve. (Courtesy of Adapco)

Fig. 3.24(b) Velocity vectors for inflow at an intake poppet valve. (Courtesy of Adapco)

The Ideal Discharge Coefficients in the 2v Honda Moped Engine

In some of my previous publications [3.16, 3.17, 3.18, 3.21], the numerical differences between the ideal and the actual discharge coefficients are presented in great detail for throttled pipes, two-stroke engine porting and four-stroke engine poppet valves. Figs 3.5 and 3.6, for outflow at plain-end and bellmouth-end pipes, further reinforce the point. The differences are considerable, illustrating the need already stressed in Sec. 3.2.5 to analyze the experimental data so as to acquire the actual discharge coefficient, Cd_a, if such data are required to be used within an engine simulation and accuracy expected from it.

For the flow at the poppet valves in the cylinder head of this 2v Honda engine, over the complete range of valve lifts, the ideal discharge coefficients, Cd_{id}, are acquired by reanalyzing the same measured data according to the approach given in Sec. 3.2.3. Those results of major interest, i.e., outflow at the exhaust valve and inflow at the intake valve, are presented in Figs. 3.25 and 3.26.

These two figures may be compared directly with those of the actual discharge coefficients, Cd_a, in Figs. 3.19 and 3.22. At first glance, there seems to be little difference for exhaust outflow, so an "error" map is produced and presented in Fig. 3.27. In precisely the fashion as

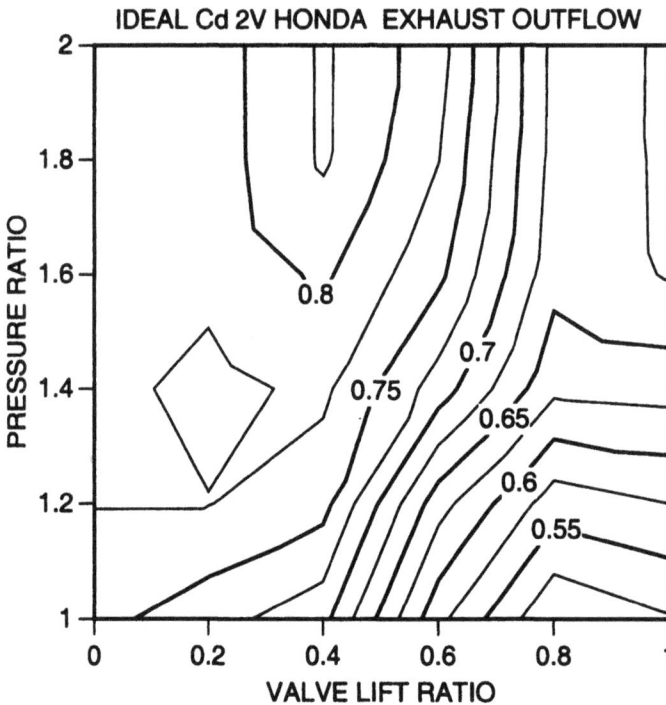

Fig. 3.25 The ideal discharge coefficients for exhaust outflow of a 2v Honda 50 cm³ engine.

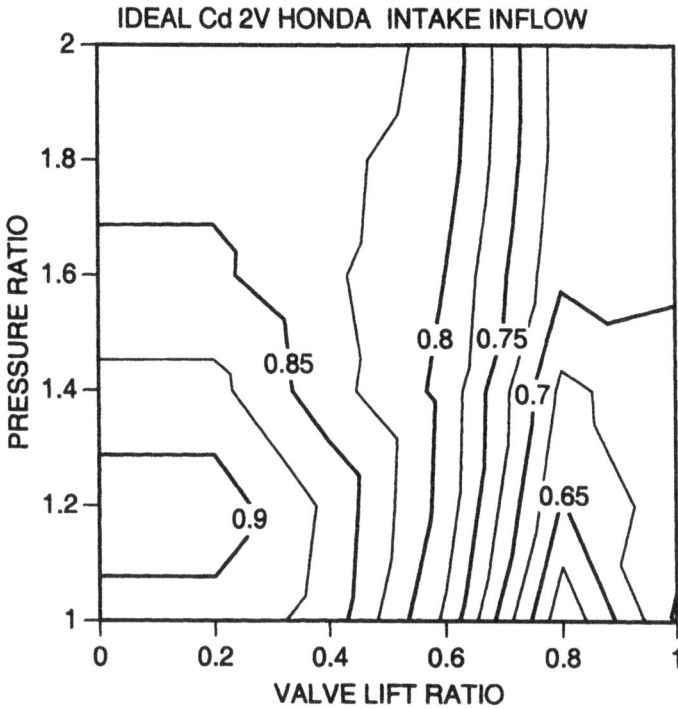

Fig. 3.26 The ideal discharge coefficients for intake inflow of a 2v Honda 50 cm³ engine.

described in Sec. 3.5.3, the Cd values at the "plane" corners of the two maps are subtracted and the difference reduced to a percentage value:

At each map point:
$$Cd_{\%} = 100 \times \left(\frac{Cd_{id} - Cd_a}{Cd_a} \right) \qquad (3.6.17)$$

The "error" for the ideal discharge coefficient is now visible. In the discussion below it is seen to be significant, bearing in mind my little homily above on the importance of accumulating 3% discrepancies during engine modeling!

The error for exhaust outflow is seen in Fig. 3.27. The location of the worst of this error is at high valve lifts and modest pressure ratios where it has the most damaging effect on simulation accuracy. The particular lift positions and pressure ratio levels at which this occurs are in the middle of the exhaust stroke when most of the exhaust pumping is taking place. The magnitude of the error is high; it can be up to 6%. The error during the exhaust blowdown process, which is at low valve lifts and high pressure ratios, can be seen to be insignificant.

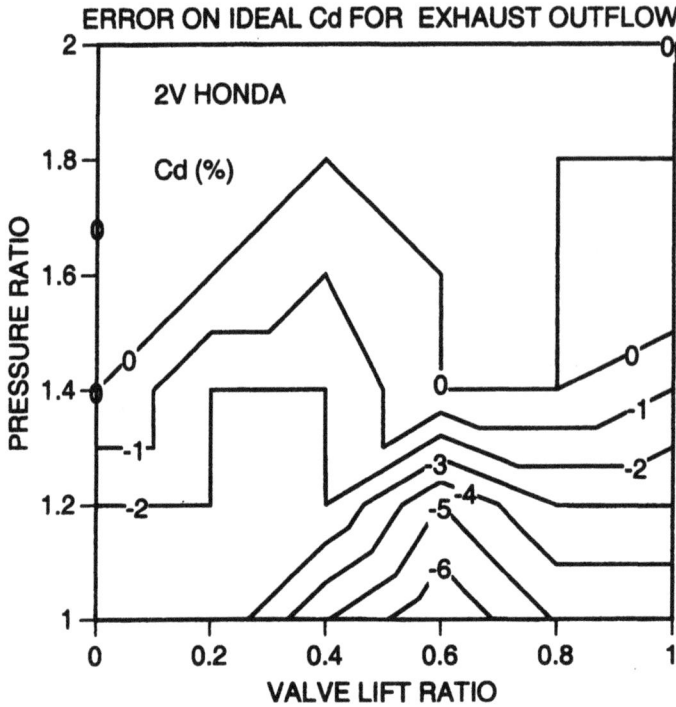

Fig. 3.27 The error between Cd_{id} and Cd_a for exhaust outflow of a 2v Honda 50 cm³ engine.

A similar error map is produced for the intake valve, by applying Eq. 3.6.17 to the two intake inflow maps given in Figs. 3.22 and 3.26. The results, as percentage errors, are plotted in Fig. 3.28.

For the intake valve, the worst error locations are at valve lifts and pressure ratios which would occur at about mid-stroke during the induction process. The valve is at high lift, the pressure ratio is modest, and the air mass flow rate is at its maximum. Here, the error is unacceptably high; it rises to 30%. As with the exhaust valve outflow process, the error at valve lift ratios less than 0.4 can also be seen to be insignificant.

From the standpoint of engine simulation, there are other awkward aspects of the traditional method of analyzing measured data which yields ideal discharge coefficients. The "error" illustrated in Figs. 3.27 and 3.28 is not consistently the negative numbers shown there. The negative numbers shown on both figures suggest that the ideal discharge coefficient, Cd_{id}, is always less than the actual discharge coefficient, Cd_a. Such is not the case. As an illustration, consider exhaust outflow at the same 2v Honda geometry given above, for the same valve lift of 4.1 mm and pressure ratio of 1.25 as used above. The actual discharge coefficient, Cd_a, deduced from the measured data is 0.5878. The ideal discharge coefficient, Cd_{id}, is 0.5772 and the isentropic discharge coefficient, Cd_{is}, is 0.6944. The 2% reduction in the ideal to the actual coefficient, seen in Fig. 3.27, is thereby illuminated. The isentropic discharge coefficient clearly has a poor correspondence with either number. If the exhaust pipe diameter were arbitrarily

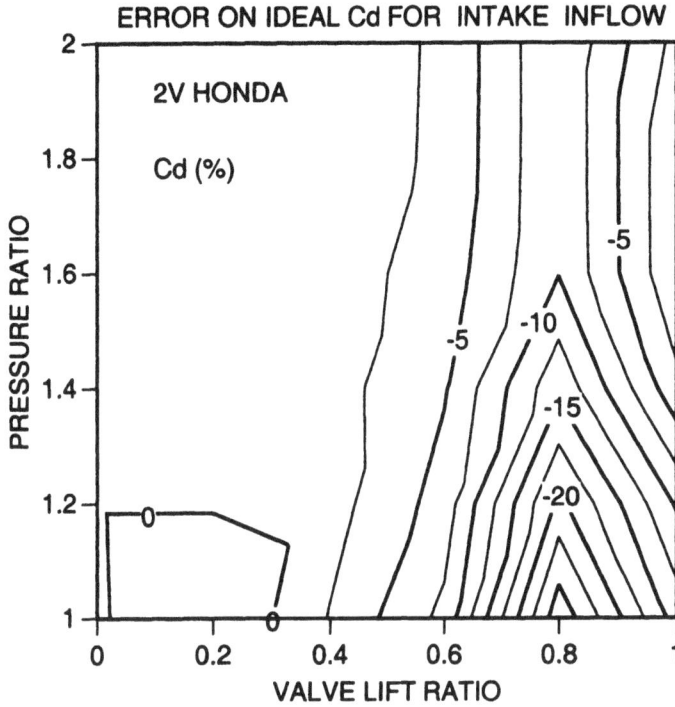

Fig. 3.28 The error between Cd_{id} and Cd_a for intake inflow of a 2v Honda 50 cm³ engine.

reduced to 17 mm, which is still larger than the inner port diameter, d_{ep}, of 16.7, and all other measured data were held constant purely for the purposes of this illustration, the several discharge coefficients, Cd_a, Cd_{id}, and Cd_{is}, would compute to 0.5803, 0.6020, and 0.6944. The isentropic value, having no analytic input from a pipe diameter, remains the same. The ideal discharge coefficient is now higher by some 3.7% than the actual discharge coefficient, which has marginally decreased. Thus, purely as a function of the pipe to port area ratio, C_m, the ideal discharge coefficient may be higher, or lower, than the actual discharge coefficient in specific experimental or geometrical circumstances. The implications from this simple example are that one cannot reexamine historical Cd data from the literature [3.1-3.14] and simply ratio them in some fashion so that they could be reused as Cd maps within an engine simulation model. One would not know whether to take those data and ratio them up, or down.

3.6.5 The Discharge Coefficients of Poppet Valves in a Two-Valve Automobile Engine

The discharge coefficients are investigated in a British Leyland 2.0-L four-cylinder automobile engine. I am indebted to Graeme Mawhinney, a doctoral student at QUB, for providing the test data reported here. The cylinder head has two valves arranged in a vertical formation as seen in Fig. 1.3. This is an engine that has been used by QUB researchers over many years, and papers on modeling gas flow through it, by the now-redundant method of characteristics, have been published previously [3.22, 3.23]. It is instructive to look at these discharge coefficients, of another two-valve head in an engine with a larger cylinder bore and larger poppet valves than the Honda moped, in order to examine the possibility that discharge coefficients may be influenced by valve size effects. The cylinder bore of this British Leyland (BL) engine is 84.5 mm and its stroke is 89 mm. This gives a cylinder swept volume of a nominal swept volume of 500 cm³, although the precise capacity is 499.1 cm³. The intake valve has inner and outer seat diameters of 36.3 mm and 39.0 mm, with no constriction at the inner port, and with a valve stem diameter of 7.4 mm. The maximum intake valve lift is 9.6 mm, giving a modified lift-diameter ratio, LD, of 0.66. The intake manifold is 30 mm in diameter. The exhaust valve has inner and outer seat diameters of 29.69 mm and 34.11 mm, with a valve stem diameter of 7.4 mm. The maximum exhaust valve lift is also 9.6 mm, giving a modified lift-diameter ratio, LD, for the exhaust valve of 0.81. The exhaust manifold is 33 mm diameter. The valve seat angles are the normal 45°. This gives pipe-to-port area ratios, C_{em} and C_{im}, of 1.3 and 0.71 for the exhaust and intake tracts, respectively. It should be noted that this is one of those engines where the intake pipe ostensibly becomes the major area restriction at high intake valve lift (see Eq. 3.6.6).

The actual discharge coefficients, Cd_a, for inflow and outflow at the intake valve are measured and presented in Figs. 3.29 and 3.30, respectively. During testing for the Cd values, the maximum valve lift is increased to 10 mm, which corresponds to a modified lift-diameter ratio, LD, of 0.69.

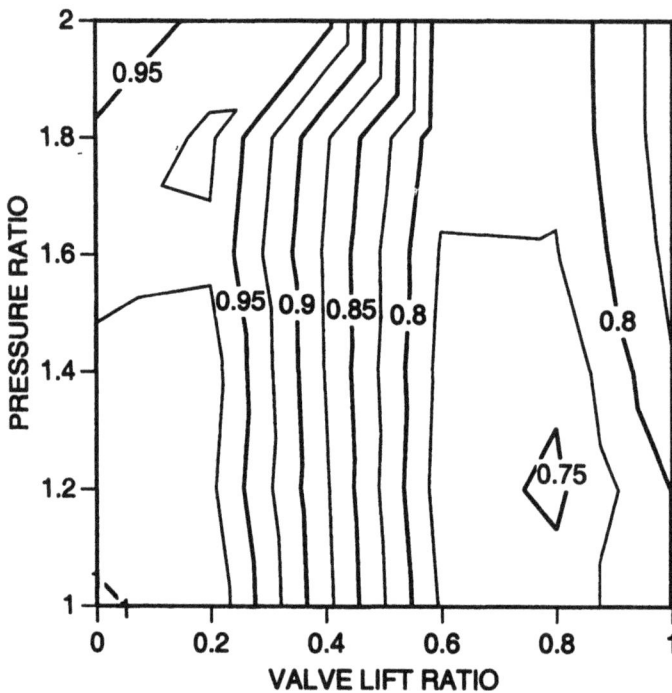

Fig. 3.29 The Cd_a maps for intake valve inflow of a 2v BL 500 cm³ cylinder.

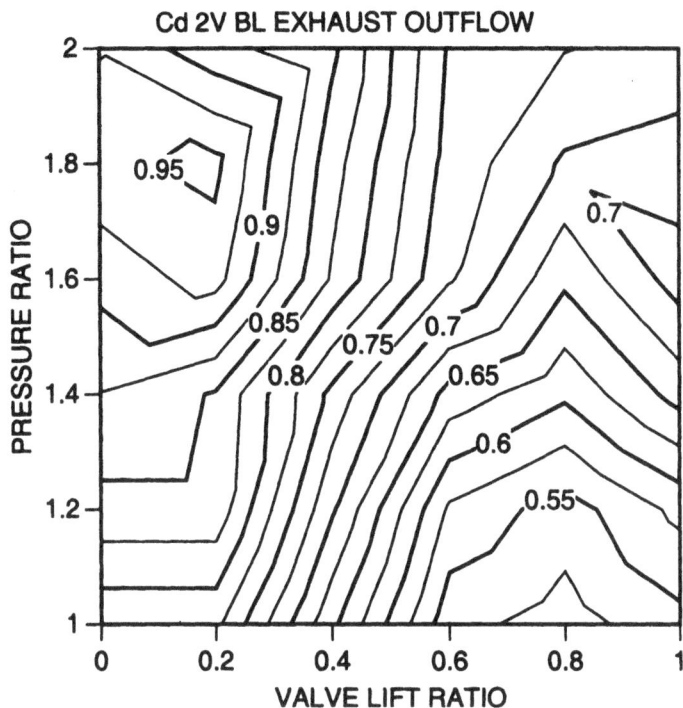

Fig. 3.30 The Cd_a maps for intake valve outflow of a 2v BL 500 cm³ cylinder.

This BL cylinder is ten times larger in capacity than the Honda moped engine discussed above in Sec. 3.6.4, but the intake valve inner seat diameter is only 1.76 times larger than that in the moped engine. The intake valve of the moped engine lifts to 5.0 mm, whereas the car engine intake valve lifts to 9.6 mm, which is a ratio of 1.92. The simplistic equation, Eq. 1.5.5, shows that the ratio of the valve throat area to the cylinder bore area should give the moped an intake inflow breathing advantage:

$$\frac{\text{Honda}}{\text{BL}} = \frac{\pi \times 20.6 \times 5.0}{\pi \times 39 \times 39} + \frac{\pi \times 36.26 \times 9.6}{\pi \times 84.5 \times 84.5} = 1.39 \tag{3.6.18}$$

This is far too simple a criterion, for the Honda is trying to breathe its best at 10000 rpm, whereas the BL car engine is only able to breathe up to 5000 rpm; bluntly, this crude empiricism is relatively meaningless as you will discover when—or if—you get to Sec. 6.1. Nevertheless, the aim of examining the discharge coefficients of a physically larger poppet valve in a cylinder with a larger bore, is achieved.

The first obvious factor seen in Fig. 3.29 is that the discharge coefficient for intake valve inflow is, apart from those at the very highest lift ratio, virtually independent of pressure ratio, i.e., it is similar to the intake valve inflow for the Honda moped in Fig. 3.22. The BL data shows Cd values of 1.0 at up to 20% lift and exhibit a large minimum area at 0.7. At a relevant pressure ratio of 1.2 at the highest valve lifts, the Honda moped valve has values about 0.8 which are superior to those for the BL engine at 0.7. The conclusion is that the intake valve inflow for the BL engine is slightly better than the Honda moped unit at lower valve lifts and similar in the higher lift regions. The Cd maps for both engines ar sufficiently similar as to warrant stating that engine size and valve size appear to have little influence on the discharge coefficients.

Assuming that intake valve outflow can be regarded as being similar to exhaust valve outflow, the close correspondence between Figs. 3.19 and 3.21 would support the same contention. The Cd map for intake valve outflow for the BL engine in Fig. 3.30 is indeed similar in both shape and magnitude to that for the Honda moped. The values in Fig. 3.30 are slightly higher for the BL engine compared to the Honda moped. As with inflow, the conclusion is that there is little difference in Cd for outflow with respect to valve size, particularly as the geometries are very different and the Cd maps are so similar; for example, one valve is some 36 mm diameter with a maximum lift of 10 mm and the other is about 21 mm diameter with a maximum lift of 5 mm.

3.6.6 The Discharge Coefficients of Poppet Valves in a Four-Valve Yamaha Motorcycle Engine

The engine used in the testing is from a FZR 600 four-cylinder Yamaha sports racing motorcycle. The cylinder head has four valves arranged in a pent-roof formation as seen in Fig. 1.6. The cylinder bore of this Yamaha engine is 62 mm and its stroke is 49.6 mm. This gives a cylinder swept volume of a nominal 150 cm^3, although the precise capacity is 149.7 cm^3. The intake valve has inner and outer seat diameters of 21.1 mm and 23.1 mm, with an inner port diameter of 20.9 mm, and with a valve stem diameter of 5 mm. The maximum intake valve lift is 7.3 mm, giving a modified lift-diameter ratio, LD, of 0.865. The intake manifold is 28.7 mm in diameter. The exhaust valve has inner and outer seat diameters of 18.7 mm and 20.7 mm, with an inner port diameter of 18.5 mm, and with a valve stem diameter of 5 mm. The maximum exhaust valve lift is also 7.3 mm, giving a modified lift-diameter ratio, LD, for the exhaust valve of 0.98. The exhaust manifold is also 28.7 mm in diameter. The valve seat angles are a conventional 45°. This gives pipe-to-port area ratios, C_{em} and C_{im}, of 1.3 and 1.0 for the exhaust and intake tracts, respectively.

A photograph of the FZR cylinder head mounted on the experimental apparatus to measure the discharge coefficients for inflow at the intake valves, is shown in Plate 3.4. This photograph corresponds to the layout sketched in Fig. 3.18, and the several static pressure and static temperature tapping points can be observed. The large flange, with bellmouth, at the intake is to facilitate re-mounting the cylinder and cylinder head, in the manner of Fig. 3.17, in order to measure the outflow discharge coefficients of the intake valves.

The discharge coefficients are measured in both flow directions for both valves, using the apparatus shown in Figs. 3.17 and 3.18. Some of these test results have been previously published [3.21].

The Discharge Coefficients of the Poppet Valves Given Free Access in a 4v Cylinder Head

To investigate the effect of allowing easier access for the gas flow to or from the cylinder, a test cylinder bore of 66 mm is used instead of the standard 62 mm dimension. In the standard

Plate 3.4 The FZR 600 Yamaha cylinder head on the C_d apparatus at QUB.

cylinder bore of 62 mm, the outer edges of the valves are 2.5 mm from the nearest cylinder wall. This means that all valves have at least an extra 2 mm clearance to an adjacent cylinder wall face, i.e., the edge clearance is increased to 4.5 mm. The results of the experiments and the analysis of the results for the actual discharge coefficients, Cd_a, are shown in Figs. 3.31 to 3.34.

These results may be compared directly with the equivalent maps for the two-valve Honda moped engine in Figs. 3.19 to 3.22, or the 2v BL car engine in Figs. 3.29 and 3.30.

A similarity pattern for exhaust outflow emerges as the experimental evidence mounts. The shape of the Cd_a map in each case is virtually unaltered, but the Yamaha engine displays the highest numbers for exhaust outflow of the series. This is particularly true above valve lift ratios of 0.4 and at pressure ratios below 1.4. Although one must bear in mind that this is a sports racing engine, that the valves are given freer access to the cylinder, and that the flow may well have been experimentally optimized on a "flow bench," the fact remains that the Cd values are higher.

For intake valve inflow, the same remarks apply. The virtual lack of dependence on pressure ratio, remarked upon for the BL engine, can be restated here. As with exhaust outflow, the Cd values for intake valve inflow on this Yamaha map are the highest shown for any poppet valve.

A more general comment can be made: Not one of these poppet valve Cd maps shows superior flow coefficients compared to the equivalent pictures for two-stroke engine ports in

373

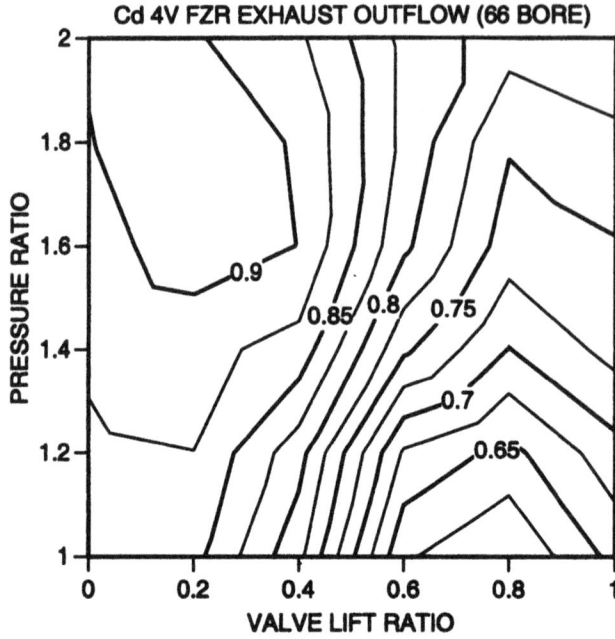

Fig. 3.31 The discharge coefficients for exhaust outflow of a FZR Yamaha engine.

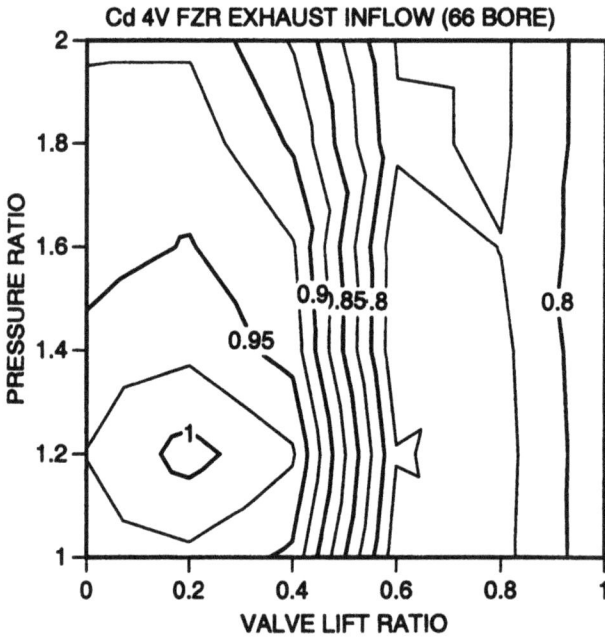

Fig. 3.32 The discharge coefficients for exhaust inflow of a FZR Yamaha engine.

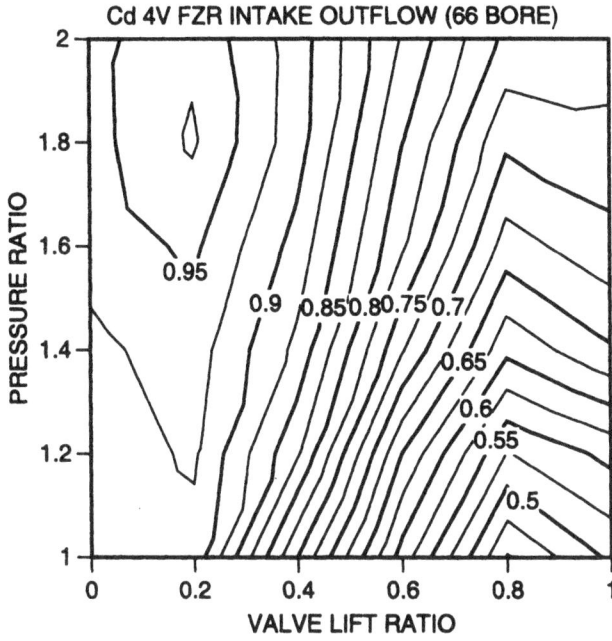

Fig. 3.33 The discharge coefficients for intake outflow of a FZR Yamaha engine.

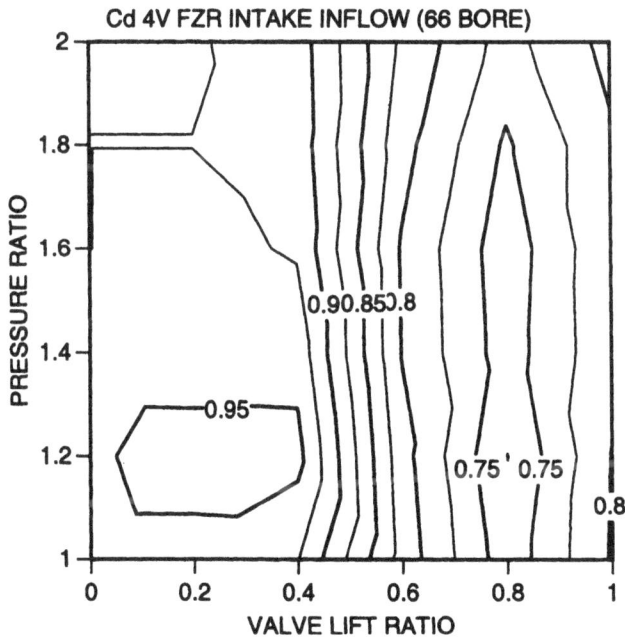

Fig. 3.34 The discharge coefficients for intake inflow of a FZR Yamaha engine.

375

Figs. 3.13 and 3.14. I remain somewhat puzzled by this finding, as I had expected that, at least, inflow at an intake poppet valve would be superior to that for a simple two-stroke engine port, simply because the latter appears to be a much sharper and cruder pipe-end restriction. Indeed, the same gas flow argument for inflow can be applied even more cogently by comparing Figs. 3.10 and 3.34, i.e., central sharp-edged orifices vis-à-vis poppet valves. On the other hand, the CFD picture in Fig. 3.24 of the inflow at an intake poppet valve is hardly one that suggests a low-loss fluid mechanic process, so perhaps I should not be so surprised!

The implications are that sleeve-valve and rotary-valve engines will flow more charge than a poppet valve at the same pressure ratio as they have the higher Cd values. As a rotary valve and a sleeve valve exposes more aperture area more rapidly than a poppet valve such engines should produce higher imep levels; indeed, the lessons of history say they did just that.

The Ideal Discharge Coefficients of the Poppet Valves Given Free Access in a 4v Cylinder Head

To complete the display of ideal discharge coefficients and their relationship to the actual values, Cd_a, and also to show that there is no unique distinction between a four-valve and a two-valve cylinder head arrangement, the ideal discharge coefficients, Cd_{id}, analyzed from the same measured data in the FZR 600 cylinder with the 66-mm bore dimension, are presented below in Figs. 3.35 and 3.36. These are for the exhaust valve outflow and the intake valve inflow.

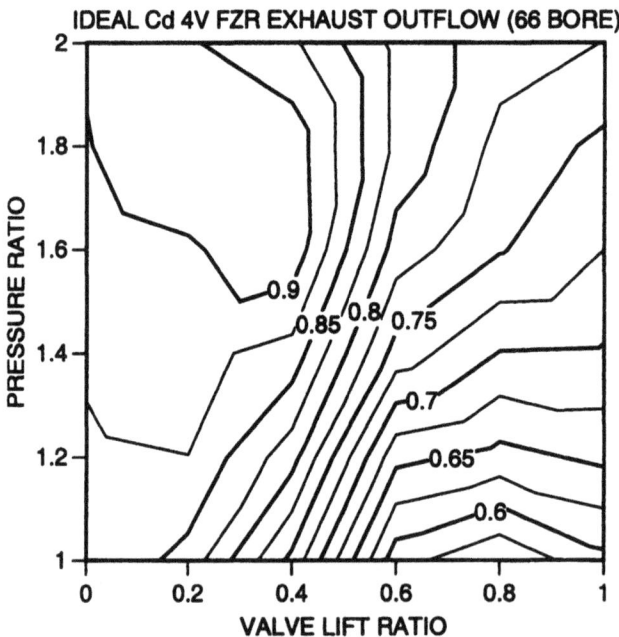

Fig. 3.35 The ideal discharge coefficients for exhaust outflow of a FZR Yamaha engine.

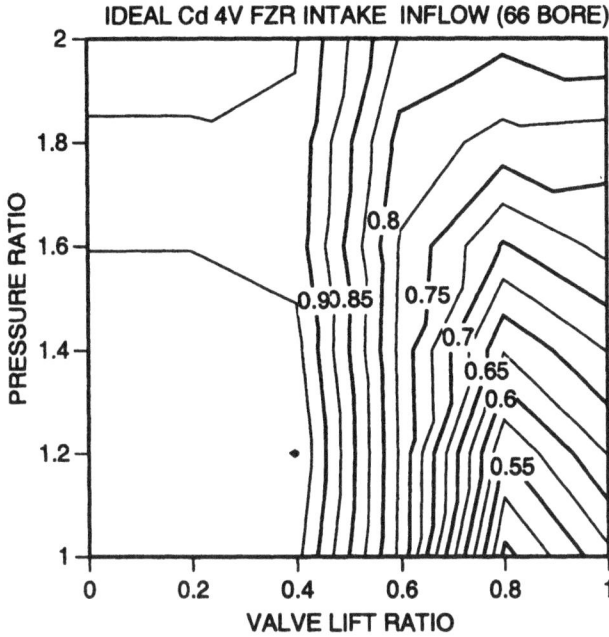

Fig. 3.36 The ideal discharge coefficients for intake inflow of a FZR Yamaha engine.

These results are very similar in magnitude and profile to the equivalent ideal Cd maps recorded in the 2v Honda moped engine shown in Figs. 3.25 and 3.26. Consequently, the basic difference between Cd_{id} and Cd_a, recorded as an error function in Figs. 3.27 and 3.28, is maintained and would appear to have no connection with either the number of intake or exhaust valves or the size of the cylinder bore.

The Discharge Coefficients of Poppet Valves Masked by the Cylinder Bore in a 4v Head

One of the major design decisions to be made is always the proximity of the edge of a poppet valve, be it exhaust or intake, to the nearest cylinder wall. If it is too close, this will mask—in effect reduce—the local valve throat flow area, A_t, at that segment of the valve circumference. To assess the impact of this on the discharge coefficients of valves, the FZR Yamaha, four-cylinder, four-valve, 600 cm³ engine is used for the experiments [3.21]. Here, by contrast with the tests reported above where the cylinder bore is increased to give the valves freer access to it, the cylinder bore is reduced to provide the opposite effect of masking the valves. In the standard cylinder bore of 62 mm, the outer edges of the valves are 2.5 mm from the nearest cylinder wall. The cylinder bore is now reduced to 59 mm, which means that all valves have some 1.5 mm less clearance to an adjacent cylinder wall face, i.e., the edge clearance of each valve is reduced to about 1.0 mm, giving the desired experimental effect of masking the valves. The discharge coefficients are measured in both flow directions for both valves, employing the apparatus shown in Figs. 3.17 and 3.18. The results of the experiments and the analysis of the results for the actual discharge coefficients, Cd_a, are shown in Figs. 3.37 to 3.40.

These results may be compared directly with the equivalent maps of the same cylinder head and valves, but where a 66 mm bore cylinder provides free air flow access, in Figs. 3.31 to 3.34. The effect of valve masking on exhaust outflow can be seen by comparing Figs. 3.37 and 3.31. It is clear for exhaust outflow that, at high valve lift, the Cd values have been reduced. The effect of valve masking on intake inflow can be observed by comparing Figs. 3.40 and 3.34. Similarly, it can be seen for intake inflow that, also at high valve lift, the masking effect lowers the Cd values.

To quantify these effects, the "masked" map in Fig. 3.37 is subtracted from the "free access" map of Fig. 3.31 and presented as a contour map of percentage change of discharge coefficient, $Cd_\%$, using the same theoretical approach of Eq. 3.6.17, but with differing input data appropriate to these two figures:

At each map point:
$$Cd_\% = 100 \times \left(\frac{Cd_{mask} - Cd_{free}}{Cd_{free}} \right) \qquad (3.6.19)$$

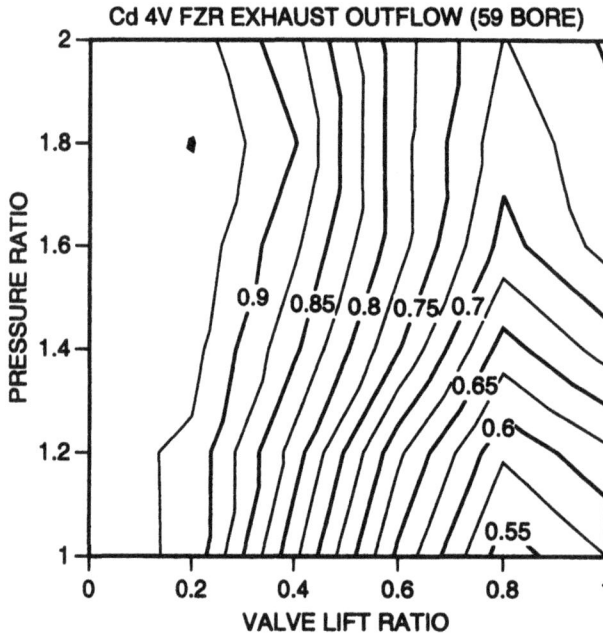

Fig. 3.37 The discharge coefficients for exhaust outflow of masked valves.

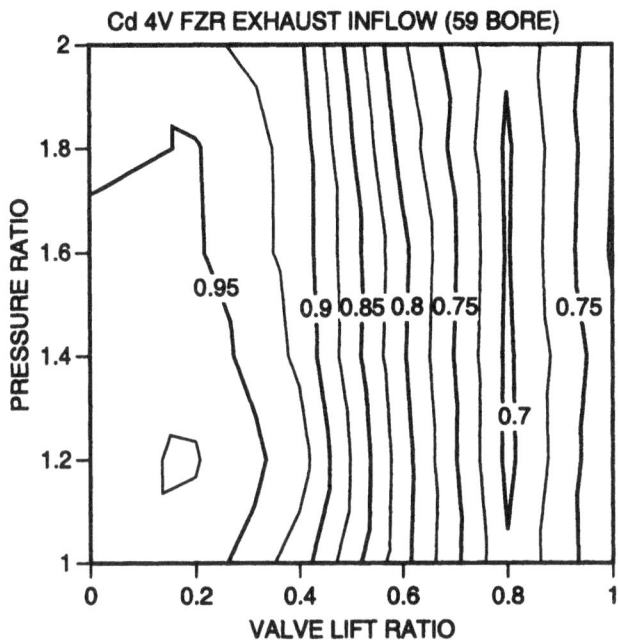

Fig. 3.38 The discharge coefficients for exhaust inflow of masked valves.

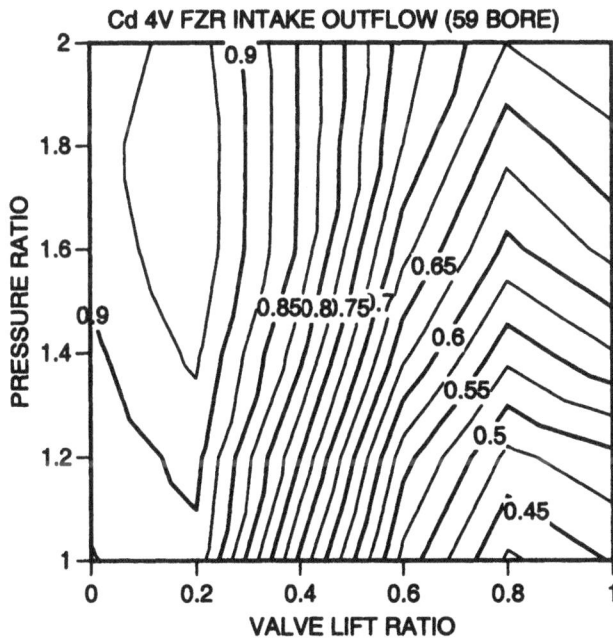

Fig. 3.39 The discharge coefficients for intake outflow of masked valves.

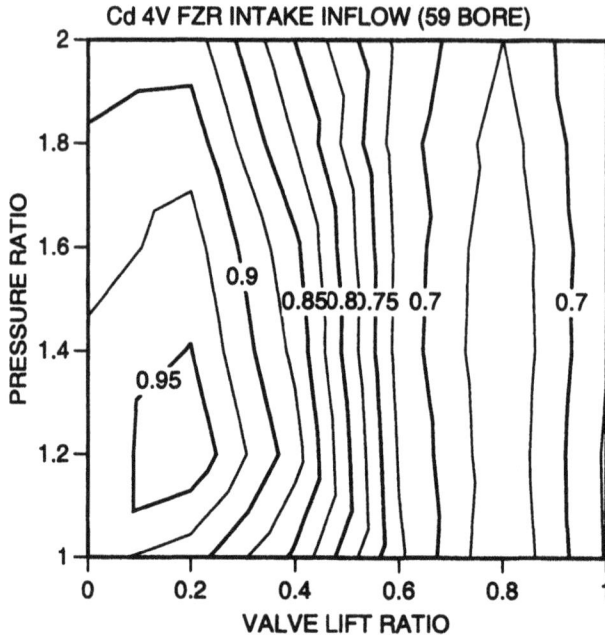

Fig. 3.40 The discharge coefficients for intake inflow of masked valves.

For exhaust outflow, the contour map of percentage change of discharge coefficient, $Cd_\%$, that results from this process is shown in Fig. 3.41. In the perverse manner that so characterizes fluid mechanics, and makes instinctive mental thinking in this area of technology such a misleading pastime, the discharge coefficient is actually improved by masking at up to half the maximum valve lift, but deteriorates by some 10% toward the highest lift and the lowest pressure ratios. Hence, gas flow from the cylinder will be improved by valve masking during a blowdown process and deteriorated during the exhaust pumping period. You would be most unwise to assume that this is a general "law" that applies in each and every design circumstance.

For intake inflow, the contour map of percentage change of discharge coefficient, $Cd_\%$, that results from applying the same theoretical reduction process to the relevant Cd contour maps presented in Figs. 3.34 and 3.40, is shown in Fig. 3.42.

In this case, the perversity of fluid mechanics is virtually absent and human instincts nearly prevail. The discharge coefficient is seen to be generally reduced by valve masking, with the maximum decrease of about 10% at high valve lift and low pressure ratio. This is precisely the region where the intake air flow rate into the engine should be at a maximum during the induction period. There is a small region at a valve lift ratio of 0.2, which just stops me from declaring that these statements are universal "laws."

The designer can now begin to see the level of compromise that is required for the poppet valves. From an existing "free access" valve design, one would have to ensure that the valve flow area, A_t, at its maximum valve lift ratio, L_r, has been increased by at least 10% if the valve

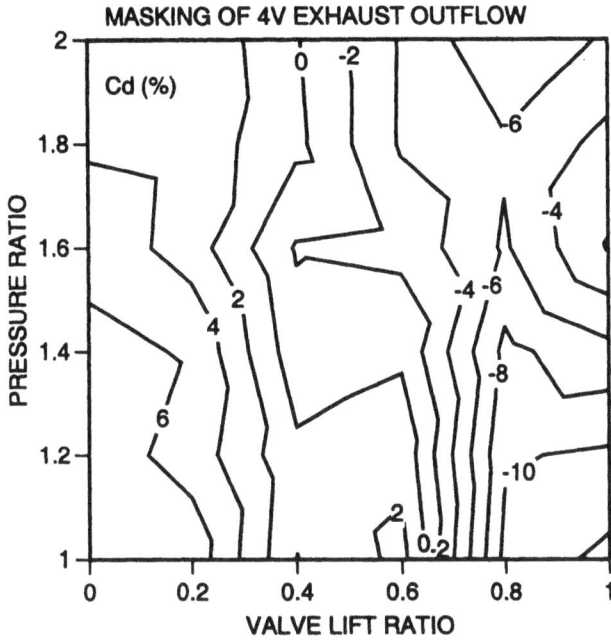

Fig. 3.41 The change in Cd caused by masking exhaust outflow in the FZR Yamaha engine.

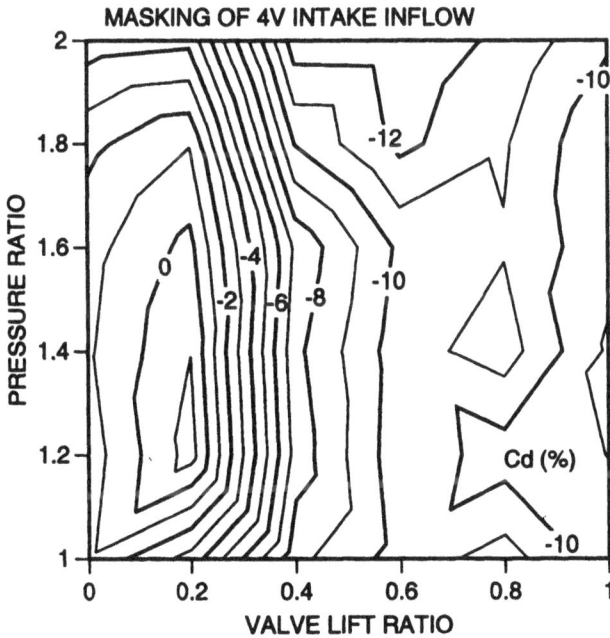

Fig. 3.42 The change in Cd caused by masking intake inflow in the FZR Yamaha engine.

381

head size is raised to the masking level described here, otherwise the potential reduction of 10% in Cd could negate the effect of incorporating the larger valve head design. If this statement is confusing, bear in mind that the masking experiment described here is not based on increasing the valve sizes, because this has the further undesirable design effect of moving the valves closer together in a four-valve head design. Moving valve edges closer together contains the potential of a further reduction of the discharge coefficient, as inter-valve masking can then occur as well as bore-edge masking. The phrase "double jeopardy" comes to mind.

3.6.7 The Discharge Coefficients of Poppet Valves in a Four-Valve 50 cm³ Moped Engine

This experiment is reported to illustrate the difficulties a designer faces when attempting to provide an alternative cylinder head design for a very small engine, namely a 50 cm³ moped unit. This cylinder head was designed to replace the two-valve design of the Honda moped discussed above in Sec. 3.6.4, with a four-valve layout. The bore and the stroke of the engine are retained at 39 mm and 41 mm, respectively. In such a small-bore cylinder, it becomes very difficult to arrange four valves around a centrally-located spark plug, even if it is just 8 mm in diameter. The designer, faced with the further limitation that one can only make valve guides and stems so small, yet retain mechanical reliability, settled for 4.5-mm thick valve stems. The result is a four-valve design with outer and inner seat diameters, d_{os} and d_{is}, of 10.9 mm and 9.9 mm, and 13.1 and 12.1 mm, for the exhaust and intake valves, respectively. The inner port diameters, d_{ip}, are 8.9 mm and 11.1 mm for the exhaust and intake ports, respectively. The maximum lift of both valves, L_v, is 3.2 mm. The exhaust and intake manifold diameters are 14.4 and 15.7, respectively, giving pipe-to-port ratios, C_{em} and C_{im}, of 1.75 and 1.19, respectively. The tract from each pair of valves to its manifold is of the conventional "Y" format, as in Plate 3.2.

The discharge coefficients are measured in both flow directions for both valves, employing the apparatus shown in Figs. 3.17 and 3.18. The results of the experiments and the analysis of the results for the actual discharge coefficients, Cd_a, are shown in Figs. 3.43 to 3.46.

These results may be compared directly with the equivalent maps of the two-valve Honda version of this cylinder head in Figs. 3.19 to 3.22, or with the Cd contours of a four-valve layout in a larger cylinder bore in Figs. 3.31 to 3.34 or Figs. 3.37 to 3.40.

The Cd maps for intake inflow, Figs. 3.45 and 3.46, are very similar to all previous 4v maps, and are at least as high in terms of the magnitude of the discharge coefficients across the entirety of the contours. Considering that the valves in this engine are "packed" into the cylinder, i.e., the potential for valve-masking effects is high, Fig. 3.46 for intake inflow must be considered to be superior to that for the masked valves of the FZR Yamaha engine in Fig. 3.40. There is no evidence here that the large (in terms of scale) valve stem diameter, or the relatively small diameter of this intake valve, which is less than half the diameter of that in the FZR Yamaha engine, has reduced in any way the magnitude of the intake inflow discharge coefficients. Indeed, the same remarks can be passed about outflow, i.e., reverse flow, at the intake valve.

Consequently, when the discharge coefficients for the exhaust valve are examined in Figs. 3.43 and 3.44, the contour maps come as rather more than a surprise. They do not resemble any of the previous maps for poppet valves, particularly in terms of profile at up to 50% maximum valve lift. Above 50% valve lift ratio, the levels and profiles of these figures are

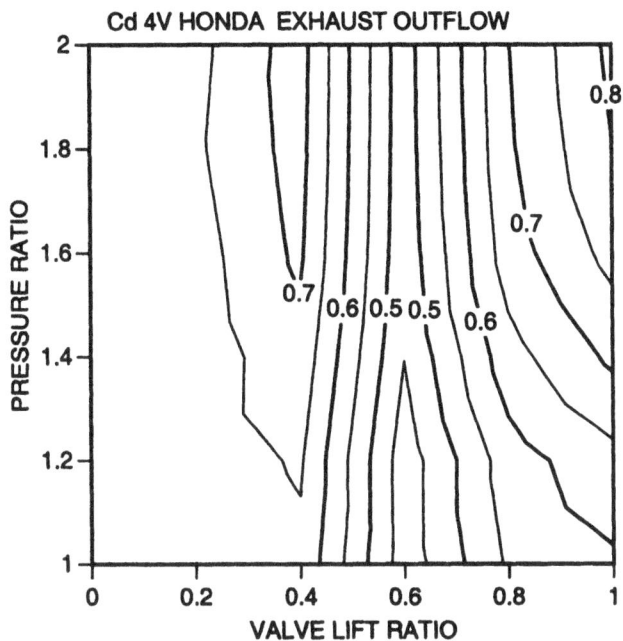

Fig. 3.43 The discharge coefficients for exhaust outflow of a 4v 50 cm³ moped engine.

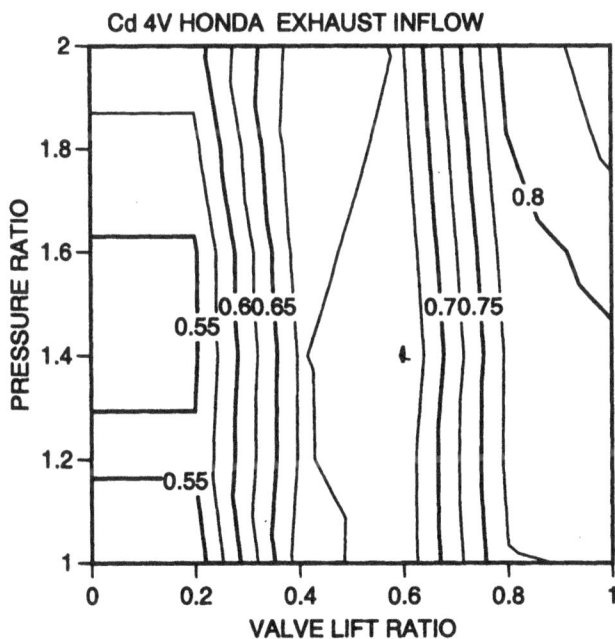

Fig. 3.44 The discharge coefficients for exhaust inflow of a 4v 50 cm³ moped engine.

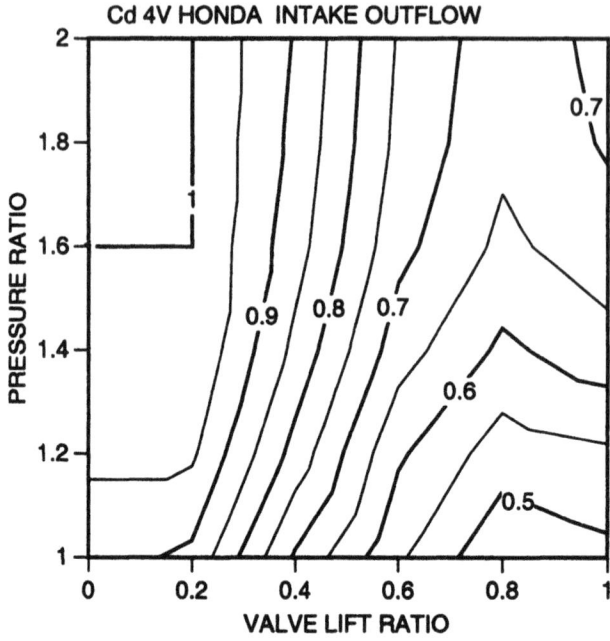

Fig. 3.45 The discharge coefficients for intake outflow of a 4v 50 cm³ moped engine.

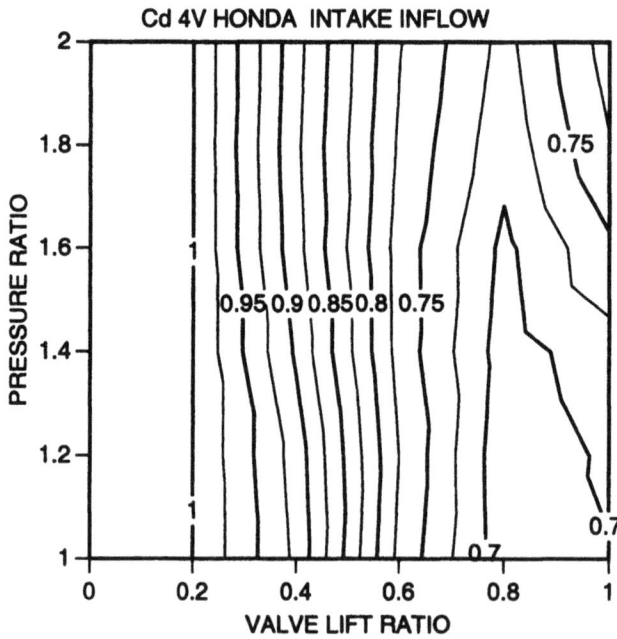

Fig. 3.46 The discharge coefficients for intake inflow of a 4v 50 cm³ moped engine.

more comparable to those of previous Cd maps. From an engine operational standpoint, the loss of 30 to 35 percentage points of mass flow rate at low exhaust valve lift will significantly reduce exhaust gas egress during blowdown, significantly increase pumping loss on the exhaust stroke, and ultimately reduce the quantity of air that can be induced during the intake stroke. It has already been noted that the effects of scale on valve size had no impact on the Cd maps for the intake valve; now, it must be concluded that it is the shape of the tract(s) leading away from the valves to the manifold at the exhaust pipe that seriously reduces the discharge coefficients for these exhaust valves. Although I have not included a drawing of the exhaust duct leading from each port to the manifold, be assured that if they had been so presented I doubt if you would observe anything unusual about the design other than the angle of the "Y," which is somewhat larger than normal.

You should take careful note of these measurements and be warned that making assumptions, viz., that all is well with the fluid mechanics of the flow from the valves to the manifolds simply because the geometry appears to be smoothly profiled, can be presumptuous.

3.7 The Discharge Coefficients of Restrictions within Engine Ducts
3.7.1 Restrictions within the Pipes of Engines
Sec. 2.12 in Chapter 2 presents a fundamental theoretical discussion on the thermodynamics and unsteady gas dynamics of the flow at a pipe discontinuity where there is a sudden change of area. The discontinuity extends to include the presence of a further restriction in pipe area at the very section where the pipe areas change. During the simulation process, where the transient behavior within the pipes and ducting of the engine is being analyzed, such a discontinuity denotes the location at which two pipes are "butted" together, giving rise to the slang nomenclature for this boundary as a "butted joint." Fig. 3.47 shows the several versions of this type of boundary, all of which are covered by the theory of Sec. 2.12. Fig. 3.48 illustrates a brief selection of such "butted joints" to be found within the ducts of engines.

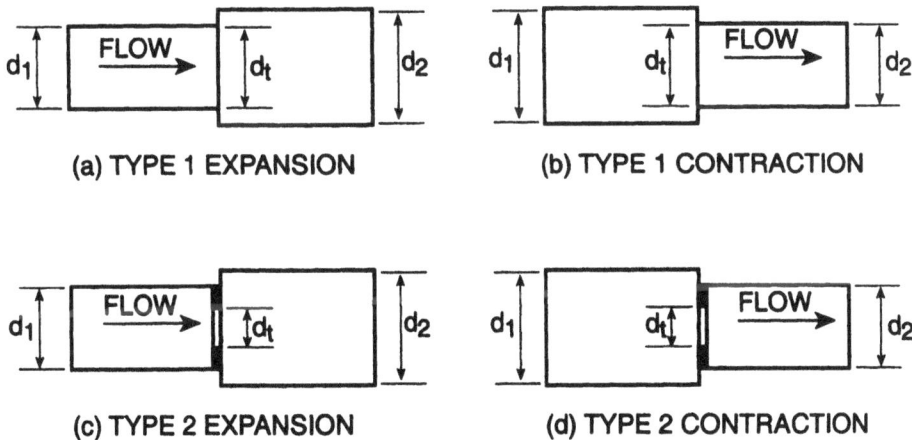

(a) TYPE 1 EXPANSION   (b) TYPE 1 CONTRACTION

(c) TYPE 2 EXPANSION   (d) TYPE 2 CONTRACTION

Fig. 3.47 The types of discontinuity at a junction between two pipes.

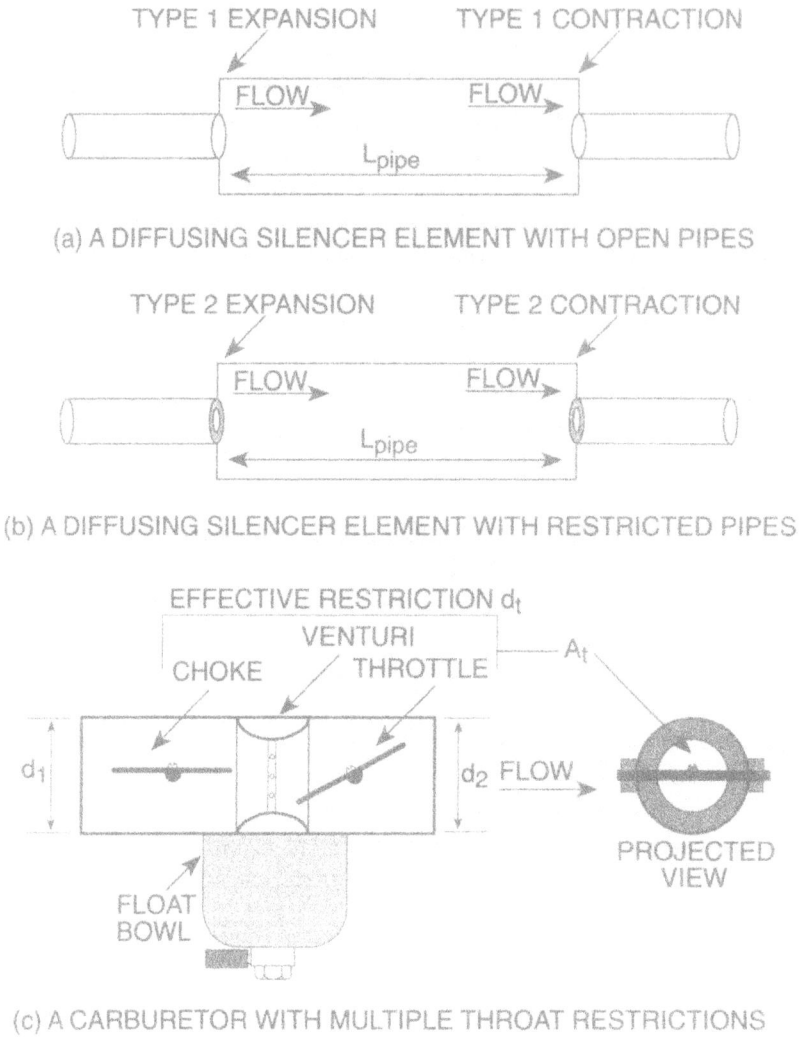

TYPE 1 EXPANSION   TYPE 1 CONTRACTION

FLOW   FLOW

L_{pipe}

(a) A DIFFUSING SILENCER ELEMENT WITH OPEN PIPES

TYPE 2 EXPANSION   TYPE 2 CONTRACTION

FLOW   FLOW

L_{pipe}

(b) A DIFFUSING SILENCER ELEMENT WITH RESTRICTED PIPES

EFFECTIVE RESTRICTION d_t

VENTURI

CHOKE   THROTTLE A_t

d_1 d_2 FLOW

PROJECTED VIEW

FLOAT BOWL

(c) A CARBURETOR WITH MULTIPLE THROAT RESTRICTIONS

Fig. 3.48 Examples of "butted joints" to be found within engine ducting.

The Type 1 Boundary with No Restriction at the Change of Section

Figs. 3.47(a) and (b) illustrate one version of this type of boundary between two pipes where the area changes from A_1 to A_2. In this sketch, the pipes are represented by the actual, or equivalent, diameters, d_1 and d_2. Figs. 3.47(a) and (b) show a Type 1 expansion or contraction, respectively, which Type 1 is defined as having no throttle or area restriction at the discontinuity. In short, the area and diameter of the throttle, A_t and d_t, are equal to the upstream area and diameter, A_1 and d_1. The theory of Sec. 2.12 copes with such an eventuality.

The pipe boundary typical of this geometry can be found as part of an exhaust silencer, such as that discussed in Chapter 7, as a diffusing silencer element. This element, which is sketched in Fig. 3.48(a), is composed of an entry and exit pipe, in between which there is a third, larger-area pipe whose length, L_{pipe}, provides multiple internal reflections. These reflections change the frequency spectrum and lower the amplitude of the transmitted noise signal to other parts of the system and, ultimately, to the atmosphere. Simply put, there is a Type 1 expansion at entry and a Type 1 contraction at exit to this diffusing silencer element. Naturally, because the flow reverses due to the internal reflections of, say, exhaust compression waves within the central pipe, the expansion at the left end of the center pipe becomes a contraction. One must incorporate all of the engine and ducting geometry within a simulation if it is to accurately predict all of the engine performance characteristics, including the exhaust and intake noise levels and spectra. An essential part of that requirement is the need for accurate Cd maps of all discontinuities, which means including the "butted joints" under discussion. The discharge coefficient, Cd, at this boundary is employed within Eq. 2.12.4 during its solution with its companion equations.

Manifestly, the Cd for a Type 1 expansion process in Fig. 3.47(a) is either unity or within a few percentage points off it. The flow during a Type 1 contraction process would appear to resemble that of inflow at a fully open port to a two-stroke engine. However, by now you are well aware that, in fluid mechanics, appearances can be deceptive.

From the theoretical standpoint of Sec. 2.12, the Type 1 discontinuity does not include the possibility of "joining" two pipes of equal area which have no restriction between them. The theory of Sec. 2.12 will arithmetically "crash" on a computer if such a numerical solution is attempted. In short, Eq. 3.7.1 formulates the definition that it is not a Type 1 butted joint.

Not a butted joint: $$A_1 = A_t = A_2 \qquad (3.7.1)$$

The Type 2 Boundary with a Restriction at the Change of Section
Fig. 3.47 (c) and (d) illustrates another version of this type of boundary between two pipes where the area changes from A_1 to A_2, as represented in the sketch by the actual, or equivalent, diameters, d_1 and d_2. The Type 2 junction is defined as having a throttle or area restriction at the discontinuity. In short, the area and diameter of the throttle, A_t and d_t, are not equal to the upstream area and diameter, A_1 and d_1 in the case of the Type 2 expansion, nor are they equal to the downstream area and diameter in the case of the Type 2 contraction in Fig. 3.47(d). Unlike the Type 1 discontinuity, the upstream and downstream areas may be identical and the theory of Sec. 2.12 can handle that eventuality:

A butted joint: if $A_1 > A_t < A_2$ then $A_1 \leq A_2$ or $A_1 \geq A_2$ $\qquad (3.7.2)$

As stated above, the discharge coefficient, Cd, at this discontinuity is inserted into Eq. 2.12.4 and is employed in the solution together with the other equations listed there. The Type 2 discontinuity is geometrically akin to the Type 1 contraction, irrespective of whether it is a Type 2 contraction or expansion, for the throat area, A_t, in both cases presents a restriction for unsteady flow from pipe 1.

Because the geometry for the Type 2 contraction, shown in Fig. 3.47 (d) is somewhat akin to that for outflow at a port in a two-stroke engine (see Figs. 3.11 or 3.12), you might imagine that the resulting Cd map might well have a similar profile to it, viz., Fig. 3.13. As the ensuing discussion here will show, fluid mechanic perversity triumphs again and the human imagination and ego suffer yet another dent. The sharper-eyed may retort that perhaps I have it wrong, as the geometry in Fig. 3.47 (c) resembles that for inflow at a port in a two-stroke engine—so perhaps the Cd map for the "butted joint" will have a similar profile to that type of boundary instead, viz., Fig. 3.14. The score, soccer fashion, now reads, *fluid mechanics-*2 *humans-*0.

The geometry of a Type 2 discontinuity can also be found in diffusing silencer elements where the entry and exit pipes are crimped, i.e., locally reduced in area at entry and exit to further enhance the internal reflections that are shuttled to and fro within the central pipe system, and decrease the magnitude of those transmitted into the exit pipe. This effect is illustrated geometrically in Fig. 3.48(b) for a diffusing silencer element and, by using an engine simulation model, on the performance and noise characteristics in Chapter 7 (see Fig. 7.31).

Within the intake system of an engine ducting, it is quite common to find even more complex "butted joints." This is illustrated in Fig. 3.48(c), using a carburetor as the example. The upstream and downstream diameters, d_1 and d_2, are often identical. The restriction between the two pipes that form the carburetor is composed of the venturi, the throttle, and the choke plate. The sketch shows the latter to be butterfly valves, but the throttle can be a slide valve and needle combination, and the choke plate may be fully retractable to one side of the intake duct. The venturi section in racing carburetors often has a quite minimal area restriction. As long as there is some area restriction at the junction between pipes 1 and 2, then it qualifies as a Type 2 contraction, or expansion, as the case may be. It is often difficult to specify the combined throat area, A_t, of such a restriction to include the choke, venturi, and throttle as shown in Fig. 3.48(c). Although there is no optimum method to do this, the simplistic way is to observe the projected throat area, A_t, of the combination of the restrictions as shown in Fig. 3.48(c). The most essential point is that one should employ that estimated throat area in the identical fashion within the measurement process for the discharge coefficient as one does within the engine simulation model. Only then will a simulation model compute the same measured mass flow for the carburetor, to quote just one example, when the state conditions across it are identical with those in the Cd experiment.

3.7.2 The Measurement of Discharge Coefficients at Restrictions within the Pipes of Engines

The apparatus for measuring discharge coefficients at area restrictions within the pipes of engines is shown in Fig. 3.49. This is basically identical to that described more completely in Sec. 3.1; only the device in the test section is different.

Between the pipes 1 and 2, is interposed the area restriction whose discharge coefficient is to be measured. It could be a carburetor, a crimped pipe, a butterfly throttle plate, or, as shown in Fig. 3.49, a series of orifice plates with differing throat areas to represent a fundamental example of the "butted joint" described theoretically in Sec. 2.12. Measurements of static pressures and temperatures across it, p_1 and T_1, and p_2 and T_2, respectively are taken at some four pipe diameters, upstream and downstream. The mass flow of air at any given experimental setting is recorded after the settling tank by an orifice plate system designed according to

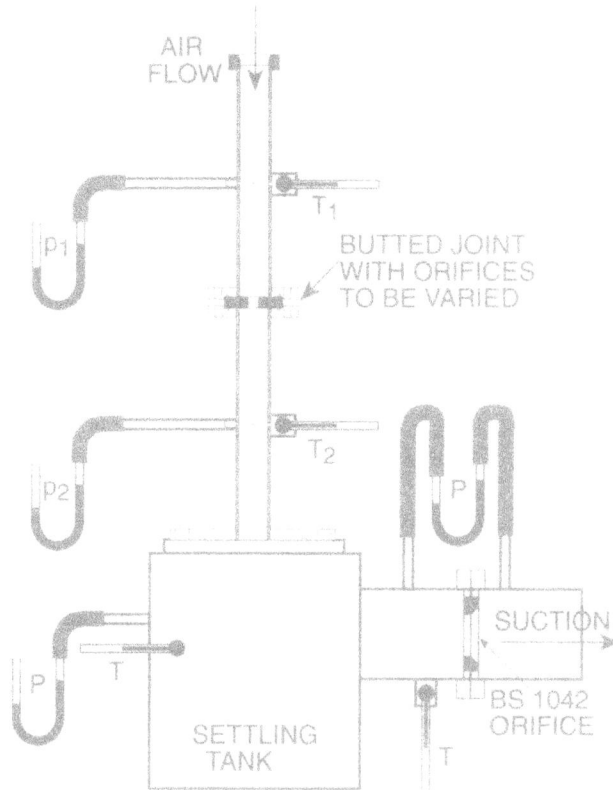

Fig. 3.49 Apparatus for measuring the discharge coefficients of "butted joints."

the British Standards Institution [3.15]. The air flow rate is altered by changing the vacuum pressure in the settling tank, thereby increasing the pressure ratio across the test section. The pressure ratio, P, for a butted joint is defined as:

$$P = \frac{\text{upstream pressure}}{\text{downstream pressure}} = \frac{p_1}{p_2} \tag{3.7.3}$$

Because there is no concept of inflow or outflow at a butted joint, irrespective of it being classified as either Type 1 or Type 2, or as a contraction or an expansion, the pressure ratio is always defined as given in Eq. 3.7.3.

To investigate the discharge coefficients of fundamental area restrictions between pipes, a series of orifices are designed and placed in the position sketched in Fig. 3.49. The orifice plates are designed as "thick" orifices and the dimensions and layout of the series of five orifices are shown in Fig. 3.50.

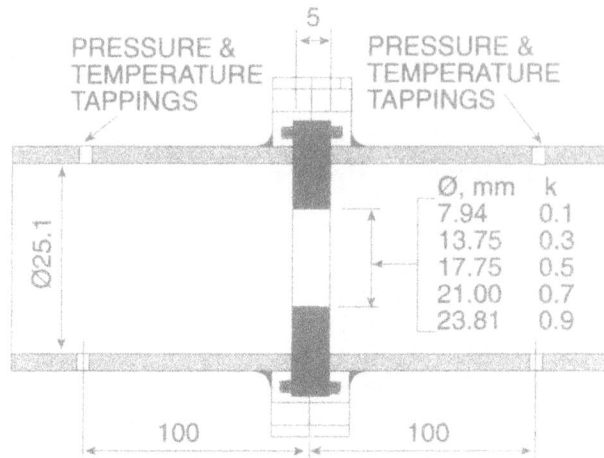

Fig. 3.50 The orifices used as fundamental examples of a "butted joint."

The upstream and downstream pipe diameters, d_1 and d_2, are identical at 25.1 mm diameter. The five orifice plates are each of 5 mm thickness and the central hole sizes are 7.94, 13,75, 17.75, 21.0, and 23.81 mm. This give area ratios, k, that are nominally 0.1, 0.3, 0.5, 0.7, and 0.9, respectively. For a butted joint, the area ratio, k, is defined as:

$$k = \frac{\text{throat area}}{\text{upstream pipe area}} = \frac{A_t}{A_1} \qquad (3.7.4)$$

In this particular case, because the upstream and downstream pipe areas are identical, the area ratio, k, is the same for flow in either direction. For many Type 2 discontinuities, where A_2 is not equal to A_1, the k value will depend on the particle flow direction at any instant during unsteady gas flow.

3.7.3 The Analysis of the Actual Discharge Coefficient for a Butted Joint

Sec. 3.2.4 describes the method for analyzing the measured data to acquire the actual discharge coefficient, Cd_a, for flow to a pipe from a cylinder using the theory of Sec. 2.16. For the analysis of any given set of experimental data, the cylinder pressure, p_1, has no "pressure wave" content, and the double iterative process, complicated as it is, has but two variables, namely, the incident pressure amplitude ratio, X_{i2}, and the value of effective throat area, A_{te}.

For the butted joint, the theory of Sec. 2.12 is now relevant and is solved on a computer using the numerical techniques detailed in that section. Because the upstream section at pipe position 1 now also contains a variable, the incident pressure amplitude ratio, X_{i1}, it follows that both this and the value of the incident pressure amplitude ratio, X_{i2}, at the downstream section 2, are varied until the calculated superposition pressures correspond precisely to the

390

measured pipe pressures at both sections, i.e., p_1 and p_2. At the same time that incident pressure amplitude ratios, X_{i1} and X_{i2}, are varied, the value of the throat area is also changed until a value of effective area, A_{te}, is arrived at which gives a theoretical mass flow rate that corresponds precisely with the measured mass flow rate, $\dot{m}_{ex}$. This triple iterative process is clearly more complex than that described in Sec. 3.2.4. The actual discharge coefficient, Cd_a, is now defined by:

$$Cd_a = \frac{\text{effective throat area}}{\text{geometrical throat area}} = \frac{A_{te}}{A_t} \qquad (3.7.5)$$

To compute the ideal discharge coefficient, Cd_{id}, for flow at the butted joint, the iterative procedure is simplified by retaining as a constant the throat area at the geometrical value, A_t, and determining the ideal mass flow rate, $\dot{m}_{id}$, at the measured overall pressure ratio, P. In other words, apart from the extra variable involved (the upstream X_{i1}), the procedure to compute the ideal discharge coefficient basically follows the approach given in Sec. 3.2.3. The ideal discharge coefficient, Cd_{id}, follows the previous definition presented in Eq. 3.2.6:

$$Cd_{id} = \frac{\text{measured mass flow rate}}{\text{ideal mass flow rate}} = \frac{\dot{m}_{ex}}{\dot{m}_{id}} \qquad (3.7.6)$$

3.7.4 The Discharge Coefficients of a Butted Joint

The five orifices described in Sec. 3.7.2 are tested and the experimental data is analyzed by the methods described above to obtain the actual discharge coefficients, Cd_a. Recall that these orifices constitute a fundamental version of a butted joint. The results are shown in Fig. 3.51, plotted as xyz contours with respect to pressure ratio, P, and area ratio, k.

It is clear, from a comparison with all of the other Cd_a maps presented in this chapter, that the discharge coefficient characteristics of a butted joint are totally different from those of an open pipe end, pipe-end orifices, two-stroke ports, or four-stroke poppet valves. Whereas all of the Cd maps for all other pipe end restrictions basically exhibit the highest Cd values at the lowest area ratios, the butted joint displays the converse characteristics. The lowest Cd values are at maximum throttling, i.e., at the smallest area ratio. At the lower area ratios, the Cd values for the butted joint have a considerable dependence on pressure ratio, but this trend decreases toward a fully open throttle value. All other pipe end restrictions, be they valves or ports, for either inflow or outflow, show very little dependence on pressure ratio at the lowest area ratios, or lowest valve lift ratios. As remarked in Sec. 3.7.1, second-guessing the results of experimental fluid mechanics can be a fruitless exercise.

For completeness, and also because it is at odds with the results from all other pipe-end restrictions, it should be reported that Blair et al. [3.18] found very little difference between the ideal and the actual discharge coefficients for a butted joint.

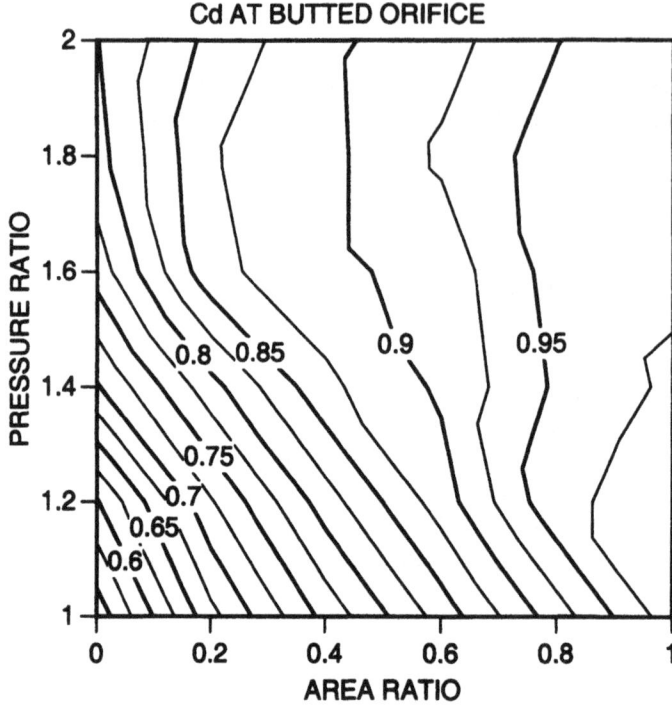

Fig. 3.51 The discharge coefficients of orifices placed within a pipe.

3.7.5 The Geometry of Throttles Used in Engine Intake Systems

Fig. 3.48 shows a carburetor with a butterfly throttle valve controlling the flow area of the intake ducting. The throttle gives the necessary variation of the delivery ratio in an Otto cycle engine at any given engine speed, in order to produce the power and torque changes desired by the user. Such engines not only employ carburetors but also EFI (electronic fuel injection) and GDI (gasoline direct injection), all of which use air-flow throttling devices ranging from the butterfly valves illustrated in Fig. 3.48 to slide valves and sliding valves. Sketches of all three throttle types are shown in Fig. 3.52.

In each case, these devices are shown to provide a "butted joint" between two round pipes of diameter, d_1 and d_2, giving upstream and downstream areas of A_1 and A_2, respectively. The aperture area at the throttle in each case is illustrated as having an area A_t. Almost without exception, the throttle is mechanically fitted within the downstream section, d_2, and although Fig. 3.52 shows the tract to be of common diameter, many carburetors have an upstream section that is larger than that downstream. In other words, d_1 is often greater than d_2, which enhances the pressure drop at the venturi in order to induce fuel from a carburetor float bowl

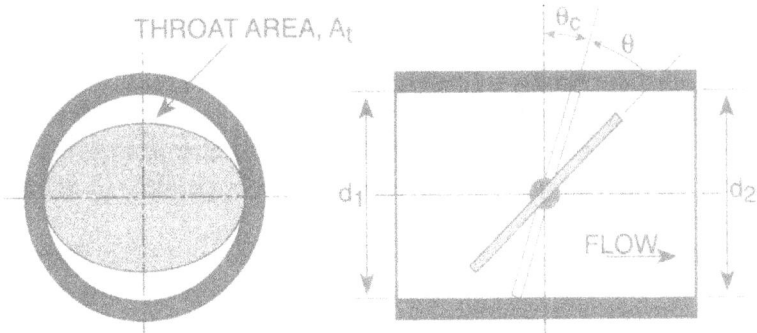

(a) BUTTERFLY THROTTLE

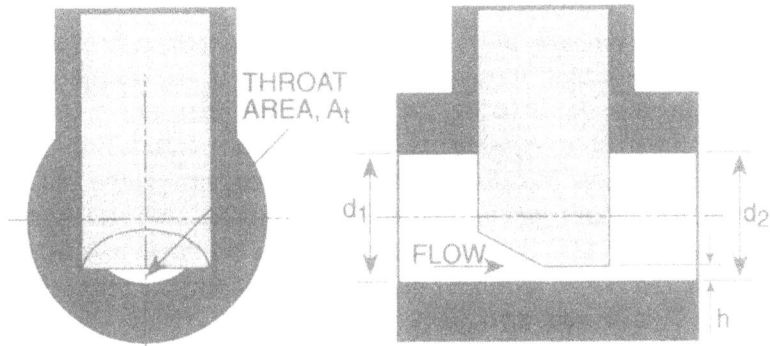

(b) SLIDE VALVE THROTTLE

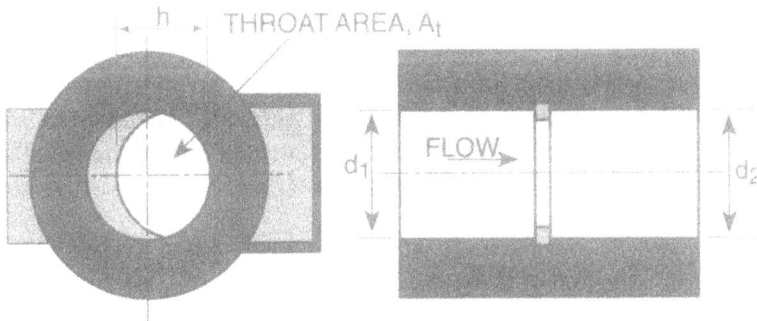

(c) SLIDING PLATE THROTTLE

Fig. 3.52 The geometry of various throttling devices.

into the air stream. Some, albeit few, throttle-body designs for EFI and GDI systems are also made with differing upstream and downstream sections. In the geometry analysis below, it is assumed in each case that the throttle is fitted in the downstream section. In general, no detailed account is taken of throttle plate thickness, or of the diameter of the spindle that holds it, or of tapered fuel needles or other main jet protrusions that can further obstruct the throat area, A_t. Thus, these aspects of physical intrusion must be assessed separately in terms of the actual design being simulated.

The engine modeler requires this geometrical analysis in order to be able to simulate carburetor throttles, or EFI and GDI throttle bodies, within an engine intake system. In particular, the modeler must insert the accurate geometry of any throttle to give meaning to the use within an engine simulation of the typical Cd map shown in Fig. 3.51.

The Butterfly Throttle

The butterfly valve is probably the most common of all throttle valves, probably because it is the cheapest to manufacture! It commonly used for the choke control as well as the air flow throttle valve, much as shown in Fig. 3.48. The throttle plate, shown in Fig. 3.52(a), sits at an angle θ_c in the closed position, which means that its obstruction profile is that of a conic section, in this case an ellipse. It is that elliptical—not circular—profile which is rotated to open the throttle, until it is fully open when it is aligned with the tract center line. The closing angle, θ_c, is typically about 15°. The flow area exposed is the throat area, A_t, at any further angular movement θ, and can be approximately expressed with quite sufficient accuracy, as:

$$A_t = \frac{\pi d_2^2}{4}\left[1 - \frac{\cos(\theta + \theta_c)}{\cos\theta_c}\right] \tag{3.7.7}$$

In a practical sense, bear in mind the above-mentioned caveat that this area must be further reduced to account for the thickness of the butterfly plate and the diameter of its holding spindle. Assuming that the thickness of the butterfly plate and its holding spindle represent a diametral obstruction of dimension s, when the obscuration exceeds that of the angled throttle plate, the aperture area, A_t, is adjusted as follows:

$$\text{if } sd_2 > \frac{\pi d_2^2 \cos(\theta + \theta_c)}{4\cos\theta_c} \text{ then } A_t = A_2 - sd_2 \tag{3.7.8}$$

In short, with the spindle and plate thickness as an obstruction to flow, the last few degrees of butterfly movement produce no further increase of throat area. Using Eq. 3.7.4, the area ratio, k, for use within the Cd map of Fig. 3.51 is defined as:

$$k = \frac{\text{throat area}}{\text{upstream pipe area}} = \frac{A_t}{A_1}$$

As an example, Eq. 3.7.7 above is solved for a 25-mm constant area tract for a butterfly valve that is fully closed at an angle θ_c of 15°. The results, at 3° intervals, are plotted in Fig. 3.53 for area ratio, k, with respect to the valve angular movement θ. Clearly, the maximum value that θ can have is 75°.

The profile can be seen to be somewhat nonlinear, and is fitted with a third order polynomial curve. The same solution for area ratio, k, is plotted in Fig. 3.54, but with respect to angle ratio C_{bt}, which is defined as:

$$C_{bt} = \frac{\theta}{90° - \theta_c} \qquad (3.7.9)$$

y = 9.0288e-4 + 4.3707e-3x + 1.7269e-4x^2 - 7.0944e-7x^3

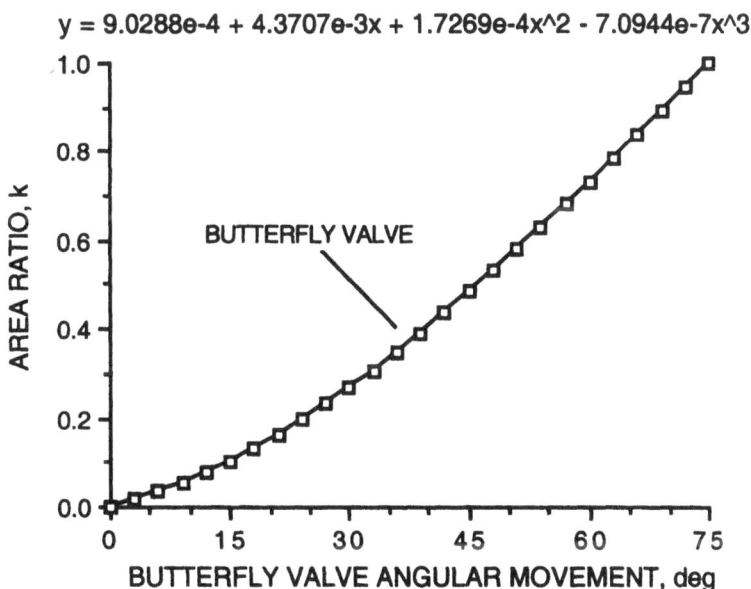

Fig. 3.53 Butterfly valve movement with respect to angle.

395

This, too, can be fitted with a third-order polynomial, an approximation that will assist those who do not wish to solve the above equations for a particular throttle geometry, but who need to acquire data for a butterfly throttle valve to insert into an engine simulation.

$$\text{Butterfly valve:} \quad k = 9.0288 \times 10^{-4} + 0.3278 C_{bt} + 0.97136 C_{bt}^2 - 0.29929 C_{bt}^3 \quad (3.7.10)$$

The Slide Valve Throttle

The slide valve throttle is shown in Fig. 3.52(b). Throughout history, almost all motor-cycle carburetors used a slide valve carburetor, although emissions legislation, which requires tight control of the air-to-fuel ratio over the entire engine speed and load range, is bringing about a change-over to EFI similar to that which occurred for automobiles two decades earlier. Nevertheless, the slide valve carburetor is still very common, not only on motorcycles, but on many low-cost industrial engines. Slide valves are also employed as throttles on multi-cylinder EFI and GDI systems because they provides a much better seal at zero and idle throttle settings compared to butterfly valves; the inter-cylinder air flow equality is thereby enhanced.

The throttle area, A_t, exposed as the slide lifts by a height, h, from the fully closed position is shown in Fig. 3.52(b). This is a segment of a circle. The throttle area, A_t, is found from the following equations, where r_2 is the radius of the downstream pipe:

$$\text{if } h < r_2 \text{ then } z = r_2 - h \qquad x = \sqrt{r_2^2 - z^2} \qquad \phi = \tan^{-1}\left(\frac{x}{z}\right) \qquad A_t = \phi r_2^2 - xz$$

$$(3.7.11)$$

$$\text{if } h = r_2 \text{ then} \qquad A_t = \frac{\pi r_2^2}{2} \qquad (3.7.12)$$

$$\text{if } h > r_2 \text{ then } z = h - r_2 \qquad x = \sqrt{r_2^2 - z^2} \qquad \phi = \pi - \tan^{-1}\left(\frac{x}{z}\right) \qquad A_t = \phi r_2^2 + xz$$

$$(3.7.13)$$

Inserting this information into Eq. 3.7.4 provides the area ratio, k, which is used as the variable on the x-axis of the Cd map shown in Fig. 3.51. The area ratio, k, is defined as:

$$k = \frac{\text{throat area}}{\text{upstream pipe area}} = \frac{A_t}{A_1}$$

A lift ratio, L_{sv}, is defined for the slide valve throttle:

$$L_{sv} = \frac{h}{d_2} \qquad (3.7.14)$$

Eqs. 3.7.11 to 3.7.13 can be solved for the slide valve throttle area, A_t, for any diameter of pipe, d_2, and plotted against the lift ratio of the slide valve from closure to full opening. If the results are plotted as area ratio, k, with respect to lift ratio, L_{sv}, such a plot is dimensionless and virtually applies to any slide valve geometry. As shown in Fig. 3.54, this approaches a linear relationship, and is much better in this regard than the butterfly valve shown plotted on the same figure, albeit against its valve angle ratio, C_{bt}, as defined by Eq. 3.7.9.

Because the relationship between throttle valve lift ratio and area ratio is dimensionless, it can be fitted with a third-order polynomial and accurately used to input data to an engine simulation. That relationship is:

Slide valve: $\quad k = -0.0048736 + 0.4421L_{sv} + 1.703L_{sv}^3 - 1.1353L_{sv}^3 \qquad (3.7.15)$

The Sliding Plate Throttle

The sliding plate throttle is shown in Fig. 3.52(c). This type of intake throttle valve is commonly found on multi-cylinder racing engines, as it allows a single-throttle plate to equally

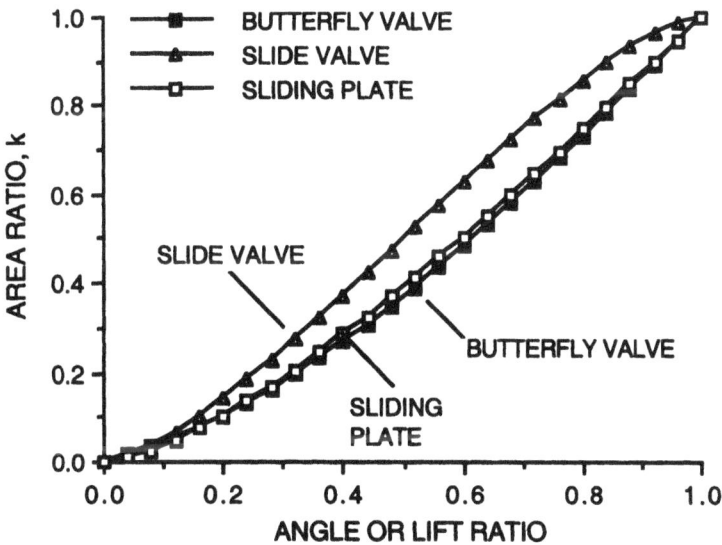

Fig. 3.54 Area ratio characteristics for several types of throttle valve.

control the intake area on all of the cylinders on that bank. At full open it permits an unobstructed tract area with an EFI or GDI system, but on a carburetor setup there may be some further throttle area reduction caused by the presence of a needle, main jet protrusion, or venturi. The comments regarding the quality of sealing at the near-zero throttle setting, and the effect of charging on cylinder-to-cylinder equality, passed above with respect to slide valves, also apply here.

The throttle area, A_t, exposed as the slide lifts by a height, h, from the fully closed position is shown in Fig. 3.52(c). This is composed of two segments of a circle. The throttle area, A_t, is found from the following equations, where r_2 is the radius of the downstream pipe:

$$\text{if } h < d_2 \text{ then } z = r_2 - \frac{h}{2} \qquad x = \sqrt{r_2^2 - z^2} \qquad \phi = \tan^{-1}\left(\frac{x}{z}\right) \qquad A_t = 2\left(\phi r_2^2 - xz\right)$$

$$(3.7.16)$$

$$\text{if } h = d_2 \text{ then } \qquad A_t = \pi r_2^2 \qquad\qquad (3.7.17)$$

Inserting this information into Eq. 3.7.4 provides the area ratio, k, which is used to index the same variable on the x-axis of the Cd map shown in Fig. 3.51. The area ratio, k, is defined as:

$$k = \frac{\text{throat area}}{\text{upstream pipe area}} = \frac{A_t}{A_1}$$

A lift ratio, L_{sp}, is defined for the sliding plate throttle:

$$L_{sp} = \frac{h}{d_2} \qquad\qquad (3.7.18)$$

If Eqs. 3.7.16 to 3.7.18 are solved for the sliding plate throttle area, A_t, for any diameter of pipe, d_2, and plotted against lift ratio of the sliding plate from closure to full opening, and if the results are plotted as area ratio, k, with respect to lift ratio, L_{sp}, such a graph is dimensionless and applies to any sliding plate geometry. As shown in Fig. 3.54, this is very similar in profile to that of a butterfly valve, which is plotted on the same figure, but is not as linear as a slide valve throttle in terms of its exposed area with respect to throttle plate movement. Like that for the slide valve, the relationship for a sliding plate between the throttle valve lift ratio and the area ratio is also dimensionless, so it can be fitted with a third-order polynomial and accurately used to input data to an engine simulation. That relationship is:

Sliding plate: $\qquad k = -0.006165 + 0.36551L_{sp} + 1.063L_{sn}^3 - 0.42515L_{sn}^3 \qquad (3.7.19)$

Application in Practice

You will recall that the polynomial equations, Eqs. 3.7.10, 3.7.15, and 3.7.19, which relate the throttle area ratio, k, to valve movement, are deduced on the assumption that the upstream area, A_1, is equal to the downstream area, A_2, where the valve is considered to be fitted. You will also recall, from Eq. 3.7.4, that the discharge coefficient, Cd, is based on, and used within an engine simulation as, the area ratio of that restriction as a function of the upstream pipe area. As a consequence, if the upstream and downstream areas are different, the numeric value resulting from an application of any of these throttle area-ratio equations must be modified by multiplying that number by the ratio of the downstream and upstream areas:

$$\text{if } A_1 \neq A_2 \text{ then } k = \frac{A_2}{A_1} \times k_{\text{from previous Eqs.}} \qquad (3.7.20)$$

The rationale behind this obscure procedure is that the throttle area ratio must be set with respect to the upstream area in order to obtain the correct Cd value from the relevant map of the particular "butted joint."

It will also be observed that the same polynomial equations, Eqs. 3.7.10, 3.7.15, and 3.7.19, contain a zero offset error in each case, i.e., when either C_{bt}, L_{sv}, or L_{sp} is zero the area ratio, k, is not zero. The calculated value of k is negative and infinitesimal but, if applied in practice into the theory of Sec. 2.12, or for the acquisition of a discharge coefficient, Cd, from the data represented graphically in Fig. 3.51, it could numerically "crash" any computation process. If you are using these theories take note that, in the highly unlikely event that you wish to investigate the effect of throttle area ratio settings below 0.005 on the performance characteristics of an engine, you must take the obvious numerical precautions!

3.8 Using the Maps of Discharge Coefficients within an Engine Simulation
3.8.1 Digitizing the Cd Maps for the Simulation Process

The xyz contour maps of discharge coefficients at all of the pipe boundaries normally found within the ducting of engines are presented in the text above. These cover plain and bellmouth pipe ends, throttled pipe ends, ports, poppet valves, and butted joints, and they cope with flow in both directions for all discontinuities, except for the butted joint where it is not necessary. In any application of the fundamental theory of Secs. 2.12, 2.16 or 2.17 within an engine simulation, it is necessary to know at any juncture the appropriate area ratio, k, or valve lift ratio, L_r, if it is a poppet valve, for the boundary being investigated. Together with the pressure ratio, P, across this boundary, it is possible to index the appropriate Cd map and deduce the relevant discharge coefficient at that instant. The effective throat area, A_{te}, of the actual geometrical throat area, A_t, is now obtained as:

$$A_{te} = C_d A_t \qquad (3.8.1)$$

Only when one is armed with this necessary information may the theory in the appropriate section be revisited and relevancy expected of the numerical solution so obtained. You will doubtless wish to know the "best" method of carrying out this exercise. Below I discuss the method that I currently use for this process. This is not presented as an optimum approach, merely one that I have found to be of the same order as the experimental accuracy that mapped the Cd values into those same xyz contours. Because I have presented the Cd values as xyz contour maps, you can digitize them relatively easily and use them within an engine simulation, either with my method given below or with whatever numerical method you concoct and consider to be a superior approach.

Sec. 3.4, in particular the discussion pertaining to Fig. 3.8, describes the method for constructing the xyz contour maps of the Cd values. Fig. 3.8 shows the outcome of the GPB method for plotting the experimental results and fitting them to "planes" at pressure ratio, P, intervals of 0.2 atm, and area ratio, k, also at intervals of 0.2. If a poppet valve is being considered, then you will recall that the valve lift ratio, L_r, replaces the area ratio on the x-axis. The y-axis always represents the pressure ratio, which has a maximum value of 2.0. Anything above this level in a simulation will definitely represent choked flow, and for the purposes of Cd assessment will be considered to be at 2.0. Hence, the x-axis runs numerically from 0 to 1 and the y-axis from 1 to 2. When the experimental Cd values are "best fitted" by both interpolation and extrapolation to these twenty-five "planes," they are actually fifty triangular surfaces as described in Sec. 3.4; the "plane" corners are transferred into the xyz contour maps that litter this chapter!

Consider the redrawing of Fig. 3.8, showing only the plane corners and the selection of a particular plane with corners numbered 1 to 4. This is presented in Fig. 3.55.

With the x-axis as area ratio and the y-axis as pressure ratio, these plane corners have x, y coordinates, i.e., (x_1, y_1), (x_2, y_2), (x_3, y_3), and (x_4, y_4), that are (1.4, 0.2), (1.4, 0.4), (1.2, 0.4), and (1.2, 0.2), respectively. For the purposes of illustrating this discussion, let it be supposed that the discharge coefficients, Cd, at these corners which are on the z-axis, have coordinates z_1, z_2, z_3 and z_4, of 0.6, 0.6, 0.5, and 0.5, respectively. A close-up view of this plane in isolation is shown in Fig. 3.56.

The splitting of the selected plane into the two triangles, referred to above and previously in Sec. 3.4, are the surfaces to which the experimental points are actually fitted, due to the geometrical fact that one normally cannot have a flat surface pass through any four points in space; actually the abnormal would be the four z coordinates I have numerically selected above! Change any one of those four numbers, e.g., one of the 0.5 values to 0.51, and you will get the point.

Somewhere within an engine simulation, consider the need to acquire the discharge coefficient for the mythical point 5 seen in Figs. 3.55 and 3.56, which happens to lie within the x and y boundaries of this particular plane. There are two ways of doing this; one is simple and the other is more complex as it uses in reverse the same GPB plotting process that originally determined the plane corners to be "best fitted" to the experimental points.

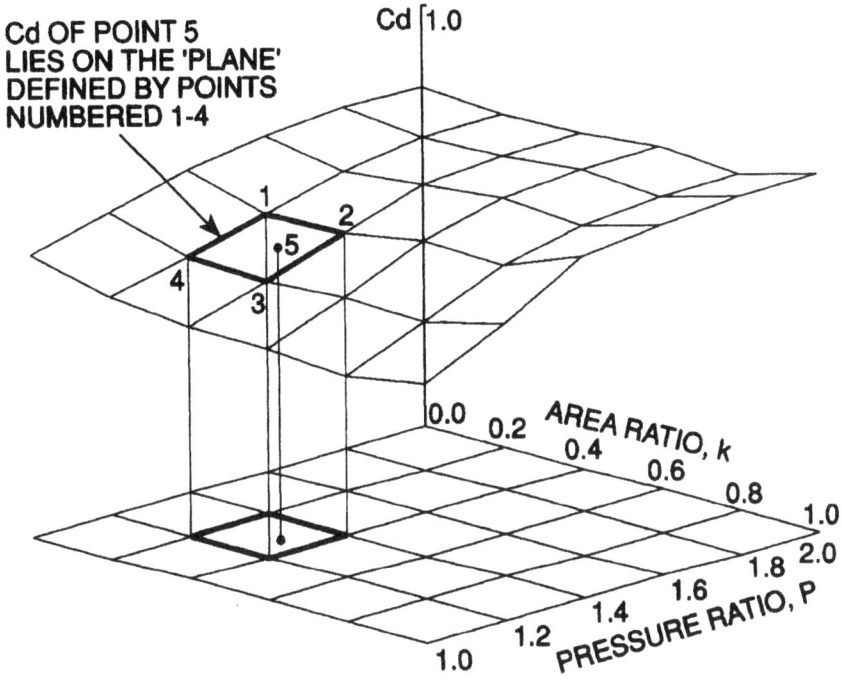

Fig. 3.55 The corners of the planes on a Cd map.

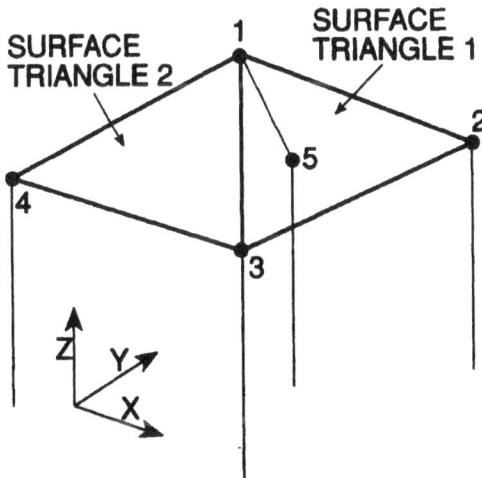

Fig. 3.56 The corners of the selected plane on the Cd map.

3.8.2 A Simple Method to Use a Digitized Cd Map

Consider point 5 in Figs. 3.55 and 3.56, which is determined to lie within the boundaries of the selected plane. The first simple approach is to assume that the Cd value of point 5 is an arithmetic average of those at the plane corners:

Simple method No. 1:
$$z_5 = \frac{z_1 + z_2 + z_3 + z_4}{4} = 0.55 \qquad (3.8.2)$$

Using the data already selected for points 1 to 4, this yields a Cd value of 0.55 for point 5, as seen in Eq. 3.8.2. If point 5 happened to lie in the middle of the plane, this outcome would not be unreasonable, but if the y coordinate (for pressure ratio) lay closer to points 1 or 2, or lay closer to points 3 or 4, the accuracy level of the value obtained by Eq. 3.8.2 would be inadequate.

The second simple method can described as the "nearest three points" approach. As the name implies, the position of point 5 is investigated to determine its nearest three neighboring corner points. By simple coordinate geometry, the lengths from point 5 to the four corners, s_{51}, s_{52}, s_{53}, and s_{54}, respectively, are found by a series of four equations, of which that for s_{51} is the first and can represent the others:

Simple method No. 2
$$s_{51} = \sqrt{(x_5 - x_1)^2 + (y_5 - y_1)^2} \qquad (3.8.3)$$

Whichever one of these lengths is the greatest—and by a simple visual inspection of Fig. 3.56 this would be corner number 4—causes that corner to be rejected from the ensuing averaging process for the Cd value of point 5:

Simple method No. 2:
$$z_5 = \frac{z_1 + z_2 + z_3}{3} = 0.567 \qquad (3.8.4)$$

As with the first approach, the usefulness of this second method is highly dependent on the location of point 5. Indeed, if that point happened to be positioned as sketched in Fig. 3.56, a prediction of 0.567 is actually worse than the 0.55 found by method No. 1. These simple methods are not at all effective unless the grid is made very considerably finer than that of the twenty-five squares shown in Fig. 3.55, which is an option you may well wish to investigate.

3.8.3 The GPB Method to Use a Digitized Cd Map

Point 5, in Figs. 3.55 and 3.56, lies on the selected plane between the four corner points 1 to 4 and within one of the two triangles defined as t_1 and t_2, which are identified as having corner points 1, 2, and 3, or 1, 3, and 4, respectively. To determine which triangle point 5 lies within, the slope of the line from point 1 to point 5, namely m_{15}, is compared with the slope of the line from point 1 to point 3, namely m_{13}:

Line slopes: $$m_{15} = \frac{y_5 - y_1}{x_5 - x_1} \qquad m_{13} = \frac{y_3 - y_1}{x_3 - x_1} \qquad (3.8.5)$$

Comparison of the slopes reveals that if slope m_{15} is greater than, or equal to, slope m_{13}, then point 5 is considered to lie within triangle number 1, else it lies within triangle number 2. Having determined the relevant triangle, the equation of the surface through the three corner points must be found. Take triangle 1 as the theoretical example of a flat surface passing through the corner points 1, 2, and 3 with coordinates (x_1, y_1, z_1), (x_2, y_2, z_2), and (x_3, y_3, z_3), respectively. The general equation of a point with coordinates, (x, y, z), that lies on the flat surface fitting those points, is given by:

Surface equation: $$ax + by + cz = 1 \qquad (3.8.6)$$

The coefficients, a, b, and c of this surface equation are found by solving the three simultaneous equations resulting from the insertion of the coordinates at the three corners into Eq. 3.8.6.

$$
\begin{aligned}
ax_1 + by_1 + cz_1 &= 1 \\
ax_2 + by_2 + cz_2 &= 1 \\
ax_3 + by_3 + cz_3 &= 1
\end{aligned}
\qquad (3.8.7)
$$

Although I could present this solution as a "one-liner" statement in matrix algebra, because this may not be universally understood, I will present the algebra of the solution in full. The complete solution for the three coefficients, a, b and c, is as follows, using a series of intermediate collection terms, p and q:

Where: $$p_1 = y_1x_2 - y_2x_1 \qquad p_2 = z_1x_2 - z_2x_1 \qquad p_3 = x_2 - x_1 \qquad (3.8.8)$$

and: $$q_1 = y_2x_3 - y_3x_2 \qquad q_2 = z_2x_3 - z_3x_2 \qquad q_3 = x_3 - x_2 \qquad (3.8.9)$$

then: $$c = \frac{p_3q_1 - p_1q_3}{p_2q_1 - p_1q_2} \qquad b = \frac{p_3 - p_2c}{p_1} \qquad a = \frac{1 - z_2c - y_2b}{x_2} \qquad (3.8.10)$$

The solution for any point lying in triangle 2 is straightforward and is found by replacing the subscripts 2 and 3 in the above equations with the subscripts 3 and 4, which represent the corner points of triangle number 2.

Consequently, because the coordinates of point 5 are known, i.e., (x_5, y_5), and the discharge coefficient for point 5, labeled as Cd_5, is found at z_5:

$$Cd_5 = z_5 = \frac{1 - ax_5 - by_5}{c} \qquad (3.8.11)$$

Consider our previous simple numeric example. In triangle number 1, with corner points 1, 2, and 3 at coordinates (x_1, y_1, z_1), (x_2, y_2, z_2), and (x_3, y_3, z_3), the unknown Cd value is required at coordinates (x_5, y_5). The example data set has numeric coordinate values of (0.2, 1.4, 0.6), (0.4, 1.4, 0.6), and (0.4, 1.2, 0.5), with the unknown at the (x, y) coordinates of (0.35, 1.3). The solution of Eqs. 3.8.8 to 3.8.11 shows that the values of the coefficients a, b, and c are 0, 5, and 10, respectively. The z_5 coordinate is the required value of Cd and, as with the simple method No. 1, Eq. 3.8.11 gives the correct answer of 0.55, whereas simple method No. 2 continues to report a higher Cd value of 0.567. If the coordinates of point 5 are moved closer to point 2, for example at (x_5, y_5) values of (0.39, 1.39), both simple methods report the same results as before, but the GPB solution provides a much more realistic Cd value of 0.595. Similarly, if the coordinates of point 5 are moved close to point 3, say at (x_5, y_5) coordinates of (0.39, 1.22), there is again no reaction from the two simple methods, but the GPB solution responds with a realistic Cd value of 0.510.

Clearly, this triangle surface solution is a simple, yet accurate, way to incorporate an entire Cd map into an engine simulation. This method is economical in terms of computer space and time, as it manages to execute this algebraic solution while holding a mere twenty-five points in a numeric array. This numeric array describes each Cd map for the relevant flow direction at its particular duct discontinuity.

3.9 Conclusions Regarding Discharge Coefficients

An engine simulation code that does not use maps of discharge coefficients to describe each and every pipe end discontinuity cannot possibly compute accurately either pressure wave formation, or mass flow rates, or thermodynamic state conditions at each location as the computation steps through an engine cycle.

Even if such Cd maps are included within an engine simulation code, if these maps of discharge coefficients do not extend to cover both flow directions over a relevant range of pressure ratios, then computation accuracy must suffer.

Even if such Cd maps are included within an engine simulation code, and cover the relevant range of pressure ratios and flow directions, if the experimental process to determine those Cd values does not include an analytic technique to determine the actual discharge coefficient, as distinct from the historical procedure which provides an isentropic or an ideal discharge coefficient, then computation accuracy cannot be expected.

In the experimental and analytic process to determine the actual discharge coefficient, it is essential to use the very same description of the geometry at that particular pipe-end discontinuity as is used within the engine simulation code, else that Cd map might just as well have been estimated instead of measured.

Unless any bank of historical data contains the original experimental data values, which can be reanalyzed to find the actual discharge coefficient, I am afraid that the usefulness of such historical data may be relegated to a guidance role only, such as the directions of future experimental work and the types of experiments and geometries to be investigated. Recall that, if an isentropic or an ideal Cd value is experimentally declared to be higher by the alteration of some geometrical parameter, then the Cd_a value is normally increased as well.

It is very unfortunate that the traditionally obtained discharge coefficient, Cd_{is} or Cd_{id}, is not the correct number as far as an engine simulation code is concerned. It is quite clear, from the paucity of the experimental data presented here, that a very considerable amount of experimental work is required to rebuild the banks of useful and relevant Cd_a data necessary to cover all of the future requirements of engine simulation. The engineers in this field, and that includes me, thought these data were all safely stored away. Next time, just to err on the safe side, we should make sure that we retain the raw experimental data as well.

References for Chapter 3

3.1   I. Fukutani, E. Watanabe, "Air Flow through Poppet Inlet Valves - Analysis of Static and Dynamic Flow," SAE Paper No. 820154, 1982.

3.2   G. Leydorf, R. Minty, and M. Fingeroot, "Design Refinement of Induction and Exhaust Systems using Steady-State Flow Bench Techniques," SAE Paper No. 720214, 1972.

3.3   K. Tanaka, "Experimental Research on Poppet Valves," *Journal of Aero. Research Institute,* Tokyo Imperial University, No. 22 and No. 27, 1926.

3.4   E.M. Nutting and G.W. Lewis, "Air Flow through Poppet Valves," NACA Report No. 24, Washington, DC, 1928.

3.5   E.S. Dennison, T.C. Kuchler, and D.W. Smith, "Experiments on the Flow of Air through Engine Valves," *Trans. ASME,* Vol. 53, OGP-53-6, 1931.

3.6   G.B. Wood, D.U. Hunter, E.S. Taylor, and C.F. Taylor, "Air Flow through Intake Valves," *Trans. SAE,* 5Q, 212, 1942.

3.7   J.D. Stanitz, R.E. Lucia, and F.L. Massel, "Steady and Intermittent Flow Coefficients of Poppet Intake Valves," NACA Report No. 1035, 1946.

3.8   L.J. Kastner, T.J. Williams, and J.B. White, "Poppet Inlet Valve Characteristics and their Influence on the Induction Process," *Proc.I.Mech.E.,* Vol. 178, No. 36, p. 955, 1963.

3.9   W.A. Woods and S.R. Khan, "An Experimental Study of Flow through Poppet Valves," *Proc.I.Mech.E.,* Vol. 180, Pt. 3N, p. 32, 1965-66.

3.10 W.A. Woods and S.R. Khan, "Discharge from a Cylinder through a Poppet Valve to an Exhaust Pipe," *Proc.I.Mech.E.,* Vol. 182, Pt. 3H, p. 126, 1967-68.

3.11 W.A. Woods, "Steady Flow Tests on Twin Poppet Valves," *Proc.I.Mech.E.,* Vol. 182, Pt. 3D, p. 32, 1967-68.

3.12 W.J.D. Annand and G.E. Roe, *Gas Flow in the Internal Combustion Engine*, Foulis, Yeovil, Somerset, 1974.

3.13 W.A. Woods and G.K. Goh, "Compressible Flow Through a Butterfly Throttle Valve in a Pipe," *Proc.I.Mech.E.*, Vol. 193, p. 237, 1979.

3.14 D.D. Agnew, "What is Limiting Engine Air Flow. Using Normalised Steady Air Flow Bench Data," SAE Motorsports Engineering Conference, Dearborn, Mich., December 1994, SAE Paper No. 942477.

3.15 BS 1042, Fluid Flow in Closed Conduits, British Standards Institution, 1981.

3.16 G.P. Blair, H.B. Lau, A. Cartwright, B.D. Raghunathan, and D.O. Mackey, "Coefficients of Discharge at the Apertures of Engines," SAE International Off-Highway Meeting, Milwaukee, Wisc., September 1995, SAE Paper No. 952138, pp. 71-85.

3.17 G.P. Blair, *Design and Simulation of Two-Stroke Engines*, R-161, Society of Automotive Engineers, Warrendale, Pa., 1996.

3.18 G.P. Blair, D. McBurney, P. McDonald, P. McKernan, and R. Fleck, "Some Fundamental Aspects of the Discharge Coefficients of Cylinder Porting and Ducting Restrictions," SAE International Congress, Detroit, Mich., SAE Paper No. 980764, 1998.

3.19 A. Jante, "Scavenging and Other Problems of Two-Stroke Cycle Spark-Ignition Engines," SAE Mid-Year Meeting, Detroit, Mich., May 1968, SAE Paper No. 680468.

3.20 T. Sakiya, *Honda Motor; the Men, the Management and the Machines*, Kodansha International, New York, 1982.

3.21 G.P. Blair and F.M.M. Drouin, "Relationship between Discharge Coefficients and Accuracy of Engine Simulation," SAE Motorsports Engineering Conference, Dearborn, Mich., December 1996, SAE Paper No. 962527, pp. 151-163.

3.22 J.F. Bingham and G.P. Blair, "An Improved Branched Pipe Model for Multi-Cylinder Automotive Engine Calculations," *Proc.I.Mech.E.*, Vol. 199, No. D1, 1985, pp. 65-77.

3.23 A.J. Blair and G.P. Blair, "Gas Flow Modelling of Valves and Manifolds in Car Engines," *Proc.I.Mech.E.*, International Conference on Computers in Engine Technology, University of Cambridge, April 1987, C11/87, pp. 131-144.

3.24 S. Nadarajah, S. Balabani, M.J. Tindal, and M. Yianneskis, "The Turbulence Structure of the Annular Non-Swirling Flow past an Axisymmetric Poppet Valve," *Proc.I.Mech.E.*, Vol. 212, Part C, pp. 455-471, 1998.

3.25 S. Nadarajah, S. Balabani, M.J. Tindal, and M. Yianneskis, "The Effect of Swirl on the Annular Flow Past an Axisymmetric Poppet Valve," *Proc.I.Mech.E.*, Vol. 212, Part C, pp. 473-484, 1998.

Chapter 4

Combustion in Four-Stroke Engines

4.0 Introduction

This chapter deals mainly with combustion in the spark-ignition engine, but a significant portion of the discussion and theory concerns compression-ignition engines. This chapter presents the subject from a practical design standpoint, and thus is not to be considered a fundamental treatise on the subject of combustion.

It would not be stretching the truth to say that combustion, and the heat transfer behavior accompanying it, are the phenomena least understood by the average engine designer. The study of combustion has always been a specialized topic which has often been treated at a mathematical and theoretical level beyond the grasp of all but the dedicated researcher. And very often, the knowledge garnered by research has not been disseminated in a manner suitable for use by the designer. This situation is changing, however. The advent of computational fluid dynamic (CFD) design packages will ultimately allow the engine designer to predict combustion behavior without requiring him or her to become a specialist in mathematics and chemistry at the same time. For those who wish to study the subject at a fundamental level, and to be made aware of the current state of the science, the References [1.2-1.3, 4.1-4.9] will help to provide a starting point for that study. There seems to be little point in repeating the fundamental theory of combustion in this book, because such theory is well covered in the literature. However, this chapter will cover certain aspects of combustion theory which are particularly applicable to the design and development of engines, and which are rarely found in the standard reference textbooks.

The first objective of this chapter is to make you aware of how the combustion process in a real engine differs from the theoretical ideal introduced in Chapter 1, and how it differs from an explosion, which is how the layperson commonly views engine combustion.

The second objective is to introduce you to the analysis of cylinder pressure records to reveal the heat release characteristics, and how such information is the most realistic available for the theoretical modeling of engine behavior. However, other possible theoretical models of combustion are presented and discussed briefly.

The third objective is to make you aware of the potential effect of various design variables, such as squish action on detonation and combustion rates, and of the means available to design cylinder heads for engines which take these factors into account.

Finally, this chapter contains a section on the latest theoretical developments in the analysis of combustion processes. In the near future, our understanding of engine combustion will be on a par with, say, unsteady gas dynamics or scavenging flow, i.e., the theory will be used to design combustion chambers for engines and the resulting engine will exhibit the designed combustion performance characteristics. I believe that we are approaching this level of understanding, but it has not yet been achieved.

4.1 The Spark-Ignition Process
4.1.1 Initiation of Ignition
It is a well-known fact that a match thrown onto some spilled gasoline in the open atmosphere will ignite the gasoline and release a considerable quantity of heat, with a significant rise in temperature. The gasoline is observed to burn as a vapor mixed with air above the remaining liquid, but rapidly vaporizing, gasoline. The procedure, for those who have witnessed it (although, on safety grounds, I am not recommending that the experiment be conducted), commences when the lighted match arrives at the vapor cloud above the liquid, and the ignition takes place with a "whoosh," apparently a major or minor "explosion," depending on the mass fraction of the gasoline that has evaporated before the match arrives. This tends to leave us with the impression that the gasoline-air mixture within the cylinder of an ic engine will "explode" upon the application of the spark at the sparking plug. Not so; that the flammability characteristics of a gasoline-air mixture is decidedly a critical phenomena should be obvious to all those who have had difficulty in starting either their lawnmower or their automobile!

What then are the requirements of an ignition process? Why does an engine fire up? When does it not fire up? The technical papers on combustion and the engineering textbooks tend to bypass such fundamental concepts, so I felt that a paragraph or two here would not be amiss, especially because more difficult concepts will only be understood against a background of some fundamental understanding.

Fig. 4.1(a) depicts an engine in which a spark has ignited the air-fuel mixture, and has produced a flame front burning its way through the mixture. For this to happen, as in the match example presented before, there had to be a gasoline *vapor* and air mixture, of the correct mass proportions, within the spark gap when that spark occurred. The energy in the spark provides a localized rise in temperature of several thousand degrees Kelvin, which causes any gasoline vapor present to be raised above its auto-ignition temperature. The auto-ignition temperature of any hydrocarbon fuel is the temperature at which the fuel now has sufficient internal energy to break its carbon-hydrogen bond structure and be oxidized to carbon dioxide and steam. In the case of gasoline, the auto-ignition temperature is about 220°C. The compression process prior to the ignition point helps to vaporize the gasoline, the maximum boiling point of which is about 200°C. The mass of gasoline vapor within the spark gap, which has begun to break down in an exothermic reaction, raises the local temperature and pressure. The reaction, if it was stoichiometric, would be as given previously in Eq. 1.6.14. The actual reaction process is much more complex than this, with the gasoline molecule breaking down in many stages to such intermediates as methane and aldehydes. Immediately after the ignition point, the initial flame front near the spark plug becomes established and heats the unburned layers of gasoline vapor-air mixture surrounding it, principally by radiation, but also by convection heat transfer as the motion of the mixture propels the mixture into the flame front. This induces further layers of mixture to reach the auto-ignition temperature, and thus the flame front moves through the combustion chamber until it arrives at the physical extremities of the chamber [4.9-4.11]. The velocity of this flame front has been recorded in engines as being between 20 and 50 m/s [4.3, 4.7]. It will be observed that this is hardly an "explosive" process, although it is sufficiently rapid to allow the engine to burn its fuel reasonably efficiently even at the highest engine speeds. In an example quoted by Kee [4.12], the flame speed, c_{fl}, in an engine of physical geometry similar to that in Fig. 4.1(a), was measured at 24.5 m/s at an engine speed, N, of 3000 rpm. The longest flame path, x_{fl}, from the spark plug to the extremity of the combustion chamber in an engine of 85 mm bore, was about 45 mm. The crankshaft rotation angle, θ_{fl}, for this flame transmission to occur is given by:

$$\theta_{fl} = \frac{6 \times x_{fl} \times N}{c_{fl}} \tag{4.1.1}$$

In the example quoted by Kee [4.12], the flame travel time to the chamber extremity is 33° crankshaft, as from Eq. 4.1.1 the value of θ_{fl} is given by:

$$\theta_{fl} = \frac{6 \times \dfrac{45}{1000} \times 3000}{24.5} = 33.06°$$

This does not mean that the combustion process is completed in 33°; what it does mean is that initiation of combustion has taken place over the entire combustion space for this particular homogeneous charge.

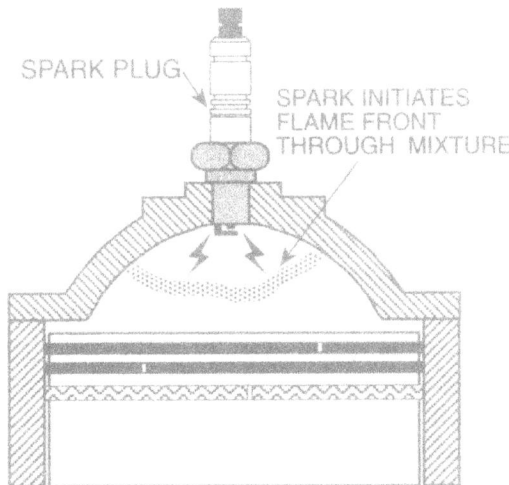

Fig. 4.1(a) Initiation of combustion in a spark-ignition engine.

409

4.1.2 Air-Fuel Mixture Limits for Flammability

The flammability of the initial flame kernel has a rather narrow window for success [4.8, 4.11]. For a flame to survive, the upper and lower values of the proportion by volume of gasoline vapor to air are 0.08 and 0.06, respectively. Because one is supplying a "cold" engine with liquid fuel—by whatever device, ranging from a carburetor to a fuel injector—the *vaporization* rate of the liquid gasoline due to the induction and compression processes is going to be highly dependent on the temperatures of the cylinder wall, the cylinder head, the piston crown, and the atmospheric air. Not surprisingly, in cold climatic conditions, it takes several compression processes to raise the local temperature sufficiently to provide the statistical probability of success. In addition, a high-energy spark may be used to assist that procedure; this has become more conventional [4.9]. At one time, ignition systems had spark characteristics of about 8 kV with a rise time of about 25 μs. Today, with "electronic" or capacitor discharge ignition systems those characteristics are more typically at 20 kV and 4μs, respectively. This current need for ever longer spark durations is being set by the demands of meeting emissions legislation whereby combustion of stoichiometric, or leaner, mixtures is essential. The higher-voltage output and the faster spark rise time ensure that sparking will take place, even if the electrodes of the spark plug are covered with liquid gasoline. Spark duration, and even multiple sparking, also assist flame kernel growth, and this is even more applicable for the ignition of a spray of liquid fuel in a stratified charging approach using direct fuel injection (GDI).

Under normal firing conditions, if the fuel vapor-air mixture becomes too lean, e.g., if the volume ratio falls below the limit of 0.06 quoted above, then a flame is prevented from growing due to an inadequate initial release of heat. When the spark occurs in a lean mixture, the mass of fuel vapor and air that is ignited in the vicinity of the spark is too small to provide an adequate release of heat to raise the temperature of the surrounding layer of unburned mixture to the auto-ignition temperature. Consequently, the flame does not develop and combustion does not take place. In this situation, intermittent misfire is the normal experience as unburned mixture forms the bulk of the cylinder contents during the succeeding exhaust and intake process and will supplement any fuel supplied by it. Needless to add, the pumping action of the exhaust stroke will send some of this unburned fuel into the exhaust duct and dramatically raise the emissions of hydrocarbons.

Under normal firing conditions, if the fuel vapor-air mixture becomes too rich, e.g., if the 0.08 volume ratio quoted above is exceeded, then the flame is prevented from growing due to insufficient mass of air present at the onset of ignition. As with the flame propagation process in the too-lean mixture, if an inadequate amount of heat is released at the critical inception point, the flame is snuffed out.

4.1.3 Effect of Scavenging Efficiency on Flammability

In the spark-ignition engine, at an idle or light-load condition, the scavenging efficiency varies with delivery ratio. As the engine load, or brake mean effective pressure, bmep, is varied by altering the throttle opening, thereby producing changes in air flow, the scavenging efficiency, SE, also changes. It will be observed later in Chapter 5 that, even for the best design of engines, the scavenging efficiency varies at light loads. Interpreting this reasonably accurately as being "charge purity" for the firing engine situation, it is clear that at light loads and low engine rotational speeds there will be a throttle position at which a considerable mass of exhaust gas present may permit ignition of a gasoline vapor-air mixture, but may not deliver complete combustion of the mixture. When the spark occurs, the mass of vapor and air that is ignited in the vicinity of the spark is too small to provide an adequate release of heat to raise the temperature of the surrounding layer of unburned mixture to the auto-ignition temperature.

Consequently, the flame does not develop and combustion does not take place. The effect is somewhat similar to the lean-burning misfire limit discussed above. Even if the mixture ignites, only a partial combustion of the totality of mixture contained within the cylinder may occur.

During the next intake process, the scavenging efficiency, SE, is raised as some of the "exhaust residual" is expelled during the succeeding exhaust stroke. Should the new SE value prove to be greater than the threshold condition for flammability, a more complete combustion will take place. To reiterate: pumping by the exhaust stroke during any intermittent or partial firing behavior ejects considerable quantities of unburned fuel and air into the exhaust duct, to the very considerable detriment of the specific fuel consumption of the engine and its emission of unburned hydrocarbons.

Active Radical Combustion

Onishi et al. [4.13] have shown that it is possible under certain conditions to ignite, and ignite with great success, an air-fuel mixture in high concentrations of residual exhaust gas, i.e., at low SE values. This has been confirmed by Ishibashi and Asai [4.14, 4.15]. Essentially, this is accomplished by raising the trapping pressure and temperature so that the exhaust residual remains sufficiently hot and active as to provide an ignition source for the air-fuel mixture distributed through it—hence the term "active radical," or AR, combustion. The ensuing combustion process is very stable, very efficient, and eliminates both the "four-stroking" skip-firing regime and the high exhaust emissions of hydrocarbons that accompany misfire due to a low SE condition, in this case in a two-stroke engine [4.15]. The concept has potential application in lean-burn four-stroke engines [4.16, 4.17]. Both AR combustion and the low-load misfire problem have fundamental connections to the "run-on" occasionally experienced when the engine ignition is switched off [4.18]. This effect must not be confused with detonation, as discussed in the next section.

4.1.4 Detonation or Abnormal Combustion

Detonation occurs in the combustion process when the advancing flame front, which is pressurizing and heating the unburned mixture ahead of it, does so at such a rate that the unburned fuel in that zone achieves its auto-ignition temperature before the arrival of the actual flame front. The result is that the unburned mixture combusts "spontaneously" and over the entire zone where the auto-ignition temperature has been achieved [4.19]. The apparent flame speed in this zone is many orders of magnitude faster than that in conventional combustion initiated by a normal flame front, with the result that the local rise of pressure and temperature is significantly sharp. This produces the characteristic "knocking" or "pinking" sound, and the local mechanical devastation that this can produce on piston crown or cylinder head can be considerable. Actually, "knocking" is the correct terminology for what is really a detonation behavior over a small portion of the combustion charge. A true detonation process would be one occurring over the entire compressed charge. However, because detonation in this strictly defined sense does not take place in the spark-ignition engine, the words "knock-ing" and "detonation" are used interchangeably in the literature, without loss of meaning, to describe the effects just discussed.

The knocking effect is related to compression ratio, because the higher the compression ratio, the smaller the clearance volume, the higher the charge density—and at equal flame speeds—the higher the heat release rate as the flame travels through the mixture. Consequently, there will be a critical level of compression ratio at which any unburned mixture in the extremities of the combustion chamber can attain the auto-ignition temperature. This effect can be alleviated by various methods, such as raising the octane rating of the gasoline used, promoting charge turbulence, squish effects, improving scavenging efficiency, stratifying the combustion process, and, perhaps more commonly, retarding the ignition or lowering the compression ratio [4.20]. Some of these techniques will be discussed later. The effect of reducing the compression ratio has been discussed earlier in Sec. 1.9.1.4.

It is clear that the combustion chamber shape has an influence on these matters. The optimum shape is a compact chamber with a low surface-to-volume ratio. The chamber should have with no nooks or crannies to trap and heat the unburned charge, and no hot protuberant piston crown with sharp-edged valve pockets containing air-fuel mixture getting hotter by the millisecond.

In the Otto engine, the presence of any hot exhaust gas residual within the combustible charge compounds the problem of designing the enging to avoid knock.

4.1.5 Homogeneous and Stratified Combustion

The conventional spark-ignition engine burns a homogeneous charge. The air-fuel mixture is supplied to the cylinder via the intake tract with some of the fuel already vaporized during its residence there, or near to the hotter intake valve. The intake tract tends to get hotter within the cylinder head, i.e., the closer it is to the combustion chamber. The remainder of the liquid fuel vaporizes during the intake and compression processes, so that by the time ignition takes place, the combustion chamber is filled with a vapor-air-exhaust gas residual mixture that is evenly distributed throughout the combustion space. This is known as a homogeneous combustion process.

If the fuel be supplied to the combustion space by some other means, such as direct in-cylinder fuel injection, all of the vaporization process will take place during the compression process. In such a case, there is a distinct possibility that, by the onset of ignition, there may be zones in the combustion space that are at differing air-fuel ratios. This is known as a stratified combustion process. This stratification may be deliberately induced, for example, to permit the local efficient burning of a small mass of air and fuel of the correct proportions to permit the use of a more open throttle setting. Such a design reduces the pumping losses of the engine and the brake specific fuel consumption is thereby improved.

It is also possible to utilize charge stratification to help alleviate detonation behavior. If the extremities of the combustion chamber contain only air, or a very lean mixture, then the possibility exists of raising engine thermal efficiency through a higher compression ratio, while lowering the potential for detonation to occur. The end-gas cannot "knock" if it contains no fuel. For example, there is no "knocking" in a diesel combustion process; there may be a rattle for other reasons, but there is no "knock."

4.1.6 Compression Ignition

This subject will be dealt with more fully in Sec. 4.3.7. Nevertheless, the fundamentals of the process are discussed here to distinguish this process clearly from the spark-ignition process described above. A sketch showing a typical combustion chamber and the fuel injector location for a direct injection (DI) diesel engine is presented in Fig. 4.1(b). The disposition of this engine with regard to the valves can be seen in Fig. 1.7.

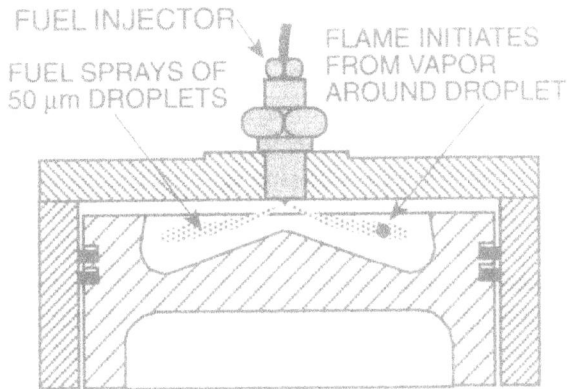

Fig. 4.1(b) Initiation of combustion in a compression-ignition engine.

In the compression-ignition process, the compression ratio employed for the engine is much higher so that the air temperature by the end of compression is significantly above the auto-ignition temperature of the fuel; this has already been described in Sec. 1.9.2. When fuel is sprayed directly into the air at this state condition and at this juncture, the fuel on the outside of the droplets begins to vaporize. When the vapor temperature has risen to the auto-ignition temperature, combustion commences between the vapor and the air in its immediate vicinity. Thus, in diesel combustion—by complete contrast with homogeneous spark-ignition—it is possible to burn one microscopic droplet of fuel in a whole "sea" of air. The great advantage of such lean combustion is that it is possible to have it occur at air-fuel ratios that are four or five times leaner than the stoichiometric mixture. The penalty is that efficient combustion cannot occur at air-fuel ratios that approach the stochiometric value, for under such conditions the fuel would be chasing through the combustion chamber searching for air. In the limited time available for combustion, it is not possible for all of the carbon component of the fuel to find the air it needs and thus there is a "rich limit" for diesel combustion. This rich limit is typically some 40% leaner than stoichiometric, or at an air-fuel ratio of about 20. If the air-fuel ratio is richer than this value, then the exhaust gas will contain considerable quantities of unburned carbon particulates, or black smoke. Today, exhaust emissions legislation specifies and limits the amount of carbon particulates that may be emitted by a diesel engine.

This effect is compounded by the heavier hydrocarbon fuel employed in diesel engines. Gasoline is typically and ideally octane, C_8H_{18}, i.e., the eighth member of the family of paraffins whose general family formula is C_nH_{2n+2}. This is much too volatile a fuel to be used in a compression-ignition engine because the ensuing rates of rise of pressure and temperature would be so rapid as to cause mechanical damage to the cylinder components. The fuel employed for most automotive diesel engines is typically and ideally dodecane, $C_{12}H_{26}$, i.e., the twelth member of the family of paraffins. Because dodecane has a more complex structure and a higher density than gasoline, it requires more energy to separate the carbon and hydrogen atoms from each other and thus it burns more slowly, giving a less rapid rate of pressure rise during a compression-ignition process than would octane. However, this more complex molecule does not readily burn to completion, hence there is a compounding effect on the production of black smoke, as described at the end of the last paragraph.

Prior to the combustion process, the fuel is injected into the air in the cylinder, which by dint of the high compression ratio is above the auto-ignition temperature of the fuel. Compression ratios, CR, of about 18 are required to accomplish this latter effect, compared to the values of between 9 and 11 normally employed for spark-ignition with gasoline. The fuel droplets must first be heated by the compressed air to vaporize the fuel, and then to raise that vapor temperature to the auto-ignition temperature. This, like all heat transfer processes, takes time. Naturally, there will be an upper limit of engine speed at which the heating effect has not been fully developed and combustion will not take place efficiently, or may not even take place at all. Thus, a diesel engine tends to have an upper speed limit for its operation, a limit not imposed on spark-ignition engines. This limit is controlled by the effectiveness of the heat transfer and mixing process for the particular type of diesel engine. For indirect injection (IDI) diesel engines for automobiles, the limit is about 4500 rpm; for direct injection (DI) engines for automobiles it is about 4000 rev/min; and for larger capacity direct injection (DI) diesel engines for trucks or buses it is about 3000 rpm. Sec. 4.3.7 contains a more extensive discussion of this topic. In that section, the considerable differences in cylinder capacity and combustion chamber geometry for these two dissimilar approaches to the generation of combustion by compression-ignition are discussed.

4.2 Heat Released by Combustion
4.2.1 The Combustion Chamber
The combustion process, discussed earlier in Sec. 1.9, is one described thermodynamically as a heat addition process in a closed system. It occurs in a chamber of varying volume proportions, the minimum value of which is the clearance volume, V_{cv}. Eq. 1.1.1 details the values of V_{cv} in terms of the swept volume, V_{sv}, to attain the requisite parameters of the geometric compression ratio, CR.

In Fig. 4.2, the piston is shown positioned at top dead center, tdc, so the clearance volume, V_{cv}, is seen to be composed of a bowl volume, V_b, and a squish band volume, V_s. The piston has a minimum clearance distance from the cylinder head which is known as the squish clearance, x_s. The areas of the piston that are covered by the squish band and the bowl are A_s and A_b, respectively. The squish action gives rise to the concept of a squish area ratio, C_{sq}, where:

$$C_{sq} = \frac{\text{area squished}}{\text{bore area}} = \frac{A_s}{\frac{\pi}{4}d_{bo}^2} = \frac{A_s}{A_s + A_b} \qquad (4.2.1)$$

The definitions above are also applicable to a compression-ignition engine, particularly the direct injection (DI) engine, but also the indirect injection (IDI) engine shown in Figs. 1.8 or 4.22.

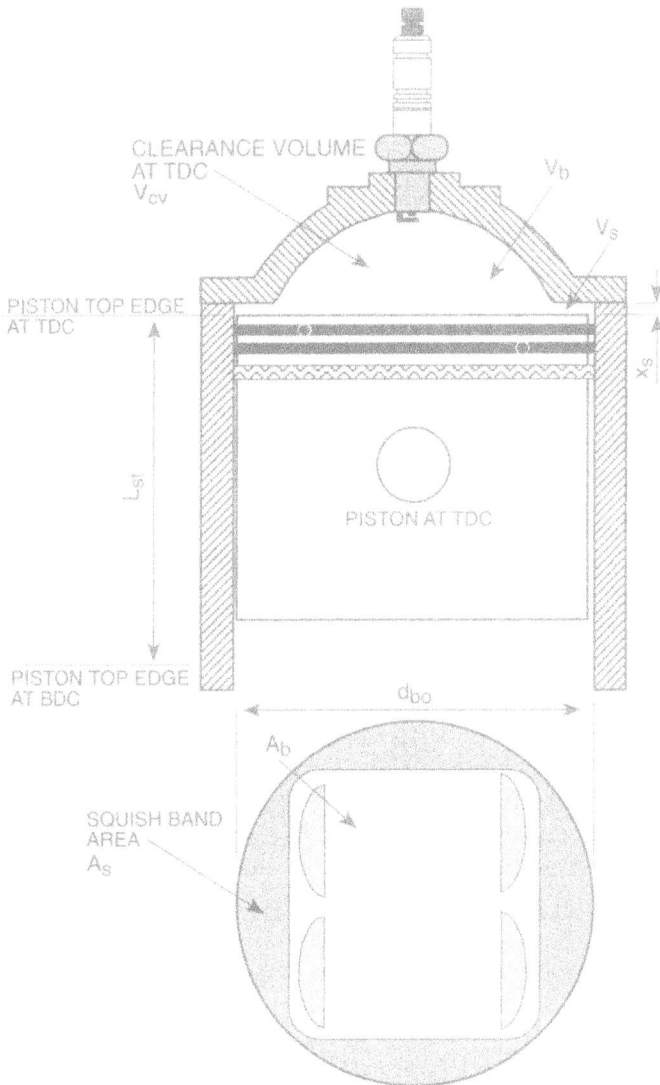

Fig. 4.2 Details of combustion chamber and cylinder geometry.

4.2.2 Heat Release Prediction from Cylinder Pressure Diagram

Sec. 1.9 contains a discussion of the theoretically ideal thermodynamic engine cycle, the Otto cycle. The combustion process is detailed as occurring at constant volume, i.e., the imaginary explosion. The reality of the situation, in Figs. 1.30 to 1.32, is that the engine pressure diagram as measured in an automobile engine shows a time-dependent combustion process. This is pointed out in Sec. 4.1.1, where the flame speed is detailed as having been measured at 24.5 m/s through a combustion chamber.

It is possible to analyze the cylinder pressure diagram and deduce from it the heat release rate for any desired period of time or crankshaft angular movement. Consider the thermodynamic system shown in Fig. 4.3, where the piston has been moved through a small crankshaft angle producing known volume changes, V_1 to V_2, for the total volume above the piston including the combustion chamber. These volumes can be found from the crankshaft, connecting rod, and cylinder geometry. The combustion process is in progress and a quantity of heat, δQ_R, has been released during this time step. A quantity of heat, δQ_L, has been lost through heat transfer to the cylinder walls and coolant at the same time. The cylinder pressure has been measured as changing from p_1 to p_2 during this time, or crankshaft angle, step. The internal energy of the cylinder gas changes from U_1 to U_2, and the temperature changes from T_1 to T_2. The work done on the piston is δW during the time interval. The First Law of Thermodynamics for this closed system states that, during this time step, such events are related by:

$$\delta Q_R - \delta Q_L = U_2 - U_1 + \delta W \qquad (4.2.2)$$

Where m is the mass of gas in the cylinder, and C_V is the specific heat at constant volume, the internal energy change is approximately given by:

$$U_2 - U_1 = mC_V(T_2 - T_1) \qquad (4.2.3)$$

The word "approximately" is used above because the value of the specific heat at constant volume is a function of temperature and also of the gas properties. During combustion, both the gas properties and the temperatures vary rapidly, for example, several hundreds of degrees Kelvin per degree crankshaft. Thus, C_V is not a constant as is evident from the discussion in Sec. 2.1.6. In any arithmetic application of Eq. 4.2.3, the value of C_V must be for some particular state condition; such as at either T_1 or T_2 and at the gas properties pertaining at state condition 1 or 2; or at the mean of those two values and at the mean of the gas properties pertaining at state condition 1 and 2. The simplest method is to employ the properties at state condition 1, because this will always be a known condition; the properties at T_2 are usually the ones being forecast and therefore are "unknowns." This caution will not be repeated throughout the thermodynamic analysis; it will be taken as understood that it is included within the theory being discussed.

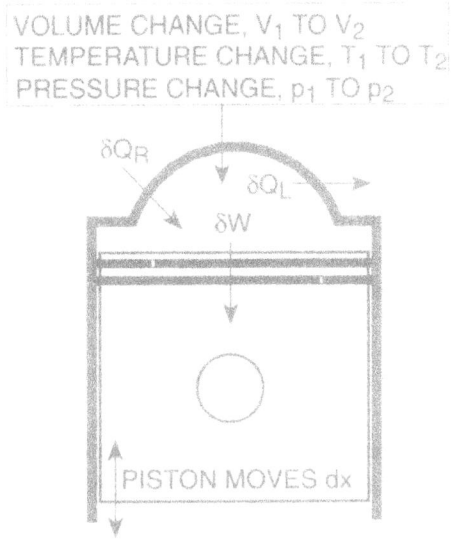

Fig. 4.3 Thermodynamic system during combustion.

The pressure and temperature values at each point can be derived through the equation of state as:

$$p_2V_2 = mRT_2 \text{ and } p_1V_1 = mRT_1 \qquad (4.2.4)$$

where R is the gas constant.

Specific heat at constant volume $\qquad C_V = \dfrac{R}{\gamma - 1}$ $\qquad\qquad$ (4.2.5)

If γ is the ratio of specific heats for the cylinder gas, Eq. 4.2.3 can be restated as:

$$U_2 - U_1 = \frac{p_2V_2 - p_1V_1}{\gamma - 1} \qquad (4.2.6)$$

The average work done on the piston during this interval is:

$$\delta W = \frac{p_1 + p_2}{2}(V_2 - V_1) \qquad (4.2.7)$$

Consequently, substituting Eqs. 4.2.6 and 4.2.7 into Eq. 4.2.2:

$$\delta Q_R - \delta Q_L = \frac{p_2 V_2 - p_1 V_1}{\gamma - 1} + \frac{p_1 + p_2}{2}(V_2 - V_1) \tag{4.2.8}$$

If the combustion process had not occurred, then the compression or expansion process would have continued in a normal fashion. The polytropic process, seen in Figs. 1.31 and 1.32, is taking place at a relationship defined by:

$$pV^n = a \text{ constant} \tag{4.2.9}$$

The polytropic index, n, is known from an analysis of the measured pressure trace as the compression (or the expansion, if post-combustion) index. Hence, p_2 would not have been achieved by combustion, but a value p_{2a} would have occurred, where:

$$p_{2a} = p_1 \left(\frac{V_1}{V_2} \right)^n \tag{4.2.10}$$

As this imaginary process is one of non-heat addition, i.e., δQ_R is zero, the First Law of Thermodynamics shown in Eq. 4.2.8 could be rewritten to calculate the heat loss, δQ_L, by substituting p_{2a} for p_2:

$$-\delta Q_L = p_1 \left\{ \frac{V_2 \left(\frac{V_1}{V_2} \right)^n - V_1}{\gamma - 1} + \frac{(V_2 - V_1)\left(\left(\frac{V_1}{V_2} \right)^n + 1 \right)}{2} \right\} \tag{4.2.11}$$

On the assumption that this heat loss behavior is that which continues during the combustion process, substitution of this value for δQ_L into Eq. 4.2.8 yields an expression for the heat released, δQ_R:

$$\delta Q_R = \left\{ p_2 - p_1 \left(\frac{V_1}{V_2} \right)^n \right\} \left(\frac{V_2}{\gamma - 1} + \frac{V_2 - V_1}{2} \right) \tag{4.2.12}$$

The value of the polytropic exponent, n, either as n_c for compression, or as n_e for expansion, is known from an analysis of the measured pressure trace taken within the cylinder, and during the analytic process can logically be applied on both sides of the tdc position. A value of the ratio of specific heats for the cylinder gas, γ, could be taken as being between 1.2 and 1.25 at the elevated temperatures found during and after combustion, and as between 1.36 and 1.38 during the pre-combustion period. The gas properties could be determined reasonably accurately from an exhaust gas analysis, and possibly even more accurately from a chemical analysis of a sample of the cylinder gas at the end of combustion. However, it will be seen from Sec. 2.1.6 that temperature has the much greater effect on the value of γ. It will be noted that the experimentally determined values of the polytropic exponents, n_c and n_e, for the compression and expansion processes for the automobile engine shown in Fig. 1.35 are 1.25 and 1.33, respectively.

Irrespective of the method chosen to determine the value of γ, it is clear from Eq. 4.2.12 that the calculated value of heat released, δQ_R, is linearly related to the value of (γ-1). To put this in context, consider the difference in the computed value of heat release caused by estimating the value of γ as either 1.20 or 1.21. The difference for this minor change of γ value is a considerable 5%. This is not new information, although some publications stress its importance more than others [4.27], and even suggest relevant formulae for the ratio of specific heats as a function of temperature. However, our thermodynamic capabilities cannot possibly approach the determination of the composition and state conditions of the cylinder gas with sufficient accuracy to compute the ratio of specific heats. Because the cylinder pressure and volume are our only concrete experimental data, and temperature is not—for we know neither cylinder mass nor its gas constant—any value of heat release so determined must be considered merely as a relative value. Fortunately, for simulation modeling purposes this is exactly what is required, so the precise selection of a value for γ is not an absolute necessity.

Rassweiler and Withrow [4.21] presented their analysis of the above theory in a slightly simpler form as:

$$\delta Q_R = \left\{ p_2 - p_1 \left(\frac{V_1}{V_2} \right)^n \right\} \left(\frac{V_2}{\gamma - 1} \right) \qquad (4.2.13)$$

The difference between using Eqs. 4.2.12 and 4.2.13 is small, and Kee [1.20] shows it to be less than 1% for crank angle steps of 1° during the analysis of experimental pressure records.

The cylinder pressure record is digitized, the values of δQ_R are determined, and the sum of all of the values of δQ_R at each crank angle step, $d\theta$, is evaluated to provide a value of the total heat released, Q_R. The total crankangle period during which heat release is discerned is defined as b, in crankshaft degree units. At any angle, θ, after the start of heat release, the summation of δQ_R to Q_R reveals the value of the mass fraction of the fuel that has been burned up until that moment; this mass fraction burned is defined as B_θ. This is found from:

Mass fraction burned
$$B_\theta = \frac{\sum\limits_{\theta=0}^{\theta=\theta} \delta Q_R}{\sum\limits_{\theta=0}^{\theta=b} \delta Q_R}$$
(4.2.14)

A definition for the heat release rate with respect to crankshaft angle, $\dot{Q}_{R_\theta}$, at each and every crank angle position is found from:

Heat release rate
$$\dot{Q}_{R_\theta} = \frac{\delta Q_{R_\theta}}{d\theta}$$
(4.2.15)

There have been many publications on heat release characteristics of engines, mostly about four-stroke engines. To mention but six, one of the original papers related to diesel engines was presented by Lyn [4.22]; others in more recent times were presented by Hayes et al. [4.23], Lancaster et al. [4.24], Martorano et al. [4.25], Daniels [4.26], and Brunt et al. [4.27]. Those who wish to compare these data with those to be found for two-stroke engines should consult another textbook [4.36].

The paper by Daniels [4.26] is particularly interesting as it reports on the use of the ionization signal at the spark plug of a spark-ignition engine to determine the heat release rate or the mass fraction burned. The conventional method is to use a pressure transducer in the engine cylinder, as I have described here. Daniels [4.26] also gives some useful numerical data on the variation of profile of the mass fraction burned with both load and equivalence ratio.

4.2.3 Heat Release from a Spark-Ignition Engine

Fig. 4.4 shows the heat release rate and mass fraction burned characteristics determined from the analysis of a cylinder pressure diagram from a conventional automobile engine that is naturally-aspirated and spark-ignited and has a four-valve pent-roof combustion chamber.

The characteristics of the combustion are listed on the right of the figure, where it is seen that the ignition timing is at 21 °btdc, the delay period is 10°, and the total burn period, b, occupies 44 degrees from the end of the delay period to the conclusion of combustion. The delay period is the period after the point of ignition and before any heat release is detected from the analysis of the cylinder pressure diagram. On Fig. 4.4, the combustion can be observed to conclude at 54° after ignition.

The heat release rate is seen to have a profile similar to an isosceles triangle, with a maximum of 108 J/deg at about 8 °atdc, corresponding approximately to the 50% value on the mass fraction burned curve. The mass fraction burned curve has an exponential profile, which has already been introduced in Sec. 1.9, but will be discussed more thoroughly in terms of measured data in Sec. 4.3.6.

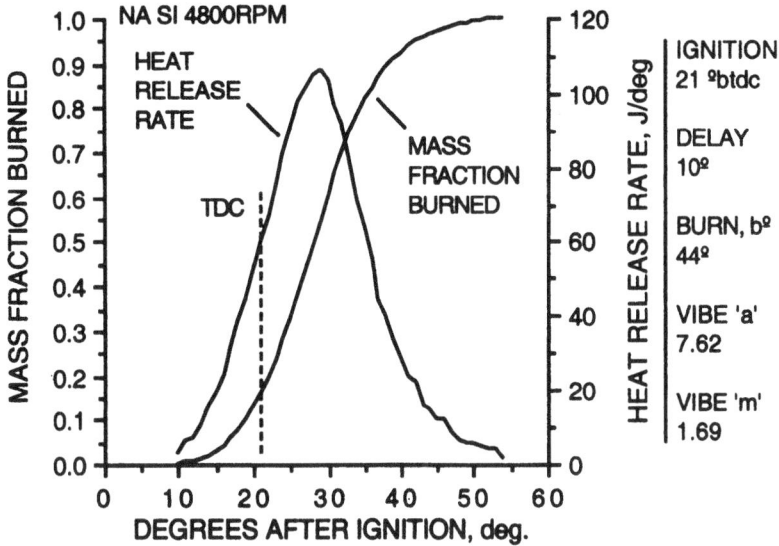

Fig. 4.4 Combustion analysis of a naturally-aspirated spark-ignition engine at 4800 rpm.

4.3 Heat Availability and Heat Transfer During the Closed Cycle

4.3.1 Properties of Fuels

Knowledge of the properties of fuels is required to assess the amount of heat available during combustion. The fuel is supplied to an engine as a liquid, which must be vaporized during the compression process in a spark-ignition engine, or during combustion in a diesel engine. Ideally, as stated earlier, the fuel for the spark-ignition engine is octane, and for the diesel unit it is dodecane. However, the real situation is more complex than that, as Tables 4.1 and 4.2 show. The properties are the hydrogen to carbon molecular ratio, H/C (or n as seen in the chemical relationship CH_n), the lower calorific value, C_{fl}, the density, ρ, and the latent heat of vaporization of the fuel, h_{vap}. If the fuel is oxygenated, such as an alcohol, or some of the reformed gasolines, then the generic formula is CH_nO_m.

Alcohols have the generic chemical formula of $C_nH_{2n+1}OH$. Methanol is the first of the alcohol family with a chemical formula of CH_3OH, and ethanol is the second with a chemical formula of C_2H_5OH.

Table 4.1 Properties of Some Gasoline Fuels Used in Engines.

Fuel	Octane C_8H_{18}	Gasoline Regular	Gasoline Premium	Gasoline Super-Unleaded	Gasoline Aviation
n and m	2.25, 0	1.95, 0	1.95, 0	1.65, 0	2.12, 0
C_{fl}, MJ/kg	44.6	42.7	43.5	43.0	43.5
density, kg/m³	690	735	755	760	720
h_{vap}, kJ/kg	400	420	420	420	400

Table 4.2 Properties of Diesel and Other Liquid Fuels Used in Engines.

Fuel	Paraffin Kerosene	Dodecane $C_{12}H_{26}$	Diesel Automotive	Alcohol Methanol	Alcohol Ethanol
n and m	1.79, 0	2.17, 0	1.81, 0	4, 1	3, 0.5
C_{fl}, MJ/kg	43.0	42.5	42.5	19.7	26.8
Density, kg/m³	800	830	835	790	790
h_{vap}, kJ/kg	350	250	300	1100	900

The tabular values show the typical properties of iso-octane, C_8H_{18}, and dodecane, $C_{12}H_{26}$. The values in Tables 4.1 and 4.2 for the gasolines labeled regular, premium, aviation, and super-unleaded, for a paraffin such as kerosene, for an automotive diesel fuel, and for the alcohols, are reasonably representative of the typical properties of such commercially available fuels. The properties of all paraffin family fuels are very dependent on the refining process, which will vary from country to country, depending on the institutional standards of each country, from refinery to refinery, and depending on the constituents and origin of the crude oil source for that particular fuel.

4.3.2 Properties of Exhaust Gas and Combustion Products

Sec. 2.1.6 shows the basic theory for computing the properties of a mixture of gases. The example chosen is the stoichiometric combustion of octane. This also continues the introduction given in Sec. 1.6.6. In both of these examples stoichiometric, i.e., chemically and ideally exact, combustion was used and so all carbon was burned to carbon dioxide and carbon monoxide was not formed. In all real combustion processes, dissociation takes place at elevated temperatures and pressures so that, even under stoichiometric conditions, free carbon monoxide and free hydrogen will be created. There are two principal dissociation reactions involved:

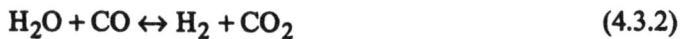

$$CO_2 \leftrightarrow CO + \frac{1}{2}O_2 \qquad (4.3.1)$$

$$H_2O + CO \leftrightarrow H_2 + CO_2 \qquad (4.3.2)$$

The latter reaction is often called the "water gas" reaction. The combustion equation to include all of these effects, and for all air-fuel ratios with a generic hydrocarbon fuel CH_nO_m, is then completely stated as:

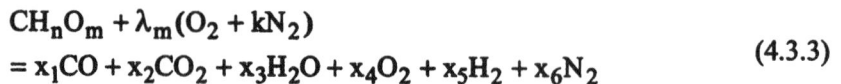

$$CH_nO_m + \lambda_m(O_2 + kN_2)$$
$$= x_1CO + x_2CO_2 + x_3H_2O + x_4O_2 + x_5H_2 + x_6N_2 \qquad (4.3.3)$$

where the value of λ_m is the molecular air-fuel ratio, and k is the nitrogen-to-oxygen molecular ratio, which is conventionally taken to be 79/21 or 3.76, as seen previously in Sec. 1.6.6. The relationship between λ_m and the actual air-to-fuel ratio, AFR, is then given by;

$$\text{AFR} = \frac{\lambda_m\left(M_{O_2} + kM_{N_2}\right)}{M_C + nM_H + mM_O} \tag{4.3.4}$$

where M_{O2}, etc., are the molecular weights of the constituent gases to be found numerically in Table 2.1.1. The molecular weight of an oxygen atom, M_O, is 15.999; of a carbon atom, M_C, 12.01; and of a hydrogen atom, M_H, 1.008. The value of λ_m can only be determined by balancing the equation for the carbon, hydrogen, oxygen, and nitrogen present in the combustion products:

Carbon balance $\qquad\qquad\qquad\qquad 1 = x_1 + x_2 \qquad\qquad\qquad\qquad\qquad$ (4.3.5)

Hydrogen balance $\qquad\qquad\qquad\qquad n = 2x_3 + 2x_5 \qquad\qquad\qquad\qquad\quad$ (4.3.6)

Oxygen balance $\qquad\qquad m + 2\lambda_m = x_1 + 2x_2 + x_3 + 2x_4 \qquad\qquad$ (4.3.7)

Nitrogen balance $\qquad\qquad\qquad\quad 2\lambda_m k = 2x_6 \qquad\qquad\qquad\qquad\qquad$ (4.3.8)

4.3.2.1 Stoichiometry and Equivalence Ratio

In the ideal case of stoichiometry, which means ignoring dissociation effects, the values of x_1, x_4, and x_5 in Eq. 4.3.3 are zero, in which case the solution of Eqs. 4.3.5 to 4.3.8 is found as:

$$x_2 = 1 \quad x_3 = \frac{n}{2} \quad \text{and} \quad \lambda_m = 1 + \frac{n}{4} - \frac{m}{2}$$

This reveals the stoichiometric air-fuel ratio, AFR_s, as:

$$\text{AFR}_s = \frac{\left(1 + \dfrac{n}{4} - \dfrac{m}{2}\right)\left(M_{O_2} + kM_{N_2}\right)}{M_C + nM_H + mM_O} \tag{4.3.9}$$

To confirm this with the previous solution in equation Eq. 1.6.15, for iso-octane:

$$AFR_s = \frac{\left(1 + \frac{2.25}{4} - 0\right)(31.99 + 3.76 \times 28.01)}{12.01 + 2.25 \times 1.008 + 0} = 15.02$$

It will be observed, using the properties of the fuels in Table 4.1, that the equivalent value for AFR_s for super-unleaded gasoline is 14.2; for dodecane it is 14.92; and for diesel fuel it is 14.42.

The concept of equivalence ratio, λ, is useful in the solution of these equations and in engine technology generally. It is defined as:

$$\lambda = \frac{AFR}{AFR_s} \tag{4.3.10}$$

The combination of Eqs. 4.3.4, 4.3.9, and 4.3.10 reveals:

$$\lambda = \frac{AFR}{AFR_s} = \frac{\lambda_m}{1 + \frac{n}{4} - \frac{m}{2}} \tag{4.3.11}$$

Obviously, the equivalence ratio is unity at stoichiometry.

4.3.2.2 Rich Mixture Combustion
In rich mixture combustion, the value of the equivalence ratio is less than unity:

$$\lambda < 1 \tag{4.3.12}$$

In the first instance, let dissociation effects be ignored. In this case, Eqs. 4.3.5 to 4.3.8 are directly soluble, and x_5 is zero. There is insufficient oxygen to burn all of the fuel, so there is no free oxygen and the value of x_4 is zero. The molecular balances become:

$$x_1 + x_2 = 1 \quad x_3 = \frac{n}{2} \quad m + 2\lambda_m = x_1 + 2x_2 + x_3 \quad x_6 = \lambda_m k$$

Hence, for the remaining unknowns x_1 and x_2:

$$x_1 = 2 - m - 2\lambda_m + \frac{n}{2} \quad x_2 = m + 2\lambda_m - \frac{n}{2} - 1 \qquad (4.3.13)$$

Consider a practical example where the equivalence ratio is 15% rich of stoichiometry for the combustion of super-unleaded gasoline, i.e., the value of n is 1.65, m is zero, and the value of λ is 0.85. From Eq. 4.3.11, the value of λ_m is found as:

$$\lambda_m = \lambda\left(1 + \frac{n}{4} - \frac{m}{2}\right) = 0.85 \times \left(1 + \frac{1.65}{4} - 0\right) = 1.20$$

Hence

$$x_1 = 2 - 0 - 2 \times 1.20 + \frac{1.65}{2} = 0.424$$

and

$$x_2 = 0 + 2 \times 1.20 - \frac{1.65}{2} - 1 = 0.576$$

also

$$x_3 = \frac{1.65}{2} = 0.825 \quad x_6 = 1.20 \times 3.76 = 4.52$$

The actual combustion equation now becomes:

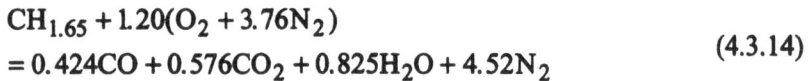

$$CH_{1.65} + 1.20(O_2 + 3.76N_2)$$
$$= 0.424CO + 0.576CO_2 + 0.825H_2O + 4.52N_2 \qquad (4.3.14)$$

The total moles of combustion products is 6.342. Dividing this into the moles of each gaseous component, gives the volumetric proportion of that particular gas. The exhaust gas composition is therefore 6.7% CO, 9.1% CO_2, and 13.0% H_2O, and the remainder is nitrogen. The emission of carbon monoxide is now significant, compared to the zero concentration ideally to be found at stoichiometry, as seen in the equivalent analysis for octane in Sec. 2.1.6; carbon dioxide and steam at stoichiometry had higher proportions by volume, 12.5% and 14.1%, respectively.

If the mixture is very rich, where the equivalence ratio, λ, is less than about 0.75, no carbon dioxide is produced, only carbon monoxide. In the solution of the above equations, this is detected by the appearance of a negative value for x_2. If this happens, x_2 is assigned a value of zero and the theory is resolved for the other coefficients. I find that this situation appears occasionally as a result of inadequately recorded experimental data submitted for the redesign of racing engines, but it is undesirable as the best power point is typically to be found at λ values around 0.85.

4.3.2.3 Lean Mixture Combustion
In lean mixture combustion, the value of the equivalence ratio is greater than unity:

$$\lambda > 1 \qquad (4.3.15)$$

In the first instance let dissociation effects be ignored. In this case Eqs. 4.3.5 to 4.3.8 are directly soluble, and x_5 is zero. There is excess oxygen to burn all of the fuel, so there is no free carbon monoxide and the value of x_1 is zero. The molecular balances become:

$$x_2 = 1 \quad x_3 = \frac{n}{2} \quad m + 2\lambda_m = 2x_2 + x_3 + 2x_4 \quad x_6 = \lambda_m k$$

Hence, for the remaining unknown, x_4:

$$x_4 = \lambda_m - 1 - \frac{n}{4} + \frac{m}{2} \qquad (4.3.16)$$

Consider a practical example where the equivalence ratio is 50% lean of stoichiometry for the combustion of diesel fuel, i.e., the value of n is 1.81, m is 0, and the value of λ is 1.5. From Eq. 4.3.11 the value of λ_m is found as:

$$\lambda_m = \lambda\left(1 + \frac{n}{4} - \frac{m}{2}\right) = 1.5 \times \left(1 + \frac{1.81}{4} - 0\right) = 2.178$$

Hence
$$x_4 = 2.178 - 1 - \frac{1.81}{4} + 0 = 0.726$$

also $\qquad x_3 = \dfrac{1.81}{2} = 0.905 \qquad x_6 = 2.178 \times 3.76 = 8.196$

The actual combustion equation now becomes:

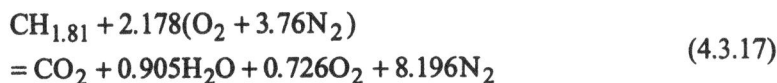

$$CH_{1.81} + 2.178(O_2 + 3.76N_2)$$
$$= CO_2 + 0.905H_2O + 0.726O_2 + 8.196N_2 \qquad (4.3.17)$$

The total moles of combustion products is 10.827. Dividing this into the moles of each gaseous component gives the volumetric proportion of that particular gas. The exhaust gas composition is therefore 9.24% CO_2, 8.4% H_2O, 6.71% O_2, and the remainder is nitrogen.

4.2.4 Effects of Dissociation

In the hot, high-pressure conditions occurring within the cylinder during combustion and expansion, gases such as carbon monoxide, carbon dioxide, steam, and oxygen will associate and dissociate in chemical equilibrium, thus changing the proportions of the components of exhaust gas. Because the above analysis is to be used to predict the composition of exhaust gas for use with simulation models, it is important to know the extent that dissociation effects will have on that composition.

If the above analysis is the precursor to complete flame travel models and to more detailed models of the thermodynamics of combustion as described by Douglas and Reid [4.28-4.30] then this section is too brief and simplistic.

Because the intention here is to describe a practical model of combustion to be included with 1D engine simulations, the need for a reasonably accurate description of exhaust gas composition assumes the higher priority. The question, therefore, is to what extent dissociation alters the composition of the combustion products from ignition until the end of expansion where the cylinder state conditions would typically finish up at a pressure of 8 to 10 bar and a temperature between 1400 and 1700 K. Consider one of the common dissociation reactions, as in Eq. 4.3.1, but extended to cope with the reality of that situation:

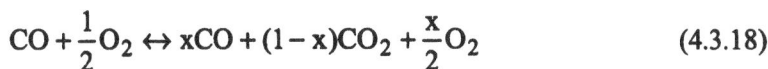

$$CO + \frac{1}{2}O_2 \leftrightarrow xCO + (1-x)CO_2 + \frac{x}{2}O_2 \qquad (4.3.18)$$

The total number of moles is then N where:

$$N = 1 + \frac{x}{2}$$

The partial pressures of the three components of the equilibrium mixture are, by Dalton's Laws, a function of their partial pressures in relation to the total pressure, p, in atmospheres:

$$p_{CO} = \frac{xp}{N} \quad p_{CO_2} = \frac{(1-x)p}{N} \quad p_{O_2} = \frac{\frac{x}{2}p}{N} \qquad (4.3.19)$$

The equilibrium constant for this reaction, K_p, is a function of temperature and is to be found described and tabulated in many standard texts on thermodynamics. As a fitted curve as a function of the temperature, T (in Kelvin units), this is as follows:

$$\log_e K_p = -68.465 + \frac{6.741}{10^2}T - \frac{2.6658}{10^5}T^2 + \frac{4.9064}{10^9}T^3 - \frac{3.3983}{10^{13}}T^4 \qquad (4.3.20)$$

For the dissociation reaction it is incorporated as:

$$K_p = \frac{p_{CO}\sqrt{p_{O_2}}}{p_{CO_2}} \qquad (4.3.21)$$

Substituting the values in Eq. 4.3.19 into Eq. 4.3.21 reveals:

$$x^3\left(1 - \frac{p}{K_p^2}\right) - 3x + 2 = 0 \qquad (4.3.22)$$

As an example let us assume that the gas at the end of expansion, when the exhaust valve is about to open, is at a pressure of 8 bar and a temperature of 1800 K. The value for Kp for this reaction at this temperature from Eq. 4.3.20 as:

As
$$\log_e K_p = -8.452$$

then
$$K_p = 2.134 \times 10^{-4}$$

The solution of Eq. 4.3.22 reveals that the value of x is 0.00224 and the volumetric proportions of the gas mixture on the right-hand side of Eq. 4.3.18 are 0.224% CO, 99.776% CO_2, and 0.112% O_2. In short, the gas composition is barely altered as a function of dissociation at these state conditions. Thus, it is not really worth considering dissociation as a factor in predicting the composition of exhaust gas. In effect, what this shows is that at stoichiometry carbon monoxide and free oxygen are present in the exhaust gas to the detriment of the proportion of carbon dioxide. Instead of containing 12.5% CO_2 by volume, the exhaust gas will contain some 0.025% CO, 12.46% CO_2, and 0.0125% O_2. This corresponds with experimental test results.

However, to make the point thoroughly, consider the above computation again, but this time as if it were at the height of the combustion process. Now the pressure would be some 40 atm and the temperature some 2500 K, and the situation would be very different. The gas composition would now be 4.3% CO, 9.36% CO_2, and 2.1% O_2. Thus, dissociation should be taken into account in any detailed model of the thermodynamics of a combustion process in order to calculate the heat release, at any instant, caused by the formation of the products.

It should be noted that, when a more complex emissions calculation is described in Appendix A4.1, a further equilibrium reaction is required. This is the so-called "water gas" reaction given above in Eq. 4.3.2. The equilibrium chemistry of that reaction and the numerical details of its equilibrium constant are given in Eqs. A4.1.15 to A4.1.17. The implementation of this reaction within a simulation is basically as described above; however, because there are now five reactions to solve in total, i.e., a carbon balance, a hydrogen balance, an oxygen balance, and the two equilibrium reactions, the arithmetic is not trivial and the Gaussian solution, so often mentioned within Chapter 2, is required yet again.

4.3.2.5 The Relationship Between Combustion and Exhaust Emissions

Chapter 5 contains a fuller discussion of this topic along with the results of engine simulations. Appendices A4.1 and A4.2 present the fundamental theoretical base for the computation of exhaust emissions within an engine simulation.

It is appropriate to point out here that the analysis of Eq. 4.3.3 shows quite clearly that rich-mixture combustion leads to an exhaust emission of carbon monoxide and unburned hydrocarbons. Because there is insufficient air present in rich mixture combustion to oxidize all of the carbon and hydrogen present, the outcome is self-evident. The effects of dissociation have been shown here to lead to both CO and hydrocarbon (HC) emissions, even if the mixture is stoichiometric.

The extent of the CO emission is shown in Appendix A4.2, and Fig. A4.4 gives a classic picture of the profile of CO emission decaying on the expansion stroke to a virtually nonexistent level from the combustion of a mixture with stoichiometric air-to-fuel ratio [4.7, 4.13].

Hydrocarbon emissions for the four-stroke engine operating at, or near, the stoichiometric mixture, are relatively low. In the modern spark-ignition engine designed to meet emissions legislation, the design work to further reduce HC emissions is now focussed on minimizing the last remaining vestiges of lubricating oil in the crevices around the piston rings. There is little profit in there for the thermodynamicist.

The creation of nitrogen oxides occurs as a function of temperature, oxygen concentration, and time within the burn zone, and this is maximized by increasing load (i.e., bmep) levels, or by combustion at the stoichiometric air-fuel ratio at any given bmep value. The creation of nitrogen oxides can be reduced by recirculating exhaust gas (as cold EGR). However, because this lowers the effective scavenging efficiency of the engine, it inevitably reduces combustion efficiency and raises the HC emissions. There is a limit to the amount of EGR that can be profitably employed; this is typically about 10% of the delivery ratio. An interesting and highly relevant further discussion on this subject is presented in Sec. 5.9.4.

For a two-stroke engine, combustion-related emissions are for the most part avoided because the combustion chamber in the cylinder head is normally a much simpler, more compact, and less mechanically cluttered design than that of the four-stroke engine. The "clutter" here refers to the poppet valves, and the less compact shape that ensues from their incorporation into a cylinder head. This produces crevices into which the flame dies or is quenched, leading to incomplete combustion in those regions. By definition, the more complex the head design, e.g., four valves rather than two, the greater is the potential for crevices in that design.

4.3.3 Heat Availability During the Closed Cycle

From the discussion on dissociation it is clear that, at the high temperatures and pressures that are the reality during a combustion process, carbon dioxide cannot be produced directly at any instant during burning. In addition, carbon monoxide is produced during the combustion process and its heat of formation is much less than that associated with the complete oxidation of the carbon to carbon dioxide. Thus, the full heat potential of the fuel is not realized or released during burning. There can be further factors that inhibit the efficiency of combustion. For example, the mixture can be so rich that there is insufficient oxygen to effect that process, even if it were to be ideal. There can be—and in a two-stroke engine there certainly will be—considerable quantities of exhaust gas residual present within the combustion chamber to inhibit the progress of the flame development and the efficiency of combustion. Thus, the complete combustion efficiency is composed of subsets related to equivalence ratio and scavenging efficiency:

$$\eta_c = C_{burn}\eta_{af}\eta_{se} \qquad (4.3.23)$$

Experimental evidence is used to determine these factors, as the theoretical chemistry of combustion is not sufficiently advanced as to be able to provide them, although this situation is improving with the passage of time [4.41-4.43]. The factors listed in Eq. 4.3.23 are:

η_{af} The relative combustion efficiency with respect to equivalence ratio, which has a maximum value of unity, usually close to the stoichiometric level for spark-ignition engines, and about 2.3 for compression-ignition engines.

η_{se} The relative combustion efficiency with respect to scavenging efficiency, which has a maximum value of unity at a value equal to or exceeding 90%. Although the term "scavenging efficiency" is used, the real criterion is the absolute trapped charge purity. For the spark-ignition engine they are virtually the same thing, but for a diesel engine they are numerically different.

C_{burn} This is unashamedly a "fudge" factor used to express the reality of all other combustion effects that are virtually inexplicable by a simple analysis, and are unrelated to fueling or charge purity. Such effects include incomplete oxidation of the hydrocarbon fuel, which always occurs even in optimum circumstances, incomplete flame travel into the corners of particular combustion chambers, weak or ineffective ignition systems, poor burning in crevices, and flame decay by quenching in most circumstances. The value of C_{burn} is between 0.85 and 0.90 and rarely changes by more than a few percentage points.

Effect of Trapped Charge Purity or Scavenging Efficiency

The experimentally determined relationship for the relative combustion efficiency related to charge purity or scavenging efficiency, SE, for spark-ignition engines is set out below:

if SE > 0.9 then $$\eta_{se} = 1.0 \tag{4.3.24}$$

if SE < 0.9 then

$$\eta_{se} = -12.558 + 70.108SE - 135.67SE^2 + 114.77SE^3 - 35.542SE^4 \tag{4.3.25}$$

For diesel engines, the relative charge purity, Π, should be in excess of 90% or else there will be a considerable emission of carbon particulates and hydrocarbons from combustion. Hence, for diesel engines the purity should be near unity or the design is in some jeopardy. Recent attempts to reduce NO_x emissions in DI diesel engines by using considerable amounts of EGR (exhaust gas recirculation) have found particulate emissions to be a major stumbling block.

Effect of Equivalence Ratio for Spark-Ignition Engines

The effect of fueling level, in terms of equivalence ratio, is best evaluated from experimental evidence. The alternative is to rely on theory, as done by Douglas and Reid [4.28-4.30] using equilibrium and dissociation theory, as outlined above, in an engine simulation using a flame propagation model. To do this accurately, so as to phase the peak power point at the correct "rich" mixture level and the peak thermal efficiency at the correct "weak" mixture level, it is necessary to employ reaction kinetics within the combustion theory. This is a complex topic well beyond the scope of this design-based text, but within the province of fundamental scientific research on combustion [4.5, 4.6].

The experimental evidence, based on many fueling loops carried out on several spark-ignition research engines at QUB reveals that the relative combustion efficiency with respect to equivalence ratio, η_{af}, is measured as:

if $0.8 < \lambda < 1.2$ $$\eta_{af} = -1.6082 + 4.6509\lambda - 2.0746\lambda^2 \tag{4.3.26}$$

Analysis of this function reveals that η_{af} has a maximum of unity at a λ value of about 1.12, i.e., 12% "weak" of stoichiometric, and produces a maximum total heat release at a λ value of about 0.875, i.e., at 14% "rich" of stoichiometric.

Effect of Equivalence Ratio for Compression-Ignition Engines

The effect of fuelling level, in terms of the trapped equivalence ratio, λ, is best evaluated from experimental evidence, as in the case of the spark-ignition engine above. The phrase "trapped equivalence ratio" is used to emphasize that the exhaust residual within a diesel engine cylinder contains significant quantities of oxygen. Any recirculated exhaust gas (EGR), by definition, will also contain the same oxygen level. On the other hand, for a correctly designed diesel engine, the scavenging efficiency should be above 90%. This limits the amount of exhaust gas residual that can be present, or that can be recirculated, to keep the exhaust smoke levels within legislated limits. The experimental evidence reveals that the relative combustion efficiency with respect to equivalence ratio, η_{af}, is measured as:

if $1.3 < \lambda < 2.3$ $\qquad$ $\eta_{af} = 0.35332 + 0.56797\lambda - 0.12472\lambda^2$ $\qquad$ (4.3.27)

if $\lambda > 2.3$ $\qquad\qquad\qquad$ $\eta_{af} = 1.0$

Analysis of this function reveals that η_{af} has a maximum of unity at a λ value of 2.3, i.e., 130% "weak" of stoichiometric. Above this equivalence ratio, the value is constant at unity. Because a diesel engine will produce peak power at a λ value of approximately 1.3, but with unacceptably high levels of black smoke emission, the equations above are sensibly applicable for equivalence ratios above 1.3.

4.3.4 Heat Transfer During the Closed Cycle

The experience at QUB is that, particularly for spark-ignition engines, the most effective and accurate method for calculating heat transfer from the cylinder during the closed cycle is that based on Annand's work [4.31-4.33] In addition, for diesel engines, the heat transfer research by Woschni [4.34] is also highly regarded as being equally effective for theoretical computation. A very full discussion of heat loss from air-cooled engines is presented in the book by Mackerle [4.35]. The logic of the Annand approach is that it separates out the convection and radiation terms, and this distinguishes it from the heat transfer theories of other researchers. Typical of the approach to the heat transfer theory proposed by Annand is his expression for the Nusselt number, **Nu**, leading to a conventional derivation for the convection heat transfer coefficient, C_h. The methodology is almost exactly that adopted for the pipe theory in Sec. 2.4. Annand recommends the following expression to connect the Reynolds and the Nusselt numbers:

$$\mathbf{Nu} = a\mathbf{Re}^{0.7} \qquad (4.3.28)$$

where the constant, a, has a value of 0.26 for a two-stroke engine and 0.49 for a four-stroke engine.

The Reynolds number is calculated as:

$$\mathbf{Re} = \frac{\rho_{cy} c_p d_{cy}}{\mu_{cy}} \qquad (4.3.29)$$

The value of cylinder bore, d_{cy}, is self-explanatory. The values of density, ρ_{cy}, mean piston velocity, c_p, and viscosity, μ_{cy}, deserve more discussion.

The prevailing cylinder pressure, p_{cy}, temperature, T_{cy}, and gas properties combine to produce the instantaneous cylinder density, ρ_{cy}:

$$\rho_{cy} = \frac{p_{cy}}{R_{cy} T_{cy}}$$

During compression, the cylinder gas will be a mixture of air, rapidly vaporizing fuel, and exhaust gas residual. During combustion, it will change from the compression gas to become exhaust gas, and during expansion it will be exhaust gas. Tracking the gas constant, R_{cy}, and the other gas properties listed in Eq. 4.3.29 at any instant during a computer simulation is straightforward.

The viscosity is that of the cylinder gas, μ_{cy}, at the instantaneous cylinder temperature, T_{cy}, but I have found that little accuracy is lost if the expression for the viscosity of air, μ_{cy}, in Eq. 2.3.11 is employed as a simplification.

The mean piston velocity is found from the dimension of the cylinder stroke, L_{st}, and the engine speed, N, in rev/min units:

$$c_p = \frac{2 L_{st} N}{60} \qquad (4.3.30)$$

Having obtained the Reynolds number, the convection heat transfer coefficient, C_h, can be extracted from the Nusselt number, as in Eq. 2.4.3:

$$C_h = \frac{C_k \, \mathbf{Nu}}{d_{cy}} \; W/m^2 K \qquad (4.3.31)$$

The parameter C_k is the thermal conductivity of the cylinder gas, and can be assumed to be identical with that of air at the instantaneous cylinder temperature, T_{cy}, and consequently may be found from Eq. 2.3.10.

Annand also considers the radiation heat transfer coefficient, C_r, to be given by:

$$C_r = 4.25 \times 10^{-9} \frac{T_{cy}^4 - T_{cw}^4}{T_{cy} - T_{cw}} \ W/m^2K \qquad (4.3.32)$$

However, the value of C_r is much less than that of C_h, to such a degree that C_r may be neglected for most engine cycle calculations, if reasonable simplification is desired. The value of T_{cw} in the above expression is the average temperature of the cylinder wall, the piston crown, and the cylinder head surfaces. The heat transfer, δQ_L, over a crankshaft angle interval $d\theta$ and a time interval dt can be deduced for the mean value of that transmitted to the total surface area exposed to the cylinder gases:

As
$$dt = \frac{d\theta}{360} \times \frac{60}{N} \qquad (4.3.33)$$

then
$$\delta Q_L = (C_h + C_r)(T_{cy} - T_{cw})A_{cw}dt \qquad (4.3.34)$$

The surface area of the cylinder, A_{cw}, is composed of:

$$A_{cw} = A_{cylinder\,liner} + A_{piston\,crown} + A_{cylinder\,head} \qquad (4.3.35)$$

It is straightforward to expand the heat transfer equation in Eq. 4.3.34 to deal with the individual components of the head or crown by assigning a surface temperature to those specific areas. This improves the accuracy of heat transfer assessment during a simulation. It should also be noted that Eq. 4.3.34 produces a "positive" value for the "loss" of heat from the cylinder, aligning it with the sign convention assigned in Eq. 4.2.2 above.

The typical values obtained from the use of the above theory are illustrated in Table 4.3. The example employed is that of an engine of 86 mm bore, 86 mm stroke, running at 4000 rpm. Various timing positions throughout the cycle are selected, and the potential state conditions of pressure, p_{cy}, (in atm units) and temperature, T_{cy}, (in °C units) are estimated to arrive at the tabulated values, based on the solution of the above equations. An estimated value of the average cylinder wall temperature, T_{cw}, is also selected for each juncture. The timing positions are in the middle of induction, at the point of ignition, at the peak of combustion for both a naturally-aspirated spark-ignition and a turbocharged compression-ignition engine, and at exhaust valve opening (release), respectively. It will be observed that the heat transfer coefficients predicted for the radiation component, C_r, are indeed very much less than that for the convection component, C_h, and might well be neglected, or indeed incorporated by a minor change to the constant "a" in the Annand model in Eq. 4.3.27—although this would defeat the very logic that makes Annand's model so effective.

Table 4.3 Heat Transfer Coefficients Using the Annand Model.

Timing Position	P_{cy} (atm)	T_{cy} (°C)	T_{cw} (°C)	Nu	Re	C_h	C_r
induction (na si)	0.6	35	150	739	34702	238	0.9
ignition (na si)	15.0	350	200	3149	275397	1733	2.9
burning (na si)	75	2300	220	2120	156507	2796	90
burning (tc ci)	150.0	2500	250	3257	289046	4349	111
release (na si)	9.0	1100	175	890	45279	869	16

It can also be seen that the heat transfer coefficients increase dramatically during combustion, but of course that is where the minimum surface area and maximum gas temperature occur during the heat transfer process, and this will have a direct influence on the total heat transferred through Eq. 4.3.34.

4.3.5 Internal Heat Loss by Fuel Vaporization
Fuel is vaporized during the closed cycle. For the spark-ignition engine vaporization normally occurs during compression and prior to combustion. For the compression-ignition unit, it occurs immediately prior to, but mostly during, combustion.

Fuel Vaporization for the Spark-Ignition Engine
Let it be assumed that the cylinder mass trapped is m_t and the scavenging efficiency is SE, with an air-fuel ratio AFR. The masses of trapped air, m_{ta}, and fuel, m_{tf}, are given by:

$$m_{ta} = m_t SE \quad \text{and} \quad m_{tf} = \frac{m_{ta}}{AFR} \tag{4.3.36}$$

If the crankshaft interval between trapping at intake valve closing and the ignition point is declared as θ_{vap}, and is the crankshaft interval over which fuel vaporization is assumed to occur linearly, then the rate of fuel vaporization with respect to crankshaft angle, $\dot{m}_{vap}$, is given by:

$$\dot{m}_{vap} = \frac{m_{tf}}{\theta_{vap}} \quad \text{kg/deg} \tag{4.3.37}$$

Consequently, the loss of heat from the cylinder contents, δQ_{vap}, for any given crankshaft interval, $d\theta$, is found by using the latent heat of vaporization of the fuel, h_{vap}. Numeric values of latent heat of vaporization of various fuels are presented in Tables 4.1 and 4.2.

$$\delta Q_{vap} = \dot{m}_{vap} h_{vap} d\theta \tag{4.3.38}$$

It should be noted that this equation provides a "positive" value for this heat loss, in a similar fashion to the application of Eq. 4.3.34. It should also be noted that the above statements cope with mixtures that are at stoichiometric or richer. If they are leaner than stoichiometric, then the analysis below for the diesel engine is more accurate. You may wish to reexamine Sec. 1.6.4 to confirm that.

Fuel Vaporization for the Compression-Ignition Engine

Let it be assumed that the cylinder mass trapped is m_t and the absolute charge purity is Π_{abs}, with a trapped air-fuel ratio of AFR_t. The masses of trapped air, m_{ta}, and fuel, m_{tf}, are given by:

$$m_{ta} = \Pi_{abs}m_t \quad \text{and} \quad m_{tf} = \frac{m_{ta}}{AFR_t} \tag{4.3.39}$$

The crankshaft angle interval over which combustion occurs is defined as $b°$. Fuel vaporization is assumed to occur as each packet of fuel, $dm_{b\theta}$, is burned over a time interval, and is related to the mass fraction burned at that juncture, B_θ, at a crankshaft angle, θ, from the onset of the combustion process. Let it be assumed that an interval of combustion occurs over a crankshaft interval, $d\theta$. The increment of fuel mass vaporized and burned during this time and crankshaft interval is given by dm_{vap}, thus:

$$dm_{vap} = dm_{b\theta} = \left(B_{\theta+d\theta} - B_\theta\right)m_{tf}h_{vap} \tag{4.3.40}$$

Consequently, the loss of heat from the cylinder contents, δQ_{vap}, for this crankshaft interval, $d\theta$, is found by using the latent heat of vaporization of the fuel, h_{vap}. A numeric value of latent heat of vaporization of a diesel fuel is presented in Table 4.2.

$$\delta Q_{vap} = dm_{vap}h_{vap} \tag{4.3.41}$$

It should be noted that this equation provides a "positive" value for this heat loss, in similar fashion to the application of Eq. 4.3.34, and that this is only soluble if the mass fraction burned profile is available as numerical information.

4.3.6 Heat Release Data for Spark-Ignition Engines

The heat release and mass fraction burned data for a naturally-aspirated spark-ignition engine as in Fig. 4.4 has already been presented and discussed. In the companion volume to this book [4.36], which is devoted to two-stroke engines, I fitted the heat release curves with a series of straight lines and triangles. This approach is very rigid when applied to experimental data, and is abandoned in favor of the more flexible method suggested by Vibe [4.37].

The Vibe Analysis of Experimental Data from a Naturally-Aspirated Spark-Ignition Engine

It is possible to analyze a mass fraction burned curve and fit a mathematical expression to the experimental data. This is often referred to as the Vibe method [4.37]. The mathematical fit is exponential with numerical coefficients, a and m, for the mass fraction burned, B_θ, at a particular crankshaft angle, θ, from the onset of heat release and the combustion persists for a total crankshaft angle duration of b°. It is expressed thus:

$$B_\theta = 1 - e^{-a\left(\frac{\theta}{b°}\right)^{m+1}} \tag{4.3.42}$$

The experimental data in Fig. 4.4 are found to be fitted with coefficients a and m of values 7.62 and 1.69, respectively, for a total burn period, b°, of 44 degrees duration. These values are printed on Fig. 4.4 at the right of the diagram, together with the recorded ignition timing at 21 °btdc, and the measured ignition delay, d°, at 10° crankshaft after the ignition point. At the end of the ignition delay, in this case at 11 °btdc, analysis of the cylinder pressure diagram shows the beginnings of heat release into the cylinder. The experimental data for the mass fraction burned are replotted in Fig. 4.5(a), and for the heat release rate data, in Fig. 4.5(b). On these diagrams, the full lines indicate the replay of the numerical values of 7.62, 1.69, and 44 for the values of a, m, and b°, respectively, into Eqs. 4.3.42, 4.2.14 and 4.2.15. There is a very good correspondence between the experimental data derived from the cylinder pressure diagram using the theory of Sec. 4.2.2 and the fit to that data by Vibe's analytic technique.

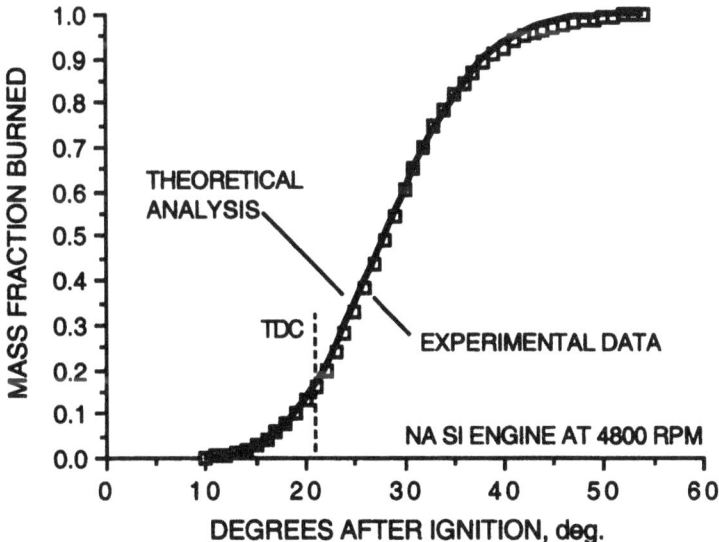

Fig. 4.5(a) Mass fraction burned in a naturally-aspirated spark-ignition engine at 4800 rpm.

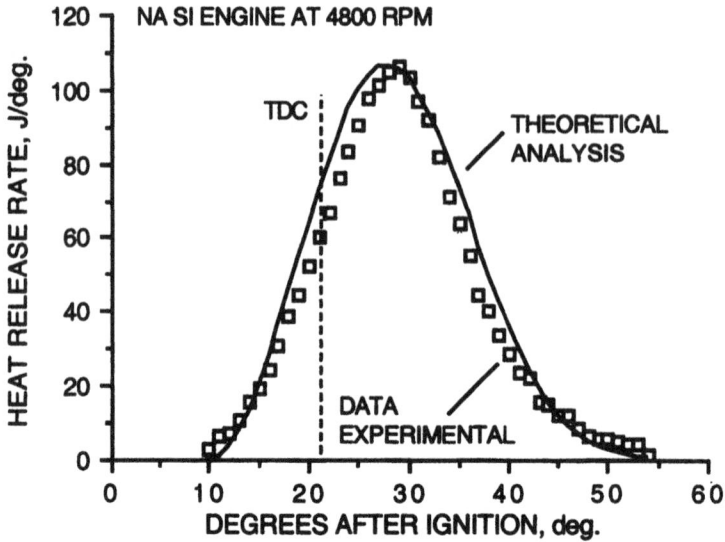

Fig. 4.5(b) Heat release rate in a naturally-aspirated spark-ignition engine at 4800 rpm.

Analysis of Cylinder Pressure Data from a Naturally-Aspirated Spark-Ignition Moped Engine

The Honda moped, which was used as an example of discharge coefficients for poppet valves in a two-valve overhead valve design in Sec. 3.6.4, is used here for the equivalent purpose of illustrating combustion in a small four-stroke engine. The cylinder head design is of the pushrod-actuated, two-valve, hemi-head type as seen in Fig. 1.2, with a compression ratio of 9.96, and the capacity of each cylinder is nearly 50 cm^3. The fuel is gasoline and the data are recorded at full load at 7500 rpm, where the bmep is measured at 7.7 bar. The data for mass fraction burned and heat release rate are determined from the cylinder pressure diagram with the theory of Sec. 4.2.2, fitted by the Vibe approach defined by Eq. 4.3.42, and presented in Fig. 4.6. The imep is found to be 10.86 bar, giving a mechanical efficiency of just 0.71 by applying Eq. 1.7.8. This is a rather low mechanical efficiency, but is typical of small-capacity four-stroke engines where the friction loss is disproportionately high.

In comparing Figs. 4.4 and 4.6, at first glance they seem similar. However, the duration of burn is very much longer at 70° crank angle in the smaller 2v engine—its capacity is nearly ten-fold smaller—compared to a 44° burn duration in the 4v automobile engine. The ignition delay in the moped engine is 17°, which is some 70% longer than the 10° of the car engine. The position of the mass fraction burned value of 50% is often used a marker of burn disposition and duration; for the moped engine this is at 17 °atdc and for the car engine it is at 7 °atdc. This confirms the later and slower burn in the much smaller moped engine cylinder. Note the values of heat release rate in Figs 4.4 and 4.6. In the moped engine heat release rate peaks at 6.2 J/deg, and in the car engine at 107 J/deg. Recall that the cylinder size ratio is about 10, so it is not surprising that the amount of heat released is proportional to the air contained in each cylinder space. The ignition timing for the slower burn process in the moped is at 27 °btdc, and for the automobile unit it is 21 °btdc, and this degree of ignition advance also bespeaks a slow combustion procedure. The Vibe coefficients of both engines are not dissimilar, giving the appearance that the exponential functions are comparable in profile when in reality they are dissimilar in extent.

Small-capacity, pushrod-actuated, two-valve, hemi-head four-stroke engines are used for many applications as industrial engines. The data in Fig. 4.6 can be used to characterize the combustion processes in such naturally-aspirated spark-ignition units using gasoline as a fuel.

Analysis of Cylinder Pressure Data from a Naturally-Aspirated Spark-Ignition Automobile Engine

The engine is a naturally-aspirated, spark-ignition engine for a current production automobile. The cylinder head design is of the four-valve, pent-roof type, as in Fig. 1.6, and the capacity of each cylinder is nearly 500 cm^3. The fuel is gasoline, and all of the data are recorded at full load. The data for mass fraction burned and heat release rate are determined from the cylinder pressure diagram using the theory of Sec. 4.2.2 and fitted by the Vibe approach defined by Eq. 4.3.42. The data at full load at 1200, 2400, 3600, and 6000 rpm are given in Figs. 4.7 to 4.10, respectively. The equivalent data output for the 4800 rpm case has already been shown in Fig. 4.4. The data on the measured ignition timing, and the recorded values for ignition delay, d°, combustion duration, b°, and the Vibe exponential coefficients, a and m, are shown to the right of each diagram.

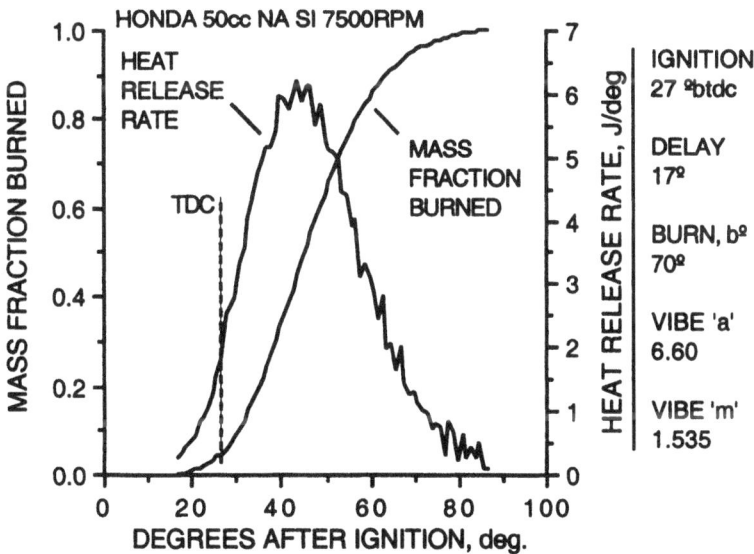

Fig. 4.6 Combustion analysis of a Honda 50 cm^3 moped engine at 7500 rpm.

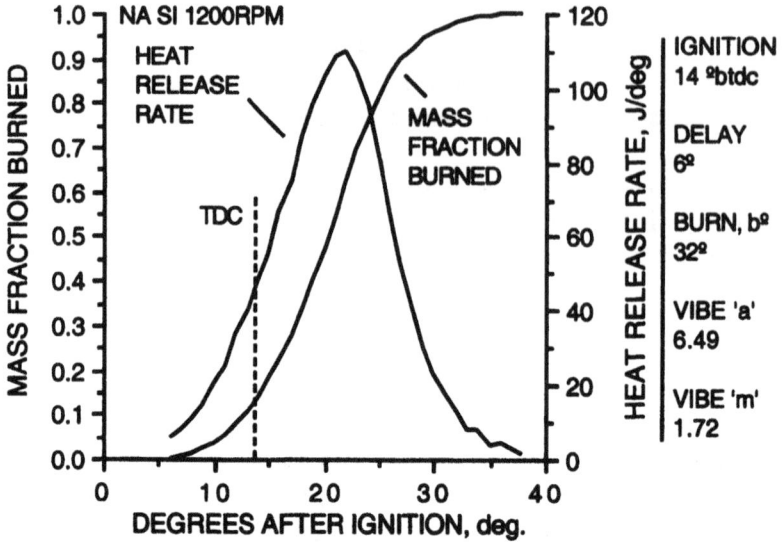

Fig. 4.7 Combustion analysis of a naturally-aspirated spark-ignition engine at 1200 rpm.

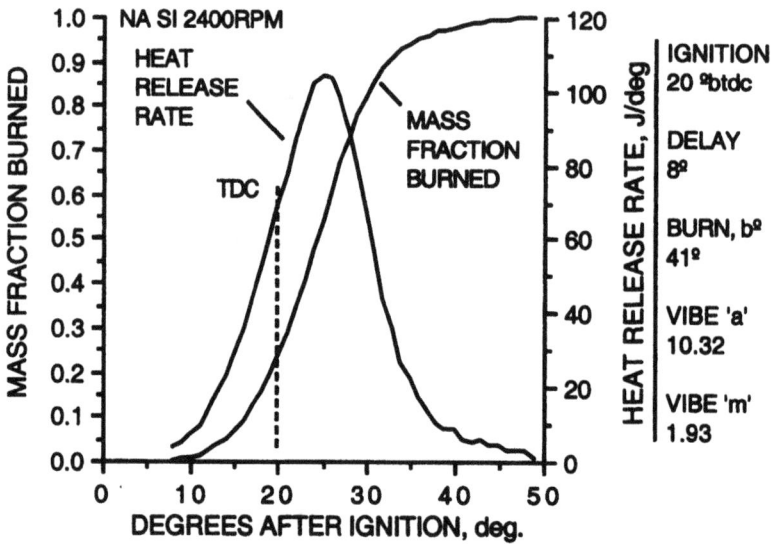

Fig. 4.8 Combustion analysis of a naturally-aspirated spark-ignition engine at 2400 rpm.

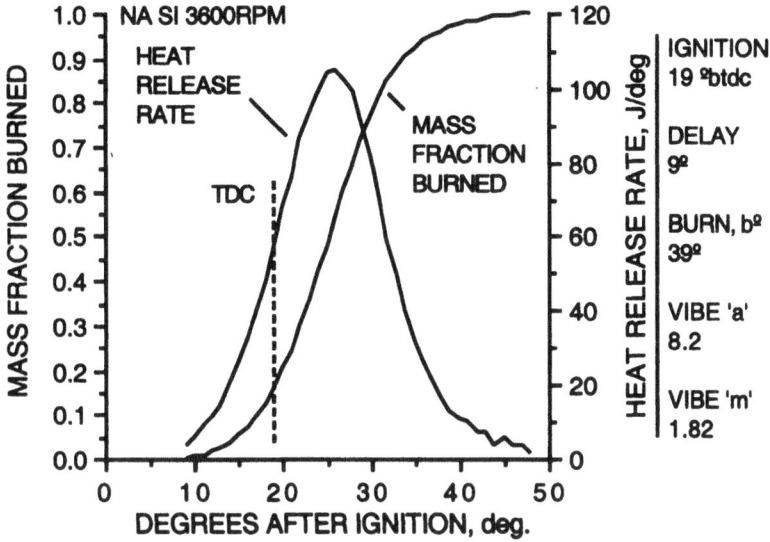

Fig. 4.9 Combustion analysis of a naturally-aspirated spark-ignition engine at 3600 rpm.

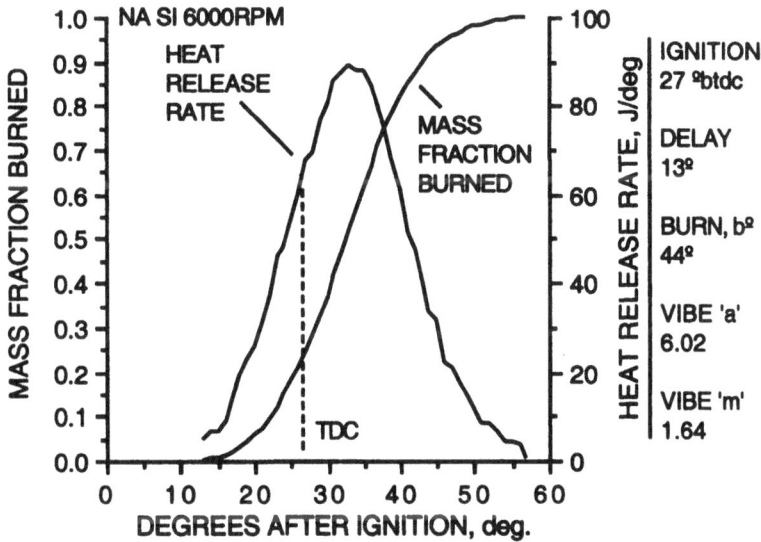

Fig. 4.10 Combustion analysis of a naturally-aspirated spark-ignition engine at 6000 rpm.

Apart from the lowest-speed case at 1200 rpm in Fig. 4.7, it can be observed that the ignition timing is progressively advanced with increasing engine speed to optimize the location of combustion around tdc, to provide the highest imep; the ignition delay increases with engine speed; and the burn duration is relatively constant with increasing engine speed, but the Vibe "a" and "m" coefficients decrease.

In the absence of measured data for an engine, where design by simulation is undertaken, these data may be used with some confidence to characterize the combustion of gasoline in a naturally-aspirated, spark-ignition engine at full load over a speed range.

Analysis of Cylinder Pressure Data from a Turbocharged Spark-Ignition Engine

The engine is a turbocharged spark-ignition engine for a current production automobile. It has a four-valve, pent-roof head design. The fuel is gasoline and all data are recorded at full load. The data for mass fraction burned and heat release rate are determined from the cylinder pressure diagram using the theory of Sec. 4.2.2 and fitted by the Vibe approach defined by Eq. 4.3.42. The data at full load at 2400, 3000, 3600, 4200, 4800, and 5400 rpm are given in Figs. 4.11 to 4.16, respectively. The data on the measured ignition timing, and the recorded values for ignition delay, $d°$, combustion duration, $b°$, and the Vibe exponential coefficients, a and m, are shown to the right of each diagram.

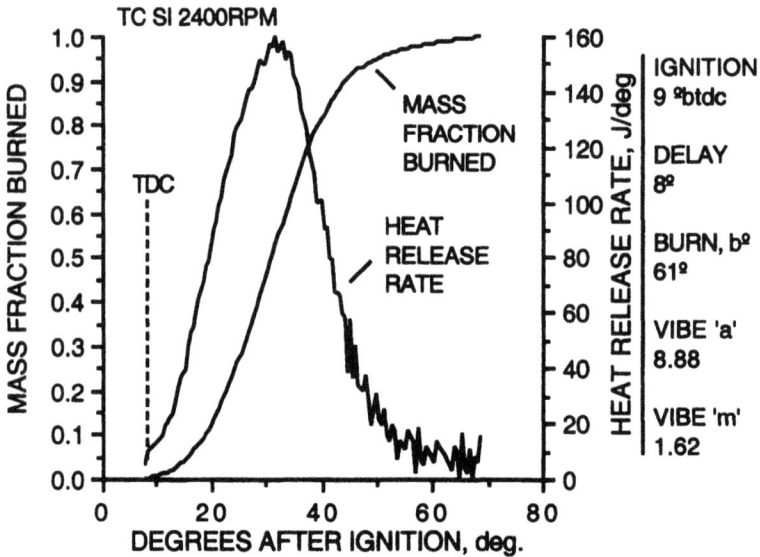

Fig. 4.11 Combustion analysis of a turbocharged spark-ignition engine at 2400 rpm.

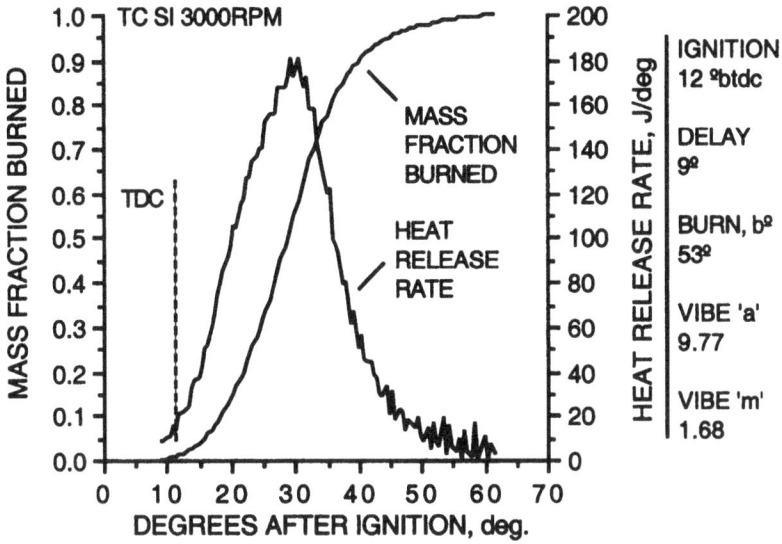

Fig. 4.12 Combustion analysis of a turbocharged spark-ignition engine at 3000 rpm.

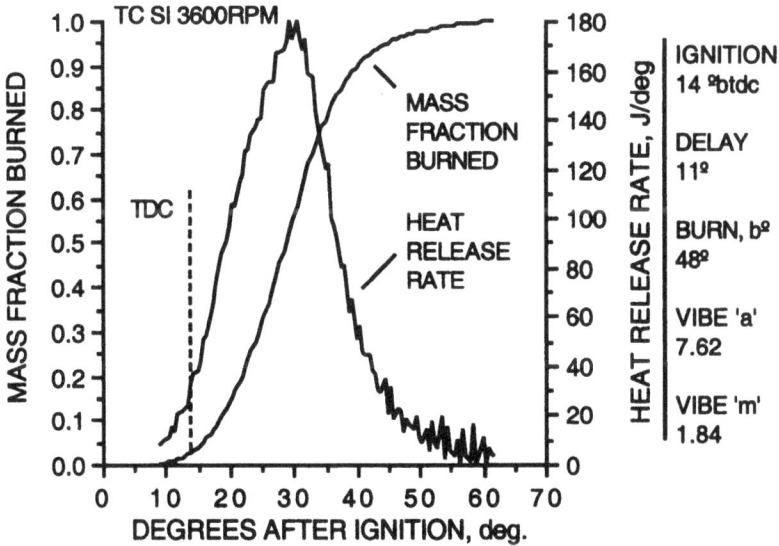

Fig. 4.13 Combustion analysis of a turbocharged spark-ignition engine at 3600 rpm.

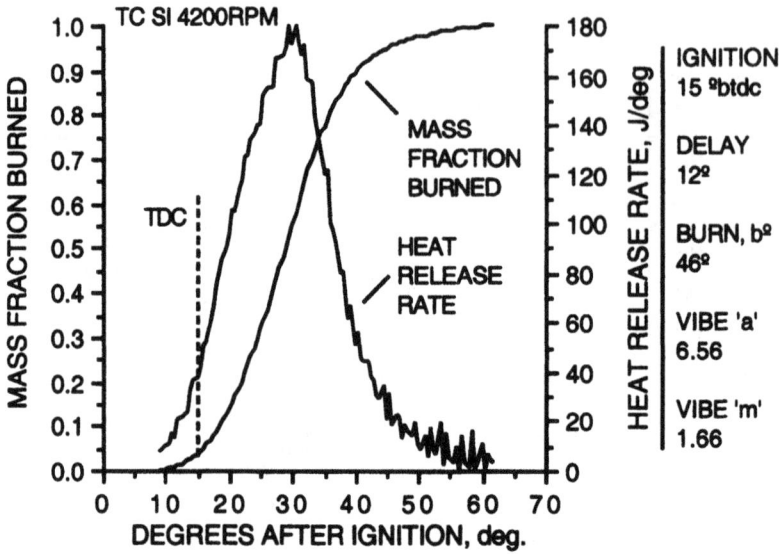

Fig. 4.14 Combustion analysis of a turbocharged spark-ignition engine at 4200 rpm.

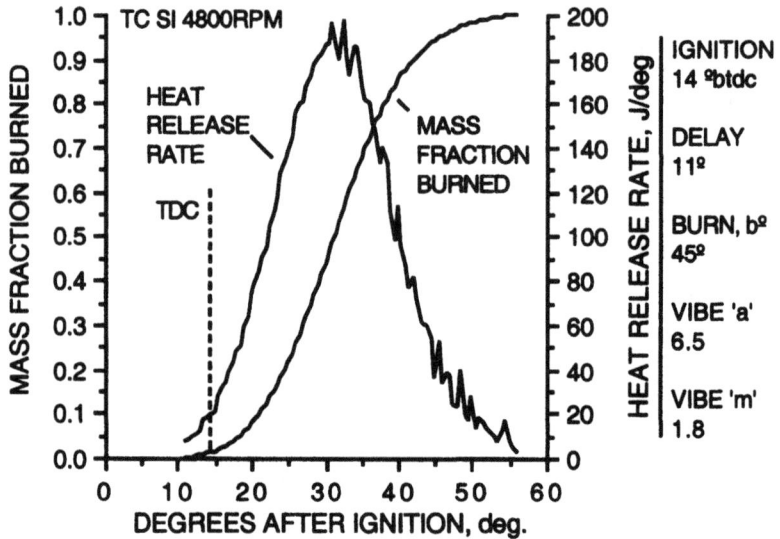

Fig. 4.15 Combustion analysis of a turbocharged spark-ignition engine at 4800 rpm.

444

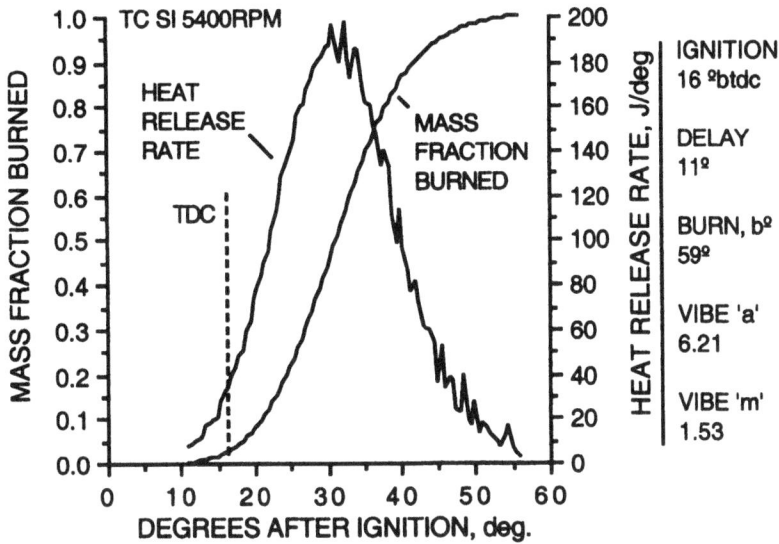

Fig. 4.16 Combustion analysis of a turbocharged spark-ignition engine at 5400 rpm.

It can be observed that the ignition timing is progressively advanced with increasing engine speed, but is considerably retarded when compared with the data for the naturally-aspirated engine in Figs. 4.7 to 4.10. The ignition delay increases somewhat with engine speed. The burn duration is relatively constant with increasing engine speed, and is longer than that for the naturally-aspirated engine. The location of the maximum rate of heat release is noticeably much later than for the naturally-aspirated engine.

If measured data are not available for such an engine, where design by simulation is undertaken, you may use these data with some confidence to characterize the combustion of gasoline in a turbocharged spark-ignition engine at full load over a speed range.

Combustion of Gasoline in Naturally-Aspirated and Turbocharged Spark-Ignition Engines

The set of figures consisting of Figs. 4.4 and Figs. 4.7 to 4.16 shows the combustion characteristics in naturally-aspirated and turbocharged spark-ignition engines. The engines are used in a production automobile from the same manufacturer. Both engines have four-valve, pent-roof cylinder heads, and have many common design features. The effect of turbocharging is to increase the delivery ratio and the mass of air and fuel trapped within the cylinder at the onset of combustion. The result is a major shift upward in the indicated mean effective pressure, imep, and power at any engine speed, produced by the turbocharged engine. The imep, assessed from the cylinder pressure diagrams, is shown in Fig. 4.17 over the engine speed range. The turbocharged (tc si) engine has imep levels between 22 and 24 bar, and the naturally-aspirated (na si) engine has imep levels from 11 to 14 bar over the operating speed range. This means that the power output at any engine speed is virtually doubled. To provide a doubling of the imep means that the work recorded by a cylinder pressure diagram will exhibit that effect. The cylinder diagrams for both engines at 3600 rpm are plotted in Fig. 4.18 to illustrate this point. The extra work produced by the turbocharged engine is quite visible on Fig. 4.18, and is numerically plotted in Fig. 4.17.

The combustion of a homogeneous charge of air and gasoline in a spark-ignition engine is always conducted at the highest possible compression ratio to obtain the maximum power and thermal efficiency (see Figs. 1.45 and 1.50) and the associated discussion. However, burning at the highest possible compression ratio means operating near the detonation limit—and doubling the mass of trapped charge by turbocharging merely exacerbates that problem. As a consequence, it comes as no surprise to find that the ignition timing of the turbocharged engine is retarded by about 6° at any engine speed, which reduces the potential for detonation to occur. The burn diagram for the turbocharged engine occurs later in the cycle, and this can be seen on all of the figures showing mass fraction burned and heat release rate. As a summary, Fig. 4.19 shows the location at which half of the fuel mass fraction has been burned, termed the "50% mass fraction burned" point. Because the cylinder sizes of the naturally-aspirated and the turbocharged engines are virtually identical, the values of heat release rate determined are in proportion to the imep levels developed as the air-fuel ratios at full load are comparable. In Fig. 4.4, the maximum heat release rate is 107 J/deg, and in Fig. 4.15 it is 198 J/deg for the turbocharged engine at the same engine speed of 4800 rpm.

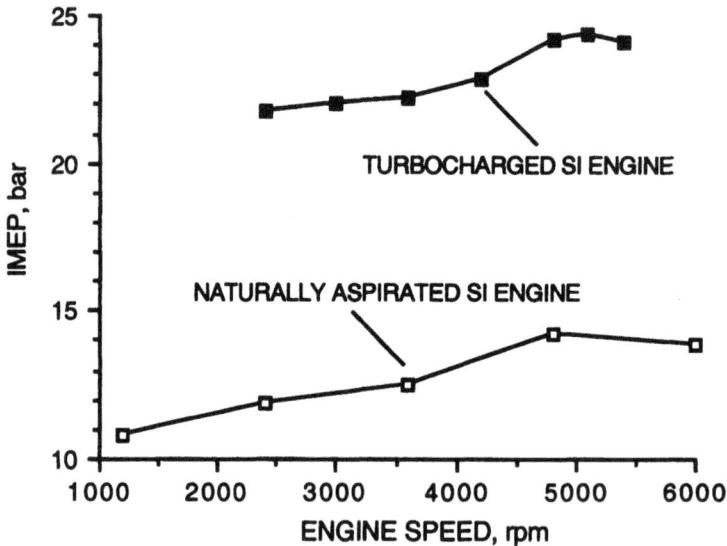

Fig. 4.17 Imep for naturally-aspirated and turbocharged spark-ignition engines.

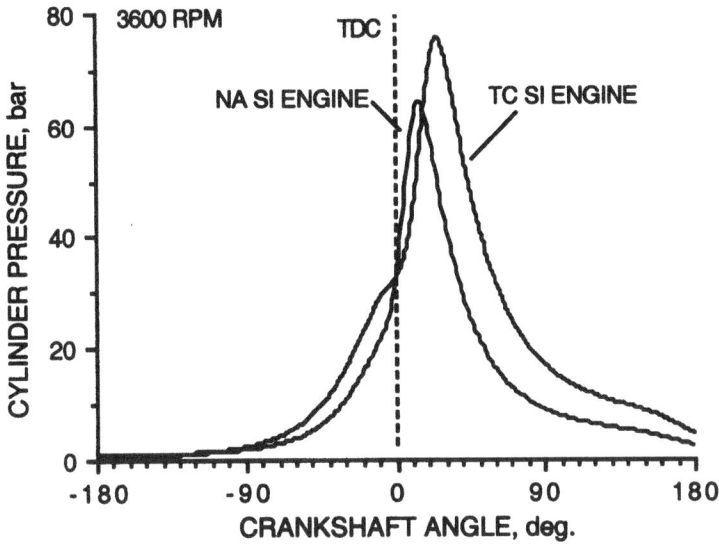

Fig. 4.18 Cylinder pressure of naturally-aspirated and turbocharged spark-ignition engines at 3600 rpm.

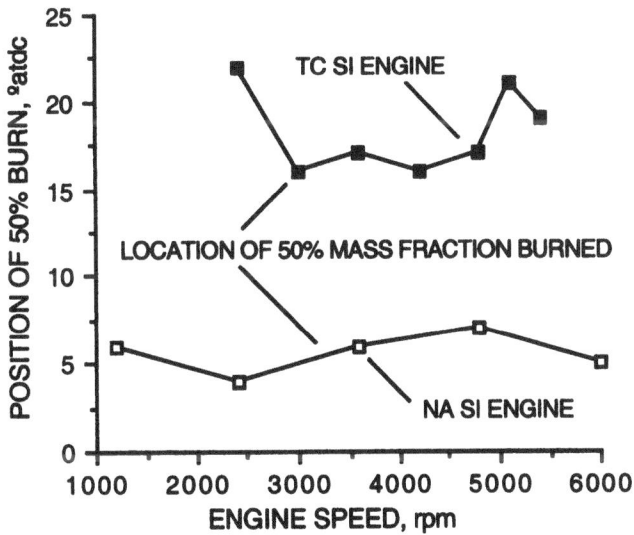

Fig. 4.19 Location of 50% mass fraction burned for naturally-aspirated and turbocharged spark-ignition engines.

It can be seen in Fig. 4.19 that, although the ignition timing of the turbocharged engine is retarded by about 6° crank angle, the location at which B_θ is 50% is retarded by some 10° or more. As observed in Fig. 4.18, this puts the peak cylinder pressure at a later position in the cycle. In the naturally-aspirated engine it is around 7 °atdc, and in the turbocharged engine it is about 17 °atdc, i.e., a 10° differential. In both cases—and virtually universally for spark-ignition combustion—the maximum rate of heat release rate occurs at the "50% mass fraction burned" point, and the heat release rate and mass fraction burned profiles are reasonably symmetrical in that their profiles are equally disposed around this mean location.

Combustion of Methanol in a Naturally-Aspirated Spark-Ignition Engine

Methanol is used widely in the racing of automobiles in the United States, but this is not common practice in the rest of the world. The alcohol family has the general chemical formula of $C_nH_{2n+1}OH$. Methanol, the first member of this alcohol family, has the chemical formula of CH_3OH. As reported in Table 4.2, methanol has a calorific value of 19.7 MJ/kg and an effective latent heat of vaporization of about 1000 kJ/kg. This means that it has about half the heat content and twice the latent heat of vaporization of gasoline. If Eq. 4.3.4 is solved for a stoichiometric case, the AFR is determined to be 6.43. Thus, for the stoichiometric combustion of methanol, compared to that of gasoline, some 2.25 times more fuel mass is required. To vaporize this in the engine means that about 4.5 times the heat energy is extracted from the air, cooling the gas on compression very considerably. The density of methanol is 790 kg/m³, which is not too dissimilar to a modern gasoline. However, at double the fuel mass flow rate in a racing engine with the same delivery ratio, the intake tracts and cylinder walls tend to be awash in liquid. Needless to add, this does nothing to improve the lubrication of such highly stressed components as the top piston ring. A further point to note is that all alcohols are hygroscopic, i.e., they absorb any water in their vicinity. This includes the water vapor in a humid atmosphere, so storage of alcohols in a vented fuel tank poses problems.

I will not report any combustion characteristics for it, however, for the sake of completeness, I will mention that the other common alcohol fuel is ethanol. The latter is also the alcohol fuel for the human engine, whereas methanol is toxic! Ethanol has been used widely for automobiles in Brazil, and is produced from biomass, e.g., sugar cane. Ethanol is the second member of the alcohol family (n is equal to 2) and has the chemical formula C_2H_5OH. As presented in Table 4.2, it has a lower calorific value of 26.8 MJ/kg and an effective latent heat of vaporization of about 900 kJ/kg. The stoichiometric AFR for ethanol is 8.95, and the density is the same as that of methanol at 790 kg/m³.

An approximate and relative computation of the maximum possible heat available during the stoichiometric combustion of ethanol, methanol and gasoline at equal delivery ratio shows:

$$\text{Ethanol heat available} = \frac{\text{calorific value}}{\text{stoichiometric AFR}} = \frac{26.8}{8.95} = 2.99 \text{ units}$$

$$\text{Methanol heat available} = \frac{\text{calorific value}}{\text{stoichiometric AFR}} = \frac{19.7}{6.43} = 3.06 \text{ units}$$

$$\text{Gasoline heat available} = \frac{\text{calorific value}}{\text{stoichiometric AFR}} = \frac{43.5}{14.7} = 2.96 \text{ units}$$

In short, these are within a few percentage points of each other with regard to heat available, with the alcohols having a slight advantage, and methanol releasing about 3% more heat than gasoline. The real advantage of methanol, in terms of producing power in spark-ignition racing engines, is that a much higher compression ratio can be used without inducing detonation. Hence, more work becomes available from the combustion process, e.g., consider Fig. 1.50 and Eq. A1.66. At an equal potential for detonation to occur, the compression ratio employed in such racing engines is around 10 for a gasoline engine and can be raised to about 15 for a methanol-burning engine. The potential relative power advantage of burning methanol is about 10% as shown by:

$$\frac{\text{methanol power}}{\text{gasoline power}} = \frac{1 - \dfrac{1}{CR^{\gamma-1}_{methanol}}}{1 - \dfrac{1}{CR^{\gamma-1}_{gasoline}}} = \frac{1 - \dfrac{1}{15^{0.4}}}{1 - \dfrac{1}{10^{0.4}}} = 1.1 \qquad (4.3.43)$$

Considering that you cannot drink methanol because it is toxic, it burns with an invisible flame so that a fire from a racing accident can be quite lethal, it provides no real power advantage over gasoline, and it requires at least double the fuel tank capacity to go the same racing miles as with gasoline, I have always totally failed to understand the American preoccupation with using methanol as a racing fuel. The concept of racing with ethanol I could at least stomach, making post-race celebrations worth attending! Casting all personal prejudice aside, the following describes how it burns.

Figs. 4.20 and 4.21 show the analysis of cylinder pressure diagrams taken from a 4.5-liter naturally-aspirated, spark-ignition racing engine at 7600 and 9200 rpm. The imep recorded from the cylinder pressure diagrams are 19.4 bar and 21.0 bar, respectively. This translates to indicated power, ihp, values of 740 and 970 hp (552 and 724 kW), respectively. The imep values are about 10% higher than could be expected from the same engine using gasoline at the appropriate compression ratio for that fuel, i.e., as predicted by Eq. 4.3.42. The profiles of the mass fraction burned and heat release rate for methanol, and their location with respect to top dead center, tdc, are very similar to those from the gasoline-burning, naturally-aspirated, spark-ignition engine. For methanol, the location of "50% mass fraction burned" is between 7 and 10 °atdc, i.e., much as for the naturally-aspirated gasoline engine. However, the ignition timing is quite advanced at 28 °btdc, with a long delay of some 15-17° before any heat is observed to be released. The data plotted in Figs. 4.20 and 4.21 indicate some irregularities in the mass fraction burned and heat release rate profiles, and it is not impossible that some of the copious quantities of liquid fuel are still vaporizing, and subsequently burning, during these latter stages of the process. The burn durations, considering the high-speed nature of this racing engine, are relatively short, and the Vibe function coefficients, a and m, are similar to those of the naturally-aspirated, spark-ignition engine burning gasoline.

General

The modeler should note that increasing the value of the Vibe coefficient, a, and decreasing the value of the coefficient m advances the rate of heat release at the onset of combustion. This point should be noted in connection with the discussion in Sec. 4.3.7 on compression-ignition engines, for this is precisely the same difference in profile of heat release rate in a diesel, compared to an Otto, engine. You may recall the discussion in Sec. 1.9. The nearer we can approach heat release at constant volume at the highest compression ratio, the higher will be the thermal efficiency and the power output.

In most situations encountered in the design by simulation of a new engine, by definition, the data on combustion rates are not available. The above data on combustion in spark-ignition engines are presented to enable you to successfully characterize virtually any combustion process within an engine simulation.

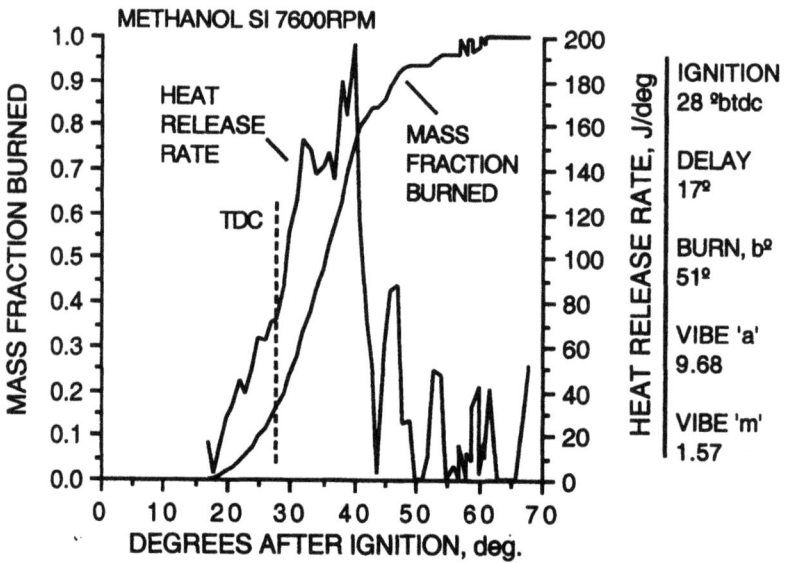

Fig. 4.20 Combustion of methanol in a naturally-aspirated spark-ignition engine at 7600 rpm.

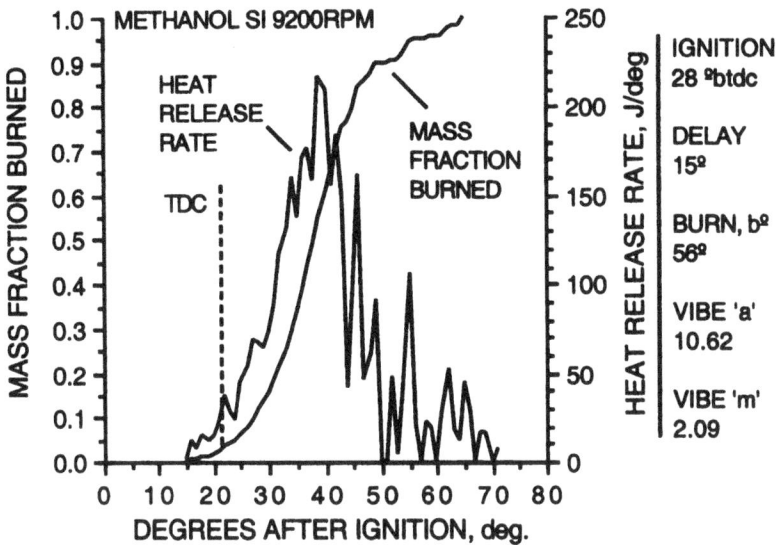

Fig. 4.21 Combustion of methanol in a naturally-aspirated spark-ignition engine at 9200 rpm.

4.3.7 Heat Release Data for Compression-Ignition Engines

In Sec. 4.1.6, it is pointed out that combustion by compression-ignition requires rapid heating of the droplets in the fuel spray, initially to vaporize the liquid, and then to promote the rise of the vapor temperature to the auto-ignition temperature. To accomplish this the in-cylinder air must be hotter than that state. The smaller the droplet—but more importantly the higher the relative velocity between fuel droplet and air—the more rapid will be the transfer of heat from air to fuel. There are two ways that this can be accomplished: (1) fast-moving fuel and slow-moving air, or (2) fast-moving air and slow-moving fuel.

The direct injection (DI) engine uses the fast-moving fuel approach by employing high-pressure injection into a bowl formed in the piston crown (see Fig. 4.1(b) or Fig. 1.7) to give the highest speed to the fuel droplets. Recent designs tend toward ever-higher injection line pressures, up to 2000 bar or more, to give more rapid motion to smaller fuel droplets. Because the air and fuel relative velocities are virtually fixed at any engine speed, there tends to be a modest maximum engine speed, typically 3000 rpm in automotive truck applications, where the cylinder size is typically about 1 liter, and some 4000 rpm in automobile applications, where the cylinder swept volume is normally around 500 cm^3. The combustion in DI engines is heavily influenced by in-cylinder and in-bowl swirl of the air with respect to the injected fuel spray(s) and droplets, and this swirl is created by suitably orientating the flow past the intake valves [4.38]. It should be noted that swirl creation in this manner reduces the discharge coefficients of the inflow process past the intake poppet valve(s); see Plate 3.3.

The indirect injection (IDI) engine uses the fast-moving air approach by employing compression-created rapid swirl in a side combustion chamber into which low-pressure fuel injection can be used to impart a moderate speed to the fuel droplets. The fuel injection line pressures are rarely higher than 250 bar, and this implies both a cheaper and a quieter injection system well suited to automobile applications. Because the swirling air flow (see Fig. 4.22) is compression-created, the relative velocity of air to fuel tends to rise linearly with engine speed, and so the high-speed limit for this engine is slightly greater than that of a DI engine. The limiting speed tends to be set by a combination of the characteristics of the mechanical/hydraulic fuel injection equipment and the combustion process. A more complete discussion of the compression-created swirl in an IDI engine is presented in Sec. 4.5.4.

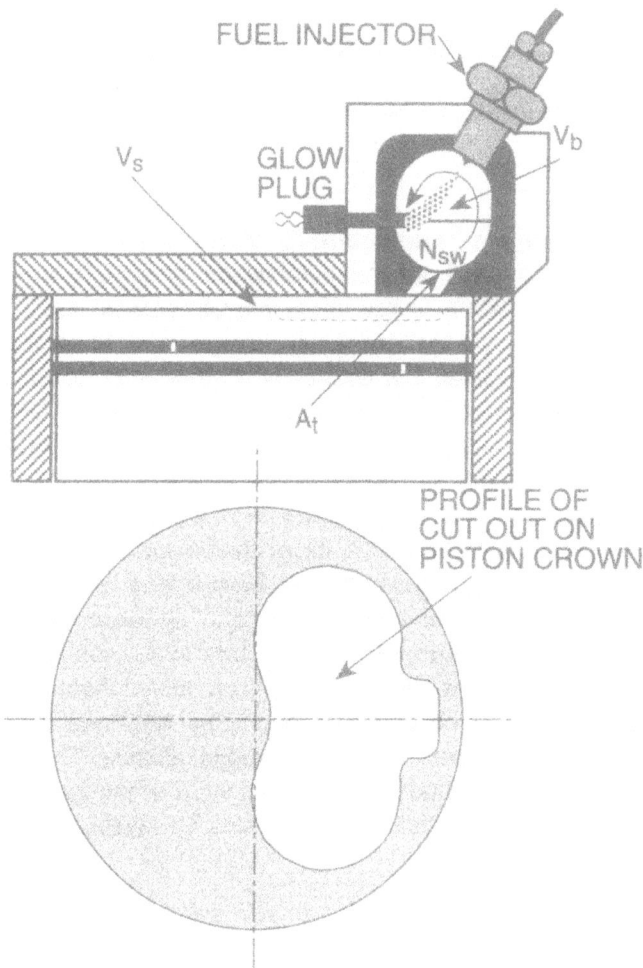

Fig. 4.22 Combustion chamber geometry of an IDI diesel engine.

The Direct Injection (DI) Diesel Engine

A sketch of a typical combustion chamber for a direct injection (DI) diesel engine is shown in Fig. 4.1(b) and Fig. 1.7. This shape is often referred to as a "Mexican hat," for obvious reasons. In practice, many other DI designs are employed, from the simple "open bowl" as shown in Fig. 4.23, to designs referred to in jargon as "squish lip," "re-entrant," "wall wetting," etc., [4.38]. A photograph of the combustion bowl in the piston of a direct-injection diesel engine is shown in Plate 4.1; this is best described as being one of the "re-entrant" types. The cut-outs seen on the piston crown provide adequate working clearances for the heads of the two poppet valves.

Combustion in direct injection diesel engines is characterized by rapid burning around the tdc position. A measured heat release rate and mass fraction burned profile from a turbo-charged DI engine are shown in Figs. 4.24 and 4.25, respectively. Apart from the high compression ratio, which in itself provides high thermal efficiency as seen in Sec. 1.9, the rapid burn around tdc approaches the ideal for combustion, which is a constant volume process. However, the rapid rates of pressure rise make it extremely noisy, hence the typical DI diesel "rattle." That the combustion process in a diesel engine is more rapid than in an Otto engine can be seen clearly by comparing the heat release rate profiles in Figs. 4.24 and 4.6 and the mass fraction burned curves in Figs. 4.25 and 4.5. In the DI diesel engine, some 40% of the fuel is burned by tdc, compared to about 20% in the Otto engine.

In Fig. 4.24 the sharper spike of heat release near the tdc position is known as "pre-mixed" combustion and the remainder as "diffusion" burning. The position of the peak of pre-mixed burning and the profile of the diffusion burning period, tend to be influenced by engine speed, as forecast in Sec. 4.1.6. With the necessity of meeting legislation on exhaust emissions, particularly those regarding the oxides of nitrogen (NO_x), which are very much a function of peak

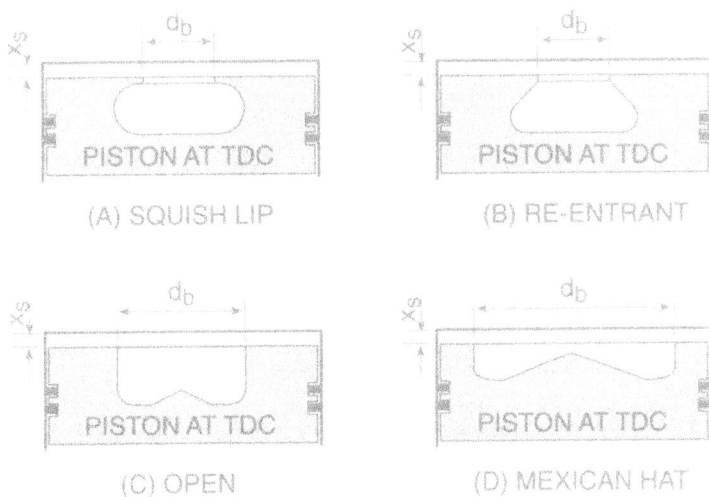

Fig. 4.23 Combustion chamber geometries of DI diesel engines.

453

cycle temperature, the pre-mixed burning phase in current diesel engines has been reduced to the point where it is often not observable. The theoretical fundamentals of creation of NO_x are discussed in Appendix A4.1.

In Figs. 4.24 and 4.25, the data for heat release rate and mass fraction burned are determined from the cylinder pressure diagram using the theory of Sec. 4.2.2, and fitted by the Vibe approach defined by Eq. 4.3.42. The experimental points are shown on these figures and the fitted Vibe exponential curve is shown as a solid line. The quality of the fit is seen to be good. Note that the x-axis of both graphs is plotted as being the crankshaft angle after the onset of the injection of the fuel, whereas in the equivalent diagrams for a spark-ignition engine, it is plotted as being the crankshaft angle after the ignition event.

To provide further information, Fig. 4.26 shows the measured heat release rate and mass fraction burned curves for a turbocharged DI truck diesel engine at 1800 rpm. The cylinder swept volume is 1 liter and the engine speed at which the cylinder pressure diagram is recorded at full load is 1800 rpm. The similarity to the data in Figs. 4.24 and 4.25 is obvious even though they are from a different engine. The details of the Vibe function, i.e., the data on the measured injection timing, and the recorded values for ignition delay, d°, the combustion duration, b°, and the Vibe exponential coefficients, a and m, are shown to the right of the diagram.

Plate 4.1 Combustion chamber in the piston of a 2v DI diesel engine.
(Courtesy of Perkins Engines)

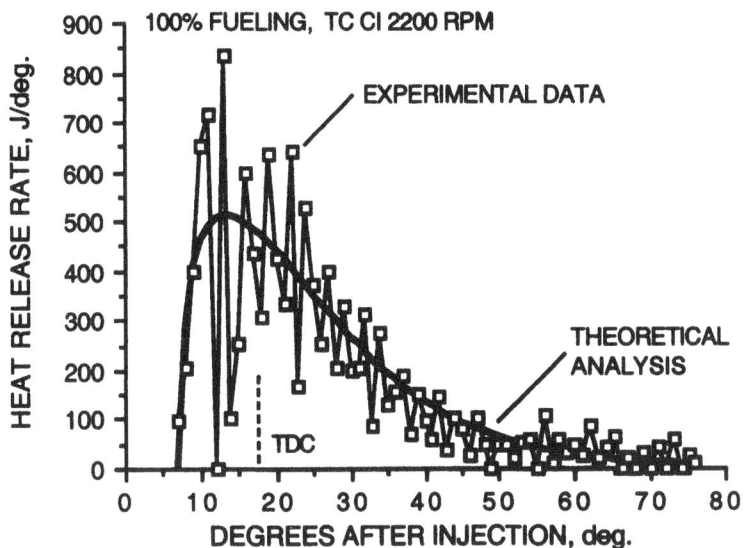

Fig. 4.24 Heat release rate in a turbocharged compression-ignition engine at 2200 rpm.

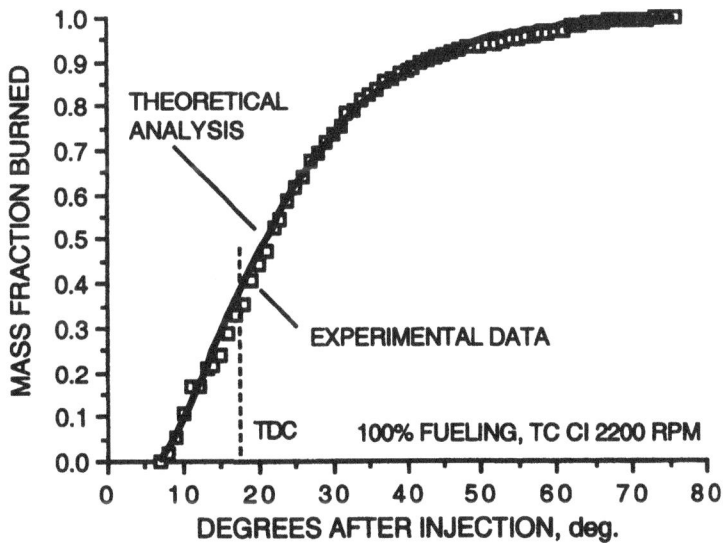

Fig. 4.25 Mass fraction burned in a turbocharged compression-ignition engine at 2200 rpm.

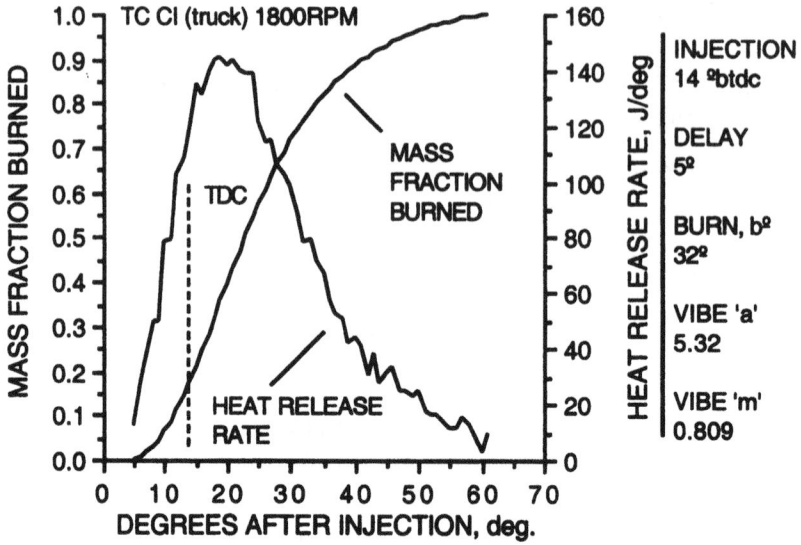

Fig. 4.26 Combustion analysis of a turbocharged truck DI diesel engine at 1800 rpm.

Effect on Combustion of Load Variation in the Direct Injection (DI) Diesel Engine

A diesel engine controls engine load by varying the quantity of fuel injected, or heat released, into each cylinder on each cycle. By contrast, the Otto engine does this by throttling the air and fuel intake. For both engines the outcome is a cylinder pressure diagram with a variable work content. Consider Fig. 4.27, which shows the cylinder pressure diagrams taken at the same engine speed of 2200 rpm in a turbocharged DI engine with a cylinder swept volume of about 1.7 liters. The several cylinder pressure records are taken at 80%, 60%, 40%, and 20% of the maximum fueling level, which is represented by the 100% diagram. The maximum cylinder pressure is about 200 bar at 100% fueling and is only 60 bar at 20% fueling. The variation in the work output, i.e., the imep, as a function of the injected fuel quantity is obvious. The relationship between equivalence ratio, λ, and fueling level for this engine is shown in Fig. 4.28. At maximum fueling, the equivalence ratio, λ, is as rich as 1.44, or an air-fuel ratio, AFR, of about 21, and at minimum fueling, λ is 3.325 and the AFR is about 49.

What is not immediately obvious from Fig. 4.27 is the reason for the variation in the p-V lines, or pressure-crank angle, p-θ lines as drawn, during compression. Because this is a turbocharged diesel engine, the energy content of the exhaust gas is reduced at the lesser fueling levels. This can be seen from the lower cylinder pressures at the exhaust valve opening point at some 60 °bbdc. With a reduced exhaust energy content entering the turbine, the compressor

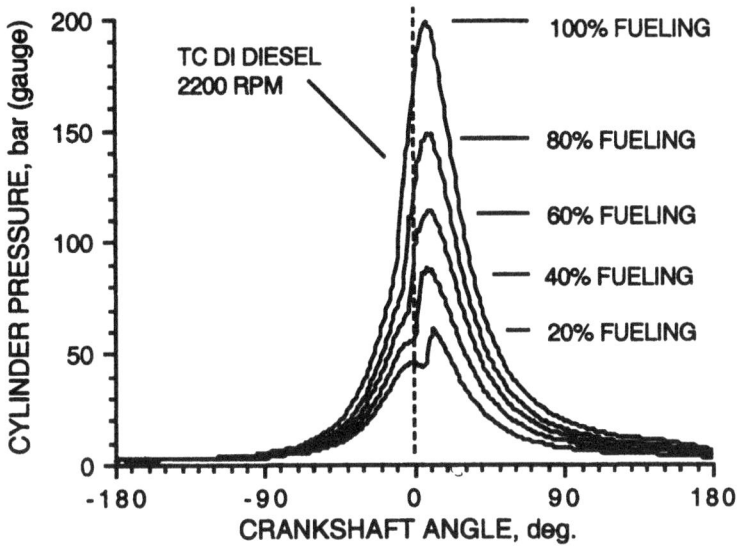

Fig. 4.27 Cylinder pressure diagrams for a DI diesel engine as a function of fueling.

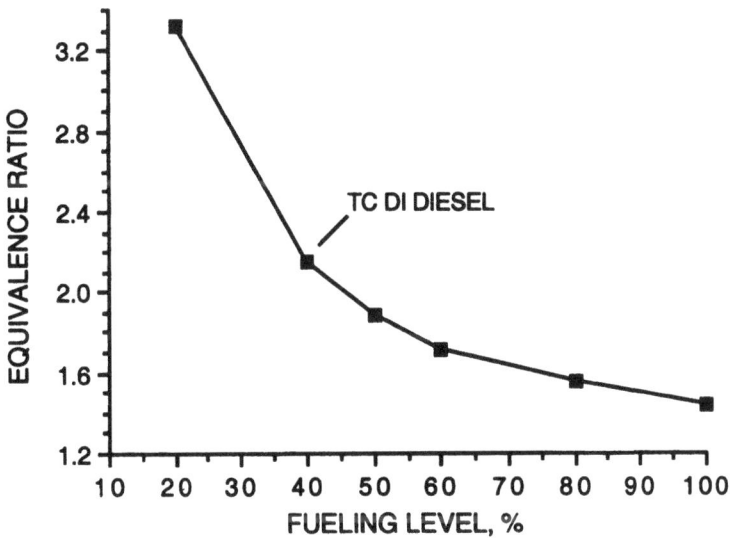

Fig. 4.28 Relationship between fueling and equivalence ratio for a DI diesel engine.

component of the turbocharger delivers less airflow at a lower pressure. The result is that the mass of air trapped within the cylinder varies with fueling, hence the observed differences in the p-θ lines during compression. Fig. 4.29, which records the measured variations in the imep and the air pressure exiting the compressor of the turbocharger, confirms this.

The indicated mean effective pressure, imep, is found from the several cylinder pressure diagrams in Fig. 4.27. That this is a high-specific-output engine can be seen from the maximum imep, which is at 27 bar; and that it is highly turbocharged can be observed from the peak blower pressure at 1.8 bar gauge or about 2.8 bar absolute. At the minimum fueling point of 20%, with an AFR of 49, the blower pressure is a mere 0.3 bar gauge, the imep is about 7 bar, and the engine would behave in a similar fashion to a naturally-aspirated engine, although a markedly different compression ratio would be used in that configuration.

To complete the set of information on the effect of fueling, Fig. 4.30 shows the measured brake specific fuel consumption (bsfc) and the emission of carbon particulates, as a Bosch smoke number [4.38, 4.57]. The minimum bsfc value is just under 200 g/kWh, which corresponds to a thermal efficiency approaching 42% and very much represents a state-of-the-art value. The Bosch smoke number is seen to approach 3 at the maximum fueling level, putting this in the "just visible" category.

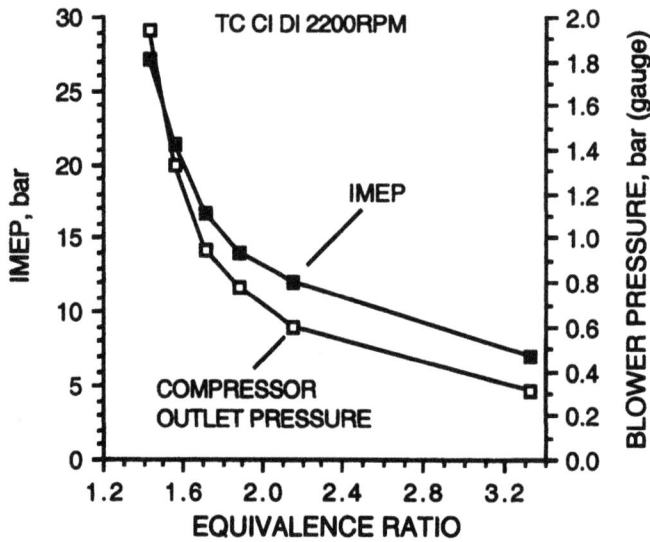

Fig. 4.29 Variation of imep and blower air pressure as a function of fueling.

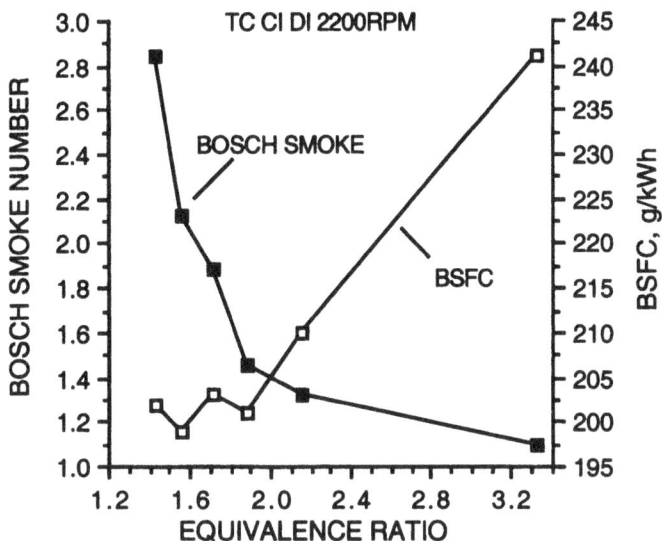

Fig. 4.30 Variation of bsfc and exhaust smoke as a function of fueling.

Finally, the mass fraction burned and heat release diagrams are shown in Figs. 4.31 to 4.36 at fueling levels of 20%, 40%, 50%, 60%, 80%, and 100%, respectively. Fig. 4.36 is actually plotted using the same data as Figs. 4.24 and 4.25. The details of the Vibe function, i.e., the data on the measured injection timing, and the recorded values for ignition delay, d°, combustion duration, b°, and the Vibe exponential coefficients, a and m, are shown to the right of each diagram. The characteristic that first catches the eye in Figs. 4.31 to 4.36 is that the combustion process becomes ever more retarded as the fueling level is increased. At the 20% fueling level it is 9° atdc before the first fuel is recorded as being burned. At the 100% fueling level it is at 11 °btdc, a difference of 18° over the fueling range. Most of the Vibe "a" coefficients are about 5 and most of the Vibe "m" coefficients are between 0.2 and 0.3. The exception to this is at the 20% fueling level where a and m are 19.6 and 2.45, respectively. The other noticeable characteristic is that the burn period varies almost linearly with the fueling level; this is shown in Fig. 4.37.

The burn period at maximum fueling is 69° but is only 31° at the 20% fueling level. The location of 50% mass fraction burned is at 3 °atdc at 100% fueling, and is at 15 °atdc at minimum fueling. The effect of the retarded combustion with increased fueling is evident in the collected cylinder pressure diagrams in Fig. 4.27, as the peak cylinder pressure occurs ever later. The ignition delay, d°, seems almost unaffected by fueling, as it remains between 7° and 9° throughout the test series.

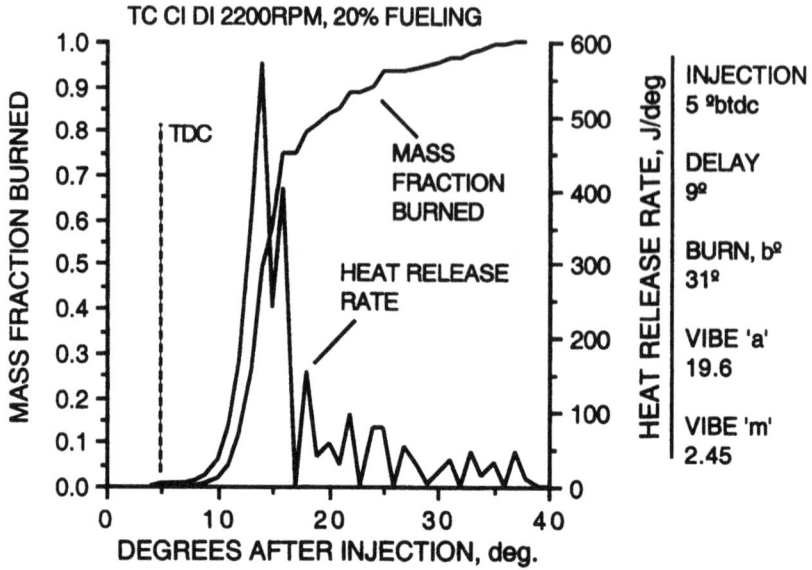

Fig. 4.31 Combustion analysis of a turbocharged DI diesel engine at 20% maximum load.

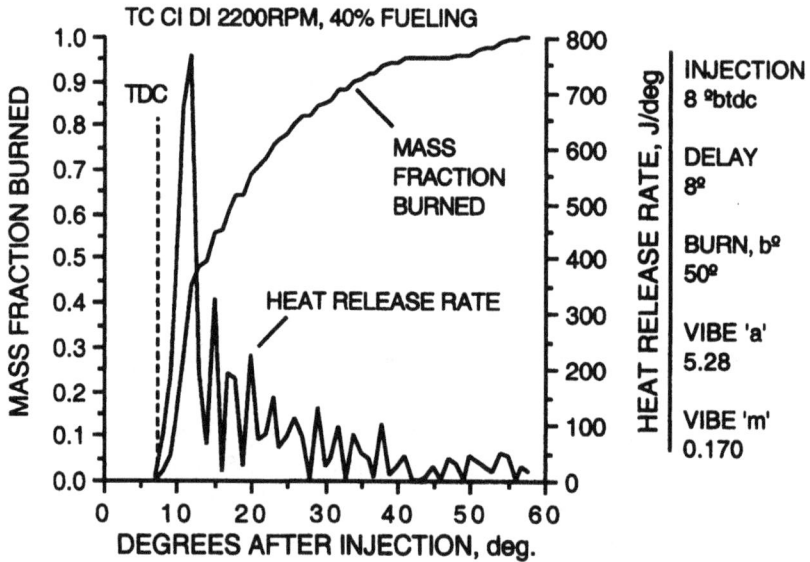

Fig. 4.32 Combustion analysis of a turbocharged DI diesel engine at 40% maximum load.

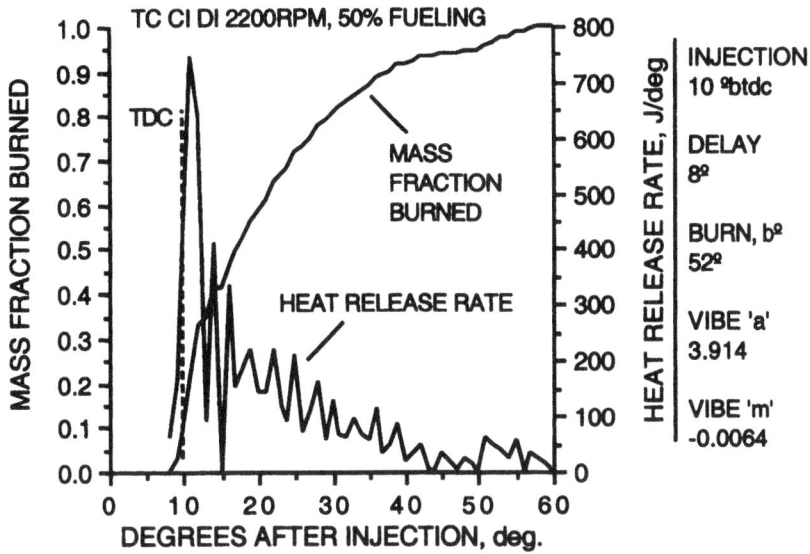

Fig. 4.33 Combustion analysis of a turbocharged DI diesel engine at 50% maximum load.

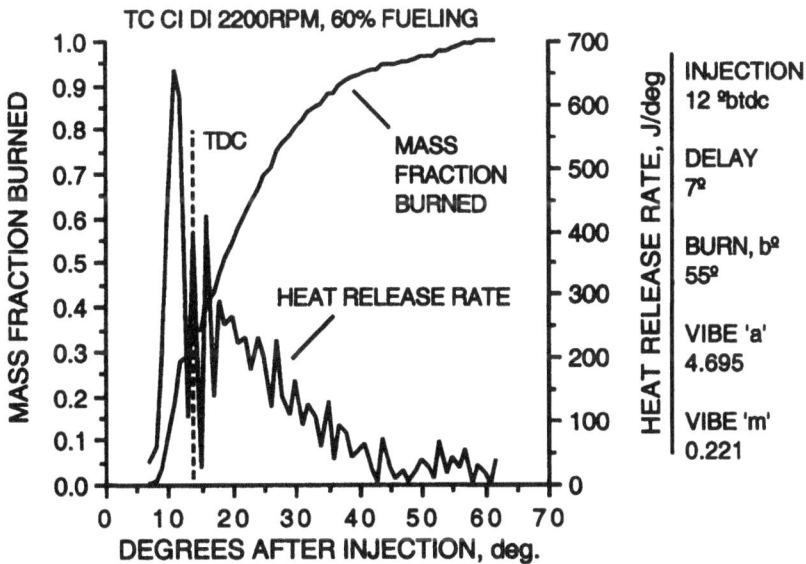

Fig. 4.34 Combustion analysis of a turbocharged DI diesel engine at 60% maximum load.

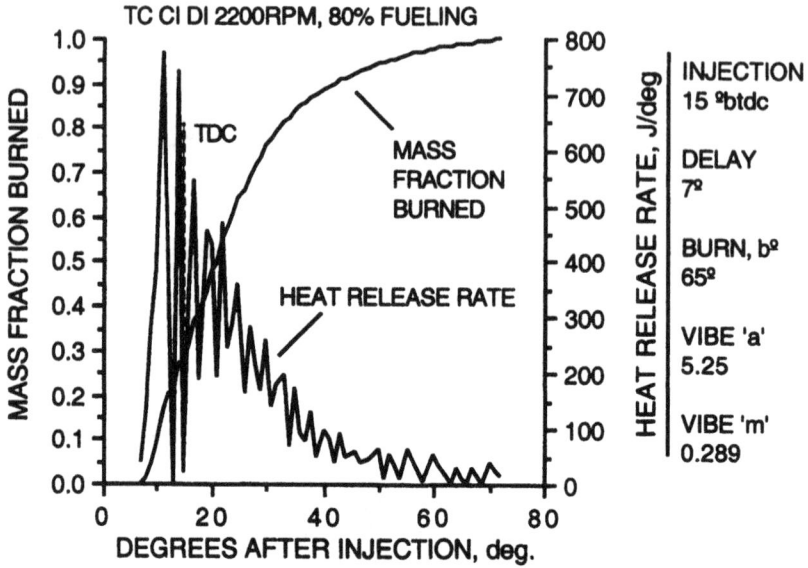

Fig. 4.35 Combustion analysis of a turbocharged DI diesel engine at 80% maximum load.

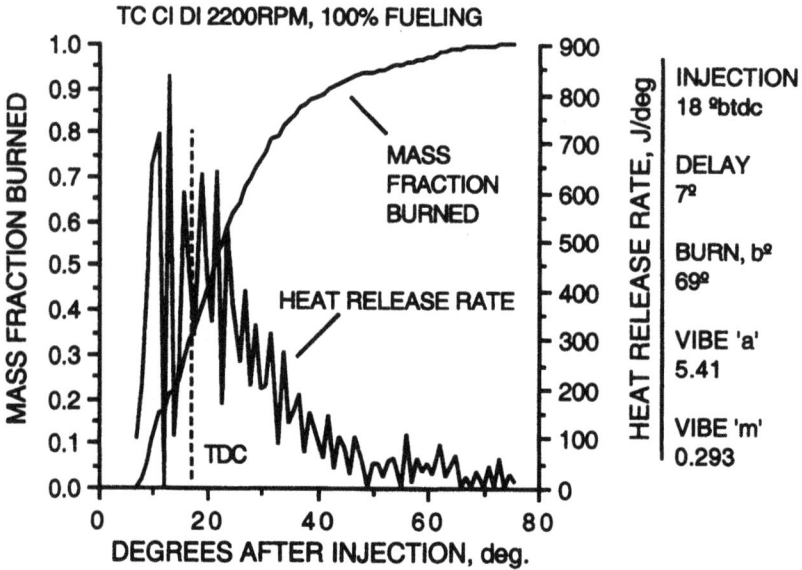

Fig. 4.36 Combustion analysis of a turbocharged DI diesel engine at maximum load.

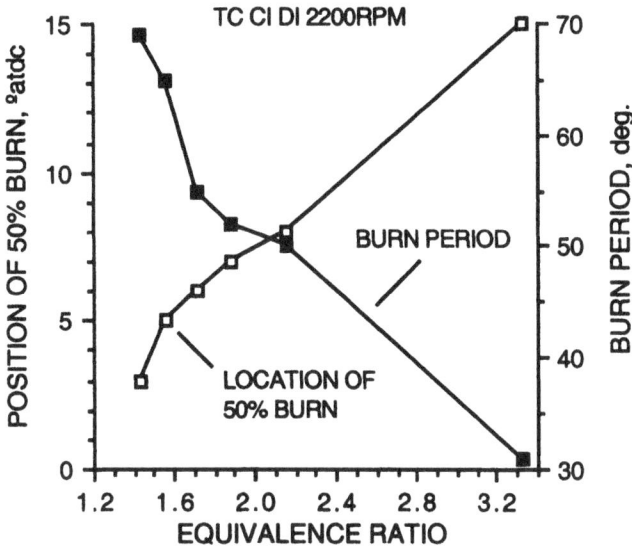

Fig. 4.37 Variation of the location of 50% mass fraction burned and the burn period as a function of fueling.

In the absence of measured data for an engine, where the engine design is conducted by simulation, the data shown to the right of Figs. 4.26 and 4.31 to 4.36 may be used with some confidence to characterize the combustion of varying amounts of diesel fuel in a turbocharged compression-ignition engine.

The Indirect Injection (IDI) Diesel Engine
A sketch of a typical combustion chamber for an indirect injection diesel engine is shown in Fig. 4.22, and in Fig. 1.8, which shows the disposition of the engine with respect to the valves. This shape, and particularly the cut-out on the piston crown, is often referred to as a Ricardo Comet design. In practice, many other IDI designs are employed [4.38].

Combustion in indirect injection diesel engines is characterized by less rapid burning around the tdc position, compared to that in engines of the DI design. To illustrate this, measured heat release rate and mass fraction burned curves are shown in Fig. 4.38. These are taken in an IDI automobile engine at full load at 4400 rpm. Comparing Fig. 4.26 for the DI engine with Fig. 4.38 for the IDI engine, the slower combustion in the IDI engine can be seen clearly. Although the Vibe exponents are very similar, the burn for the IDI engine takes 79° compared to just 56° for the DI. The peak of the heat release rate curve for the IDI engine is at 14 °atdc compared to 7 °atdc for the DI engine. Although the initial burn is very similar in both cases, the throat between the two segments of the clearance volume delays the flow from the inner, higher-pressure region as the piston moves away on the power stroke. Note that the heat release rate for the IDI engine rises sharply just before tdc, and nearly as sharply as in the DI engine in Fig. 4.26, and both are much more rapid than in a spark-ignition engine (as exemplified by Fig. 4.4). Nevertheless, the mass fraction burned by tdc is roughly similar for all of these engines, i.e., about 0.15—an interesting feature.

By definition, the deviation of IDI combustion from the approximation to constant volume combustion provided by the DI engine, is less efficient and in practice leads to a drop of some 10-15% in the thermal efficiency of the IDI unit. However, the IDI unit is somewhat less "noisy" as the rates of pressure rise within the cylinder are reduced. A further loss of efficiency occurs due to the air pumping loss engendered by the piston motion into, and during the power stroke from, the combustion chamber where the fuel is injected. Due to this pressure drop, and hence temperature drop, into the combustion bowl during compression, it is necessary to employ a glow plug (heater) to assist starting the engine.

You may well ask, "What is the purpose of this engine if its thermal efficiency is so compromised?" The answer lies in the high-speed swirl created through the throat of area A_t into the side chamber (see Fig. 4.22). The rotational speed of swirl, N_{sw}, is designed to be some 20 to 25 times the rotational speed of the engine, and is ensured by designing the area of the throat, A_t, to be some 1.0 to 1.5% of the piston area, A_p. At top dead center the disposition of the clearance volume, V_{cv}, into the volume within the bowl, V_b, and the volume above the piston, V_s, is about equal, i.e., V_s roughly equals V_b. The theory behind this matter will be discussed in much greater detail in Sec. 4.5.4.

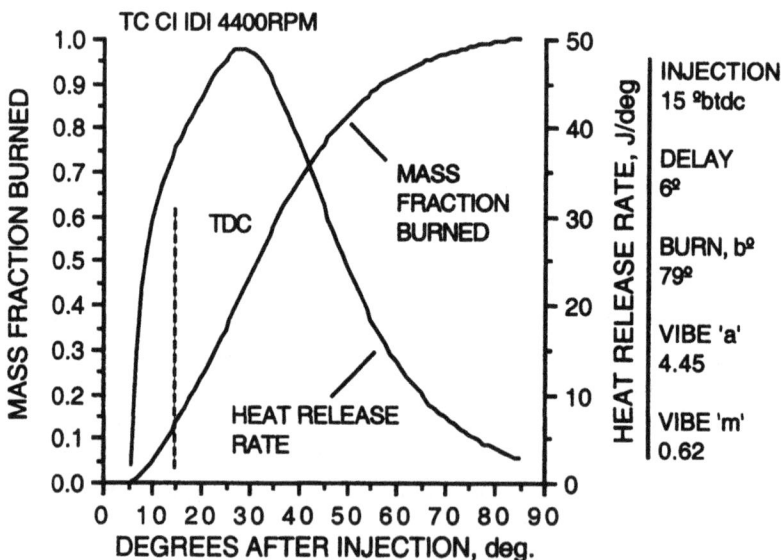

Fig. 4.38 Combustion analysis of a turbocharged IDI diesel engine at 4400 rpm.

The compression ratio employed in the IDI engine is higher than in the DI engine, due to the air temperature drop through the throat into the combustion bowl where the fuel is sprayed. A four-stroke IDI diesel engine has a geometric compression ratio, CR, between 20 and 22 depending on the air supercharge (or turbocharge) pressure level. The purpose of the shape of the cut-out on the piston crown is to assist in the mixing of the burning charge emanating from the combustion bowl with the rest of the air above the piston. In effect, combustion in the bowl takes place at air-fuel ratios approaching stoichiometric, and then "torches" through the throat into the main chamber to continue as "lean" combustion.

The result is a high-speed engine with a high specific output, that provides less noise and less visible black smoke (soot) than the DI unit, and that can run in automobile applications to engine speeds in excess of 4500 rpm. Furthermore, because these features lead to a lesser combustion pressure and temperature rise around the tdc point, and a burn in the inner chamber that occurs in a much richer mixture, the IDI engine has much lower NO_x emissions than either the DI diesel engine or many a spark-ignition unit.

As with much of the previous data presented, when an engine is designed by simulation without access to known burn-rate data, the information shown at the right of Fig. 4.38 can be confidently used to characterize the combustion process within an IDI compression-ignition engine.

4.4 Theoretical Modeling the Closed Cycle

For the engine designer who wishes to predict by simulation the performance characteristics of the engine, it is clear that the fundamental theory already given in Chapter 1 on the thermodynamics of engine cycles, in Chapter 2 on applied thermodynamics and unsteady gas flow, and in Chapter 3 on discharge coefficients permits the formulation of a computer model that will calculate the thermodynamic and gas-dynamic behavior of an engine at any throttle opening and any rotational speed. Much of this theory pertains to the open cycle when either, or both, the exhaust or intake valve(s) are open. The remaining portion of the required computer analysis must cover the closed cycle period when the combustion phase takes place, and this requires more detailed thermodynamics than is presented in Chapter 1. However, it is pertinent in this chapter on combustion to point out the various possibilities that exist for formulating such a model. There are four main methods possible. Each of these will be discussed in turn:

(a) A simple heat release or mass fraction burned model.

(b) A more complex heat release or mass fraction burned model.

(c) A one-dimensional flame propagation model for homogeneous combustion in spark ignition engines which should include fuel droplet vaporization for stratified and homogeneous combustion in direct injection (GDI) spark-ignition engines and in all compression-ignition engines.

(d) As in (c), but a three-dimensional combustion model for spark-ignition and compression-ignition engines.

465

4.4.1 The Simplest Closed Cycle Model for an Engine Simulation

One of the simplest models of engine combustion is formulated through the use of the heat release analysis presented in Sec. 4.2.2, but in reverse. For example, Eq. 4.2.12 would become a vehicle for the prediction of the pressure, p_2, at the conclusion of any incremental step in crankshaft angle, if all of the other parameters were known as input data. Equally, a Vibe function for the mass fraction burned, with appropriate exponential coefficients, as in Eq. 4.3.24, may be replayed as an alternative to a heat release rate function. This relies on the assumption of data values for such factors as the combustion efficiency and the polytropic exponents during compression and expansion. At the same time, a profile must be assumed for the heat release during the combustion phase as well as a delay period before that heat release commences.

At The Queen's University of Belfast, heat release models have been used for many years, beginning with the original publication on the subject [4.39], and then some seventeen years later [4.10]. That this is found to be reasonably effective is evident from the accuracy of correlation between measurement and calculation of engine performance characteristics described in those publications.

The heat release characteristics of homogeneously charged spark-ignition engines have been found to be quite similar, which is not too surprising, as virtually all of the two-stroke engines seen in the QUB publications have somewhat similar combustion chambers. This is not the case for four-stroke engines, where the combustion chamber shape is mainly dictated by the poppet valve mechanisms involved, and of which the side-valve engine is a very special case.

One of the simplifying assumptions in the use of a simple heat release model for combustion is that the prediction of the heat loss to the cylinder walls and coolant is encapsulated in the selection of the polytropic exponents for the compression and expansion processes. This will be regarded by some as an unnecessary and potentially inaccurate simplification. However, because all models of heat transfer are more or less based on empirical forms, the use of experimentally determined polytropic exponents could actually be regarded as a more realistic assumption. This is particularly true for the air-cooled two-stroke engine, of which so many engine types are similar in construction. For four-stroke engines, the polytropic exponents are not universally constant. The key to success is to have a complete map of the polytropic indices of compression and expansion covering the entire speed and load range of any engine. Unfortunately, such information rarely exists, and much of that which is published is very contradictory.

The modeler who wishes to use this simple approach, in the absence of better experimental evidence from a particular engine, could have some confidence in using polytropic indices of 1.25 and 1.35 for the compression and expansion phases, respectively, and 1.2 for the numerical value of the ratio of specific heats, γ, in Eq. 4.2.12, together with a value between 0.85 and 0.9 for the overall combustion efficiency, h_c.

4.4.2 A Closed Cycle Model within Engine Simulations

A slightly more complex approach, but one that is much more complete and of greater accuracy, is to use all of the theory provided above on heat transfer, fuel vaporization, heat release rates, or mass fraction burned behavior, and solve the First Law of Thermodynamics as expressed earlier in Eq. 4.2.8 at every step in a computation, but necessarily extended to include vaporization of fuel in the manner described in Sec. 4.3.5.

$$\delta Q_R - \delta Q_L - \delta Q_{vap} = \frac{p_2 V_2 - p_1 V_1}{\gamma - 1} + \frac{p_1 + p_2}{2}(V_2 - V_1)$$

All of the physical geometry of the engine will be known, as will physical parameters such as all surface areas and their temperatures. The open cycle model that is used, for example, that described in Chapter 2, must give the initial masses, purities and state conditions of the cylinder contents at the onset of the closed cycle. The air-fuel ratio will be known, so that either the mass of fuel to be injected, or to be vaporized during compression, can be computed. A heat release rate, or mass fraction burned, profile must be assumed so that at any juncture during combustion the heat to be released into the combustion chamber can be determined. The heat transfer to or from the cylinder walls can be found at any juncture using the Annand model of Sec. 4.3.4.

The theory of Sec. 2.1.6, at any point thereafter in the closed cycle, will yield the properties of the cylinder gas at any instant, during compression and before combustion, during combustion, and in the final expansion process prior to release at exhaust valve or port opening; recall that sleeve valve engines use ports and not valves! The closed cycle commences at intake valve closure on the compression stroke, i.e., the trapping point.

The computation time step is dt, corresponding to a crankshaft angle, $d\theta$. The notation as used in Sec. 4.2.2 is reused, and Fig. 4.3 is still applicable. The crankshaft is at θ_1 degrees and moves to θ_2. All properties and values at state position 1 are known, and the new cylinder volume, V_2, is known from its geometry. Using the theory of Sec. 2.1.6, it is assumed that the gas properties at state position 1 will persist for the small time step, dt, as a function of the temperature, T_1, and purity, Π_1, at the beginning of the time step. These will be updated at the end of that time step as a function of the temperature, T_2, and purity, Π_2, to become those at the commencement of the next time step.

Let us deal with each phase of the closed cycle in turn to acquire each of the numerical values of the terms of Eq. 4.2.8. The basic solution of Eq. 4.2.8 is for the new pressure, p_2:

From Sec. 2.1.3

$$G_6 = \frac{\gamma + 1}{\gamma - 1}$$

then
$$p_2 = \frac{2(\delta Q_R - \delta Q_L - \delta Q_{vap}) + p_1(G_6 V_1 - V_2)}{G_6 V_2 - V_1} \qquad (4.4.1)$$

Consequently, from the state equation:

$$T_2 = \frac{p_2 V_2}{m_2 R}$$
(4.4.2)

and from mass continuity, if a fuel mass, dm, is added by vaporization:

$$m_2 = m_1 + dm$$
(4.4.3)

The heat transfer term, δQ_L, is determined from Sec. 4.3.4.

Compression in Spark-Ignition Engines (Eqs. 4.3.36 to 38)

$$dm = \dot{m}_{vap} d\theta \quad \delta Q_{vap} = \dot{m}_{vap} h_{vap} d\theta \quad \delta Q_R = 0$$

Compression in Compression-Ignition Engines

$$dm = 0 \quad \delta Q_{vap} = 0 \quad \delta Q_R = 0$$

Combustion in Spark-Ignition Engines (Eqs. 4.3.23 to 27)

$$dm = 0 \quad \delta Q_{vap} = 0 \quad \delta Q_R = \eta_c \frac{\dot{Q}_{R\theta_1} + \dot{Q}_{R\theta_2}}{2} d\theta$$

Combustion in Compression-Ignition Engines (Eqs. 4.3.23 to 41)

$$dm = dm_{b\theta} = dm_{vap} \quad \delta Q_{vap} = dm_{vap} h_{vap} \quad \delta Q_R = \eta_c \frac{\dot{Q}_{R\theta_1} + \dot{Q}_{R\theta_2}}{2} d\theta$$

Expansion in Spark-Ignition and Compression-Ignition Engines

$$dm = 0 \quad \delta Q_{vap} = 0 \quad \delta Q_R = 0$$

In the above equations where the heat release rates at the beginning and end of the time steps, $\dot{Q}_{R\theta_1}$ and $\dot{Q}_{R\theta_2}$, are used to acquire the heat release term, this may be replaced by the equivalent expressions using the mass fraction burned approach for the fuel quantity consumed in the time period. Using the same symbolism as before, where q is the crankshaft angle from the onset of burn, the new expression is:

$$\delta Q_R = \eta_c \left(B_{\theta + d\theta} - B_\theta \right) m_{tf} C_{fl}$$
(4.4.4)

Gas Purity throughout the Closed Cycle

It is essential to trace the gas purity for, together with temperature, this provides the essential information on the gas properties at any instant. At the commencement of the closed cycle, at trapping, the scavenging efficiency is SE_t and the purity is Π_t. The latter is composed of the purity of the air, Π_a, and the exhaust gas, Π_{ex}. It should be reemphasized that the absolute charge purity of exhaust gas is not necessarily zero for, if it is lean combustion in a spark-ignition engine, or combustion in any diesel engine, the exhaust gas will contain oxygen. (Sec. 1.6.4 covers this point.) Therefore the relative charge purity, Π_{ex}, of exhaust gas is given by:

$$\text{if } \lambda < 1 \qquad \Pi_{ex} = 0 \qquad\qquad (4.4.5)$$

and
$$\text{if } \lambda \geq 1 \qquad \Pi_{ex} = \frac{\lambda - 1}{\lambda} \qquad\qquad (4.4.6)$$

The actual constituents of the gas mixture are dictated by the theory in Sec. 4.3.2, and are fed into the fundamental theory in Sec. 2.1.6 to calculate the actual gas properties of the ensuing mixture.

During compression, the gas purity is constant and is dictated by:

$$\Pi_t = SE_t \Pi_a + (1 - SE_t)\Pi_{ex} \qquad\qquad (4.4.7)$$

During combustion, the gas purity at an interval θ from the onset of burn is dictated by:

$$\Pi_\theta = (1 - B_\theta)\Pi_t + B_\theta \Pi_{ex} \qquad\qquad (4.4.8)$$

During expansion, the gas purity is constant and is Π_{ex}.

Some may wonder why the purity of air, Π_a, is not simply expressed as unity in the above equations. If the engine is running with EGR, then the purity of the incoming air is no longer unity, which must be taken into account by the simulation.

A More Accurate Combustion Model in Two Zones

The theory presented here shows a single-zone combustion model. A simple extension to burning in two zones is given in Appendix A4.2. Arguably, what is presented there is merely a more accurate single-zone model. This is not accidental, because the computation of any combustion process that is based on heat release data (from a heat release analysis such as that by Rassweiler and Withrow), or on a mass fraction burned curve (in the Vibe fashion), must theoretically replay that approach within a simulation in precisely the same manner as the data was experimentally gathered. The experimental data are referred to, and analyzed with reference to, a single zone, i.e., the entire combustion chamber. Replaying the data in a computer simulation in any other way makes the end result totally inaccurate—or "theoretically meaningless," using a better choice of words to describe the lack of mathematical logic. Today, all GPB engine simulation software uses a two-zone burn model with an emissions computation included, as shown in Appendix A4.2.

4.4.3 A One-Dimensional Model of Flame Propagation in Spark-Ignition Engines

One of the simplest one-dimensional models of flame propagation in spark-ignition engines was proposed by Blizard and Keck [4.2], and is of the eddy entrainment type. This model was used by Douglas and Reid [4.28-4.30] at QUB. In essence, the procedure used in this model is to predict the mass fraction burned curves as seen throughout this chapter, and then apply equilibrium and dissociation thermodynamics to the in-cylinder process.

The model is based on the propagation of a flame as shown in Fig. 4.1, and as already discussed in Sec. 4.1.1. The model assumes that the flame front entrains the cylinder mass at a velocity that is controlled by the in-cylinder turbulence. The mass is entrained at a rate controlled by the flame speed, c_{fl}, which is a function of both the laminar flame speed, c_{lf}, and the turbulence velocity, c_{trb}.

The assumptions made in this model are:

(a) The flame velocity is the sum of the laminar and turbulence velocities.

(b) The flame forms a portion of a sphere centered on the spark plug.

(c) The thermodynamic state of the unburned mass that has been entrained is identical to that of fresh charge that is not yet entrained.

(d) The heat loss from the combustion chamber is to be predicted by convection and radiation heat transfer equations based on the relative surface areas and thermodynamic states of burned and unburned gases. There is no heat transfer between the two zones.

(e) The mass fraction of entrained gas burned at any given time after its entrainment is to be estimated by an exponential relationship.

Clearly, a principal contributor to the turbulence present is squish velocity or tumble, which is discussed further in Sec. 4.5. The theoretical procedure progresses by the use of complex empirical equations for the various values of laminar and turbulent flame speed, all of which are determined from fundamental experiments in engines or combustion bombs [4.1, 4.4, 4.5].

It is clear from this brief description of the turbulent flame propagation model that it is much more complex than the heat release model posed in Secs. 4.4.1 and 4.4.2. Because the physical geometry of the clearance volume must be specified precisely, and all of the chemistry of the reaction process followed, the calculation requires considerably more computer time. By using this theoretical approach, the use of empirically determined coefficients, particularly for factors relating to turbulence, has increased greatly over my initial proposal of a simple heat release model to simulate the combustion process. Due to the need for such estimated coefficients it is somewhat questionable if the overall accuracy of the calculation has been greatly improved, although the results presented by Douglas and Reid [4.28-4.30] are impressive. There is no doubt that valuable understanding is gained, in that the user obtains from the computer calculation data on such important factors as exhaust gas emissions and flame duration. However, this type of calculation is probably more logical when applied in three-dimensional form and allied to a more general CFD calculation for the gas behavior throughout the cylinder leading up to the point of ignition. This is discussed briefly in the next section.

4.4.4 Three-Dimensional Combustion Models for Engines

From the previous comments it is clearly necessary that reliance on empirically determined factors for heat transfer and turbulence behavior, which refer to the combustion chamber as a whole, will have to be exchanged for a more microscopic examination of the entire system if the accuracy of the calculation is to be improved. This is possible through the use of a combustion model in conjunction with a computational fluid dynamics model of the gas flow behavior within the chamber. Computational fluid dynamics, or CFD, was presented to you in Chapter 3, where it was shown to be a powerful tool to aid in the understanding of gas flow into a pipe and past valves in the cylinder head.

That the technology is moving toward providing the microscopic in-cylinder gas-dynamic and thermodynamic information is seen in many publications, such as in the References [4.41-4.43]. The further references in those publications will form a reading list for those who wish to pursue more study of this subject. Fig. 4.39(a) shows the in-cylinder velocity vectors in an open-chamber DI diesel engine toward the conclusion of the compression stroke; however, this also represents all of the thermodynamic properties of the charge at all points just before

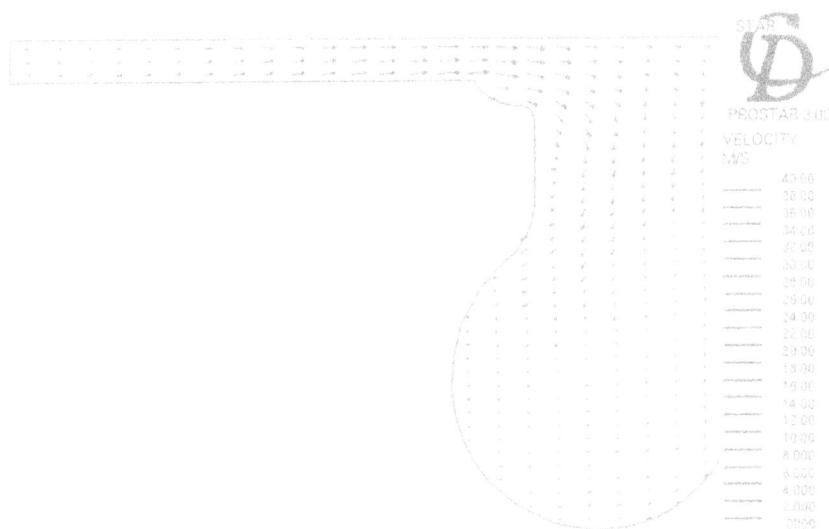

Fig. 4.39(a) An elevation view obtained by a CFD computation of squish flow in a DI diesel engine. (Courtesy of Adapco)

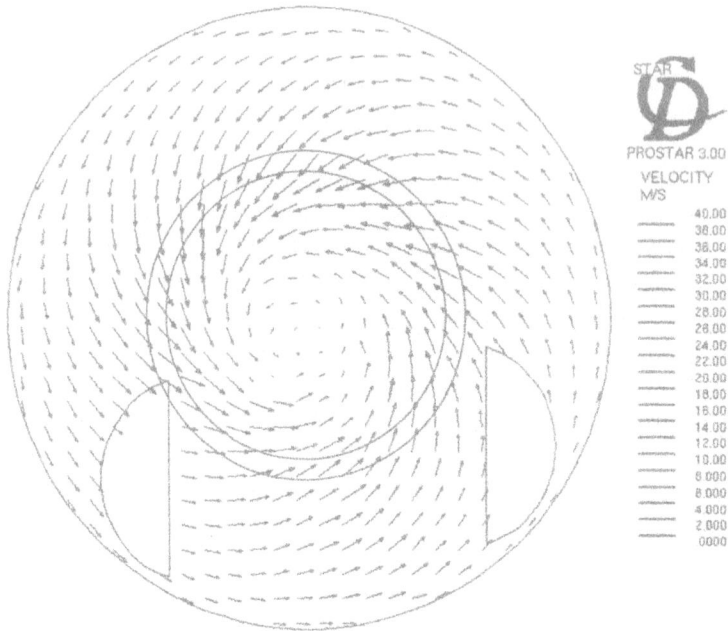

Fig. 4.39(b) A plan view obtained by a CFD computation of squish flow in a DI diesel engine. (Courtesy of Adapco)

ignition. This means that the prediction of heat transfer effects at each time step in the calculation will take place at the individual calculation mesh level, rather than by empiricism for the chamber as a whole, as in the theory in preceding sections. For example, should any one surface or side of the combustion bowl be hotter than another, the computation will predict the heat transfer in this microscopic manner giving new values and directions for the motion of the cylinder charge. This will affect the resulting fuel vaporization and combustion behavior.

Such a calculation can be extended to include the chemistry of the subsequent combustion process. Examples of this have been published by Amsden et al. [4.41], and Fig. 4.40 is an example of their theoretical predictions for a direct-injection, stratified-charge, spark-ignition engine. Fig. 4.40 illustrates a section through the combustion bowl, the flat-topped piston, and cylinder head. Reading down from top to bottom, at 28° btdc, the figure shows the spray droplets, gas particle velocity vectors, isotherms, turbulent kinetic energy contours, equivalence ratio contours, and the octane mass fraction contours. The Amsden et al. paper [4.41] goes on to show the ensuing combustion of the charge. This form of combustion calculation is clearly superior to any of the (a)-(c) models previously discussed, as the combustion process is now being theoretically examined at the correct "microscopic" level. No matter how much computer time such calculations require, they will become the norm in the future, for computers are becoming ever more powerful, ever more compact, faster, and less costly with the passage of time. Having said that, something better than the conventional κ-ε models for turbulence [4.41] will have to be concocted to improve their computational accuracy in that area, for there is no universal satisfaction with that approach.

For those interested in a similar combustion computation for diesel engines the paper by Aneja and Abraham [4.43] should be examined together with the further references that it contains. This paper gives illustrations similar to that shown in Fig. 4.40, but in an open-chamber, direct-injection compression-ignition engine.

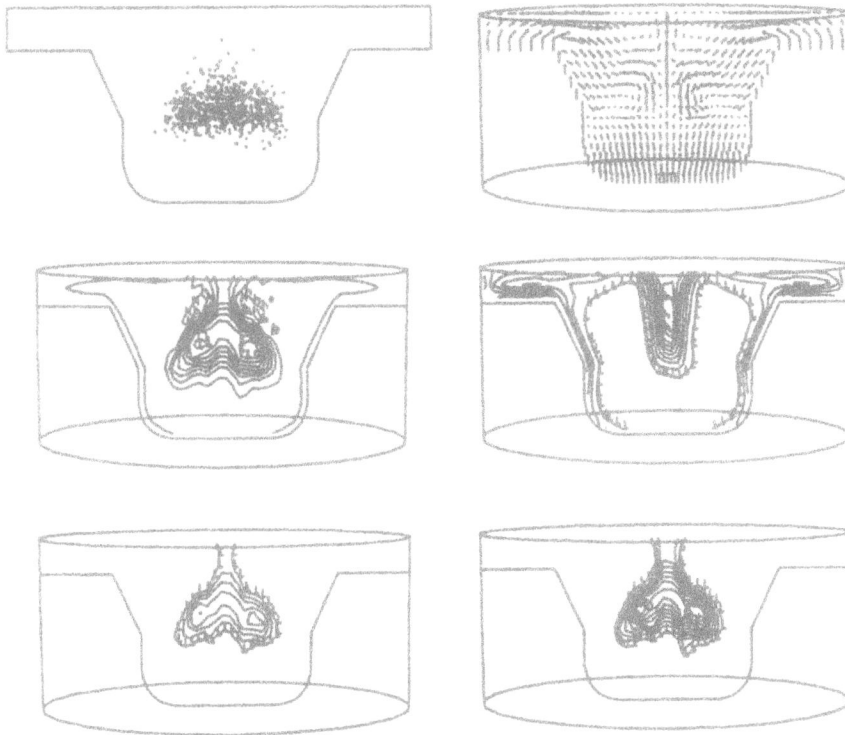

Fig. 4.40 Detailed calculation of gasoline combustion in a fuel-injected SI engine (from Ref. [4.41]).

4.4.5 Chaos and Combustion

From the literature, it would appear that the most sophisticated combustion theories, allied to the most advanced computational analysis in three dimensions, are still not capable of the design of a combustion chamber—and I use the word "design" in its true sense, i.e., to mean that the input data are prepared, the theory is used and predicts the outcome in terms of cylinder pressure with respect to time at all loads and speeds within the operating range of the engine, including the cyclic variability of the cylinder pressure at each of the loads and speeds. The combustion chamber is constructed based on the geometry of the input data, and cylinder pressures are measured for the engine at the same loads and speeds. The measurements and the calculations coincide to some acceptable degree. This is a design process, and my point is that I am unaware of this being accomplished to date.

My opinion, based on the research of others, is that a combustion process is chaotic [4.59-4.61] because the controlling parameter for flame propagation and convection heat transfer is turbulence, and turbulence is a classic example of chaos [4.62-4.63]. The theorists of chaos show that there is order even in chaos, and one element that has been highlighted as a predictable phenomenon is turbulence. Because no one is deliriously happy with the current microscopic turbulence models, and the κ-ε model is good evidence of that attitude, perhaps the time has come to step back from the microscopic and deal with the reality of the chaos of turbulence using this new approach. If that were accomplished, then perhaps the chemistry of the reactions would not need to be so detailed, and the physical geometry could be made even more detailed, within a computation model. The outcome of such a model might be a more accurate prediction of the cylinder pressure at all engine speeds and loads and its cyclic variability at each condition. In short, perhaps the relatively simple approach of Douglas and Reid [4.28-4.30] could be usefully extended to include a model of turbulence based on the reality of its chaos. The end product might well be the all-important first step toward true design of combustion chambers.

4.5 Squish Behavior in Engines

As already stated in Sec. 4.1.4, detonation is an undesirable feature of spark-ignition engine combustion, particularly when the designer attempts to operate the engine at a compression ratio that is too high. As described in Appendix A1.1, Sections A1.1.8, the highest thermal efficiency is obtained at the highest possible compression ratio. Any technique that assists the engine to run reliably at a high compression ratio on a given fuel must be studied thoroughly and, if possible, designed into the engine and experimentally tested. Tumble action of the intake charge within the chamber is often highlighted as being one such technique [1.34]. Squish action is another technique that provides added turbulence both within the combustion chamber, and in the end zones of the squish band where detonation is most prevalent.

Figs. 4.39(a) and (b) show the computation by CFD of such a process in a DI diesel engine with a bowl much as shown in Plate 4.1 and a helical intake port promoting swirl much as shown in Plate 3.3. The arrows represent the velocity vectors in both magnitude and direction. The bore and stroke of this engine are 80 and 93 mm, respectively, and the compression ratio is 20.4. The diameters at the squish edge and the bowl entrance are 42 and 38 mm, respectively. The minimum squish clearance is 0.78 mm. The computation is conducted at 2000 rev/min. The views shown in Fig. 4.39 are snapshots of the velocity vectors at 20 °btdc. In Fig. 4.39(a), which is an elevation view over half the cylinder bore and the bowl, the squished flow is very visible, with a maximum velocity of 39 m/s at the edge of the squish band. In Fig. 4.39(b), which is a plan view over the cylinder bore, where the edges of two of the valves, the squish edge on the piston and the inner bowl entrance, are shown in overlay, the squish flow is observed to compound, and enhance, the residue of the in-cylinder swirling flow engendered by the directioning of the ports leading up to the intake valves. The radial component of the velocity vector can be considered to be that due to the squish effect, and the axial component that due to the swirling flow. From Fig. 4.39(b), as the squish and swirl flow are approximately at a mutual 45° from the resultant vector direction, swirl and squish velocity values of 27.5 m/s are obtained. The swirl rotation rate, N_{sw}, at the edge of the bowl is then found to be given by:

$$N_{sw} = \frac{c_{sw}}{\pi d_b} = \frac{27.5}{\pi \times 0.0375} = 233.4 \text{ rev/s}$$

The engine speed in the simulation is 2000 rev/min, or 33.33 rev/s, so the swirl ratio, C_{sw}, is defined by:

$$C_{sw} = \frac{\text{swirl rotation rate}}{\text{engine rotation rate}} = \frac{233.4}{33.3} = 7.0$$

A value of 7.0 for a swirl ratio in a direct-injection diesel engine is regarded as conventional. Further discussion of this topic is presented in Sec. 4.5.3. In-cylinder swirl and swirl ratio are discussed even more extensively in connection with IDI diesel engines in Sec. 4.5.4.

A photograph of squish action producing a tumbling motion in the combustion space, visually enhanced by smoke in a motored QUB-type deflector piston two-stroke engine, is shown in Plate 4.2. This photograph, albeit in a two-stroke engine, illustrates (a) the type of squish action that could be expected in a side-valve, four-stroke engine where much of the compressed charge is shuttled sideways into the clearance volume where combustion will occur and (b) the swirling flow in a bowl caused by a similar squishing action through the throat leading to the combustion space in the IDI diesel engine sketched in Fig. 4.22.

You may be a little puzzled right now as to the difference in definition between "tumbling" and "swirling" flow. "Swirling" is loosely defined as flow that is spinning concentrically with the axis of the cylinder, and "tumbling" is equally loosely defined as in-cylinder flow that is rotating at right angles to the cylinder axis. I use the word "loosely" in connection with these definitions to stop you from writing to me and telling me that the so-called "swirling" flow seen in an IDI combustion bowl in Fig. 4.22, should be called "tumbling" flow by that definition. Such are the inconsistencies that pervade the jargon-laden world of internal combustion engines!

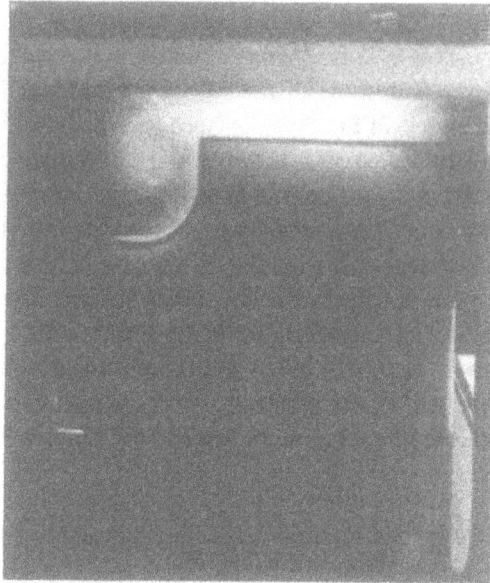

Plate 4.2 The vigorous squish action in a QUB-type cross-scavenged engine at the end of the compression stroke.

4.5.1 A Simple Theoretical Analysis of Squish Velocity

From Fig. 4.40, it is clear that the use of a CFD model permits the accurate calculation of squish velocity characteristics within the cylinder. This insight into the squish action is excellent, and this is a calculation method that is likely to be in common use by most designers in the not-too-distant future. However, much work in correlating the results with measurements will be required before total confidence is placed in the accuracy of this model [4.44]. Because a CFD calculation uses much time on even the fastest of supercomputers or work stations, the designer needs a more basic solution for everyday use. Indeed, to save both supercomputer and designer time, the CFD user will always need some simpler guidance tool to narrow down the many design options in order to reduce both the number of runs and the tedious, time-consuming process of generating space mesh for a CFD package. The analytical approach presented below is intended to fulfill that need. This approach is particularly helpful because within it, the cylinder heads of the conventional four-stroke spark-ignition engine fall into several discrete categories, as do the bowl geometries of conventional compression-ignition units, so the designer can rapidly analyze a considerable variety of combustion chamber shapes.

The squish characteristics of the spark-ignition engine, not counting the side-valve engine, are relatively modest compared to those of compression-ignition engines. This is because the valves in a spark-ignition engine tend to occupy a considerable proportion of the bore area and so a high squish design can compromise their effective flow area. The cylinder heads of all spark-ignited overhead-valve engines tend to fall into the four classifications, (a)-(d), shown in Fig. 4.41.

476

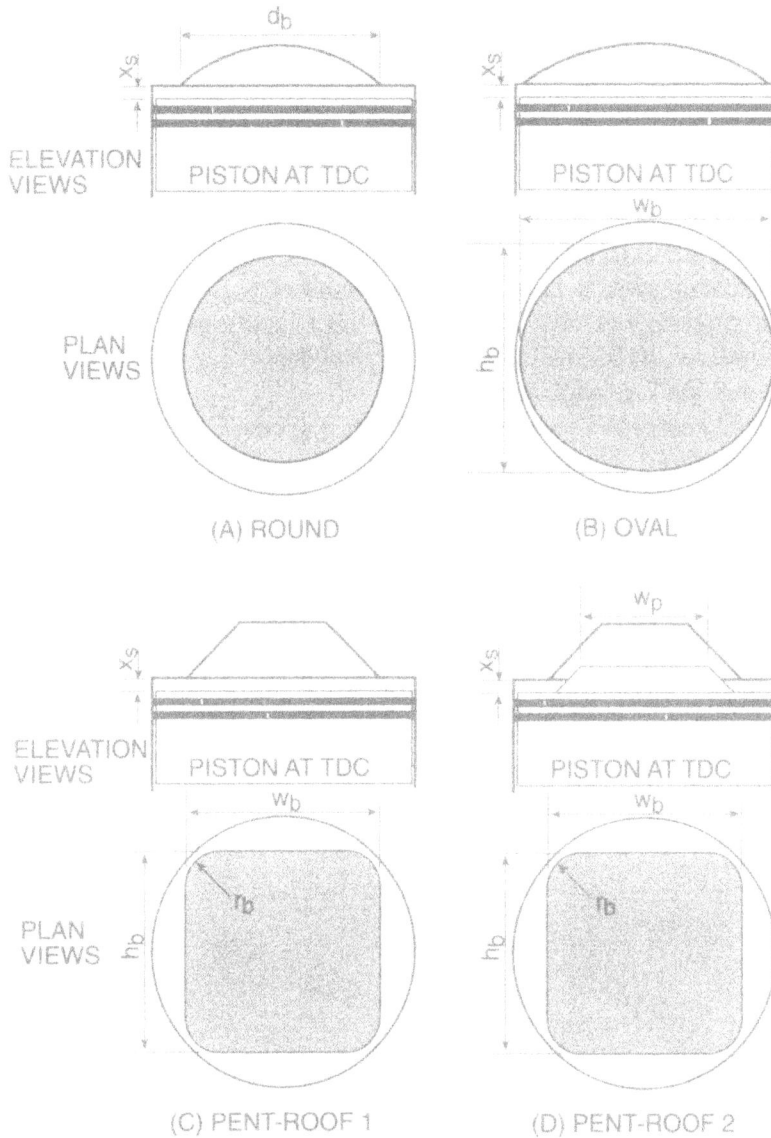

Fig. 4.41 The squish geometry of cylinder heads on spark-ignited engines.

The shaded areas for the four head shapes shown in Fig. 4.41 are the non-squished areas, i.e., the areas of the open bowls facing the piston crown. In Fig. 4.41(a) and (b), the head shape is typical of two-valve designs as in Fig. 1.2. In Fig. 4.41(c) and (d), the head shape is typical of four-valve designs as in Fig. 1.6, where in some cases, as in (d), the pent-roof on the piston protrudes into what might be loosely termed "the combustion space."

477

The squish characteristics of the DI compression-ignition engine, the shape of which is normally similar to that sketched in Fig. 1.7 or illustrated in Plate 4.1, and which has already been presented in geometric detail in Fig. 4.23, shows the entrance to the bowl as being normally circular with a diameter, d_b, enclosing the non-squished area of the piston.

There have been several attempts to produce a simple analysis of squish velocity, often with theory that is more empirically based than fundamentally thermodynamic. One of the useful papers that has been widely quoted in this area is that by Fitzgeorge and Allison [4.45]. Experimental measurements of such phenomena are becoming more authoritative with the advent of instrumentation such as laser doppler velocimetry (LDV). The paper by Fansler [4.46] is an excellent example of what is possible by this accurate and non-intrusive measurement technique. However, the following theoretical procedure is one that is quite justifiable in thermodynamic terms, yet is remarkably simple.

Figs. 4.2 and 4.42 represent a compression process inducing a squished flow between the piston and the cylinder head. The process commences at trapping, i.e., intake valve closure. From Sec. 1.6.4, the value of trapped mass, m_t, is known, and is based on an assumed value for the trapped charge pressure and temperature, p_t and T_t. At this juncture, the mass will be evenly distributed between the volume subtended by the squish band, V_{st}, and the volume subtended by the bowl, V_{bt}. The actual values of V_{s1} and V_{b1} at any particular piston position, as shown in Fig. 4.42, are a matter of geometry based on the parameters illustrated in Fig. 4.2. From such input parameters as squish area ratio, C_{sq}, and bowl diameter, d_b, the values of squish area, A_s, and bowl area, A_b, are calculated from Eq. 4.2.1.

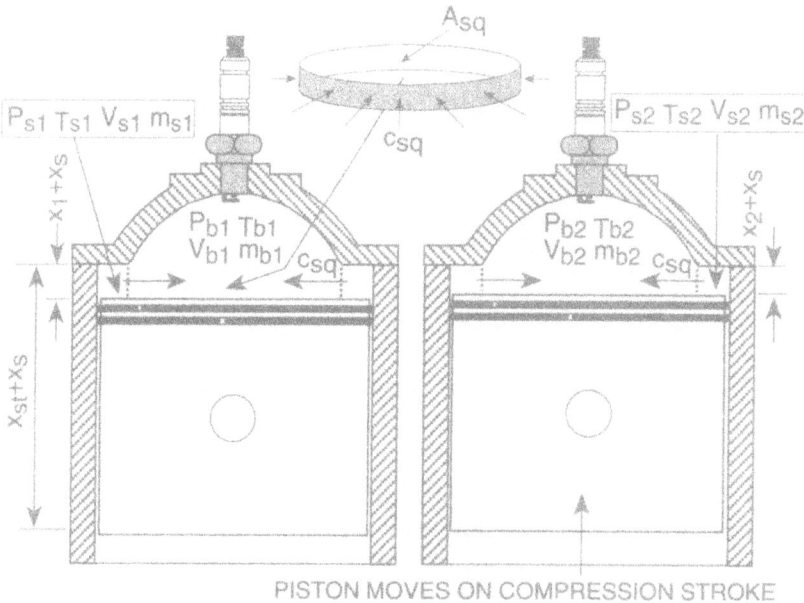

Fig. 4.42 A simple theoretical model of squish behaviour.

For a particular piston position, as shown in Fig. 4.42, the values of thermodynamic state and volumes are known before an incremental piston movement takes place.

The fundamental thesis behind the calculation is that all state conditions equalize during any incremental piston movement. Thus, the values of cylinder pressure and temperature are p_{c1} and T_{c1}, respectively, at the commencement of any time step dt, which will give an incremental piston movement dx, an incremental crank angle movement $d\theta$, and a change of cylinder volume dV.

However, if equalization has taken place from the previous time step, then:

$$p_{c1} = p_{s1} = p_{b1} \quad T_{c1} = T_{s1} = T_{b1} \quad \rho_{c1} = \rho_{s1} = \rho_{b1}$$

and
$$V_{c1} = V_{s1} + V_{b1} \quad m_{c1} = m_t = m_{s1} + m_{b1}$$

For the next time step, the compression process will occur in a polytropic manner with an exponent, n, as discussed in Sec. 4.2.2. If the process is considered as ideal, i.e., isentropic, the exponent is γ, the ratio of specific heats for the cylinder gas. Because of the multiplicity of trapping conditions and polytropic compression behavior, it is more logical for any calculation to consider engine cylinders to be analyzed on a basis of equality, in which case, for sheer simplicity, one should assumed that the process commences with an air-filled cylinder at bdc and pretend that the intake valve is already closed!

Therefore, the compression process is assumed to be isentropic, with air as the working fluid, and the trapping conditions are assumed to be at the reference pressure and temperature of 1 atm and 20°C, in which case the value of γ is 1.4.

The initial cylinder density is given by:

$$\rho_t = \frac{p_t}{RT_t} = \frac{101325}{287 \times 293} = 1.205 \ \text{kg/m}^3$$

No matter what degree of simplicity or complexity we choose for the application of the theory, the value of the individual compression behavior in the squish and bowl volumes, as well as the overall macroscopic values, follows:

$$\frac{p_{s2}}{p_{s1}} = \left(\frac{V_{s1}}{V_{s2}}\right)^{\gamma} \quad \frac{p_{b2}}{p_{b1}} = \left(\frac{V_{b1}}{V_{b2}}\right)^{\gamma} \quad \frac{p_{c2}}{p_{c1}} = \left(\frac{V_{c1}}{V_{c2}}\right)^{\gamma}$$

and
$$\frac{\rho_{c2}}{\rho_{c1}} = \frac{V_{c1}}{V_{c2}}$$

If a squish action takes place, or by definition, if:

$$A_{sq} < A_b$$

then the value of squish pressure, p_{s2}, is greater than either p_{c2} or p_{b2}, but:

$$p_{s2} > p_{c2} > p_{b2}$$

The squish pressure ratio, P_{sq}, causing gas flow to take place, is found from:

$$P_{sq} = \frac{p_{s2}}{p_{b2}} \tag{4.5.1}$$

At this point, consider that a gas flow process takes place within the time step so that the pressures equalize in the squish band and the bowl, and are equal to the average cylinder pressure. This implies movement of mass from the squish band to the bowl, so that the mass distributions at each end of the time step are proportional to the volumes, as follows:

$$m_{s1} = m_t \frac{V_{s1}}{V_{c1}} \qquad m_{s2} = m_t \frac{V_{s2}}{V_{c2}} \tag{4.5.2}$$

During the course of the compression analysis, the mass in the squish band is considered to be the original, and equalized in the manner above, value of m_{s1}, so the incremental mass squished, dm_{sq}, is given by:

$$dm_{sq} = m_{s1} - m_{s2} = m_t \left(\frac{V_{s1}}{V_{c1}} - \frac{V_{s2}}{V_{c2}} \right) \tag{4.5.3}$$

This incremental movement of mass occurring in time dt, is further evaluated by the continuity equation to determine the squish velocity, c_{sq}:

By continuity

$$dm_{sq} = \dot{m}_{sq} dt \tag{4.5.4}$$

where $$\dot{m}_{sq} = \rho_{c2} A_{sq} c_{sq} \tag{4.5.5}$$

and as

$$dt = \frac{d\theta}{360 \times rps}$$ (4.5.6)

then combining Eqs. 4.5.3 to 4.5.6 to resolve the value of squish velocity, c_{sq}, at angle, $\theta+d\theta$:

$$c_{sq} = \frac{V_{c2}}{A_{sq}}\left(\frac{V_{s1}}{V_{c1}} - \frac{V_{s2}}{V_{c2}}\right)\frac{360rps}{d\theta}$$ (4.5.7)

The value of the squish flow area, A_{sq}, which requires careful definition, is shown sketched for diesel and Otto engines in Figs. 4.23 and 4.41, respectively. It is the annular area between the bowl and the squish band formed as the "curtain" in Fig. 4.42, the length of which is the line separating the squish band from the bowl, and the height of which is the mean of the piston movement from x_1+x_s to x_2+x_s during the time interval of the calculation.

From Eq. 4.5.7 it can be seen that the value of squish velocity is independent of the gas properties selected for the computation, but the value of squish pressure ratio will be dependent on them.

At any juncture during the computation the squish band and bowl volumes, V_s and V_b, are found as follows:

Bore area

$$A_{bo} = \frac{\pi}{4}d_{bo}^2$$ (4.5.8)

Squish area

$$A_s = C_{sq}A_{bo}$$ (4.5.9)

Bowl area

$$A_b = A_{bo} - A_s = (1 - C_{sq})A_{bo}$$ (4.5.10)

Bowl volume

$$V_{b1} = A_b(x_1 + x_s) \quad \text{or} \quad V_{b2} = A_b(x_2 + x_s)$$ (4.5.11)

Squish band volume

$$V_{s1} = A_s(x_1 + x_s) \quad \text{or} \quad V_{s2} = A_s(x_2 + x_s)$$ (4.5.12)

The remaining factor to be evaluated is the value of A_{sq}, the squish flow area through which the incremental mass dm_{sq} will pass during the time step dt. This will have a numerical value that depends on the type of cylinder head employed, and is irrespective of the parameters previously mentioned. For example, consider the cases of the basic cylinder head types shown in Figs. 4.23 and 4.41.

Direct Injection Compression-Ignition Engines, As in Fig. 4.23(a)-(d)

All of the bowls are considered to be round with a diameter d_b. The following is the analysis for the squish flow area, A_{sq}. The flow will be radial through the circumference of the squish band, and if the flow is considered to take place midway through the incremental step, then:

$$A_b = \frac{\pi}{4}d_b^2 \qquad A_s = A_{bo} - A_b \qquad C_{sq} = \frac{A_s}{A_{bo}} \qquad (4.5.13)$$

$$A_{sq} = \pi d_b \left(x_s + \frac{x_1 + x_2}{2} \right) \qquad (4.5.14)$$

Spark-Ignition Engine, As in Fig. 4.23(a)

This is a typical two-valve hemi-head design. As with the diesel engine, the bowl is considered to be round with a diameter d_b. The following is the analysis for the squish flow area, A_{sq}. The flow will be radial through the circumference of the squish band and, if the flow is considered to take place midway through the incremental step, then:

$$A_b = \frac{\pi}{4}d_b^2 \qquad A_s = A_{bo} - A_b \qquad C_{sq} = \frac{A_s}{A_{bo}} \qquad (4.5.15)$$

$$A_{sq} = \pi d_b \left(x_s + \frac{x_1 + x_2}{2} \right) \qquad (4.5.16)$$

Spark-Ignition Engine, As in Fig. 4.23(b)

This is also a typical two-valve hemi-head design. The bowl is considered to be an ellipse with a width w_b, and a height h_b. The following is the analysis for the squish flow area, A_{sq}. The flow will be radial through the circumference of the squish band, which is elliptical in shape and, if the flow is considered to take place midway through the incremental step, then:

$$A_b = \frac{\pi}{4}\left(w_b + h_b \right) \qquad A_s = A_{bo} - A_b \qquad C_{sq} = \frac{A_s}{A_{bo}} \qquad (4.5.17)$$

$$A_{sq} = \frac{3\pi}{4}\left(w_b + h_b \right)\left(x_s + \frac{x_1 + x_2}{2} \right) \qquad (4.5.18)$$

Spark-Ignition Engine, As in Fig. 4.23(c)

This is a typical four-valve design where the combustion chamber is usually within the cylinder head; however, to meet the required compression ratio, some may be in the form of a thin matching dish set in the piston crown. The bowl is considered to be a rectangle with a width w_b, a height, h_b, and corner radii r_b. The following is the analysis for the squish flow area, A_{sq}. The flow will be radial through the circumference of the squish band, which is rectangular in shape and, if the flow is considered to take place midway through the incremental step, then:

$$A_b = w_b \times h_b - \left(4 - \pi\right)r_b^2 \qquad A_s = A_{bo} - A_b \qquad C_{sq} = \frac{A_s}{A_{bo}} \qquad (4.5.19)$$

$$A_{sq} = \left(2w_b + 2h_b - 8r_b + 2\pi r_b\right)\left(x_s + \frac{x_1 + x_2}{2}\right) \qquad (4.5.20)$$

Spark-Ignition Engine, As in Fig. 4.23(d)

This is another typical four-valve layout where the combustion chamber is within the cylinder head but, to meet the required compression ratio, the pent-roof on the piston crown protrudes into the cylinder head. The bowl is considered to be a rectangle with a width w_b, a height h_b, and corner radii r_b. In a conventional pent-roof design, this radius is normally minimal or even non-existent. The width of the pent-roof within the combustion chamber is w_p. The following is the analysis for the squish flow area, A_{sq}. The flow will be radial through the circumference of the squish band, which is rectangular in shape and, if the flow is considered to take place midway through the incremental step, then:

$$A_b = w_p \times h_b \qquad A_s = A_{bo} - A_b \qquad C_{sq} = \frac{A_s}{A_{bo}} \qquad (4.5.21)$$

$$A_{sq} = \left(2w_p + 2h_b\right)\left(x_s + \frac{x_1 + x_2}{2}\right) \qquad (4.5.22)$$

Kinetic Energy Associated with Squish Flow

It is also possible to determine the turbulence energy induced by this flow, on the assumption that it is related to the kinetic energy created. The incremental kinetic energy value at each time step is dKE_{sq}, where:

$$dKE_{sq} = dm_{sq}\frac{c_{sq}^2}{2} \qquad (4.5.23)$$

The total value of turbulence kinetic energy squished, KE_{sq}, is then summed for all of the calculation increments over the compression process from trapping to tdc.

4.5.2 Evaluation of Squish Velocity for Spark-Ignition Engines

The equations in Sec. 4.5.1 are programmed into software. All of the combustion chamber types shown in Figs. 4.23 and 4.41 can be handled by this program. The operator is prompted to type in the chamber type by name, either "round," "oval," "pent-roof 1," or "pent-roof 2."

A typical output from the program is shown in Fig. 4.43, where the output values for squish velocity and kinetic energy are plotted with respect to crank angle position from bdc to tdc.

The input data for this calculation are for a "round" chamber in a spark-ignition engine with a bore and stroke of 86 mm and a connecting rod that is 165 mm long. The engine speed is 5000 rev/min. The compression ratio is 10. The bowl diameter, d_b, is 70 mm and the squish clearance is 1.4 mm. This gives a squish area ratio, C_{sq}, of 0.337. It can be seen that the maximum squish velocity is 15.4 m/s and the cumulative kinetic energy is 6.53 mJ. Both maxima occur at 10 °btdc, which coincides closely—and most usefully as far as combustion enhancement is concerned—with the end of the ignition delay reported in most of the data shown for spark-ignition engines in Figs. 4.7 to 4.21. If the squish clearance, x_s, is reduced to 1.2 mm, or raised to 1.6 mm, the outcome is that the maximum squish velocity is raised to 17.3 m/s or lowered to 13.8 m/s, respectively. Because a 1.2 mm minimum head clearance would be mechanically regarded as a "narrow" squish clearance value, and 1.6 mm as "generous", it can be seen that the value of maximum squish velocity is not overly sensitive to these design changes.

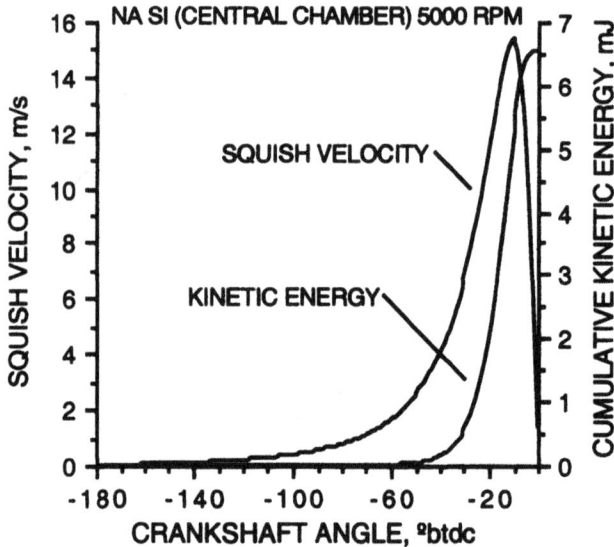

Fig. 4.43 Squish velocity and kinetic energy during compression in an si engine.

If the basic engine geometry is retained, but an "oval" chamber of width w_b and a height h_b of 70 and 76 mm, respectively, is substituted for the "round" bowl, the squish area ratio is reduced to 0.281. The result is a maximum squish velocity of 8.2 m/s, still maximized at 10 °btdc. The cumulative kinetic energy is reduced by a factor of four to 1.54 mJ. This is a significant reduction in the value of maximum squish velocity for a relatively small change in squish area ratio. Clearly, squish area ratio, C_{sq}, and the chamber shape as determined through squish flow area, A_{sq}, have a more profound effect on squish velocity than squish clearance, x_s.

Let us keep the same basic engine geometry, but use a "pent-roof 1" chamber of width w_b and height h_b of 62 and 62 mm, respectively, and a corner radius, r_b, of 5 mm. The squish area ratio is increased slightly to 0.342. The result is a maximum squish velocity of 14.3 m/s, still maximized at 10 °btdc. The cumulative kinetic energy is at 5.73 mJ, representing a small reduction in the value of maximum squish velocity compared to the "round" chamber. Although the squish area ratio is higher for the pent-roof chamber, the squish flow area, A_{sq}, found by Eq. 4.5.20 is greater than that determined by Eq. 4.5.16. The outcome is a lower maximum squish velocity, c_{sq}.

Let us now replace the "pent-roof 1" chamber with a "pent-roof 2" clearance volume of width w_p and height h_b of 48 and 62 mm, respectively, and the same corner radius, r_b, of 5 mm. The squish area ratio is increased significantly to 0.488. It is assumed the same squish clearance of 1.4 mm is retained over the protruding pent-roof. The result is a maximum squish velocity of 22.2 m/s which is 50% greater than that of "pent-roof 1," but still maximized at 10 °btdc. The cumulative kinetic energy is also raised considerably to 19.69 mJ. Not only is the squish area ratio higher for this second pent-roof chamber, but the squish flow area, A_{sq}, found by Eq. 4.5.22 is also less than that determined by Eq. 4.5.20. The outcome is a much higher maximum squish velocity, c_{sq}. It is clear that a protruding pent-roof chamber is quite capable of raising the squish velocity very effectively. Equally importantly, due to the inclination of the pent-roof sides on the piston crown, this inserts the squished flow more effectively into the combustion chamber by directing it toward the advancing flame front.

4.5.3 Evaluation of Squish Velocity for DI Compression-Ignition Engines

The input data for this calculation is for the "round" chamber as seen in Fig. 4.41, but with dimensions appropriate to the bowls of the several types of DI combustion systems shown in Fig. 4.23. The same engine with a bore and stroke of 85 mm and 88 mm, respectively, and a connecting rod that is 165 mm long, is elected as an example. The engine speed is 3000 rev/min. The compression ratio is 18. The squish clearance is 1.25 mm. Various bowl diameters, d_b, are selected which are representative of "squish-lip," "re-entrant," "open," and "mexican hat" designs at 41 mm, 44 mm, 47 mm, and 60 mm, respectively. The computation software is run, and the maximum squish velocities recorded for each design are plotted in Fig. 4.44 with respect to their computed squish area ratios.

The outcome, as with all previous "round" chambers, is that the maximum squish velocity is found at the highest squish area ratio. It will be noted that even the most open chamber in a compression-ignition engine, i.e., that of a "mexican hat," has a higher squish area ratio than seen in any of the spark-ignition engine examples discussed above.

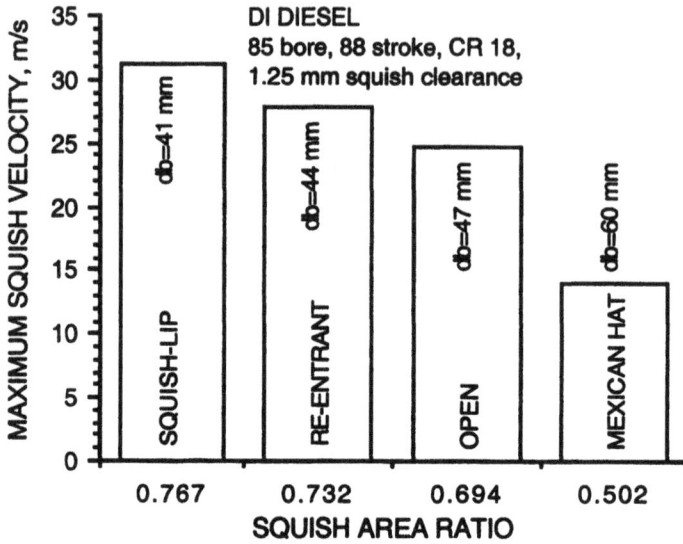

Fig. 4.44 Squish velocities for various designs of DI diesel combustion chambers.

Although in most open-chamber diesel engines it is true that the in-cylinder swirl created by the induction system has the most pronounced effect on fuel vaporization and combustion, the very existence of a "squish-lip" design, and these results, emphasizes the extra significance of squish effects on high-speed combustion, particularly for those power units used in automobiles.

At the very beginning of this section, the results of a complex CFD computation of squish action in a DI bowl are presented in Fig. 4.39. The peak velocity at the edge of the squish band, noted as 39 m/s, is a compound of the squish action and the induced swirl. The radial component can be considered to be due to the squish component, and this is computed by the CFD as 27.5 m/s. If the same geometrical data for that DI diesel engine are inserted into the simpler theory presented here, a maximum squish velocity of 26.5 m/s, located at 9 °btdc, is obtained. By the CFD, the maximum value of squish (+swirl) velocity is located at 10 °btdc. The correlation between the simple and the complex approaches seems reasonable.

4.5.4 Evaluation of Squish Velocity for IDI Compression-Ignition Engines

The squish effect in the indirect-injection (IDI) diesel engine is a special case of the theory shown in Sec. 4.5.1. In Figs. 1.8 and 4.22, the compression process pushes the air from the main chamber above the piston through a throat of area A_t into the side chamber. The throat area is represented by a diameter d_t quite correctly if it is circular in cross section, or as an equivalent diametral value if it is not round:

$$d_t = \sqrt{\frac{4A_t}{\pi}} \qquad (4.5.24)$$

The squish area ratio, using a suitably modified version of Eq. 4.5.13, becomes:

$$A_s = A_{bo} - A_t \qquad C_{sq} = \frac{A_s}{A_{bo}} \qquad (4.5.25)$$

The squish area ratio, C_{sq}, typically found in IDI engines is 0.99, very considerably higher than anything discussed for spark-ignition or DI compression-ignition engines. It is also clear that the basic theory in Sec. 4.5.1 now needs to be modified because the bowl volume, V_b, is no longer a variable as before, but is a constant throughout the compression process. The actual value of the bowl volume is determined from the clearance volume, V_{cv}, and a factor known as the bowl volume ratio, C_{bv}. This is defined as:

Bowl volume ratio $\qquad C_{bv} = \dfrac{\text{bowl volume}}{\text{clearance volume}} = \dfrac{V_b}{V_{cv}} \qquad (4.5.26)$

The clearance volume is determined in the usual manner from the cylinder swept volume, V_{sv}, and the compression ratio, CR:

$$V_{cv} = \frac{V_{sv}}{CR - 1} \qquad (4.5.27)$$

The values of bowl volume ratio, C_{bv}, to be found in IDI diesel engines range from 0.4 to 0.6, with 0.5 being the almost universal choice found in practice. Hence, the bowl and squish volumes, V_b and V_s, are found from:

$$V_b = C_{bv} V_{cv} \qquad\qquad V_s = V_{cv} - V_b \qquad (4.5.28)$$

Due to the very small volumes involved, and the high compression ratio, the volume in the throat is significant and is included within any theory as part of the squish volume above the piston. The simple assumption, based on conventional geometry, is that the length of the throat is 1.5 diameters:

$$L_t = 1.5d_t \qquad V_t = A_t L_t = \frac{3\pi d_t^2 L_t}{8} \qquad (4.5.29)$$

This provides a means of determining the effective squish height, x_s, as:

$$x_s = \frac{V_s - V_t}{A_{bo}} \qquad (4.5.30)$$

The theoretical solution for the squished velocity, listed as c_{sq} in Eq. 4.5.7, must be modified to take into account the fact that the bowl volume, V_b, is no longer considered as a variable, but as a constant. The major differences are straightforward and are inserted into the same theory, Eqs. 4.5.1 to 4.5.7, to determine the throat velocity, c_t, into the IDI bowl.

$$V_{b2} = V_{b1} = V_b \qquad p_{b2} = p_{b1} \qquad A_{sq} = A_t \qquad c_t = c_{sq} \qquad (4.5.31)$$

The shape of the bowl is normally spherical, or so nearly spherical that it can be considered to be such with little loss of computational accuracy. The bowl diameter can be found from:

Bowl diameter $$\qquad d_b = \sqrt[3]{\frac{6V_b}{\pi}} \qquad (4.5.31)$$

The bowl and its throat are shown in Fig. 4.45. The outside edge of the throat is assumed, as is normal, to be tangential to the outer radius of the bowl, r_b. The jet of squished gas is at velocity c_t, and the center line is assumed to arrive into the bowl, and act on the spinning bowl contents, at a radius r_j, where the rotational velocity at that point is c_j. The entire contents of the bowl are assumed to be spinning in a solid-body rotation at an angular velocity ω_b.

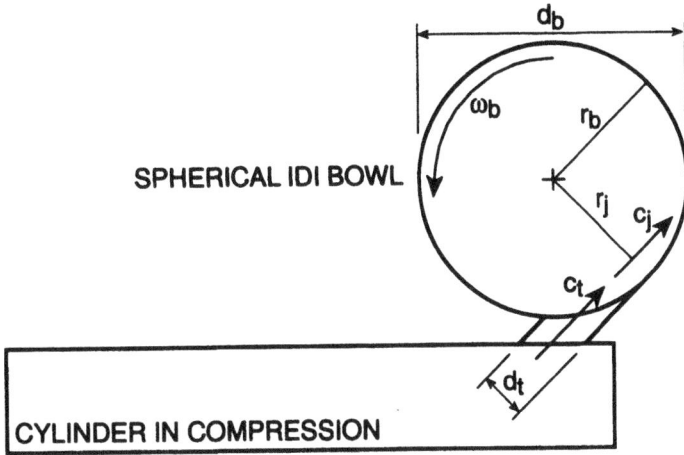

Fig. 4.45 The throat and bowl geometry of an IDI diesel combustion chamber.

The energy, dE_b, required to change the rotation of this spinning "gas" flywheel from an angular velocity ω_{b1} to ω_{b2} is given by:

$$dE_b = dE_s - dE_f = \frac{I_{b2}\omega_{b2}^2 - I_{b1}\omega_{b1}^2}{2} \qquad (4.5.32)$$

where I_{b2} and I_{b1} are the moments of inertia of the gas flywheel before and after the event. The moment of inertia of a spherical flywheel of mass m_b and diameter d_b is given by:

$$I_b = \frac{m_b d_b^2}{10} \qquad (4.5.33)$$

The moments of inertia can be calculated as the mass of the gas in the bowl at the start of any time increment, m_{b1}, is a known quantity, and at the mass at the conclusion of the time increment, m_{b2}, from Eq. 4.5.7 is:

$$m_{b2} = m_{b1} + dm_{sq} \qquad (4.5.34)$$

In Eq. 4.5.32, there are two further energy terms, dE_s and dE_f, which are, respectively, the energy supplied by the squished gas jet and that lost by the gas flywheel due to friction with the walls of the bowl. The energy supplied by the gas jet at velocity, c_t, is kinetic energy, which will act over the time increment of the process, dt:

$$dE_s = \frac{dm_{sq}c_t^2}{2}$$ (4.5.35)

The energy lost due to friction may be assessed by calculating the shear stress at the wall of the bowl, τ_b, due to a friction factor, C_f:

$$\tau_b = \frac{C_f \rho_b c_b^2}{2}$$ (4.5.36)

The friction force, F_b, at the surface of the bowl is given by:

$$F_b = \tau_b A_{sb} = \tau_b \left(\pi d_b^2 \right)$$ (4.5.37)

where A_{sb}, is the surface area of the sphere that is the bowl.

The work done against this force during the time increment dt, is the energy lost due to friction, dE_f, by the gas flywheel during that period:

$$dE_f = F_b c_b dt = F_b \varpi_b r_b dt$$ (4.5.38)

Friction energy loss may be assessed using the bowl angular velocity at the commencement of any time step, or the calculation may be reiterated with a mean value found after a first computation step. Either method yields essentially the same end result.

The contents of Eqs. 4.5.33 to 4.5.38 are inserted into Eq. 4.5.32, and the angular velocity of the spinning gas in the bowl, ϖ_{b2}, determined at the conclusion of the time step:

Bowl swirl velocity $$\varpi_{b2} = \sqrt{\frac{\dfrac{20\left(dE_s - dE_f\right)}{d_b^2} + m_{b1}\varpi_{b1}^2}{m_{b2}}}$$ (4.5.39)

490

It is considered conventional to quote the swirl as a swirl ratio, C_{sw}, which is defined as follows in terms of the engine speed, N, in rev/min units:

$$\text{Swirl ratio} \qquad C_{sw} = \frac{\text{rotational speed of gas in bowl}}{\text{rotational speed of engine}} = \frac{\varpi_b}{2\pi} \times \frac{60}{N} \qquad (4.5.40)$$

It is realized that, by assuming that all bowl surface gas is moving at velocity c_b, the energy loss due to friction is somewhat overestimated. On the other hand, as can be seen in Fig. 4.22, the bowl normally contains a glow plug (see Fig. 4.22), which produces a considerable aerodynamic drag, so this apparent overestimation is not illogical. It would be quite appropriate to use a friction factor, C_f, of 0.004. It would also be quite sensible to use a discharge coefficient, C_d, for the air velocity entering the bowl through the throat. Considering the experimental results quoted in Chapter 3, the following relationship provides a reasonably realistic approximation for the effective throat velocity, $c_{t_{eff}}$, to be inserted into the energy equation, Eq. 4.5.35:

$$c_{t_{eff}} = C_d c_t = \left(0.75 + \frac{c_t}{20} \right) c_t \qquad (4.5.41)$$

This theoretical process has been investigated before by researchers such as Ajakaiye and Dent [4.47] and Somerville et al. [4.48], to name but two, and for those interested in the topic, these papers and the further references in them will prove helpful. The theoretical solution of the swirling flow can be effected by using either the equations of conservation of energy, as I have done above, or the equations describing the conservation of angular momentum [4.47]. However, the theoretical "nub" of the matter is an ability to compute with some realism a value for the time-varying, effective squish velocity through the throat into the IDI bowl. Some (unreferenced) publications have been noticeably ineffective in this vital regard.

The Design of an IDI Diesel Combustion Chamber to Optimize the Swirl Characteristics
Consider an IDI diesel engine with a bore and stroke of 85 and 88 mm, respectively, with a connecting rod length of 165 mm, and a compression ratio of 21. As an initial guess, let us assume a bowl volume ratio, C_{bv}, of 0.5 and a throat diameter, d_t, of 9 mm, which by Eq. 4.5.25 gives a squish area ratio, C_{sq}, of 0.989. The data for the engine, the bowl volume ratio and the compression ratio, show that the swept volume, V_{sv}, is 499.4 cm^3, the clearance volume, V_{cv}, is 25 cm^3, and the bowl volume, V_b, is 12.5 cm^3. The bowl diameter, d_b, is found from Eq. 4.5.31 to be 28.79 mm. The above computation procedure is run from bdc to tdc on the compression stroke at 4000 rev/min with the initial state conditions at bdc as assumed previously at 1 atm and 20°C. The results for swirl ratio, C_{sw}, and (effective) throat velocity, c_t, are plotted with respect to crankshaft angle in Fig. 4.46.

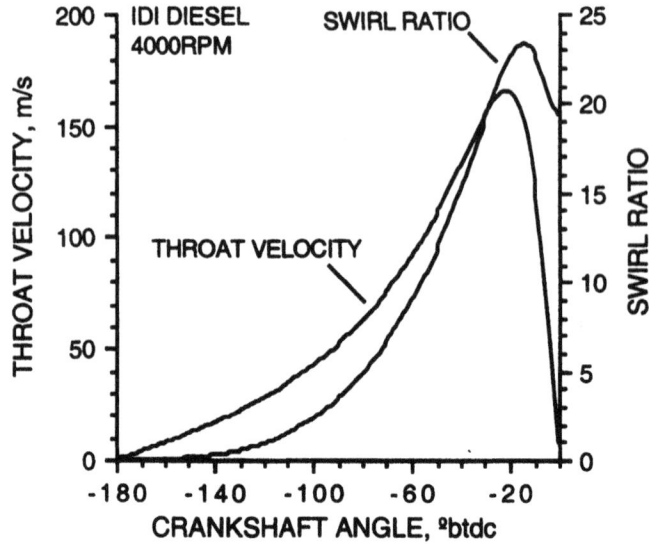

Fig. 4.46 The throat velocity and swirl ratio in an IDI diesel combustion bowl.

It can be seen that the swirl ratio is a maximum at 14 °btdc, which is almost precisely the location of the start of injection as observed from the measured data for an IDI diesel engine in Fig. 4.38. The swirl ratio, C_{sw}, has a maximum value of 23.3. The throat velocity, c_t, has a maximum value of 165.5 m/s and peaks at 21 °btdc. It can be seen that this is a very much higher squish velocity, which maximizes earlier before tdc than any observed for the open chambers of either spark-ignition or compression-ignition engines discussed in Sec. 4.5.3.

If the engine speed is varied from 1500 to 4500 rev/min, a conventional speed range for an IDI diesel car engine, the computation output as plotted in Fig. 4.47, shows that the throat velocity varies almost linearly with engine speed from 60 to 190 m/s. Both the throat velocity and the swirl ratio are plotted as the maximum values at their respective crank angle positions. The swirl ratio does vary, but to a lesser degree, over this speed range from 21.2 to 23.7, providing almost constant mixing and vaporization of the injected fuel—one of the profound advantages of the IDI diesel engine. By comparison, in a DI diesel engine, the swirl ratio induced by intake flow tends to vary almost linearly with engine speed.

If the throat diameter, d_t, is varied from 8.0 to 9.5 mm, a range of squish area ratios, C_{sq}, from 0.991 to 0.988 is obtained, which from a design standpoint is apparently a very narrow range. The computation is repeated at 4000 rev/min, and the results plotted in Fig. 4.48. The throat velocity and the swirl ratio are plotted as the maximum values at their respective crank angle positions. The effect on both throat velocity and swirl ratio is considerable as they change

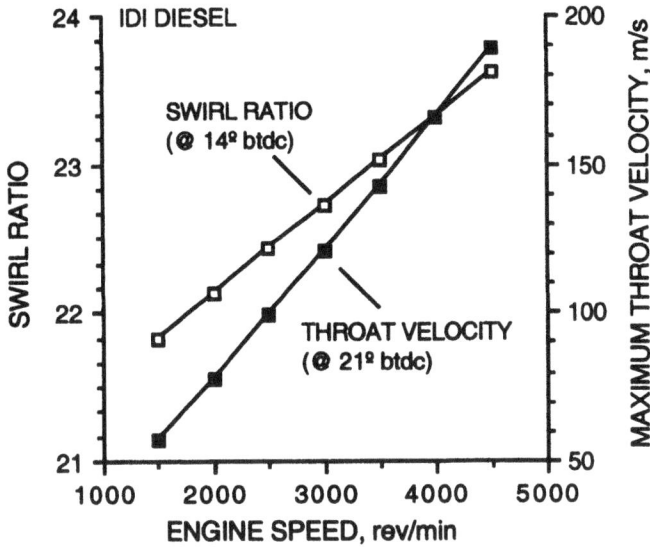

Fig. 4.47 The throat velocity and swirl ratio in an IDI bowl as a function of engine speed.

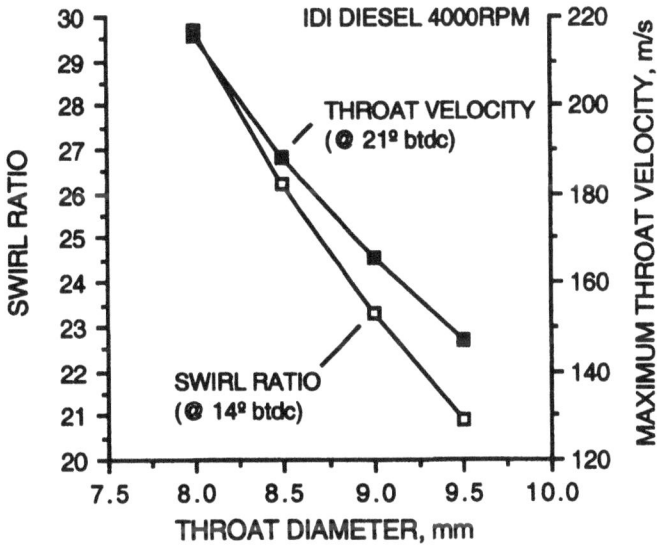

Fig. 4.48 The throat velocity and swirl ratio in an IDI bowl as a function of throat diameter.

from 247 to 147 m/s, and from 29.7 to 21.6, respectively. Clearly the size, i.e., the area, A_t, of the throat is an extremely sensitive design parameter even from a manufacturing standpoint; it must be machined with some accuracy to ensure inter-cylinder combustion equality. The swirl ratio changes by about 3 units for every 0.5 mm difference in diameter.

If the compression ratio, CR, is changed from 19 to 23, and all other data are retained at the "standard" values, the computation at 4000 rev/min, the results of which are plotted in Fig. 4.49, shows that CR is not a hugely important parameter. The swirl ratio decreases with compression ratio and varies from 23.61 to 23.05 over the range examined; however, this is an insignificant decrease in design terms. The throat velocity drops from 176 m/s to 157 m/s, apparently exhibiting a greater effect not seen in the swirl ratio, but the smaller bowl size and the lower spin rate reduces the friction loss within it.

If the bowl volume ratio, C_{bv}, is changed within the "standard" data set to run from 0.4 to 0.6, and the computation is repeated at 4000 rev/min, the calculation results, plotted in Fig. 4.50, show that this factor also has a significant design influence. The swirl ratio increases with bowl volume ratio, i.e., with a larger bowl volume and a smaller squish band volume, from 19.8 to 26.6. The throat velocity is increased by the stronger squish action from 130 m/s to 204 m/s.

In summary, the most sensitive design parameters with regard to swirl ratio, in decreasing order of significance, are throat area, bowl volume ratio, engine speed, and compression ratio. The designer of such a combustion chamber must realize that this is definitely not an area where experience alone will suffice, but where some form of theoretical design treatment is absolutely essential. Failure to include some form of theoretical design treatment will result in an extensive, and even potentially inconclusive, experimental process on the test-bed, if the lessons on the relative significance of the various design parameters are ignored. The computation process given here is a very good place to begin the design procedure. From that provisional geometrical database, a CFD process can finalize the design with fewer iterations. Because that CFD process can extend to compute the droplet vaporization and combustion process (viz., Fig. 4.40 and Ref. [4.43]), the designer is urged to take those final steps.

4.5.5 General Remarks on Design to Include Squish Effects
Design of Squish Effects for Spark-Ignition Engines

The design message here is that high squish velocities lead to rapid burning characteristics, and that rapid burning approaches the thermodynamic ideal of constant volume combustion. However, there is a price to be paid for this, evidenced by more rapid rates of pressure rise, which will lead to an engine with more vibration and noise emanating from the combustion process. Nevertheless, the designer has available theoretical tools, as presented above, to tailor this effect to the best possible advantage for any particular design.

One of the side benefits of squish action is the possible reduction of detonation effects. The squish effect gives high turbulence characteristics in the end zones and, by inducing locally high squish velocities in the squish band, this increases the convection coefficients for

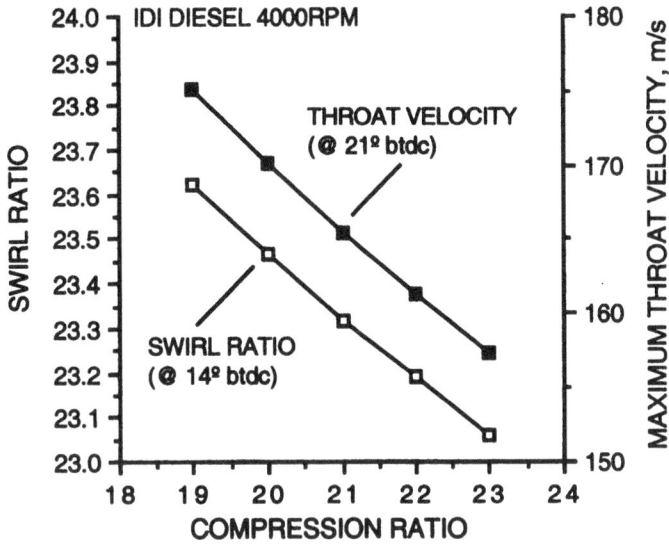

Fig. 4.49 The throat velocity and swirl ratio in an IDI bowl as a function of compression ratio.

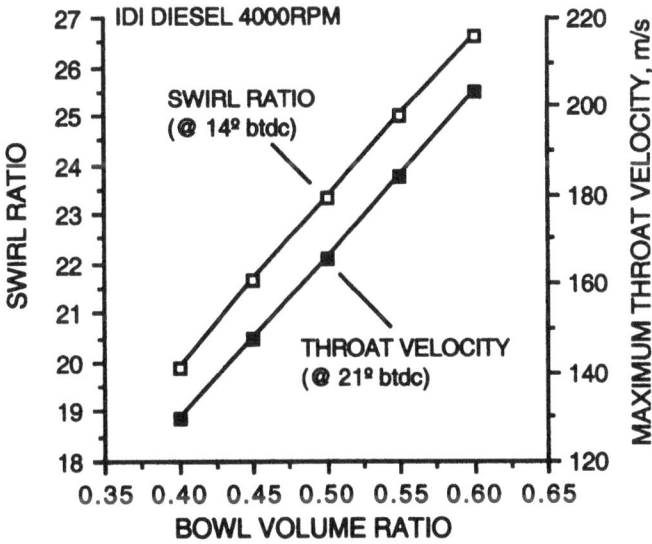

Fig. 4.50 The throat velocity and swirl ratio in an IDI bowl as a function of bowl volume ratio.

heat transfer. Should the cylinder walls be colder than the squished charge, the end zone gas temperature can be reduced to the point where detonation is avoided, even under conditions of high bmep and compression ratio. For high-performance engines, such as those used for racing, the design of squish action must be carried out by a judicious combination of theory and experimentation. A useful starting point for the design of gasoline-fueled spark-ignition engines is to keep the maximum squish velocity between 15 and 20 m/s, perhaps up to 25 m/s for methanol-fueled units, at the peak power engine speed. If the value is higher than this, the mass trapped in the end zones of the squish band may be sufficiently large and, with the faster flame front velocities engendered by a too-rapid squish action, may still induce detonation in a worst-case scenario or slow end-zone burning at best.

For combustion of other fuels, such as kerosene or natural gas, which are not noted for having naturally high flame-speed capabilities, the creation of turbulence by squish action will speed up the combustion process. For such fuels in a spark-ignition engine, squish velocities greater than 30 m/s could be advantageous to assist with the combustion of fuels that are notoriously slow burning. As with most design procedures, compromise is required, and that compromise is different depending on the performance requirements of the engine, and its fuel, over the entire speed and load range. On the other hand, it would be remiss of me not to point out that catalyst-based ignition systems have been shown to be very effective for the combustion of lower-grade paraffins such as kerosene, and also for natural gas, and should be investigated [4.58].

Design of Squish Effects for Spark-Ignition Engines with Stratified Combustion

A stratified charge is one in which the air-fuel ratio is not the same throughout the chamber at the point of combustion. This can be achieved by, for example, direct injection of fuel into the cylinder at a point sufficiently close to ignition that the vaporization process takes place as combustion commences or proceeds. By definition, diesel combustion is stratified combustion. Also by definition, the combustion of a directly injected charge of gasoline (GDI) in a spark-ignition engine is potentially a stratified process—depending on the timing of the fuel injection.

Stratified charge burning is that in which the corners of the chamber contain air only, or a very lean mixture. The engine can be run in "lean-burn, high-compression" mode without real concern for detonation problems. The homogeneous charge engine will always, under equality of test conditions, produce the highest specific power output because virtually all of the air in the cylinder can be burned with the fuel. This is not the case with a stratified-charge burning mechanism, thus the aim of the designer is to raise the air utilization value at full load to as high a level as possible, with 90% regarded as a good target value.

All such stratified combustion processes require a vigorous in-cylinder air motion to assist the mixing of air with the fuel and vaporization of the mixture, particularly at, or near, the full-load points. However, at low speed and light load, a GDI engine may not respond favorably to a vigorous in-cylinder air motion emanating from a too-strong squish action, as it can insert so generous a supply of air into the flame kernel that the local mixture becomes so lean that it is extinguished.

Design of Squish Effects for Compression-Ignition Engines

The use of squish action to enhance diesel combustion is common practice, particularly for DI diesel combustion. Of course, as shown in Sec. 4.5.4, IDI diesel combustion is a special case and it could be argued that it is one of exceptional squish behavior, in that some 50% of the entire trapped mass is squeezed into a side combustion chamber by a squish area ratio exceeding 98%!

However, for a DI diesel combustion as shown in Fig. 4.40, where the bowl has a more "normal" squish area ratio of some 50 or 60%, and with a high compression ratio, with a typical value of 17 or 18, an effective squish action must be incorporated by design, and illustrated in Sec. 4.5.3. The design message here is that the higher the compression ratio, the more severely the squish effect must be applied to acquire the requisite values of squish velocity and squish kinetic energy, and an enhancement of in-cylinder swirl into the open bowl. Thus, what is an extreme and unlikely design prospect for the spark-ignition combustion of gasoline becomes an effective and logical set of bowl geometry values for a DI diesel engine.

References for Chapter 4

4.1   B. Lewis and G. Von Elbe, *Combustion, Flames and Explosions of Gases*, Academic Press, 1961.

4.2   N.C. Blizard and J.C. Keck, "Experimental and Theoretical Investigation of Turbulent Burning Model for Internal Combustion Engines," *SAE Transactions*, Paper No. 740191.

4.3   T. Obokata, N. Hanada, and T. Kurabayashi, "Velocity and Turbulence Measurements in a Combustion Chamber of SI Engine under Motored and Firing Conditions by LDA with Fibre-Optic Pick-up," Society of Automotive Engineers, International Congress and Exposition, Detroit, Mich., February 23-27, 1987, SAE Paper No. 870166.

4.4   W.G. Agnew, "Fifty Years of Combustion Research at General Motors," *Prog. Energy Combust. Sci.*, Vol. 4, pp. 115-155, 1978.

4.5   R.J. Tabaczynski, "Turbulence and Turbulent Combustion in Spark-Ignition Engines," *Prog. Energy Combust. Sci.*, Vol. 2, p. 143, 1977.

4.6   R.J. Tabaczynski, S.D. Hires, and J.M. Novak, "The Prediction of Ignition Delay and Combustion Intervals for a Homogeneous Charge, Spark-ignition Engine," Society of Automotive Engineers, International Congress and Exposition, Detroit, Mich., February, 1978, SAE Paper No. 780232.

4.7   J.B. Heywood, *Internal Combustion Engine Fundamentals*, McGraw-Hill, New York, 1989.

4.8   R. Herweg, Ph. Begleris, A. Zettlitz, and G.F.W. Zeigler, "Flow Field Effects on Flame Kernel Formation in a Spark-Ignition Engine," Society of Automotive Engineers, International Fuels and Lubricants Meeting and Exposition, Portland, Ore., October 10-13, 1988, SAE Paper No. 881639.

4.9   M.S. Hancock, D.J. Buckingham, and M.R. Belmont, "The Influence of Arc Parameters on Combustion in a Spark-Ignition Engine," Society of Automotive Engineers, International Congress and Exposition, Detroit, Mich., February 24-28, 1986, SAE Paper No. 860321.

4.10   G.F.W. Zeigler, A. Zettlitz, P. Meinhardt, R. Herweg, R. Maly, and W. Pfister, "Cycle Resolved Two-Dimensional Flame Visualization in a Spark-Ignition Engine," Society of Automotive Engineers, International Fuels and Lubricants Meeting and Exposition, Portland, Ore., October 10-13, 1988, SAE Paper No. 881634.

4.11   P.O. Witze, M.J. Hall, and J.S. Wallace, "Fiber-Optic Instrumented Spark Plug for Measuring Early Flame Development in Spark-ignition Engines," Society of Automotive Engineers, International Fuels and Lubricants Meeting and Exposition, Portland, Ore., October 10-13, 1988, SAE Paper No. 881638.

4.12   R.J. Kee, "Stratified Charging of a Cross-Scavenged Two-Stroke Cycle Engine," Doctoral Thesis, The Queen's University of Belfast, October, 1988.

4.13   S. Onishi, S.H. Jo, K. Shoda, P.D. Jo, and S. Kato, "Active Thermo-Atmosphere Combustion (ATAC) - A New Combustion Process for Internal Combustion Engines," Society of Automotive Engineers, International Congress and Exposition, Detroit, Mich., February, 1979, SAE Paper No. 790501.

4.14   Y. Ishibashi and M. Asai, "Improving the Exhaust Emissions of Two-Stroke Engines by Applying the Activated Radical Combustion," Society of Automotive Engineers, International Congress and Exposition, Detroit, Mich., February, 1996, SAE Paper No. 960742.

4.15   Y. Ishibashi and M. Asai, "A Low-Pressure Pneumatic Direct Injection Two-Stroke Engine by Active Radical Combustion Concept," Society of Automotive Engineers, International Congress and Exposition, Detroit, Mich., February, 1998, SAE Paper No. 980757.

4.16   P.M. Najt, and D.E. Foster, "Compression-Ignited Homogeneous Charge Combustion," Society of Automotive Engineers, International Congress and Exposition, Detroit, Mich., February, 1983, SAE Paper No. 830264.

4.17   L.A. Gusak, "High Chemical Activity of Incomplete Combustion Products and a Method of Pre-Chamber Torch Ignition for Avalanche Activation of Combustion in Internal Combustion Engines," Society of Automotive Engineers, International Congress and Exposition, Detroit, Mich., February, 1975, SAE Paper No. 750890.

4.18   W.S. Affleck. P.E. Bright, and A. Fish, "Run-On in Gasoline Engines: A Chemical Description of Some Effects of Fuel Composition," *Combustion and Flame*, 12, p. 307-317.

4.19   G.T. Kalghatgi, P. Snowdon, and C.R. McDonald, "Studies of Knock in a Spark-Ignition Engine with CARS; Temperature Measurements and Using Different Fuels," Society of Automotive Engineers, International Congress and Exposition, Detroit, Mich., February, 1995, SAE Paper No. 950690.

4.20   A. Lööf, K. Rydquist, and S. Strömberg, "Influence of Gas Exchange and Volumetric Efficiency on Knock Behavior in a Spark-Ignition Engine," Society of Automotive Engineers, International Congress and Exposition, Detroit, Mich., February, 1998, SAE Paper No. 980894.

4.21   G.M. Rassweiler and L. Withrow, "Motion Pictures of Engine Flames Correlated with Pressure Cards," SAE Paper No. 800131, 1980 (originally presented in 1938).

4.22   W.T. Lyn, "Calculations of the Effect of Rate of Heat Release on the Shape of Cylinder Pressure Diagram and Cycle Efficiency," *Proc.I.Mech.E*, No. 1, 1960-61, pp. 34-37.

4.23   T.K. Hayes, L.D. Savage, and S.C. Sorensen, "Cylinder Pressure Data Acquisition and Heat Release Analysis on a Personal Computer," Society of Automotive Engineers, International Congress and Exposition, Detroit, Mich., February 24-28, 1986, SAE Paper No. 860029.

4.24   D.R. Lancaster, R.B. Krieger, and J.H. Lienisch, "Measurement and Analysis of Engine Pressure Data," *Trans. SAE*, Vol. 84, p. 155, 1975, SAE Paper No. 750026.

4.25   L. Martorano, G. Chiantini, and P. Nesti, "Heat Release Analysis for a Two-Spark-ignition Engine," International Conference on the Small Internal Combustion Engine, Paper C372/026, Institution of Mechanical Engineers, London, 4-5 April, 1989.

4.26   C.F. Daniels, "The Comparison of Mass Fraction Burned Obtained from the Cylinder Pressure Signal and Spark Plug Ion Signal," Society of Automotive Engineers, International Congress and Exposition, Detroit, Mich., February, 1998, SAE Paper No. 980140.

4.27   M.J. Brunt, H. Rai, and A.L. Armitage, "The Calculation of Heat Release Energy from Engine Cylinder Pressure Data," Society of Automotive Engineers, International Congress and Exposition, Detroit, Mich., February, 1998, SAE Paper No. 981052.

4.28   M.G.O. Reid, "Combustion Modelling for Two-Stroke Cycle Engines," Doctoral Thesis, The Queen's University of Belfast, May, 1993.

4.29   M.G.O. Reid and R. Douglas "A Closed Cycle Model with Particular Reference to Two-Stroke Cycle Engines," Society of Automotive Engineers International Off-Highway & Powerplant Congress, Milwaukee, Wisc., September 1991, SAE Paper No. 911847.

4.30   M.G.O. Reid and R. Douglas "Quasi-Dimensional Modelling of Combustion in a Two-Stroke Cycle Spark-ignition Engine," Society of Automotive Engineers International Off-Highway & Powerplant Congress, Milwaukee, Wisc., September 1994, SAE Paper No. 941680.

4.31   W.J.D. Annand, "Heat Transfer in the Cylinders of Reciprocating Internal Combustion Engines," *Proc.I.Mech.E.*, Vol. 177, p. 973, 1963.

4.32   W.J.D. Annand, and T.H. Ma, "Instantaneous Heat Transfer Rates to the Cylinder Heat Surface of a Small Compression-ignition Engine," *Proc.I.Mech.E.*, Vol. 185, p. 976, 1970-71.

4.33 W.J.D. Annand and D. Pinfold, "Heat Transfer in the Cylinder of a Motored Recip-rocating Engine," Society of Automotive Engineers, International Congress and Exposition, Detroit, Mich., February, 1980, SAE Paper No. 800457.

4.34 G. Woschni, "A Universally Applicable Equation for the Instantaneous Heat Trans-fer Coefficient in the Internal Combustion Engine," SAE Paper No. 670931, Novem-ber, 1967.

4.35 J. Mackerle, *Air-Cooled Automotive Engines*, Charles Griffin, London, 1972.

4.36 G.P. Blair, *Design and Simulation of Two-Stroke Engines*, R-161, Society of Auto-motive Engineers, Warrendale, Pa., March, 1996, p. 623.

4.37 I.I. Vibe, "Brennverlauf und Kreisprozeb von Verbrennungs-Motoren," VEB Technik Berlin, 1970.

4.38 L.R.C. Lilly (Editor), *Diesel Engine Reference Book*, Butterworths, London, 1984.

4.39 G.P. Blair, "Prediction of Two-Cycle Engine Performance Characteristics," SAE Paper No. 760645, Society of Automotive Engineers, Warrendale, Pa., 1976.

4.40 G.P. Blair, "Correlation of Measured and Calculated Performance Characteristics of Motorcycle Engines," 5th Graz Two-Wheeler Symposium, Graz, Austria, April 1993, p. 5-16.

4.41 A.A. Amsden, J.D. Ramshaw, P.J. O'Rourke, and J.K. Dukowicz, "KIVA: A Com-puter Program for Two- and Three-Dimensional Fluid Flows with Chemical Reac-tions and Fuel Sprays," Los Alamos National Laboratory Report, LA-102045-MS, 1985.

4.42 A.A. Amsden, T.D. Butler, P.J. O'Rourke, and J.D. Ramshaw, "KIVA–A Compre-hensive Model for 2-D and 3-D Engine Simulations," Society of Automotive Engi-neers, International Congress and Exposition, Detroit, Mich., February, 1985, SAE Paper No. 850554.

4.43 R. Aneja and J. Abraham, "Comparisons of Computed and Measured Results of Combustion in a Diesel Engine," Society of Automotive Engineers, International Congress and Exposition, Detroit, Mich., February, 1998, SAE Paper No. 980786.

4.44 J.F. O'Connor and N.R. McKinley, "Comparison of Intake Port CFD Flow Pre-dictions using Automatic Brick Meshing with Swirl Laser Sheet and LDA Flow Measurements," Society of Automotive Engineers, International Congress and Exposition, Detroit, Mich., February, 1998, SAE Paper No. 980129.

4.45 D. Fitzgeorge and J.L. Allison, "Air Swirl in a Road-Vehicle Diesel Engine," *Proc.I.Mech.E.*, Vol. 4, p. 151, 1962-63.

4.46 T.D. Fansler, "Laser Velocimetry Measurements of Swirl and Squish Flows in an Engine with a Cylindrical Piston Bowl," Society of Automotive Engineers, Inter-national Congress and Exposition, Detroit, Mich., February, 1985, SAE Paper No. 850124.

4.47 B.A. Ajakaiye and J.C. Dent, "A Correlation for Air Velocities in Cylindrical Pre-Chambers of Diesel Engines," *Trans. ASME*, Journal of Engineering for Power, Vol. 103, p. 499-504, July 1981.

4.48 B.J. Somerville, S.J. Charlton, S.A. McGregor, and B. Nasseri, "A Study of Air Motion in an IDI Passenger Car Diesel Engine," I.Mech.E. Conference, C465/002, 1993, p.37-44.

4.49 D.A. Hamrin and J.B. Heywood "Modelling of Engine-Out Hydrocarbon Emissions for Prototype Production Engines," Society of Automotive Engineers, International Congress and Exposition, Detroit, Mich., February, 1995, SAE Paper No. 950984.

4.50 R.W. Schefer, R.D. Matthews, N.P. Ceransky, and R.F. Swayer, "Measurement of NO and NO_2 in Combustion Systems," Paper 73-31 presented at the Fall meeting, Western States Section of the Combustion Institute, El Segundo, Calif., October 1973.

4.51 Ya.B. Zeldovitch, P. Ya. Sadovnikov, and D.A. Frank-Kamenetskii, "Oxidation of Nitrogen in Combustion," (transl. by M. Shelef), Academy of Sciences of USSR, Institute of Chemical Physics, Moscow-Leningrad, 1947.

4.52 V. Sulakov and G. Merker, "Nitrogen Oxidizing in Modeling of Diesel Engine Operation," Society of Automotive Engineers, International Congress and Exposition, Detroit, Mich., February, 1995, SAE Paper No. 952063.

4.53 National Standards Reference Data System, Table of Recommended Rate Constants for Chemical Reactions Occuring in Combustion, QD502 /27

4.54 G.A. Lavoie, J.B. Heywood, and J.C. Keck, *Combustion Science Technology*, 1, 313, 1970.

4.55 G. De Soete, *Rev. Institut Français du Pétrole*, 27, 913, 1972.

4.56 V.S. Engleman, W. Bartok, J.P. Longwell, R.B. Edleman, Fourteenth Symposium (International) on Combustion, p. 755, The Combustion Institute, 1973.

4.57 *Bosch Automotive Handbook*, Robert Bosch, Stuttgart, 4th Edition, October 1996 (distributed by SAE).

4.58 M.A. Cherry, R.J. Morrisset, and N.J. Beck, "Extending Lean Limit with Mass-Timed Compression Ignition using a Catalytic Plasma Torch," Society of Automotive Engineers, SAE Paper No. 921556.

4.59 J.W. Daily, "Cycle to Cycle Variations; A Chaotic Process," Society of Automotive Engineers, SAE Paper No. 870165.

4.60 J.C. Kantor, "A Dynamical Instability of Spark-Ignited Engines," *Science*, pp. 1233-1235, June 15, 1984.

4.61 L. Chew, R. Hoekstra, J.F. Nayfeh, and J. Navedo, "Chaos Analysis of In-Cylinder Pressure Measurement," Society of Automotive Engineers, SAE Paper No. 942486.

4.62 J. Gleick, *Chaos, Making a New Science*, Penguin Books, New York, 1987.

4.63 H-O Peitgen, H. Jürgens, and D. Saupe, *Chaos and Fractals, New Frontiers of Science*, Springer-Verlag, Berlin, 1992.

Appendix A4.1 Exhaust Emissions
The Combustion Process

The combination of Eqs. 4.3.3 and 4.3.4 permits the determination of the molecular composition of the products of combustion for any hydrocarbon fuel, CH_nO_m, at any given air-fuel ratio, AFR.

The mass ratio of any given component gas "G" within the total, ε_G, is found with respect to the total molecular weight of the combustion products, M_c:

$$M_c = \frac{x_1 M_{CO} + x_2 M_{CO_2} + x_3 M_{H_2O} + +x_4 M_{O_2} + x_5 M_{H_2} + x_6 M_{N_2} + x_7 M_{CH_4}}{x_1 + x_2 + x_3 + x_4 + x_5 + x_6 + x_7}$$

(A4.1.1)

$$\varepsilon_G = \frac{x_G M_G}{M_c}$$

(A4.1.2)

Hence, if the engine power output is $\dot{W}$ (in kW units), the delivery ratio is DR, the trapping efficiency is TE, and the engine speed is N (in units of rev/min), many of the brake specific pollutant emission figures from combustion can be determined from an actual engine simulation, or from simple design estimations, in the following manner. Using information from Eqs. 1.6.4 and 1.6.8, the total mass of air and fuel trapped during each cycle within the engine, m_{cy}, is given by:

$$m_{cy} = TE \times DR \times m_{dref} \times \left(1 + \frac{1}{AFR}\right)$$

(A4.1.3)

The mass of gas pollutant "G" produced per hour, is therefore:

$$\dot{m}_G = 60 \times \frac{N}{2} \times m_{cy} \times \varepsilon_G \quad kg/h$$

(A4.1.4)

and the brake specific pollutant rate for gas "G," bsG, is found as:

$$bsG = \frac{\dot{m}_G}{\dot{W}} \quad kg/kWh$$

(A4.1.5)

Carbon Monoxide Emissions

Carbon monoxide is obtained only from the combustion source, so the brake specific pollutant rate for CO is found as:

$$bsCO = \frac{\dot{m}_{CO}}{\dot{W}} \quad kg/kWh \quad (A4.1.6)$$

Combustion Derived Hydrocarbon Emissions

The brake specific pollutant rates for these are found as:

$$bsHC_{comb} = \frac{\dot{m}_{CH_4}}{\dot{W}} \quad kg/kWh \quad (A4.1.7)$$

For the simple two-stroke engine hydrocarbon emissions are but a minor contributor compared to those from scavenge losses, if scavenging is indeed being conducted by air-containing fuel. Reference [4.36] contains a full discussion of this topic.

For the four-stroke spark-ignition engine, "scavenging losses" can occur if the engine has a valve overlap period of any significance. It must never be assumed that its trapping efficiency is always unity. If this engine has direct fuel injection (GDI) then, even if TE is less than unity, the short-circuited loss will be air, but not fuel.

For the four-stroke compression-ignition engine, any "scavenging losses" will be of air only and while the air trapping efficiency, TE, can be—and often is—less than 100%, the fuel trapping efficiency, TE_f, is unity.

The computation of combustion-derived hydrocarbons is a very difficult subject from a theoretical standpoint, and the chemistry is not only complex but highly dependent on the mechanism of flame propagation and decay, by quenching or otherwise at the walls or in the corners of the chamber [4.7]. The situation is compounded by the burning of lubricating oil from the cylinder walls, the valve stems, and the piston ring crevices. Although the literature contains many publications on this topic, even some of the most authoritative sources turn to what can only be described as semi-empiricism in tackling the subject from a design stand-point [4.49].

Scavenge Derived Hydrocarbon Emissions

The mass of charge through the inefficiency of scavenging is found as:

$$\dot{m}_{HCscav} = 60 \times \frac{N}{2} \times \frac{DR \times (1 - TE) \times m_{dref}}{AFR} \quad kg/h \quad (A4.1.8)$$

Consequently these devolve to a brake specific pollutant rate as:

$$bsHC_{scav} = \frac{\dot{m}_{HC_{scav}}}{\dot{W}} \ kg/kWh \qquad (A4.1.9)$$

Total Hydrocarbon Emissions

The total hydrocarbon emission rate is then given by the sum of that in Eq. A4.1.7 and Eq. 4.1.9 as bsHC:

$$bsHC = bsHC_{comb} + bsHC_{scav} \qquad (A4.1.10)$$

Emission of Oxides of Nitrogen

Extensive field tests have shown that nitric oxide, NO, is the predominant nitrogen oxide emitted by combustion devices; this is to be seen in recent investigations by Schefer et al. [4.50] and Zeldovich et al. [4.51].

The two principle sources of NO in the combustion of conventional fuels are oxidation of atmospheric nitrogen (molecular N_2) and to a lesser extent oxidation of nitrogen-containing compounds in the fuel (fuel nitrogen).

The mechanism of NO formation from atmospheric nitrogen has been extensively studied by several prominent researchers. It is generally accepted that in combustion of lean and near stoichiometric air-fuel mixtures, the principle reactions governing formation of NO from molecular nitrogen are those proposed by Zeldovitch [4.51]. The reactions in compression-ignition engines are considered to be similar [4.52].

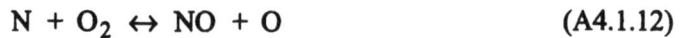

$$O + N_2 \leftrightarrow NO + N \qquad (A4.1.11)$$

$$N + O_2 \leftrightarrow NO + O \qquad (A4.1.12)$$

The forward and reverse rate constants for these reactions have been measured in numerous experimental studies, and kinetic details for this model have been sourced from NSRDS [4.53]

Lavoie et al. [4.54] have suggested that the reaction described by,

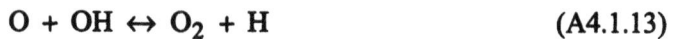

$$O + OH \leftrightarrow O_2 + H \qquad (A4.1.13)$$

should also be included. This suggestion should not be disregarded, as the argument for including this reaction with the Zeldovitch equations above is that during rich and near-stoichiometric air-fuel ratios this third, rate-limiting reaction prevails. If this model is used to predict the formation rate of NO in the cylinder of a compression-ignition engine, where the richest trapped air-fuel ratio will be approximately 20, then the NO rate model may be able to exclude this third, rate-limiting reaction.

The NO formation rate is much slower than the combustion rate, and most of the NO is formed after the completion of the combustion due to the high temperatures present in the combustion zone. Therefore, Eqs. A4.1.11 and 4.1.12 are decoupled from the combustion model.

It was reported [4.54, 4.53] that measured values of the NO formed in the post-flame front zone are greater than those predicted by the reaction kinetics. Several models have been formulated to account for this situation, which is due to "prompt NO" formation of NO.

The reaction kinetics, which are an integral part of the model, are generally formulated under ideal laboratory conditions whereby the combustion occurs in a shock tube. Clearly, however, this is somewhat remote from the closed cycle combustion taking place in the spark-ignition or the compression-ignition engine.

The temperature that is calculated in the cylinder as an outcome of the increase in pressure corresponding to the period of burn is the average in-cylinder temperature. This temperature is the product of two distinct zones in the cylinder, namely the burn zone, composed of the current products of combustion and the unburned zone, composed of the remaining air and exhaust gas residual. De Soete [4.55], proposed that a greater rate of NO formation would be recorded in the post-flame zone, the temperature of which has been raised by the passing flame. To emulate this, the NO_x model uses the average temperature, T_b, in the burn zone. As described previously, the flame packet contains air, fuel, and exhaust gas residual. This packet mass varies with time step, and its position during the heat release process. At the conclusion of the burning of each packet during a computation time step, normally about one degree crankshaft, the burn zone has new values of mass, volume, temperature, and mass concentrations of oxygen and nitrogen. The NO rate formation model may now be generally described as follows:

$$\frac{dNO}{dt} = k \times F \left(m_{b\,O_2}, m_{b\,N_2}, V_b, T_b \right) \qquad (A4.1.14)$$

where V_b is the volume of the burn zone, T_b is the temperature in the burn zone, and k is a rate limiting constant. Once the formation of NO is determined as a function of time, the formation of NO in any given time-step of an engine simulation is determined, and the summation of that mass increment over the combustion period gives the total mass formation of the oxides of nitrogen.

In practice, the execution of this model within a computer simulation is not quite so straightforward, as it is necessary to solve the equilibrium and dissociation behavior within the burn zone. The amount of free oxygen within the burn zone is a function of the local temperature and pressure, as has been discussed in Sec. 4.3.2. Two equilibrium reactions must be followed closely, the first being that for the carbon monoxide, given in Eq. 4.3.18, and the second for the so-called "water-gas" reaction:

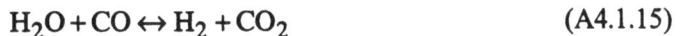

$$H_2O + CO \leftrightarrow H_2 + CO_2 \qquad (A4.1.15)$$

It is clear that this can only be solved through the use of a two-zone burning model, as described in Appendix A4.2. There, the results of a computer simulation of an engine which predicts NO formation using the above theory are presented.

The "Water Gas" Reaction

The equilibrium constant, K_p, for this reaction introduced above is a function of temperature, and is described and tabulated in many standard texts on thermodynamics. As a fitted curve, as a function of the temperature, T (in Kelvin units), K_p is evaluated as follows:

$$\log_e K_p = -5.858 + \frac{8.4348}{10^3} T - \frac{3.4973}{10^6} T^2 + \frac{6.5595}{10^{10}} T^3 - \frac{4.5914}{10^{14}} T^4 \quad (A4.1.16)$$

For the dissociation reaction it is incorporated as:

$$K_p = \frac{P_{H_2} P_{CO_2}}{P_{H_2O} P_{CO}} \quad (A4.1.17)$$

General

The use of these theories to predict pollutant gas emission rates is illustrated in Appendix A4.2 and in Chapters 5 and 6. In particular, this theory is employed to show the effect of changing engine geometry on either the performance or the exhaust emission characteristics.

Appendix A4.2 A Simple Two-Zone Combustion Model
The Combustion Process

Sec. 4.4.2 details a single zone combustion model. The physical parameters for the theoretical solution of a combustion process in two zones are shown in the sketch in Fig. A4.1.

The assumption that pervades this approach is that the pressure in the unburned and burned zones is equal at the beginning and end of a time step in a computation, which is represented by a time interval, dt, or a crankshaft interval, dθ. The piston movement produces volume variations from V_1 to V_2 during this period, and so the mean cylinder conditions of pressure and temperature change from p_1 to p_2, and T_1 to T_2, respectively. The total cylinder mass, m_c, is constant, but the masses in the unburned and burned zones, m_b and m_u, change with respect to the increment of mass fraction burned, dB, during this time interval, thus:

$$dB = B_2 - B_1 = B_{\theta_1 + d\theta} - B_{\theta_1} \tag{A4.2.1}$$

and
$$m_{b2} = m_{b1} + dB \times m_c \tag{A4.2.2}$$

$$m_{u2} = m_{u1} - dB \times m_c \tag{A4.2.3}$$

also
$$m_c = m_{u1} + m_{b1} \quad \text{and} \quad m_c = m_{u2} + m_{b2} \tag{A4.2.4}$$

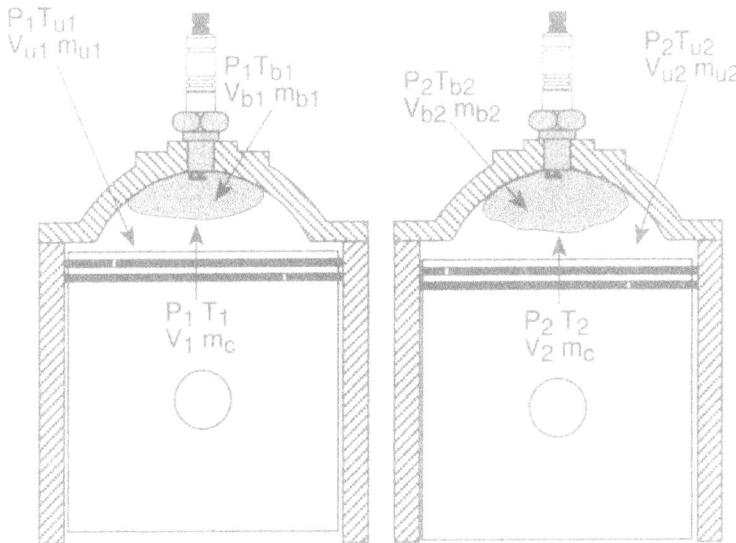

Fig. A4.1 Simple theoretical model for two-zone combustion.

The purities in both the burned and unburned zones are known. That in the burned zone is zero and at the initial temperature, T_{b1}, the theory of Sec. 2.2.6 can be used to determine the gas properties at that temperature with respect to the air-fuel ratio and the particular hydrocarbon fuel being used. This provides the numerical values of gas constant, R_b, specific heat at constant volume, C_{vb}, and the ratio of specific heats, γ_b. The purity in the unburned zone is that at the trapping condition, Π_t. In the unburned zone, at temperature T_{u1}, the properties of gas constant, R_u, specific heat at constant volume, C_{vu}, and the ratio of specific heats, γ_u, may also be determined from Sec. 2.1.6. The average properties of the entire cylinder space at the commencement of the time step, and those assumed to prevail during the time step, may be found as:

$$R_1 = B_1 R_b - (1 - B_1) R_u \qquad (A4.2.5)$$

$$C_{v1} = B_1 C_{vb} - (1 - B_1) C_{vu} \qquad (A4.2.6)$$

$$\gamma_1 = \frac{C_{p1}}{C_{v1}} = \frac{C_{v1} + R_1}{C_{v1}} = 1 + \frac{R_1}{C_{v1}} \qquad (A4.2.7)$$

Using heat release data, or a mass fraction burned approach, as described in Sec. 4.4.2, the overall behavior of the entire cylinder space may be found as before, using Eq. 4.4.1, or even more simply as below if the time interval is sufficiently short. A sufficiently short time interval is defined as $1°$ crankshaft angle.

$$T_2 = \frac{\delta Q_R - \delta Q_L + m_c C_{v1} T_1 - p_1(V_2 - V_1)}{m_c C_{v1}} \qquad (A4.2.8)$$

and

$$P_2 = \frac{m_c R_1 T_2}{V_2} \qquad (A4.2.9)$$

A simple solution for the properties within the two zones is only possible if some assumption is made regarding inter-zone heat transfer. Without such an assumption, it is not possible to determine the individual volumes within each zone, and hence the individual zone temperatures cannot be determined, as the thermodynamic equation of state must be satisfied for each of them, thus:

$$T_{u2} = \frac{p_2 V_{u2}}{m_{u2} R_u} \quad \text{and} \quad T_{b2} = \frac{p_2 V_{b2}}{m_{b2} R_b} \qquad (A4.2.10)$$

Some researchers [4.27-4.29] have employed the assumption that there is zero heat transfer between the zones during combustion. This is clearly unrealistic, for if there is no heat transfer between the burned and unburned zones then detonation would not occur. The first simple assumption that can be made to instill some realism into the solution is that the process in the unburned zone is adiabatic. At first glance this also appears very naive, but this is not the case as an alternate way of stating an adiabatic assumption is that the unburned zone is gaining as much heat from the burned zone as it is losing to the surfaces of the cylinder and the piston crown. The more this assumption is examined, in terms of the relative masses, volumes, surface areas, and temperatures of the two zones, the more logical it becomes as a useful theoretical starting point. Using this assumption, the solution is straightforward:

$$\frac{p_2}{p_1} = \left(\frac{\rho_2}{\rho_1}\right)^{\gamma_u} = \left(\frac{m_{u2}}{m_{u1}} \times \frac{V_{u1}}{V_{u2}}\right)^{\gamma_u} \tag{A4.2.11}$$

The volume of the unburned zone, V_{u2}, is the only unknown in the above equation. Consequently, the volume of the burned zone is:

$$V_{b2} = V_2 - V_{u2} \tag{A4.2.12}$$

and the temperatures in the two zones may now be found, using Eq. A4.2.10.

This is arguably a more accurate solution for the single-zone theory described in Sec. 4.4.2. There, the average gas properties for the entire cylinder space are determined as in Eqs. A4.2.5 to A4.2.7, but the individual properties of the trapped air and the burned gas are determined less realistically using the average cylinder temperature, T_1, instead of the zone temperatures, T_{u1} and T_{b1}, employed here. In the example given below, it is clear that a single zone solution will induce errors in the estimation of the cylinder pressure, which translates to discrepancies in imep and power. Although these discrepancies are not negligible, they are relatively small. However, only a two-zone model will permit a calculation of the proportions of the cylinder gas constituents with respect to time. This means that exhaust emissions can be computed and, because the properties of the exhaust gas are known in some detail, this means improved accuracy in tracking the unsteady gas flow in the ducting.

The Use of This Theory within an Engine Simulation

In Chapters 5 to 7, a four-cylinder 2.0-liter automobile engine is frequently employed as a design example for the computer simulations. The geometrical data for the engine and its ducting are given in Sec. 5.9. In this section the data for the so-called "standard" engine are employed. The fuel for the simulation is unleaded gasoline with a stoichiometric air-fuel ratio of 14.3. We have available a measured burn diagram for a four-valve, naturally-aspirated spark-ignition car engine at 4800 rev/min, as shown in Fig. 4.4. The pertinent results of a computer simulation of the complete engine at 4800 rev/min, with air-fuel ratios of 12, 13, 14, 15, and 16, are shown in Figs. A4.2 to A4.4. The simulation shows that the calculated delivery ratio is

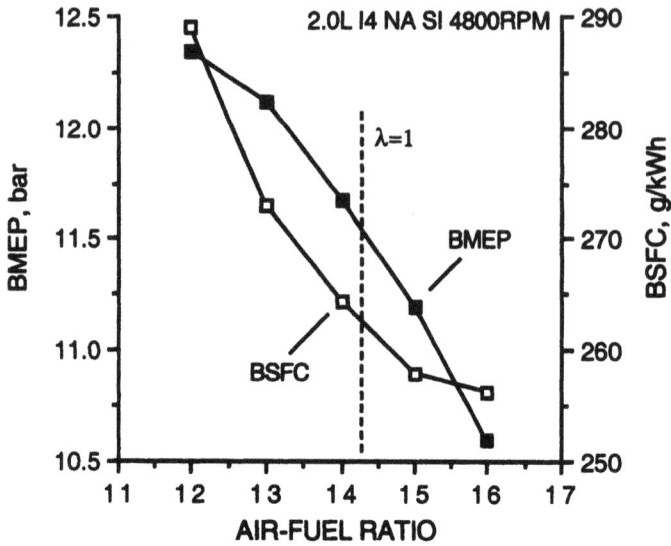

*Fig. A4.2 Brake mean effective pressure and brake specific fuel
consumption at 4800 rev/min.*

almost precisely 1.0 for these several computations at the various fueling levels. The trapping efficiency, TE, is unity throughout the calculation series. Thus, from Appendix A4.1, the bypassed hydrocarbon emissions are zero and, as you will remember, the combustion hydrocarbons are not calculated.

Fig. A4.2 shows that the engine torque, shown here as brake mean effective pressure, bmep, falls with a near-linear behavior from the richest setting, where the air-fuel ratio is 12, and the equivalence ratio, λ, is 0.84, to the leanest setting where the air-fuel ratio is 16, and the equivalence ratio, λ, is 1.12. The maximum bmep is 12.3 bar and the minimum is 10.6 bar. On the same figure is also plotted the brake specific fuel consumption. This improves with leaning the mixture, from a value of 290 g/kWh to 256 g/kWh when the equivalence ratio, λ, is 1.12, and it levels off at that juncture.

Fig. A4.3 shows the minor changes in scavenging and charging efficiency due to fueling; these are the averages of those in each of the four cylinders, but inter-cylinder variability is negligible. It can be seen that leaning off the mixture does marginally improve scavenging efficiency. I will say now that this is due to a reduced blowback into the intake tract at the onset of the valve overlap period; however, you will have to wait until Chapter 5 for a further insight into the pressure wave action involved. There is a 2% improvement in charging efficiency from the same gas dynamic origins.

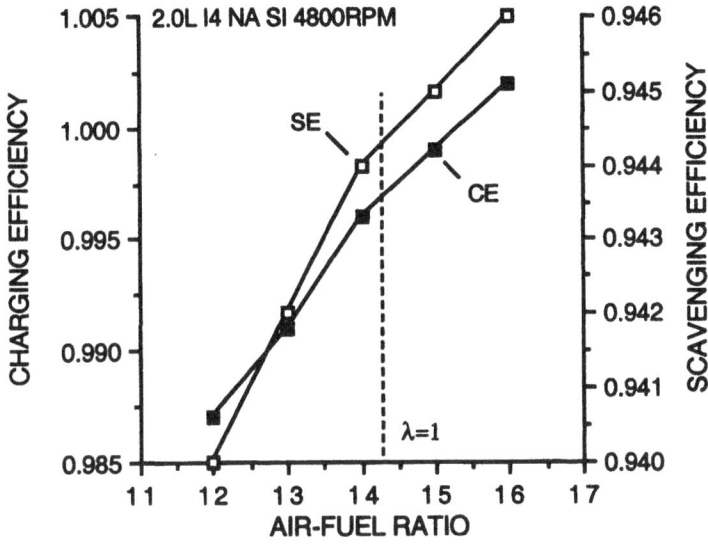

Fig. A4.3 Scavenging and charging efficiencies with varying fueling at 4800 rev/min.

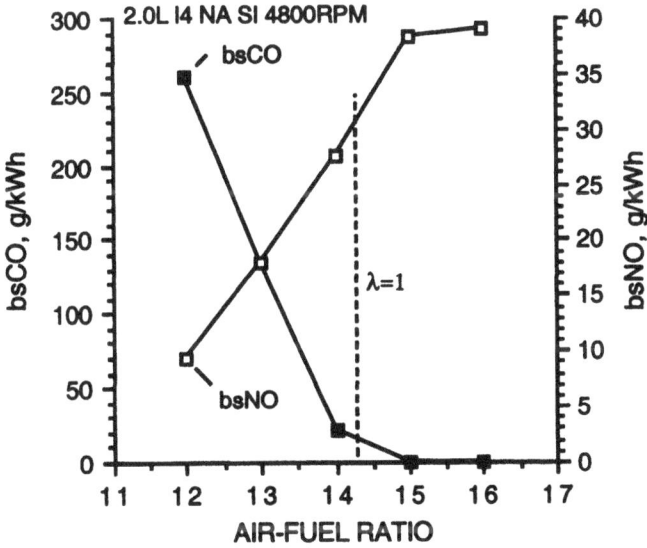

Fig. A4.4 Carbon monoxide and nitrogen oxide emissions with varying fueling at 4800 rev/min.

Fig. A4.4 shows a plot of the computed brake specific emissions of carbon monoxide and NO_x, as a nitric oxide equivalent. The traditional fall-off in bsCO toward the stoichiometric point can be observed, as is the increase in bsNO toward it [4.7, 4.36].

The results of this simulation will come as no surprise to those familiar with measuring such data, although the level of bsNO emissions at full load and a stoichiometric mixture at 4800 rev/min for a four-stroke, spark-ignition automobile engine is a data value rarely seen published as it is embarrassingly high. The equivalent data for a two-stroke engine, both measured and calculated, have been published [4.37], and indicate that the peak value around the stoichiometric region is only 10 g/kWh.

The Burned and Unburned Zone Temperatures

Fig. A4.5 shows the effect on zone temperatures of a two-zone burn process within this simulation of the 2.0-liter automobile engine. Fig. A4.5 shows the in-cylinder zone temperatures as predicted by the two-zone combustion model. The mean cylinder temperature, and the temperatures in the burned and unburned zones, are indicated on the figure, which is plotted around the tdc period. The rapid rise of temperature in the burned zone to nearly 2600°C is clearly visible, and this peaks at about 5 °atdc. The mean cylinder temperature peaks at about 2250°C, but this occurs at some 22 °atdc. The temperature in the unburned zone has a shallow

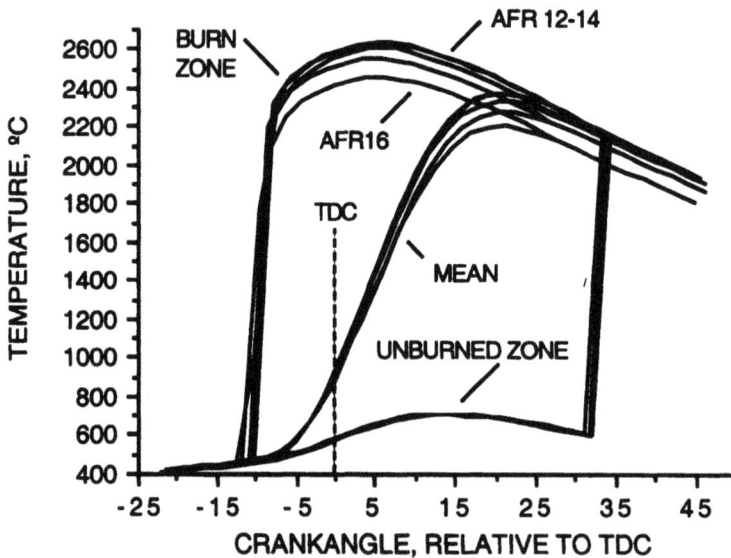

Fig. A4.5 Cylinder temperatures in a simple two-zone burn model.

rise to a peak at about 15 °atdc, whereas the spike at 33 °atdc indicates the engulfment of, and disappearance of, the unburned zone by the combustion process. The maximum burn zone temperatures recorded during the simulations are shown in Fig. A4.6, with the highest value shown to be 2650°C at an AFR of 13; this is not at the stoichiometric point, i.e., where λ is unity. The maximum temperature in the unburned zone has only a small variation over the fueling range, but does peak at 708°C. Interestingly, in a simple two-stroke engine the maximum temperature in the unburned zone is shown to be some 150°C higher [4.37].

The Formation of Nitric Oxide (NO)

The formation of oxides of nitrogen is very dependent on temperature and on the oxygen content at any juncture. In Fig. A4.6, as captured from Fig. A4.5, the two-zone burn model shows the peak temperatures in both the burned and the unburned zones, with respect to the fueling changes. The equilibrium and dissociation chemistry and the reaction kinetics, described in Sec. 4.2.4 and these Appendix A4.1, provide the basis of the time- and location-related gas constituents.

The Zeldovich [4.51] approach to the computation of NO formation, as in Eq. A4.1.14, takes all of these factors into account. The profile of the calculated NO formation, shown in Fig. A4.4 with respect to air-fuel ratio, is quite conventional. That the bsNO peaks at a leaner

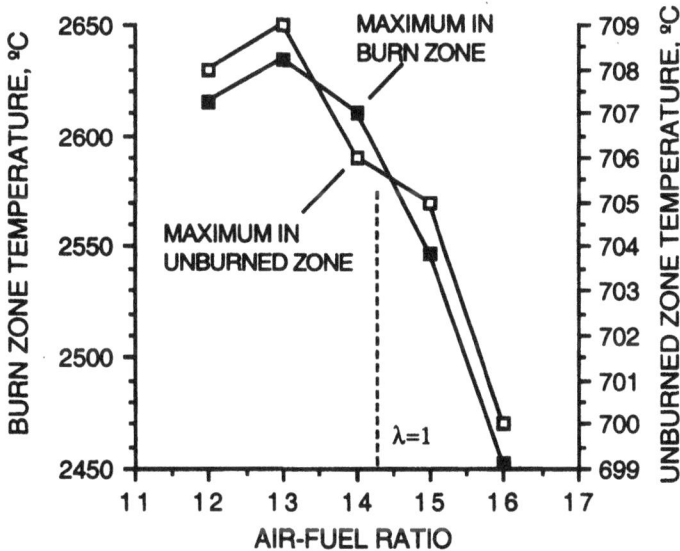

Fig. A4.6 The effect of fueling on peak temperatures in the two zones.

513

AFR than the AFR that gives the maximum burn zone temperature, may seem surprising. The reason is, whereas the maximum burn zone temperature is decreasing toward the stoichiometric air-fuel ratio, the oxygen concentration therein is rising (see Fig. A4.8).

The formation of nitric oxide, NO, is time-related, as evidenced by Eq. A4.1.14. This is shown clearly by the results of the simulation of the automobile engine in Fig. A4.7. The formation of nitric oxide, NO, is shown plotted with respect to crank angle during and after the combustion period. The highest rates of formation are at the air-fuel ratios of 14 and 15, which bracket the stoichiometric value. In addition, they occur at about 20 °atdc, which just happens to coincide with the peak of the mean cylinder temperature. This is but coincidence, as the real formation is taking place within the higher temperature of the burn zone where there is oxygen present in increasing quantities as the mixture is leaned off. Clearly visible in Fig. A4.7, the mass of NO present within the cylinder rises and falls on both sides of its maximum, indicating that the rates of formation operate in both directions.

The Time Related Equilibrium and Dissociation Behavior

As discussed in Appendix A4.1, it is necessary to solve simultaneously the dissociation reactions for CO/CO_2 and the "water gas" reaction. The results of such a computation have great relevance, not only for the NO formation model discussed above, but also for the exhaust gas properties presented in Sec. 2.1.6. The time-related cylinder gas properties expand on the information given by the relatively simple chemistry found at the beginning of Sec. 4.3.2. Exhaust gas is ultimately the gas that occupies the burn zone at the point of exhaust valve opening, for at that juncture the combustion period is over and the "burn zone" covers the entire volume of the cylinder.

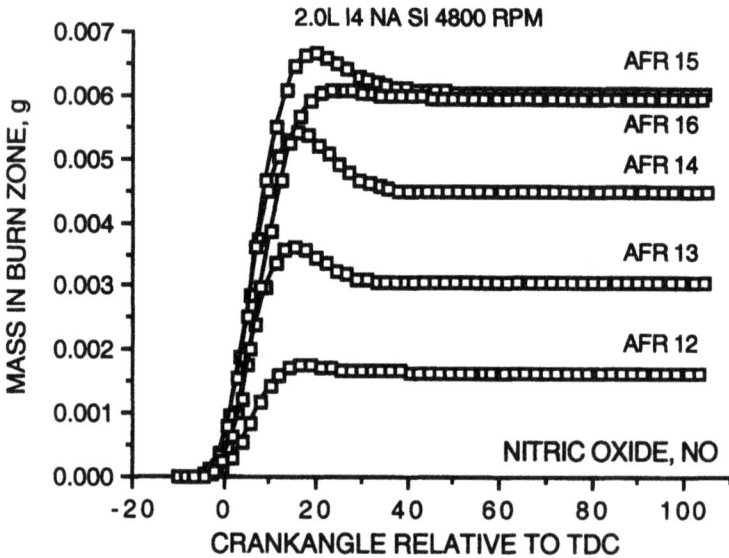

Fig. A4.7 Mass of nitric oxide (NO) in cylinder burn zone.

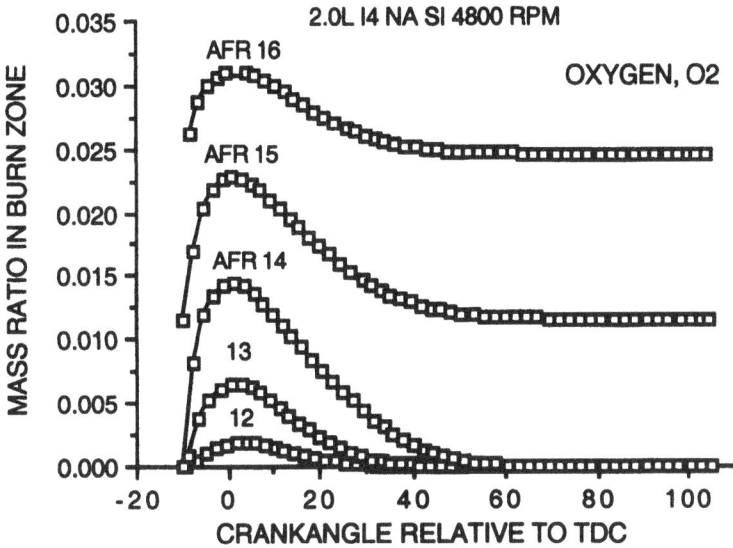

Fig. A4.8 Mass of oxygen in cylinder burn zone.

The simulation of the car engine at 4800 rpm, which provides the data for Figs. A4.2 to A4.5, is queried further to yield the time histories of the mass concentrations of some of the other gases within the burn zone. These are extracted and plotted in Figs. A4.8 to A4.12, showing the mass concentration ratios for the following gases: oxygen, carbon monoxide, carbon dioxide, hydrogen, and steam., i.e., O_2, CO, CO_2, H_2, and H_2O, respectively.

Oxygen and Its Effect on NO Formation

From the discussion above on the formation of nitric oxide, Fig. A4.8 is the most relevant. It shows the mass concentration of oxygen which, together with the nitrogen present, yields the nitric oxide. At equal burn zone temperatures and total mass, the rate of formation of nitric oxide will be almost directly proportional to the mass concentration of oxygen. It is interesting to note that the mass concentrations of oxygen at air-fuel ratios of 12 and 14 are 0.0019 and 0.0145, respectively, which differ by a factor of 7.6. The masses of NO created at the same fueling levels and 0.0035 and 0.0108 grams, respectively, which only differ by a factor of 3. The exponential effect of burn zone temperature is now clarified, for the peak burn zone temperatures at AFR values of 12 and 14 are 2615 and 2610°C, respectively, seemingly almost identical, yet the exponential functions involved produce a factor of 2.5 on NO formation for this 5° temperature differential.

Fig. A4.8 shows that the dissociation of oxygen is dependent on temperature and mixture strength. However, at any given fueling level, by 50 °atdc, which is about 50° crankshaft before the exhaust valve opens, the mass concentration of oxygen has become stabilized at an appropriate level. The leaner is the fueling, the greater is the oxygen content of the exhaust gas, although it is only significant for air-fuel ratios that are leaner than stoichiometric. The

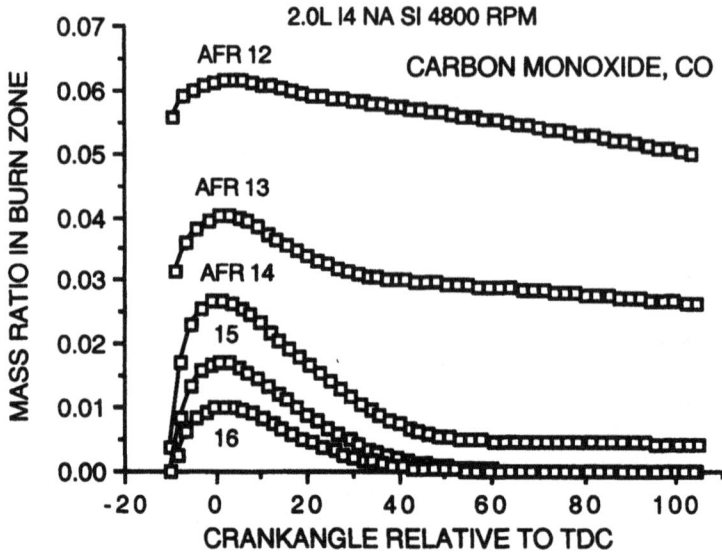

Fig. A4.9 Mass ratio of carbon monoxide in cylinder burn zone.

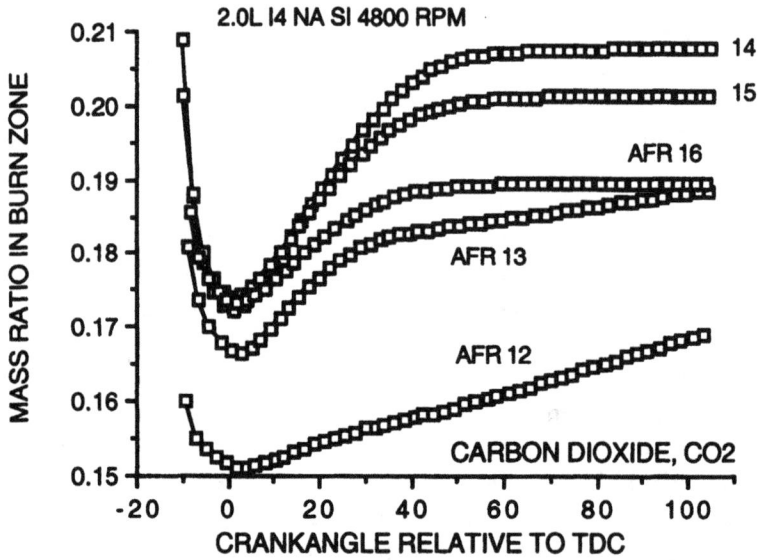

Fig. A4.10 Mass ratio of carbon dioxide in cylinder burn zone.

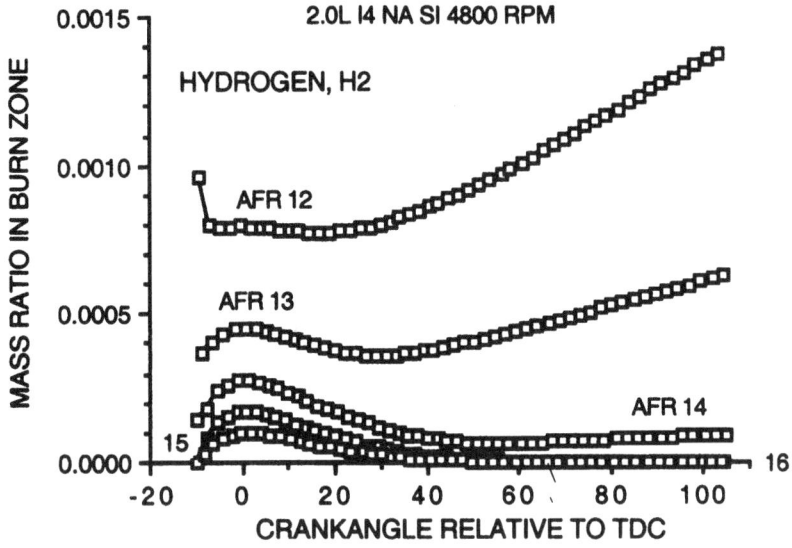

Fig. A4.11 Mass ratio of hydrogen in cylinder burn zone.

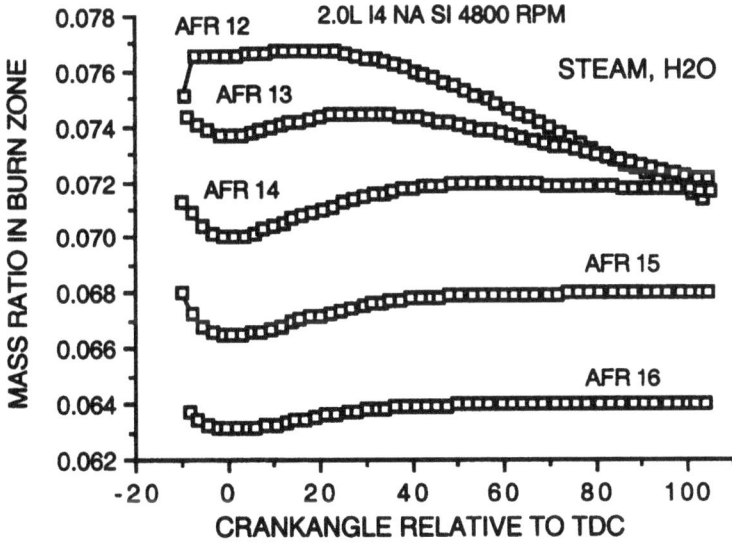

Fig. A4.12 Mass ratio of steam in cylinder burn zone.

simple chemistry of Eq. 4.3.3 already supplies that information but, without the temperature-dependent data from equilibrium and dissociation chemistry, computation of time-dependent data such as NO formation is not possible for mixtures that are richer than stoichiometric.

Carbon Monoxide

Simple chemistry dictates that the richer the fueling, the greater the CO content, and Fig. A4.9 confirms this. Dissociation at high temperature temporarily increases this characteristic during combustion. At air-fuel ratios near the stoichiometric value, the behavior is not unlike that of oxygen, i.e., stabilization has occurred well in advance of the exhaust valve opening at 105 °atdc. However, at rich mixture combustion, the dissociation behavior continues right up to exhaust opening, although the variation in real terms is quite small. Note the behavior at the richest AFR of 12: The CO mass ratio peaks at the highest burn zone temperature and then dwindles to about 5% as the temperature falls. There should be a corresponding "reverse image" for CO_2 formation, and this is seen in Fig. A4.10 at the same rich air-fuel ratio.

Carbon Dioxide

The time-related behavior of CO_2 for air-fuel ratios from 12 to 16 is shown in Fig. A4.10. The process is the reverse of that for carbon monoxide and oxygen, in that carbon dioxide is reformed by association as the temperature falls with expansion on the power stroke, as previously illustrated in Fig. A4.5. In short, the higher the temperature, the greater the dissociation of carbon dioxide. Again, as with carbon monoxide, the richest mixture combustion is the slowest to reform with the passage of time, and is barely complete as the exhaust valve opens and freezes the process. It should be noted that the closer the AFR is to stoichiometric during combustion, the higher is the carbon dioxide mass concentration, which means that the most efficient extraction of heat has been obtained from the fuel. This reinforces the measurement-based empirical relationship given in Eq. 4.3.26 for the relative combustion efficiency with respect to fueling for homogeneous combustion in a spark-ignition engine.

Hydrogen

This is probably the most complex reaction behavior of the series, for it can be seen in Fig. A4.11 that hydrogen formation is dissimilar on either side of the stoichiometric fueling level, which is at an AFR of 14.3 for this particular (unleaded gasoline) fuel. Near-stoichiometric, or richer than the stoichiometric value, where λ is unity, the presence of hydrogen is more pronounced than at lean mixture burning.

At rich mixtures, and at higher temperatures in the burn zone, more hydrogen is created leaving more free oxygen. As the cylinder contents become cooler—if 1800°C can be described as "cooler"—then the reduced availability of oxygen means that there is less oxygen around to combine with the hydrogen to form steam, and hence the free hydrogen content of the embryo exhaust gas is increased. The steam content of the richer mixtures falls by some 7% by exhaust valve opening which also reduces the amount of heat liberated because the hydrogen remains "unburned." At the richest AFR of 12, the process is not completed by exhaust valve opening at 105 °atdc. The proportions involved are tiny; a mere 0.15% of the cylinder mass is free hydrogen, even at the richest mixture combustion. Nevertheless, the computation depends on such complete chemistry to provide its accurate deduction of the oxygen mass concentration; without this level of accuracy, the prediction of nitric oxide (NO) formation inevitably becomes incorrect.

Steam

The remarks passed above regarding steam formation can be observed through Fig. A4.12 to be accurate. At lean mixtures, the steam content dips slightly at the highest in-cylinder temperatures, but remains relatively constant. At the richer mixtures, with the behavior at an AFR of 12 being the most extreme, the steam content maximizes at the highest temperatures and tails off with reducing temperature toward exhaust valve opening, losing some 7% of mass content as it does so, and freeing hydrogen as it proceeds. In short, much as the CO and CO_2 graphs are "reverse images" of each other, so are the free hydrogen and steam characteristics.

General

As discussed in Appendix A4.1, it is evident that this combustion model, using both equilibrium chemistry and reaction kinetics, is absolutely necessary to provide temperature and mass concentration data for all of the relevant gas species in the burned zone so that the mass emissions of oxides of nitrogen—indeed of all exhaust gases—can be predicted. Without this theoretical procedure, the properties of the exhaust gas, either within the exhaust ducting, or retained unscavenged within the cylinder, or blown back into an intake duct, cannot be accurately determined, with the result that the accuracy of the simulation is undermined.

Chapter 5

Computer Modeling of Four-Stroke Engines

5.0 Introduction

In the first four chapters of this book, you were introduced to the four-stroke engine; the unsteady nature of gas flow in, through, and out of such an engine; and the combustion process in such an engine. In the course of describing the combustion process, the characteristics of burning air and fuel in either the spark-ignition or the compression-ignition mode were explained. For each topic, theoretical models were described which can be solved, or used to index measured data, on a digital computer. The purpose of this chapter is to bring together those separate models and thereby illustrate the effectiveness of a complete simulation of the four-stroke engine. A complete simulation is capable of predicting all of the pressure, temperature, and volume variations with time within an engine of specified geometry, and of calculating the resulting performance characteristics of power, torque, fuel consumption, and air flow; most of the exhaust emission components; and the intake and exhaust noise levels.

Because of the (almost) infinite number of conceivable combinations of intake and exhaust geometry that can be used with the various types of engine geometry, it is impossible to show here how the computer simulation can cope with virtually every eventuality. Instead, the engine simulation is used here to illustrate many of the points made in earlier chapters regarding the design of various types of engines, and to highlight further issues that are not commonly found in the literature, but that have a significant influence on the design of four-stroke engines.

Perhaps one of the most useful aspects of an engine simulation is that it allows the designer to imagine the unimaginable, by allowing him or her to see on a computer screen the temporal variations of pressure, volume, and gas flow rate that take place during the engine cycle. The parametric changes observed in this visual manner, together with their net effect on power output and fuel consumption, provide the designer with much food for thought. Computer output is traditionally in the form of numbers, with graphs of the analysis normally being created after the computation process. In the era when such analysis was conducted on mainframe computers, the designer acquired the output as a package of printout and (rarely) pictorial information, and then pondered over it for its significance. This "post-mortem" approach, regarded at the time as a miracle of modern technology, was much less effective as a design aid than the kind of computer output that is available today. Today, the desktop microcomputer presents the same information in the same time span, but permits the designer to watch the engine "run" on the computer screen in a slow-motion mode. It is from such pictorial informa-

521

tion that a designer conceives of future improvements in, say, power or fuel consumption, or exhaust emissions. However, in the midst of all of this sophisticated computer output, don't forget that it is the human brain that synthesizes, whereas the computer merely analyzes. Also note that, with the pace of development of stand-alone computer systems, in terms of performance for cost, comments such as those made above become virtually out of date as soon as they are typed into the computer!

This chapter attempts to illustrate, through the use of computer models, the operational behavior, in very fine detail, of both spark-ignition and compression-ignition engines. This is often done by predicting data on thermodynamic properties throughout the engine, data that, even at this point in history, still cannot be measured. For instance, the measurement of temperature as a function of time has always posed a serious problem in terms of instrumentation, yet a computer model has no difficulty in calculating this state condition anywhere within the engine at intervals of one degree crankshaft while the engine is turning at 10,000 rpm. Because temperature cannot be measured with the same time-dependence, there is no means of telling the absolute accuracy of this prediction, although some technological progress in this area has been made recently [5.1-5.4]. On the other hand, the measurement of pressure as a function of time is a practical instrumentation exercise, although it must be conducted with great care so that measurements and calculations can be effectively compared.

By viewing the pictorial output of the simulation, you should be more able to comprehend the ramifications of changing design variables on an engine, such as valve timings and areas, or compression ratios, or the dimensions of the exhaust or intake ducting. Actually, this topic is discussed in even greater detail in Chapter 6.

The simulations presented in this chapter have been selected to show the effects of exhaust and intake tuning, or the lack of it, on the ensuing performance characteristics. The simulations are applied to several single-cylinder and multi-cylinder engines. In the latter, the effect of coupling the ducting to the cylinders of such engines reflects on the performance levels attained. Several factors were considered in selecting the engines for simulation, the foremost being that the simulation would illustrate some design principle. In each case, that design principle is noted below the description of the engine. Another, almost nostalgic, factor was the historical value. I have been very personally acquainted with the design and tuning of most of these engines, and the design lessons they contain should be passed on to you before they get lost and unrecorded in the mists of time.

The first five engines to be simulated here are all naturally-aspirated, spark-ignition engines.

The first engine to be simulated does not use exhaust or intake system tuning to achieve its performance characteristics. This engine is a hand-held unit for a weed trimmer, with simple intake and exhaust silencers of a form that can only be described as "boxes" or plenums. This type of engine is sold for applications where compactness and light weight are crucial; very small "boxes" are used for the intake and exhaust silencers. This uniquely-small four-stroke engine is manufactured by Ryobi in the United States to attain low hydrocarbon emissions in hand-held tool applications [5.5-5.7]. It has a swept volume of just 26 cm^3, and produces a target bmep of some 4 bar at 7000 rpm. The design lesson here is the basic behavior of the simple, untuned four-stroke engine.

The second engine to be simulated uses all of the tuning effects it can acquire from the intake and exhaust ducting attached to the cylinder. It was used by privateers in World Championship Grand Prix motorcycle racing up to about 1975, and then the target bmep was around 14 bar at engine speeds up to 7500 rpm. This single-cylinder engine, labeled as a G50 model, was manufactured by Matchless, and later by Seeley, in England, and has a swept volume of 499 cm^3. The design lesson here is the fundamental behavior of exhaust and intake tuning, and its effect on the output of a single-cylinder, high-performance engine.

The third engine, like the second, is a highly tuned unit, and is currently still used for World Superbike motorcycle racing. The target bmep is around 12 bar at engine speeds up to 12,000 rpm. This engine is a 90° V-twin unit. Labeled as a 916SP model, it is manufactured by Ducati in Italy, and has a swept volume of 955 cm^3 [5.8]. The design lesson here is the tuning of an asymmetrically-firing engine.

The fourth engine is the Nissan Infiniti unit used for the Indy Racing League (IRL) in the United States. It burns methanol, unlike the previous three engines which consume gasoline. This engine, a 90° V8 unit, has a swept volume of 4000 cm^3 and produces 700 bhp around 10,000 rpm. The design lesson here is the difference in the performance characteristics of symmetrically-firing and asymmetrically-firing multi-cylinder engines.

The fifth and sixth engines to be simulated are both in-line four-cylinder (I4) units, each with a total swept volume of about 2000 cm^3. One is a naturally-aspirated spark-ignition (na si) unit for a sports car, and the other is a turbocharged direct-injection (tc ci) diesel unit for a passenger-car application. The design lessons here are on the differences in the full-load performance characteristics of diesel and gasoline automobile engines, on effective intake tuning over a wide speed range in a sports car engine, and on the behavior of diesel and gasoline engines at the low-speed and light-load urban driving conditions where low exhaust emissions and low fuel consumption assume the highest design priorities.

5.1 Structure of a Computer Model

As indicated first in Sec. 2.18, the key elements required of any computer program to model a four-stroke engine are:

(i)   The physical geometry of each cylinder of the engine, so that all of the valve or port flow areas, and the cylinder volumes are known at any crankshaft angle during the rotation of the engine for several revolutions at a desired engine speed.

(ii)   A model of the unsteady gas flow in the inlet and exhaust ducts of the engine.

(iii)   A model of the unsteady gas flow at the ends of each of the inlet and exhaust ducts of the engine where they encounter either cylinders or the atmosphere, or branches, expansions, contractions, or restrictions within that ducting. This must include comprehensive mapping of the discharge coefficients of the flow in both directions at all discontinuities in the ducting.

(iv)   A model of the thermodynamic and gas-dynamic behavior within each cylinder of the engine while the valves or ports are open, i.e., during the open cycle period.

(v)   A model of the thermodynamic behavior within each cylinder of the engine while the valves or ports are closed, i.e., during the closed cycle period.

(vi) A model of the scavenge process during the valve overlap period, so that the proportion of fresh charge retained within the cylinder can be predicted. In this book, a "constant mixing process" is assumed for the four-stroke engine [5.14].

(vii) A model of the engine friction characteristics, so that the predicted values of the indicated performance characteristics may be converted to measured engine output data, such as brake mean effective pressure, power, or brake specific fuel consumption.

Virtually all of these elements have already been described and discussed in earlier chapters; here they are brought together to describe a complete engine. Only item (vii) remains to be addressed within this chapter.

The software code for the thermodynamic and gas dynamic segments of an engine simulation program is very lengthy, running to many tens of thousand of lines. It is normally written in Fortran, a fast number-crunching code. The preparers for the data input to the simulation code are written in a software code such as Microsoft™ QuickBasic for the Macintosh™, Microsoft™ VisualBasic for the PC™, or TrueBasic™, the latter of which is a cross-platform language for either the PC™ or the Macintosh™. These three software codes permit a highly visual data input procedure, with the cylinders or the valves or the ducting of the engine appearing as moving entities on the computer screen. This allows even the tyro user to enter numeric data for the ensuing engine simulation with little doubt as to the physical meaning of the data parameter that is required.

Before introducing you to the considerable array of design information that comes from such computer simulations, two items remain to be discussed: (1) the acquisition of the physical geometry of the engine, which is required as input data for such a simulation, and (2) the friction characteristics to be assigned to a particular engine geometry, which allow the computations detailing the indicated performance characteristics to be translated into the brake performance behavior required of any design, as first discussed in Secs. 1.5 and 1.6.

5.2 Physical Geometry Required for an Engine Model

Recall now that detailed descriptions of engine geometry were presented throughout Chapter 1. Compression ratio is defined in Sec. 1.4.2. The theory describing the relationship between the connecting-rod and the crankshaft, which determines the piston position at any instant during the crank rotation for each cylinder of an engine, is presented in Sec. 1.4.3.

5.2.1 Cylinder Head Geometry

The geometries of the cylinder head, its apertures, and its valving, are discussed at length in Sec. 1.5, and, for any of the engines to be simulated within this chapter, this information is presented in a sketch format. Fig. 5.1, which is for the Ryobi 26 cm^3 hand-held power-tool engine, is a typical example of such a sketch.

This sketch provides a simple visualization of the valve layout in this case, as it depicts an engine with two vertical valves. The symbolism on the sketch follows the nomenclature introduced in Sec. 1.5 in that the engine has one valve, n_v, for intake and exhaust. The intake valve has inner and outer seat diameters, d_{is} and d_{os}, of 9.3 and 10.8 mm, respectively. The exhaust

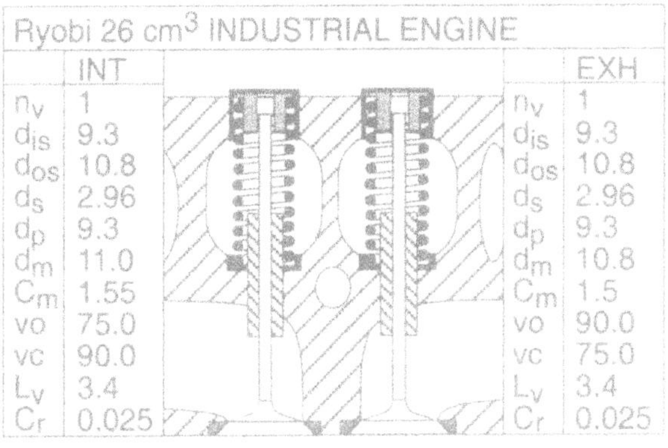

Fig. 5.1 The cylinder head and valve geometry for the Ryobi 26 cm³ engine.

valve also has inner and outer seat diameters of 9.3 and 10.8 mm, respectively. The intake valve stem, d_s, has a diameter of 2.96 mm, as has the exhaust valve stem. The intake port at the valve, d_p, has a diameter of 9.3 mm, as has the port at the exhaust. The intake port at the manifold, d_m, has a diameter of 11 mm, whereas the exhaust port at the head face has a diameter of 10.8 mm. The manifold-to-port area ratio on the intake side, C_m, is 1.55, whereas on the exhaust side, it is 1.5. The valve opening and closing events, vo and vc, are at 75 °btdc and 90° abdc for the intake valve, whereas they are at 90 °bbdc and 75 °atdc for the exhaust valve. The ramp lift ratio for the intake valve, C_r, is 0.025, i.e., 2.5%, and is the same for both the intake and exhaust valves. The procedure for creating the valve lift geometry—either as input data for the simulation of a new engine design or to mimic an existing cylinder head and valve lift geometry—is as given in Sec. 1.5.

For the remaining engines to be discussed within this chapter, a sketch similar to Fig. 5.1 will be used to detail each cylinder head geometry, thereby eliminating the need for substantive commentary such as that presented above.

5.2.2 The Intake and Exhaust Ducting

The intake and exhaust ducting attached to the cylinder head of the single-cylinder Ryobi 26 cm³ engine is sketched in Fig. 5.2. Observe that this is composed of a two-box exhaust silencer and a single-box intake muffler with twin intake pipes.

The lengths and diameters of each pipe in the entire system and the volumes of the silencer boxes are needed for the simulation. The surface areas of the three muffler boxes are also needed, but the sketch is already so complex that they are not included. The pipes of the ducting are meshed as described in Sec. 2.18.2, based on selecting a mesh length L in the pipe containing the hottest gas, obviously an exhaust pipe. As described in Sec. 2.18.9, the mesh length in the exhaust, L, is normally selected to be 20 mm, which typically permits a simulation to proceed in suitably small crank-angle steps of about 1° at 5000 rpm. By the same token,

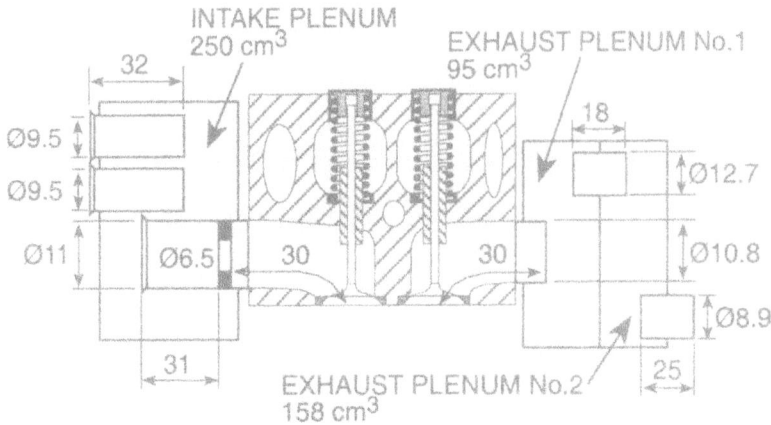

Fig. 5.2 The intake and exhaust ducting of the Ryobi 26 cm³ engine.

the mesh length for the intake system is normally selected to be about 12 mm. However, as is reasonably obvious from the discussion in Sec. 2.18.9, the minimum number of meshes required in any one pipe between the two ends is three in total, i.e., a left- and right-hand mesh and just one central mesh. Because one of the Ryobi exhaust pipes is just 18 mm long, perforce a 6 mm mesh length is used in the exhaust pipe. All this means is that the calculation will proceed in time steps of about 0.25 crank degrees at 5000 rpm, so the computation takes longer, but is inherently more accurate!

Sketches similar to that in Fig. 5.2 are presented for all of the other engines described in this chapter, thereby eliminating the need for much written commentary in each case.

5.2.3 The Cylinder Firing Order for a V-Twin Engine

For the single-cylinder engine, the firing order is simplistic; the engine fires at 0 °atdc and again at 720 °atdc. However, this fundamental principle also applies to the multi-cylinder layout because that is always the declared firing order for cylinder number 1, and the firing of all other cylinders is referenced with respect to the rotation of cylinder number 1.

In a multi-cylinder engine, where intake and exhaust ducting for each cylinder are connected to the cylinder head, and pressure waves are generated into them by the opening and closing of the valves, it is manifestly vital that the exact sequence of these events is relayed to the simulation. In a multi-cylinder arrangement this can become surprisingly complicated, particularly because there is no universally accepted system for numbering the engine cylinders, or for inferring precise information about the timing events of the engine simply by knowing the so-called "firing order."

Consider the structure of the Ducati 955 cm³ V-twin racing engine. The simple sketch in Fig. 5.3 shows that the cylinders are arranged at 90° to each other in a vee formation.

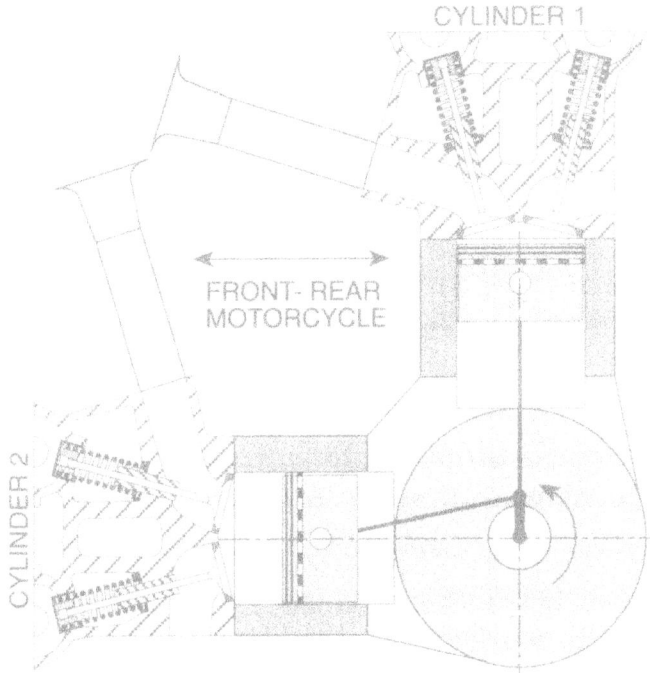

Fig. 5.3 The cylinder layout of the Ducati 955 cm³ V-twin engine.

Cylinder number 1 is shown at tdc during combustion. Because the cylinders are disposed at a mutual 90°, and the connecting rods are on a common crankpin, cylinder number 2 is clearly at 90 °btdc *on one of its strokes*. It may be at 90 °btdc on the compression stroke, or it may be at 90 °btdc on the exhaust stroke. As it happens, for the Ducati engine it is at 90 °btdc on the exhaust stroke, so cylinder number 2 will fire when cylinder number 1 is at 450 °atdc on its 720° cycle. This is the precise information that an engine simulation requires in order to correctly compute the performance characteristics of a multi-cylinder engine. The simplistic information that the cylinder firing order is 1-2-1 is totally inadequate because both of the firing scenarios postulated above have a 1-2-1 firing order.

If required, the simulation could equally well simulate the alternative firing scenario posed for the engine, i.e., that in which the firing position for cylinder number 2 is at tdc when cylinder number 1 is at 90 °atdc. Because this is a mere 90° crankshaft after cylinder number 1 has fired and is still on its power stroke, the simulation would: (a) correctly predict the ensuing performance characteristics, but (b) would be quite oblivious to the horrendous vibration as experienced by the rear end of the motorcyclist, not to mention the "white knuckles" on the handlebars, in the event that this power unit was ever installed within a motorcycle frame!

The data input to the simulation for the cylinder firing sequence in the Ducati is presented in Fig. 5.4, together with that of other, more complex, engines described further below. The salient number of 450° appears beside cylinder number 2, and 0° beside the reference cylinder (which is always cylinder number 1).

5.2.4 Numbering the Cylinders of a Multi-Cylinder Engine

The cylinders of a multi-cylinder engine are numbered according to the German standard DIN 73021 [5.9]. The cylinder numbering process for a selection of two-, four-, and eight-cylinder engines is shown in Fig. 5.4, where the power output shaft for each engine is seen at the bottom of the cylinder grouping. Observe in Fig. 5.3 that the V-twin Ducati engine adheres strictly to this notation.

The Double-Plane Crankshaft

American readers will now become acutely aware that this is not how they are used to numbering the cylinders of their traditional V8. This is unfortunate, but in order to perform an effective simulation of this type of engine, you will have to face the irrefutable logic that precision in this matter is essential. An example of such an engine is shown in Plate 5.1. This is the 1998 version of the V8 unit used in the General Motors' Chevrolet Camaro and Pontiac Firebird. It has two valves per cylinder, and the valves are pushrod operated. The exhaust

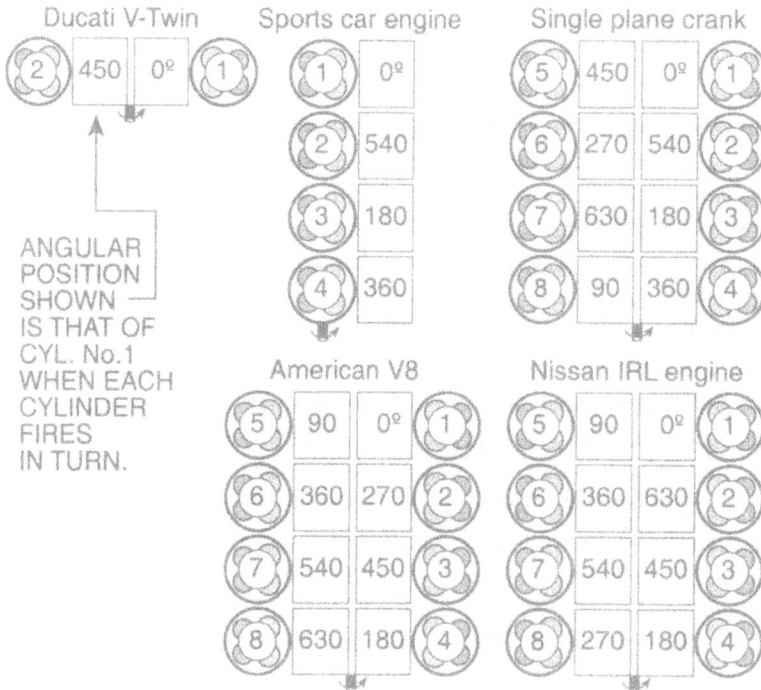

Fig. 5.4 Simulation data input for cylinder numbering and firing sequence.

Plate 5.1 The traditional "American V8" engine used in the Camaro.
(Courtesy of General Motors)

manifold is a compact collector near the cylinder head, and the intake manifold has a long, separate, tuned intake runner for each cylinder. In the notation of DIN 73021, the cylinder nearest to you on Plate 5.1 is cylinder number 1.

Consider the engine shown in Fig. 5.4 to be an "American V8." This is an engine with a double-plane crankshaft, as shown in Fig. 5.5(a), with the cylinder banks mutually set at 90°, just as in the Ducati motorcycle.

Ignore the fact that the title on the sketch reads "Nissan Infiniti," because an "American V8" uses the same double-plane crankshaft. The cylinders are numbered as decreed by DIN 73021 [5.9]. The logic of this system is that a V8 layout is simply a "straight-eight" engine, i.e., an I8 configuration where, as in Fig. 5.4, cylinders numbered 5-8 are relocated alongside cylinders numbered 1-4, and the power output end of the crankshaft is the end that is farthest away from cylinder number 1. Each pair of cylinders, such as 1 and 5, 2 and 6, 3 and 7, or 4 and 8, will behave like, and have the same firing options as, the Ducati V-twin. For the "American V8" unit, and for the first three pairs of cylinder named, it can be seen that the alternative Ducati firing scheme, which is 90° apart, is employed, whereas cylinders 4 and 8 fire as in the Ducati engine. Beside each cylinder is the all-important number required by the simulation, i.e., the crankshaft angle position for cylinder number 1 when each cylinder fires in turn. For the "American V8," this produces a cylinder firing order of 1-5-4-2-6-3-7-8. The Oldsmobile Aurora engine, also used for IRL racing, employs this firing sequence in a 90° V8 unit.

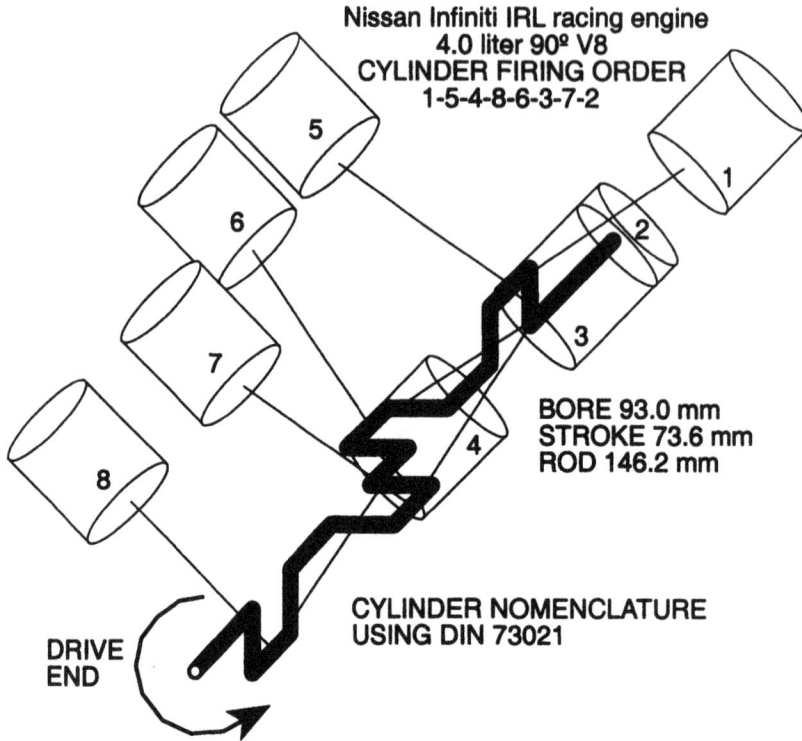

Fig. 5.5(a) The double-plane crankshaft of a typical "American V8" engine.

A cutaway drawing of the Aurora engine as used in the production Oldsmobile car is shown in Plate 5.2. Observe that this is a dohc unit with 4 valves per cylinder in a classic pent-roof cylinder head design. In the notation of DIN 73021, the cylinder nearest to you on Plate 5.2 is cylinder number 5. However, when such an engine gets developed for the Indy Racing League, and is installed in the racing car shown in Plate 5.3, it looks a little less benign with some 700 bhp on tap at 10,500 rpm. The four-into-one collector exhaust system is clearly visible. In the notation of DIN 73021, reading from left to right on Plate 5.3, the cylinder numbers are 4, 3, 2, and 1.

The Nissan Infiniti 90° V8 engine, as named in Fig. 5.5(a), has the same double-plane crankshaft, but uses the Ducati firing option for the 2-6 cylinder pair rather than the 4-8 pair, so the crank angle firing information as listed in Fig. 5.4 is somewhat different. The cylinder firing order is also different; it now becomes 1-5-4-8-6-3-7-2. The crankshaft and cylinder

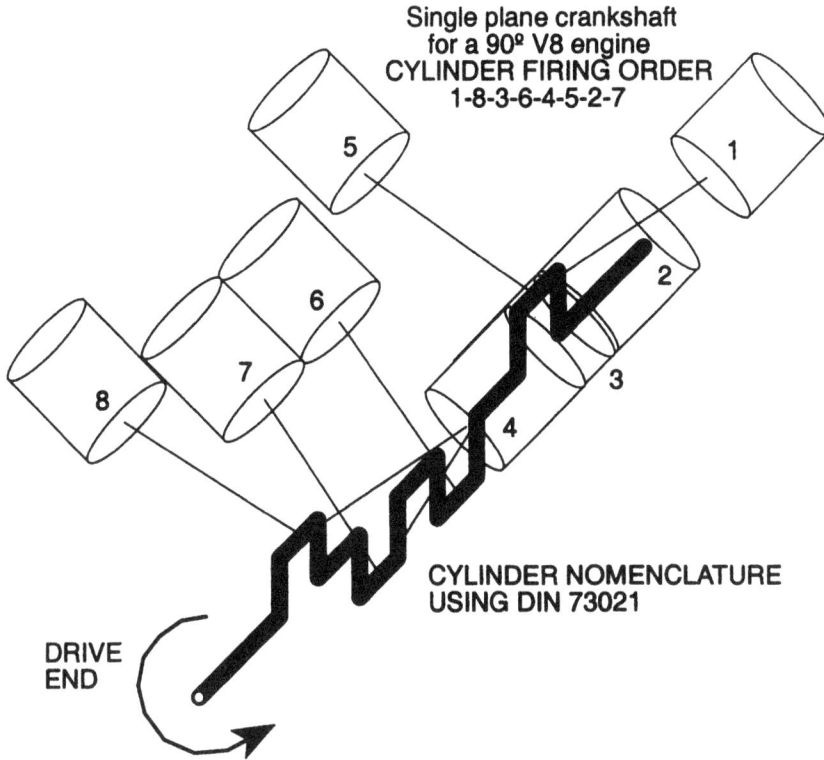

Fig. 5.5(b) The single-plane crankshaft for a V8 or an I4 engine.

components of the Nissan Infiniti engine for IRL racing, including this double-plane crankshaft, can be seen in Plate 5.4. The layout of the cylinders and the crankshaft in this photograph is the same as that shown in Fig. 5.4.

The Effect of the Inter-Bank Angle on Cylinder Firing

It now becomes clear that the inter-bank angle of the cylinders is another critical geometrical feature in determining the angular firing sequence of the cylinders of an engine with a vee layout. Imagine that the inter-bank angle of the cylinders of the Nissan engine was widened to 120°, from 90°, but that the same crankshaft was employed. In Fig. 5.4, the angles of cylinders numbered 1-4 would be unchanged, but those of cylinders 5-8 would be extended by 30°, namely to 120°, 390°, 570°, and 300°, respectively. The point I wish to make here is that the cylinder firing order would be quite unchanged at 1-5-4-8-6-3-7-2. Hence, as far as the engine

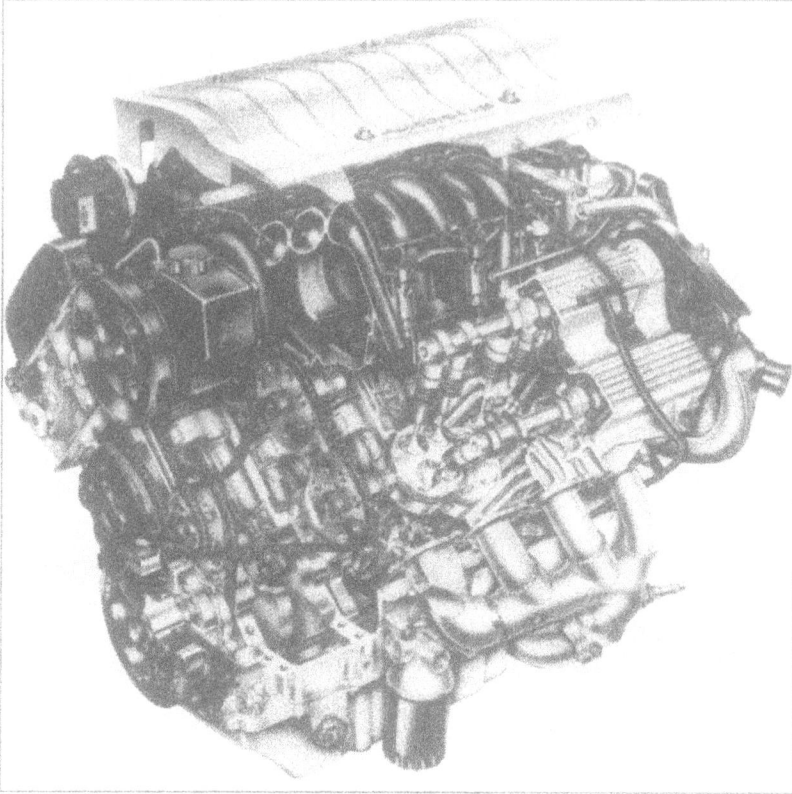

Plate 5.2 The V8 engine used in the Oldsmobile Aurora car. (Courtesy of General Motors)

modeler is concerned, the firing order of an engine is only part of the data needed for an engine simulation. The requisite accuracy is only obtained by inserting the crankshaft angular sequence in which each cylinder fires, with those angular intervals specified precisely.

The Single-Plane Crankshaft

The use of a double-plane crankshaft as shown in Fig. 5.5(a) gives the uneven firing sequences shown in Fig. 5.4 for the Nissan Infiniti engine or typical "American V8." In short, if the exhaust pipes emanating from each bank of cylinders 1-4, or 5-8, are coupled into a single collector to an outlet pipe, as is quite conventional, then exhaust pressure pulses arrive at the 4-into-1 branch at unequal crankshaft angle intervals. From the standpoint of exhaust tuning of a high-performance engine, it is reasonably obvious that this is undesirable—although precisely how undesirable will be illustrated in Sec. 5.8. An alternative solution is to use a single-plane crankshaft as shown in Fig. 5.5(b). If the rotation of each set of "V-twin" cylinders is followed, it can be seen that it is possible to have the cylinders in each bank fire at 180° intervals to each other, giving the angular sequence for the "single-plane crank" engine as shown in Fig. 5.4, with a firing order of 1-8-3-6-4-5-2-7. The exhaust pulses would arrive at

Plate 5.3 The Aurora V8 engine used in an IRL racing car. (Courtesy of Hans Hermann)

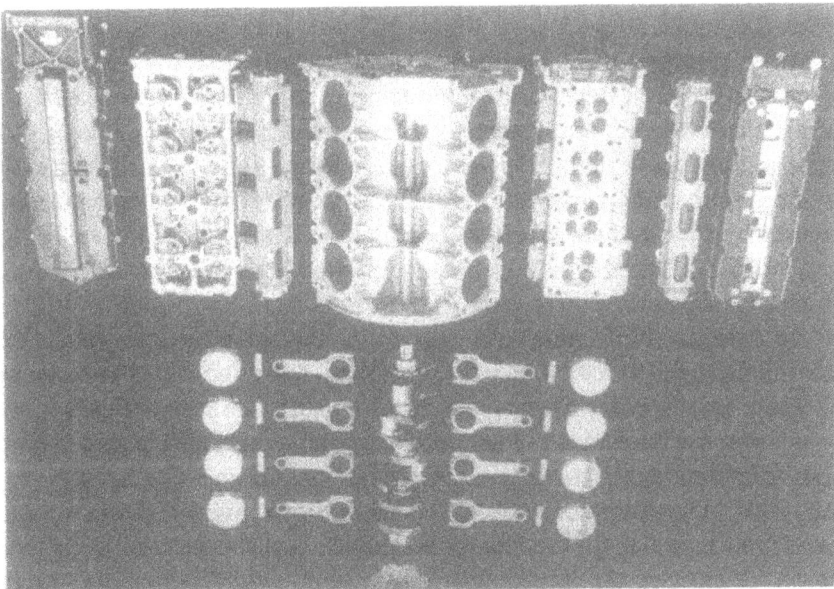

*Plate 5.4 The crank and cylinder components of the Nissan Infiniti for IRL racing.
(Courtesy of Nissan)*

equal angular intervals at a 4-into-1 collector exhaust system from the cylinders on each bank. Some of the V8 engines used in FIA Formula 1 racing had just such a crankshaft layout. Later in this chapter, in Sec. 5.8, the effect of using a single-plane crankshaft in a high-performance V8 engine is described.

The In-Line Engine

Also shown in Fig. 5.4 is an in-line four-cylinder engine labeled as a "sports car engine," i.e., an I4 layout using a single-plane crankshaft as shown in Fig. 5.5(b). This layout is almost universally employed for I4 engines in the automotive industry, and is common for both the sports car engine and the diesel car engines introduced above, the simulation of which will be described in great detail in Secs. 5.9 and 5.10. It can be seen in Fig. 5.4 that the angular sequence of the cylinder firing of the I4 engine is the same as for cylinders 1-4 of the V8 unit, giving a firing order of 1-3-4-2.

By now you will realize that, should you have the software to simulate engines on a computer, you must provide more extensive information on cylinder firing intervals than the simplistic "firing-order," because the angular intervals between them may be unequal.

5.3 Mechanical Friction Losses of Four-Stroke Engines

The internal components of the four-stroke engine reciprocate and rotate. This means that friction opposes every motion, at the expense of work from the crankshaft. The four-stroke engine contains many more components than the simple two-stroke engine, and this factor alone implies that the 4-stroke unit has proportionately greater friction losses. The extent to which the number of components is greater in a four-stroke engine can be seen in Plates 5.4 and 5.5. In Plate 5.4, which shows the V8 Nissan Infiniti engine, the components seen are the main crankshaft, connecting rod, and pistons of the engine. If this were a two-stroke V8 engine, these would be all of the rotating and reciprocating elements needed for the engine [5.14]. However, this same four-stroke unit, the Nissan Infiniti IRL racing engine, requires the further components shown in Plate 5.5, but these are for just one bank of four cylinders. In addition, an equal number of components would be required for the second bank of cylinders. Camshafts must be driven and all valves pushed by their cams against the spring forces and through the valve guides, hence this, too, raises the friction loss of the four-stroke by comparison with the simple two-stroke power unit.

In short, the designs of all four-stroke engines contain construction details that are relevant variables as far as the friction losses are concerned. Although it is possible to establish a fundamental theoretical approach to this topic, this is outside the scope of this text, which uses thermodynamics and gas dynamics to obtain information about performance characteristics. The best I can offer you here is a series of empirical relationships that have been shown to correlate quite well with experimental observations on various types of four-stroke engines. The situation is more complex for the four-stroke engine than for the two-stroke unit. This is because four-stroke engines are sometimes manufactured with "frictionless" bearings, i.e., rolling-element bearings using balls, rollers, or needle rollers, but are more conventionally constructed with hydrodynamic bearings, i.e., the oil pressure-fed plain bearings as seen in virtually all four-stroke production units. The friction characteristics of any engine are related to the type of lubrication. For example, crankshafts supported by rolling-element bearings

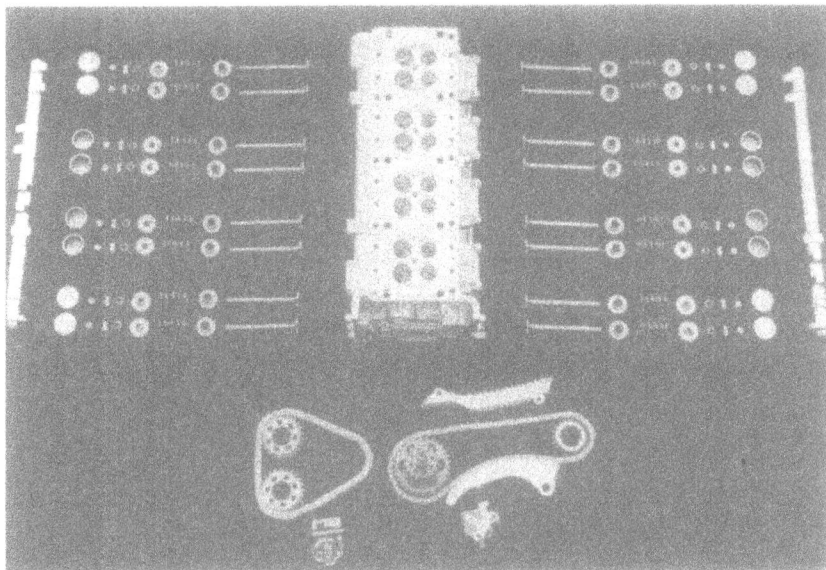

Plate 5.5 The cylinder head components of the Nissan Infiniti for IRL racing.
(Courtesy of Nissan)

normally employ a dry sump system in which the oil is scavenged from the crankcase sump and returned directly to a separate oil tank, often via an oil cooler. High-performance racing engines with pressure-fed plain bearings function largely in the same manner. The most common automotive method of lubrication is the wet sump approach, in which the oil lies below the rotating crankshaft in the crankcase sump. From there, it is pumped through the engine bearing shells, camshaft bearings, etc., and is returned to the sump via galleries, usually by gravity.

Four-stroke engines can be divided into several classes based on their friction characteristics, i.e., those with rolling element bearings or those with plain bearings, and those with spark-ignition or compression-ignition combustion. Virtually all compression-ignition four-stroke engines use plain bearings and, due to their larger and heavier components designed to withstand the higher cylinder pressures, tend to have relatively greater friction losses. All of the input and output data in the empirical equations quoted below are in strict SI units: the friction mean effective pressure, fmep, is in Pa units; the engine stroke, L_{st}, is in meters; and the engine speed, N, is in revolutions per minute.

It can be seen that the equations below relating friction mean effective pressure to the variables listed above are of a linear format [5.10-5.13]:

$$fmep = a + bL_{st}N$$

The terms a and b are constants. It is interesting to note, and it would be supported by both fundamental theory and experiment on lubrication and friction, that the value of a is effectively zero for rolling element bearings, hence the term "frictionless" for systems containing this type of bearing. It is also interesting to note that the second term in these equations, which combines stroke and engine speed, is piston speed in all but name.

Spark-Ignition Engines with Rolling Element Bearings

This classification covers virtually all small engines used in industrial units such as hand-held power tools and, normally, outboard and motorcycle engines. From the experimental evidence, the small industrial engines, i.e., those of cylinder capacity less than 100 cm^3, appear to have somewhat higher friction characteristics. In this regard, I have never satisfactorily resolved whether the data bank of information that provides Eq. 5.3.1 also incorporates the energy loss associated with the cooling fan normally employed on such engines.

Industrial engines ($V_{sv} > 50$ cm^3) $\qquad$ fmep = $275 L_{st} N$ $\qquad$ (5.3.1)

Industrial engines ($V_{sv} < 50$ cm^3) $\qquad$ fmep = $350 L_{st} N$ $\qquad$ (5.3.2)

Motorcycles, etc. $\qquad$ fmep = $250 L_{st} N$ $\qquad$ (5.3.3)

Spark-Ignition Engines with Plain Bearings

This type of engine is normally found in automotive units, which can include industrial engines, generating sets, motorcycles, and outboards. Because the engine swept volume decreases, the value of the constant a increases.

Engines ($V_{sv} > 500$ cm^3) $\qquad$ fmep = $100000 + 350 L_{st} N$ $\qquad$ (5.3.4)

Engines ($V_{sv} < 500$ cm^3) $\qquad$ fmep = $100000 + 100 \left(500 - V_{sv} \right) + 350 L_{st} N$ $\qquad$ (5.3.5)

Compression-Ignition Engines with Plain Bearings

This type of engine is normally found in automobiles, and in trucks, buses, and generating sets in which fuel is injected directly into the cylinder and the air supply to the engine is provided by a supercharger or a turbocharger. Here, there appears to be two classifications, where that including the automobile has somewhat lesser friction characteristics because the engine is more lightly constructed and runs to a higher piston speed. The dividing line between the two appears to be at a stroke length of 100 mm. However, diesel engines have higher compression and combustion pressure loadings than spark-ignition engines, and because the bearing and piston ring designs must cope with this loading, they are proportionately greater in size or number. This gives higher friction characteristics for this type of engine [5.12, 5.13].

Automobiles $\qquad$ fmep = $150000 + 400 L_{st} N$ $\qquad$ (5.3.6)

Trucks $\qquad$ fmep = $153000 + 420 L_{st} N$ $\qquad$ (5.3.7)

5.4 The Thermodynamic and Gas Dynamic Engine Simulation

Virtually everything presented in this text up until now has been leading you to this section of this chapter. The physical geometry of the engine and its valving and ducting, the firing sequence of the cylinders, the mechanical friction characteristics of the engine, the thermodynamic and gas dynamic theory in Chapter 2, the structure of a simulation model in Chapter 2, the mapping of discharge coefficients in Chapter 3, the theory of combustion and heat transfer in Chapter 4, and the properties of fuels in Chapter 4, are now, in this section, all linked together to simulate an engine.

For each of the engines to be simulated, the relevant data bank is read by the simulation, and at a selected engine speed, the computation is run for about nine or ten complete engine cycles until an equilibrium, of sorts, is achieved. Just as a real engine never produces the same power or torque, or breathes the same air mass, on each successive cycle for each cylinder, due to the variations in time and phase of the gas properties within it, neither does a simulation with any pretensions of accuracy. However, a simulation is marginally more cyclically stable than a real engine because the statistical combustion variations are rarely included within such a model. Perhaps the time has come to apply chaos theory to this aspect of an engine simulation [5.22].

On the assumption that equilibrium is reached after some nine or ten cycles of computation, the output data for that cycle are collected and stored as being representative of the engine performance characteristics. The calculated data range from the indicated and brake performance characteristics to the temporal changes of the leftward and rightward wave pressures at every mesh location within the ducting. An accurate prediction of the performance characteristics of the engine, based on such basic input data, makes the simulation a most powerful design tool. To be able to interrogate the output data in terms of gas motion and properties at every location, particularly for data that cannot yet be measured owing to technological limitations, gives the engine designer the engineering equivalent of the scientist's microscope, i.e., the designer is provided with the vital insight needed before making the next design decision. Let us now examine the results of the simulation of each of the engines I have selected.

One final caveat before we go on: Recall now the discussions about scavenging efficiency and charge purity in Secs. 1.6.4 and 2.18.9. In the following graphs of simulation output, the word "purity" should be interpreted precisely as defined by Eqs. 1.6.10 and 1.6.11, namely as the mass ratio of air to the total mass present, where that air is defined as being supplied air and is not a combustion residue. Exhaust gas is defined as being the products of combustion. In short, the purity of exhaust gas is zero, and that of air is unity.

5.5 The Ryobi 26 cm^3 Hand-Held Power Tool Engine

A cutaway sketch of this engine is shown in Fig. 5.6, where it is seen to have two vertical valves operated by pushrods actuated by a single cam. I am indebted to Gene Rickard of Ryobi for providing this informative artwork. By definition, such mass-production engines must be designed up to selected performance criteria and down to a minimum manufacturing cost. The design methodologies for some of the performance criteria, and for minimizing manufacturing cost have been described extensively elsewhere [5.5-5.7].

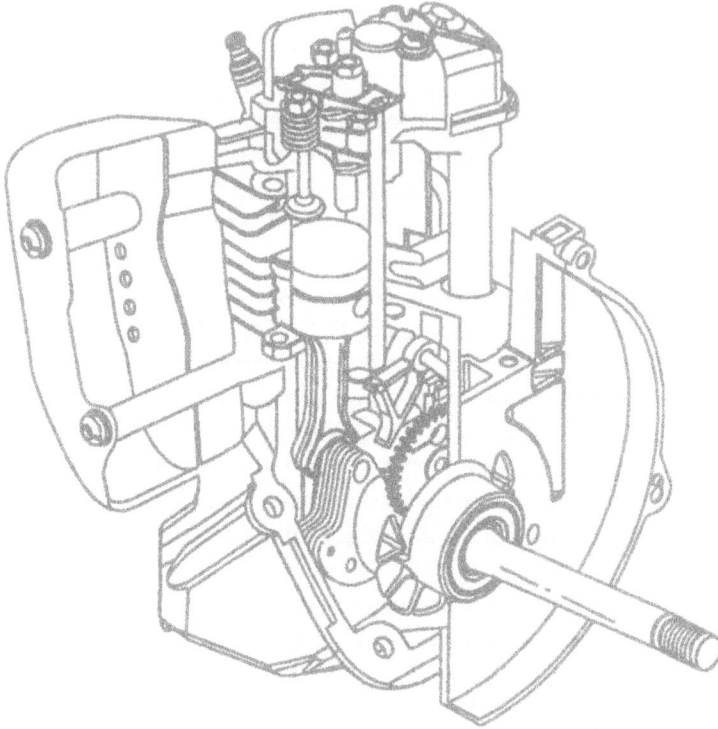

Fig. 5.6 The Ryobi 26 cm³ hand-held power tool engine.

5.5.1 Input Data for the Simulation of the Ryobi Engine

This engine has bore, stroke, and connecting rod dimensions of 32, 32.6, and 58.2 mm, respectively. The compression ratio is 8.5. The total swept volume of the engine is 26.2 cm³. The cylinder head and ducting geometries are as presented in Figs. 5.1 and 5.2. During the computer simulation, the following assumptions are made: the combustion of unleaded gasoline occurs at a constant air-fuel ratio of 13.5; the combustion process is as shown in Fig. 4.6 for the (somewhat similar) 50 cm³ Honda 2v moped engine; the discharge coefficient maps for the valves, pipe ends, and carburetor throttle/venturi are as presented in Figs. 3.37 to 3.40, 3.9 to 3.10, and 3.51, respectively; the discharge coefficients for unthrottled bellmouth-ended pipes are as quoted in Eqs. 3.3.1 to 3.3.4; the skin temperatures of the cylinder head, piston crown, and cylinder liner are 250, 200, and 150°C, respectively; the skin temperatures of the exhaust duct surfaces range from 450 to 350°C; and the engine breathes to and from the atmosphere at 1.01325 bar and 25°C.

5.5.2 The Performance Characteristics of the Ryobi Engine

Comparisons of the measured [5.6] and the calculated power and torque curves are shown in Figs. 5.7 and 5.8, respectively. Above 6000 rpm, the correlation is very good; below 6000 rpm, the deviation is entirely due to the assumption of a constant air-fuel ratio, AFR, which, in

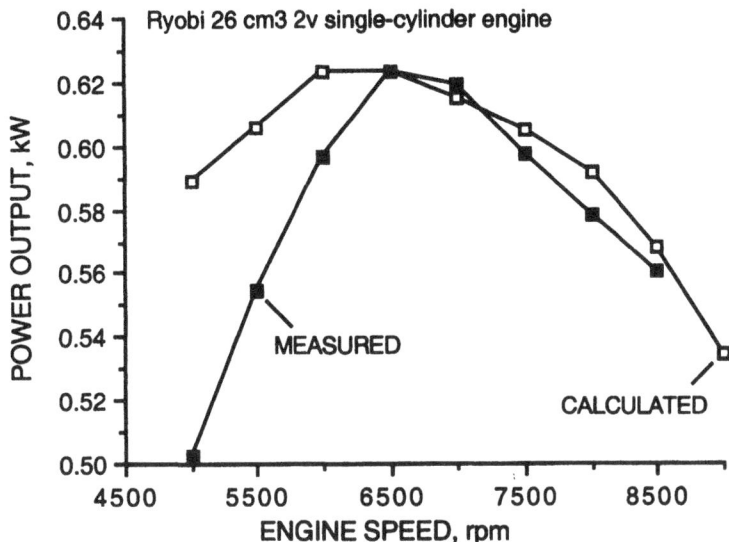

Fig. 5.7 Measured and calculated power output of the Ryobi 26 cm³ engine.

reality, the diaphragm carburetor on the engine would be quite unable to supply. Because the deviation from a constant AFR is not known, making assumptions about the values of the AFR at the lower engine speeds is a fruitless and misleading exercise.

The calculated airflow and trapped charge purity characteristics are shown in Fig. 5.9. The air flow, presented as delivery ratio, is seen to fall linearly with engine speed from 0.72 to 0.55, as it would for any engine that is simply pumping air without the benefit of any intake or exhaust system tuning. The trapped charge purity also drops over the same speed range from 0.88 to 0.84, and these values indicate levels of exhaust gas retention somewhat similar to those for a simple, equivalent two-stroke engine operating at the same torque level [5.14].

The calculated fuel consumption, as bsfc, and carbon monoxide emissions, as bsCO, are shown in Fig. 5.10. Considering the doubt regarding the low-speed air-fuel ratios, the minimum bsfc is probably at 6500 rpm at 520 g/kWh, which would be somewhat similar to a well-optimized, small, two-stroke engine aimed at the same market segment; the CO emissions levels would also be somewhat similar. The engine-out hydrocarbon emissions from the combustion process of the four-stroke engine would be considerably superior to the combined scavenging and combustion inefficiencies of the simple, conventional two-stroke engine [5.14, 5.23].

The mean effective pressure characteristics are shown in Fig. 5.11. The indicated mean effective pressure, imep, behavior tracks the linear delivery ratio characteristic over the speed range, as does the pumping mean effective pressure, pmep. The friction mean effective pressure, computed using Eq. 5.3.5, is also a linear function. Thus, from the theory of Sec. 1.6 and Eq. 1.7.9, the brake mean effective pressure, bmep, is also a reasonably linear function. For such a small-capacity, low-output engine, it is interesting to note that the combination of fmep

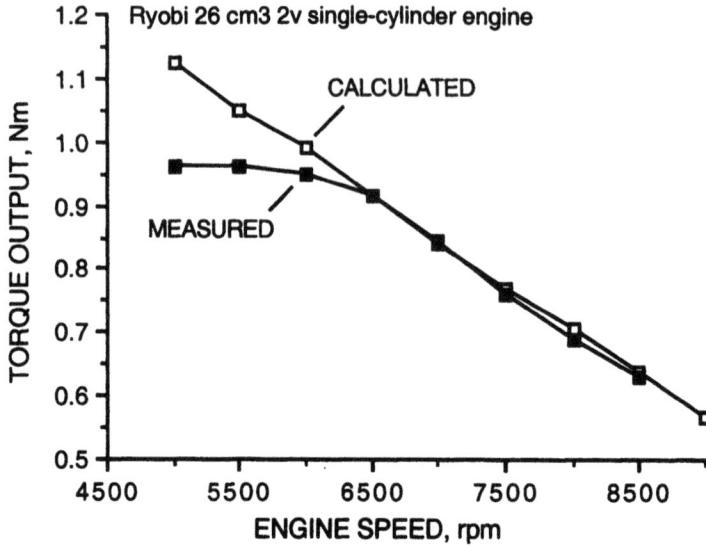

Fig. 5.8 Measured and calculated torque output of the Ryobi 26 cm³ engine.

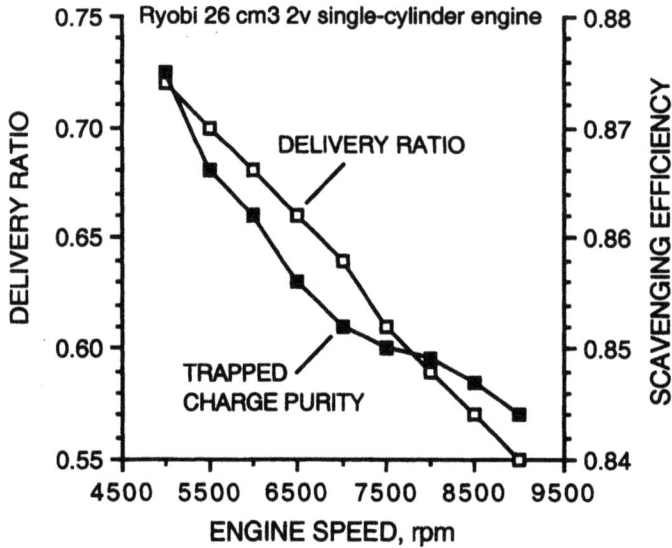

Fig. 5.9 Calculated airflow and charge purity of the Ryobi 26 cm³ engine.

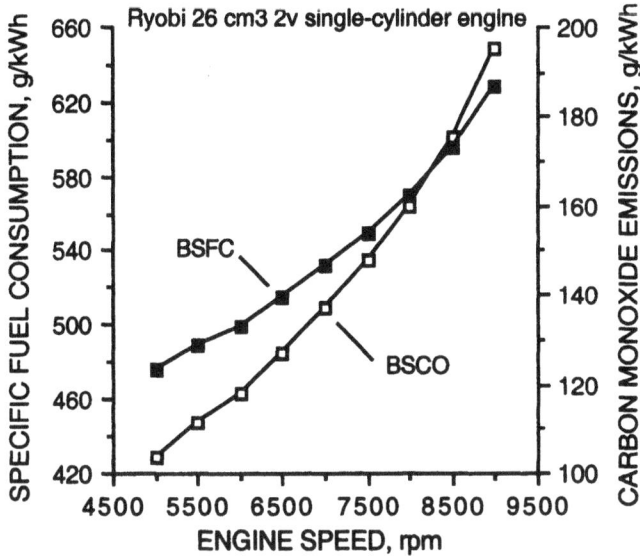

Fig. 5.10 Calculated fuel consumption and CO emissions of the Ryobi engine.

and pmep exceeds the bmep at the higher engine speeds. The imep is only 7.0 bar at 9000 rpm, which is some 50% less than the higher-performance engines discussed later in this chapter. Because the intake airflow is only moderate, the imep attained is also moderate, and it is not dissimilar to the low-speed throttled performance levels of the sports car engine discussed in Sec. 5.9.4 and seen in Fig. 5.70.

In Fig. 5.12, the mechanical efficiency varies from 0.7 to 0.45 over the speed range, which helps to explain the indifferently low bsfc levels in Fig. 5.10.

5.5.3 The Thermodynamics within the Ryobi Engine Cylinder

A further interrogation of the simulation output reveals the origins of these performance characteristics. The cylinder pressure and temperatures are shown in Figs. 5.13 and 5.14, respectively.

The peak cylinder pressure is seen to rise to 26.6 atm, with the peak at 21.6 °atdc. The burn zone temperature rises to a peak of 2142°C at 12 °atdc. The mean temperature peak is 1778°C at 38.4 °atdc, whereas that in the unburned zone is 583°C at 21.6 °atdc, coinciding with the location of maximum pressure.

Due to the scale of the maxima in Figs. 5.13 and 5.14, it is difficult to observe pressure and temperature effects during the open cycle. These graphs are therefore repeated in Figs. 5.15 and 5.16, and are supplemented by the data in the intake and exhaust ducting just behind each valve.

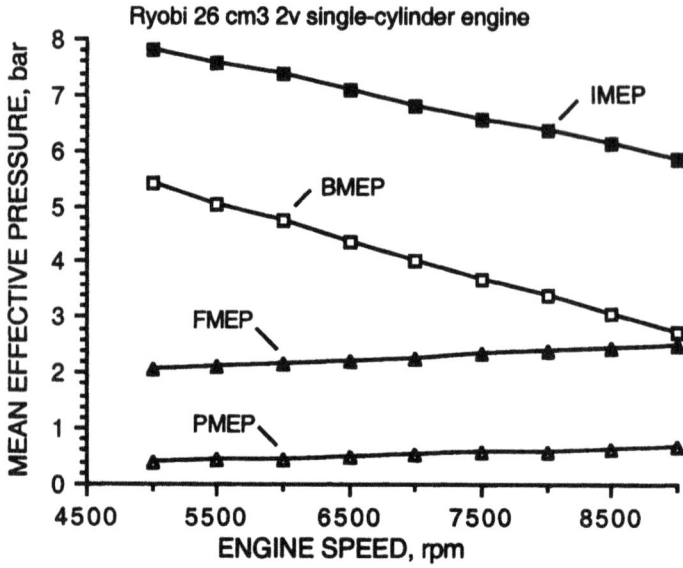

Fig. 5.11 Calculated mean effective pressures of the Ryobi engine.

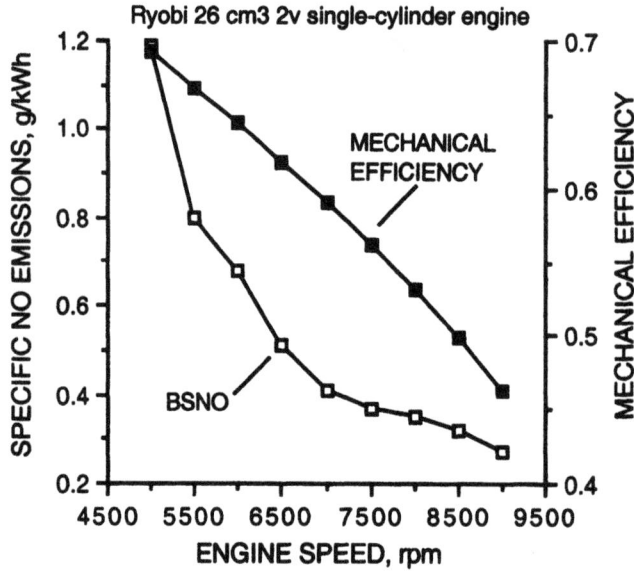

Fig. 5.12 Calculated mechanical efficiency and bsNO emissions of the Ryobi engine.

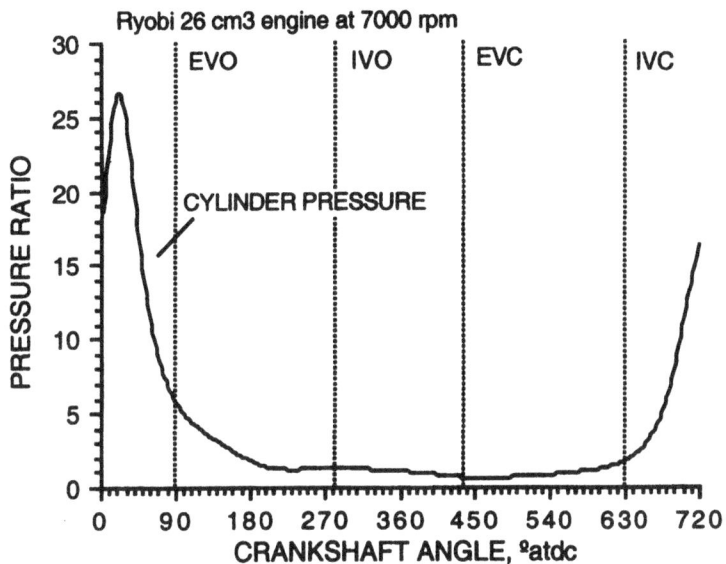

Fig. 5.13 Calculated cylinder pressure at 7000 rpm in the Ryobi engine.

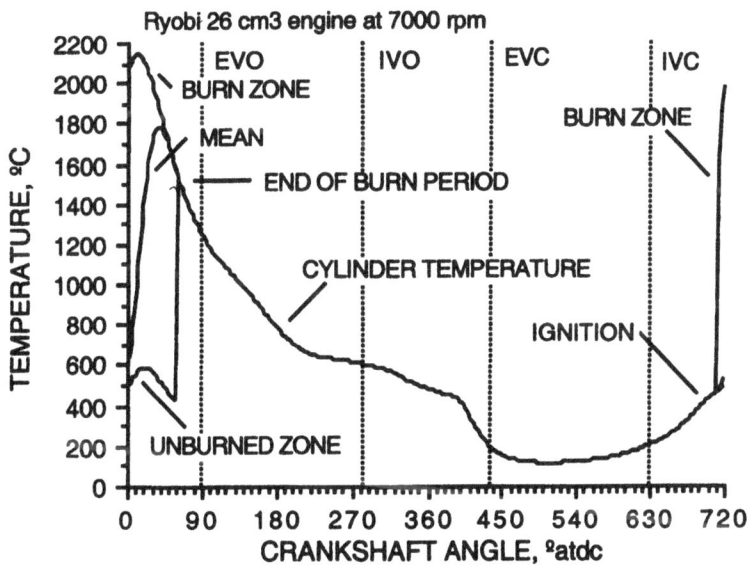

Fig. 5.14 Calculated cylinder temperatures at 7000 rpm in the Ryobi engine.

5.5.4 The Pressures and Temperatures within the Ryobi Engine Cylinder and Ducting

In Fig. 5.15, it is now possible to observe the bulge of cylinder pressure on the exhaust stroke between 180 and 360 °atdc. This is the exhaust stroke half of the pumping loop. The suction segment of the pumping loop is shown on the intake stroke from 360 to 540 °atdc, and the combination gives a hefty pmep value of 0.53 bar, using the theory of Eq. 1.6.30. (To refresh your memory on "pumping loops," reexamine Fig. 1.33.) Although the valve timings, and in particular the valve overlap periods, appear to be over-generous, with a ramp lift coefficient, C_r, of just 0.025, the valves lift and close very slowly. This can be seen in Fig. 5.15, where the exhaust pressure trace rises for a full 90° before it peaks at bdc, then the remaining outflow on the exhaust stroke continues to tdc. The cylinder pressure during suction is strong and falls to 0.58 atm at 460 °atdc, a value that is 100 °atdc on the suction stroke, and coincides with the maximum piston speed around the mid-stroke point. The pressure in the intake duct is just as strong, for it falls to 0.711 atm at the same position. There is no evidence of any effective intake ramming, or exhaust tuning, as the very short intake and exhaust ducts give reflections almost immediately superposed on each other. These high-frequency pressure signals are clearly visible on the intake duct pressure trace from 570 to 720 °atdc.

In Fig. 5.16, the exhaust temperature rises rapidly from exhaust valve opening, evo, to equal the cylinder temperature, with a peak of 993°C at 137 °atdc, and tracks the cylinder temperature closely by outflow until the next tdc at 360 °atdc. It then falls to, and stabilizes around, 475°C until the next exhaust event. From a "simulation transducer" installed in the middle of the outlet pipe, it is possible to estimate the mean exhaust gas outlet temperature. The simulation logs it at 380°C, which forms part of the overall engine design criteria owing to the fire-hazard legislations for weed trimmers and chainsaws. Interestingly, this is about 60°C higher than in an equivalent two-stroke engine [5.14].

By comparing Figs 5.15 and 5.16, a much more significant event is seen to occur on both the cylinder and intake temperature traces. From intake valve opening, ivo, until tdc, the cylinder pressure exceeds the intake duct pressure and backflow takes place into it from the cylinder. The temperature in the intake duct rises rapidly until it equals the cylinder temperature at about 500°C. The thermal shock across the exhaust valve consists of a hot pulse varying from about 500 to 1000°C; that across the intake valve varies from 100 to 500°C, and it is arguably just as severe. The extent of this backflow can be more thoroughly observed in Figs. 5.17 and 5.18, where the local gas particle velocities and charge purities are plotted.

5.5.5 The Particle Velocities and Charge Purities within the Ryobi Engine Cylinder and Ducting

In Fig. 5.17, the gas particle velocities in the intake and exhaust ducts behind each valve are plotted as Mach number, where a positive value indicates intake inflow and exhaust outflow. The Mach number for gas particle velocity is defined in Eq. 2.1.23. On the exhaust trace,

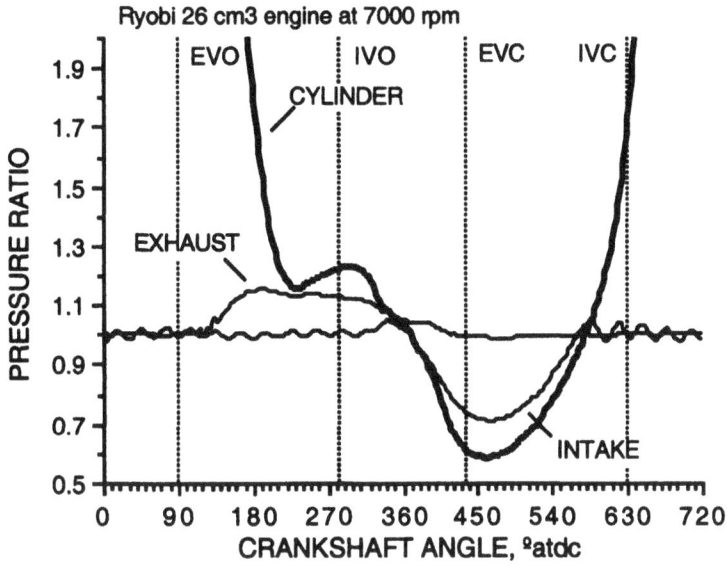

Fig. 5.15 Cylinder, exhaust, and intake pressures at 7000 rpm in the Ryobi engine.

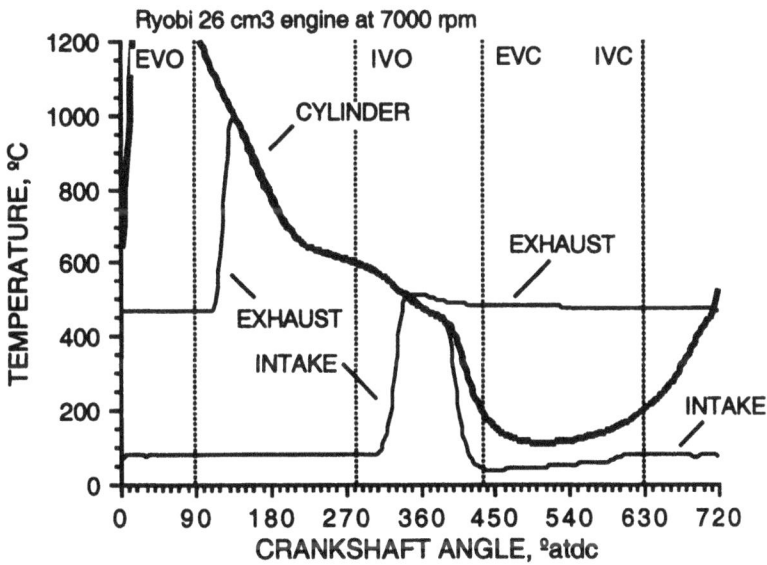

Fig. 5.16 Cylinder, exhaust, and intake temperatures at 7000 rpm in the Ryobi engine.

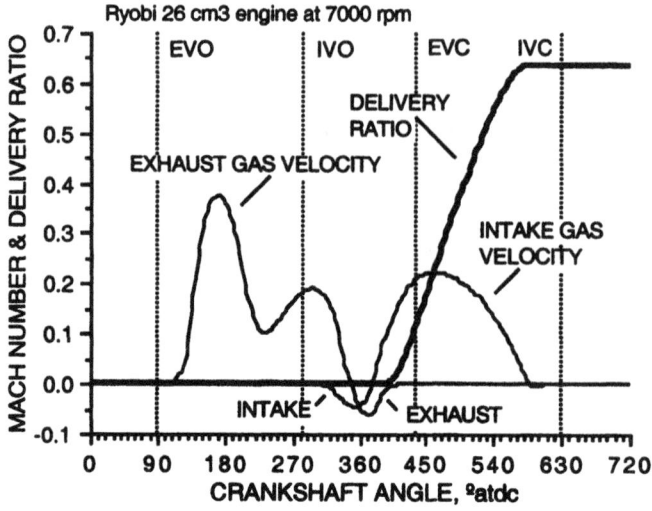

Fig. 5.17 Exhaust and intake particle velocities and DR at 7000 rpm in the Ryobi engine.

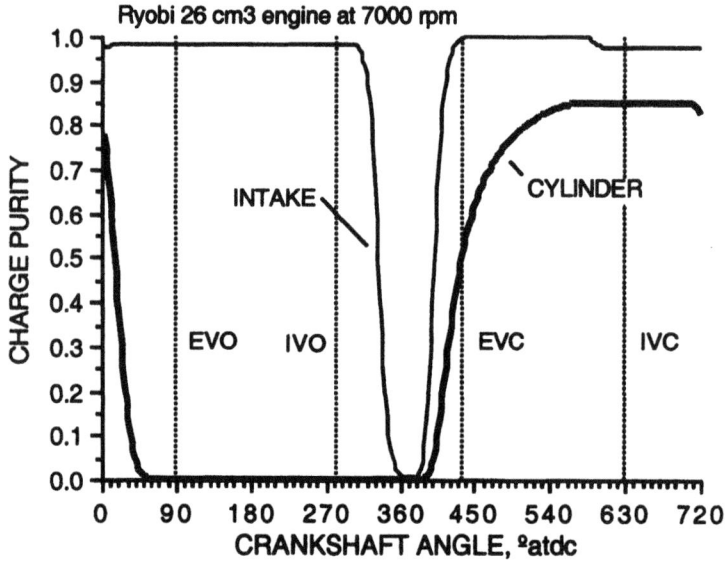

Fig. 5.18 Cylinder and intake charge purity at 7000 rpm in the Ryobi engine.

the particle velocities associated with the outgoing exhaust pulse start at 115 °atdc and continue until about 240 °atdc, followed by those from the exhaust stroke, until the next tdc point at 360° atdc. The peak Mach number associated with the main exhaust pulse is 0.373 at 173 °atdc. Exhaust backflow into the cylinder is evident from 352 °atdc until 411 °atdc, which has a negative effect on the amount of air that can be induced. Intake backflow from the cylinder is seen to occur from 312 °atdc until 376 °atdc. The intake airflow mass quantity is plotted on the same diagram as delivery ratio, DR, and from this it is clear that the cylinder does not begin filling with air until all of this backflow has ceased and that which has been pushed back into the intake tract has been reingested.

In Fig. 5.18, where the charge purities in the intake duct and in the cylinder are plotted, the extent of the intake backflow is clearly observed because it is so extensive. The purity in the intake duct behind the intake valve falls to zero around the tdc point, and it is 75° later, i.e., at 435 °atdc, before the last remnants of the cylinder backflow are reingested into the cylinder. The first whiff of fresh air to enter the cylinder is indicated by the rise of cylinder purity above zero at 387 °atdc, an event that coincides with the beginning of the DR curve in Fig. 5.17.

Note in Fig. 5.17 the small amount of backflow from the cylinder as the intake valve is closing. This barely registers on the intake gas particle velocity from the cylinder at 600 °atdc, and it has no noticeable effect on the DR curve, which is virtually flat at 0.64. However, the cylinder gas has a purity of 0.86 at this juncture, and when some of it empties back into the intake tract, the local intake duct purity drops from 1.0 to 0.976. Because that gas is trapped behind the closed intake valve until the next time it opens at 285 °atdc, it remains at that purity level, apart from some minor diffusion mixing with the gas in the neighboring (mesh) spaces.

5.5.6 The Noise Characteristics of the Ryobi Engine

As will be clarified in Chapter 7, it is possible to compute the gas-created noise from the intake and exhaust systems at a "microphone location" point in space beyond the outlets of the intake and exhaust ducting. Noise legislation is in force for all engines fitted to hand-held power tools, so an engine simulation that can compute the trade-off between generated noise and power output is of great assistance to the designer. For the record, and as part of a later discussion in Chapter 7, the simulation output for the Ryobi engine for these parameters is plotted in Fig. 5.19. The computational microphone is placed at 1.0 m from each intake and exhaust outlet. The total, i.e., the combined, noise level in dBA units is plotted from 4500 to 9000 rpm, and is seen to rise to 76.5 dBA. The exhaust is the noisier of the two sources.

5.6 The Matchless (Seeley) 496 cm³ Racing Motorcycle Engine

This single-cylinder, air-cooled engine, called the G50 model, was manufactured by Matchless, and later by Seeley, in England, and has a swept volume of 499 cm³. This engine was originally conceived by Jack Williams [6.17] following his development of the very successful

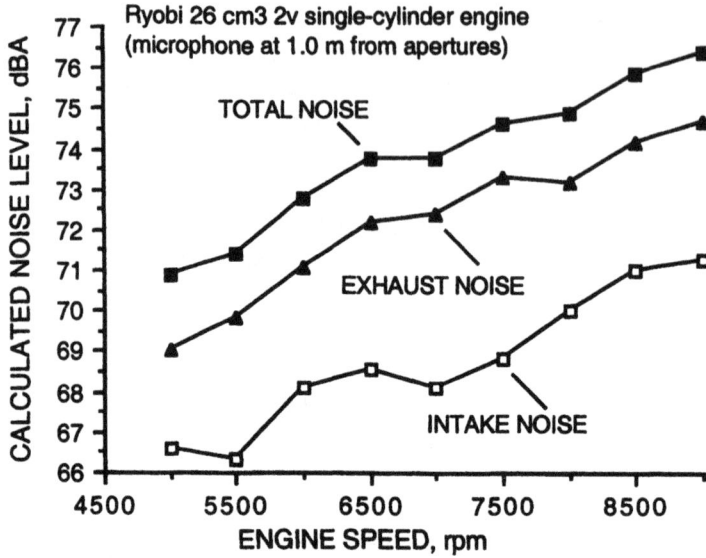

Fig. 5.19 Exhaust and intake noise characteristics of the Ryobi engine.

350 cm³ engine, called the 7R model, and manufactured under the AJS label. AJS and Matchless were both marque names of Associated Motorcycles. The 7R AJS and the G50 Matchless had an sohc layout with two valves per cylinder. The camshaft was driven by chain from the crankshaft, and both valves, as in the Norton in Plate 1.2, were controlled by hairpin valve springs. In the 7R or the G50 the valves were operated by roller-ended rocker arms from the single, central camshaft. A cutaway drawing of the 7R AJS engine is shown in Plate 5.6. Apart from the bore-stroke ratio, this is virtually indistinguishable from the G50 Matchless. The exhaust valve had a sodium-cooled stem, and its thicker diameter, larger than for a conventional valve, can be observed in Plate 5.6. The ignition system, in those pre-electronic days, consisted of a gear-driven Lucas magneto and a contact-breaker.

I first became closely acquainted with the G50 engine when its manufacture had been taken over by Colin Seeley at Erith in Kent, England. One of my students at Queen's University, Brian Steenson, under the QUB and Irish Racing Motorcyles banner, rode a 500 Seeley with this engine installed, and finished in second place at the 1969 Ulster Grand Prix to the great Giacomo Agostini, the World Champion, riding a three-cylinder MV Agusta. Brian Steenson was the only rider still on the same lap as Agostini when the race concluded, and was a long way ahead of the third place finisher, Malcolm Uphill, himself no mean rider. Brian Steenson is shown doing just that in Plate 5.7, and, from the scuff marks on the side of the fairing, you may guess just how far over it was heeled in the corners and then wonder how far away that might have been from the hedges, ditches, and stone walls that lined the Dundrod circuit. In those days the Ulster Grand Prix, a real road race, was a World Championship event, so the entry list was literally a "who's who" of the great riders of the day.

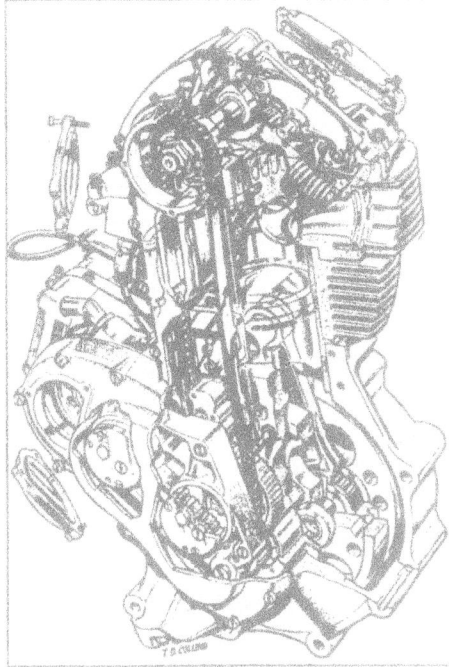

Plate 5.6 The AJS model 7R road-racing motorcycle engine. (Courtesy of Classic Bike)

Plate 5.7 Brian Steenson on the G50 Seeley with the long exhaust pipe.
(Courtesy of Rowland White)

5.6.1 Input Data for the Simulation of the G50 Engine

The engine has bore, stroke, and connecting rod dimensions of 90, 78, and 161.9 mm, respectively. The compression ratio is 11.2. The general layout of the 2v cylinder head and combustion chamber is as shown in Fig. 1.2, although the operation of the valves is by a single overhead camshaft. The cylinder head and ducting geometries are presented in Figs. 5.20 and 5.21. Note the diametral dimension of the thick stem of the sodium-cooled exhaust valve and the rather modest valve timings used in the 1960s compared to those in the Ducati or the Nissan engine in the 1990s (see Figs. 5.36 and 5.49). During the computer simulation, the following assumptions are made: the combustion of aviation gasoline occurs at a constant air-fuel ratio of 12.0, which corresponds to a λ value of 0.85; the combustion process is as shown in Fig. 4.10 for a somewhat similar engine; the discharge coefficient maps for the valves and pipe ends are as presented in Figs. 3.37 to 3.40 and 3.9 to 3.10, respectively; the discharge coefficients for unthrottled bellmouth-ended pipes are as quoted in Eqs. 3.3.1 to 3.3.4; the skin temperatures of the cylinder head, piston crown, and cylinder liner are 300, 250, and 150°C, respectively; the skin temperatures of the exhaust duct surfaces range from 450 to 350°C; and the engine breathes to and from the atmosphere at 1.01325 bar and 20°C.

5.6.2 The Performance Characteristics of the G50 Engine with the Short Exhaust Pipe

Jack Williams, the G50 designer, was the epitome of the perfect English gentleman and was a man I regarded as a friend; he stayed as a guest in my house in Belfast on at least one occasion. It was he who gave me the original copy of the measured power and torque curves shown in Fig. 5.22. He also gave me the engine dimensions at that time—it would have been about 1970—but they got lost long ago in the swirling mists of my inefficient filing system! An acknowledgment is due to Mr. Ron Lewis, of East Dulwich, London, for supplying the data on intake ducting, the short, standard exhaust system seen in Fig. 5.21, and the cylinder head geometry tabulated in Fig. 5.20; there was a quid pro quo, but of that perhaps another day.

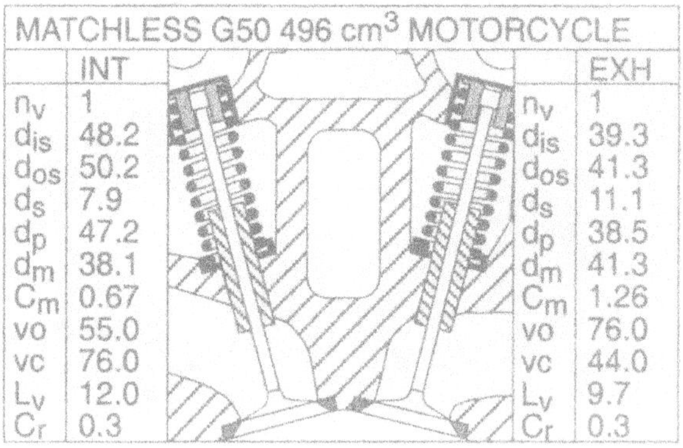

Fig. 5.20 The cylinder head and valve geometry for the G50 496 cm³ engine.

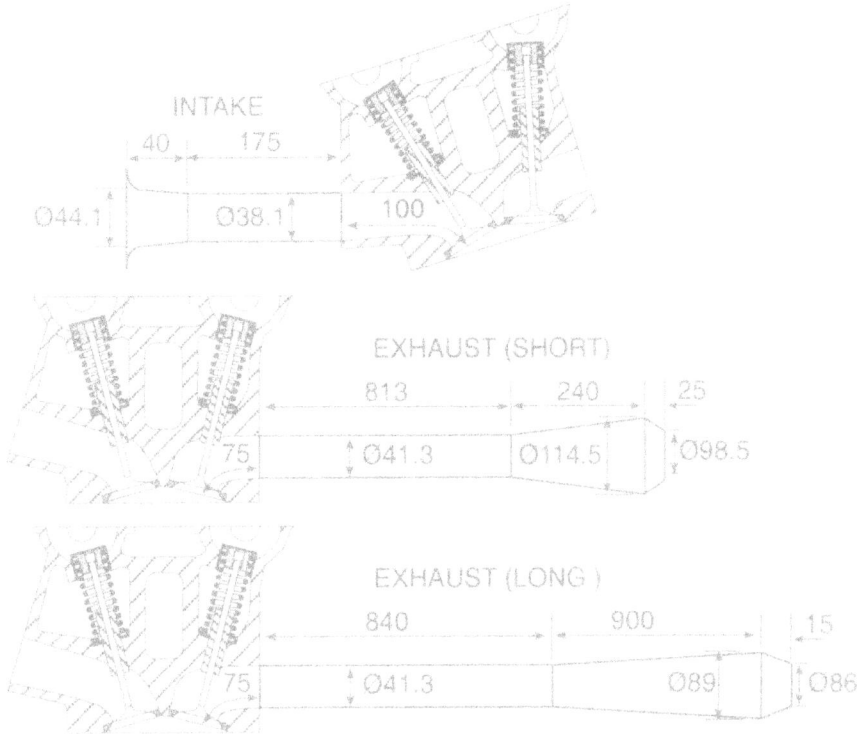

Fig. 5.21 The intake and exhaust ducting of the G50 496 cm³ engine.

Two exhaust systems are shown Fig. 5.21, one labeled as "short" and the other "long." The exhaust duct in both cases consists of a straight pipe followed by a diffuser, and then a short nozzle, which was colloquially referred to in the racing world back then as the "reverse-cone" part of the "megaphone." It is the diffuser, for obvious reasons, that is referred to as the "megaphone." The short exhaust megaphone, a mere 240 mm long, was the standard production G50 system, and almost certainly was the exhaust system used when the measured power and torque curves shown in Fig. 5.22 were recorded by Jack Williams.

Also shown in Fig. 5.22 are the power and torque results of the simulation of the standard G50 engine. The correlation between measurement and calculation is very good. The units of power and torque in this discussion are left in the original British units, for historical reasons.

5.6.3 Pressure Diagrams for the G50 with the Short Exhaust Pipe

Fig. 5.23 shows the computed superposition pressure diagrams at the peak power point of 7000 rpm. These are shown in the intake ducting at the intake valve, in the exhaust ducting at the exhaust valve, and within the cylinder. If measurements had been taken in the actual engine, the experimental set-up would have looked rather similar to that shown in Plate 5.8,

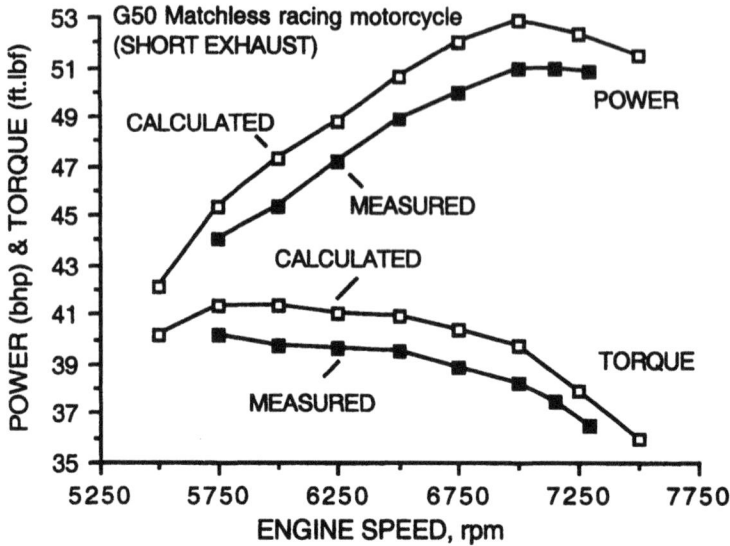

Fig. 5.22 The power and torque output of the production G50 496 cm³ engine.

although that engine is a 500 cm³ BSA single-cylinder engine for motocross racing [5.15, 5.26]. As you can see, placing transducers close to the intake and exhaust valves in their respective ducts is mechanically difficult, but the simulation can easily output the superposition traces at any physical position in order to compare measurement and calculation at those locations. For our purposes, the events closer to the valves are of the greater interest.

In terms of the pressure diagrams in Fig. 5.23, you may wish to refresh your memory on the fundamental action of an intake ramming process by looking at Sec. 2.19.2 (ii), and on the behavior of an exhaust megaphone by looking at Sec. 2.19.7. For the G50 engine in Fig. 5.23, the exhaust pressure is seen to peak at 2.3 atm and to create a strong suction wave by reflection from the megaphone, so that in the valve overlap period that pressure drops to 0.3 atm. This drags the cylinder pressure down to 0.6 atm by the time the exhaust valve closes.

In Fig. 5.23, the intake ramming process is clearly visible by the reflection of the intake suction wave at the bellmouth end of the intake tract. This traps the induced air almost until the intake valve closes, by which point the trapping pressure is about 2.0 atm. Of equal effectiveness is the arrival of a subsidiary and multiple reflection of that same ramming wave at intake valve opening, which blocks the cylinder from excessively blowing exhaust gas back into the intake tract. I will not discuss intake ramming to any greater extent at this stage, for a very detailed study of this topic is presented in Sec. 6.3. Consequently, you may wish to review the remarks passed in this chapter on intake ramming for this, and the other, engines discussed herein.

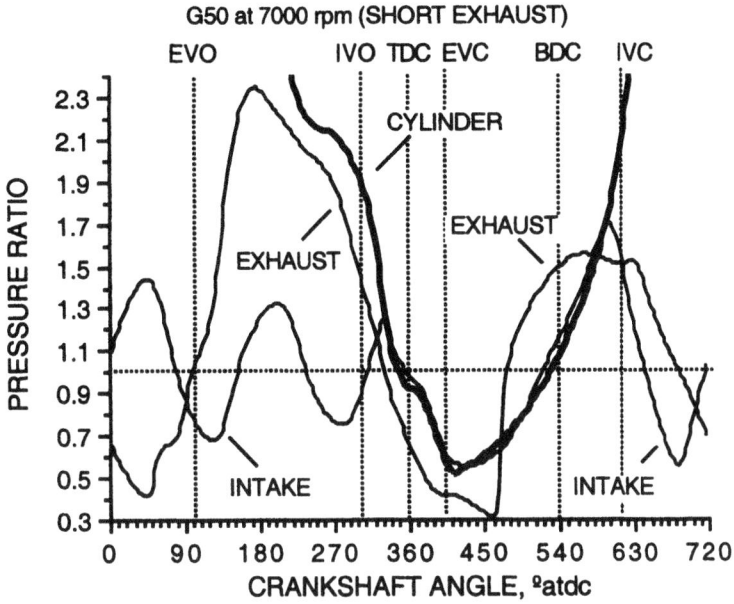

Fig. 5.23 Cylinder, exhaust, and intake pressures at 7000 rpm in the standard G50 engine.

Plate 5.8 Cylinder, exhaust, and intake pressure transducers in a 500 BSA engine.

5.6.4 The Performance Characteristics of the G50 Engine with the Long Exhaust Pipe

As Brian Steenson was racing this G50-engined Seeley out of QUB for Mick Mooney and Ronnie Conn of Irish Racing Motorcycles, I designed the long exhaust pipe shown in Fig. 5.21 using the more limited theoretical techniques available to me at that time [5.15]. This was very successfully raced in 1969, for example, it received a win at the Southern 100 event in the Isle of Man and the podium place at the Ulster Grand Prix already mentioned above (see Plate 5.7). Then, racing teams were not as fancy, and the team was not as extensive, as they are today. In the Isle of Man, I remember being the one, and only, on-grid mechanic for Brian Steenson, and the stop-watch wielder and lap-scorer, while Hubert Gibson (Ray McCullough's mechanic [5.14]) chalked up the signal board. I suspect that there are not too many professors who have acted in this capacity for their students at motorcycle road races!

The computed performance characteristics of the G50 engine with the long exhaust pipe, together with those of the short exhaust pipe for comparison, are presented in Figs. 5.24 to 5.29.

From Fig. 5.24, it is clear that the long exhaust pipe provided a higher peak power with more power above 6750 rpm, and less below, that engine speed. Using a five-speed gearbox, this gave a distinct advantage to the engine fitted with the long exhaust pipe.

The origins of this extra power are not obvious in Fig. 5.25, where the charging efficiency, CE, is seen to be very similar for both pipes throughout the speed range, with some gain for the

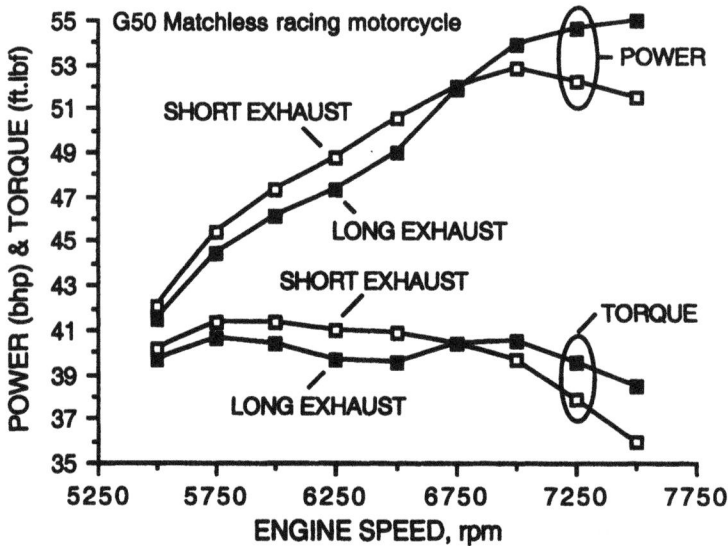

Fig. 5.24 The power and torque output of the G50 496 cm³ engine.

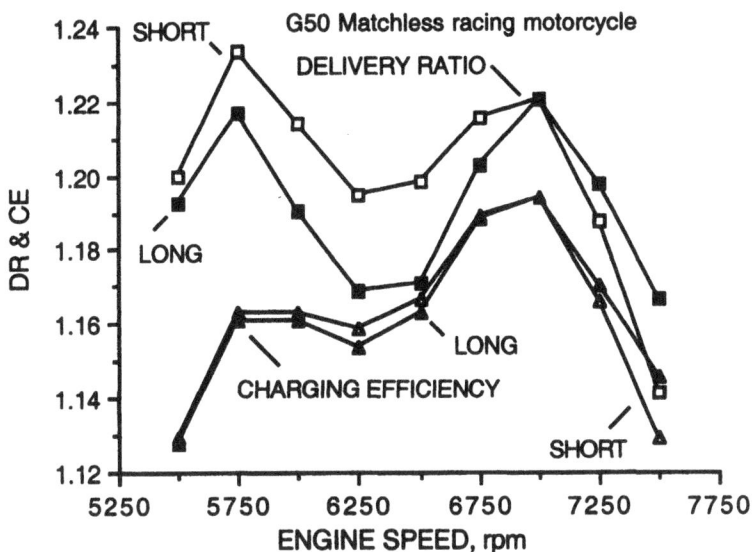

Fig. 5.25 Delivery ratio and charging efficiency of the G50 496 cm³ engine.

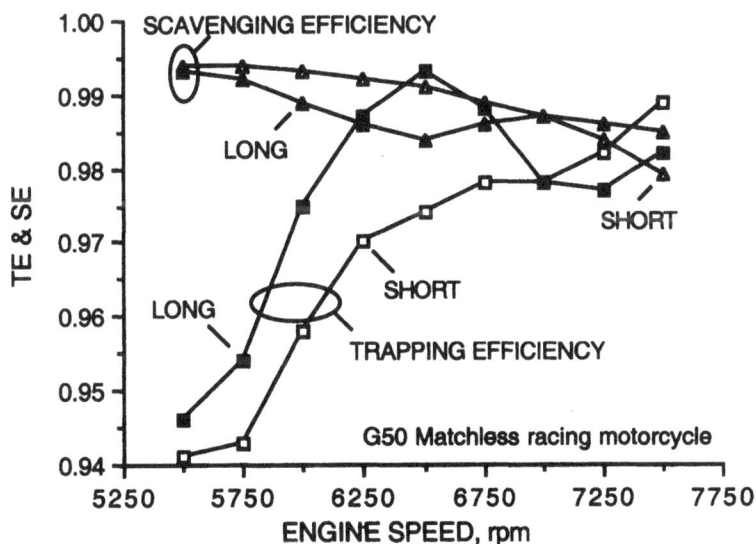

Fig. 5.26 Trapping and scavenging efficiencies of the G50 496 cm³ engine.

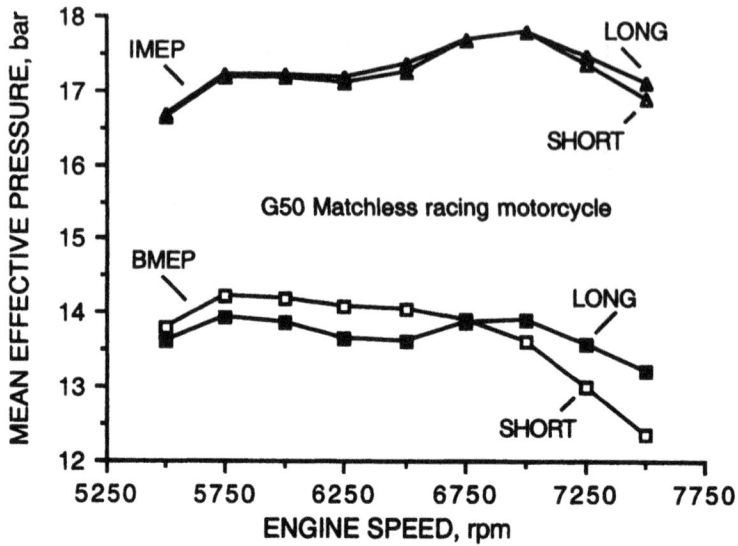

Fig. 5.27 Mean effective pressures of the G50 496 cm³ engine.

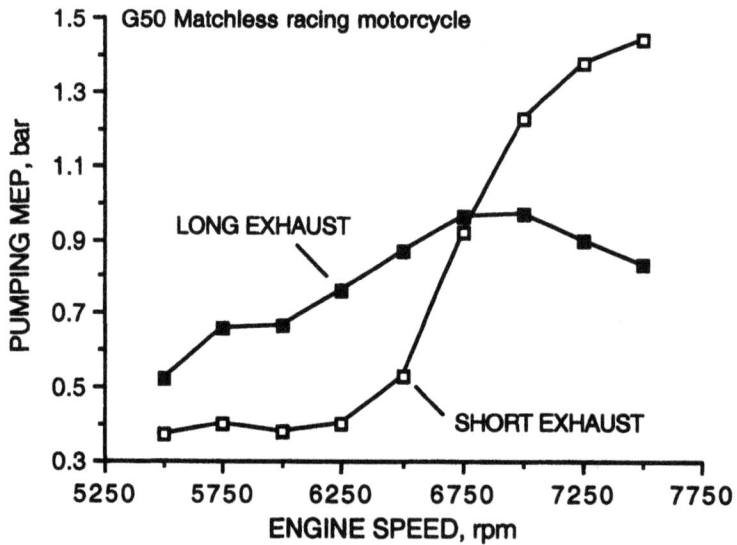

Fig. 5.28 The pumping losses of the G50 496 cm³ engine.

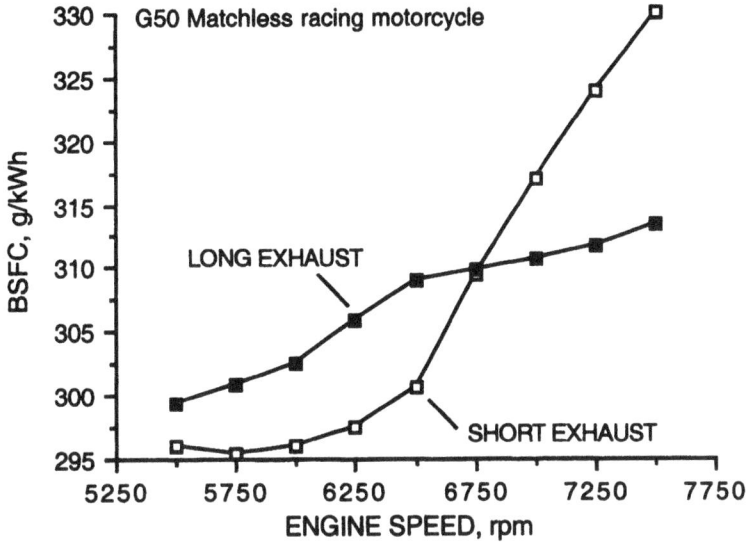

Fig. 5.29 The specific fuel consumption of the G50 496 cm³ engine.

long pipe above 7000 rpm. The delivery ratio, DR, on the other hand, is considerably superior for the short pipe below 7000 rpm, so the trapping efficiency of the longer pipe has a profound influence on the all-important mass of trapped air, i.e., as CE.

This is confirmed in Fig. 5.26, where the trapping efficiency of the longer exhaust pipe is noticeably superior below 7000 rpm and little different above that. The trapped charge purities for the two exhaust pipes, presented as scavenging efficiency, SE, are equally high and are virtually indistinguishable from each other.

Because indicated mean effective pressure, imep, is virtually directly related to charging efficiency, the imep characteristics plotted in Fig. 5.27 for the two exhaust pipes are almost identical, with the long exhaust pipe possessing only a minor advantage above 7000 rpm.

Yet the bmep, which is directly related to torque, and is plotted on Fig. 5.27, mimics the torque separation exhibited by the two exhaust systems, above and below 6750 rpm, already noted in Fig. 5.24. The long pipe is much better at high engine speeds.

The explanation for the bmep outputs lies in Fig. 5.28, which plots the pumping losses for the two exhaust pipes as pumping mean effective pressure, pmep. The long exhaust pipe has a somewhat higher pumping loss than the short exhaust pipe below 6750 rpm, but a significantly lower pmep value above that speed.

The overall outcome, already noted for power and torque in Fig. 5.24, and bmep in Fig. 5.27, is completed by the plots of brake specific fuel consumption in Fig. 5.29, where the long exhaust pipe is seen to possess a significant advantage above 6750 rpm in the peak power zone.

All of the above test data could have been measured, but the explanation for the curious disparities in the performance characteristics caused by the fitting of these two exhaust systems would not have been forthcoming. Clearly, pressure traces should be measured in the cylinder and in the exhaust pipes, so that the source of these performance differences may be determined. These pressure diagrams are shown in Figs. 5.30 and 5.31 at 7000 rpm for the short and the long exhaust pipe, respectively.

5.6.5 Pressure Diagram Comparisons for the G50 with the Long and Short Exhaust Pipes

In Figs. 5.30 and 5.31, the simulation output for the superposition pressure in the exhaust duct at the exhaust valve is represented by the heavy line marked "exhaust." This is plotted from exhaust valve opening to exhaust valve closing, which makes Fig. 5.30 an expanded version of what was already presented as Fig. 5.23. Recall that a pressure transducer measures superposition pressure only. If these pressure diagrams had been recorded, way back then in 1969, and the above measured performance characteristics confirmed for both exhaust systems, I presume that I would have stared at them in something approaching disbelief. The suction component during the valve overlap period appears similar for both exhaust pipes. However, the main exhaust pulse for the short exhaust pipe is at a conventional 2.3 atm, whereas for the long exhaust pipe it has a double hump with the highest pressure at a mere 1.5 atm. The "measured" cylinder and exhaust pressure traces have, if anything, further confused the situation. This will require another troll of the simulation output before the mystery is finally solved.

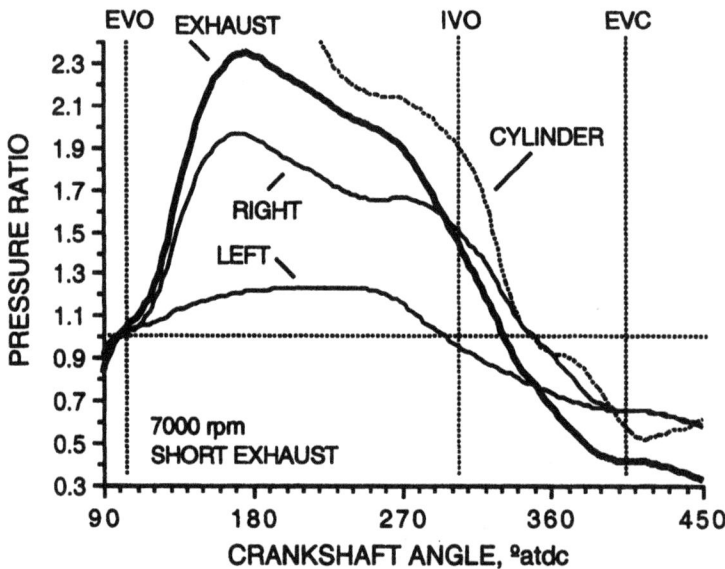

Fig. 5.30 Cylinder and exhaust pressures at 7000 rpm for the G50 short exhaust pipe.

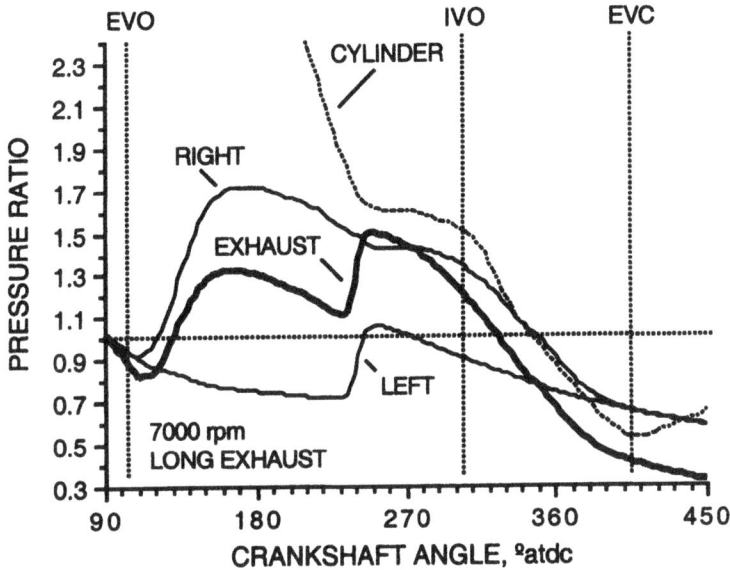

Fig. 5.31 Cylinder and exhaust pressures at 7000 rpm for the G50 long exhaust pipe.

Figs. 5.30 and 5.31 show plots of the computed rightward and leftward pressure waves, the combination of which provides the superposition pressure. These reveal the two classic methods of exhaust tuning a high-performance engine.

In Fig. 5.30, for the short exhaust pipe, the leftward wave during the exhaust stroke opposes particle outflow, but gives a large-amplitude rightward wave of 1.95 atm. This larger exhaust pulse propagates toward the diffuser and will reflect as a stronger suction wave to aid the lowering of the cylinder pressure during the valve overlap period. The use of a subsidiary compression wave reflection to combine with an outgoing exhaust pulse is known as "resonance." The penalty to be paid for opposing the exhaust outflow is an increased pumping loss during the exhaust stroke (see Fig. 5.28).

In Fig. 5.31, for the long exhaust pipe, the opposite effect occurs, manifested as a leftward suction reflection during most of the exhaust stroke. This helps extract exhaust gas from the cylinder, lowers the cylinder pressure, and thereby reduces the pumping loss. This is at the expense of creating a lower amplitude rightward exhaust pulse of 1.7 atm. The rightward waves, the real exhaust pulses, are seen to be similar in profile when either exhaust pipe is installed, but that for the short exhaust pipe is greater by 0.25 atm. The origins of the grossly dissimilar superposition exhaust pressure traces are now revealed. The extent to which the pumping loss is increased with the short exhaust pipe can be gauged from the cylinder pressure at intake valve opening on the two pressure traces. With the short exhaust pipe it is about 1.9 atm, whereas with the long exhaust pipe it is just 1.5 atm, and its entire profile is much lower for most of the exhaust stroke.

5.6.6. Particle Velocity Comparisons for the G50 with the Long and Short Exhaust Pipes

Further information from the simulation, in the form of exhaust gas particle velocities, gives more detail on the gas flow around the cylinder at 7000 rpm. In Fig. 5.32, the gas particle velocities, as Mach number in the exhaust duct at the exhaust valve, are plotted for both exhaust pipes. The effect of the leftward suction reflection provided by the long exhaust pipe during the exhaust stroke is seen to double the outgoing particle velocity during this period. This implies the virtual doubling of the mass of exhaust gas that is being extracted without the cylinder expending the extra work to pump it out, as is the case with the short exhaust pipe. This effect continues until by intake valve opening the short pipe is exhibiting the higher exhaust gas particle velocity simply because the cylinder *has* to pump it out.

5.6.7 Temperature and Charge Purity Comparisons for the Long and Short Exhaust Pipes

The cylinder pressure is visibly higher at intake valve opening for the short exhaust pipe, which must involve backflow from the cylinder into the intake tract, and so it proves in Fig. 5.33. The temperatures and charge purities in the intake duct at the intake valve at 7000 rpm are plotted in Fig. 5.33. The extra backflow of exhaust gas with the short exhaust pipe causes the intake gas temperature to increase by a further 200°C, and the local charge purity to deteriorate from 0.93 to 0.66. More blowback into the intake tract always means that less air is ingested when normal induction commences. Because this is the cross-over engine speed, i.e., the long pipe performs better above this point, the delivery ratio will deteriorate for the short exhaust pipe, whereas the tuning of the long exhaust pipe becomes ever more effective.

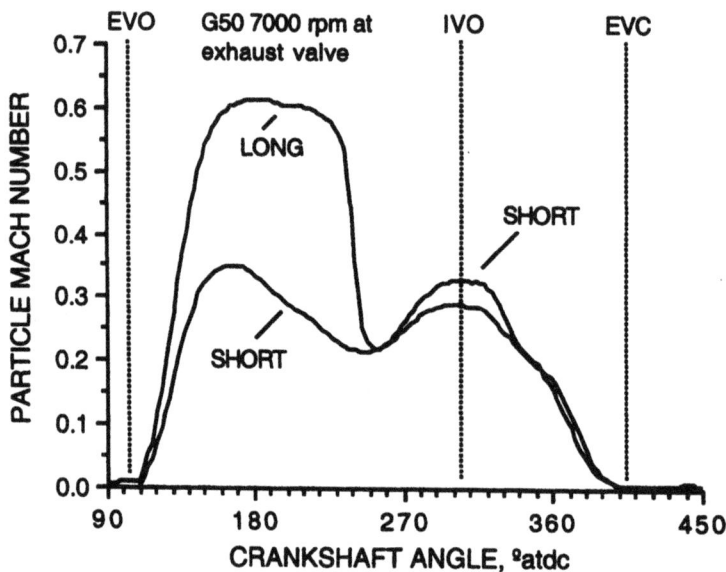

Fig. 5.32 Exhaust gas particle velocity at 7000 rpm in the G50 engine.

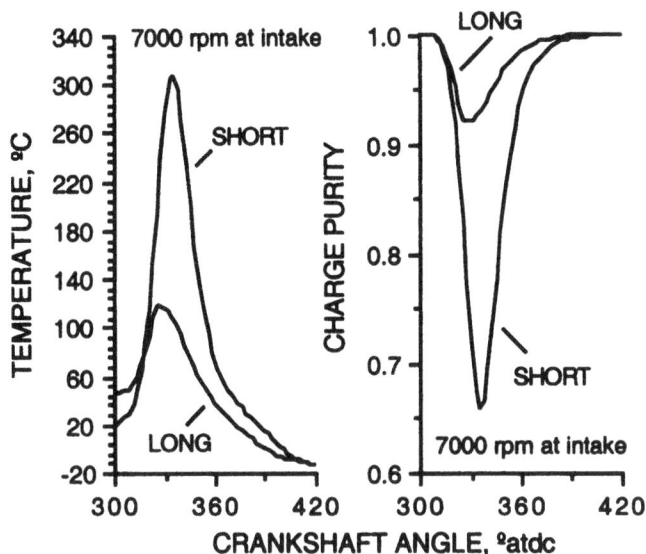

Fig. 5.33 Intake temperature and charge purity at 7000 rpm in the G50 engine.

5.6.8 Racing Experience with the Two Exhaust Pipes on the G50

When the long exhaust pipe was first fitted to the G50 Seeley, Brian Steenson reported that he needed to raise the transmission gearing for the engine, i.e., he found that he could typically pull a final drive sprocket with one less tooth on it. This was to be expected, for I had assumed that the long exhaust pipe would produce more power at the same, or perhaps slightly lower, engine speed. Once the carburetion was optimized, I was even more surprised to be told that not only was the engine pulling a higher final gear ratio, but it was pulling it to 7400 rpm rather than 7200 rpm. In short, the engine power was noticeably increased with the long pipe and the peak power was located at a higher engine speed.

The Unusual Noise Characteristics with the Long Pipe on the G50

This factual information on power at a higher engine speed was at odds with another curiosity that arose: the noise produced by the engine when fitted with the long pipe. This had a flat, almost toneless, note, which made the engine sound as if it was less powerful, and lower revving, than the same G50 Seeley with the standard short exhaust pipe. The standard pipe had the sharp, classic bark of the British racing single. I should add that all of this took place in an era before the rules dictated that racing motorcycles be fitted with reasonably effective mufflers. Out of curiosity, I completed this nostalgic exercise by asking the engine simulation to predict the exhaust noise characteristics of the short and long exhaust pipes along with the performance characteristics. In each case, the computational microphone was placed at 1.0 m from either the intake bellmouth or the exhaust pipe outlet. Fig. 5.34 shows plots of the intake and exhaust noise levels, in dBA-weighted units, over the engine speed range. Fig. 5.35 shows the dBA-weighted frequency spectra at 7000 rpm, allowing us to assess the tonal quality of the noise produced by the two exhaust pipes.

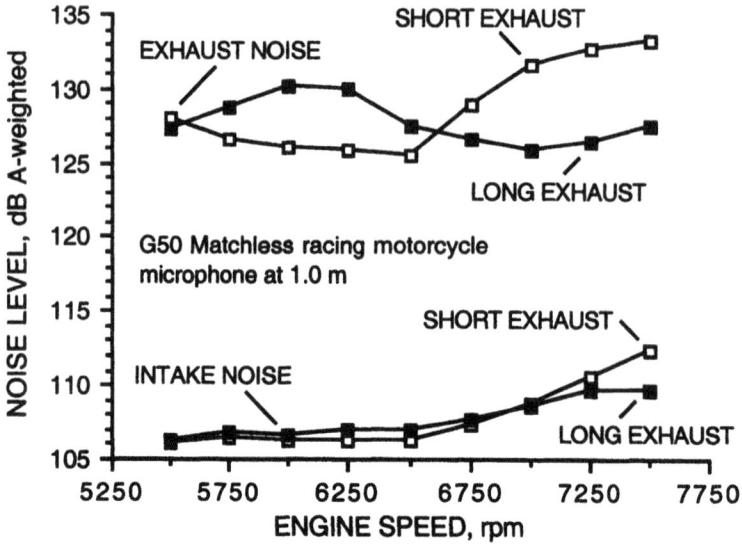

Fig. 5.34 The noise characteristics of the G50 496 cm³ engine.

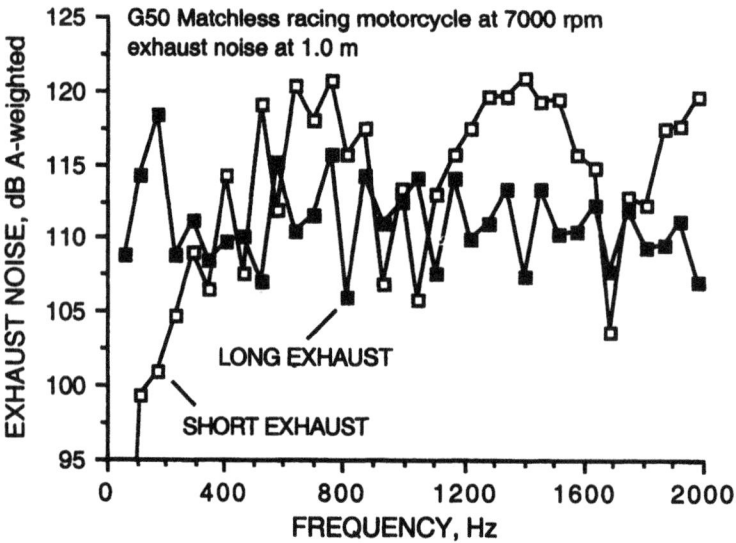

Fig. 5.35 The exhaust noise spectra at 7000 rpm of the G50 496 cm³ engine.

As you can see, the computation has reproduced the evidence that my own ears heard a quarter of a century ago. In Fig. 5.34, it can be seen that the short exhaust pipe gets noisier as it moves up the engine speed range through the peak power point. Its tonal quality, as seen in Fig. 5.35, exhibits equally noisy frequency peaks at 700, 1400, and 2100 Hz, which gives it the sharp "bark" referred to above. On the other hand, the long exhaust pipe produces its maximum exhaust noise at 6000 rpm, but this decreases thereafter as it proceeds up the speed range to peak power. You may imagine how strange that sounded as it accelerated toward you through the gears. Actually, considering the Doppler effect, it sounded even worse as it went past! The tonal quality of the long pipe shows an almost flat frequency profile with a peak around a low frequency point at 150 Hz.

The Low Fuel Consumption of the G50 When Racing with the Long Pipe

Also of interest are the intake noise levels plotted in Fig. 5.34. The engine fitted with the short exhaust pipe has the highest intake noise at 7500 rpm, even though it is breathing a lower delivery ratio at that speed. The answer to this conundrum lies in Fig. 5.33, where the backflow is seen to be greater in the engine with the short exhaust pipe. The cylinder-to-intake backflow is a compression pressure wave which travels to the bellmouth end of the intake pipe, spits out a little air and not a little fuel, and sends a little "exhaust" compression wave out into the atmosphere as noise! It is the spitback of the fuel at the intake pipe end, caused by the short exhaust pipe, which is perhaps of even more interest because, from Fig. 5.29, the brake specific fuel consumption is observed to be superior at the highest engine speeds for the long exhaust pipe. The reduced liquid fuel spitback further improves the inherent good fuel economy characteristics of the G50 Seeley with the long exhaust pipe. This was indeed the experience of all who raced with it. Tommy Robb, another Ulsterman [5.16], rode one to fourth place in the 500cc World Road-Racing Championship of 1970, and often passed comment on how little petrol (gasoline) it used.

5.7 The Ducati 955 cm^3 Racing Motorcycle Engine

This engine is used in World Superbike racing, which is a class based on production machines. This class permits the competition of twin-cylinder engines of up to 1000 cm^3, against four-cylinder machines with capacities up to 750 cm^3. The engines in this—indeed, in all Fédération Internationale Motocycliste (FIM) racing—are required to use exhaust silencers and meet reasonably stringent noise regulations to protect the ears of the race-going public. By contrast, the engines of Formula 1 cars are unsilenced, and going to a Formula 1 race can be an aurally-painful experience!

The Superbike class provides some of the most exciting and hardest-fought motorcycle racing ever seen. The action photograph in Plate 5.9 shows Carl Fogarty riding a Ducati V-twin in Northern Ireland to win the North-West 200 race of 1993. I remember being mightily

Plate 5.9 Carl Fogarty on a Ducati V-twin winning the 1993 North-West 200.
(Courtesy of Bill McLeod)

impressed by the manner of the win and the remarkable noise that accompanied it—and I might add that this was not the first time that Carl Fogarty had ridden a motorcycle in Ulster. I can remember him as a six-year-old riding a mini-bike around the paddock at the Ulster Grand Prix where his father, George, a Lancastrian, was competing, as he did so regularly at Irish road races.

The Ducati 916SP model engine is a 90° V-twin and has been raced in this class with capacities of 916, 955, and, more recently, 995 cm^3, such as in 1998, when Carl Fogarty rode one to become World Superbike Champion for the third time on a Ducati. The cylinder layout is shown in Fig. 5.3 and the asymmetrical firing order of its two cylinders, already discussed in detail in Sec. 5.2.3, is shown in Fig. 5.4. The objective of this further discussion is to demonstrate the gas dynamic and thermodynamic implications in the design and tuning of such a high-performance engine.

5.7.1 Input Data for the Simulation of the Ducati Engine

The engine has bore, stroke, and connecting rod dimensions of 96, 66, and 126 mm, respectively. The compression ratio is 11.5. The total swept volume of the engine is 955.5 cm^3. Although the engine valve operation is unique to the Ducati and is desmodromic, the general layout of the cylinder head and combustion chamber is as shown in Fig. 1.6. The four-valve, pent-roof cylinder head and ducting geometries are presented in Figs. 5.36(b) and 5.37.

Fig. 5.36(a) contains a sketch describing desmodromic valve operation where the valves are positively pushed and pulled by separate cams. Ducati has applied several versions of this

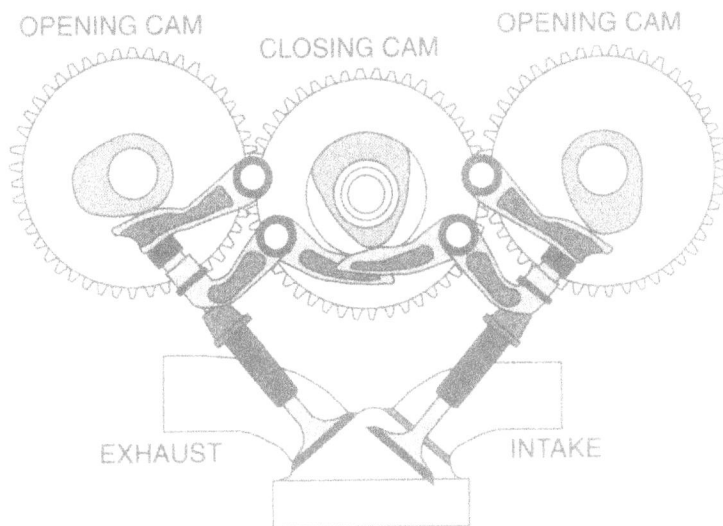

Fig. 5.36(a) The desmodromic valve operation of a Ducati engine.

DUCATI 955 cm^3 90° V-TWIN MOTORCYCLE				
	INT			EXH
n_v	2		n_v	2
d_{is}	35.0		d_{is}	29.0
d_{os}	37.0		d_{os}	31.0
d_s	7.0		d_s	7.0
d_p	34.5		d_p	28.5
d_m	46.0		d_m	43.0
C_m	0.93		C_m	1.21
vo	70.0		vo	90.0
vc	90.0		vc	65.0
L_v	12.0		L_v	11.0
C_r	0.25		C_r	0.25

Fig. 5.36(b) The cylinder head and valve geometry for the Ducati 995 cm^3 engine.

technique, and the one sketched here represents one such variation on the desmodromic valve theme. The positive control of valve motion permits the engine to run to high engine speeds without any fear of valve "bounce," which is always a concern when the valve contact with the cam is spring-controlled. Irrespective of the type of spring being used, e.g., a hairpin type, as in Plate 1.2 for the Norton, or a coil spring or a pneumatic spring, at some speed of operation, a natural frequency will exist at which the valves will tend to float off the cam and perhaps contact the piston crown, or even each other. The success of Ducati in Superbike racing suggests that they have perfected this system of valve control, so I am always a little surprised that the technique has not been applied by others, such as in Formula 1 car racing. In the current

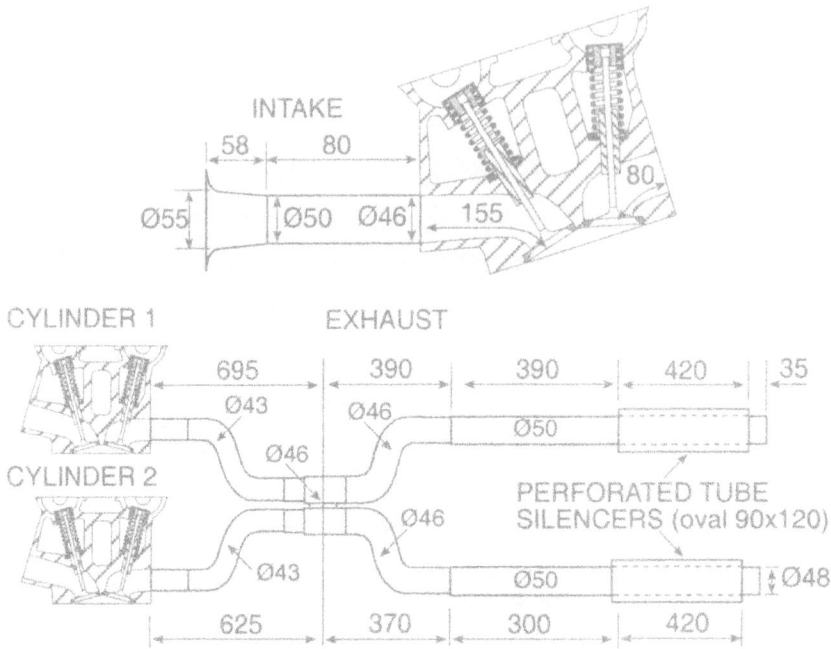

Fig. 5.37 The intake and exhaust ducting of the Ducati 995 cm³ engine.

Ducati engines, the bulky timing-gear drive seen in Fig. 5.36(a) is replaced by more compact pinions rotated by rubber-toothed timing belts. Otherwise, the two-valve design in Fig. 5.36(a) is very similar to that in a motorcycle I currently own, the 944 cm³, 90° V-twin, ST2 model Ducati, which I air on the country roads of Ulster on, sadly all too few, sunny summer days.

In Fig. 5.37, the exhaust system can best be described as a variation on the 2-1-2 design theme described by Roe [5.17]. In this unusual configuration, the four-way branch is best modeled as two connected tee-junctions, the cross-connection being a very short (46 mm diameter) pipe. The interconnection between the two pipes is actually an elliptical hole with an area of 1662 mm², which is equivalent in area to a 46 mm diameter pipe.

During the computer simulation, the following assumptions are made: the combustion of aviation gasoline occurs at a constant air-fuel ratio of 12.0, which corresponds to a λ value of 0.85; the combustion process is assumed to be as shown in Fig. 4.10 for a 4v na si engine; the discharge coefficient maps for the valves and pipe ends are assumed to be as presented in Figs. 3.33 to 3.36 and 3.9 to 3.10, respectively; the discharge coefficients for unthrottled bellmouth-ended pipes are as quoted in Eqs. 3.3.1 to 3.3.4; the skin temperatures of the cylinder head, piston crown, and cylinder liner are 300, 250, and 150°C, respectively; the skin temperatures of the exhaust duct surfaces range from 500 to 300°C; and the engine breathes to and from the atmosphere at 1.01325 bar and 25°C.

An acknowledgment is due to Mr. Steve Wynne, of Stockport, England, for supplying the data on the intake and exhaust ducting of the 916SP model seen in Fig. 5.37 and some of the data on cylinder head geometry tabulated in Fig. 5.36(b). More information on the engine, the cylinder head and duct geometry and the engine output performance data has been published by Boretti [5.8].

5.7.2 The Performance Characteristics of the Ducati Engine

The computed performance characteristics of torque and power are shown in Fig. 5.38. These are plotted as ratios from the maxima as this is the only way they can be compared with the measured data published by Boretti et al. [5.8]. Naturally, these are purely comparative, as I have no means of knowing whether the engine ducting or cylinder head geometry are precisely the same as those used by Boretti, as he gives only very limited geometric information in his paper. However, in the context of a discussion on designing a V-twin engine with asymmetrical firing, this is not overly important. Even though the exhaust system is semi-silenced, this is a high-performance engine because the peak bmep level is greater than 14 bar. A noticeable dip in the torque (or bmep) curve is evident at 9500 rpm. The issue to be discussed here is the reason for this dip. Is it caused by the asymmetrical firing? If so, have the development engineers exacerbated it, or alleviated it, by their ducting design?

You will see in Fig. 5.37 that the developers have opted for a duct that connects the exhaust down-pipes from each cylinder relatively closely to each cylinder head. By making this cross-connection for the primary exhaust pipes, the exhaust pressure pulses will transmit

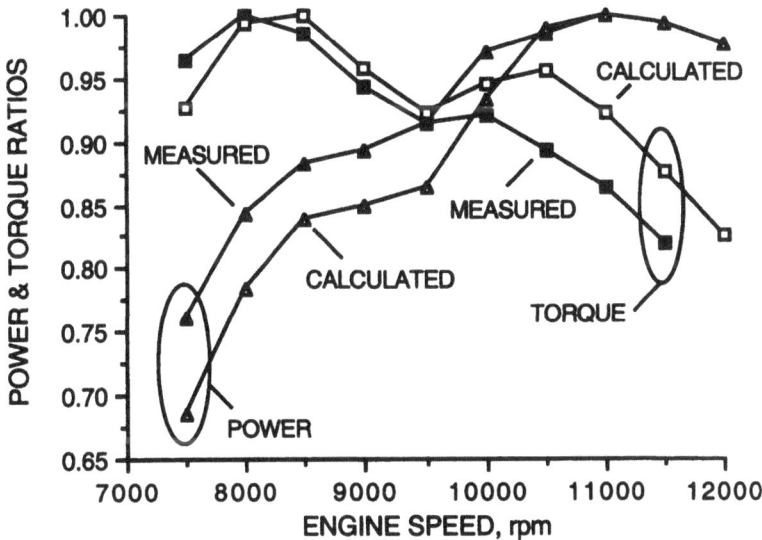

Fig. 5.38 The power and torque characteristics of the Ducati 995 cm^3 engine.

at the junction into the other pipes, so the pressure-time histories at the exhaust valves of each cylinder cannot be identical at all or any engine speed. The question is, how significantly does this affect performance characteristics? The further information we have is that the primary pipes are unequal in length, as are the remaining pipes after the junction up to the absorption mufflers. For a definition of absorption mufflers and the method for incorporating these into a simulation, consult Sec. 7.6.2 and Fig. 7.20.

The simulation calculates what would be difficult, but not impossible, to measure, namely, the individual breathing, charging, and trapping characteristics of the cylinders. These are shown in Figs. 5.39 and 5.40.

From Figs. 5.39 and 5.40, it is quite evident that the torque dip at 9500 rpm affects both cylinders because there are airflow, charge purity, and charging efficiency peaks at 8500 and 10,500 rpm. The high trapping efficiency at 9500 rpm ramps up a relatively high charging efficiency from a delivery ratio that has fallen by 10%, so that the hole in the torque curve ends up as shallow as it is. This is the overall effect for both cylinders, but it is also seen that cylinder number 2 is the least favored of the two cylinders as far as charging efficiency and scavenging efficiency, i.e., imep, is concerned. The worst zone within the speed range of this indifferent inter-cylinder behavior is precisely where it is least required, i.e., at 9500 rpm. What causes these effects?

5.7.3 Time-Varying Properties within the Ducati Engine and Ducting

Figs. 5.41 to 5.43 show the temporal variations at 9500 rpm of pressure, particle velocity, and temperature around both cylinders. Fig. 5.41 shows the pressures within cylinder number 1 and the pressures at the exhaust and intake valves within their separate ducts. Fig. 5.42

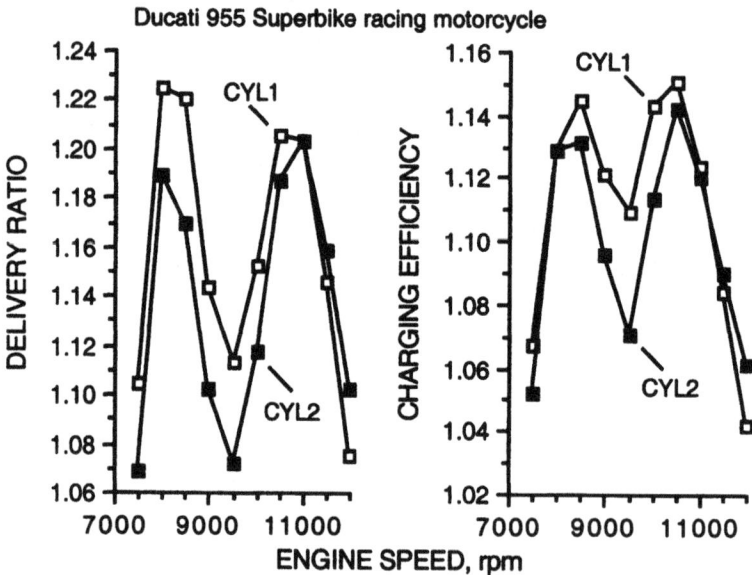

Fig. 5.39 Airflow and charging characteristics of the Ducati 995 cm³ engine.

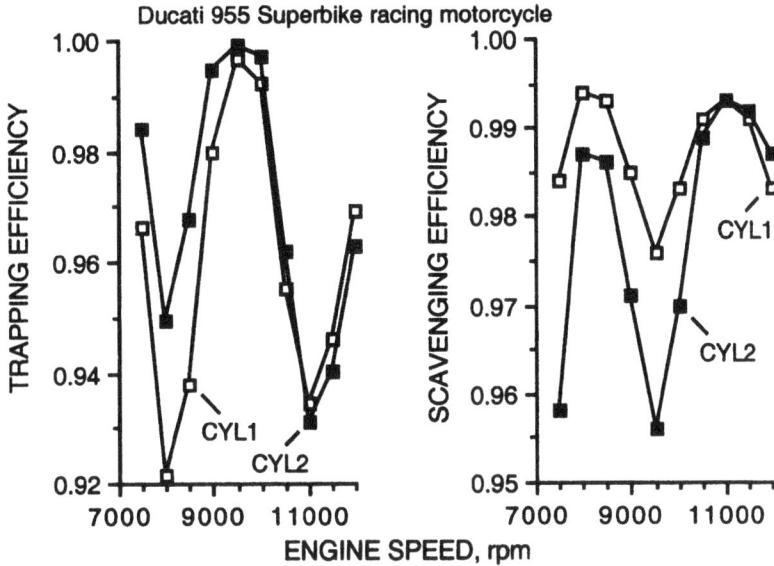

Fig. 5.40 Trapping and scavenging characteristics of the Ducati 995 cm³ engine.

shows the equivalent diagram for cylinder number 2. As forecast earlier, these exhaust pressure diagrams are quite dissimilar, and that for cylinder number 2 seriously opposes both the removal of exhaust gas toward the end of its exhaust stroke and air inflow at the start of the intake stroke. These awkward exhaust events are at precisely the speed where the intake ramming is at its weakest, because the multiple reflection of the primary intake ramming wave arrives during the valve overlap period as an expansion wave below atmospheric pressure.

The outcome for gas particle flow and the local intake tract gas properties can be seen in Fig. 5.43. The first of the three diagrams in this figure shows the gas particle velocities, as Mach number, behind the exhaust and intake valves for both cylinders during the valve overlap period at 9500 rpm. Exhaust outflow continues until about 55° before the exhaust valve closes, when the flow reverses and sends exhaust gas back into the cylinder to displace air. In the second diagram, the intake particle flow commences with backflow, increasing the local gas temperature very considerably. In the third diagram, the local intake tract charge purity drops to 0.6 for cylinder number 1 and below 0.35 for cylinder number 2. Although the naturally weaker ramming effect at this engine speed is always going to present tuning problems, these are exacerbated by the exhaust system giving no assistance to either cylinder, to the detriment of the charging of cylinder number 2, as already observed in Figs. 5.39 and 5.40.

Contrast this behavior with that under well-tuned conditions at 10,500 rpm, as represented by Figs. 5.44 to 5.46.

In Fig. 5.45, although cylinder number 2 still has that awkward exhaust reflection at the end of the exhaust stroke, the intake tuning is now optimized due to the presence of a compression wave at the beginning of the valve overlap period. The exhaust system gives increasingly strong suction on both cylinders down to 0.4 atm, from before tdc to exhaust valve closing.

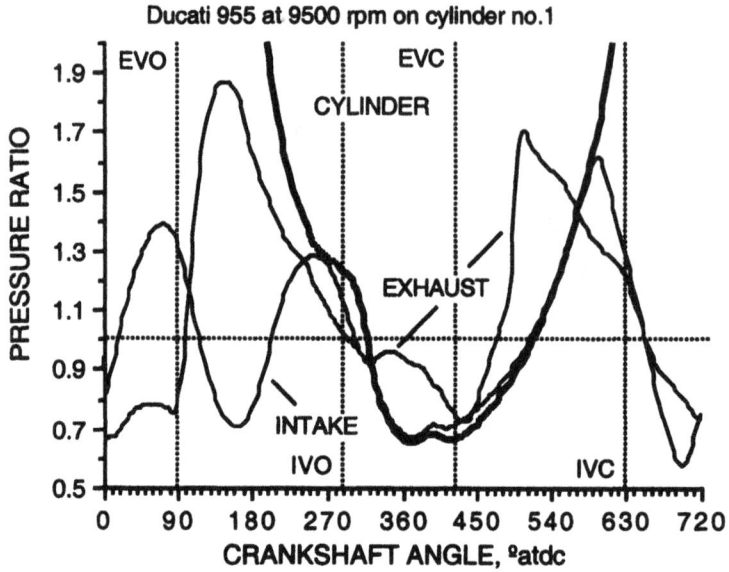

Fig. 5.41 Ducati: cylinder, exhaust, and intake pressures at 9500 rpm in cylinder number. 1.

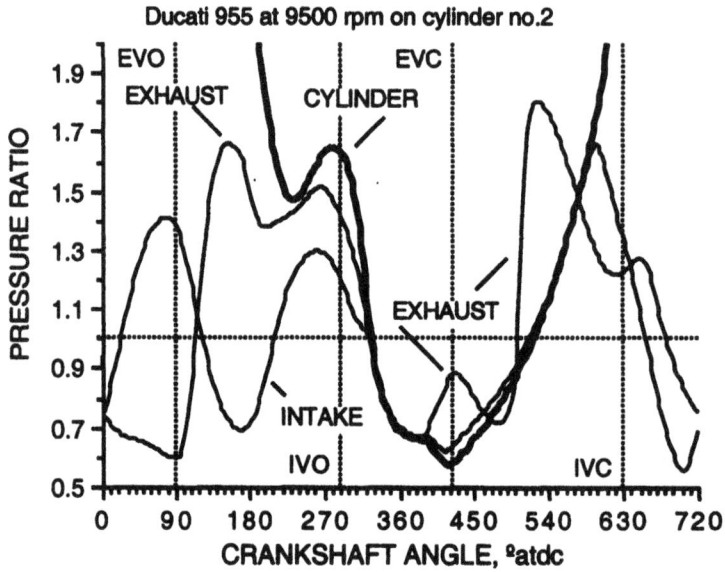

Fig. 5.42 Ducati: cylinder, exhaust, and intake pressures at 9500 rpm in cylinder number 2.

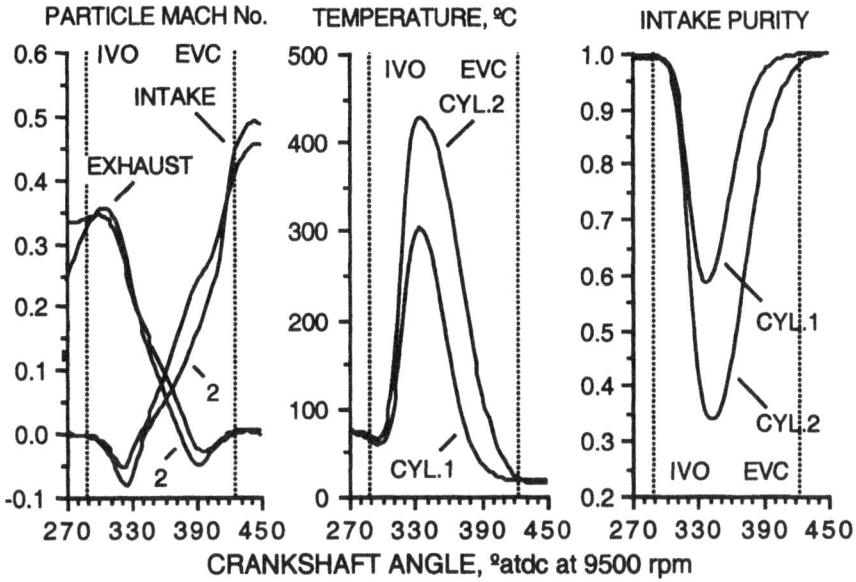

Fig. 5.43 Ducati: intake velocity, temperature, and charge purity at 9500 rpm in both cylinders.

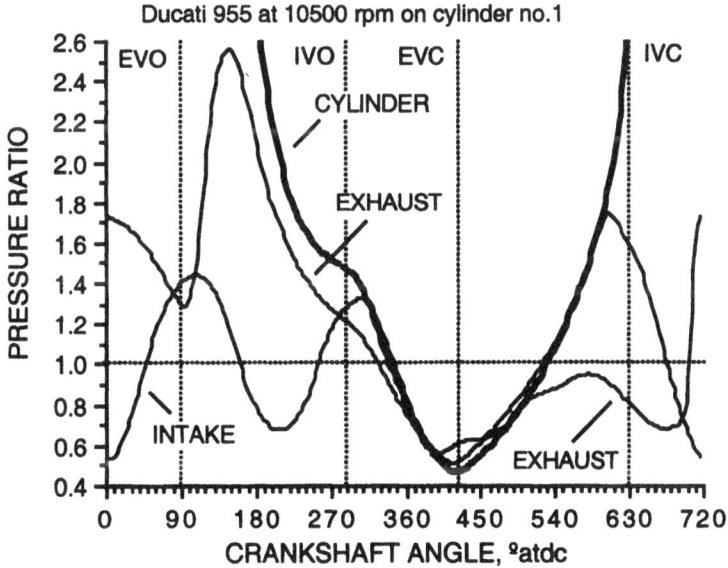

Fig. 5.44 Ducati: cylinder, exhaust, and intake pressures at 10,500 rpm in cylinder number 1.

Ducati 955 at 10500 rpm on cylinder no.2

EVO   IVO   EVC   IVC

CYLINDER

EXHAUST

EXHAUST

INTAKE

PRESSURE RATIO

CRANKSHAFT ANGLE, ºatdc

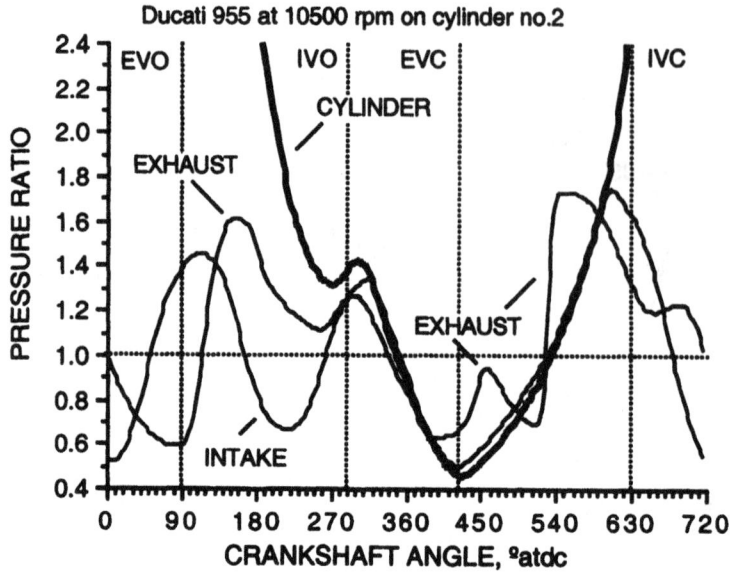

Fig. 5.45 Ducati: cylinder, exhaust, and intake pressures at 10,500 rpm in cylinder number 2.

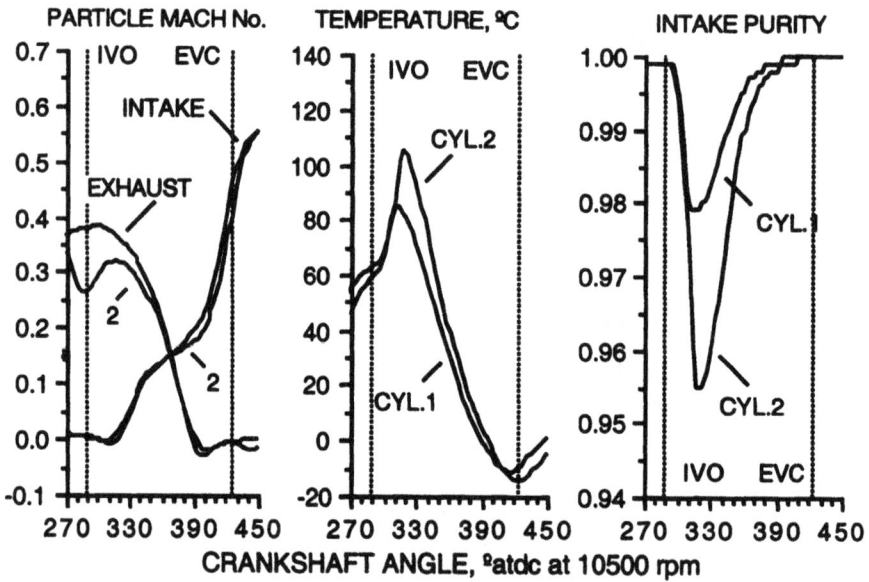

PARTICLE MACH No.   TEMPERATURE, ºC   INTAKE PURITY

IVO   EVC

INTAKE

EXHAUST

2

2

IVO   EVC

CYL.2

CYL.1

CYL.

CYL.2

IVO   EVC

CRANKSHAFT ANGLE, ºatdc at 10500 rpm

Fig. 5.46 Ducati: intake velocity, temperature and charge purity at 10,500 rpm in both cylinders.

In Fig. 5.46, the outcome is seen to be very little backflow for either cylinder. The gas temperature in the intake tract barely climbs by 20-40°C, and even poor cylinder number 2 only sees some 5% drop in charge purity behind the intake valve. The outcome is a high charging efficiency for both cylinders (see Fig. 5.39), with just a 1% difference between the two cylinders. This may seem insignificant, but for cylinder number 2 it translates into a 1 kW drop in power output that it is shy of cylinder number 1; by such tiny margins one can narrowly finish in second place!

It will be observed in Figs. 5.38 and 5.39 that there is another torque and airflow peak at 8500 rpm. Figs. 5.47 and 5.48 are presented to show the reason for this lower-speed peak: It is caused by intake ramming. For both of the cylinders represented in these graphs, you can see the multiple reflection of the primary intake ramming wave. It returns in the middle of the valve overlap period, assisting in the prevention of cylinder backflow into the intake tract, i.e., that which causes such problems at 9500 rpm. Here, at 8500 rpm, the exhaust system is also quite well-tuned by providing suction on the cylinder throughout the valve overlap period, although cylinder number 2 still sees that awkward reflection toward the conclusion of its power stroke, which still increases its pumping loss over that of cylinder number 1.

5.7.4 Some Logic Regarding Tuning a High-Performance V-Twin Engine

Connecting together the exhaust primary pipes of an asymmetrically firing engine will always provide potentially awkward, and always dissimilar, exhaust tuning effects on each cylinder. If effective intake ramming is used to enhance performance, then an airflow trough can be expected in the middle of the speed range. The main objective of the designer and

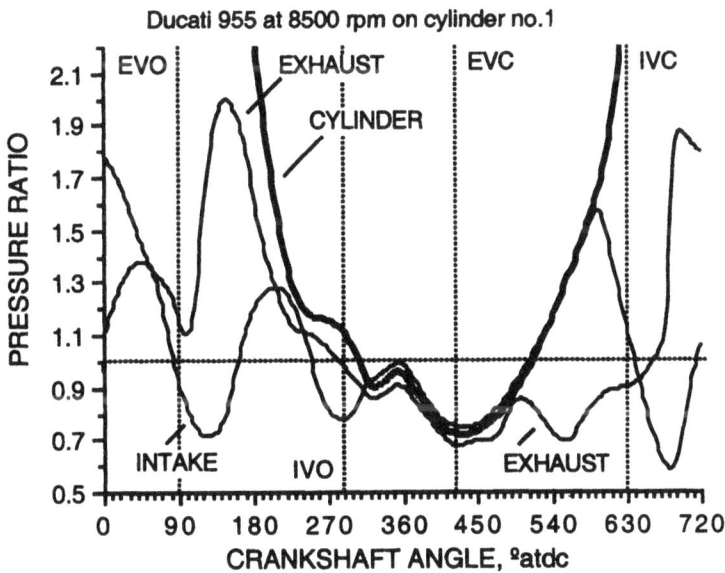

Fig. 5.47 Ducati: cylinder, exhaust, and intake pressures at 8500 rpm in cylinder number 1.

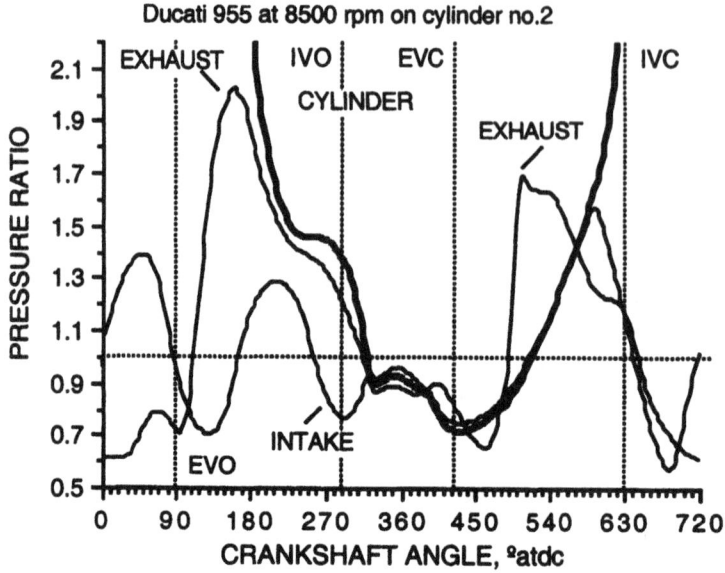

Fig. 5.48 Ducati: cylinder, exhaust, and intake pressures at 8500 rpm in cylinder number 2.

developer is to ensure that the engine speeds that give the worst pressure wave tuning in the exhaust and the intake systems do not coincide. For this, an accurate engine simulation is invaluable in that it provides design insights where technology cannot yet provide measurements.

5.8 The Nissan Infiniti 4000 cm³ Car Engine for the Indy Racing League (IRL)

This engine is used for IRL oval track racing in the United States. I was involved with Hans Hermann, of Hans Hermann Engineering, in the initial design by simulation for the performance characteristics of this engine. Hans Hermann continues to use this simulation for the ongoing development of the engine to enhance its competitiveness using the minimum number of manufacturing iterations. I am indebted to him for the geometrical data on the current engine design and its measured performance characteristics. A photograph of this engine appears in Plate 5.10 and also in Plate 5.11 as installed in a car driven by Roberto Guerrero at an IRL race in Texas in 1998, where he led the race for many laps before finishing fourth.

The objective here is to continue the discussion begun with the Ducati, namely, that of the design of high-performance engines with asymmetrical firing characteristics. The IRL racing rules include the use of a double-plane crankshaft as shown in Fig. 5.5(a), which is based on

Plate 5.10 The Nissan Infiniti racing engine for Indy Racing League (IRL). (Courtesy of Nissan)

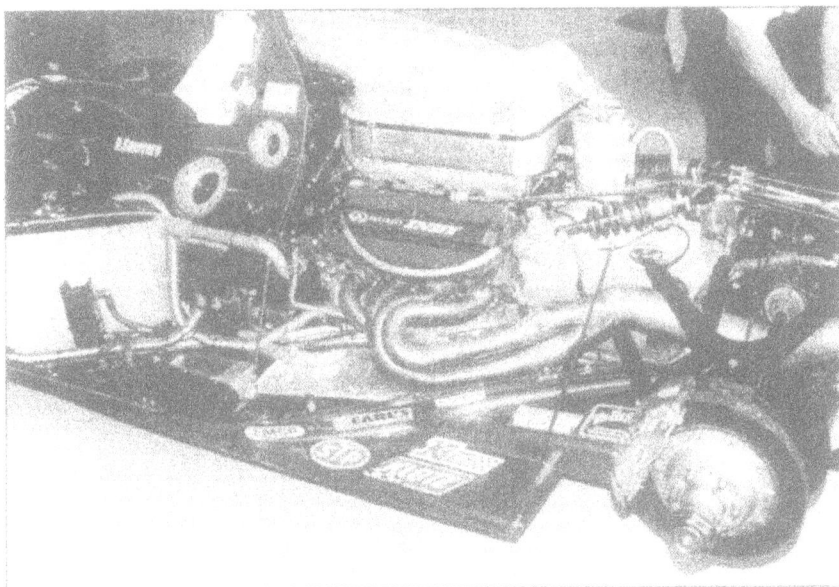

Plate 5.11 The Nissan Infiniti racing engine installed in an IRL car. (Courtesy of Nissan)

the Nissan Infiniti engine. The outcome is the cylinder firing characteristics discussed in Sec. 5.2.4 and enumerated in Fig. 5.4. The other engine that is currently racing in 1998, in the IRL, is the Oldsmobile Aurora, shown in Plates 5.2 and 5.3. This has the cylinder firing characteristics listed in Fig. 5.4 for an "American V8." Other IRL racing rules of design significance are that the maximum permitted engine speed is 10,500 rpm and the fuel used must be methanol.

5.8.1 Input Data for the Simulation of the Nissan Infiniti Engine

The engine has bore, stroke, and connecting rod dimensions of 93, 73.6, and 146.2 mm, respectively. Because the fuel burned is methanol, not gasoline, the compression ratio is higher at 14.5. The total swept volume of the engine is 3999.7 cm^3. The engine valves are operated by double overhead camshafts, and the general layout of the cylinder head and combustion chamber is as shown in Fig. 1.6. The four-valve pent-roof cylinder head and ducting geometries are presented in Figs. 5.49 and 5.50. The exhaust system is of the 4-1 collector type very commonly seen on racecar engines. Fig. 5.50 presents a sketch of cylinders numbered 5-8, and the intake and exhaust ducting for cylinders 1-4 is simply a mirror image of this set-up. The branch angles at each five-way branch can best be described as follows: the cylinder primary pipes are at a mutual 15° to each other and the outlet pipe is symmetrically disposed facing the four primary pipes. Many of the internal and external features of this engine, as discussed above, can be seen in Plates 5.4, 5.5, 5.9, and 5.10.

During the computer simulation, the following assumptions are made: the combustion of methanol occurs at a constant air-fuel ratio of 5.47, which corresponds to a λ value of 0.85; the combustion process is assumed to be as shown in Fig. 4.21 for a 4v na si engine using methanol; the discharge coefficient maps for the valves and pipe ends are assumed to be as presented in Figs. 3.37 to 3.40 and 3.9 to 3.10, respectively; the discharge coefficients for unthrottled bellmouth-ended pipes are as quoted in Eqs. 3.3.1 to 3.3.4; the skin temperatures of the cylinder head, piston crown, and cylinder liner are 250, 200, and 150°C, respectively; the skin

Nissan 4000 cm³ V8 IRL racing engine				
	INT			**EXH**
n_v	2		n_v	2
d_{is}	36.0		d_{is}	30.7
d_{os}	37.6		d_{os}	32.7
d_s	7.0		d_s	7.0
d_p	35.0		d_p	29.0
d_m	46.6		d_m	47.9
C_m	0.93		C_m	1.45
vo	58.0		vo	81.0
vc	90.0		vc	53.0
L_v	12.8		L_v	11.3
C_r	0.2		C_r	0.263

Fig. 5.49 The cylinder head and valve geometry for the Nissan Infiniti IRL engine.

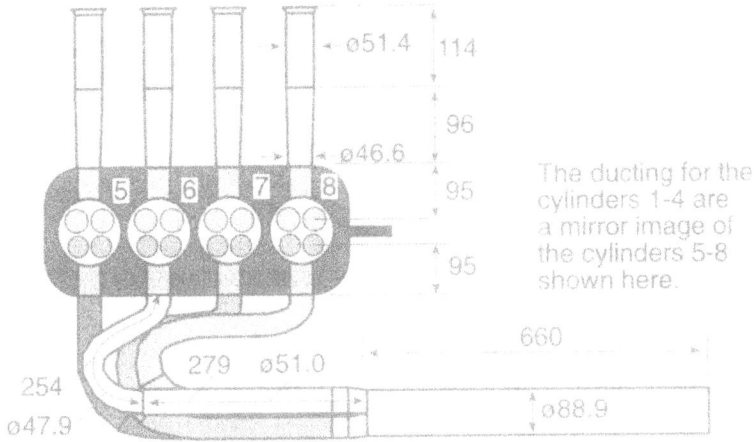

Fig. 5.50 The intake and exhaust ducting of the Nissan Infiniti IRL engine.

temperatures of the exhaust duct surfaces range from 500 to 400°C; the engine breathes to and from the atmosphere at 1.01325 bar and 25°C.

5.8.2 The Performance Characteristics of the Nissan Infiniti IRL Engine

The measured and computed performance characteristics of power and torque (as bmep) are shown in Figs. 5.51 and 5.52.

The correlation between measurement and calculation is seen to be very good, particularly so because the design computation is carried out prior to engine construction! The discrepancy in the bmep computation at 9000 rpm is more apparent than real due to the plotting scale used for Fig. 5.52; it amounts to a difference of 0.3 bar out of 15.2 bar, an error of only about 2%.

The computation yields data on air and fuel flow rates, which are shown in Fig. 5.53. The airflow, plotted as delivery ratio, is in the same 1.1-1.2 bailiwick as that of the Ducati and G50 engine (see Figs. 5.39 and 5.25), whereas the bmep is about 1.5 bar higher, because the consumption of methanol permits a compression ratio some 3 ratios higher. This confirms the discussion presented in Sec. 4.3.6, which elicited the simple empirical relationship, Eq. 4.3.43, that is seen here to be very relevant.

The profile of the airflow curve in Fig. 5.53 should look familiar; it has delivery ratio peaks around 8500 and 10,500 rpm, with a trough at 9500 rpm, just like the Ducati engine in Fig. 5.39. This similarity is reinforced by Fig. 5.54, a graph of the cylinder, exhaust, and intake port pressures at 10,000 rpm for cylinder number 1. In this figure, the intake pressure wave profile for the Nissan engine is very similar to that observed for the Ducati engine at 10,500 rpm, in Fig. 5.44. The ducting geometries for the two engines are shown in Figs. 5.37 and 5.50, and you can see there that the total lengths of the intake tracts for the Ducati and the Nissan

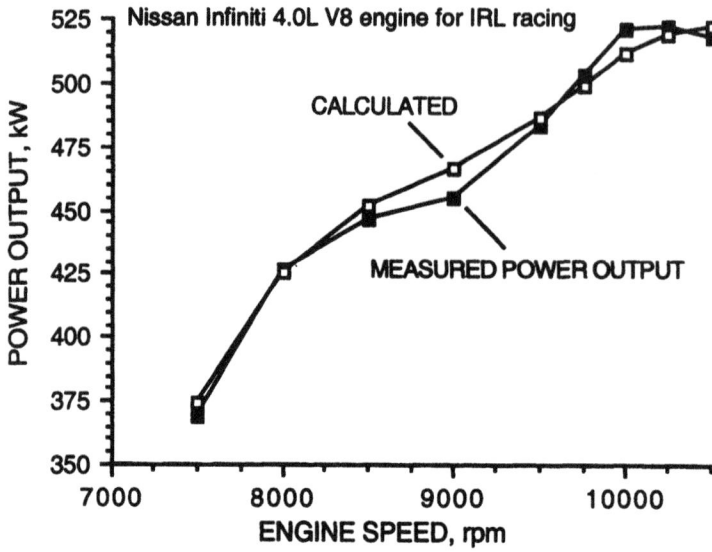

Fig. 5.51 Measured and calculated power characteristics of the Nissan Infiniti engine.

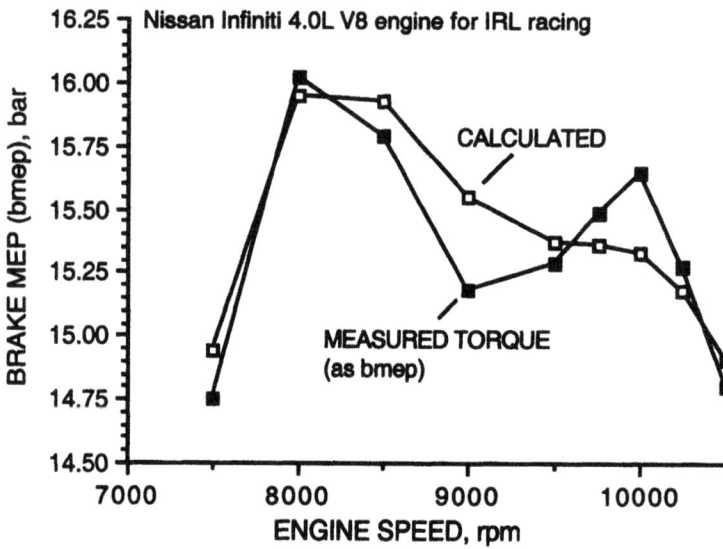

Fig. 5.52 Measured and calculated torque characteristics of the Nissan Infiniti engine.

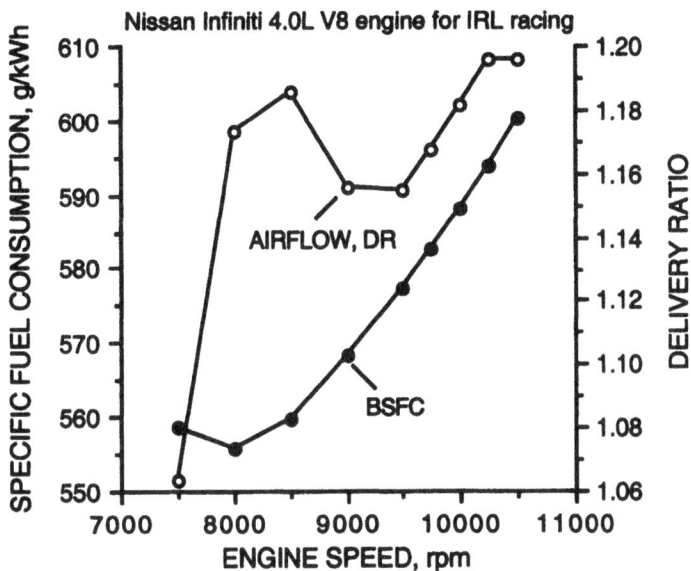

Fig. 5.53 Calculated airflow and fuel consumption for the Nissan Infiniti engine.

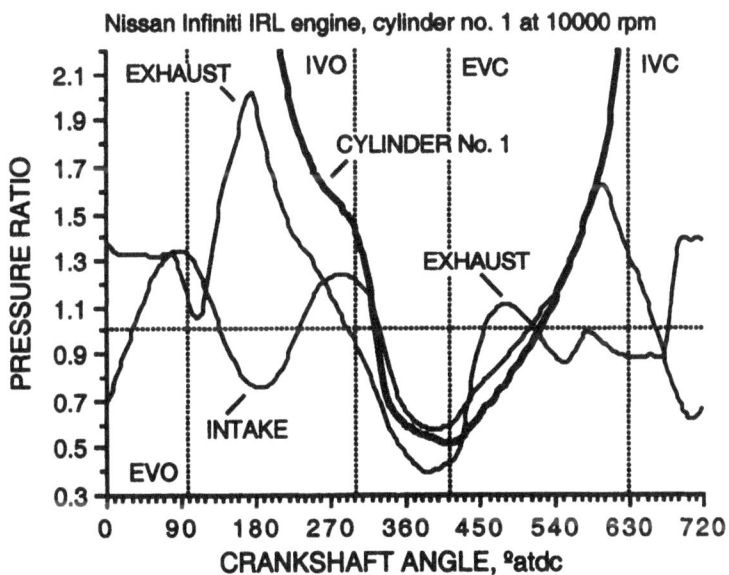

Fig. 5.54 Nissan: cylinder, exhaust, and intake pressures at 10,000 rpm in cylinder number 1.

engine are almost the same at 293 and 305 mm, respectively. I will remind you again of this geometrical coincidence when intake ramming is discussed in detail in Sec. 6.3.

In Fig. 5.54, the exhaust system for the Nissan V8 is shown to provide even stronger suction around the valve overlap period compared to that shown for the Ducati V-twin in Fig. 5.44. This is the satisfactory situation in cylinder number 1, but in the further discussion below some other cylinders are shown to be not so well served.

Note in Fig. 5.53 the level of the brake specific fuel consumption, which has a minimum of about 555 g/kWh. Contrast this with the minimum of 300 g/kWh for the gasoline-burning Ducati engine and that for the Matchless G50 engine observed in Fig. 5.29. Methanol-burning engines certainly know how to drink the alcohol!

5.8.3 The Effect of Asymmetrical Firing Due to the Double-Plane Crankshaft

The engine simulation records the singular events for each cylinder of the Nissan engine. As you will recall, this means that there are dissimilar intercylinder angular intervals on each bank of the vee layout due to the regulation double-plane crankshaft. Figs. 5.55 and 5.56 show the airflow, as DR, and brake mean effective pressure, bmep, data for each cylinder at 9000 and 10,000 rpm.

It can be observed from these figures that the cylinders numbered 3 on the first bank, and 5 on the second bank, breathe less, and produce less torque, than their fellows. While cylinders numbered 1, 2, and 4 on the first bank, and cylinders numbered 6, 7, and 8 on the second bank,

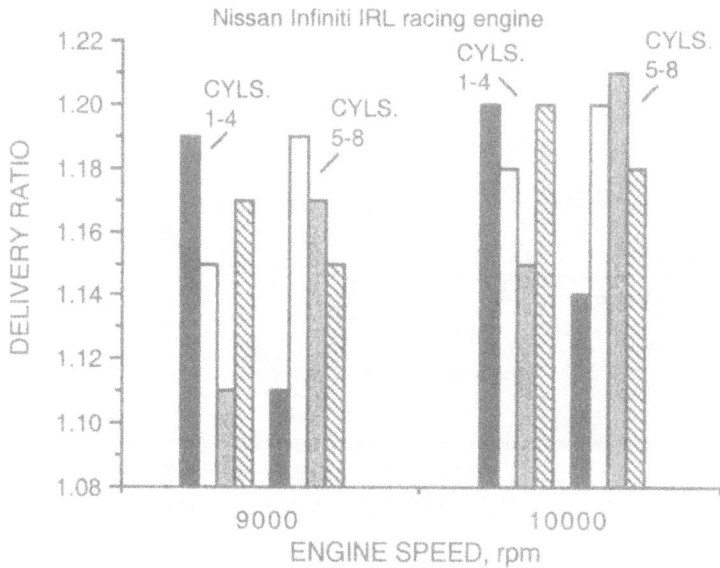

Fig. 5.55 Nissan: inter-cylinder variations of airflow at 9000 and 10,000 rpm.

580

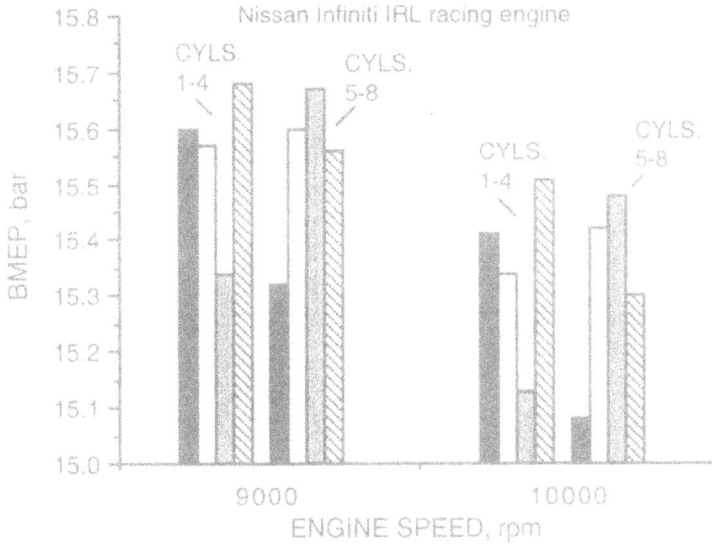

Fig. 5.56 Nissan: intercylinder variations of torque at 9000 and 10,000 rpm.

are not precisely equal in terms of airflow or torque attained, they are superior to the weaker cylinders already noted. This relative performance of the cylinders on either bank is maintained at 9000 and 10,000 rpm. The reason, as usual, is to be found in the pressure profiles facing the open exhaust valve during the valve overlap period. These profiles are shown in Figs. 5.57 and 5.58, as computed at 10,000 rpm for cylinders numbered 1 and 3, and 5 and 7, respectively.

It is quite clear, in Fig. 5.57, that the weakest cylinder, numbered 3, has a reduced suction pressure during the valve overlap period compared to a stronger cylinder, numbered 1. This situation is repeated in Fig. 5.58 for the weakest cylinder, numbered 5, compared to the strongest cylinder on the other bank, numbered 7. The two weak cylinders, 3 and 5, have almost identical pressure profiles for the main exhaust outflow after release at evo, as do the strong cylinders, 1 and 7. The question remains as to the extent of the performance loss due to these effects, and this can be answered by modeling the engine with symmetrical cylinder timing and firing, i.e., as if it had a single-plane crankshaft.

5.8.4 The Effect of Symmetrical Firing Provided by a Single-Plane Crankshaft

The engine is re-modeled in the simulation with a few changes to the input database. The firing order and intercylinder angular variations for the double-plane Nissan crankshaft shown in Figs. 5.4 and 5.5(a) are replaced by the single-plane data shown in Fig. 5.4 and 5.5(b). Otherwise, the data are identical to those above. Now the cylinders fire evenly at 180° intervals on each bank. Each cylinder, although I do not display it here, breathes the "same"

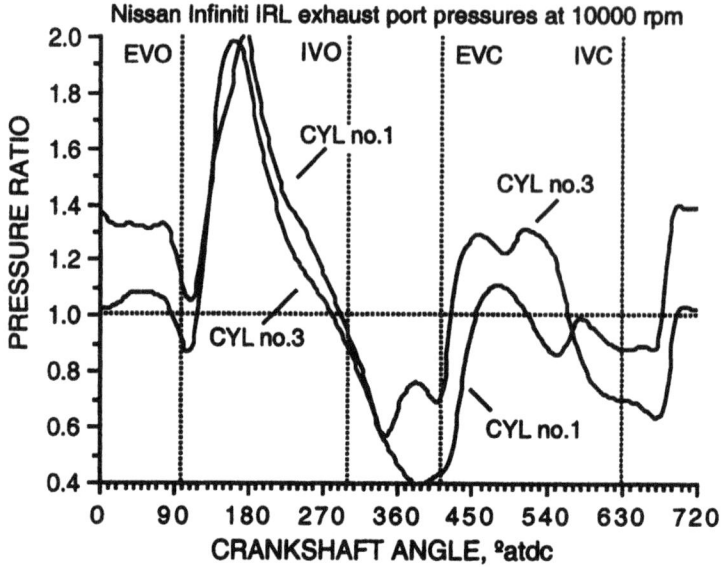

Fig. 5.57 Nissan: exhaust pressures for cylinders numbered 1 and 3 at 10,000 rpm.

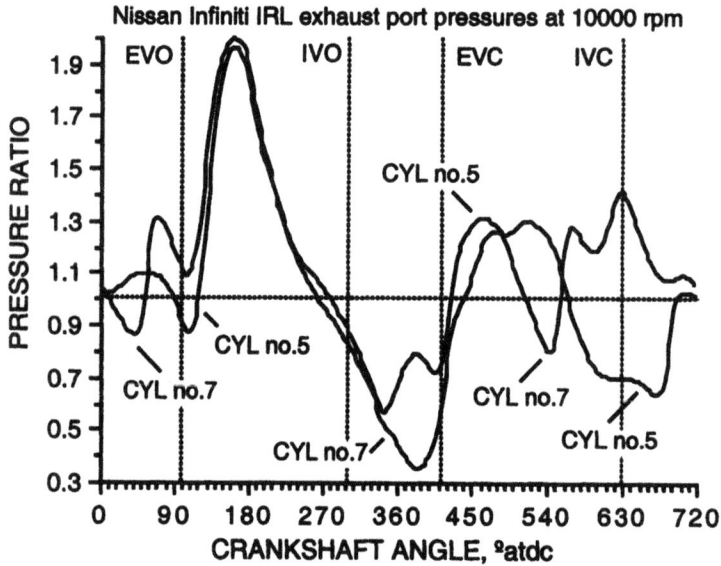

Fig. 5.58 Nissan: exhaust pressures for cylinders numbered 5 and 7 at 10,000 rpm.

DR value at all speeds, produces the "same" torque at all speeds, and exhibits the "same" pressure diagrams at all speeds. I have place quotation marks around the word "same" in each case to show that it is a correct statement, within the normal limits expected of arithmetic computation. The ensuing performance characteristics for airflow, power, and torque are shown in Figs. 5.59 and 5.60, where they are compared with the output data obtained with the standard double-plane crankshaft.

In Fig. 5.59, the overall increase in airflow caused by using a single-plane crankshaft is seen to be significant. If you compare these results with the maximum values for DR observed in Fig. 5.55, it is clear that all cylinders with a single-plane crankshaft have been brought up to the same level as the best cylinder in Fig. 5.55. This translates into the torque, as bmep, plotted in Fig. 5.60 where all cylinders approach the maximum produced by the best cylinder as shown in Fig. 5.56. The gain in power output by using a single-plane crankshaft is also plotted in Fig. 5.60. In oval-track racing, an IRL engine typically "pencils" between 9500 and 10500 rpm during a (normal) transmission gear change, and this is where the symmetrical firing would pay the greatest dividends—if it were legal within the racing regulations to use it. There is a power advantage for the single-plane crankshaft design, all other data being equal, of over 6 kW, i.e., about 9 bhp. As always, this could well make the difference between winning and losing!

You might have imagined that a single-plane crankshaft would confer an even greater power advantage than the 1.3% predicted for it, and you may well be correct, as that engine

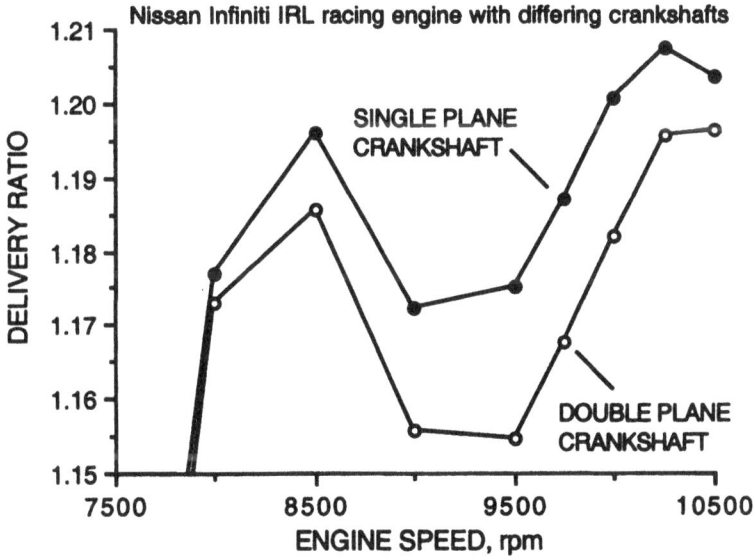

Fig. 5.59 Nissan: airflow for a single- and double-plane crankshaft

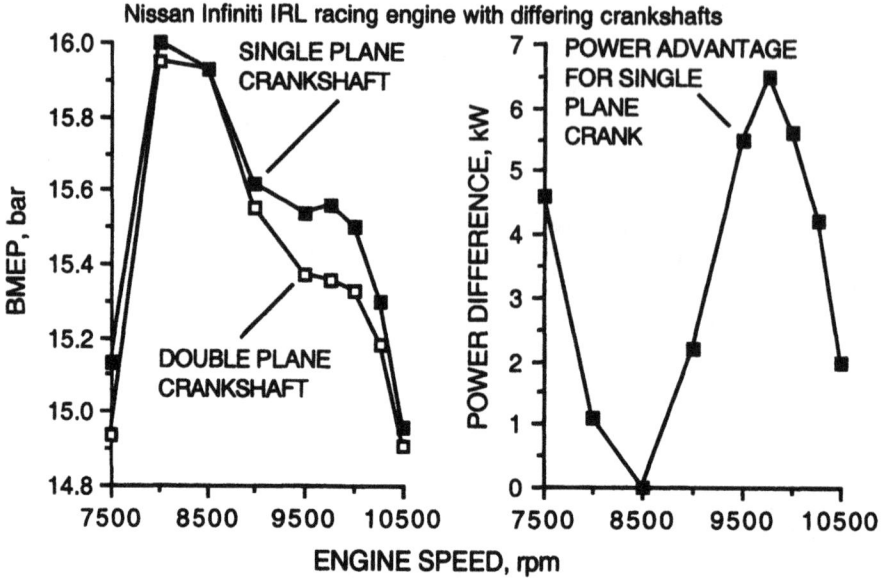

Fig. 5.60 Nissan: power and torque for a single- and double-plane crankshaft.

can be re-optimized to take more advantage of symmetrical firing. Nevertheless, what saves the double-plane crankshaft design in the present analysis is the strength of the exhaust tuning from the collector exhaust system. The effectiveness of this type of system in controlling the transmission of pressure waves, which can detune other cylinders in that bank, is demonstrated in Sec. 6.4.6

Asymmetrical Firing Provides Difficult Inter-Cylinder Fueling Control

Recall that all of the simulation output presented here uses data which decree that the air-fuel ratio to each cylinder is identical. In practice, this provides a simple fueling control mechanism for an evenly-breathing engine with a single-plane crankshaft; however, if similarly employed with a double-plane crankshaft, this could present even greater disparities in power and torque than shown in Fig. 5.60, because some cylinders will operate up to 10% richer than their fellows. To verify this statement, you could examine the numerical effects of Eq. 4.3.26 on a mixture that is 10% richer than the optimum for maximum power.

Some Logic Regarding Tuning a Vee Engine with a Double-Plane Crankshaft

In the context of the design of an IRL engine with its mandatory double-plane crankshaft, or any engine in which the cylinder firing is asymmetrical, the implication is that either not all primary pipes to an exhaust collector on each bank should be of equal length, or that some other form of branching of the exhaust pipes from each bank of cylinders should be investigated. One method to accomplish this, and to totally prevent cylinder-cylinder cross-talk via the exhaust system, is to install a separate exhaust pipe per cylinder on a traditional "American V8." Although bulky, this is a not-illogical concept—and, like most things in life, it has all been done before, as shown in Plate 5.12. This is a photograph I took of just such a layout; it was at a car race near Dublin (Ireland, not Ohio!) in 1965. However, I make no guarantee that the lengths of tuned exhaust pipe shown in this plate are optimized.

5.9 Automobiles: a 2000 cm³ Four-Cylinder Sports Car Engine

This engine is an in-line, four-cylinder (I4) unit, with a total swept volume approaching 2000 cm^3. It is a naturally-aspirated spark-ignition (na si) engine for a sports car, and the design lessons to be learned from the simulation concern the dual requirements of relatively high-performance characteristics and the necessity to meet exhaust emissions legislation under urban driving conditions. The geometrical data for the simulation do not match precisely any particular car in current production, although the general layout and ethos would be similar to a car I currently own, the Mazda MX-5 1998 model, also known as the Miata in the United States, and the Eunos Roadster in Japan [5.18].

Plate 5.12 An exhaust pipe for each cylinder on an "American V8."

5.9.1 Input Data for the Simulation of the Sports Car Engine

The engine has bore, stroke, and connecting rod dimensions of 86, 86, and 165 mm, respectively, which makes it very similar to that used in the earlier study with simple Otto cycle thermodynamics, presented in Sec. 1.9.1. The compression ratio is 10.3. The total swept volume of the engine is 1998.2 cm^3. The engine valves are operated by double over-head camshafts. The four-valve, pent-roof cylinder head and ducting geometries are presented in Figs. 5.61 to 5.63.

The firing order is dictated by the use of a single-plane crankshaft, as shown in Fig. 5.5(b), resulting in a 1-3-4-2 firing order, and the angular firing sequences listed for that unit in Fig. 5.4.

Fig. 5.61 The cylinder head and valve geometry for the sports car engine.

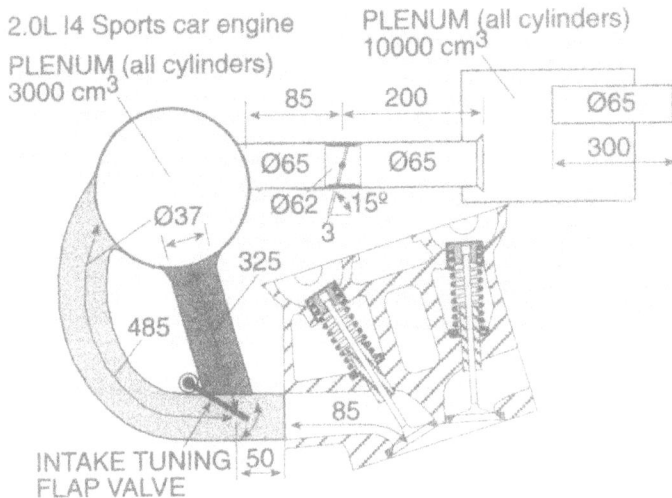

Fig. 5.62 The intake ducting of the sports car engine.

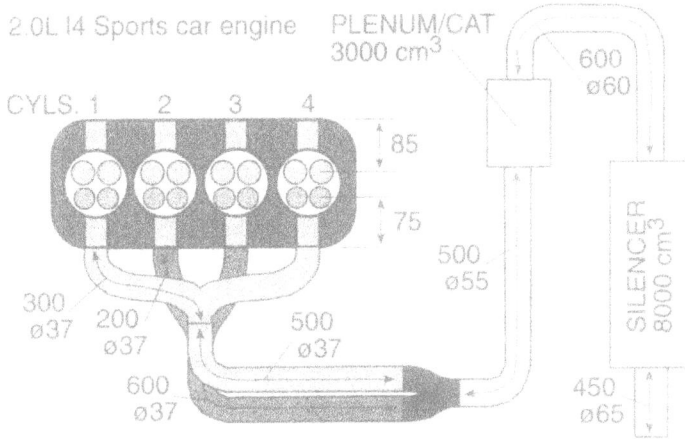

Fig. 5.63 The exhaust ducting of the sports car engine.

During the computer simulation at wide open throttle, the following assumptions are made: the combustion of unleaded gasoline occurs at a constant air-fuel ratio of 14, which corresponds to a λ value of 0.92; the combustion process is assumed to be as shown in Fig. 4.4 and Figs. 4.7 to 4.10 at each relevant engine speed for a 4v na si engine with gasoline; the discharge coefficient maps for the valves and pipe ends are assumed to be as presented in Figs. 3.37 to 3.40 and 3.9 to 3.10, respectively; the discharge coefficient map for the butterfly throttle body are assumed to be as shown in Fig. 3.51; the discharge coefficients for unthrottled bellmouth-ended pipes are as quoted in Eqs. 3.3.1 to 3.3.4; the skin temperatures of the cylinder head, piston crown, and cylinder liner are 240, 240, and 150°C, respectively; the skin temperatures of the exhaust duct surfaces range from 600 to 400°C; and the engine breathes to and from the atmosphere at 1.01325 bar and 25°C.

Intake Ducting for the Sports Car

The intake ducting sketched in Fig. 5.62 shows a dual-speed intake ramming design. There are long and short intake tract lengths which are indexed by a flap valve that is electronically controlled to change from the long to the short tract as the engine accelerates past a speed threshold. There are many possible designs that can be used to implement dual-length intake tract ramming, of which Fig. 5.62 is but one. The patent literature will yield many more, for those who wish to study further the myriad geometrical possibilities. As in Fig. 3.52(a), the intake throttling is controlled by a butterfly throttle, an elliptical plate that is 3 mm thick. In its fully closed position, the butterfly is at an angle, θ_c, of 15° in a duct segment of diameter, d_2, of 62 mm, and is bounded either side by a duct of diameter, d_1, of 65 mm. The theory regarding the geometry of this butterfly throttle, and others, is discussed in detail in Sec. 3.7.5. The intake silencer plenum has a silencer tube, which is the subject of a more extensive study in Sec. 7.6.7.

Exhaust Ducting for the Sports Car

The exhaust system is of the 4-2-1 collector type very commonly seen on production car engines. It is shown in Fig. 5.63, attached to the cylinders numbered 1-4 as sketched. In the cylinders, which fire at 360° intervals, the exhaust primary pipes are closely coupled near the cylinder head, whereas the final cross-connection to a single outlet pipe is at 875 mm from any one exhaust valve. The objective of this design is to prevent cross-talk between the cylinders during the valve overlap period when the engine speed is in the region of peak torque. The branch angles at the three-way branch can best be described as follows: for cylinders numbered 1 and 4, the primary pipes are at 45° to each other, and the outlet pipe is symmetrically disposed facing them; for cylinders numbered 2 and 3, the primary pipes are at 30° to each other, and the outlet pipe is symmetrically disposed facing them; the two down-pipes arrive at the final three-way branch at 30° to each other, and the outlet pipe to the catalyst is symmetrically disposed facing them.

The exhaust ducting contains a "plenum" holding a catalyst and, further down the system, an exhaust muffler that is treated here as a plenum. This allows the performance, but not the noise, characteristics to be accurately assessed. As with the intake silencer, I will not dwell further on the geometry of the exhaust muffler because exhaust silencing in general, and the assessment of performance and noise characteristics, are explained in detail in Chapter 7.

5.9.2 The Full-Load Performance Characteristics of the Sports Car Engine

Actually, you have encountered this sports car engine, and a simulation of it, earlier in this book. In Appendix A4.2, on two-zone combustion, this same engine provides the input data at 4800 rpm used to obtain the detailed insights into combustion presented in Figs. A4.2 to A4.12.

Here, the simulation is conducted at full-load over the speed range from 1600 rpm to 6500 rpm, with the input data as specified above. The simulation is executed twice, once for each of the two intake tract length geometries shown in Fig. 5.62. The full-load performance characteristics are shown in Figs. 5.64 to 5.67.

The full-load airflow characteristics, as delivery ratio, are shown in Fig. 5.64. The delivery ratio for the long intake tract peaks at 4500 rpm, and that of the short intake tract reaches a maximum at 5500 rpm. Over much of the usable speed range from 2500 rpm to 6500 rpm, the airflow has delivery ratio values above 90%, with most above 100%, which should translate into a sports car engine with a healthy torque curve. The cross-over point for the airflow profiles for the two different intake tract lengths is just above 5000 rpm. This should be reflected in the torque curves.

The torque (as bmep) characteristics are shown in Fig. 5.65, with the ensuing power outputs given in Fig. 5.66. The cross-over point is still at 5200 rpm. The peak bmep is near 13 bar at 4500 rpm, and peak power is 110 kW (148 bhp) at 6000-6500 rpm. The power curve is quite "fat" with a 2000 rpm spread above 100 kW. The torque curve is quite flat above 2500 rpm

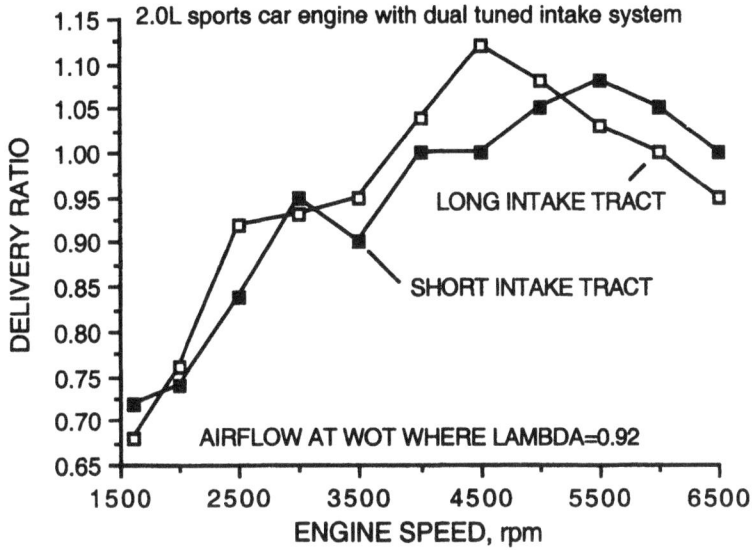

Fig. 5.64 The intake airflow characteristics of the sports car engine.

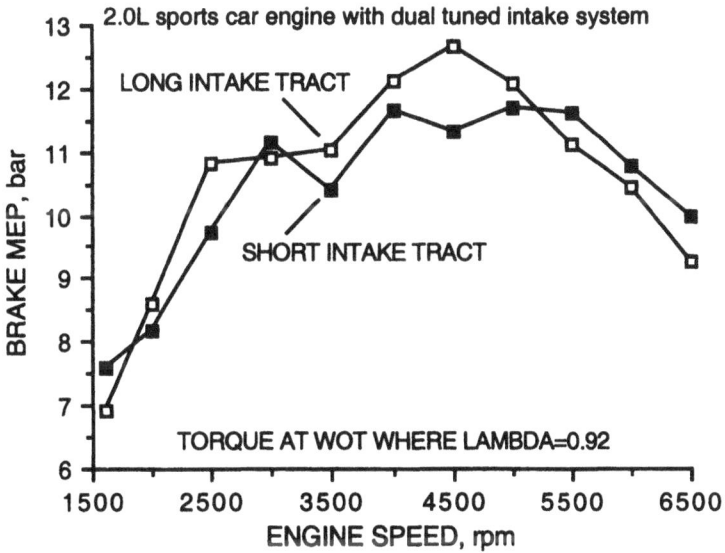

Fig. 5.65 The torque output characteristics of the sports car engine.

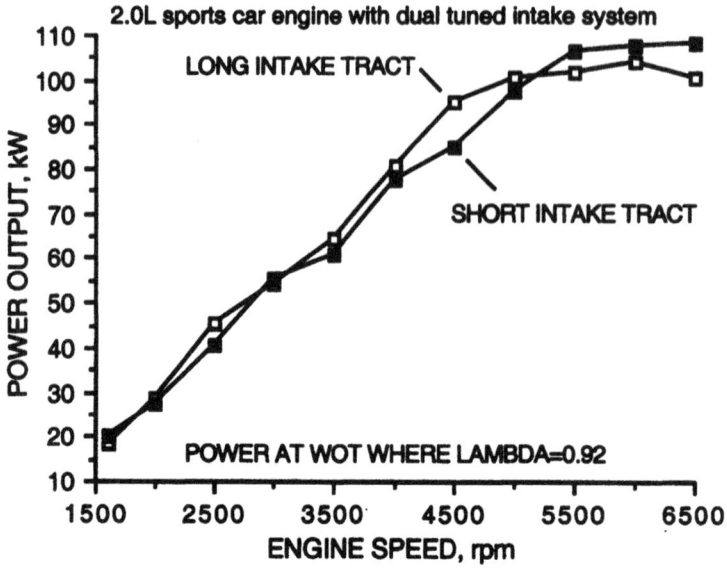

Fig. 5.66 The power output characteristics of the sports car engine.

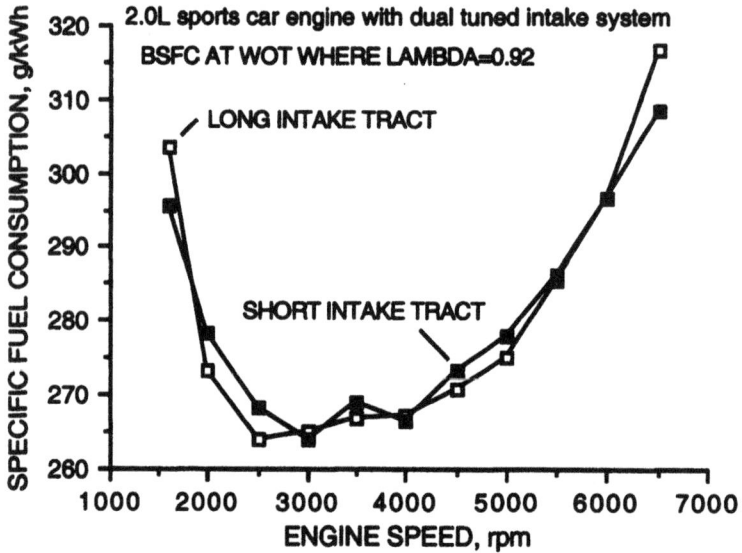

Fig. 5.67 The fuel consumption characteristics of the sports car engine.

where the bmep level exceeds 11 bar. Apart from a single point at 3000 rpm, there is a considerable gain in torque and power using the dual-length approach to intake ramming. The crossover point may seem high at 5200 rpm, but that is precisely the engine speed employed in the Mazda MX-5 control system [5.18].

With dual-length intake ramming, there is even a small gain in brake specific fuel consumption, bsfc, as shown in Fig. 5.67, but the minimum value of 265 g/kWh is fairly typical of a well-optimized 4v, pent-roof head, automobile engine at full-load.

The full-load performance characteristics in Figs. 5.64 to 5.67 are very typical of the 2.0-liter "sports car" or "GT," i.e., the uprated saloon (sedan) car, engines sold in Japan and Europe for automobiles with a higher performance specification. Nevertheless, they have to meet the same exhaust emissions laws as any other vehicle, so it is not surprising that the valve overlap periods are not as extensive as in a racing engine. The valve overlap period lasts for 80° in the sports car engine compared to 113° in the Nissan racing engine. On any graphical plot showing crankshaft angle, the ivo and evc points are visibly closer to each other than for the racing engine.

5.9.3 Pressure Diagrams at Full-Load in the Sports Car Engine

Pressure diagrams for the sports car engine are presented in Figs. 5.68 and 5.69, where the cylinder and the exhaust and intake port pressures are shown at full-load at 4500 rpm with the long intake tract, and at 5500 rpm with the short intake tract, respectively.

The engine speed and ducting combinations are selected to coincide with a maximum airflow point in each case. The intake ramming seen in Figs. 5.68 and 5.69 appears similar in that there is strong primary ramming giving trapping up to 2.0 atm at intake valve closing, and there is a pronounced secondary ramming wave at intake valve opening. The exhaust system is well-matched in that the cylinder pressure is reduced to atmospheric pressure by intake valve opening, or even before it in Fig. 5.68, and this bespeaks an exhaust system that is assisting scavenging by reducing the pumping loss on the exhaust stroke.

The effect of the cylinder-coupled exhaust system can be seen, for example in Fig. 5.69, where the pulse observed around 540° atdc on cylinder number 1 emanates from the exhaust outflow process of cylinder number 4. This appears too large in terms of amplitude in this superposition pressure plot, but it is reflecting at a closed exhaust valve, so the particle velocity is zero. Throughout the overlap valve period, the exhaust system maintains a sub-atmospheric pressure at the exhaust valve, assisting with cylinder scavenging, but at the expense of some bypassed fuel which contributes to the amount of unburned hydrocarbon emissions at full-load. Of course, this is only true if the engine employs homogeneous charging, i.e., by carburetion, or single- or multi-point EFI in the intake system; with gasoline direct-injection (GDI), there would be bypassed air, but not fuel [5.19]. Assuming this sports car engine uses EFI, the bypassed specific HC emissions, bsHC, are at 2.2 g/kWh for the situation depicted in Fig. 5.68 and 2.1 g/kWh for that depicted in Fig. 5.69. Considering the relatively high-performance nature of this engine, the trapping efficiencies are satisfactorily high at 99.2% and 99.0%, respectively. Further exhaust emissions data are presented in Fig. 5.81.

Full load at 4500 rpm with long intake duct: cylinder no.1

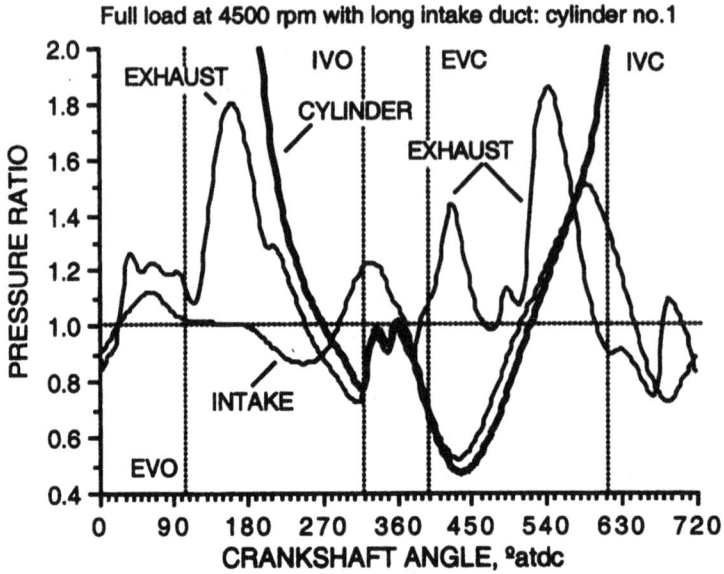

Fig. 5.68 Sports car, long intake tract: cylinder, exhaust, and intake pressures at 4500 rpm.

Full load at 5500 rpm with short intake duct: cylinder no.1

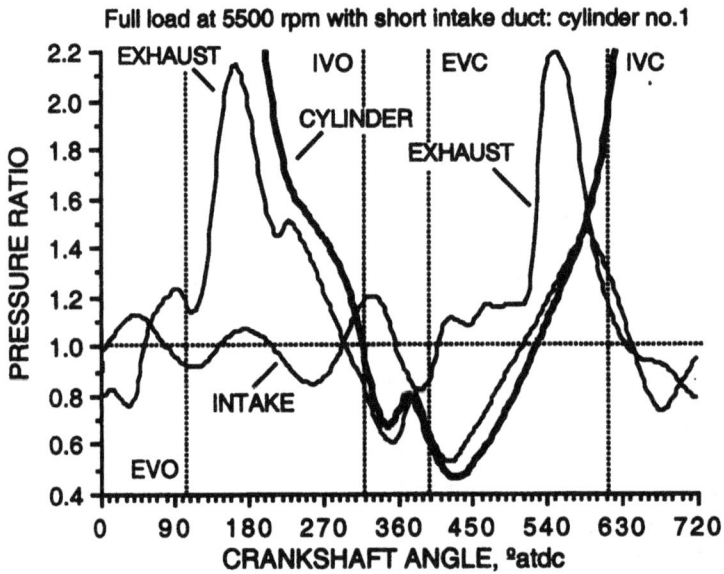

Fig. 5.69 Sports car, short intake tract: cylinder, exhaust, and intake pressures at 5500 rpm.

5.9.4 The Part-Load Performance Characteristics of the Sports Car Engine

To gain some understanding of the behavior of an automobile engine during an exhaust emissions test on the urban driving cycles on the statute books of Japan, the United States, or Europe, a torque-speed mapping point of 2.6 bar bmep at 1600 rpm is often employed during the R&D phase of a new or prototype unit. The simulation can be used to investigate the engine behavior at this part-load condition by closing the butterfly throttle seen in Fig. 5.62, and by reducing the air-fuel ratio to the stoichiometric level so that the three-way catalyst may function at maximum efficiency, until the required torque of 41.5 Nm is attained at 1600 rpm. The air-fuel ratio for a stoichiometric mixture of this unleaded gasoline is 14.2. Using the theory in Sec. 3.7.5, and the geometric data in Fig. 5.62, the area exposed by the now-angled butterfly throttle plate is found to be equivalent to 84.95 mm² when the butterfly plate has moved by 5.2° from a closing angle, θ_c, of 15° to an angle, θ, of 20.2°.

The simulation provides the performance characteristics at the test point. These are shown in Fig. 5.70 in the row marked "na si" (for naturally-aspirated spark-ignition). To provide the requisite performance of 2.61 bar bmep at 1600 rpm, which corresponds to a power output of 7.0 kW, the delivery ratio is seen to decrease to 0.38 and the brake specific fuel consumption, bsfc, to increase to 437 g/kWh. The imep is 4.39 bar, while the friction consumes, as fmep, 1.48 bar, and the pumping loss, as pmep, consumes 0.30 bar. By comparison, at full-load at 1600 rpm, the bsfc is better at 303 g/kWh, while the pumping mean effective pressure, pmep, is a mere 0.02 bar. Clearly, under these urban driving conditions, the throttling necessary to achieve the required part-load torque and power increases the pumping losses, lowers the mechanical efficiency from 0.822 to 0.595, and deteriorates the specific fuel consumption rate by 44%.

5.9.5 Pressure and Temperature at Part-Load in the Sports Car Engine

The reasons behind these performance characteristics are to be found in the detail provided by the simulation for the temporal variations of pressure and temperature within the engine. Figs. 5.71 and 5.72 show plots of the pressure and temperature histories in the cylinder and in the exhaust and intake ducts at the valves.

unit	rpm	afr	λ	power	bmep	DR	bsfc	imep	pmep	fmep
NA SI	1600	14.2	1.00	7.0	2.61	0.38	437.4	4.39	0.30	1.48
TC CI	1600	44.0	3.05	7.0	2.61	0.87	327.2	4.15	0.03	1.51

Fig. 5.70 Part-load performance of the sports car and turbo-diesel car engines.

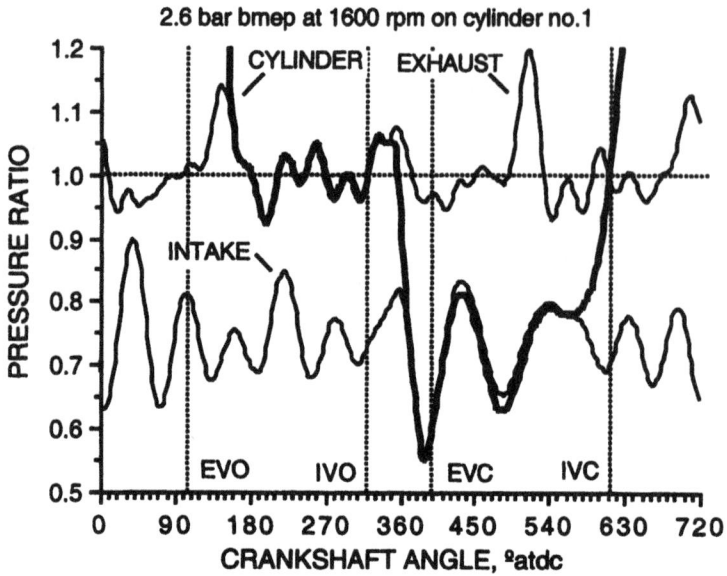

Fig. 5.71 Sports car, part-load: cylinder, exhaust, and intake pressures at 1600 rpm.

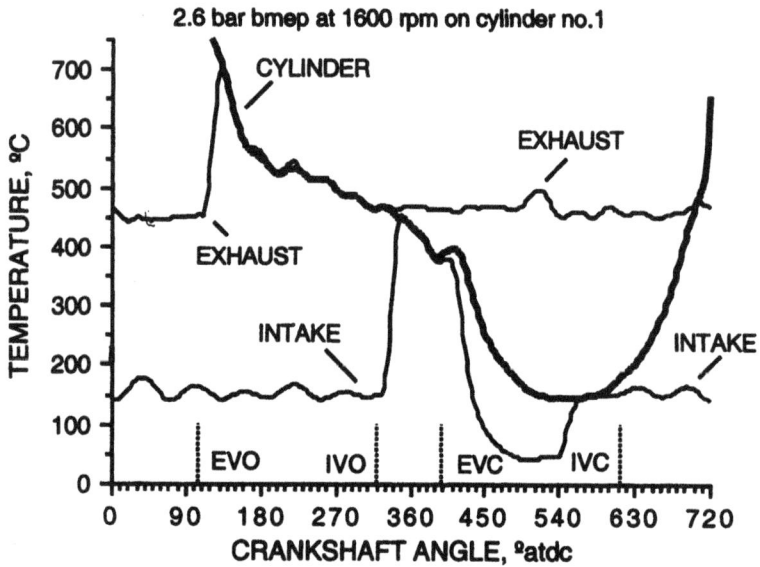

Fig. 5.72 Sports car, part-load: cylinder, exhaust, and intake temperatures at 1600 rpm.

In Fig. 5.71, the pressure in the cylinder is seen to be controlled by the exhaust system pressure during the exhaust stroke, and by the intake system during the intake stroke. The exhaust system pressure signal oscillates around atmospheric pressure, as that is basically its reference pressure. The intake pressure signal does likewise, but its reference pressure has been reduced to that of the 3-liter intake plenum, i.e., to about 0.72 atm. This plenum is filled from the atmosphere through the considerable restriction of the throttle-body because the butterfly throttle plate has opened by a mere 5.2° from its fully-closed position. The intake pressure signal is seen to oscillate around 0.72 atm. During the overlap period, exhaust gas blowback occurs in two stages. From intake valve opening to tdc, and beyond, the cylinder pressure exceeds the intake pressure, so blowback of exhaust gas into the intake duct can be expected. From tdc to exhaust valve closing, the exhaust system pressure exceeds the cylinder pressure, so exhaust gas must be reingested into the cylinder. As far as pumping losses are concerned, the pmep component normally seen during the exhaust stroke is almost zero because the mean cylinder pressure during this period is close to the atmospheric value. During the intake stroke component of the pumping loop, the mean cylinder pressure approximates to about 0.7 atm, revealing the pmep magnitude reported as 0.3 bar in Fig. 5.70. To ensure that you fully understand this pumping loss, you may wish to reexamine Sec. 1.6.9 and Fig. 1.33.

In Fig. 5.72, the exhaust blowback, forecast by the pressure signals in Fig. 5.71, can be seen to have an important influence on the temperatures in the intake duct and within the cylinder. At intake valve opening and in the intake duct at the intake valve, the temperature of the local charge is seen to rise to meet that of the exhaust gas within the cylinder, and remain there until the exhaust valve has closed. It then falls to about 40°C while some fresh air is ingested, but rises again to 150°C as further cylinder charge is blown into the intake tract in the run-up from bdc, at 540 °atdc, to trapping at intake valve closure. The minimum cylinder charge temperature never drops below 150°C. The exhaust gas in the exhaust duct sits around 450°C, but rises briefly to near 700°C as each new blowdown phase takes place between exhaust valve opening and the bdc location at 180 °atdc.

5.9.6 The Gas Movements at Part-Load in the Sports Car Engine

The cylinder-intake and cylinder-exhaust pressure differentials provide the local gas particle velocities in the intake and exhaust ducts at their respective valves. These are plotted in Fig. 5.73 as Mach number, where exhaust outflow and intake inflow are shown as positive quantities and backflow as negative, by definition.

The exhaust particle flow shows the initial exhaust outflow from evo to bdc, with further outflow by piston pumping on the exhaust stroke from bdc to near tdc. However, from then on, it is backflow into the cylinder until the exhaust valve closes.

For the intake duct, the process commences with backflow from the cylinder, virtually until the evc point. Then the normal intake procedure occurs until bdc at 540 °atdc. This is followed by backflow of the cylinder charge until the intake valve closes. The effect of these gas movements can be seen through the dynamic airflow characteristic, which is also plotted on Fig. 5.73, as delivery ratio.

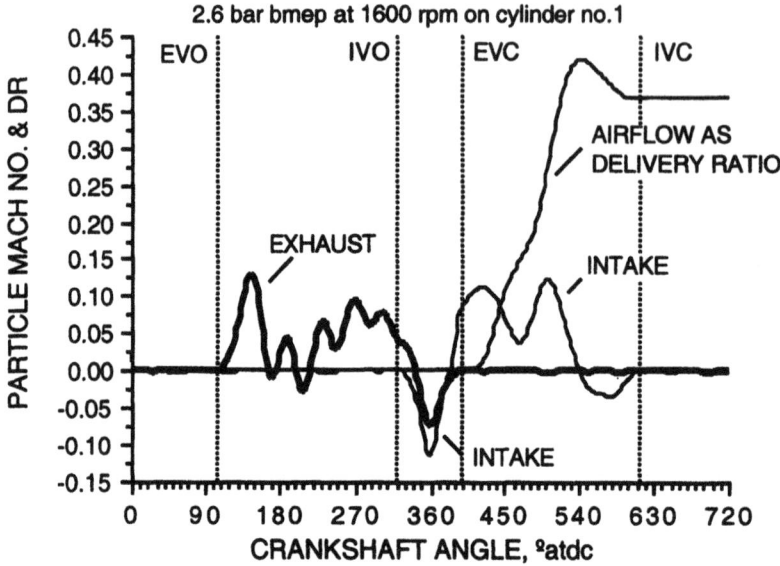

Fig. 5.73 Sports car, part-load: exhaust and intake particle velocities at 1600 rpm.

The delivery ratio curve records the first appearance of air within the cylinder at 20° after the local tdc point, or 380 °atdc, and it rises to a DR value of 0.418 before falling again to 0.368 by intake valve closing. There is some minor inter-cylinder variability here due to the exhaust system tuning. For example, for cylinder number 2, the maximum and trapped DR are 0.432 and 0.379, respectively; for cylinder number 3, they are 0.427 and 0.374: and for cylinder number 4, they are 0.428 and 0.375.

The pressure differentials and the particle velocities at the apertures of the cylinders have an ongoing effect on the local charge purity behind the intake valve and within the cylinder. These purities are plotted in Fig. 5.74 for the part-load condition at 1600 rpm; the traces are recorded for cylinder number 1, but are very similar for the other three cylinders. The charge purity trace at the intake valve shows the extent of the backflow of the exhaust gas. Never unity in the first place, the intake purity drops to zero by tdc and remains at that level as even more exhaust gas goes by. The first whiff of fresh air at the intake valve appears by exhaust valve closing, some 40° after that tdc point, and rises to unity by 500 °atdc, i.e., at 40° before bdc at the bottom of the intake stroke. Because backflow from the cylinder recommences at the beginning of the compression stroke, the intake tract purity falls until it is equal to the maximum in the cylinder, which is 0.68, an event that occurs some 30° before the intake valve closes. Being trapped at that location behind the closed intake valve, the charge purity remains at about 0.68-0.7 until the valve opens again. The maximum charge purity in the cylinder coincides with the maximum point on the dynamic DR curve in Fig. 5.73, which is to be expected because this juncture signals the reemergence of backflow from the cylinder. Because no more fresh air enters the cylinder on that cycle, the cylinder purity remains constant after that, until combustion takes place.

2.6 bar bmep at 1600 rpm on cylinder no.1

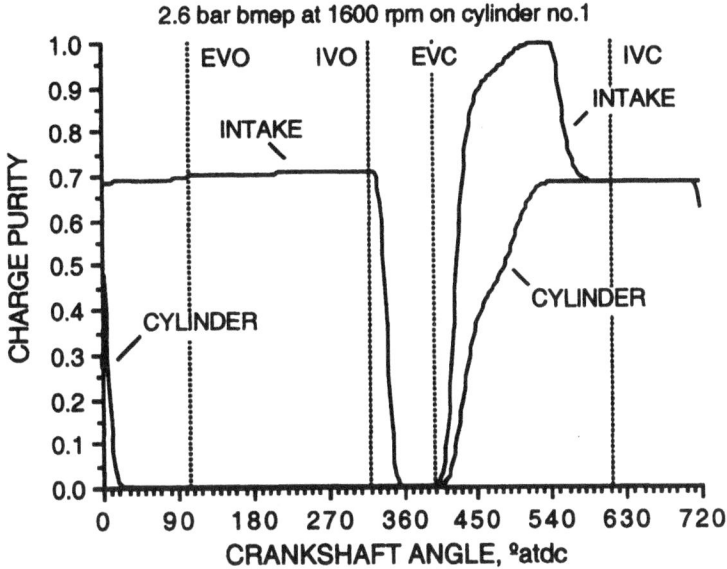

Fig. 5.74 Sports car, part-load: cylinder and intake charge purities at 1600 rpm.

5.9.7 The Implications for Combustion Efficiency of the Part-Load Trapped Charge Purity

In Fig. 5.74, the trapped charge purity of 0.68 is at a very interesting level. In many engines, exhaust gas is cooled and recirculated (EGR) into the intake tract as a means of reducing the exhaust emissions of oxides of nitrogen. This has an added benefit in that the throttle may need to be opened a little further to permit the necessary airflow, i.e., a DR value of 0.38, which is still required to provide the necessary imep of about 4.39 bar. Consequently, the ensuing pmep may be a little less than 0.3 bar, and so the mechanical efficiency and brake specific fuel consumption can be slightly improved. Whatever the minor trade-offs may be, a bmep of 2.6 bar is still required to give the requisite power of 7.0 kW to satisfy the emissions mapping point criterion. However, if too much EGR is inserted into the intake tract, the combustion quality deteriorates rapidly and high hydrocarbon emissions ensue. The typical EGR limit is often quoted as being no greater than that equivalent to 10% of the fresh intake airflow, i.e., 10% EGR. If the cylinder purity is already at 0.68, the addition of 10% EGR with a purity of zero could well cause the cylinder charge purity to drop below 0.6. This is precisely the value often quoted for two-stroke engines as being the lower threshold of cyclic combustion, i.e., the misfire limit at part-load known as "four-stroking" [5.14], and could explain why "10% EGR" is conventionally regarded as the limit for exhaust gas recirculation in four-stroke automobile engines.

Compare Figs. 5.72 to 5.74 for this sports car engine at 2.6 bar bmep with the equivalent figures for the Ryobi engine at 4 bar bmep, Figs. 5.16 to 5.18, and you will see that they are not dissimilar.

5.9.8 Thermodynamic Mayhem is Not a Joking Matter

As you may know, I have been as involved with the design of two-stroke engines [5.14] as with four-stroke engines. Many of my "friends," who live and design exclusively in a four-stroke world and know "nothing" about two-stroke engines, have been known to pass disparaging remarks, indeed even jest, about the low-speed and part-load performance characteristics of the simple Clerk two-stroke! Perhaps after they have read the above paragraphs, and have absorbed the horrifying truth about the thermodynamic and gas dynamic mayhem surrounding the cylinder of a four-stroke Otto engine at part-load, they may be less inclined to be so humorous!

5.10 Automobiles: a 2000 cm³ Four-Cylinder Turbocharged Diesel Engine

This engine is an in-line, four-cylinder (I4) unit, with a total swept volume of about 2000 cm³. It is a turbocharged, direct-injection, compression-ignition (tc di ci), engine for a car, and the design lessons from the simulation concern the same dual requirements as for the sports car, i.e., for relatively high performance characteristics and the ability to meet exhaust emissions legislation under urban driving conditions. The geometrical data for the simulation do not match any diesel car engine in current production, but the general design concept for this engine is similar to that of engines manufactured by Volkswagen/Audi, BMW or Fiat in Europe. It is interesting to note that diesel engines are now fitted in some 40% of the cars on European roads. The models most recently introduced into the marketplace are four-valve units that are turbocharged and inter-cooled with direct-injection. Generalities on the design of these engines can be found in Sec. 1.3.2.

5.10.1 Input Data for the Simulation of the Diesel Car Engine

The engine has bore, stroke, and connecting rod dimensions of 84, 90, and 180 mm, respectively. As discussed in Sec. 1.8.1, diesel engines tend to be under-square by design. The general layout of the cylinder head and combustion bowl is shown in Fig. 1.7. The compression ratio is 18. The total swept volume of the engine is 1995 cm³. This engine has been designed to have the same swept volume capacity as the sports car engine so that the performance characteristics of torque and power output are directly comparable. The engine valves are vertical and operated by double overhead camshafts (dohc). The four-valve cylinder head and ducting geometries are presented in Figs. 5.75 and 5.76.

The firing order is dictated by the use of a single-plane crankshaft as shown in Fig. 5.5(b), resulting in a 1-3-4-2 firing order and, consequently, the same angular firing sequences as listed for the "sports car engine" in Fig. 5.4.

During the computer simulation at full load, the following assumptions are made: the combustion of diesel fuel occurs at a constant air-fuel ratio of 22, which corresponds to a λ value of 1.525; the combustion process is assumed to be as shown in Fig. 4.36; the discharge coefficient maps for the valves and pipe ends are assumed to be as presented in Figs. 3.37 to 3.40 and 3.9 to 3.10, respectively; the discharge coefficients for unthrottled bellmouth-ended pipes are as quoted in Eqs. 3.3.1 to 3.3.4; the skin temperatures of the cylinder head, piston crown, and cylinder liner are 300, 250, and 150°C, respectively; the skin temperatures of the exhaust duct surfaces range from 450 to 300°C; and the engine breathes to and from the atmosphere at 1.01325 bar and 25°C.

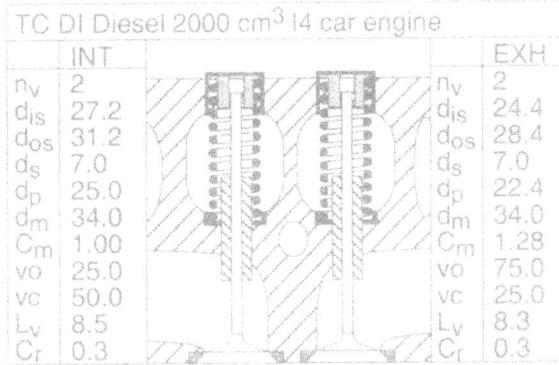

Fig. 5.75 The cylinder head and valve geometry for the turbo-diesel car engine.

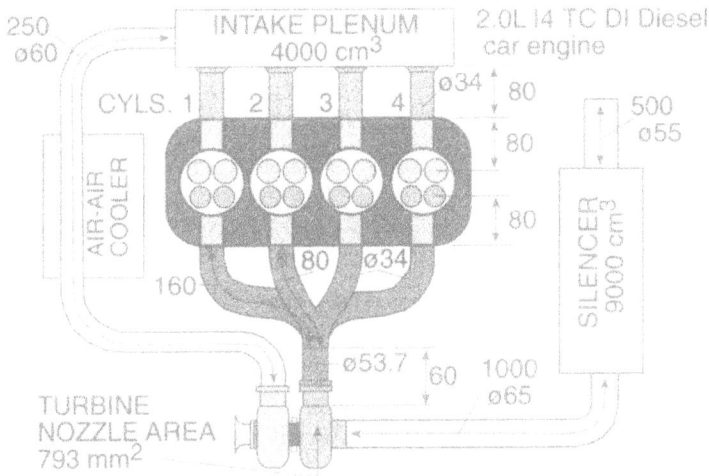

Fig. 5.76 The intake and exhaust ducting of the turbo-diesel car engine.

Exhaust and Intake Ducting of the Diesel Car Engine

The exhaust primary pipes from each cylinder are closely branched near the cylinder head, and the outlet pipe mates with the single-entry turbine to face the nozzle ring. The branch angles at this five-way branch can best be described as follows: the cylinder primary pipes are at a mutual 22.5° to each other and the outlet pipe is symmetrically disposed facing the four primary pipes. The turbine nozzle ring has a total area of 793 mm², which is 35% of the area of the 53.7 mm diameter entry pipe. Downstream of the turbocharger, the exhaust enters a silencer. The compressor of the inward-radial-flow turbocharger compresses the air through an intercooler into an intake plenum above the short intake pipes to each cylinder. Taking into consideration that the lengths of the intake pipes are very short, the intake valves close at 50 °abdc, and the maximum engine speed is 4000 rpm, it is clear that intake ramming is not a feature included in the design of this particular turbo-diesel car engine.

Modeling the Turbocharger

The turbocharger has a waste gate [5.20] to control the maximum compressor boost pressure to a pressure ratio of about 2.0. At full-load fueling, this realizes a boost pressure ratio of about 1.3 bar at 1600 rpm, which rises with engine speed until the waste gate begins to control the maximum boost pressure at the higher engine speeds. In this simulation, the turbocharger is modeled in the simple manner recommended by Watson and Janota [5.24], but not as completely as that so elegantly described by one of the late Neil Watson's students, Smith [5.25], for whom I had the honor of being the external examiner for his Ph.D.

For the basic design and theory of operation of turbochargers, I recommend the book by Watson and Janota [5.24]. However, to assist you further with some basic understanding, consider the shape of the turbocharger shown in Fig. 5.76, which can be seen to have the same profile as in the photograph in Plate 5.13. This shows a turbocharger mounted on a Volvo Penta diesel engine for marine applications, with the same orientation to the intake and exhaust system as in Fig. 5.76. Within the compressor casing on the left and the turbine casing on the right are the rotors of the compressor and turbine. A typical example of such rotors is shown in Plate 5.14, again with the same orientation as in Plate 5.13 and Fig. 5.76. For obvious reasons, the flow path of the compressor rotor is known as outward radial flow, and that of the turbine as inward radial flow. As far as the exhaust system is concerned, the restriction that the turbine nozzle ring and the rotor pose to the exiting exhaust pulses is quite evident. On a personal note, I spent three years as a doctoral student attempting to track the transmission and reflection of, and work available from, exhaust pressure waves passing through the turbines of such turbochargers [5.27].

The overall design and appearance of a four-cylinder, turbocharged, direct-injection diesel engine is shown in Plate 5.15. The car diesel engine of Fig. 5.76 would look much like this photograph. The turbocharger can be seen to be connected to the exhaust system in a compact manner similar to that shown in Fig. 5.76, although in Plate 5.15 the turbine is at the left, and the compressor to the right, on the photograph. The flat plate inter-cooler and air plenum can be seen mounted on top of the cylinder head.

5.10.2 The Full-Load Performance Characteristics of the Diesel Car Engine

The full-load performance characteristics of this engine are presented in Figs. 5.77 to 5.81, which, respectively, show the airflow, torque, power output, brake specific fuel consumption, and brake specific emissions of nitrogen oxide as a function of engine speed. These figures also show the performance characteristics of the sports car engine so that you may compare the behavior of the naturally-aspirated spark-ignition Otto cycle engine with that of the turbocharged, direct-injection Diesel cycle engine as a power unit installed in an automobile. This discussion began in Sec. 1.9, where the fundamental thermodynamic theory was presented and numerical examples were given.

Plate 5.13 A turbocharger on a Volvo Penta marine diesel engine.
(Courtesy of Volvo Penta)

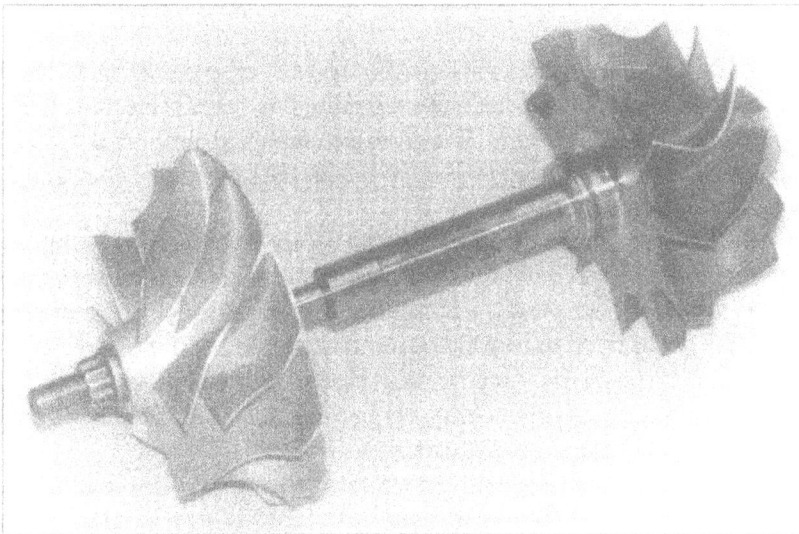

Plate 5.14 Rotors of an outward radial flow compressor and an inward radial flow turbine.

Plate 5.15 A four-cylinder, turbocharged, direct-injection diesel engine.
(Courtesy of Perkins Engines)

The sports car data, previously shown in Figs. 5.64 to 5.67, are plotted here as the composite performance characteristics, i.e., when the intake ramming flap valve is electronically activated as the engine speed approaches 5200 rpm. This gives the combined performance characteristics for the engine in speed bands from 1600 to 5000 rpm with the long intake duct, and from 5500 to 6500 rpm with the short intake tract.

In Fig. 5.77, the airflow into the diesel car engine is seen to be very much higher than for the sports-car engine. The delivery ratio rises to above 1.7, driven to this level by the compressor boost pressure, which is also plotted on the same figure (as an absolute pressure in bar units). The waste gate is observed to be fully operational above 3000 rpm.

The effect of this airflow on the torque output, plotted as brake mean effective pressure, bmep, is seen in Fig. 5.78. The maximum torque is 14.5 bar at 3000 rpm, and the entire torque curve is much higher than for the sports car engine.

The torque-speed curves translate into the power curves given in Fig. 5.79, where the more extensive speed range of the sports car engine gives the higher maximum power output. The difference in the maxima is about 20 kW, i.e., the sports car engine has a 22% power advantage. Nevertheless, the users of the two vehicles, when accelerating in the medium speed

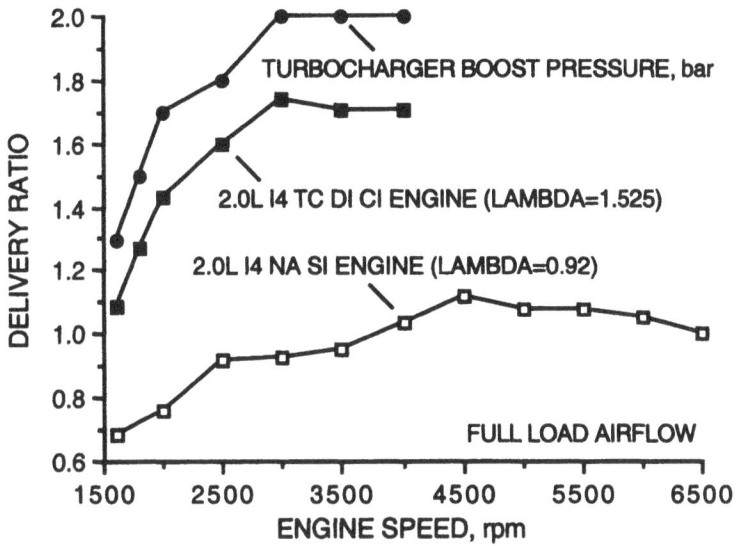

Fig. 5.77 The intake airflow characteristics of the turbo-diesel car engine.

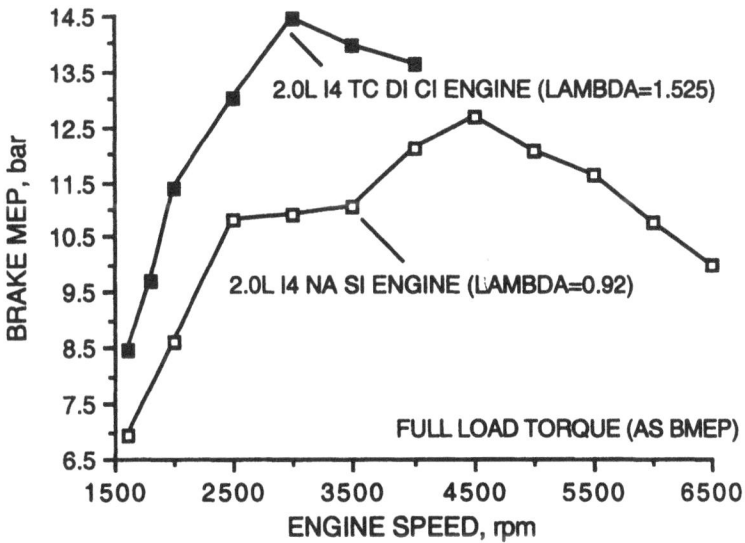

Fig. 5.78 The torque output characteristics of the turbo-diesel car engine.

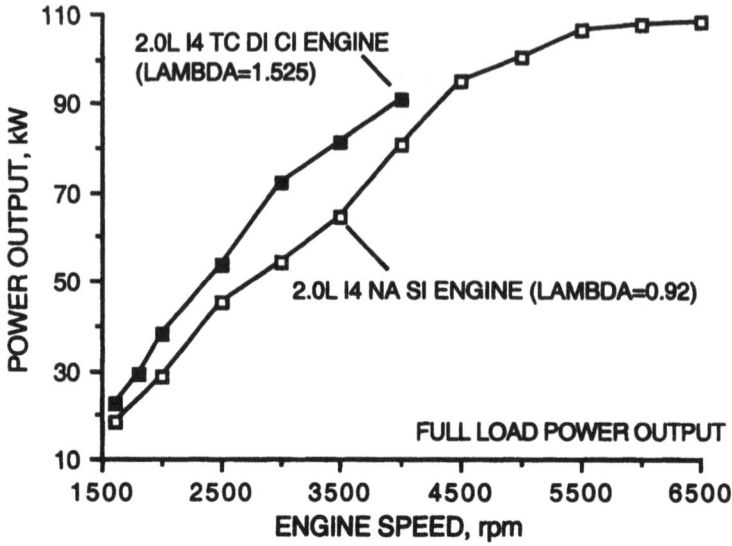

Fig. 5.79 The power output characteristics of the turbo-diesel car engine.

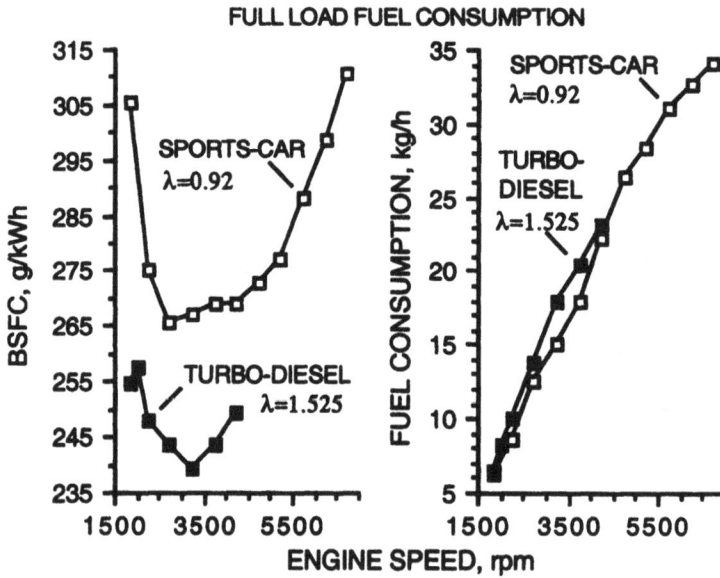

Fig. 5.80 The fuel consumption characteristics of the turbo-diesel car engine.

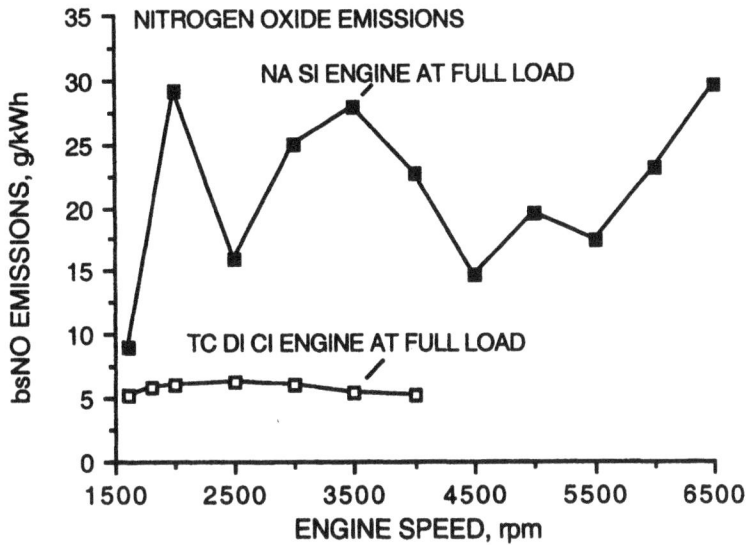

Fig. 5.81 The nitrogen oxide emission characteristics of the turbo-diesel car engine.

range of each engine, would experience a more significant thrust from the diesel engine compared to that from the gasoline engine. In short, the engine speed of the sports-car engine must be raised above 4000 rpm before its performance advantage is realized.

The fuel consumption rates are plotted in Fig. 5.80, where the fundamental thermodynamic advantage of the Diesel cycle becomes apparent, in this case through a combination of the higher compression ratio and the recouping of some of the exhaust gas energy in the turbocharger. The minimum bsfc of the diesel car engine is at 238 g/kWh, whereas that of the sports car engine is at 264 g/kWh. Thus, the diesel engine has a specific fuel consumption advantage of some 11%.

The brake specific exhaust emissions of nitrogen oxides, bsNO, at full-load are plotted in Fig. 5.81. The diesel engine is seen to have very much lower levels of NO emissions than the sports-car unit; this is because the Otto cycle unit emits some six times more than the Diesel cycle engine. As reported in Appendix A4.1, the emissions of nitrogen oxides are a function of temperature and the oxygen concentration during combustion. The diesel car engine at full-load is operating at an equivalence ratio of 1.525, so there are considerable quantities of free oxygen available throughout the combustion process. Although the sports car engine at full-load is operating very slightly rich of stoichiometric, at an equivalence ratio, λ, of 0.92, or an air-fuel ratio, AFR, of 14, and nominally no free oxygen is available, free oxygen is released during this combustion process due to dissociation, as shown in Fig. A4.8. The temperature in the burn zone during combustion has a very profound—in Appendix A4.1 it is correctly described as exponential—influence on NO formation, so Fig. 5.82, which shows the in-cylinder temperatures during combustion, is germane to the discussion. Although the cylinder pressure in the tc ci car engine is very much higher than in the na si sports-car unit, the mass ratio of trapped charge is even higher, which makes the peak in-cylinder temperatures (K) 50% higher in the gasoline engine. Hence, because temperature has a greater influence on NO formation than the local mass concentration of oxygen in the burn zone (the influence of temperature is exponential), the Otto engine creates higher NO emissions.

Fig. 5.82, which is plotted at the engine speed at peak torque for both engines, shows that the unburned zone temperatures are higher in the turbo-diesel engine compared to those in the sports car unit. This is to be expected, especially because the diesel engine needs the higher compression ratio to effect compression-ignition. At the ignition point in the spark-ignition unit, the air temperature is 457°C, whereas it is 617°C for the diesel engine. For the diesel engine, note that the location of the peak burn zone temperature is at tdc, but the location of the peak mean temperature is much retarded at 14.8 °atdc with a value of 1649°C. It is instructive to compare these data with those in Fig. 1.70, which are plotted with predictions from much simpler thermodynamics but for a similar type of engine.

5.10.3 Pressure Diagrams at Full-Load in the Diesel Car Engine

Fig. 5.83 shows the pressures in the cylinder, and in the exhaust and intake ducting at the valves, at the peak torque speed of 3000 rpm. The diesel engine has an even narrower valve overlap period than the sports car engine—you may recall the discussion on the mechanical necessity for this in Sec. 1.3.2—and this is visible by the close spacing of the valve event

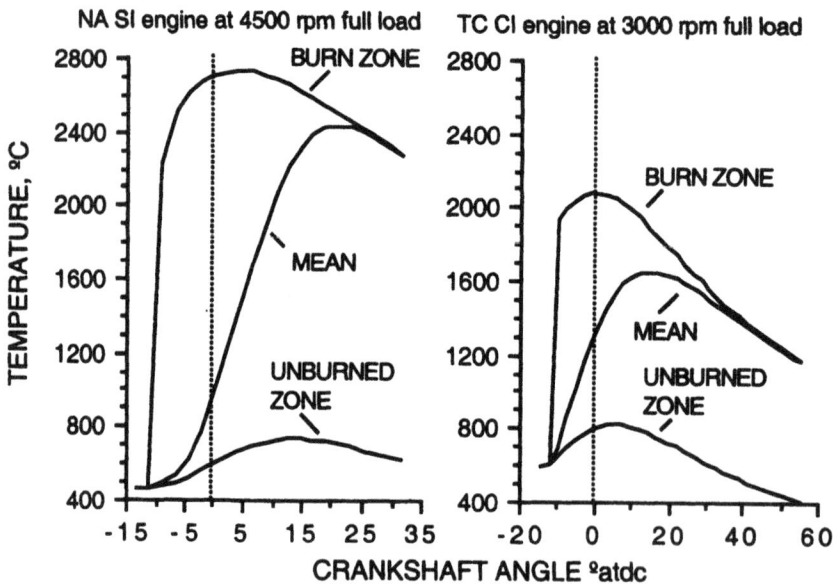

Fig. 5.82 In-cylinder temperatures for the sports car and turbo-diesel car engines.

606

markers on the diagram. The five-way exhaust branch ends at the restriction of the turbine nozzle ring, so compression wave reflections of every firing are propagated to every end of the short upstream ducting. As each cylinder fires at 180° intervals, exhaust pulse reflections from the nozzle ring can be seen in Fig. 5.83 to arrive at the exhaust valve of cylinder number 1. Consequently, because the cylinder firing order is 1-3-4-2, the reflection appearing at the exhaust valve during the overlap period emanates from cylinder number 3. In a naturally-aspirated engine, this would have a disastrous effect on the ensuing intake flow; however, in this case, because the compressor boost pressure at the intake valve exceeds both the cylinder and the exhaust duct pressure, the intake flow is unaffected. There is no exhaust tuning, yet the cylinder pressure falls rapidly to 1.2 atm before rising on the remainder of the exhaust stroke to nearly 2.0 atm at tdc, as pumping work is expended by the piston. The piston is literally pushed down the intake stroke by the compressor boost pressure as the cylinder pressure remains above 1.8 atm. Here, the outcome is a net gain of pumping work, pmep, of -0.55 bar. The imep at 3000 rpm is 16.48 bar. The friction work, as fmep, in this diesel unit at 3000 rpm is somewhat higher, at 2.58 bar, than an equivalent Otto engine. Hence, the bmep is 14.45 bar and the mechanical efficiency is 0.877.

In Fig. 5.83, the oscillation on the intake pressure trace cannot be due to conventional intake ramming because the intake duct is made deliberately short to eliminate any such ramming effects. This oscillation comes from the pulsations in the intake plenum and the duct leading through the intercooler from the compressor. The outcome is seen to very effective, as one compression wave neatly straddles the overlap period and another arrives in time to provide modest ramming before the intake valve closes early; recall that ivc is at a modest 50° abdc. The rapid oscillation on the intake trace at, and after, the intake valve has closed is the

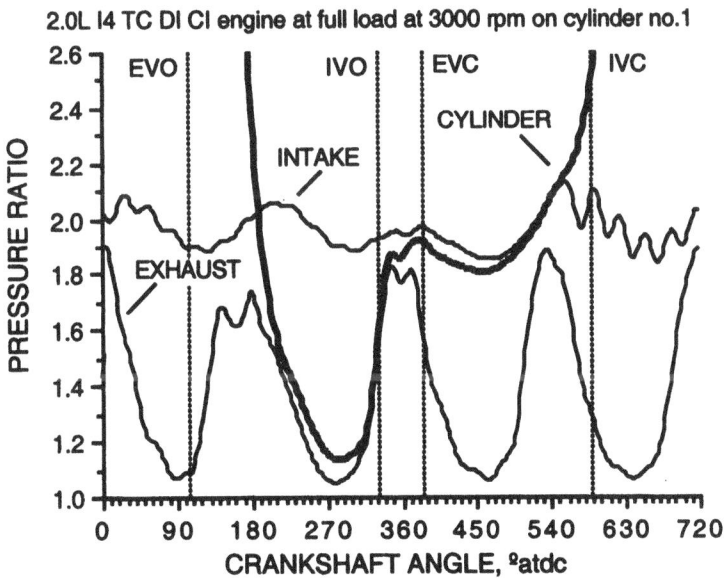

Fig. 5.83 Turbo-diesel, full-load: cylinder, exhaust, and intake pressures at 3000 rpm.

"conventional" ramming reflection process, i.e., to-and-fro the open and closed ends in this very short intake pipe at an engine speed of just 3000 rpm.

5.10.4 Temperature Diagrams at Full-Load in the Diesel Car Engine

In Fig. 5.84, the large expansion ratio of a compression-ignition engine rapidly causes the temperature, and also the pressure, to drop on the expansion stroke so that by mid-exhaust stroke the gas temperature in the cylinder and exhaust duct is a mere 350°C. As a comparison, glance back at Fig. 5.14, which shows that in the spark-ignition Ryobi engine at the same juncture the cylinder temperature is still above 600°C. During induction, the cylinder temperature falls to 100°C, just a fraction hotter than the incoming air at around 79°C. This air may be intercooled, but it has been compressed to a pressure ratio approaching 2.0 before leaving the compressor at 92°C, and has passed over many hot surfaces before it reaches the intake valve!

5.10.5 Particle Velocity Diagrams at Full-Load in the Diesel Car Engine

Fig. 5.85 shows the gas particle velocities, as Mach number, in the exhaust and intake ducting at the valves, at the peak torque speed of 3000 rpm. The lack of intake and exhaust backflow, already predicted from the favorable cylinder-exhaust and cylinder-intake pressure differentials in Fig. 5.84, is confirmed in Fig. 5.85. The exhaust pulse outflow occurs cleanly in two stages, at release and during the exhaust stroke, and continues smoothly until the exhaust valve closes. The intake inflow commences at intake valve opening and shows only a meager amount of backflow near intake valve closure. Plotted on the same figure, this is registered as a 3% drop in DR. This occasions an almost indiscernable rise in the local gas temperature at the intake valve, as seen in Fig. 5.84.

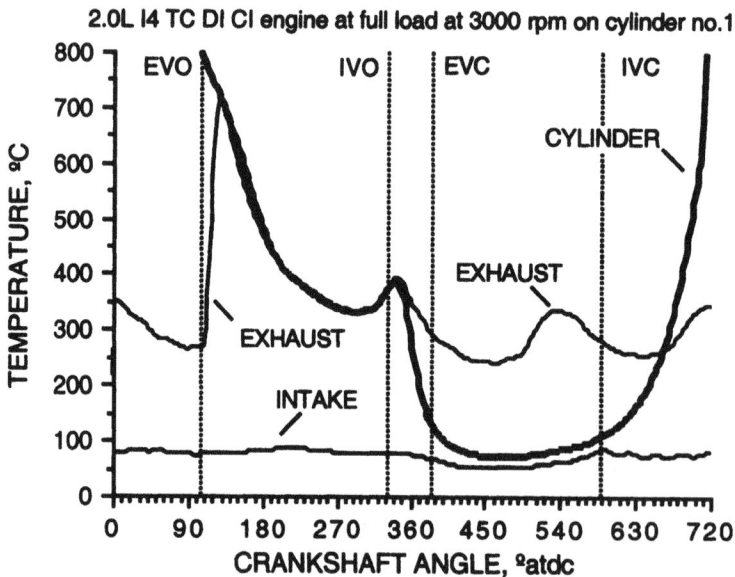

Fig. 5.84 Turbo-diesel, full-load: cylinder, exhaust, and intake temperatures at 3000 rpm.

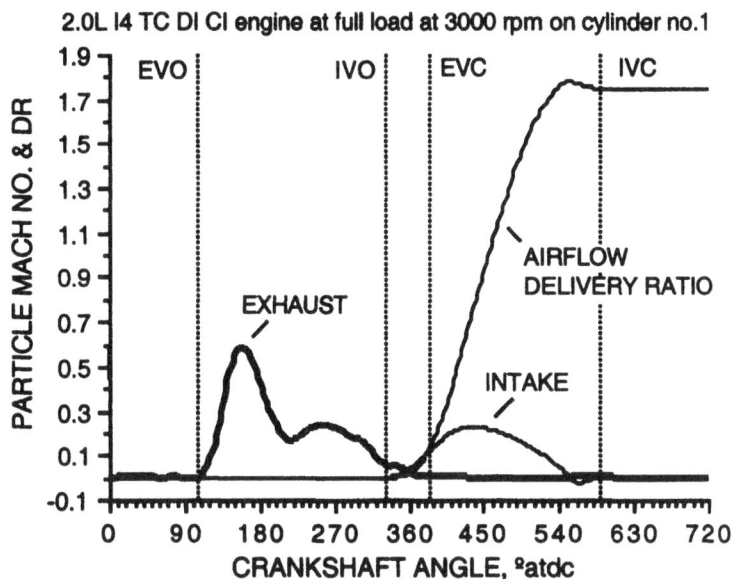

Fig. 5.85 Turbo-diesel, full-load: airflow, and exhaust, and intake particle velocities at 3000 rpm.

5.10.6 Charge Purities at Full-Load in the Diesel Car Engine

Fig. 5.86 shows the in-cylinder scavenging efficiency and the charge purities in the intake and exhaust ducting at the valves. This figure shows the effect of gas movements provided by the flow directions and particle velocities that move air and combustion products in the manner plotted in Fig. 5.85. The immediate access of air to the cylinder at intake valve opening raises the cylinder purity above zero and this continues to rise until it reaches 0.980. The minor backflow into the intake tract at ivc lowers the local purity in the intake ducting to 0.990. Some of the intake air gets into the exhaust system, even with a lesser valve overlap period, where it raises the local exhaust gas purity from zero to 0.161 before decaying slightly by diffusion behind the closed exhaust valve.

The more comprehensive charging of a turbocharged diesel engine cylinder compared to that of a naturally-aspirated Otto cycle engine as represented by the sports-car unit, and which are already recorded through the comparative performance characteristics, is explained more clearly by the simulation output of pressures, temperatures, particle velocities, and charge purities in the cylinder and in the surrounding ducting.

5.10.7 The Part-Load Performance Characteristics of the Diesel Car Engine

The behavior of the sports car engine during an exhaust emissions test is represented by the torque-speed mapping point of 2.6 bar bmep at 1600 rpm. The diesel-engined car must meet its exhaust emissions targets at the very same mapping point. The simulation exercise to determine the behavior of the turbo-diesel car engine at the same part-load condition is carried out by reducing the fueling to the injectors until the identical torque-speed point is attained at 41.5 Nm at 1600 rpm.

2.0L I4 TC DI CI engine at full load at 3000 rpm on cylinder no.1

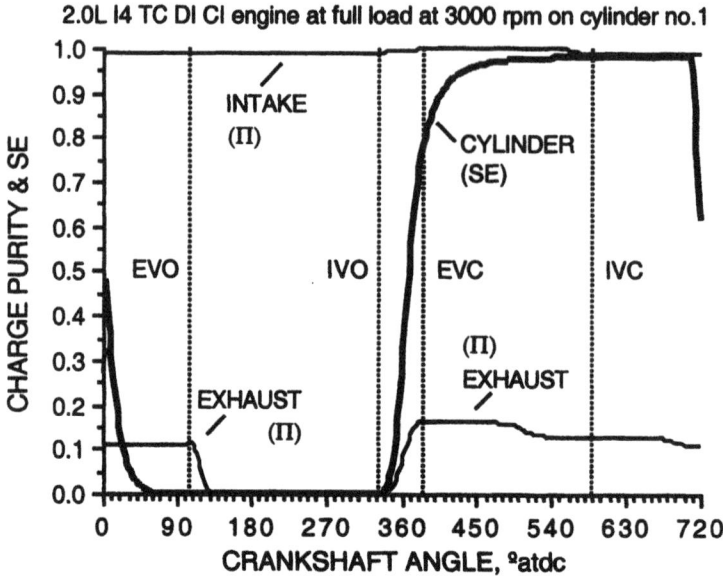

Fig. 5.86 Turbo-diesel, full-load: cylinder, exhaust, and intake charge purity at 3000 rpm.

The simulation provides the performance characteristics at the test point. These are shown in Fig. 5.70 in the row marked "TC CI" (for turbocharged, compression-ignition). To provide the requisite performance of 2.61 bar bmep at 1600 rpm, which corresponds to a power output of 7.0 kW, the fueling has been reduced to an air-fuel ratio, AFR, of 44, which corresponds to an equivalence ratio, λ, of 3.05. Naturally, the exhaust gas energy released is now much less than at a full-load condition. The compressor of the turbocharger unit receives a reduced turbine work output, hence the boost pressure drops significantly to a mere 1.05 bar absolute. Although the intake ducting is unthrottled, the delivery ratio, DR, is correspondingly lowered to 0.87 from the full-load fueling value of 1.09. The lack of intake throttling has a major influence on the pumping losses as represented by the pumping mean effective pressure, pmep. At part-load for the diesel engine the pmep value in Fig. 5.70 is 0.03 bar. By contrast, the pmep value for the sports car engine is 0.3 bar. The cry of "congratulations" at this point may be somewhat stifled by the knowledge that at full-load for the diesel engine, the pmep value at this speed, -0.2 bar (i.e., a negative value), represented a work gain!

The friction consumes, as fmep, 1.51 bar, so an imep of 4.15 bar is required to attain the necessary output bmep of 2.61 bar. The outcome is that the part-load brake specific fuel consumption, bsfc, increases to 327.2 g/kWh from a full-load value of 252.7 g/kWh. Although this represents a 30% increase in specific fuel consumption from full-load to part-load for the diesel engine, as seen in Fig. 5.70, it still leaves the naturally-aspirated Otto cycle sports car engine at a 34% disadvantage compared to the turbocharged Diesel cycle car engine.

5.10.8 Pressure and Temperature Diagrams at Part-Load in the Diesel Car Engine

The reasons behind these performance characteristics are to be found in the detail provided by the simulation for the temporal variations of pressure, temperature, etc., within the engine. Figs. 5.87 and 5.88 show plots of the pressure and temperature histories in the cylinder and at the exhaust and intake ducts at the valves.

Fig. 5.87 shows the pressures in the cylinder, and in the exhaust and intake ducting at the valves, at this part-load condition at 1600 rpm. Compare this with the equivalent plot for the sports car engine in Fig. 5.71. In Fig. 5.87, there is no intake throttling, but a mild supercharge exists, and so the intake pressure signal oscillates around 1.03 bar. The previously-noted favorable intake oscillation continues to assist in the prevention of backflow of the cylinder contents into the intake duct. Because the cylinder pressure exceeds that in the intake duct from ivo until tdc, there will be some backflow, and more will occur from bdc to ivc because a similar negative pressure drop exists during that period. The amplitude of the exhaust pressure oscillations is very modest—it does not exceed 1.2 atm—which explains why the exhaust energy recovered in the turbine is only sufficient to maintain the compressor boost pressure at a mere 0.05 atm above the atmospheric pressure. At intake valve opening, an exhaust pressure oscillation is observed which obstructs the exhaust outflow during the final stages of the exhaust stroke, and which raises the pumping loss; this compression pressure wave emanates from the release of cylinder number 3.

In Fig. 5.88, which shows temperature at the same locations, the cylinder and exhaust duct temperatures are seen to be very low compared to those in the equivalent gasoline engine

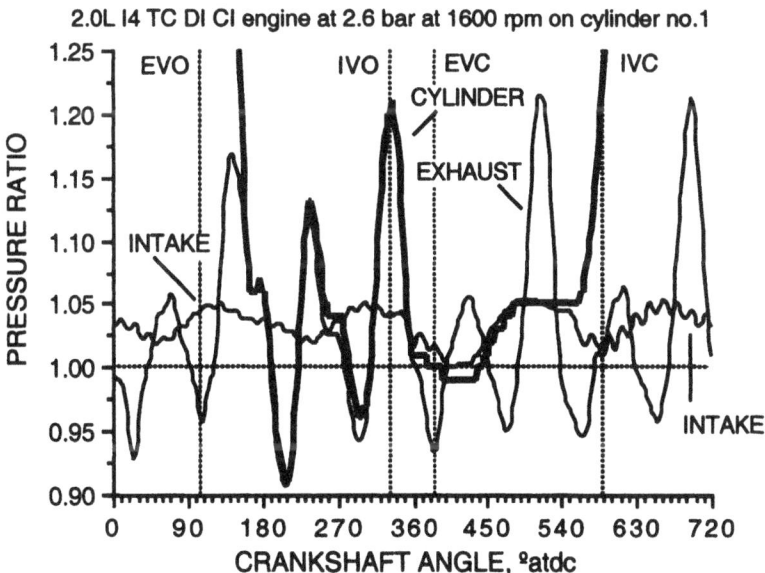

Fig. 5.87 Diesel car, part-load: cylinder, exhaust, and intake pressures at 1600 rpm.

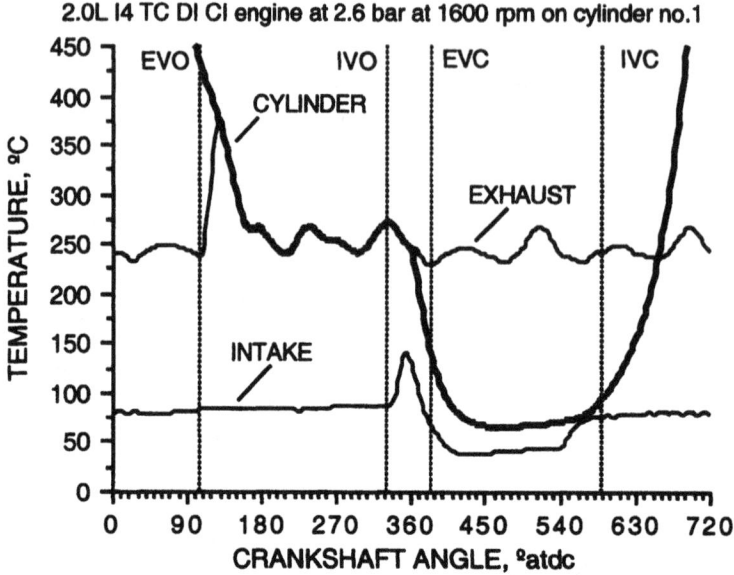

Fig. 5.88 Diesel car, part-load: cylinder, exhaust, and intake temperatures at 1600 rpm.

diagrams in Fig. 5.72. The mean exhaust gas temperature in the diesel engine is a mere 250°C, compared to 450°C in the sports car engine at the same part-load condition. This low temperature presents real problems in the application of a catalyst at part-load in a diesel engine because a catalyst requires a certain minimum exhaust gas temperature to "light off" and become effective. The temperature in the intake duct rises at intake valve opening due to cylinder backflow, but only to a mere 75°C, a small rise compared to the major event seen in Fig. 5.72 for the sports car engine.

5.10.9 Particle Velocities and Charge Purities at Part-Load in the Diesel Car Engine

Figs. 5.89 and 5.90 show the gas particle velocities and charge purities, respectively, in the cylinder, and in the exhaust and intake ducting at the valves, at this part-load mapping point at 1600 rpm.

The data in Fig. 5.89 confirm that intake backflow occurs. It is barely discernable on the plot at intake valve opening, but is clearly visible from bdc to ivc. If you compare this information with that in Fig. 5.73 for the gasoline engine, you will observe the stark contrasts. The particle velocity trace in the exhaust duct at the exhaust valve illustrates the paucity of the mass released. There is so little outflow that there are two backflow periods before the exhaust stroke commences more formally at 250 °atdc. The contents of the cylinder are then dispatched by the piston pumping action and outflow continues smoothly until the exhaust valve closes. The dynamic airflow curve, as DR, rises smoothly from tdc until some backflow commences around, and after, the bdc location.

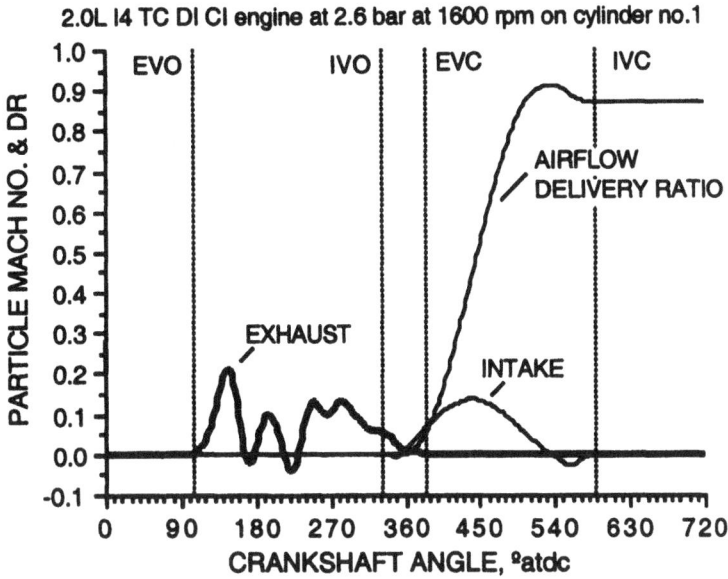

Fig. 5.89 Diesel car, part-load: cylinder, exhaust, and intake particle velocities at 1600 rpm.

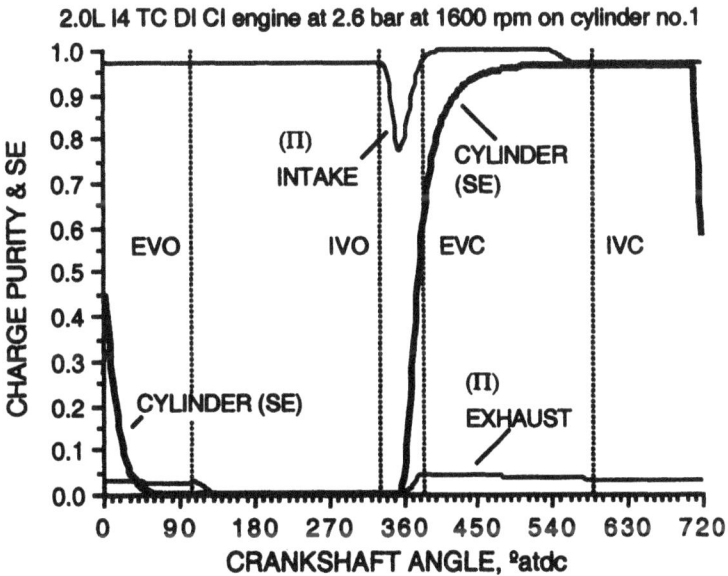

Fig. 5.90 Diesel car, part-load: cylinder, exhaust, and intake purities at 1600 rpm.

In Fig. 5.90, the cylinder purity is seen to rise as smoothly as the DR curve. Fresh air enters the cylinder by tdc. The intake duct purity dips briefly down to 0.75, due to cylinder backflow during the valve overlap period, but recovers to unity by the time the exhaust valve closes. However, these are minor hiccups compared to the grim events seen in the throttled spark-ignition engine, i.e., as plotted in Fig. 5.74. A small amount of the intake air gets into the exhaust system, as noted on the exhaust duct purity trace in Fig. 5.90.

5.10.10 Some Overall Comparisons of Diesel and Gasoline Automobile Engines

I suspect that most of you will find the comparisons of the part-load performance characteristics of the sports car and the turbo-diesel car engines rather surprising—apart from the data on brake specific fuel consumption, because most of you will anticipate that an unthrottled diesel engine has the higher thermodynamic efficiency. It is a well-known experimental fact that diesel engines have a 20-40% fuel economy advantage under urban driving conditions, of which I am sure you are already aware. This is reinforced by the simulation output data in Fig. 5.70.

However, under urban driving conditions, as many of you will have experienced, the modern gasoline engine makes a passable imitation of a good sewing machine. On the other hand, the diesel car engine, and especially turbocharged direct-injection engines equipped with a high-pressure injection system, are noisy, with an objectionable rattle that clearly originates from both the combustion and the injection processes. Many of these diesel car engines must be encapsulated in order to meet the twin objectives of customer satisfaction and noise legislation. Consequently, the reality of the thermodynamic and gas dynamic inefficiencies of the gasoline engine—or the "mayhem," as I referred to it earlier—is in stark contrast with the evidence collected by our ears.

There are other myths and legends regarding the fuel consumption rates of gasoline and diesel car engines under normal highway driving conditions that require closer examination. Evaluating the actual fuel consumption rates for the two types of vehicles is not quite as simple as noting the relative brake specific fuel consumption data shown in Fig. 5.80. It is necessary to multiply the power output data in Fig. 5.79 by the brake specific fuel consumption data in Fig. 5.80 to get the actual fuel consumption rates of the sports car and the turbo-diesel car engines. The actual fuel consumption rates are plotted in the right-hand panel of Fig. 5.80. Assuming equality of vehicle, in a full-load acceleration scenario from 1600 to 4000 rpm, the diesel car would be faster than the sports car, but would consume more fuel, maybe even up to 10% more fuel. The average fuel consumption during this test on the diesel car is probably about 15 kg/h. On the other hand, if the same full-load acceleration test for the sports car were repeated, but in its as-designed performance zone from 4000 to 6500 rpm, the sports car would be much faster than the diesel car, and would now consume more gasoline fuel. The average fuel consumption during this test on the sports car is probably about 29 kg/h, i.e., roughly twice that of the diesel car.

For those who drive at maximum velocity on a highway—and there are those on German autobahns who do—the maximum power output of the two types of vehicles would provide the sports car with a 10 mph speed advantage over the turbo-diesel car, that is, 130 mph vis-à-vis 120 mph. This statement is made on the grounds that maximum vehicle speed is basically related to the cube root of the power ratio, which is given in Fig. 5.79. If this is the case, the data on Fig. 5.80 show that the sports car would consume gasoline at a rate of 33.4 kg/h, which is a 42% greater mass of fuel per hour than the diesel car consumes at its maximum velocity.

At the 7.0 kW part-load point, the sports car engine consumes 3.06 kg/h of gasoline. That is 11 times less fuel than at the full-load point at maximum engine speed. The diesel car, at the same 7.0 kW condition, consumes diesel fuel at the rate of 2.29 kg/h, but this rises to 22.4 kg/h at maximum velocity; that is a ratio of 10. In short, under equality of urban driving conditions, the diesel car uses 25% less fuel. However, as can be seen from this discussion, it is only under conditions of equal load, power, and speed that any data on brake specific fuel consumption become relevant in a debate on the fuel consumption rate of an automobile.

Recall that the exhaust emissions of carbon dioxide, CO_2, are often described as "greenhouse gas emissions," and are directly related to the fuel consumption rate. The above discussion therefore highlights the simple fact that the quantity of such emissions is absolutely related to the manner in which the vehicle is driven, and is only partially related to the engine test data for brake specific fuel consumption. The current political scenario is that Europe, the United States, and other developed countries have volunteered, at environmental conferences such as those held in Rio de Janeiro and Kyoto, to reduce CO_2 emissions from all sources by a significant proportion by the turn of the twenty-first century. The automotive industry can look forward to being pressed to produce ever-more-fuel-efficient vehicles, as well as to satisfy all existing legislation regarding emissions of noise, soot, particulates, hydrocarbons, carbon monoxide, nitrogen oxides, etc. Considering the above discussion on the reality of the on-road fuel consumption rates of cars fitted with gasoline and diesel engines, it would appear that there is considerably more to be gained by educating the car driver's right foot than there is by further legislative harassment of the engine designer!

High-Performance Diesel Engines

Although the discussion above shows that turbocharged diesel engines have very good specific power performance characteristics, there is a general impression in the world at large that the gasoline engine will always outperform the diesel engine in any field where competition or racing is the issue. In general, this is correct, but the performance gap is narrowing, and, where the rules of competition are logically framed—inasmuch as any rules for racing ever can be so framed—the diesel engine has been known to prevail.

One example is offshore racing, an example of which is shown in Plate 5.16. The competition rules of the UIM (Union Internationale Motonautique) for Class 1 Offshore Racing permit the use of two engines in a boat, either two naturally-aspirated, spark-ignition gasoline engines, each of 8.2-liter capacity, or two turbocharged diesel engines, each of 11.7-liter capacity. There are some caveats on those rules regarding intake restrictions for the gasoline engines and maximum permitted boost pressure for the diesel engines.

Plate 5.16 Offshore Racing. (Courtesy of Mercury Marine and Jeff Girardi, Freeze Frame Video)

Many of the gasoline engines raced in this class are V8 engines and produce about 1100 bhp (820 kW). The Seatek diesel engines also compete most successfully in offshore racing with I6 engines that have bore and stroke dimensions of 127 and 135 mm, respectively. With a capacity of 10.3 liters, they also produce 1100 bhp at 3300 rpm. One of these engines is shown in Plate 5.17. The twin turbochargers can be seen at the upper left of the photograph at the cylinder head level, and the seawater intercooler at the front. This engine is very compact with maximum length, height, and width dimensions of 1508, 1058, and 730 mm, respectively. It weighs 810 kg and has a best point specific fuel consumption of 265 g/kWh. These are excellent bulk, and power performance, characteristics by any standards.

Seatek in Italy produces alternative versions of this engine for other marine applications such as high-performance navy patrol boats and similar military vehicles. These have lesser, but equally outstanding, performance characteristics, but also require endurance and longevity. The engine shown in Plate 5.18 has the same capacity as the racing engine, but with a single turbocharger produces 463 kW (620 bhp) at 3100 rpm, and has a best point specific fuel consumption of 210 g/kWh. This engine is also compact with maximum length, height, and width dimensions of 1680, 1070, and 720 mm, respectively, and weighs 800 kg. The large (sea) water-cooled intercooler can be seen to occupy the entire side of the engine.

Plate 5.17 Seatek diesel racing engine of 10.3-liter capacity and 1100 bhp output.
(Courtesy of Seatek SpA)

From these examples, and also from the performance characteristics of the latest turbo-charged direct-injection diesel engines now being manufactured and installed in automobiles in Europe, the image of the diesel engine as being "slow" compared to its spark-ignition counterpart is becoming rapidly dispelled.

5.11 Concluding Remarks

The simulation is shown to predict, with equal levels of accuracy, the performance characteristics of a wide range of engines, from a 1-hp single-cylinder unit, to a 700-hp V8 engine. Clearly, the design by simulation of four-stroke engines has been greatly enhanced by the use of the theoretical methods outlined in Chapters 1 to 4. This is evident from the examples provided here, which show not only the accuracy of correlation with experiments, but also the extent of the detailed information that is forthcoming on the temporal changes in gas properties and state conditions throughout the engine. This greatly assists the designer in comprehending the effects on performance characteristics of alterations in the physical geometry of the engine and its ducting.

617

Plate 5.18 Seatek diesel navy engine of 10.3-liter capacity and 630 bhp output.
(Courtesy of Seatek SpA)

References for Chapter 5

5.1   R.J. Kee, P.T. McEntee, G.P. Blair, T. Fickenscher, J. Hölzer, and R. Douglas, "Validation of Two-Stroke Engine Simulation by a Transient Test Method," Small Engine Technology Conference, Yokohama, Japan, 27-31 October 1997, pp. 441-449.

5.2   R.J. Kee, P.G. O'Reilly, R. Fleck, and P.T. McEntee, "Measurement of Exhaust Gas Temperature in a High Performance Two-Stroke Engine," SAE Motorsports Engineering Conference, Dearborn, Mich., November, 1998, SAE Paper No. 983072.

5.3   J. A. Caton, "Comparisons of Thermocouple, Time-Averaged and Mass-Averaged Exhaust Gas Temperatures for a Spark-Ignited Engine," SAE International Automotive Congress, Detroit, Mich., February, 1983, SAE Paper No. 820050.

5.4   W. Bauer, C. Tam, and J.B. Heywood, "Fast Gas Temperature Measurement by Velocity of Sound for IC Engine Applications," SAE Paper No. 972826, 1997.

5.5   K. Kurihara and J. G. Conley, "Design Considerations for Overhead Valve Train in Small High-Speed 4-Cycle Engines," Small Engine Technology Conference, Yokohama, Japan, 27-31 October 1997, pp. 203-208.

5.6   J.G. Conley, J.K. Olsen, K. Kurihara, Y. Imagawa, and G. Rickard, "The New Ryobi 26.2 cc, OHV, 4-Stroke Engine for Hand-Held Power Applications," SAE International Off-Highway & Powerplant Congress, Indianapolis, Ind., August 26-28, 1996, SAE Paper No. 961728, pp. 1-12.

5.7   J.G. Conley, J.K. Olsen, K. Kurihara, G. Rickard, and H. Hermann, "The Development of a Durable Cost Effective, Overhead Valve Train for Application to Small, 4-Cycle Engines," SAE International Off-Highway & Powerplant Congress, India-napolis, Ind., August 26-28, 1996, SAE Paper No. 961729, pp. 13-26.

5.8   A.A. Boretti, G. Cantore, E. Mattarelli, and F. Preziosi, "Experimental and Compu-tational Analysis of a High Performance Motorcycle Engine," SAE Motorsports Engineering Conference, Dearborn, Mich., November, 1996, SAE Paper No. 962526.

5.9   *Automotive Handbook,* 4th Edition, Robert Bosch GmbH, Stuttgart, 1996, p. 383. (distributed by SAE).

5.10 K.J. Patton, R.G. Nitschke, and J.B. Heywood, "Development and Evaluation of a Friction Model for Spark-Ignition Engines," SAE International Automotive Con-gress, Detroit, Mich., February, 1989, SAE Paper No. 890836.

5.11 W.L. Brown, "The Caterpillar IMEP Meter and Engine Friction," SAE International Automotive Congress, Detroit, Mich., 1973, SAE Paper No. 730150.

5.12 B.W. Millington and E.R. Hartles, "Friction Losses in Diesel Engines," SAE Inter-national Automotive Congress, Detroit, Mich., 1968, SAE Paper No. 680590.

5.13 S.K. Chen and P. Flynn, "Development of a Compression-Ignition Research Engine," SAE International Automotive Congress, Detroit, Mich., 1965, SAE Paper No. 650733.

5.14 G.P. Blair, *Design and Simulation of Two-Stroke Engines*, R-161, Society of Auto-motive Engineers, Warrendale, Pa., 1996, p. 623.

5.15 G.P. Blair and J.H. McConnell, "Unsteady Gas Flow Through a Four-Stroke Cycle Motorcycle Engine," SAE Farm, Construction and Industrial Machinery Meeting, Milwaukee, Wisc., September 1974, SAE Paper No. 740736, p. 20.

5.16 T. Robb, *From TT to Tokyo*, Courier-Herald Printers, Douglas, Isle of Man, 1974, p. 118.

5.17 W.J.D. Annand and G.E. Roe, *Gas Flow in the Internal Combustion Engine*, Foulis, Yeovil, Somerset, 1974.

5.18 *Automotive Engineering International*, Society of Automotive Engineers, Warrendale, Pa., April 1998, pp. 34-40.

5.19 T. Kume, Y. Iwamoto, K. Iida, K. Akishino, and H. Ando, "Combustion Control Technologies for Direct Injection SI Engine," SAE International Automotive Con-gress, Detroit, Mich., 1996, SAE Paper No. 960600.

5.20 G.P. Blair, S.J. Kirkpatrick, and R. Fleck, "Experimental Evaluation of a 1D Model-ling Code for a Pipe Containing Gas of Varying Properties," SAE International Con-gress, Detroit, Mich., February 28-March 3, 1995, SAE Paper No. 950275, p. 14.

5.21 G.P. Blair, S.J. Kirkpatrick, D.O. Mackey, and R. Fleck, "Experimental Evaluation of a 1D Modelling Code for a Pipe System Containing Area Discontinuities," SAE International Congress, Detroit, Mich., February 28-March 3, 1995, SAE Paper No. 950276, p. 16.

5.22 J. Gleick, *Chaos, Making a New Science*, Penguin Books, New York, 1987.

5.23 R.G. Kenny, D.J. Thornhill, G. Cunningham, and G.P. Blair, "Reducing Exhaust Hydrocarbon Emissions from a Small Low Cost Two-Stroke Engine," SAE International Off-Highway & Powerplant Congress, Milwaukee, Wisc., September 1998, SAE Paper No. 982013.

5.24 N. Watson and M.S. Janota, *Turbocharging the Internal Combustion Engine*, Macmillan Press, London, 1982, p. 607.

5.25 L.A. Smith, "Prediction of Air Mass Flow Rate in Turbocharged Four-Stroke Diesel Engines," PhD Thesis, Imperial College, London, March 1990, p. 468.

5.26 G.P. Blair, "Correlation of Measured and Calculated Performance Characteristics of Motorcycle Engines," Funfe Zweiradtagung, Technische Universität, Graz, Austria, 22-23 April 1993, pp. 5-16.

5.27 G.P. Blair, "Unsteady Flow Characteristics of Inward Radial Flow Turbines," PhD Thesis, The Queen's University of Belfast, May 1962.

Chapter 6

Empirical Assistance for the Designer of Four-Stroke Engines

6.0 Introduction

The first five chapters of this book introduced you to the four-stroke engine, unsteady gas dynamics, discharge coefficients, combustion, and the synthesis of these topics into computer modeling of the complete engine. Having just finished reading Chapter 5, you may therefore feel somewhat bemused; if so, you are forgiven. In Chapter 5, you were assured that modeling reveals the "inside story" on engine behavior, and I hope that is how it was presented. You were also made aware that the engine simulations presented in Chapter 5 are only of practical use in a design mode if the designer produces real input data in the form of banks of numbers to be used as input data files for the modeling software.

Some of you may feel that running the simulation will be as simple as opening up the computer program, inserting the numbers for the engine and geometry, as the mood takes you, and letting the computer inform you of the outcome regarding the pressure wave action and the performance characteristics. Of course, this is an option, but I, and possibly also your employer—if you are a designer—would not consider this to be the most rapid approach to optimize an engine design to meet a particular need. Depending on the complexity of the geometry of the engine, there could be several hundred different data values or labels that could be changed within the engine simulation to effect a single calculation, and the statistical probability of achieving an optimized engine design by an inspired selection of those numbers is about on a par with your chance of picking a winning racehorse. Or, with this approach, you may feel so overwhelmed at the range of choices available for the numeric value of any one data parameter, that you could spend more time trying to estimate a sensible value for it than using the engine simulation for the predictive task.

What is required is some empirical guidance regarding the data values to be used in engine simulations in particular and engine design in general—and this is precisely what I hope to provide in this chapter. The empirical guidance given here will range from the simplistic to the complex. Actually, what will be regarded as simplistic by some will be considered inconceivable by others.

There are some academic purists who will react adversely to the revelation that this textbook was written by a university professor, emeritus or otherwise, that it contains empirical guidance, and that it may be read by undergraduate students. My contrary view is that the design and creation of engineering involves decisions that are based just as much on past experience as they are on theoretical analyses. Thus, those whose past experience ranges from limited to non-existent, such as undergraduate engineering students, require guidance and advice, both in lectures and in textbooks, from those with that know-how. Of course, there is a danger that any empirical guidance given will be regarded as the final word on the design, the "quick-fix" that we all so cherish for cutting corners to get to the solution of a problem. Let a word of caution be sounded in that case, for empiricism can never provide an exact answer to a design question.

The literature on the four-stroke engine is not brimming with practical advice for designers and tuners of high-performance engines on such vital matters as the disposition and sizing of intake and exhaust valves [5.1-5.5]. A good deal has been written on the disposition and layout of the intake and exhaust ducting for such engines, but very little pragmatic advice has been provided on the physical dimensions of the ducting necessary to optimize target performance characteristics at a given engine speed [5.1-5.9].

What is needed is a set of empirical design criteria for the valving and ducting of the four-stroke engine that is as extensive as that available to cope with the more complex geometries, gas dynamics, and scavenging of the two-stroke engine [5.10]. The apparent simplicity of the four-stroke engine compared to the two-stroke engine may well have kept generations of engineers from formulating logical empirical design guidelines for the four-stroke engine. I propose to rectify this situation and show the relevance of such empirical criteria to the design of four-stroke engines. The empirical criteria presented here are particularly oriented toward the high-performance, naturally-aspirated, spark-ignition engine; however, the basic approach can be adapted to all types of four-stroke engines.

Armed with some empirical advice, the designer can commence the simulation process with a well-matched engine design where the valving and ducting are already part-optimized with respect to a set of target performance characteristics. Hence, the empirical advice offers a more logical and less time-consuming route for the selection and insertion of geometrical data into an engine simulation model, for either the initial design stages of a new engine or the ongoing development of an existing unit [5.10-5.16].

The empiricism is aimed at two areas:

(i) The sizing of the intake and exhaust valves and the dimensioning of the manifolding at the cylinder head, to achieve the required breathing characteristics for the engine.

(ii) The dimensioning of the intake and exhaust ducting, to achieve the required tuning characteristics for the engine.

6.1 Empiricism for the Design of the Cylinder Head
6.1.1 Time-Area Relationships of the Engine Valving

Sec. 1.5 contains a detailed discussion of the geometry of a cylinder head, giving the theory of the gas flow area created by a lifting poppet valve(s). Sec. 1.5 also contains the definitions to be quoted here regarding pipe-to-port area ratios, C_m, and explains the use of a discharge coefficient, C_d, to account for the fact that a valve is lifted somewhat higher than is apparently necessary.

The poppet valves of the engine lift and expose the annular flow areas, A_{et} and A_{it}, as illustrated in Fig. 1.14 and defined in Sec. 1.5.2. If these areas are plotted with respect to time, for example, or crankshaft angle, θ, at any given engine speed, N, the result is a graph of the type shown in Fig. 6.1.

Consider the intake pumping period, i.e., the intake stroke, as an example of the air inflow behavior. At any instant on the intake stroke, during a crankshaft angle increment $d\theta$, the uncovered annular valve flow area is A_i with a discharge coefficient C_d, and through this gap flows air at particle velocity c and density ρ. The result is an inflowing mass increment of air, dm, to the cylinder during the equivalent time increment dt, where,

$$dm = C_d \rho c A_i dt = C_d \rho c A_i \frac{dt}{d\theta} d\theta = \frac{C_d \rho c A_i}{6N} d\theta \qquad (6.1.1)$$

The total mass flow of air, M, that would flow into the cylinder by pumping during the intake stroke from top to bottom dead center, i.e., tdc to bdc, is given by,

$$M = \int_{tdc}^{bdc} dm = \int_{tdc}^{bdc} \frac{C_d \rho c A_i}{6N} d\theta \qquad (6.1.2)$$

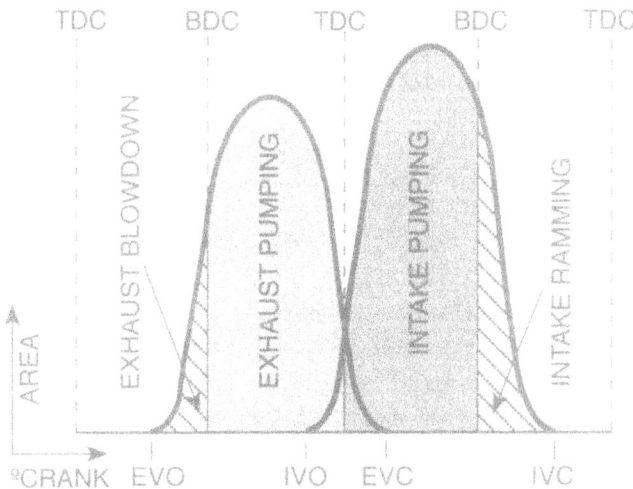

Fig. 6.1 The blowdown, pumping, and intake ramming time-areas.

The delivery ratio, DR, that this incoming mass flow will provide is,

$$DR = \frac{M}{\rho_{ref}V_{sv}} = \int\limits_{tdc}^{bdc} \frac{C_d\rho cA_i}{6N\rho_{ref}V_{sv}} d\theta \qquad (6.1.3)$$

where the reference density is ρ_{ref} and the swept volume of any one cylinder is V_{sv}.

In any high-performance engine, because the time-varying profiles of discharge coefficient, particle velocity, and density will be approximately similar, and with other parameters in Eq. 6.1.3 being constants,

$$DR \propto \int\limits_{tdc}^{bdc} \frac{A_i dt}{V_{sv}} \propto \int\limits_{tdc}^{bdc} \frac{A_i}{6NV_{sv}} d\theta \propto \Phi_{ip} \qquad (6.1.4)$$

Thus, the amount of air ingested during the pumping stroke of any engine must be related to the "time-area per unit swept volume" during the "intake pumping period," as illustrated in Fig. 6.1, and this parameter is defined as Φ_{ip}. From the data provided in the numeric examples in Chapter 5, you are already aware that delivery ratio has a strong correlation with the torque produced by the engine, i.e., by its brake mean effective pressure, bmep.

6.1.2 Time-Area Relationship for Intake Ramming
The flow of air into the cylinder is further assisted by pressure wave ramming during a period from bottom dead center to intake valve closing. Clearly, the intake valve(s) must be kept open, with the correct area function and for the correct period, for this to be optimized. As illustrated in Chapter 5, and more extensively discussed in Sec. 6.3, inadequacy leads to ineffective ramming, and an excess results in backflow of hard-won air returned into the intake tract. The "time-area per unit swept volume" during this juncture, i.e., that for intake ramming, Φ_{ir}, as drawn in Fig. 6.1, acts as a marker of the quality of a well-matched design.

$$\Phi_{ir} = \int\limits_{bdc}^{ivc} \frac{A_i dt}{V_{sv}} = \int\limits_{bdc}^{ivc} \frac{A_i}{6NV_{sv}} d\theta \qquad (6.1.5)$$

6.1.3 Time-Area Relationship for Exhaust Blowdown
The same argument applies to the "exhaust blowdown period" as sketched in Fig. 6.1. If the correct proportion of the total mass of the cylinder contents is not lost from the cylinder prior to bdc from the opening of the exhaust valve, then too much is left within the cylinder, which must be pumped from the cylinder by the piston. This causes an excess of pumping work which will (a) deteriorate the engine performance, and/or (b) potentially not be removed, but be pushed into the intake tract, thereby reducing the mass of air that can be ingested during the intake process. The deleterious effects of high pumping losses and cylinder backflow are

well illustrated in Chapter 5. On the other hand, opening the exhaust valve unnecessarily early merely reduces the effective work done on the piston during the power stroke. The relevant time-area is that defined for the exhaust blowdown period, Φ_{bd}, where,

$$\Phi_{bd} = \int_{evo}^{bdc} \frac{A_e dt}{V_{sv}} = \int_{evo}^{bdc} \frac{A_e}{6NV_{sv}} d\theta \qquad (6.1.6)$$

6.1.4 Time-Area Relationship for Exhaust Pumping

The removal of the correct proportion of the cylinder mass by blowdown, followed by pumping from bdc to tdc, is essential for success in the upcoming intake stroke and the limitation of pumping work on both pumping strokes. So, the time-area concept is extended to the exhaust pumping stroke itself, where the "time-area per unit swept volume" for exhaust pumping is defined as Φ_{ep},

$$\Phi_{ep} = \int_{bdc}^{tdc} \frac{A_e dt}{V_{sv}} = \int_{bdc}^{tdc} \frac{A_e}{6NV_{sv}} d\theta \qquad (6.1.7)$$

The area and period for the exhaust pumping computation is shown in Fig. 6.1. Because the proportion of cylinder mass that must be removed by exhaust pumping is somewhat similar to that ingested as delivery ratio during intake pumping, it should come as no surprise if the numeric values for Φ_{ep} and Φ_{ip} found in actual engines are somewhat similar.

6.1.5 Time-Area Relationship for the Valve Overlap Period

If the exhaust tuning is to be effective during the valve overlap period, then both valves must be held open for the correct period of time and with an adequate amount of valve flow area exposed. The amount of cylinder mass that can be extracted through the exhaust valve during this juncture, or the amount of air that can be ingested through the intake valve prior to the onset of the real intake stroke, is characterized by the "time-area per unit swept volume" for each valve for the valve overlap period. These time-areas are illustrated in Figs. 6.2 and 6.3, and are defined for the exhaust and intake valves as Φ_{eo} and Φ_{io}, respectively.

$$\Phi_{eo} = \int_{ivo}^{evc} \frac{A_e dt}{V_{sv}} = \int_{ivo}^{evc} \frac{A_e}{6NV_{sv}} d\theta \qquad (6.1.8)$$

$$\Phi_{io} = \int_{ivo}^{evc} \frac{A_i dt}{V_{sv}} = \int_{ivo}^{evc} \frac{A_i}{6NV_{sv}} d\theta \qquad (6.1.9)$$

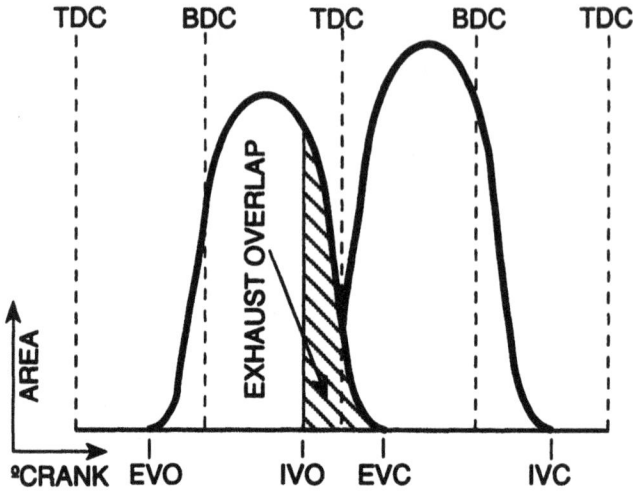

Fig. 6.2 The time-area for the exhaust valve during the overlap period.

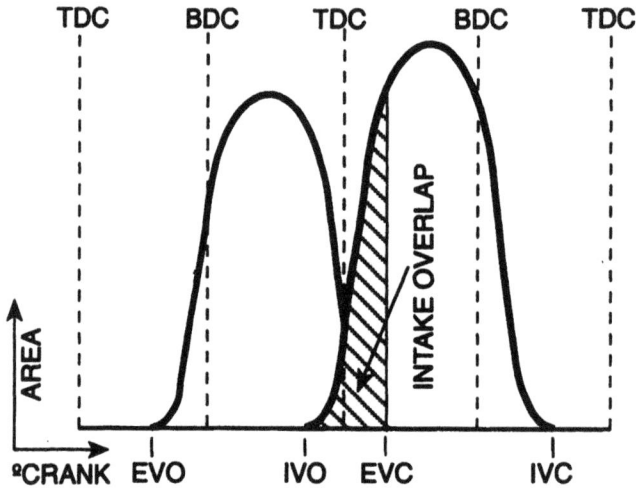

Fig. 6.3 The time-area for the intake valve during the overlap period.

6.1.6 Time-Area as a Pseudo-Dimensionless Number

Although the units of "time-area per unit swept volume" are s/m, this is in effect a pseudo-dimensionless number for an engine. Eq. 6.1.4, shows that the air ingested during the intake pumping stroke into any one cylinder is its delivery ratio, which is a dimensionless number, and this is proportional to the "time-area per unit swept volume." Thus, because all high-performance engines have somewhat similar bmep levels at the engine speed for peak torque or peak power, they should exhibit similar "time-area per unit swept volume" characteristics at that point, irrespective of their total capacity or number of cylinders. In short, a "time-area per unit swept volume" computation for any engine, for any aperture, over any angular period at a particular engine speed, provides a number that is related to dimensionless mass flow, so that dissimilar engines, running at dissimilar speeds, but at equivalent bmep, i.e., charging, levels, will have approximately similar Φ values for that same angular period. In this context, it should be obvious that the swept volume, V_{sv}, to be employed in Eqs. 6.1.3 to 6.1.9 is the swept volume of one cylinder of the engine.

6.1.7 Measurements of Time-Areas from Real Engines

To reiterate, if the above theoretical procedures are to be applicable in a design process, then real high-performance engines should exhibit somewhat similar time-area characteristics, particularly at the engine speed for peak power, where it is presumed that effective intake and exhaust tuning is occurring. Most high-performance, naturally aspirated, spark-ignition, engines operating on gasoline will attain bmep levels at the peak power speed point of between 13 and 15 bar. A selection of such engines is analyzed using software I have prepared, based on the theory of Sec. 1.5 together with that shown above. The ensuing examination of the cylinder and cylinder head geometry for the time-area characteristics of these engines is presented in Fig. 6.4. There are eight engines in total, each one competitively famous in its own right and in its own day. Due to confidentiality, the description for each engine simply details swept volume (liters), layout and number of cylinders ('I' is in-line and 'V' is vee-formation), number of valves, and the application (F1 is Formula 1, F2 is Formula 2, SRM is Supersports Motorcycle, SRC is Sports Racing Car, ITC is Touring Car, and OPB is Offshore Powerboat). Needless to say, all of these engines arrived at their particular design and development point through a plethora of methodologies, ranging from the application of the best applied science available, to nothing more subtle than painstaking black art, or "back-of-envelope" development based on the experiences of the designer's lifetime. Nevertheless, the time-areas, Φ, shown in Fig. 6.4 have a quite remarkable degree of commonality considering the variety of applications from cars to boats, maximum engine speeds ranging from 6500 to 16,500 rpm, and bore-stroke ratios ranging from 1.1 to greater than 2.0.

The last engine in the table in Fig. 6.4 is rather interesting. It is an 8.2-liter, V8, 2v ohv engine for offshore powerboat racing. The blowdown time-area is seen to be apparently excessive, yet the exhaust pumping time-area is too small, compared to those of the other four-valve engines listed. Because it has just two valves, it is not possible to raise the exhaust pumping time-area any higher, so compensation with extra blowdown time-area is required. The valves in this engine are also pushrod-actuated. Old-fashioned, you say? This engine produces well over 900 bhp on gasoline.

A second series of engines is analyzed and presented. This consists of the engines featured in Chapter 5, the Matchless G50 Grand Prix racing motorcycle, the Ducati 955 Superbike racing motorcycle, the Nissan Infiniti IRL racing car, the "sports car," the "turbo-diesel car," and the Ryobi hand-held power tool unit. The results of the analysis for the time-areas of these engines are shown in Fig. 6.5.

The time-areas of the Nissan and Ducati engines compare well with those listed in Fig. 6.4. In each time-area category, the time-areas for the sohc G50 racing engine are somewhat less

Time area per unit swept volume in units of s/m x 10^4						
ENGINE	Φ_{eb}	Φ_{ep}	Φ_{lp}	Φ_{lr}	Φ_{eo}	Φ_{lo}
2.0L I4 4v F2	15.7	97.5	114.5	17.3	25.3	29.6
3.5L V12 4v F1	15.6	105.0	130.3	18.2	37.6	46.5
0.6L I4 4v SRM	17.5	115.4	139.6	16.9	38.7	47.1
3.0L V10 4v F1	13.7	113.0	129.0	15.6	38.4	45.5
3.5L V8 4v F1	15.8	94.0	112.0	17.5	23.6	29.0
4.5L V8 4v SRC	11.5	87.3	126.9	11.6	27.3	40.7
2.0L I4 4v ITC	14.9	121.1	138.8	20.4	33.8	39.0
8.2L V8 2v OPB	18.2	90.4	117.4	22.2	21.9	30.2

Fig. 6.4 The time-areas of some high-performance engines.

than those of the dohc Nissan and Ducati race units. That is not too surprising, as the G50 is a competition engine from a previous technical generation and its specific power level at 110 hp/liter is some 40% less than that of most of the engines listed in Fig. 6.4. The time-areas for the diesel car are also interesting. Although the data for intake ramming and the valve overlap periods are, by necessity, low, those for the main pumping and blowdown periods reflect the proportionately larger mass flows of gas through this turbocharged engine; from Fig. 5.77, it has a delivery ratio of 1.7, which is some 50% more than any of the naturally-aspirated units listed in Fig. 6.4.

It is instructive to plot the data from Figs. 6.4 and 6.5 for all of these engines, excepting those from the diesel car and the Ryobi unit on the grounds that these are special cases containing potentially misleading information on the effects of either pressure-charging, or disproportionately high friction losses, respectively. The time-area data are plotted with respect to the

Time area per unit swept volume in units of s/m x 10^4						
ENGINE	Φ_{eb}	Φ_{ep}	Φ_{lp}	Φ_{lr}	Φ_{eo}	Φ_{lo}
G50 Matchless	9.85	75.6	120.8	14.6	18.3	28.5
Ducati 955	14.5	89.5	119.4	18.5	35.9	47.7
Nissan IRL	13.2	96.5	131.4	20.0	27.7	34.7
Sports-car	9.85	92.0	102.4	10.5	11.4	13.1
Diesel car	17.2	120.9	132.2	6.0	5.2	6.4
Ryobi strimmer	8.80	89.5	89.6	8.5	37.3	37.9

Fig. 6.5 The time-areas of the engines from Chapter 5.

bmep attained at the engine speed for peak power, which is precisely the same engine speed at which the time-area data are gathered. The results for the time-areas of exhaust blowdown, exhaust pumping, intake pumping, intake ramming, exhaust overlap and intake overlap, are shown in Figs. 6.6 to 6.11, respectively.

The scatter observed on the several figures is not only considerable, but is to be expected. Some engines have a lesser blowdown, but a higher exhaust pumping characteristic. Others have a considerable intake pumping capacity with a reduced intake ramming behavior. Such combinations, together with the variable quality of the intake and exhaust tuning in each case, produce the observed scatter in Figs. 6.6 to 6.11. Nevertheless, it is possible to deduce a set of simple fits to this scatter, and these are shown as a line, with its matching linear equation, on each of these plots. It is interesting to note that the least scatter is observed in Figs. 6.6 and 6.9, i.e., those for exhaust blowdown and intake ramming, respectively. This probably reflects the very close attention that is conventionally paid by each engine developer to the location of exhaust valve opening and intake valve closing.

It easy to conceive of using the data reduction, seen in Figs. 6.6 to 6.11, in reverse as a design process and orient a particular engine design to have minimalized valve timing periods, while still achieving the requisite time-areas. In short, you can take as a logical starting point the mean best thinking behind the data for all those successful engines. These relationships for the values of exhaust blowdown, Φ_{eb}, exhaust pumping, Φ_{ep}, intake pumping, Φ_{ip}, intake ramming, Φ_{ir}, exhaust overlap, Φ_{eo}, and intake overlap, Φ_{io}, are a good first approximation for a design or a development process. They are extracted from the figures as the

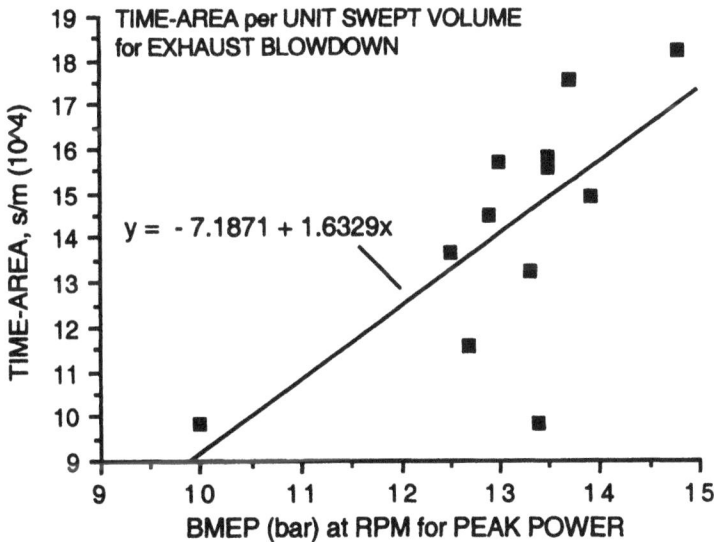

Fig. 6.6 The time-area relationship with bmep of exhaust blowdown.

required time-areas at the engine speed for peak power, N, where the required torque at that speed is expressed by the brake mean effective pressure, $bmep_N$, in bar units. The time-areas are in units of s/m.

Exhaust blowdown: $\Phi_{eb} = \left(-7.1871 + 1.6329 \times bmep_N\right) \times 10^{-4}$ s/m   (6.1.10)

Exhaust pumping: $\Phi_{ep} = \left(74.822 + 1.7775 \times bmep_N\right) \times 10^{-4}$ s/m   (6.1.11)

Intake pumping: $\Phi_{ip} = \left(57.78 + 5.02 \times bmep_N\right) \times 10^{-4}$ s/m   (6.1.12)

Intake ramming: $\Phi_{ir} = \left(-14.527 + 2.4022 \times bmep_N\right) \times 10^{-4}$ s/m   (6.1.13)

Exhaust overlap: $\Phi_{eo} = \left(-11.363 + 3.0296 \times bmep_N\right) \times 10^{-4}$ s/m   (6.1.14)

Intake overlap: $\Phi_{io} = \left(-17.985 + 4.1185 \times bmep_N\right) \times 10^{-4}$ s/m   (6.1.15)

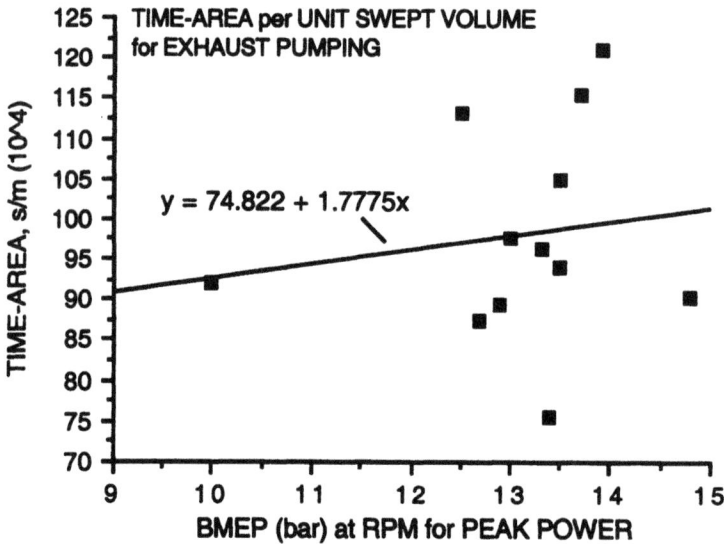

Fig. 6.7 The time-area relationship with bmep of exhaust pumping.

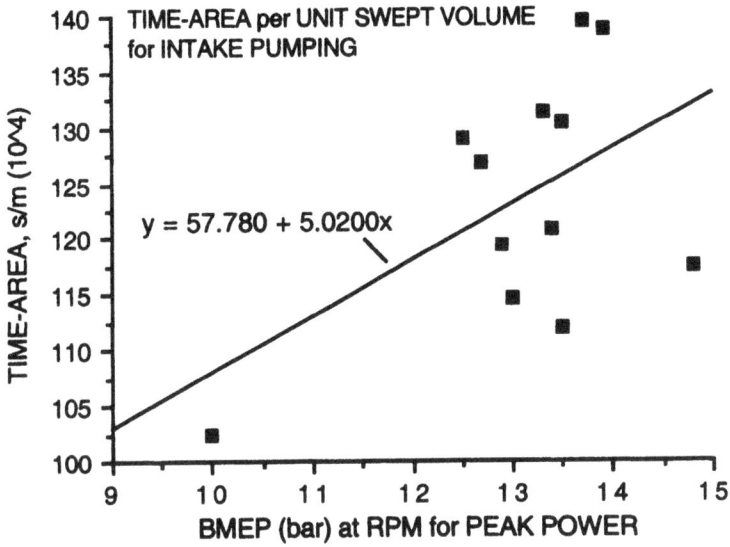

Fig. 6.8 The time-area relationship with bmep of intake pumping.

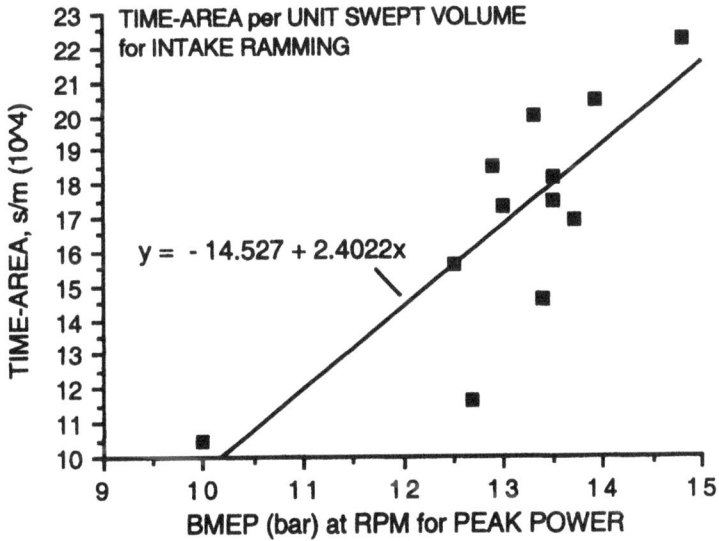

Fig. 6.9 The time-area relationship with bmep of intake ramming.

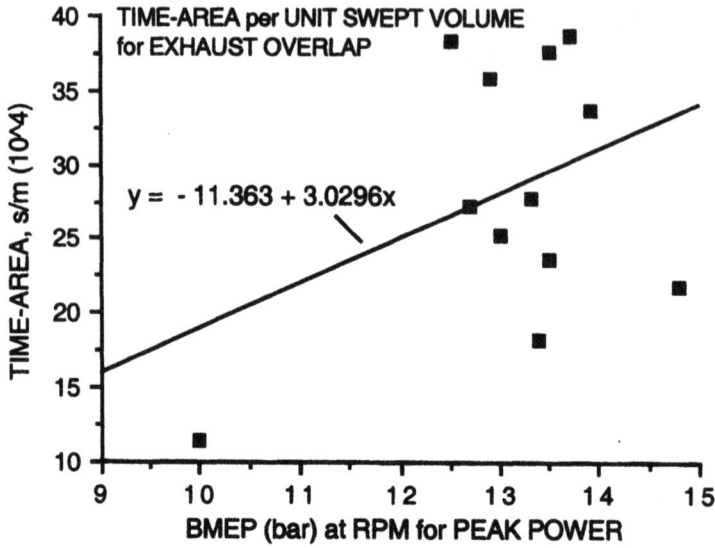

Fig. 6.10 The time-area relationship with bmep of exhaust overlap.

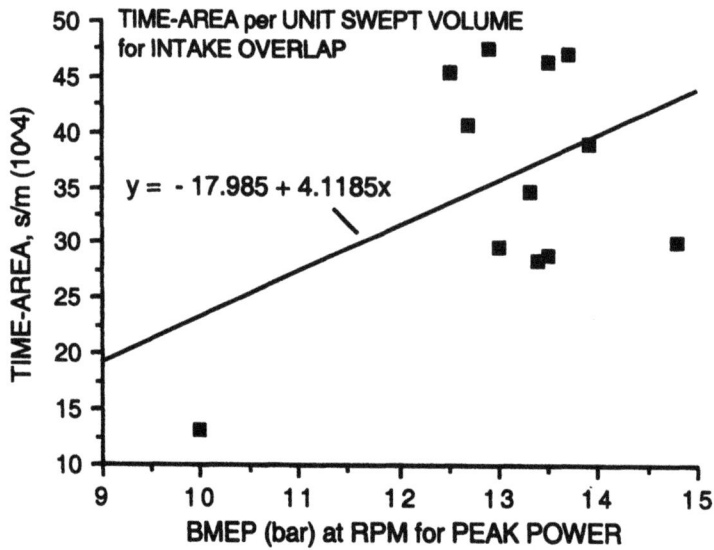

Fig. 6.11 The time-area relationship with bmep of intake overlap.

Having spent a lifetime with undergraduate students, I know that they just love being presented with what appears to be a panacea. I must therefore repeat a word of caution to those who will immediately analyze a particular engine and compare the time-area numbers with either the data in Figs. 6.4 and 6.5, or with the numbers predicted by Eqs. 6.1.10 to 6.1.15. Although I encourage you to make such comparisons, if they are to be relevant, the flow areas for the valves must be determined from the theory in Sec. 1.5.2 using Eqs. 1.5.6 to 1.5.14, and not by the simplistic Eq. 1.5.5.

6.1.8 Measurements of Valve Lift from Engines

From examining Eq. 1.5.15 and reading the associated discussion, you will see that it is possible in any given cylinder head configuration to insert values of C_d for an intake or exhaust valve until equality is attained between the valve annular flow area, A_t, and the minimum port area, A_p. If this exercise is conducted at the measured value of maximum valve lift for each of the engines listed in Figs. 6.4 and 6.5, then those values C_d of can become design criteria in reverse. Figs. 6.12 and 6.13, show the results of such an analysis for this same set of engines. The commonality of the ensuing data for C_d for either valve is quite remarkable, with the singular exception of the G50 Matchless engine.

In the 1960s when the G50 engine was raced, the engineers of the day apparently did not appreciate the gas flow advantages of lifting the poppet valves much beyond the nominal maximum, i.e., 25% of the inner seat diameter, a ratio that appears in Figs. 6.12 and 6.13 with a value of unity. The effect of high valve lift on the actual discharge coefficients is emphasized in Chapter 3, and is easily seen by a reexamination of, say, Figs. 3.37 and 3.40.

Required C_d to equate the measured maximum valve lift		
ENGINE	**intake valve C_d**	**exhaust valve C_d**
2.0L I4 4v F2	0.71	0.62
3.5L V12 4v F1	0.67	0.52
0.6L I4 4v SRM	0.72	0.64
3.0L V10 4v F1	0.77	0.64
3.5L V8 4v F1	0.71	0.58
4.5L V8 4v SRC	0.72	0.67
2.0L I4 4v ITC	0.67	0.61
8.2L V8 2v OPB	0.69	0.50

Fig. 6.12 The Cd at maximum valve lift of some high-performance engines.

Required C_d to equate the measured maximum valve lift		
ENGINE	intake valve C_d	exhaust valve C_d
G50 Matchless	0.990	0.960
Ducati 955	0.714	0.632
Nissan IRL	0.661	0.601
Sports-car	0.726	0.624
Diesel car	0.722	0.640
Ryobi strimmer	0.689	0.689

Fig. 6.13 The Cd at maximum valve lift of engines from Chapter 5.

The C_d data for the maximum lift of the intake and exhaust valves, from all of the engines in Figs. 6.12 and 6.13, are plotted with respect to the bmep attained by each engine, $bmep_N$, and are presented in Figs. 6.14 and 6.15, respectively. The exceptional data for the G50 engine are not plotted, nor are those for the diesel car engine or the Ryobi power-tool, for the singular reasons already expressed. The scatter observed in both figures appears to be considerable, but the range of y-scale values is not very extensive. The collective wisdom of the engineers who selected these maximum valve lifts is statistically found on the two lines representing the data mean on each diagram. The slope of the line is negative, indicating what Figs. 3.37 and 3.40 have already confirmed: that higher bmep levels are more readily attained with higher valve lifts. The engine designer, developer, and modeler, can use this information at any stage from initial design to ongoing R&D, either to create new maximum valve lift data, or to check the compatibility of existing data. The C_d relationships for maximum valve lift, to be employed within Eq. 1.5.15, can be expressed as a function of the brake mean effective pressure, $bmep_N$, required at the engine speed for maximum power output. The $bmep_N$ quoted below is in bar units.

Intake valve: $$C_d = 0.86871 - 0.0012452 \times bmep_N \qquad (6.1.16)$$

Exhaust valve: $$C_d = 0.88401 - 0.0021435 \times bmep_N \qquad (6.1.17)$$

It is fairly obvious that Eqs. 6.1.16 and 6.1.17 can be used in conjunction with Eq. 1.5.15 to determine the value of maximum valve lift, L_v, of the intake and exhaust valves for any new engine design or during the development of an existing power unit. In this context, even the simplistic selection, from Figs. 6.12 to 6.13, of a C_d value of 0.7 for the intake valve, and of 0.6 for the exhaust valve, would suffice for an initial estimate of the maximum valve lift in each case.

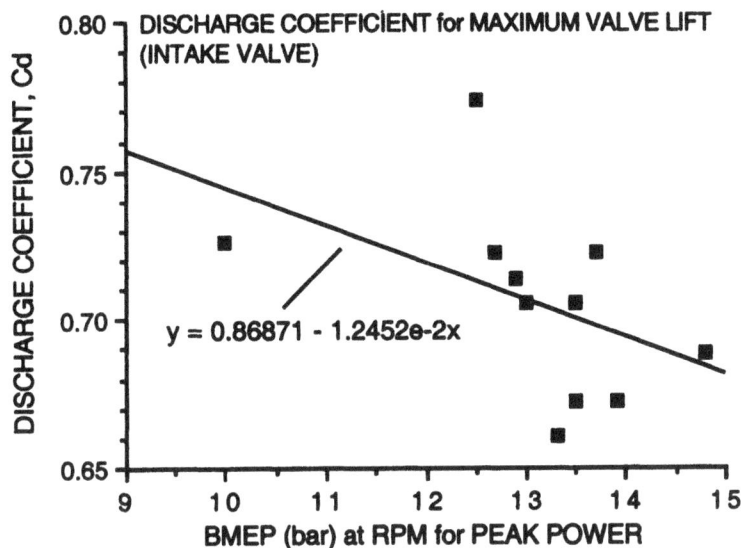

Fig. 6.14 The C_d relationship with bmep of intake valve maximum lift.

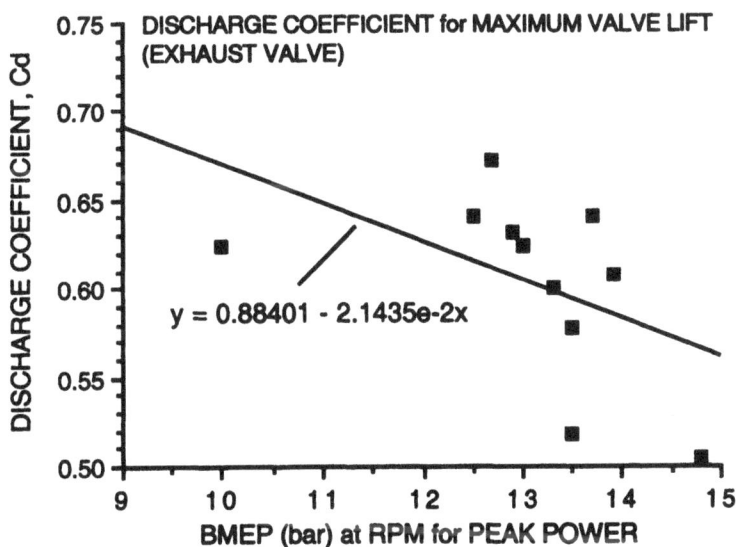

Fig. 6.15 The C_d relationship with bmep of exhaust valve maximum lift.

6.1.9 Manifold-to-Port Area Ratios from Engines

Sec. 1.5.1 discusses and defines manifold-to-port area ratios. Eqs. 1.5.1 and 1.5.2 detail the theoretical nature of the important parameter of manifold-to-port area ratio, C_m, for both the intake and exhaust system connections to the cylinder head. Analyses of the measured geometrical data for the engines shown in Figs. 6.4 and 6.5 are given in Figs. 6.16 and 6.17.

The data for the intake manifold-to-port area ratios, C_{im}, for the various engines reveal that the disparate designers of these engines thought much alike. For a design criteria, it would appear that a value for C_{im} approaching unity would be a logical starting point.

The G50 engine is a notable exception to this trend, for the data reveal that its intake duct, like its maximum intake valve lift, could have profited the engine if they were enlarged. As always, hindsight is 20/20. However, just maybe Jack Williams [6.17] knew all this. The diameter of the intake manifold, d_m, of the G50 is 38 mm (see Fig. 5.20), and the largest carburetor that Jack Williams could obtain from Amal in England, and fit to that manifold, was the 38 mm (1.5 in.) GP version.

ENGINE	Measured manifold-port area ratio	
	exhaust C_{em}	intake C_{im}
2.0L I4 4v F2	1.26	0.84
3.5L V12 4v F1	1.60	0.90
0.6L I4 4v SRM	1.40	1.00
3.0L V10 4v F1	1.40	0.90
3.5L V8 4v F1	1.60	0.90
4.5L V8 4v SRC	1.40	1.10
2.0L I4 4v ITC	1.20	1.00
8.2L V8 2v OPB	1.55	1.05

Fig. 6.16 The manifold-to-port area ratios of some high-performance engines.

ENGINE	Measured manifold-port area ratio	
	exhaust C_{em}	intake C_{im}
G50 Matchless	1.260	0.670
Ducati 955	1.210	0.930
Nissan IRL	1.448	0.925
Sports-car	1.300	1.005
Diesel car	1.280	1.000
Ryobi strimmer	1.500	1.550

Fig. 6.17 The manifold-to-port area ratios of engines from Chapter 5.

The like thinking of the disparate designers, evident from the intake manifold-to-port area ratios, does not extend to the exhaust manifold-to-port area ratio, C_{em}, of which the values for the several engines in Figs. 6.16 and 6.17 range from 1.2 to 1.6.

To observe the degree of scatter for C_{im} and C_{em}, the data for the engines in Figs. 6.16 and 6.17 are plotted with respect to the bmep attained by each engine, $bmep_N$, and are presented in Figs. 6.18 and 6.19, respectively. The mean of the data plotted in each figure is represented by a line and its associated linear equation.

In Fig. 6.18, the intake manifold-to-port area ratio has an almost flat characteristic with respect to the bmep expected at peak power. One could almost assume it to be a constant at 0.96. For the record, however, the function with respect to the brake mean effective pressure, $bmep_N$, required at the engine speed for maximum power output, is set out below. The $bmep_N$ quoted is in bar units.

Intake manifold:
$$C_{im} = 0.94801 + 0.00084783 \times bmep_N \qquad (6.1.18)$$

The equivalent data for the exhaust manifold are shown in Fig. 6.19, where the scatter, at first sight, appears random. However, the statistical mean as represented by the line and its associated function, teases some logic from the data. The simple meaning of the positive slope is that the higher the expected bmep, the larger the diameter of the exhaust manifold and the primary pipe fitted to it. The logic is impeccable. In general, as the bmep rises, the trapped cylinder mass and the magnitude of the cylinder pressure at release also rise. The outcome,

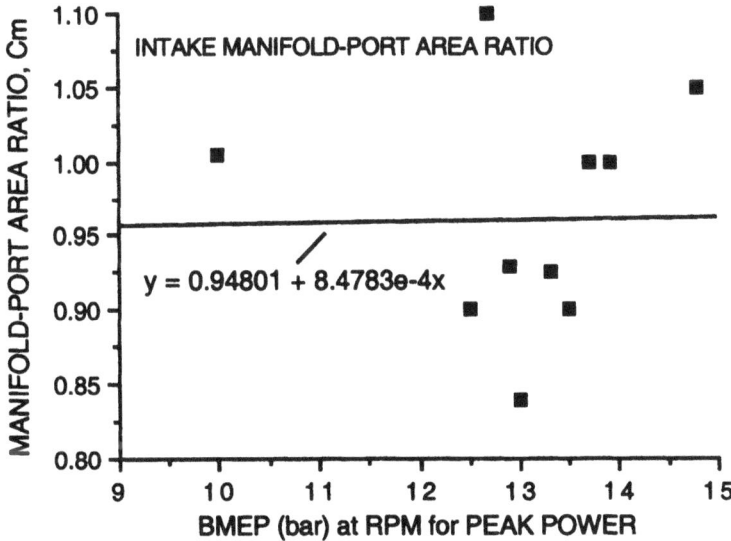

Fig. 6.18 The manifold-to-port area ratio relationship with bmep of the intake system.

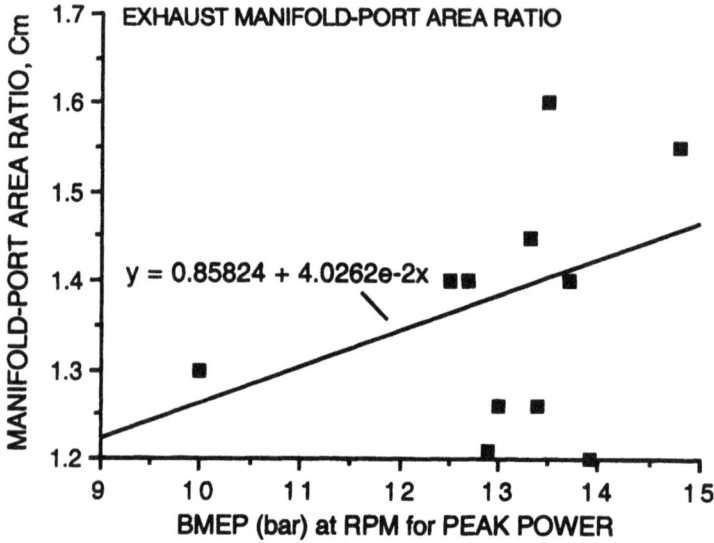

Fig. 6.19 The manifold-to-port area ratio relationship with bmep of the exhaust system.

with any given manifold diameter, is that the peak pressure of the exhaust pulse will increase. Recall from the theory in Sec. 2.1.5 and the measurements in Sec. 2.19.2 that large-amplitude compression waves become steep-fronted more easily and lose amplitude and energy. Hence, if a high bmep is expected, the exhaust manifold (pipe) diameter should be suitably enlarged to retain that amplitude and energy. The energy is needed for tuning, by reflection at some appropriate expansion of area, such as that at an open end, a collector pipe, or a megaphone. In practice, for naturally-aspirated, spark-ignition, high-performance engines burning gasoline, the maximum exhaust pulse amplitude that fits this criterion is about 2 atm. The "average" collective wisdom seen in Fig. 6.19 is oriented toward this value, and Sec. 6.4.6 describes a well-optimized engine that produces just such an exhaust pulse amplitude.

The function, shown in Fig. 6.19, for the exhaust manifold-to-port area ratio, with respect to the brake mean effective pressure, $bmep_N$, required at the engine speed for maximum power output, is set out below. The $bmep_N$ quoted is in bar units.

Exhaust manifold: $C_{em} = 0.85824 + 0.040262 \times bmep_N$ (6.1.19)

This equation provides an interesting prediction for those who design small industrial engines, such as the Ryobi trimmer unit, where the output bmep often lies between 4 and 7 bar. From Eq. 6.1.19, the value of C_{em} falls to unity when the target bmep is 3.52 bar. Recall that a not-illogical guess for C_{im} is also unity. Hence, it is quite possible that the exhaust and intake manifolds of many industrial (na si) engines in current production may well be too large.

6.2 The Relevance of Empiricism for the Design of the Cylinder Head

In Sec. 6.1, some empiricism is introduced regarding the design of the cylinder head and its components. In order to determine the relevance of the several empirical factors that have been introduced through Eqs. 6.1.4 to 6.1.19, it is proposed to simulate a gasoline-burning, naturally-aspirated, spark-ignition, high-performance engine and change the dimensions within the cylinder head.

Naturally, it would be preferable to carry out the investigation by experimental methods as well as by simulation, so that the relevance of the empiricism may be established by both fact and theory. However, apart from the financial cost involved, which would be quite astronomical, the time required to execute an experimental program as extensive as that exhibited below in the text, might be equally astronomical. I think that the correlation of this simulation with experimental data is already sufficiently credible as to warrant taking its computational conclusions seriously (viz., Chapter 5 and noted references [5.9-5.10, 5.16, 5.18-5.20]).

The engine selected for the simulation exercise is the seventh on the list in Fig. 6.4, described as "2.0L I4 4v ITC," and used for saloon (sedan) car racing in Europe. The engine has a measured power output approaching 200 kW (268 bhp) at 8400 rpm.

6.2.1 Input Data for the Simulation of the ITC Car Engine

The engine has bore, stroke, and connecting rod dimensions of 88, 82, and 148 mm, respectively. The compression ratio is 12. The total swept volume of the engine is 1995 cm^3. The engine valves are operated by double overhead camshafts. The four-valve, pent-roof cylinder head and ducting geometries are presented in Figs. 6.20 and 6.21.

The firing order is dictated by the use of a single-plane crankshaft, as shown in Fig. 5.5(b), resulting in a 1-3-4-2 firing order and the same angular firing sequences as listed for the "sports car" in Fig. 5.4.

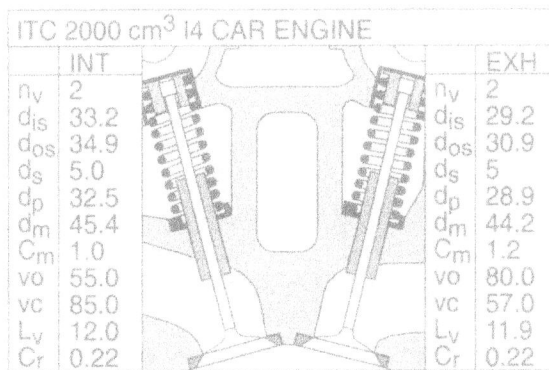

Fig. 6.20 The cylinder head geometry of the ITC car engine.

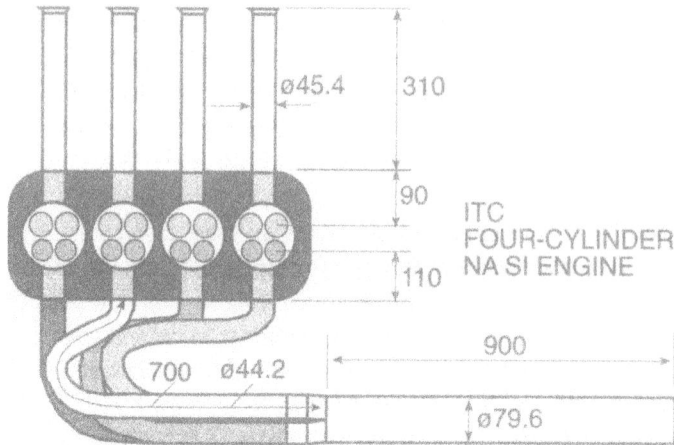

Fig. 6.21 The intake and exhaust ducting of the ITC car engine.

During the computer simulation at wide open throttle, the following assumptions are made: the combustion of aviation gasoline occurs at a constant air-fuel ratio of 12, which corresponds to a λ value of 0.846; the combustion process is assumed to be as shown in Fig. 4.10 for a reasonably similar 4v na si engine with gasoline; the discharge coefficient maps for the valves and pipe ends are assumed to be as presented in Figs. 3.37 to 3.40 and 3.9 to 3.10, respectively; the discharge coefficients for unthrottled bellmouth-ended pipes are as quoted in Eqs. 3.3.1 to 3.3.4; the skin temperatures of the cylinder head, piston crown, and cylinder liner are 300, 200, and 150°C, respectively; the skin temperatures of the exhaust duct surfaces range from 500 to 300°C; and the engine breathes to and from the atmosphere at 1.014 bar and 25°C.

Intake and Exhaust Ducting of the ITC Car Engine

The intake ducting, sketched in Fig. 6.21, shows a simple intake ramming design. Because the simulation is at full-throttle, and the throttle-plate and EFI system do not obstruct the intake tracts, it is not necessary to include these in the simulation geometry. The exhaust system is of the 4-1 collector type very commonly seen on racing car engines. It is shown in Fig. 5.63, attached to the cylinders numbered 1-4 as sketched. The branch angles at the five-way branch can best be described as follows: for cylinders numbered 1-4, the primary pipes are at a mutual 15° to each other, and the final outlet pipe is symmetrically disposed facing them.

6.2.1 Overall Engine Performance Characteristics of the ITC Engine

Before beginning the computational experiments, which involves altering empirical factors on this engine, the performance characteristics of the standard engine are presented so that you may observe the behavior of a well-optimized high-performance engine. The engine has such excellent performance characteristics, and the exhaust and intake tuning is so well optimized, that any change to the engine geometry in the theoretical experiments is more likely to deteriorate the performance of the engine rather than improve it!

The basic performance characteristics predicted by the simulation, from 4500 to 8500 rpm, are shown in Figs. 6.22 to 6.24. In Fig. 6.22, it can be seen that the peak power performance approaches 200 kW and the bmep is greater than 13 bar over the critical racing power band from 5500 to 8500 rpm. The peak torque shows a very impressive bmep value of 15 bar at 7500 rpm, which bespeaks good air breathing at high engine speed despite a bore-stroke ratio of a mere 1.07, and the mean piston speed peaks at 23.2 m/s.

In Fig. 6.23, the origins of this high torque characteristic are evident. The delivery ratio is between 1.15 and 1.27 over the operating power band. Its double peak profile has been seen before, viz., Fig. 5.25, Fig. 5.39, and Fig. 5.53, and is a characteristic of good intake tuning, as you will see in Sec. 6.3. Although the trapping efficiency falls off to 95% in this area, this still leaves the charging efficiency peaking near 1.2. The trapped charge purity is mostly above 98%, but falls off to 90-95% in the off-power-peak zones.

Fig. 6.24, shows the brake specific fuel consumption, bsfc. Because the air-fuel ratio is set at 12, the best point bsfc is at 278 g/kWh, but tails off to 295 g/kWh at peak power. This deterioration owes much to the rising pumping loss in the power band and to the decreasing trapping efficiency. This profile for pumping loss is also familiar from Fig. 5.28, the profile for another highly-tuned engine.

6.2.2 The Pressure Wave Action at the Valves at 8000 rpm in the ITC Engine

All data regarding pressure waves and other time-related data given in the figures are at cylinder number 1. Because the engine ducting and firing order are arranged symmetrically, all cylinders behave in a virtually identical fashion. The data shown in Figs. 6.25 and 6.26 are extracted from the computation at 8000 rpm.

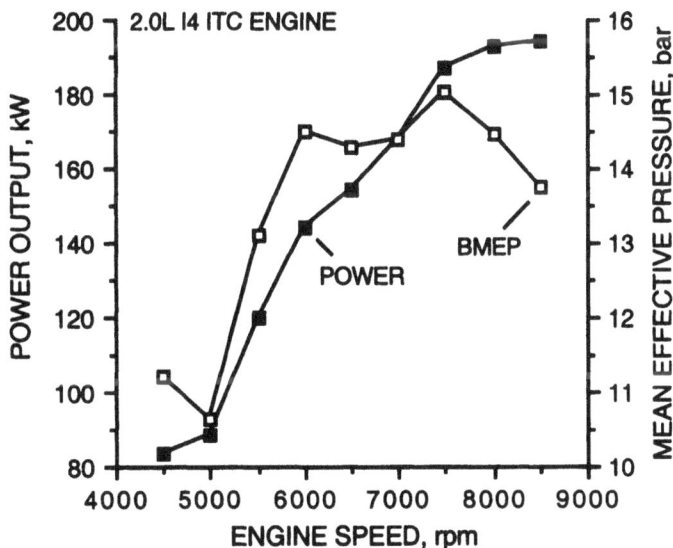

Fig. 6.22 The power and torque output of the ITC car engine.

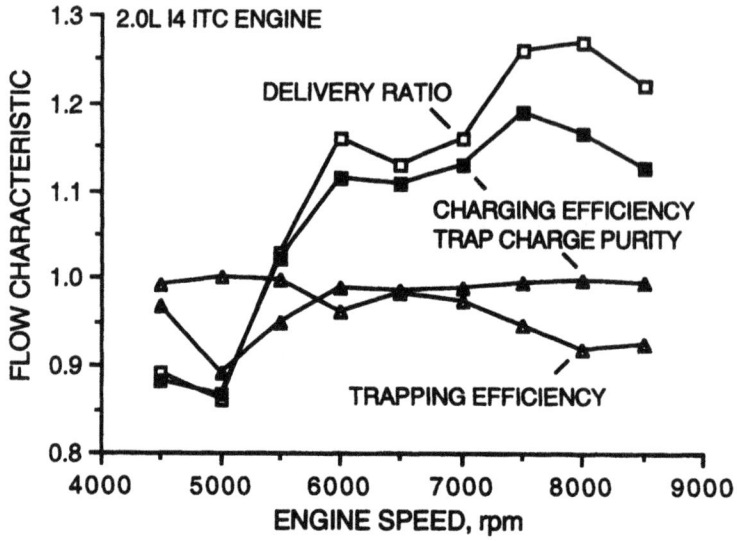

Fig. 6.23 The charging and trapping characteristics of the ITC car engine.

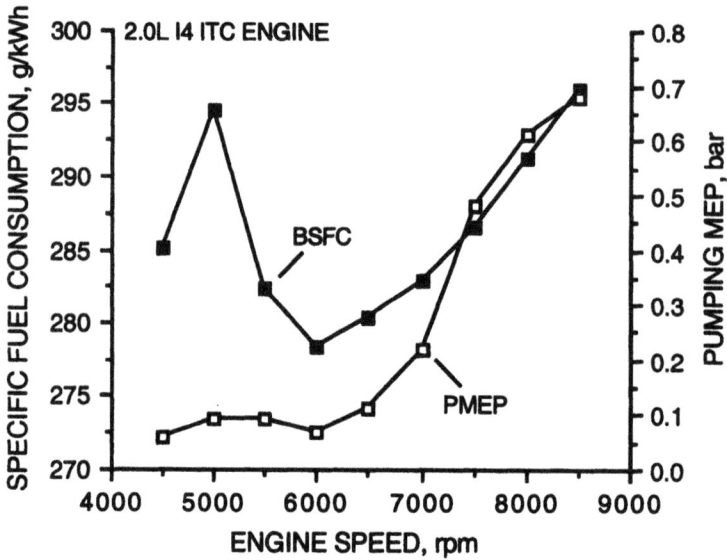

Fig. 6.24 The fuel consumption and pumping losses of the ITC car engine.

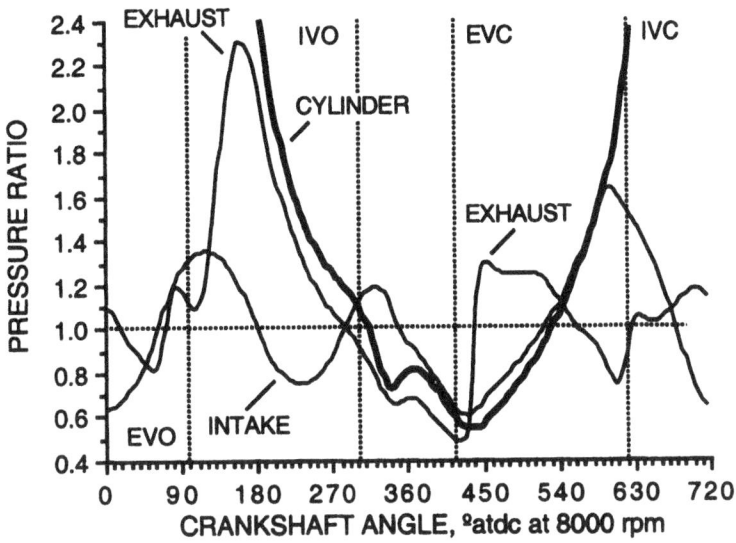

Fig. 6.25 Cylinder, exhaust, and intake pressure diagrams at 8000 rpm in the ITC engine.

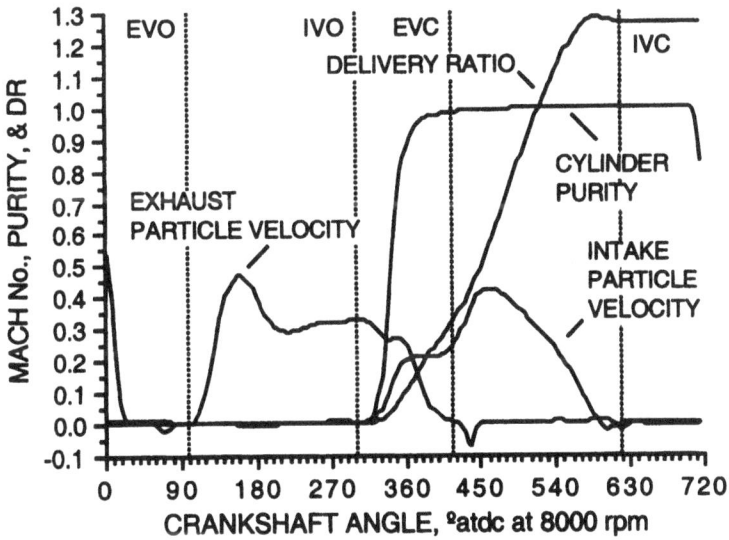

Fig. 6.26 Particle velocities, purity, and DR at 8000 rpm in the ITC engine.

Fig. 6.25 shows the predicted pressure ratios in the cylinder and in the intake and exhaust ducts behind the valves. The pressure ratios in the ducts are the superposition pressures, i.e., as they would be recorded by a pressure transducer if one were placed at these locations. The cylinder pressure is seen to fall below the atmospheric pressure even before tdc, due to the strong suction pressure emanating from the exhaust system. This extends throughout the entire valve overlap period. The particle flow direction is clearly from intake through the cylinder to the exhaust duct. A contributing factor here is the second-order wave reflection in the intake tract which coincides with the beginning of the overlap period and forces the flow inward. If this second-order intake pulsation had been phased differently, some exhaust backflow into the intake tract could have inevitably occurred; you will find this to be a recurring theme emphasized in Sec. 6.3.

The strength of the exhaust suction is significant, down to 0.5 atm, and this induces considerable exhaust particle flow from the cylinder, and intake flow into it. This is best illustrated in Fig. 6.26, where the particle velocities, as Mach number, are drawn for the entire engine cycle. On the intake duct trace, there is no sign of cylinder backflow into the intake tract, except just before the intake valve closes as the intake ramming wave is beaten by the rising cylinder pressure. On the exhaust duct trace, the exhaust stroke is most visible, for the double hump on the velocity profile shows firstly the exhaust pulse formation, and secondly the pumping action of the piston allied to the exhaust wave action. In Fig. 6.25, only the mildest of pressure bulges around tdc hints at the piston pumping action, such is the suction strength of the exhaust wave action. This suction extracts considerable quantities of fresh charge into the exhaust duct, as the exhaust duct velocity profile displays, hence the loss of trapping efficiency observed in Fig. 6.23. The unburned HC emission that this occasions is calculated at greater than 20 g/kWh! In Fig. 6.26, the purity within the cylinder rises above zero almost as soon as the intake valve opens. Because the flow directions during the valve overlap period are always from intake to cylinder to exhaust, the cylinder airflow, plotted as DR in Fig. 6.26, commences before tdc and rises to about 1.3 before tailing off with some backflow. This tuning is probably as good as, if not better than, any of the engines discussed in Chapter 5. Indeed, pressure diagrams that appear to be similar to Fig. 6.25 have been seen in Chapter 5 as Figs. 5.23, 5.44, and 5.54. Indeed, by now you should have in your head a series of mental pictures of the pressure diagram profiles that constitute good intake and exhaust tuning.

6.2.3 The Effect of Time-Areas on the Performance of the ITC Engine

It is very difficult to change any particular time-area element for a valve, or valves, in isolation. A reduction or increase for any one element has somewhat of a ripple effect on the others. Nevertheless, cylinder head geometry parameters are so changed, time-area shifts accomplished, and the outcome on engine performance is reported below.

The Effect of Changing the Exhaust Blowdown Time-Area

The data for the exhaust valve timing of the engine are changed in two experiments by delaying the opening of the exhaust valve by 5° and 10°, respectively. This alters the "time-area per unit swept volume," Φ, as presented in Fig. 6.27, and several similar figures. For convenience, all time-area values, Φ, on these figures are printed as multiplied by 10^4, in the same manner as in Figs. 6.4 and 6.5. It can be seen that reducing the blowdown time area, Φ_{eb}, causes a small ripple effect on other exhaust-related time-area parameters, but this is unavoidable in the context of logical input data preparation. Nevertheless, the basic effect of reducing blowdown time-area by about 16% and 30% is achieved.

The simulations are run, the output data are analyzed, and the outcome is presented in Fig. 6.28 as a power gain or loss at any particular engine speed. A mixture of good and bad news is evident. Some gains are noted below 7000 rpm, but losses predominate above that engine speed. These changes indicate that blowdown, while important, is not as hugely significant as might be imagined. In other words, the exhaust valve opening timing is important, but as long as the blowdown, Φ_{eb}, value remains high, i.e., greater than 10×10^{-4} s/m, the engine should still perform well, even if not quite at an optimum. It should be noted that this engine already has one of the highest values for the exhaust pumping time-area, Φ_{ep}, and so has some margin of error on blowdown with which to compensate on the exhaust stroke. This can be seen in more detail when the exhaust pumping time-area is reduced.

ENGINE	Φ_{eb}	Φ_{ep}	Φ_{ip}	Φ_{ir}	Φ_{eo}	Φ_{io}
standard	14.9	121.1	138.8	20.4	33.8	39.0
-5° evo	12.8	120.8	138.8	20.4	34.4	39.0
-10° evo	10.5	119.8	138.8	20.4	34.6	39.0

Fig. 6.27 Reducing the blowdown time-area in the ITC engine.

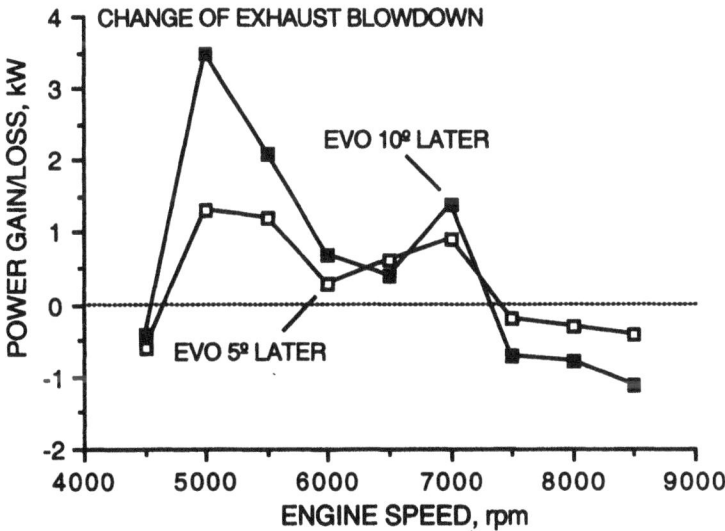

Fig. 6.28 Effect of reducing the blowdown time-area in the ITC engine.

The Effect of Changing the Exhaust Pumping Time-Area

Each exhaust valve is reduced in seat diameter by 1 mm. The valve timings are retained, but now each valve lifts by 0.4 mm less, in order to retain the similarity in lift/diameter through the use of the same C_d values for each valve. As shown in Fig. 6.29, the effect of this is a 7% reduction in exhaust pumping time-area.

The result is shown in Fig. 6.30, as power gain or loss over the entire speed range. Power loss is evident in the racing power band from 7000 to 8500 rpm, but the reduced value of 112.9 $\times 10^{-4}$ for Φ_{ep} is still not particularly low in the context of some of the other engines listed in Figs. 6.4 and 6.5.

The Effect of Changing Intake Pumping Time-Area

Each intake valve is reduced in seat diameter by 1 mm. The intake valve timings are retained, but now the valve lifts are each 0.4 mm less, in order to retain the similarity in valve lift/diameter using the same C_d criteria in each case. As shown in Fig. 6.31, the effect of this is a 6.4% reduction in intake pumping time-area, Φ_{ip}.

ENGINE	Φ_{eb}	Φ_{ep}	Φ_{ip}	Φ_{ir}	Φ_{eo}	Φ_{io}
standard	14.9	121.1	138.8	20.4	33.8	39.0
-1 mm d_{ev}	13.8	112.9	138.8	20.4	31.5	39.0

Fig. 6.29 Reducing the exhaust pumping time-area in the ITC engine.

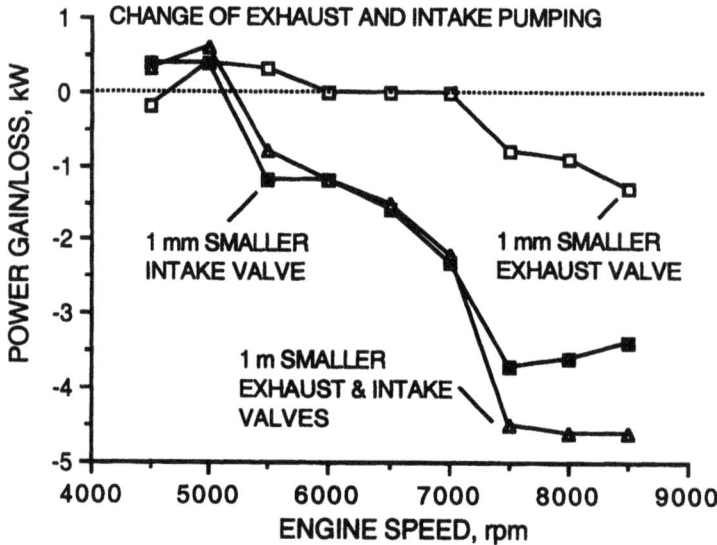

Fig. 6.30 Effect of reducing the pumping time-areas in the ITC engine.

The result on power gain or loss can be seen in Fig. 6.30. Power loss is evident in the racing power band from 7000 to 8500 rpm and is a more significant loss of power than that for the same diametral change of the exhaust valve. The power loss for the intake valve is about 2.5%, so intake pumping time-area is a sensitive parameter.

The Effect of Changing Exhaust and Intake Pumping Time-Area

If the computation is run with both intake and exhaust valves reduced in diameter by 1 mm, the effects on time-areas for Φ_{ep} and Φ_{ip} are as shown in Fig. 6.31. The effect on power change is graphed in Fig. 6.30. Now a significant power loss has developed over most of the speed range, and the effect of size change for each valve is seen to compound the power loss in the peak power zone.

The Effect of Changing Intake Ramming Time-Area

Of the engines shown in Figs. 6.4 and 6.5, it can be seen that the "standard" ITC engine has one of the higher values for intake ramming time-area as its plotted position lies above the "mean" line on Fig. 6.9. In Fig. 6.26, some delivery ratio and particle loss is seen from the cylinder near the intake valve closing position. Hence in two computation experiments, the intake valve closing timing, ivc, is reduced first by 5° and then by 10°. The effect on intake ramming time-area, Φ_{ir}, is shown in Fig. 6.32.

Fig. 6.33 shows the outcome on power gain or loss over the entire speed range. Significant power gains are evident in both cases at all but the highest speed point of 8500 rpm. This provides some evidence as to the relevance and applicability of the proposed design criteria for intake ramming time-area.

ENGINE	Φ_{eb}	Φ_{ep}	Φ_{ip}	Φ_{ir}	Φ_{eo}	Φ_{io}
standard	14.9	121.1	138.8	20.4	33.8	39.0
-1 mm d_{iv}	14.9	121.1	129.9	19.1	33.8	36.5
-1 mm d_{ev} & d_{iv}	13.8	112.9	129.9	19.1	31.5	36.5

Fig. 6.31 Reducing the intake pumping time-area in the ITC engine.

ENGINE	Φ_{eb}	Φ_{ep}	Φ_{ip}	Φ_{ir}	Φ_{eo}	Φ_{io}
standard	14.9	121.1	138.8	20.4	33.8	39.0
-5° ivc	14.9	121.1	138.3	17.6	33.8	39.5
-10° ivc	14.9	121.1	137.6	15.0	33.8	40.0

Fig. 6.32 Reducing the intake ramming time-area in the ITC engine.

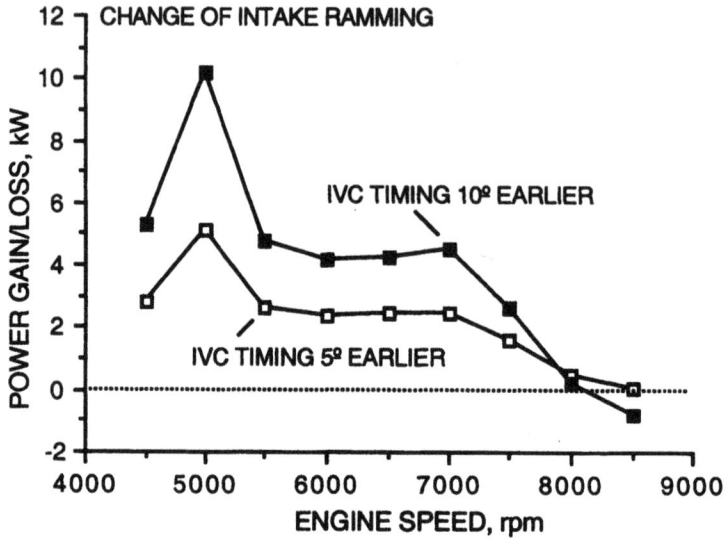

Fig. 6.33 Effect of reducing the intake ramming time-area in the ITC engine.

The Effect of Changing Valve Overlap Time-Area

The intake valve opening timing and the exhaust valve closing timing are advanced and retarded by 5°, respectively, to reduce the valve overlap period by 10° crankshaft total. This is repeated in a second experiment for a step change of 10° for each valve timing, giving a 20° total reduction in valve overlap. The effect on time-areas for these two computation experiments is seen in Fig. 6.34. The unavoidable ripple effect on other time-area parameters is again evident, but the basic effect of reducing Φ_{eo} and Φ_{io} is attained, with reductions of about 19% and 36% in the two experiments, respectively.

The effects on power gain or loss over the entire speed range can be seen in Fig. 6.35. Significant power loss is evident in the racing power band from 7000 to 8500 rpm, particularly for the larger of the two reductions of the Φ_{eo} and the Φ_{io} time-areas. The singular gain at 5000 rpm is probably not significant in terms of racing.

ENGINE	Φ_{eb}	Φ_{ep}	Φ_{ip}	Φ_{ir}	Φ_{eo}	Φ_{io}
standard	14.9	121.1	138.8	20.4	33.8	39.0
-10° overlap	15.1	119.5	136.8	20.6	27.4	31.7
-20° overlap	15.2	117.5	134.5	20.9	21.4	24.8

Fig. 6.34 Reducing the valve overlap time-areas in the ITC engine.

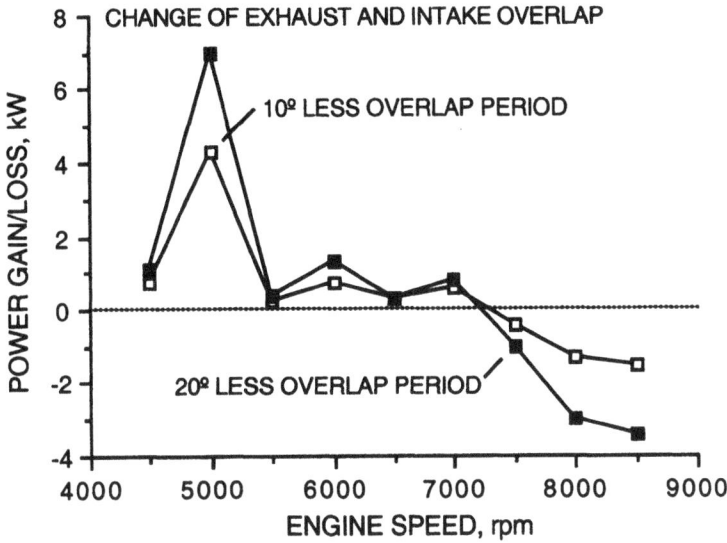

Fig. 6.35 Effect of reducing the valve overlap time-areas in the ITC engine.

The Effect of Changing the Intake Manifold-to-Port Area Ratio

In Fig. 6.18, which represents the collective thinking of several designers of high-performance engines, the manifold-to-port area ratio, C_{im}, as defined by Eq. 1.5.1, is seen to be virtually constant at about 0.96. Using the "standard" data set, which happens to have a manifold-to-port area ratio, C_{im}, of 1.0, this value is set to 0.8, 0.9, and 1.1 in three separate calculations over the speed range. Naturally, the "standard" intake duct diameter of 45.4 mm, as seen in Fig. 6.21, is adjusted for each run to match the three differing manifold diameters required. For the record, these diameters are 40.6, 43.1, and 47.6 mm, respectively. The results of the computations, in terms of power gain or loss with respect to the "standard" data, are shown in Fig. 6.36.

It is clear that, although a C_{im} value of 1.0 is a very acceptable design criteria, C_{im} values of 0.8 and 0.9 are better in all but the peak power zone, and a value of 0.9 would most likely be the winner of any race between them. It is also clear that a C_{im} value of 1.1 is unacceptable. Thus, a logical data starting point for simulation is 0.96, or that provided by Eq. 6.1.18, with the more detailed engine simulation model being used to determine in any given set of circumstances the optimum value of C_{im}.

The Effect of Changing the Exhaust Manifold-to-Port Area Ratio

On Fig. 6.19 the designer's selection of the manifold-to-port area ratio, C_{em}, as defined by Eq. 1.5.2, is seen to vary from 1.2 to 1.6. At the bmep level of this engine at peak power, Eq. 6.1.19 would select 1.4 as the correct choice for C_{em}. Using the "standard" data set, which

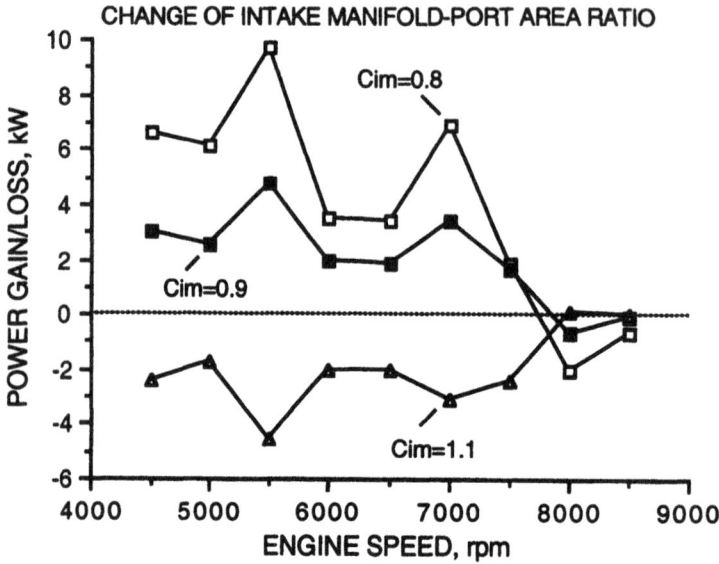

Fig. 6.36 Effect of changing the intake manifold-to-port area ratio in the ITC engine.

happens to have a manifold-to-port area ratio, C_{em}, of 1.2, the computation experiments set this value to 1.3, 1.4, 1.5, and 1.6 in four separate calculation runs over the speed range. Naturally, the "standard" exhaust duct diameter of 44.2 mm, as seen in Fig. 6.21, is adjusted for each run to match the four differing manifold diameters required. For the record, these diameters are 46, 47.7, 49.4, and 51.0 mm, respectively. The final tailpipe diameter, i.e., after the collector, is retained at 79.6 mm in all four calculations, although the later discussion in Sec. 6.4.6 throws some doubt on the validity of this decision. The results of the computations, in terms of power gain or loss with respect to the "standard" data, are shown in Fig. 6.37.

In terms of maximum power, it is obvious that the best value for C_{em} is 1.6, but only in circumstances where the engine requires a 1000 rpm power band. For example, in American oval circuit racing, such as Indy Racing League, there is not much need for a power band much wider than 1000 rpm. In European track racing, a working power band of at least 2500 rpm is needed, so the trough in the power curve displayed in Fig. 6.37 at 6500 rpm would be quite unacceptable. Thus, the selection of the correct exhaust manifold-to-port area ratio is very much dependent on the application.

If breadth of power band is part of the design criterion, then a C_{em} value of 1.2 is clearly the best choice; however, the value of 1.4, as predicted by Eq. 6.1.19 would not be too disastrous as a design starting point.

On the other hand, and on a purely practical note, the smaller the diameter of the header pipes of a collector system, the easier it is to manufacture, the higher is the quality of retention of constant diameter in the bends in the pipe runs to the collector, and the smaller is the interpipe branch angle at the collector. These are logical design considerations that must be taken into account, along with the more formal parameters investigated, if the maximum tuning function of a collector, or any, exhaust system is to be retained.

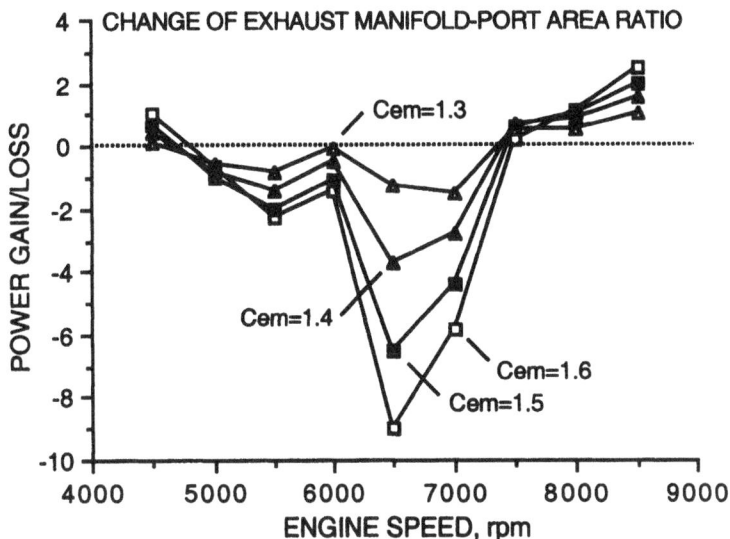

Fig. 6.37 Effect of changing the exhaust manifold-to-port area ratio in the ITC engine.

The Effect of Not Quite Getting It Right

Throughout the preceding sections, it is clear that certain factors give rise to more performance penalties than others. Imagine a designer working initially on the "standard" engine, i.e., before it is built. Conventional wisdom is employed, black art is consulted, tables of past history in the race-shop are examined, and maybe an insufficient, yet still considerable, number of data sets are inserted into an engine simulation model. Without employing the empiricism outlined above, it would not be unreasonable of that designer to start out with an ITC prototype with both valves 1 mm smaller, with a 10° less total overlap period, with an exhaust manifold-to-port area ratio of 1.6, and with an intake manifold-to-port area ratio of 1.1, compared to our "standard" set of data. Let it be assumed that all other data and the same exhaust and intake length dimensions are otherwise retained as "standard." If these data are inserted into the engine simulation, the effect on engine performance, as power gain or loss compared to the "standard" data, is as shown in Fig. 6.38.

The initial result is bad enough, with an average loss of some 5 kW throughout the power band; however, the longer-term design implications are worse. The overlap timing can be adjusted fairly simply with a new design for the camshaft. The valves can be enlarged, assuming they have not been too closely spaced initially, but this means that a new cylinder head casting will be needed, and that any previous, careful optimization of the water-coolant flow may well have come to naught. By sheer good fortune, the cylinder head casting now contains port-to-manifold dimensions that incorporate improved, but not optimized, C_m values. This may still entail new cylinder head castings in order to cure any holes in the power curve, but even that option assumes that the designer is aware of the underlying theory described here that governs the origins of the problems.

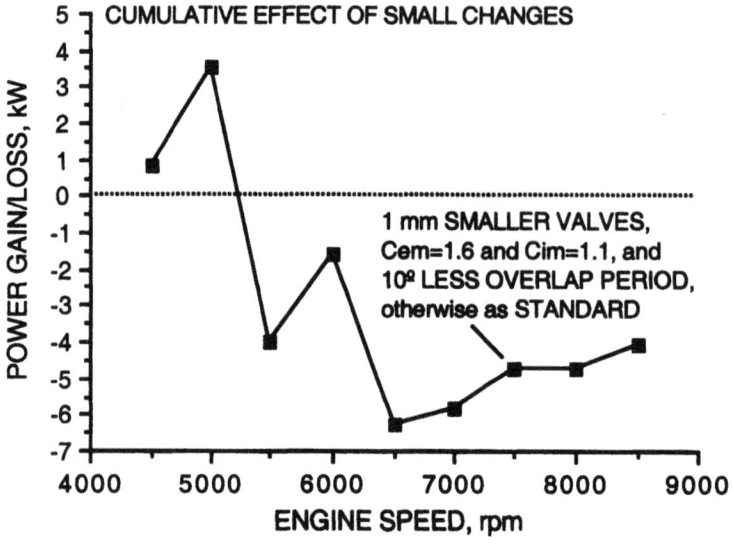

Fig. 6.38 Effect of under-optimization on the performance of the ITC engine.

As in this simple example, a lack of inital optimization in the design of a high-performance engine is clearly punitive in terms of both time and money, not to mention races narrowly lost.

The Relevance of the Empirical Criteria for Cylinder Head Design

The computational experiments confirm that the criteria expressed in Eqs. 6.1.4 to 6.1.19 are relevant. The correlation of the empiricism with the simulated outcome is not perfect, for if it were, there would be no need for either simulation or experimentation. Empiricism, by definition, is but a guide to simplify the path of data selection for the real design process by simulation and, ultimately, the development process on the dynamometer test-bed through logic-created experiments.

6.3 Empiricism for the Optimization of Intake System Tuning

A Computation to Examine the Fundamentals of Intake System Tuning

Intake ramming is clearly a vital issue for engine designers. Although an adequate intake ramming time-area gives the requisite breathing space to allow it to occur, a ramming pressure wave reflection must arrive at the intake valve at an appropriate juncture in order to accomplish this. This optimum phasing is provided by the length of the intake duct. That designers take this seriously can be seen from the illustration in Plate 6.1, which shows the upper end of the intake ducting of a V8 engine used for offshore powerboat racing. The disposition of each tract indicates that the designer has ensured that all of the intake tracts are of equal length. The EFI injectors are installed just below each intake pipe, and the common-pressure gasoline supply rail is visible running along the nearest bank of four cylinders.

Plate 6.1 The intake ducts and bellmouths of a V8 engine.
(Courtesy of Innovation Marine)

In the previous computation experiments on the ITC engine, several parameters within the cylinder head of the engine were changed and the effects on the engine performance characteristics were calculated. It is clear from Figs. 6.25 and 6.26, which are based on the "standard" input data, that the intake and the exhaust system are both well-optimized in terms of tuning characteristics. Therefore, if the intake tract length were to be changed and the simulated outcome observed in terms of change of power output, or delivery ratio, etc., it would be very difficult to come to absolute conclusions on the subject of intake tuning only, on the grounds that the interference of the exhaust system would cloud the issue. To surmount this problem, a single-cylinder test engine is created with an exhaust system that has a neutral, but effective, exhaust tuning behavior. Note that a similar logic issue will be faced in Sec. 6.4, where the fundamentals of exhaust tuning are examined and the problem is surmounted in a similar manner.

6.3.1 A Test Engine to Examine the Fundamentals of Intake System Tuning

A single-cylinder engine is created using one cylinder of the ITC unit, which means that the cylinder head data shown in Fig. 6.20 are still applicable as input data. I will refer to this engine as the "ITC" single. The intake and exhaust ducting are shown in Fig. 6.39, where it can be seen that the exhaust duct is but a 100 mm stub pipe, i.e., an effective—but extremely noisy—exhaust system with totally neutral tuning characteristics. Regarding the noise, I should know because I sat beside four of these on a Gipsy Major engine in the De Havilland Chipmunk aeroplane, RAF No. WZ845, in which I was taught to fly!

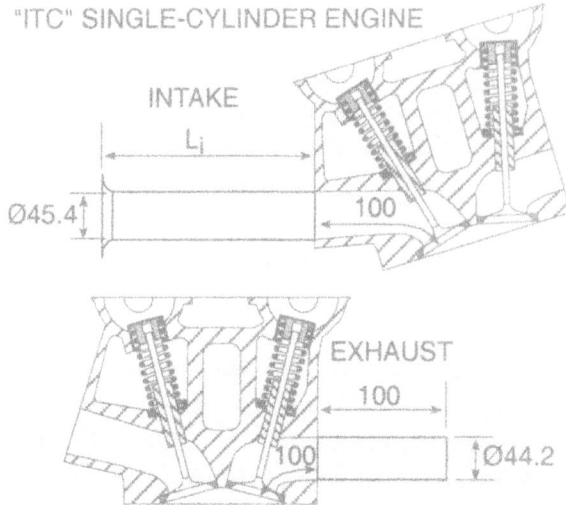

Fig. 6.39 The intake and exhaust ducting of the single-cylinder "ITC" test engine.

During the computer simulation at wide open throttle the following assumptions are made: the combustion of unleaded gasoline occurs at a constant air-fuel ratio of 12.5, which corresponds to a λ value of 0.881; the combustion process is as shown in Fig. 4.10 for a 4v na si engine with gasoline; the discharge coefficient maps for the valves and pipe ends are assumed to be as presented in Figs. 3.37 to 3.40 and 3.9-3.10, respectively; the discharge coefficients for unthrottled bellmouth-ended pipes are as quoted in Eqs. 3.3.1 to 3.3.4; the skin temperatures of the cylinder head, piston crown, and cylinder liner are 250, 200, and 150°C, respectively; the skin temperature of the stub exhaust duct surface is at 500°C; and the engine breathes to and from the atmosphere at 1.0133 bar and 25°C.

In Fig. 6.39, the intake system is composed of a duct with a constant diameter of 45.4 mm. It has a fixed length of 100 mm within the cylinder head and a variable length of duct, L_i, a length that is changed over a wide range in the computations. The total intake tract length, L_{it}, is simply related to the two length dimensions shown on Fig. 6.39.

$$L_{it} = 100 + L_i \quad \text{mm}$$

As the length, L_i, is varied in nine steps of 25 mm from 200 to 400 mm, the total intake tract length, L_{it}, is changed to 300, 325, 350, 375, 400, 425, 450, 475, and 500 mm, respectively. The engine speed range is investigated in 250 rpm steps from 4000 to 8750 rpm, i.e., in twenty individual computations for each of the nine pipe lengths, giving a total of 180 sets of engine simulation output data for torque, airflow, power, etc. You can become somewhat confused as you attempt to infer logic from this amount of simulation output. As a simple initial example of this confusion, consider Fig. 6.40, in which the prediction from the simulation of the delivery ratio of the engine with just three of the intake tract lengths, L_{it}, namely 300, 400 and 500 mm, is plotted.

For each of the three tract lengths there are three airflow peaks and three troughs. The peaks and troughs of the 300 and 400 mm intake tract lengths are, more or less, at the same engine speeds, but are very different from those of the airflow curve for the 500 mm tract length. The maximum delivery ratio is greater than 1.2, so the engine is well-tuned by its intake system, and the power and torque are well up to the standards of the four-cylinder ITC engine. Indeed, the basic shape of the DR curve for the four-cylinder ITC engine is not dissimilar to that for the "ITC" single, but the collector exhaust system clearly contributes another 10% more airflow above 7500 rpm.

However, as seen in Fig. 6.40, where a 300 mm and a 400 mm intake tract gives, more or less, the same answer in terms of peak airflow at an engine speed between 7500 and 7750 rpm, and the locations and magnitudes of the other peaks and troughs are quite similar, there appears to be little hope of assigning simple empirical logic to an intake ramming scenario.

6.3.2 The Dimensionless Intake Ramming Factor, C_{ir}

If the x-axis of the diagram, i.e., engine speed, is replaced by a dimensionless number in the form of an intake ramming factor, C_{ir}, a different picture emerges. The intake ramming factor, C_{ir}, is defined in terms of the engine speed, N, and the local reference speed of sound in the intake duct, a_0, as:

Intake ramming factor
$$C_{ir} = \frac{N \times L_{it}}{a_0}$$
(6.3.1)

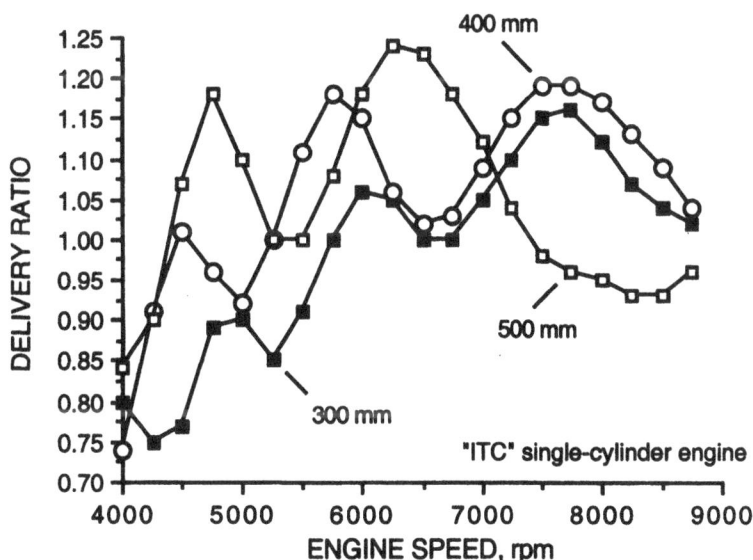

Fig. 6.40 Variation of airflow with intake tract length and engine speed.

The units of engine speed, N, are in rpm. The units of the intake tract length, L_{it}, are in mm. The units of the speed of sound, a_0, are in m/s. The origins of the right-hand side of Eq. 6.3.1 can be found in the formation of Eq. 6.4.3, where it can be seen to be related to the crankshaft angle, i.e., the time at a given engine speed required for a wave and its reflection to traverse a duct.

The engine speed axis in Fig. 6.40 is replaced by the intake ramming factor, C_{ir}. In this data reduction, the reference speed of sound, a_0, is arbitrarily assumed to be 350 m/s. With these corrections, the data of Fig. 6.40 are replotted in Fig. 6.41. To be truly dimensionless, and not a little pedantic, the numeric value displayed on the x-axis of Fig. 6.41 would need to be divided by 60,000; you may forget this number, as long as you remember the units required for the use of an intake ramming factor as defined for Eq. 6.3.1.

In Fig. 6.41, order is seen to be restored from the chaos of Fig. 6.40, which could have been so much worse if the other six pipe lengths had been added to it. Now that some logic is available, the delivery ratio characteristics with respect to engine speed, i.e., as the dimensionless intake ramming factor, for all nine pipe length factors are plotted in Fig. 6.42.

The same coding symbolism for the 300, 400, and 500 mm tract lengths, as used in Figs. 6.40 and 6.41, is repeated in Fig. 6.42. The peaks and troughs of the airflow for each of the nine intake tracts are now seen to virtually coincide. The ramming peaks, C_{rp}, are observed to be four in number and are labeled as C_{rp1}, C_{rp2}, C_{rp3}, and C_{rp4}. The ramming troughs, C_{rt}, are seen to be five in number and are labeled as C_{rt1}, C_{rt2}, C_{rt3}, C_{rt4}, and C_{rt5}. The ramming peak C_{rp1} has the highest airflow rate, and is regarded as the primary ramming peak. The numeric values of the intake ramming factor, C_{ir}, at the four ramming peaks, C_{rp1} to C_{rp4}, are found from Fig. 6.42 as 8900, 6600, 5150, and 4150, respectively. The numeric values of the intake ramming factor, C_{ir}, at the five ramming troughs, C_{rt1} to C_{rt5}, are found from Fig. 6.42 as 12,000, 7600, 5700, 4500, and 3650, respectively.

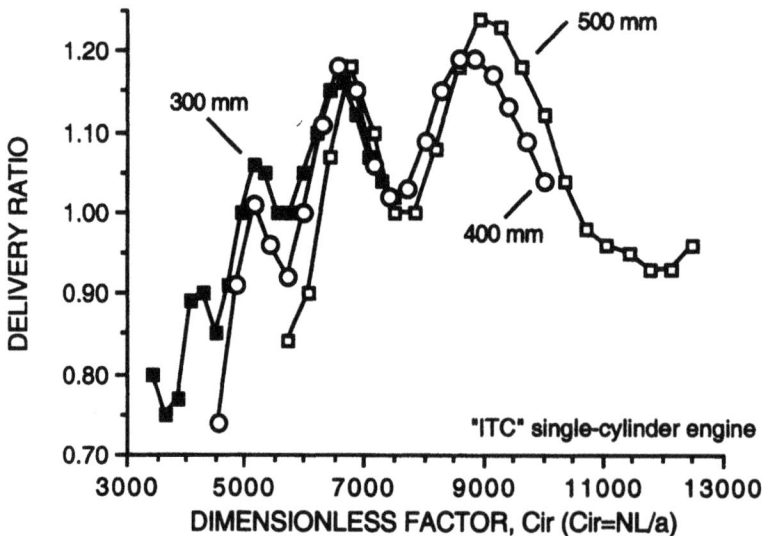

Fig. 6.41 Variation of airflow with the intake ramming factor in the "ITC" single.

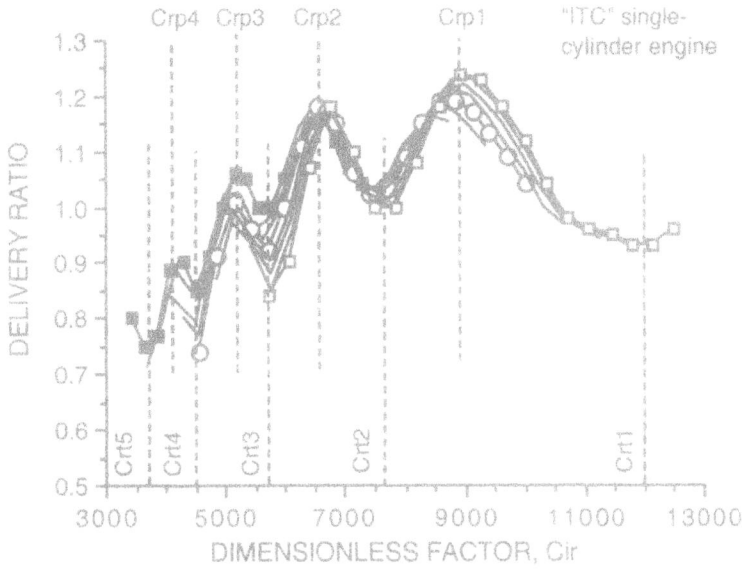

Fig. 6.42 Variation of airflow for all pipe lengths with the intake ramming factor.

6.3.3 Empirical Design Using a Dimensionless Intake Ramming Factor, C_{ir}

The C_{ir} criterion may now be used in reverse as a design process. Let us assume that intake ramming is to be optimized at a given engine speed, N_p, which is presumably selected either for peak power or for peak torque. A reasonable guess for the reference temperature, T_0, in the intake tract is always that in the atmosphere from which the engine breathes, or the best estimate of that in an intake plenum, should that be the air source, as is the case with the sports car in Sec. 5.9. The reference speed of sound in the intake duct air, a_0, at a reference temperature, T_0, is found from Eq. 2.1.10, as:

$$a_0 = \sqrt{\gamma R T_0}$$

Eq. 6.3.1 is translated and used to select the intake duct length, L_{it}, in terms of the intake ramming peak identified at the dimensionless intake ramming factor, C_{rp1}, which peak exhibits the highest delivery ratio in Fig. 6.42.

Optimum intake length: $$L_{it} = \frac{a_0 \times C_{rp1}}{N_p} \qquad (6.3.2)$$

The remaining intake ramming peaks and troughs logically follow the combined selections of the peak ramming speed, N_p, and the intake tract length, L_{it}. The engine speeds for the four ramming peaks are, N_p, N_{rp2}, N_{rp3}, and N_{rp2}, and the engine speeds for the five ramming troughs are, N_{rt1}, N_{rt2}, N_{rt3}, N_{rt4}, and N_{rt5}. These can be found from:

$$N_{rp1} = N_p \quad N_{rp2} = \frac{a_0 C_{rp2}}{L_{it}} \quad N_{rp3} = \frac{a_0 C_{rp3}}{L_{it}} \quad N_{rp4} = \frac{a_0 C_{rp4}}{L_{it}} \tag{6.3.4}$$

$$N_{rt1} = \frac{a_0 C_{rt1}}{L_{it}} \quad N_{rt2} = \frac{a_0 C_{rt2}}{L_{it}} \quad N_{rt3} = \frac{a_0 C_{rt3}}{L_{it}} \quad N_{rt4} = \frac{a_0 C_{rt4}}{L_{it}} \quad N_{rt5} = \frac{a_0 C_{rt5}}{L_{it}}$$

$$\tag{6.3.5}$$

6.3.4 The Relevance of Empiricism Using a Dimensionless Intake Ramming Factor, C_{ir}

The Ducati 955 Engine

Consider the case of the Ducati 955 motorcycle engine discussed in Sec. 5.7, which has a total intake tract length, L_{it}, of 293 mm. In Fig. 5.39, there are delivery ratio peaks at 10,750 rpm and 8250 rpm, with an observable DR trough at 9500 rpm and others that can be assumed to be at 7000 rpm, or below, and 12,000 rpm, or above. In short, for the Ducati engine, the known information is that N_p is 11,000, and N_{rp2} is 8000, whereas N_{rt1} is 12,000, N_{rt2} is 9500, and N_{rt3} is 7000. Even though the engine speed range, a full 5000 rpm, is very extensive for a high-performance engine, other ramming peaks cannot be seen, and even the locations of the upper and lower troughs are assumptions rather than precise observations.

Assuming a reference temperature, T_0, of 25°C, which from Eq. 2.1.10 gives a reference speed of sound, a_0, of 346 m/s, the application of Eqs. 6.3.2 to 6.3.5 predicts that the intake tract length, L_{it}, is 286 mm, that the ramming peaks are at 10,750, 7972, 6221, and 5013 rpm, respectively, and that the ramming troughs are at 14,494, 9180, 6885, 5435, and 4409 rpm, respectively.

The intake tract length is predicted with a precision of 97.6% and the main peaks and troughs are located quite accurately.

The Nissan Infiniti Engine

The case of the Nissan engine discussed in Sec. 5.8 is so similar to that of the Ducati, in terms of the engine speed that locates the peak intake ramming between 10,250 and 10,500 rpm, that I need not discuss it here, except to say that the application of Eqs. 6.3.2 to 6.3.5 predicts an intake tract length, L_{it}, of 293 mm, whereas it measures 305 mm, with the main peaks and troughs fairly accurately located.

6.3.5 The Awkward Case Observed in Fig. 6.40

Fig. 6.40 shows that there are two intake tract lengths, of 300 and 400 mm, that provide identical ramming peaks around 7500 rpm. It seems inexplicable that the application of Eqs. 6.3.2 to 6.3.5 can unravel this dichotomy. Before attempting to do so, the delivery ratio curves for the two pipe lengths, recorded for the "ITC" single, are extracted from Fig. 6.40 and replotted in Fig. 6.43. A small change has been made to the shorter pipe length of 300 mm: the computations are repeated for Fig. 6.43 with a length of 310 mm for the L_{it} dimension, which you will note barely affects the airflow, but ultimately simplifies the ensuing argument.

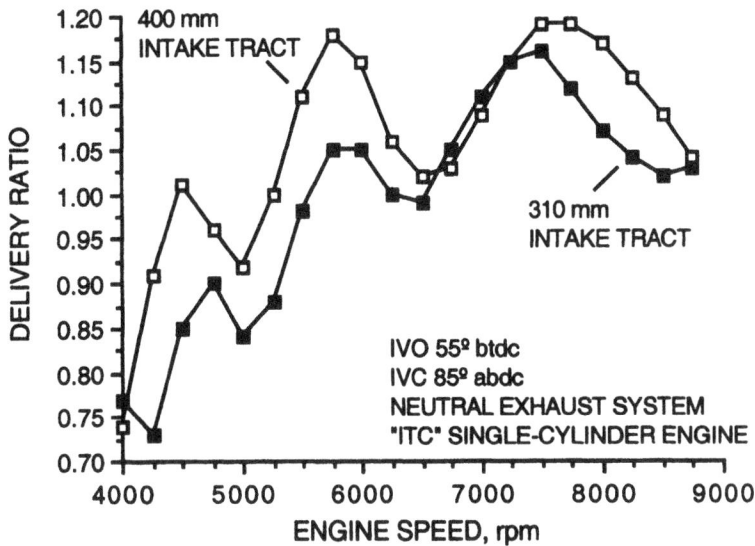

Fig. 6.43 The variation of airflow for intake tract lengths of 310 and 400 mm.

In Fig. 6.43, the maximum airflow peak for both pipe lengths is around 7500 rpm, with lesser ramming peaks at 5750 and 4500 rpm. Troughs exist above 8750 rpm and at 6500, 5000, and around 4000 rpm, again for both pipe lengths. However, the 400-mm pipe length gives superior airflow throughout the speed range.

These computations are repeated for the "ITC" single engine, but with reduced intake valve timings so that the intake valve opens at 40 °btdc and closes at 65 °abdc, compared to the "standard" timing of ivo at 55 °btdc and ivc at 85 °abdc. The airflow predictions from these calculations are plotted in Fig. 6.44. Around 7000 rpm, the situation is reversed in that the shorter pipe length of 310 mm is observed to give the higher airflow. At 6000 rpm, and below, the longer, 400-mm, intake pipe remains superior. The dichotomy observed in Fig. 6.40 has, apparently, been compounded, and it seems an even more remote possibility that Eqs. 6.3.1 to 6.3.5 can help explain the behavior involved. As has been the case throughout this book, the simulation output of the time-varying properties within the engine again provides the answers to apparent conundrums.

One matter, which I could prove using many more figures, is clarified through Figs. 6.43 to 6.44: The engine speeds for the ramming peaks and troughs are virtually unaffected by the different valve timings, so there is apparently at least some consistency in an otherwise confusing world!

Intake Pressure Diagrams at 7500 rpm with the "Standard" Intake Valve Timing

Figs. 6.45 and 6.46 show plots, at 7500 rpm, of the pressures in the cylinder and in the intake and exhaust ducts at the valves, when the two intake tract lengths of 310 and 400 mm are used as input data parameters to the simulation of the "ITC" single. Valve timings are "standard," i.e., as in Fig. 6.20. In these pressure plots, with the stub exhaust system fitted to the engine, the almost totally neutral tuning characteristic is very evident.

Fig. 6.45 shows that, with the 310 mm intake tract, the primary suction pulse drops to 0.6 atm at 400 °atdc and its ramming wave reflection exceeds the cylinder pressure until the intake valve is nearly closed. The ramming wave reflection propagates to and fro the bellmouth-open and the valve-closed ends for six traverses until it provides a compression pulse just as the intake valve opens again. A pressure diagram like this for optimized intake ramming has been seen before, in Fig. 5.23 for the G50 Matchless engine. The similarity I wish to highlight here is the number of multiple reflections of the intake ramming wave, i.e., three compression/expansion oscillations are visible from the six wave traverses of the intake tract.

Fig. 6.46 shows that, with the 400 mm intake tract, the primary suction pulse also falls to 0.6 atm at 400 °atdc and its ramming wave reflection equals the cylinder pressure until the intake valve is virtually closed. The ramming wave reflection propagates to and fro the bellmouth-open and the valve-closed ends for four traverses until it, too, provides a compression pulse just as the intake valve opens again. You have seen diagrams like this before at an optimum intake ramming speed, in Figs. 5.44 and 5.45 for the Ducati, and in Fig. 5.54 for the Nissan Infiniti engine. Here, the similarity I wish to highlight is the number of multiple reflections of the intake ramming wave, i.e., two compression/expansion oscillations are visible making the four traverses of the intake tract. The key issue, for both intake tract lengths, is that they provide good primary intake ramming and a secondary compression wave that coincides with the intake valve opening to prevent cylinder blowback. The 400 mm intake tract is known to give more airflow.

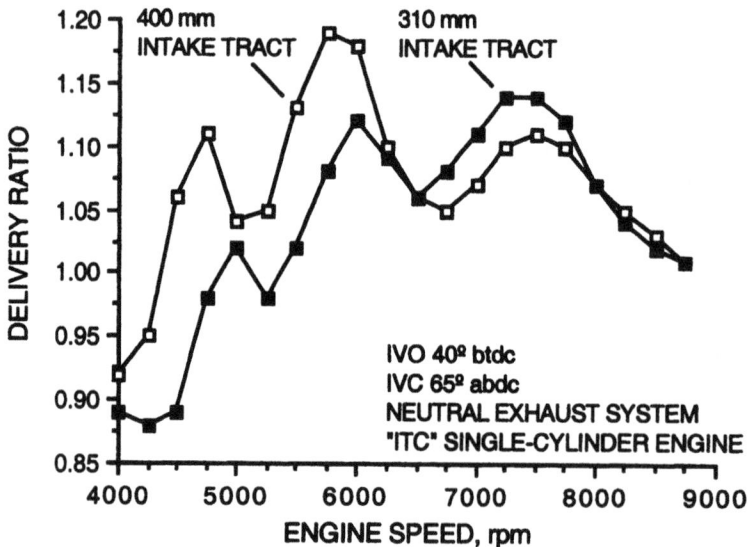

Fig. 6.44 Variation of DR with narrower valve timings for the two intake tracts.

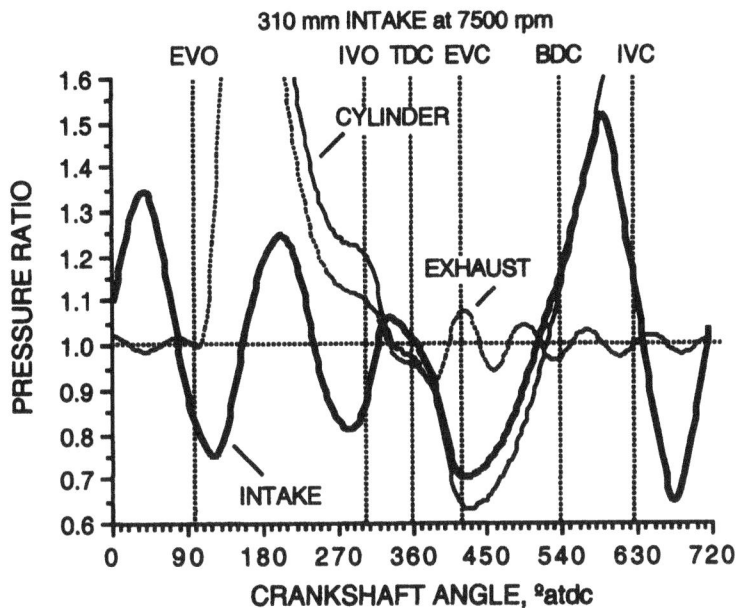

Fig. 6.45 Pressure diagrams at 7500 rpm for the 310 mm intake tract.

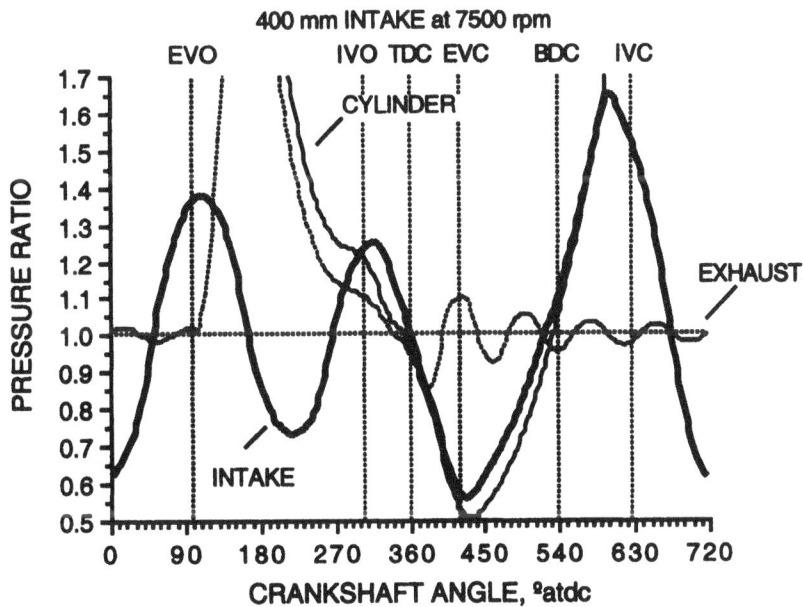

Fig. 6.46 Pressure diagrams at 7500 rpm for the 400 mm intake tract.

661

Figs. 6.45 and 6.46 are plotted with superposition diagrams, which have been shown previously to sometimes give potentially misleading information. Thus, to reinforce the issue, Figs. 6.47 and 6.48 have been prepared, which supplement the superposition information with the leftward and rightward pressure waves composing the superposition pressure trace at the intake valve. The valve is at the left-hand end of the duct.

As shown in Fig. 6.47, with a 310 mm intake tract, the suction (rightward) wave is now seen to be broader, deeper, and later than might be imagined from the superposition trace. Naturally, because this reflects at a bellmouth end, it gives rise to a much longer ramming (leftward) wave than is hinted at by the superposition signal. As the valve begins closing, and from bdc onward, this reflects as a compression (rightward) wave of increasing amplitude compounded by a little cylinder backflow. It is the final narrower segment of this (rightward) reflection that traverses to and fro the intake tract, changing sign at the open end and echoing off the closed end at the valve.

As shown in Fig. 6.48, with a 400 mm intake tract, the same process occurs, except the to-and-fro travel of the secondary ramming reflection along the intake tract occurs only twice. As the pressure reflections traverse the intake tract, wave energy is lost by friction and through the non-isentropic behavior of inflow and outflow at the open bellmouth end. Hence, as observed in Figs. 6.46 and 6.48 with the 400 mm tract, compared to Figs. 6.45 and 6.47 with a 310 mm intake duct, the amplitude of the secondary ramming wave at intake valve opening is higher, which more comprehensively prevents cylinder backflow and encourages intake inflow.

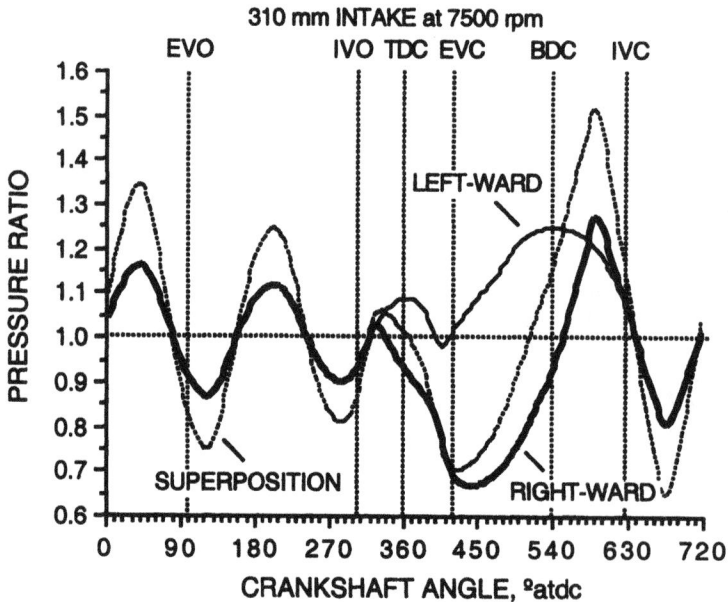

Fig. 6.47 Right and left pressure waves at 7500 rpm for the 310 mm intake tract.

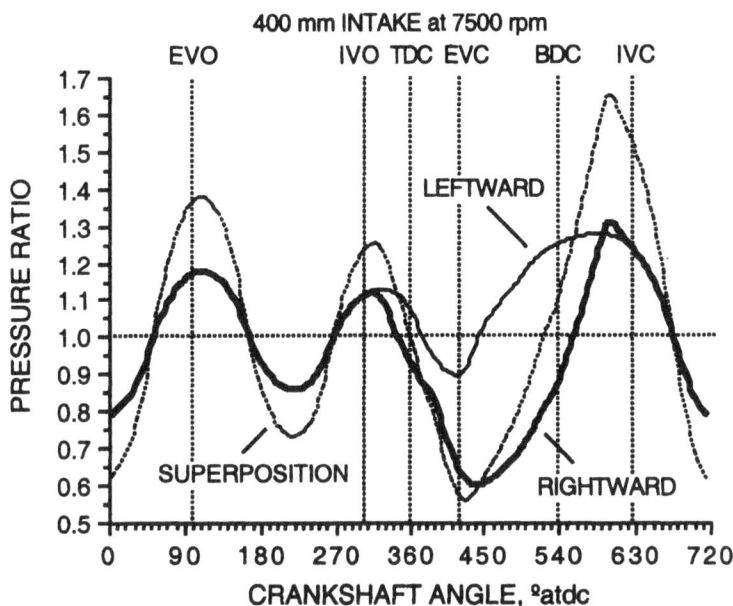

Fig. 6.48 Right and left pressure waves at 7500 rpm for the 400 mm intake tract.

The effect on the airflow particle velocities for the two intake tract lengths, 310 and 400 mm, at an engine speed of 7500 rpm, is shown in Figs. 6.49 and 6.50. In these figures, the gas particle velocity at the intake valve is plotted as Mach number, as defined in Eq. 2.1.23. Both figures exhibit good intake airflow characteristics, although that for the 400 mm intake tract is marginally superior, as Fig. 6.43 shows for the overall DR values. As shown in Fig. 6.50, with the 400 mm intake tract, inflow begins immediately as the intake valve opens. As shown in Fig. 6.49, with the 310 mm intake tract, inflow is delayed by some 20° crankshaft. Although cylinder backflow is observed in both figures during primary ramming toward intake valve closing, that for the 310 mm intake tract is greater, as the primary ramming wave seen in Fig. 6.47 arrives a fraction too early due to its shorter tract length. The outcome is a somewhat higher delivery ratio at 7500 rpm for a 400 mm intake tract when the valve timings are as "standard" for the ITC single engine.

Intake Pressure Diagrams at 7500 rpm with the Narrower Intake Valve Timings

As shown in Fig. 6.44, at 7500 rpm and with narrower intake valve timings, these airflow characteristics are reversed, i.e., the shorter, 310 mm intake tract has a higher delivery ratio than the longer, 400 mm intake duct. The pressure and particle velocity diagrams at 7500 rpm, shown in Figs. 6.51 to 6.53, reveal the origins of this effect.

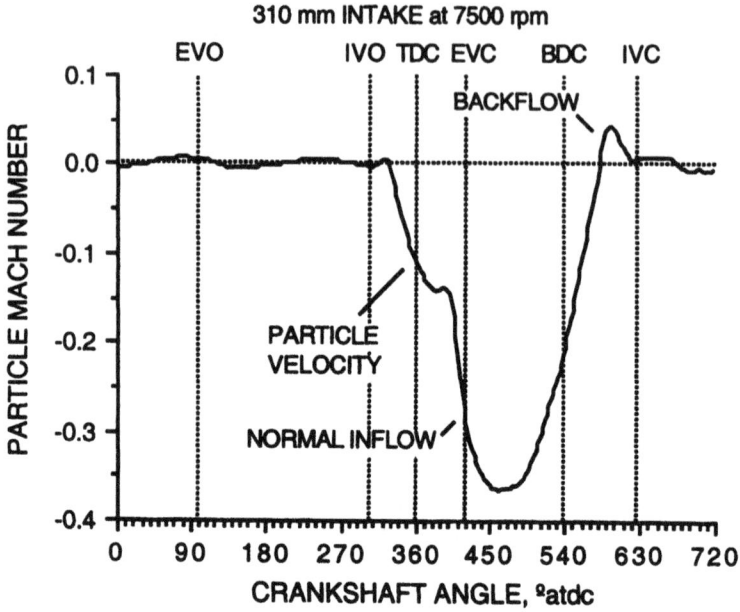

Fig. 6.49 Particle velocity at 7500 rpm for the 310 mm intake tract.

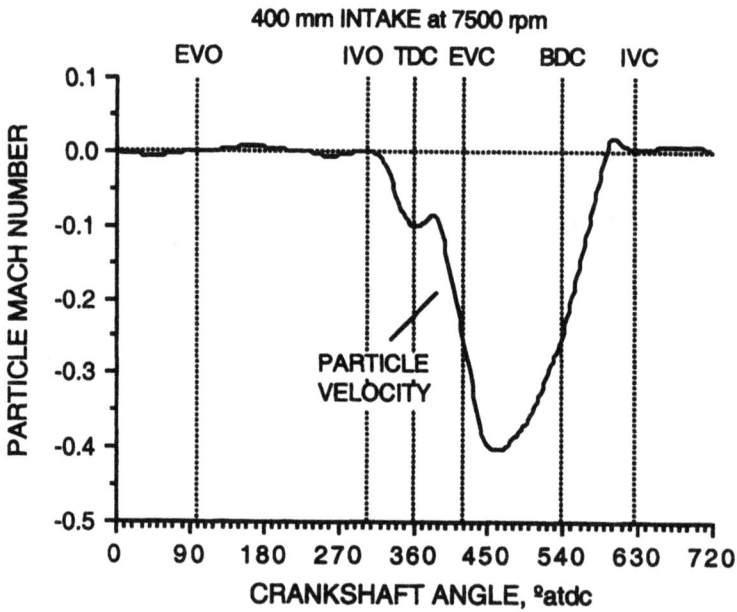

Fig. 6.50 Particle velocity at 7500 rpm for the 400 mm intake tract.

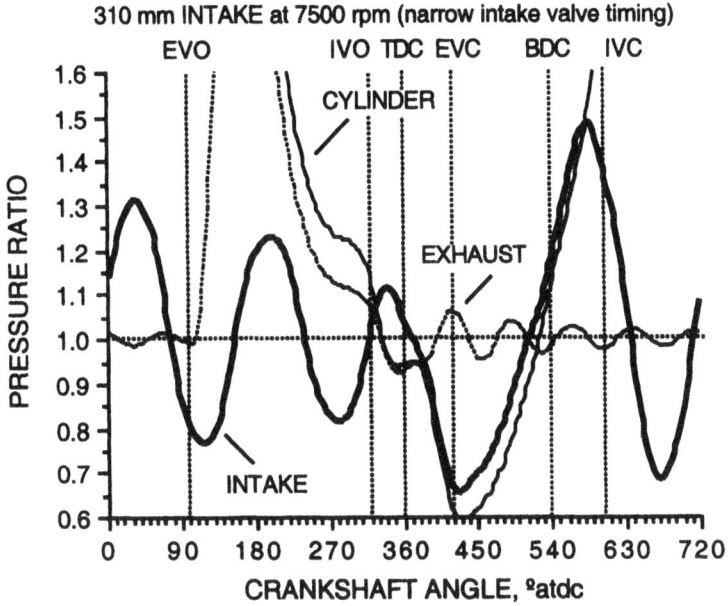

Fig. 6.51 Pressure diagrams for the 310 mm intake tract with narrower valve timings.

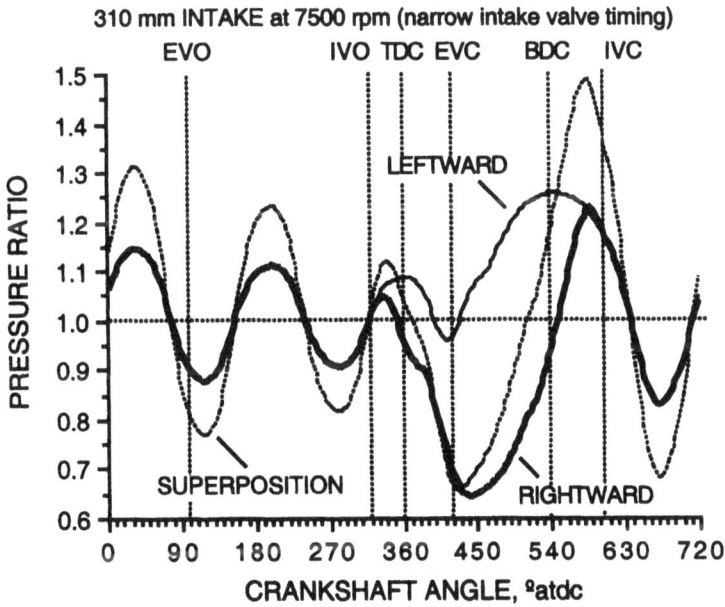

Fig. 6.52 Right/left waves for the 310 mm intake tract with narrower valve timings.

665

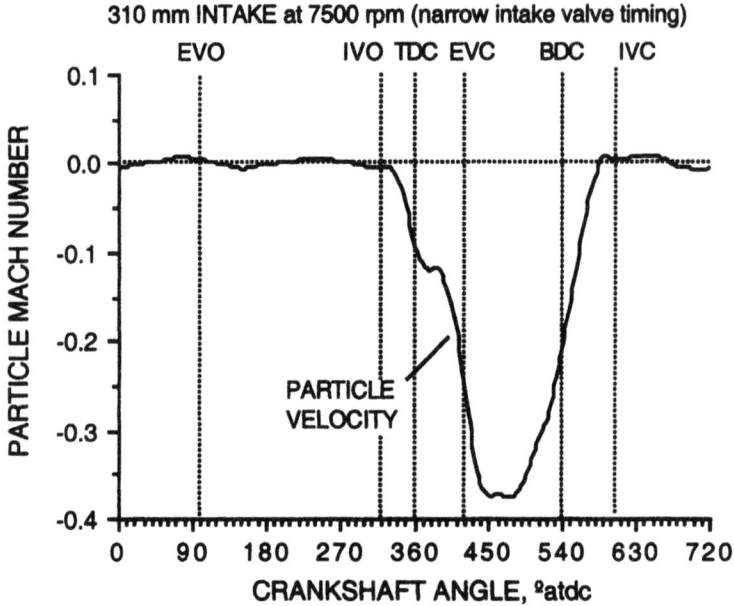

Fig. 6.53 Particle velocity for the 310 mm intake tract with narrower valve timings.

By comparing Figs. 6.51 and 6.45, it can be seen that with a 310 mm intake tract and with a narrower intake valve timing, the primary ramming pulse arrives at the intake valve at a more optimum juncture and prevents cylinder backflow in a positive manner. This is confirmed in Fig. 6.53, where no cylinder backflow occurs in the period before the intake valve closes, and all of the air that is breathed in is trapped. As shown in Fig. 6.52, the secondary ramming wave is not only stronger, but arrives at (a now later) intake valve opening and permits the air inflow to commence directly from that point onward. This, too, is confirmed by the particle velocity history in Fig. 6.53. As seen in Fig. 6.44, the overall outcome for a 310 mm intake duct produces a marginally higher DR value at 7500 rpm than that for the 400 mm intake tract. Furthermore, with narrower valve timings, the shorter tract length produces more airflow at its next lower ramming speed peak, i.e., at 6000 rpm, than it gave with the "standard" valve timings.

The Peaks and Troughs in the Airflow Characteristics
The Peaks
As seen in Fig. 6.43, the 400 mm intake tract has a second intake ramming peak at 5750 rpm. In Fig. 6.54, the pressures in the cylinder and in the intake and exhaust ducts at the valves are plotted at 5750 rpm, where the same intake tract length of 400 mm is used as input data to the simulation of the "ITC" single with the "standard" valve timings. It can be seen that there are now three double traverses of the intake duct by the secondary ramming wave before one of them coincides with intake valve opening. This is identical to the picture presented in

Fig. 6.45, or in Fig. 6.51, for a 310 mm intake tract at 7500 rpm. In short, the intake ramming peaks for the 400 mm long intake tract, which are at 7500, 5750, and 4750 rpm, are the first, second, and third ramming peaks; they are coincident with those defined by the dimensionless intake ramming factors as C_{rp1}, C_{rp2}, and C_{rp3}, in Eq. 6.3.4. On the other hand, for the 310 mm intake tract in Figs. 6.43 or 6.44, the intake ramming peaks at 7500, 6000, and 5000 rpm are the second, third, and fourth ramming peaks, as they are coincident with those defined by the dimensionless intake ramming factors, C_{rp2}, C_{rp3}, and C_{rp4}, in Eq. 6.3.4.

The Troughs

As shown in Fig. 6.43, the 400 mm intake tract has a ramming trough at 6500 rpm. In Fig. 6.55, the pressures in the cylinder and in the intake and exhaust ducts at the valves are plotted at 6500 rpm, in a simulation of the "ITC" single with "standard" valve timings and the 400 mm intake tract. It can be seen that there are now five traverses of the intake duct by the secondary ramming wave before a suction wave reflection coincides with the valve overlap period. Without the compression wave back-up that distinguishes good ramming at 5750 and 7500 rpm, considerable backflow occurs into the intake duct to disrupt the intake inflow process. The corresponding intake particle velocity diagram, Fig. 6.56, at the same engine speed of 6500 rpm, reinforces this statement as the backflow between intake valve opening and tdc is considerable and normal inflow does not commence until 380 °atdc, i.e., 20° after the local tdc point; even then it will commence by regurgitation of the exhaust gas backflow. The delivery ratio is deteriorated by the 18% drop from the surrounding peaks, as seen in Fig. 6.43.

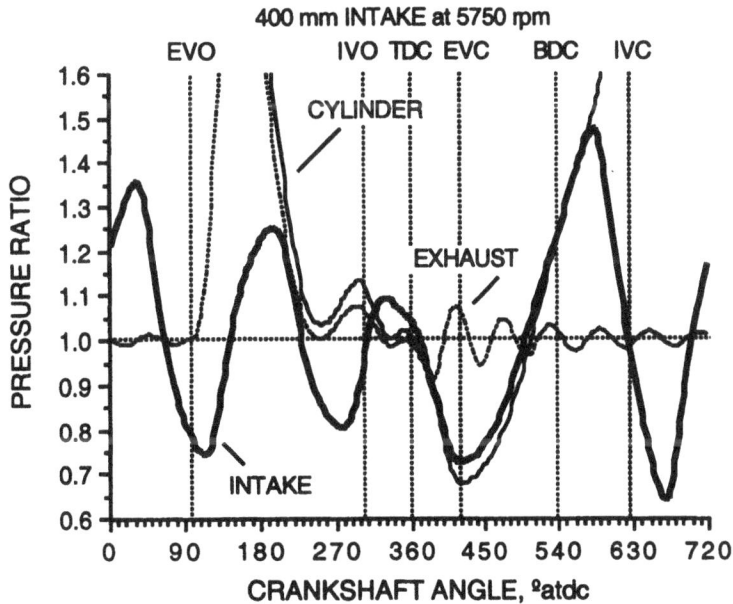

Fig. 6.54 Pressure diagrams at 5750 rpm for the 400 mm intake tract.

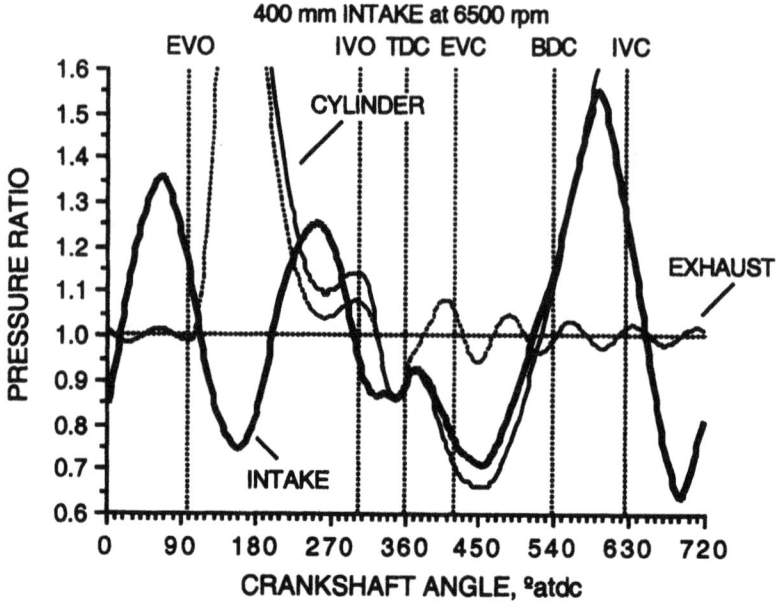

Fig. 6.55 Pressure diagrams at 6500 rpm for the 400 mm intake tract.

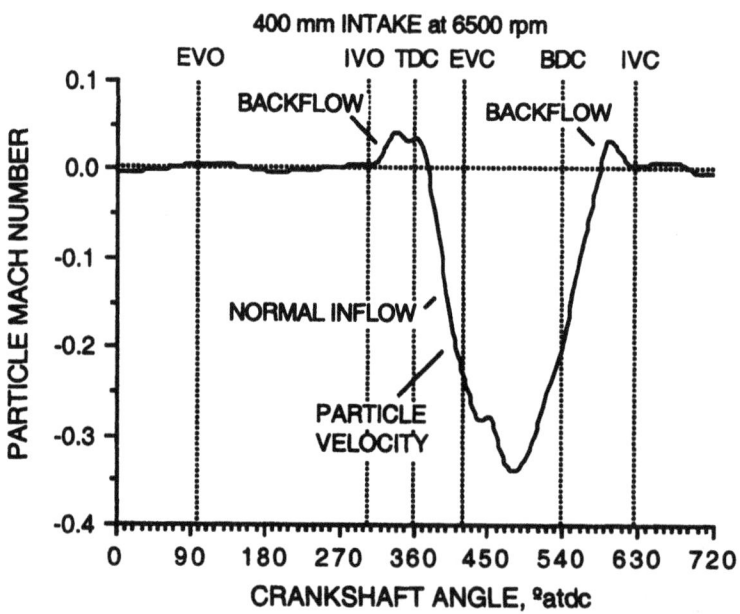

Fig. 6.56 Particle velocities at 6500 rpm for the 400 mm intake tract.

This ramming trough, i.e., the one that is located at an engine speed lower than that of the primary ramming peak, coincides with that defined as the second ramming trough, that defined by the dimensionless intake ramming factor as C_{rt2}.

Conclusions on the Awkward Case Observed in Fig. 6.40

Consider the empirical design of intake ramming for the "ITC" single engine and, by definition, for the ITC multi-cylinder engine, as both have intake duct lengths, L_{it}, of 400 mm. Using Eq. 6.3.2, with a reference speed of sound in air, a_0, at 25°C, of 346 m/s, the optimum intake ramming length, L_{it}, at 7500 rpm is given by:

$$\text{ITC intake length:} \quad L_{it} = \frac{a_0 \times C_{rp1}}{N_p} = \frac{346 \times 8900}{7500} = 410.6 \text{ mm} \tag{6.3.6}$$

This amounts to a length error of 2.6%, which is quite acceptable for an empirical solution for that data value. The application of the associated ramming factor equations, Eqs. 6.3.3 and 6.3.4, shows that, using a 410.6 mm intake length, the main intake ramming peaks are predicted at 7500, 5562, 4340, and 3497 rpm, respectively. They are observed on Fig. 6.43 as 7500, 5750, and 4500 rpm. The troughs are predicted as being at 10,112, 6404, 4803, 3792, and 3076 rpm, respectively. They are observed on Fig. 6.43 as being above 8750 rpm, at 6500 and 5000, and below 4000 rpm, respectively. The correlation is reasonably good—but then it ought to be because these same data contributed to the logic used in forming a dimensionless intake ramming factor!

Consider now the second case where an intake ramming design is required at 7500 rpm, but is optimized at that speed by the second ramming peak, which is known to be provided by an intake duct length, L_{it}, of 310 mm. In this case the application of Eq. 6.3.2 becomes:

$$\text{Intake length:} \quad L_{it} = \frac{a_0 \times C_{rp2}}{N_p} = \frac{346 \times 6600}{7500} = 304.5 \text{ mm} \tag{6.3.7}$$

This amounts to an error of 1.8% in the prediction of this shorter intake tract length, which is an acceptable error for an empiricism. Using the associated ramming factor equations, Eqs. 6.3.3 and 6.3.4, with a 304.5 mm intake length, the main intake ramming peaks are predicted at 7500, 5852, and 4716 rpm, respectively. They are observed on Fig. 6.43 as 7500, 5875, and 4750 rpm. The troughs are predicted as being at 8636, 6477, 5114, and 4148 rpm, respectively. They are observed on Fig. 6.43 as being at 8500, 6500, 5000, and 4250 rpm, respectively. If anything, the correlation for the engine speeds for these peaks and troughs is better than with the 400 mm intake tract length.

The basic conundrum is solved and the empirical relationships for intake ramming are shown to be relevant as design procedures. Only one niggling worry remains: When should an intake tract be designed to be optimized for intake ramming at the second peak, C_{rp2}, rather than at the primary peak, C_{rp1}? Although Figs. 6.51 to 6.53 have provided enlightenment, these questions are best answered by another look at that old classic, the G50 Matchless.

6.3.6 The Intake Ramming of the G50 Matchless

The intake ducting for the G50 is shown in Fig. 5.21, where the intake duct length, L_{it}, is seen to be 315 mm. From Fig. 5.20, the intake timings of this two-valve engine are known to be relatively narrow and the time-areas in Fig. 6.5 confirm that the intake pumping and ramming time areas, Φ_{ip} and Φ_{ir}, are among the lowest shown for any of the engines in either Figs. 6.4 or 6.5. In Fig. 5.25, the airflow ramming peaks are at 7000 and 5750 rpm, whereas the only observable DR trough on the narrow power band is at 6250 rpm. From Fig. 5.23, at 7000 rpm, the intake tuning is seen to be optimized for the second ramming peak, i.e., there are three double traverses of the intake tract by the reflection of the primary ramming wave before the arrival of a compression reflection at intake valve opening. Using Eq. 6.3.2 in the same manner as described above, the two choices for the intake tract length, L_{it1} and L_{it2}, can be calculated, with a reference speed of sound in air, a_0, at 15°C, of 340 m/s, i.e., taking into account the cooler British climate where the G50 was raced!

$$L_{it1} = \frac{a_0 \times C_{rp1}}{N_p} = \frac{340 \times 8900}{7000} = 432.3 \text{ mm} \quad L_{it2} = \frac{a_0 \times C_{rp2}}{N_p} = \frac{340 \times 6600}{7000} = 320.6 \text{ mm}$$

$$(6.3.8)$$

The shorter length, L_{it2}, corresponds well with the measured value of 315 mm, with a 1.8% error. Using the associated ramming factor equations, Eqs. 6.3.3 and 6.3.4, with a 320.6 mm intake length, the main intake ramming peaks are predicted at 7000, 5462, and 4402 rpm, respectively. The troughs are predicted as being at 8061, 6045, 4773, and 3871 rpm, respectively. This correlates reasonably well with the DR profile presented in Fig. 5.25.

Although I have not shown the computations here, the DR and—almost by definition—the power and torque levels of the engine are lowered considerably when the intake tract length is lengthened to 432 mm within the simulation of the G50 engine. I would not be at all surprised if Jack Williams [6.17] knew that as well!

6.3.7 The Cross-Over Point in Intake Ramming Design

Where, then, lies the cross-over point for the design of an intake tract length using the secondary, C_{rp2}, factor, rather than the primary, C_{rp1}, factor? The honest answer is that it is a grey area. Based on the evidence above, such as from the G50 Matchless, and the example of the 310 mm intake pipe on the "ITC" single engine with the narrower intake valve timings, the only tip I can pass on to you is to design for an intake tract length that is based on the intake ramming dimensionless factor, C_{rp2}, if either of the following criteria are satisfied:

(i)   The engine is a two-valve design and the intake valve closing timing, ivc, is less than 75 °abdc.

(ii) The engine is a four-valve design and the intake ramming time-area, Φ_{ir}, is less than 0.00015 s/m at the peak power point.

My best advice is always to calculate both intake tract lengths, L_{it1} and L_{it2}, insert both data values into an accurate simulation model of the engine in question, and, by varying each length around its respective datum, let the engine simulation provide the solution to this "grey-area" problem.

6.4 Empiricism for the Optimization of Exhaust System Tuning

6.4.1 The Input Data to the "ITC" Single Engine to Examine Exhaust System Tuning

To investigate exhaust system tuning, the "ITC" single-cylinder test engine is employed with intake and exhaust ducting as shown in Fig. 6.57. It can be seen that the total intake duct length is the same as in the ITC four-cylinder engine. During the computer simulation at wide open throttle, the same assumptions are made within the simulation as in Sec. 6.3.

As shown in Fig. 6.57, the exhaust system is composed of a duct that has a constant diameter of 44.2 mm, a fixed duct length of 100 mm, and a variable duct length, L_e, which is changed over a wide range in the computations. The total intake tract length, L_{et}, is simply related to the two length dimensions shown in Fig. 6.57.

$$L_{et} = 100 + L_e \quad \text{mm} \tag{6.4.1}$$

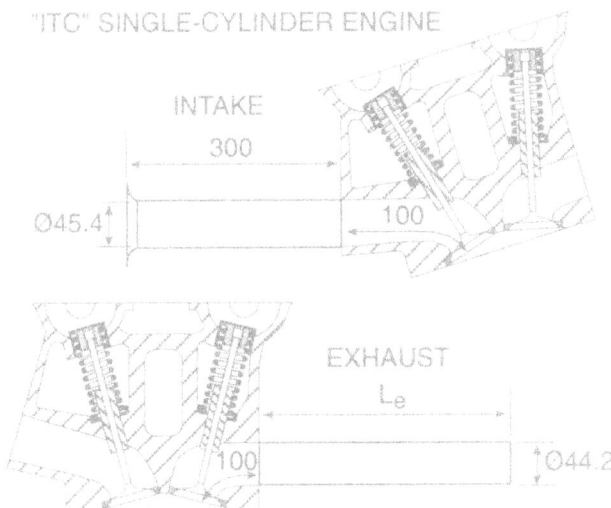

Fig. 6.57 The ducting of the "ITC" single to investigate exhaust tuning.

As the length, L_e, is varied in seven steps of 50 mm from 600 to 900 mm, the total exhaust tract length, L_{et}, is changed to 700, 750, 800, 850, 900, 950, and 1000 mm. The engine speed range is investigated in 500 rpm steps from 4000 to 9000 rpm, i.e., in eleven individual computations for each of the seven pipe lengths, giving a total of seventy-seven sets of engine simulation output data for torque, airflow, power, etc. As with the intake investigation, you can become somewhat bemused as you attempt to infer logic from this amount of simulation output.

6.4.2 The Fundamentals of Exhaust System Tuning

To set the scene for this investigation, the performance characteristics of a straight exhaust pipe of length, L_{et}, of 800 mm are compared with (a) those of the stub exhaust pipe where the intake duct length, L_{it}, is 400 mm and (b) with those of the 4-1 collector exhaust system of the ITC four-cylinder engine. In this simulation of the ITC four-cylinder engine, the assumptions regarding fueling and combustion are made identical to those for the other two engines to make the comparisons more relevant, although the ensuing performance characteristics that are computed are not quite the same as shown in Figs. 6.22 to 6.23.

Figs. 6.58 to 6.60 show plots comparing the performance characteristics for delivery ratio, charging efficiency, and torque output, respectively. Torque output is shown as brake mean effective pressure, bmep.

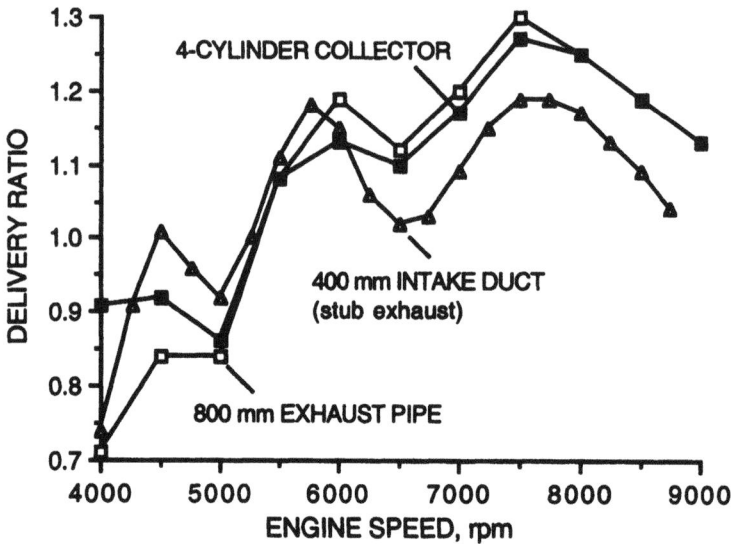

Fig. 6.58 Airflow with a stub, a straight pipe, and a collector exhaust system.

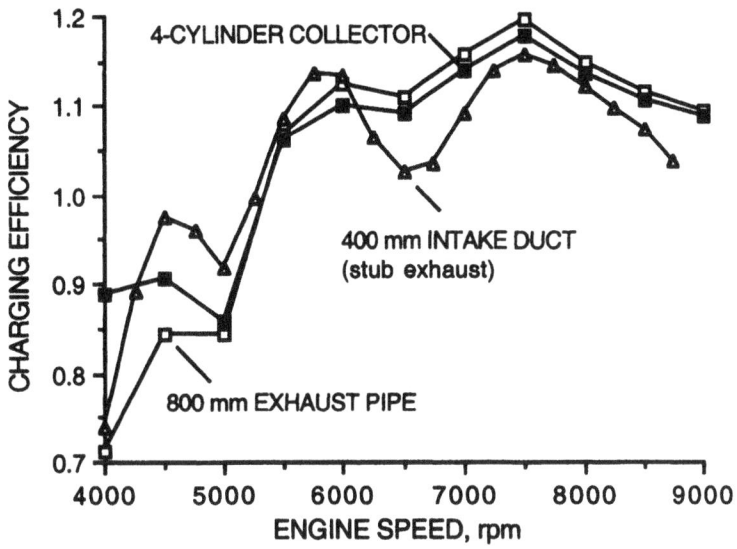

Fig. 6.59 Charging efficiency with a stub, a straight pipe, and a collector exhaust system.

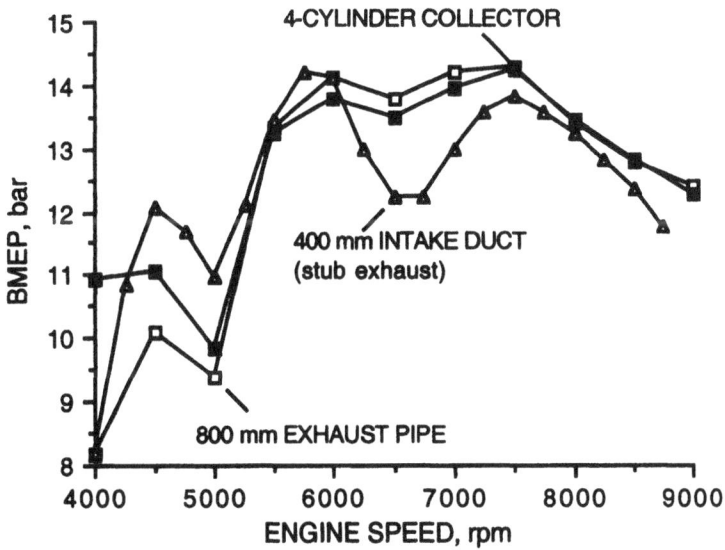

Fig. 6.60 Torque output with a stub, a straight pipe, and a collector exhaust system.

The "ITC" single fitted with the stub exhaust and the 400 mm tuned intake produces, in all three figures, the triple peaks associated with the ramming peaks and troughs discussed in Sec. 6.3. In the same figures, the action of the straight pipe and the collector exhaust systems is observed to enhance and extend the air flow and torque over the upper end of the speed range. The most important task of the optimized exhaust system is to fill in the trough left by the intake ramming process in the middle of the airflow and torque curves. Upon examining Figs. 5.38 and 5.39, for the Ducati 955 engine, you can see that its tuned exhaust system attempts to accomplish this same task, but apparently a gain of lower-speed torque is preferred to a higher peak power, and the ramming trough at 9500 rpm is not so well filled.

In Figs. 6.58 to 6.60, the straight exhaust pipe and the collector system are seen to perform in a very similar manner. The straight pipe has a length, L_{et}, of 800 mm. In the collector system, as sketched in Fig. 6.21, the primary exhaust pipe length, L_e, is seen to be 700 mm, with a manifold tract length of 110 mm, giving a total length, L_{et}, of 810 mm to the five-way branch at the collector. Because the branch location and the tailpipe design create a sort of "mini-atmosphere," the similar performance characteristics seen in Figs. 6.58 to 6.60 should not come as a total surprise. The characteristics of this "mini-atmosphere" and the tuning action of the branch and the collector tailpipe are discussed in Sec. 6.4.6.

In Figs. 6.58 to 6.60, perhaps the surprise is the level of performance tuning available from a simple, straight exhaust pipe. Apart from the speed region below 5000 rpm, where an exhaust stub pipe is superior, the torque and power available from the straight pipe are as good as, if not better than, a 4-1 collector system. However, I must caution you against assuming generalities on the basis of this single piece of information. Nevertheless, for engines for which there is no exhaust silencing requirement, or for those with asymmetrical firing, the use of a separate pipe per cylinder may be well worth investigating, assuming that their undeniable extra bulk can be packaged in a vehicle (see Plate 5.12).

It is against this backdrop that the following discussion on the exhaust tuning of a high-performance, naturally aspirated, spark-ignition engine is conducted. The first priority for any tuning is to optimize the intake system; an explanation of how to do this was presented in Sec. 6.3. The second priority for tuning is to design an exhaust system to eliminate the ramming trough(s) that it leaves behind and enhance the ramming peaks that it provides.

6.4.3 The Effect of Changing the Exhaust Pipe Length on the "ITC" Single

As described in Sec. 6.4.1, the exhaust pipe on the "ITC" single is straight and has a varying length, L_{et}, ranging from 700 to 1000 mm. The effect of these several exhaust pipes on the simulated performance characteristics of airflow and torque are shown in Figs. 6.61 to 6.62. Airflow is plotted as delivery ratio. The torque output is shown as brake mean effective pressure, bmep, and the y-axis of the plot is curtailed to illustrate more clearly the bmep variations in the area of the intake ramming trough at 6500 rpm.

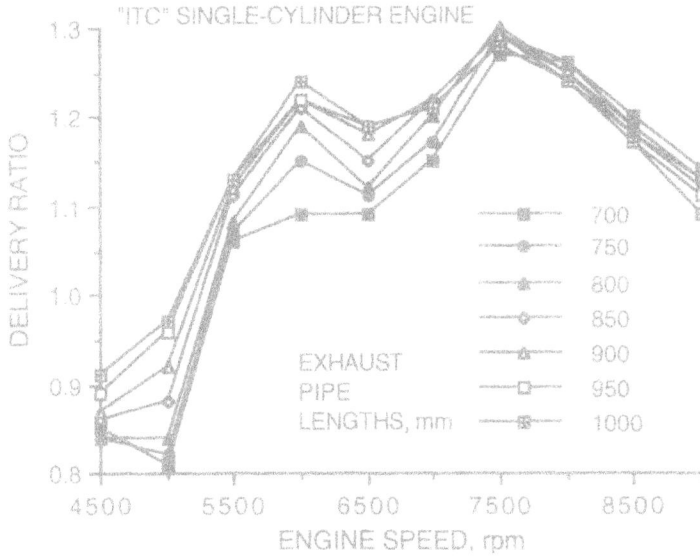

Fig. 6.61 Airflow with differing exhaust pipe lengths on the "ITC" single.

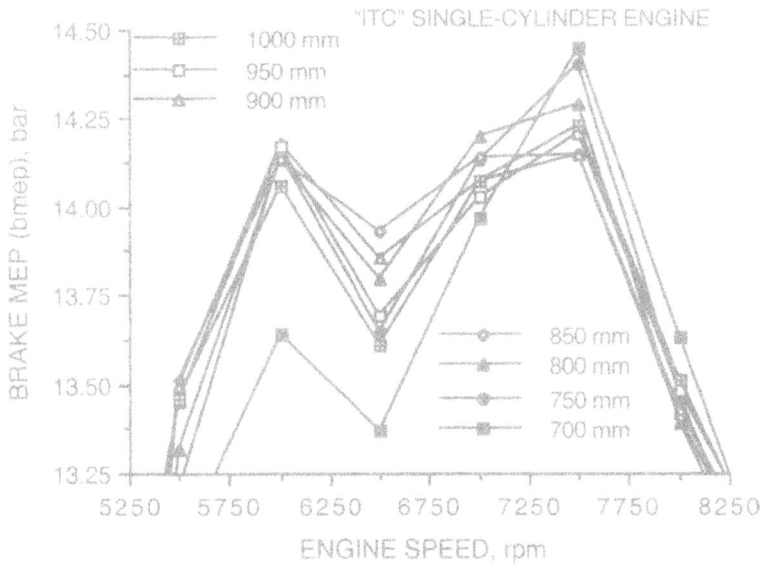

Fig. 6.62 Torque output with differing exhaust pipe lengths on the "ITC" single.

675

From the delivery ratio, in Fig. 6.61, it is quite clear that the shortest, 700 mm, exhaust pipe gives the lowest air flow at 6500 rpm and the highest at 7500-8000 rpm. The translation to torque, in Fig. 6.62, shows the same trends. If the highest maximum power is the aim, and the remainder of the power curve is of no concern, then this is the best exhaust pipe length.

As shown in Fig. 6.61, the longest, 1000 mm, exhaust pipe gives the highest air flow below 6500 rpm and the lowest above 7500-8000. However, as seen in Fig. 6.62, this does not translate into torque output, for it is the second worst performer in the torque hole at 6500 rpm and the torque peak at 6000 rpm. Clearly, only pumping or trapping losses can account for this behavior.

The exhaust pipe lengths with the best overall performance characteristics, either in terms of airflow or torque, are those with 800, 850, and 900 mm lengths. In particular, these comments apply to the 6500 rpm region where the exhaust pipe tuning must be good to enable the engine to recover from the intake ramming trough at that speed. The exhaust pipe with the worst performance characteristics in this regard is the 700 mm exhaust pipe. In order to understand the tuning mechanisms involved, the temporal variations of pressure, purity, particle velocity, etc. provided by the simulation at 6500 rpm are examined for (one of) the "good" and the "bad" exhaust pipes, i.e., the 900 mm and the 700 mm exhaust pipes, respectively.

6.4.4 Good and Bad Exhaust Tuning at 6500 rpm in the "ITC" Single

In Figs. 6.63 and 6.64, the pressures in the cylinder and in the exhaust duct at the exhaust valve, at 6500 rpm, are plotted for the 700 and 900 mm straight exhaust pipes, respectively.

Also drawn are the leftward and rightward pressure waves in the exhaust, the combination of which produces the superposition trace. The basic action of a tuned exhaust pipe is evident in both figures, in that the exhaust pulse has traversed the pipe and reflected at the open end to provide an expansion wave at the exhaust valve during the valve overlap period. This suction wave in the 700 mm pipe has arrived a little earlier than in the 900 mm pipe, but this is to be expected from the shorter pipe. The outcome, in terms of the ensuing intake flow, is very different. The problem is not the usual one, i.e., cylinder backflow into the intake tract reducing the airflow mass inhaled during the ensuing intake stroke, but is backflow from the exhaust duct into the cylinder. This can be seen, albeit not too clearly, in Fig. 6.63, where the exhaust pressure exceeds the cylinder pressure after tdc. By comparison, in Fig. 6.64, for the 900 mm duct, the exhaust pressure is always less than the cylinder pressure. To clarify this further, Fig. 6.65 is drawn for the valve overlap period only and shows the intake, cylinder, and exhaust pressures at 6500 rpm for both exhaust pipes.

At this speed, the intake ramming is in its "trough" mode, so the intake duct pressure is under atmospheric pressure. Both exhaust systems also provide sub-atmospheric pressure wave reflections at the cylinder, but that for the 700 mm pipe has arrived too early, and, between tdc and exhaust valve closing, is greater than either the intake or the cylinder pressure. For the 700 mm

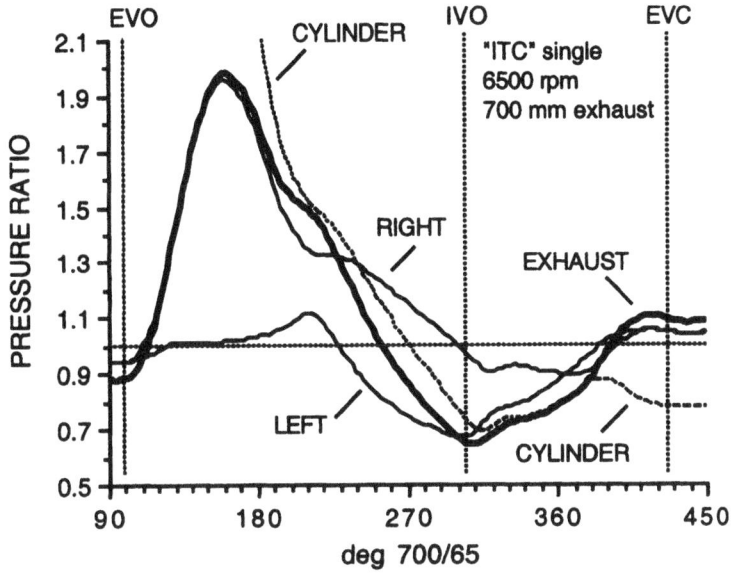

Fig. 6.63 Pressure diagrams at 6500 rpm for the 700 mm exhaust pipe.

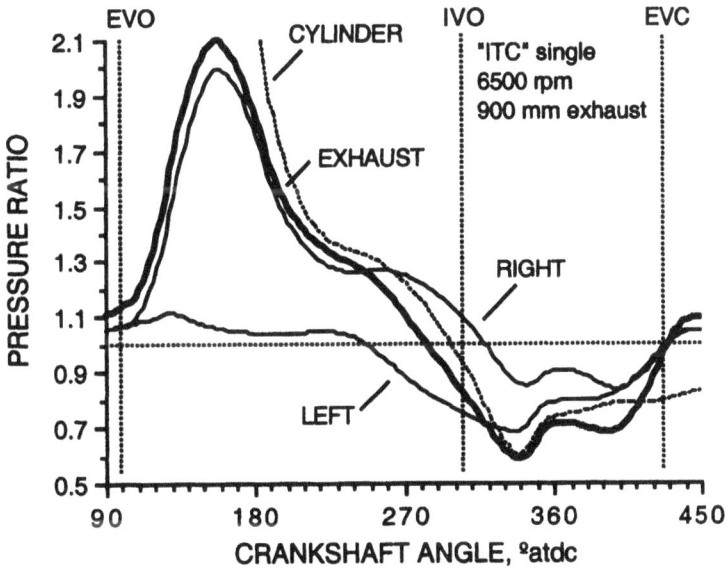

Fig. 6.64 Pressure diagrams at 6500 rpm for the 900 mm exhaust pipe.

677

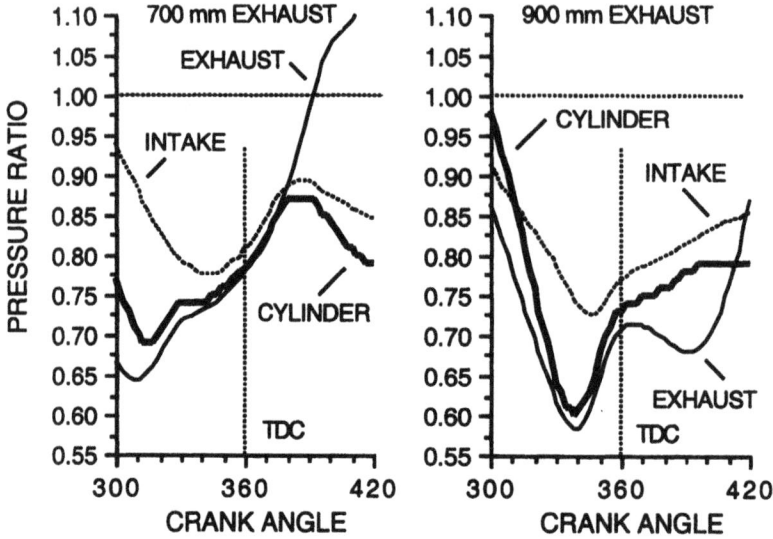

Fig. 6.65 Intake, cylinder, and exhaust pressures at 6500 rpm for both exhaust pipes.

pipe, the consequence is that exhaust gas flows back into the cylinder from tdc onward. For the 900 mm pipe, the phasing is better and this backflow is absent. This is confirmed in Fig. 6.66, which shows plots equivalent to those in Fig. 6.65, but for the gas particle velocities in the intake and exhaust ducts at the valves.

As shown in Fig. 6.66, for the 700 mm pipe, the exhaust outflow at the exhaust valve ceases at tdc and becomes backflow until evc, whereas for the 900 mm exhaust pipe it is almost continuously outflow during the entirety of the valve overlap period. At the intake valve, for both pipes, there is no evidence of cylinder backflow at any stage in the process, and, by the time the exhaust valve closes, the intake air particle velocities are almost equal. However, during valve overlap, the intake airflow process starts off much more rapidly with the 900 mm exhaust pipe, so that by evc considerably more air mass has been ingested. This point is made more obvious by the plot of dynamic cylinder air mass, as delivery ratio, shown in Fig. 6.67.

The early start to cylinder inflow given by the 900 mm exhaust pipe is confirmed in Fig. 6.67, for by exhaust valve closing the delivery ratio is 0.280, whereas it is 0.201 with the 700 mm exhaust pipe because the cylinder cannot remove the exhaust backflow. The DR advantage for the 900 mm exhaust pipe is maintained as it rises to a peak of 1.234 and finally falls to 1.178. The equivalent DR numbers for the 700 mm exhaust pipe are a peak of 1.127 and a concluding value of 1.088. The 8% EGR permanently ingested by the 700 mm exhaust pipe between tdc and evc can be seen in both the peak DR and the final trapped values.

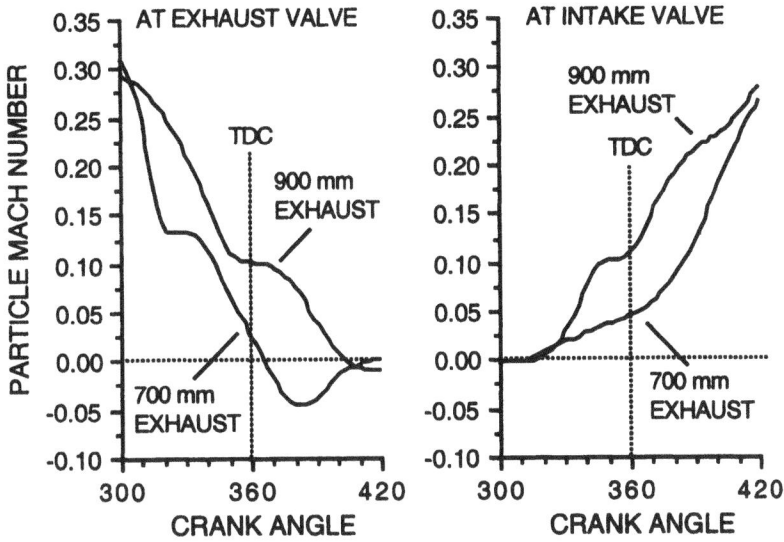

Fig. 6.66 Intake and exhaust particle velocities at 6500 rpm for both exhaust pipes.

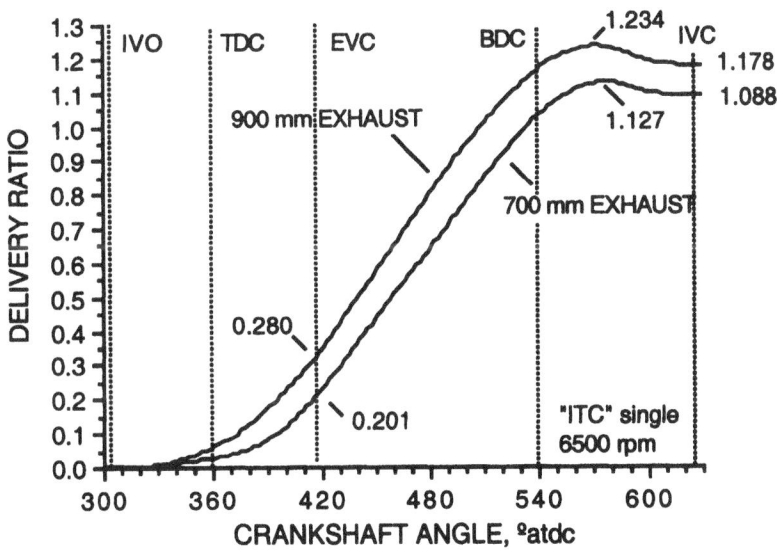

Fig. 6.67 Delivery ratio at 6500 rpm for both exhaust pipes.

679

In terms of pressure wave phasing, the reasons that a 700 mm exhaust pipe provides indifferent exhaust tuning and 800, 850, or 900 mm pipes provide better tuning, are now understood. But can empiricism provide a simple theory to calculate the "correct" exhaust pipe length at a desired engine speed? If so, could this allow us to limit the amount of data inserted into, and analyzed by, the engine simulation?

6.4.5 Empirical Theory for the Length of the Primary Exhaust Pipe

This tuned length, L_{et}, from the exhaust valve to a major expansion in area such as the atmosphere, a plenum, a branch, or a branched collector junction, clearly phases the exhaust tuning in the manner described in Sec. 6.4.4.

Empiricism for the Primary Exhaust Length Based on a Fundamental Approach

It should be possible to deduce a simple tuning formula for this length. Let us use the same, simple estimation technique that proved so effective on the single pulse rig, in Sec. 2.19.2(i). The time t equivalent to a crankshaft angle θ, taken for a double traverse of any pipe of length L, at an engine speed N, at an average propagation velocity a, is given by:

$$t = \frac{2 \times L}{1000 \times a} = \frac{\theta}{360} \times \frac{60}{N} \quad s \qquad (6.4.2)$$

For convenience, the units of length are in mm, engine speed is in rpm, and propagation velocity is in m/s.

From Fig. 6.63, it is clear that the peak of the exhaust pulse passed the exhaust valve at 161 °atdc and returned at 301 °atdc, giving a return angle, θ_{er}, of 140° crank angle. The pipe length, L_{et}, is 700 mm and the engine speed is 6500 rpm.

From Fig. 6.64, it can be seen that the peak of the exhaust pulse passed the exhaust valve at 159 °atdc and returned at 335 °atdc, giving a return angle, θ_{er}, of 176° crank angle. The pipe length, L_{et}, is 900 mm and the engine speed is still 6500 rpm.

Eq. 6.4.2 can be reworked in terms of the variables in this case to read as an empirical relationship for this pipe length:

$$L_{et} = \frac{500 a_e \theta_{er}}{6N} \quad mm \qquad (6.4.3)$$

Estimating a mean temperature, T, of 535°C deduced from the computed temperature time-history at 6500 rpm at the exhaust transducer, and assigning values for the ratio of specific heats, γ, at 1.27 and a gas constant, R, at 300 J/kgK for the exhaust gas, gives a reference speed of sound, a_e, as,

$$a_e = \sqrt{\gamma RT} = \sqrt{1.27 \times 300 \times 808} = 555 \quad m/s \qquad (6.4.4)$$

The simple theory behind the use of the reference speed of sound as the average propagation velocity is that the compression exhaust pulse travels supersonically and the expansion reflection propagates subsonically, so the average normally approximates to the reference speed of sound (viz., a previous discussion in Sec. 2.1.4).

Taking the known data for the two pipes, and inserting them into Eq. 6.4.3, gives a prediction for the pipe lengths, L_{et}, which are known to be 700 and 900 mm, respectively.

$$L_{et} = \frac{500 \times 555 \times 140}{6 \times 6500} = 996 \text{ mm} \qquad L_{et} = \frac{500 \times 555 \times 176}{6 \times 6500} = 1252 \text{ mm} \qquad (6.4.5)$$

The gross disparities are hardly an inspirational outcome for empiricism, as the values are about 14% too long, respectively, and are particularly useless because it is already known that a pipe length of 1000 mm is inferior to most other pipe lengths from a tuning standpoint (see Fig. 6.62 at 6500 rpm).

What is wrong with our solution? It worked perfectly well in Sec. 2.19.2(i). The only parameter that is not a known fact in Eq. 6.4.5 is the average propagation velocity, a_e. The correct answers to Eq. 6.4.5 are known, so what is the "correct" value of the average propagation velocity, a_e? Forcing the above solution to fit the facts shows:

$$L_{et} = \frac{500 \times "390" \times 140}{6 \times 6500} = 700 \text{ mm} \qquad L_{et} = \frac{500 \times "399" \times 176}{6 \times 6500} = 900 \text{ mm} \qquad (6.4.6)$$

From Eq. 6.4.6, the average propagation velocity, a_e, would appear to be between 390 and 399 m/s, and taking 395 m/s as a rough average, this corresponds to an exhaust gas temperature, calculated as:

$$T_e = \frac{a_e^2}{\gamma R} = \frac{395^2}{1.25 \times 300} = 416 \text{ K or } 143°C \qquad (6.4.7)$$

Because the exhaust gas in the exhaust duct is at least 530°C for most of this wave propagation process, the very low temperature predicted by Eq. 6.4.7, implying a very low average propagation velocity, makes very little sense. As always, when puzzled by an unsteady gas dynamic phenomenon, I ask the simulation to plot out relevant pressure diagrams and I study them. The answers to this particular baffling conundrum are provided by Fig. 6.68.

Empiricism for the Primary Exhaust Length Based on the Facts

In the GPB simulation software, the facility exists to place "transducers" anywhere within the engine and its ducting. During the simulation of the 900 mm exhaust pipe, three transducers are placed within the ducting, one at the exhaust valve in the duct, another at the mid-section of the pipe, and the third at the very end of the pipe. The exhaust valve is at the left-hand end of the duct and the open end is at the right-hand end. The leftward and rightward waves at each transducer are recorded and plotted in Fig. 6.68. The rightward moving waves are all compression waves, and the steepening of a compression wave, first discussed in Sec. 2.1.5, is seen to occur. At the transducer at the exhaust valve end, the wave peak occurs at 160 °atdc and reaches the end of the pipe at 180 °atdc, a mere 20° crank angle later. Due to wave superposition and interference at the end, the peak of the suction reflection, by definition a leftward moving wave, does not leave until 200 °atdc. It reaches the mid-point of the pipe at about 278 °atdc and arrives back at the exhaust valve by 330 °atdc. The return journey, including the end delay, has taken 145° crankshaft, which is more than seven times longer than the exhaust (compression) pulse took to cover the same distance from the engine. This explains the slow (average) propagation velocity already noted in Eq. 6.4.6. What causes this effect? It has already been described in Sec. 2.2.2, where a compression wave, in superposition through an oppositely moving expansion wave, is seen to accelerate, and the expansion wave is seen to be retarded. This phenomenon is labeled there as "wave interference during superposition." The simulation keeps track of this phenomenon with no difficulty—indeed that is its very purpose. However, the lesson from Fig. 6.68 is that any attempt to empirically formulate this effect, by some logic based on an average reference acoustic velocity, is inevitably doomed to failure.

The optimum exhaust tuning of most four-stroke engines will be provided by pressure diagrams such as that shown in Fig. 6.64, in that particular case for the 900 mm pipe on the "ITC" single at 6500 rpm. This information should be packaged in empiricism based on Eq. 6.4.3, but with some adjustment for differing wave propagation velocities based on a

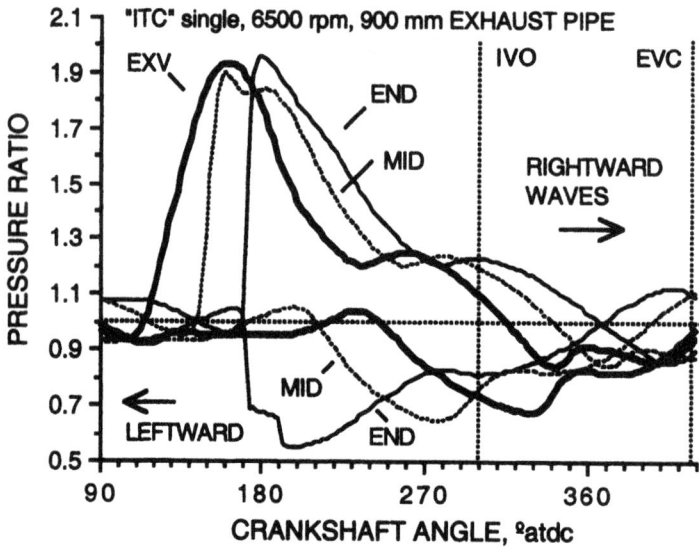

Fig. 6.68 Exhaust wave propagation in the 900 mm pipe at 6500 rpm.

mean exhaust gas temperature, T_{ex}, as measured, say, by a thermocouple in the middle of an exhaust pipe. The exhaust gas temperature, T_{ex}, is considered to be measured in °C units. The empirical adjustment of Eq. 6.4.3 provides a correct outcome for this known situation.

$$L_{et} = \frac{500 a_e \theta_{er}}{6N} = \frac{\text{constant}}{N} \times \frac{a_{ex}}{a_{ref}} = \frac{\text{constant}}{N} \times \sqrt{\frac{(T_{ex} + 273)}{(535 + 273)}} = 900 \quad \text{mm} \quad (6.4.8)$$

All constants are lumped together to become a single exhaust tuning factor, C_{et}, giving an empirical relationship for the tuned exhaust pipe length, L_{et}, as follows:

$$L_{et} = \frac{C_{et} \sqrt{(T_{ex} + 273)}}{N} \quad \text{mm} \quad (6.4.9)$$

where the value of the exhaust tuning factor, C_{et}, is numerically provided by the known facts regarding the 900 mm exhaust pipe:

$$C_{et} = \frac{L_{et} N}{\sqrt{(T_{ex} + 273)}} = \frac{900 \times 6500}{\sqrt{808}} = 2.058 \times 10^5 \quad (6.4.10)$$

For example, at the same exhaust gas temperature, T_{ex}, of 535°C, using Eq. 6.4.9 in reverse would predict for the ITC engine that:
(i) An exhaust pipe with an L_{et} of 1000 mm would be best tuned at 5850 rpm.
(ii) An exhaust pipe with an L_{et} of 700 mm would be best tuned at 8357 rpm.
(iii) An exhaust pipe with an L_{et} of 780 mm would be best tuned at 7500 rpm.
Other predictions with ramifications for other engines are:
(iv) An exhaust pipe with an L_{et} of 810 mm would be best tuned at 7220 rpm.
(v) An exhaust pipe with an L_{et} of 740 mm would be best tuned at 8000 rpm.
(vi) An exhaust pipe with an L_{et} of 650 mm would be best tuned at 9000 rpm.
There is some evidence, from Figs. 6.61 and 6.62, that this replay of Eq. 6.4.9 giving answers (i)-(iii) provides a reasonably accurate set of results based on the evidence provided by the "ITC" single test engine.

In the case of the exhaust primary length to the collector in the four-cylinder ITC engine, where peak torque is at 7500 rpm and the intake ramming trough is at 6500 rpm, answer (iv) provides reasonable correlation with the facts.

In the case of the two exhaust primary lengths to the first branch of the ducting of the Ducati engine, which are 705 and 775 mm, as seen in Fig. 5.37, answer (v) provides reasonable correlation with the facts. There is some evidence that the primary exhaust tuning is optimized to boost the 8000 rpm torque peak. With these longer exhaust pipes, the exhaust suction reflections arrive a little too late to be truly effective at the higher engine speeds, particularly at the intake ramming trough at 9500 rpm. This view is supported by Figs. 5.41 to 5.45.

Answer (vi) provides reasonable correlation with the facts for the eight-cylinder Nissan Infiniti IRL racing engine. The exhaust primary pipes to the collector are shown in Fig. 5.50 as being 628 mm in length, and the intake ramming trough is observed in Fig. 5.52 to be at 9000 rpm.

A length of 900 mm also correlates well with that of the short exhaust for the G50 Matchless engine, seen in Fig. 5.21. The G50 has its intake ramming trough at 6500 rpm, as seen in Fig. 5.25.

It is incumbent on me to remind those ever-desirous of a universal quick-fix solution, viz., the ubiquitous undergraduate student, of which species I was a prime example forty years ago, that the exhaust primary pipe tuning "formula," Eq. 6.4.9, is not as universally applicable as the intake tuning "relationships," Eqs. 6.3.2 to 6.3.8. The basic reason is that the "wave interference during superposition" effects referred to above, and discussed in Sec. 2.2.2, are very dependent on the amplitudes of the compression and expansion pressure waves within the exhaust pipe system. These amplitudes are, in turn, dependent on the trapped charge mass within the engine—the bmep, if you will—and also on the exhaust manifold-to-port area ratio, C_{em}. Although in most high-performance, naturally-aspirated engines these factors are reasonably common, as seen in Fig. 6.19, they are not constants. My best advice, as similarly given for dimensioning the intake system by empiricism, is to calculate the primary exhaust tract lengths, L_{et}, to alleviate the intake ramming trough, then insert that data value into an accurate simulation model of the engine in question. Subsequently, by varying that length around its datum, let the simulation provide the answer to the "wave-interference during superposition" problem.

Concluding Remarks on the Primary Exhaust Pipe Length

There has been enough evidence presented that the intake and exhaust tuning should be designed to work together to provide high torque over a wide speed range. The intake tuning will give two major torque and airflow peaks at two engine speeds with a trough in the middle. The major ramming peak should be located near the required engine speed for maximum power. The primary exhaust tuning should be coordinated in order to alleviate the (inevitable) intake ramming trough, and not supplement one of the ramming peaks, unless a very narrow power band is acceptable.

6.4.6 The Tuning Behavior of a Branched Collector Exhaust System

The result of exhaust pressure wave reflection in the primary exhaust pipe, in terms of its effect on the cylinder at the exhaust valve, is discussed in Sec. 6.4.5. What should be the design approach for the multi-cylinder engine? One such system in common use is the collector system, examples of which are seen in Fig. 6.21 for the ITC engine and in Fig. 5.50 for the

Nissan engine. Photographs of such exhaust systems for the Oldsmobile Aurora and the Nissan Infiniti engines used in the Indy Racing League are presented in Plates 5.3 and 5.11.

Another such system is shown in Plate 6.2. This is for a V8 engine used in offshore racing where the regulations for boat engines stipulate that all parts of the exhaust system must be water-cooled. Every primary exhaust pipe, and the final collector tailpipe, is double-skinned with water pumped between the metal layers, which explains the water-pipe connections from the primary pipes to the tailpipe. The manufacture of the junction of the double-skinned primary pipes, and the collector, is made much more difficult by this legal necessity, and this complexity is better illustrated in Plate 6.3. Each primary pipe must have the same, optimized length if the engine is to perform well. The separate water-cooling connections for each pipe at the exhaust manifold, at the collector, and onward to the tailpipe, are now evident. This is clearly not an inexpensive component, and is obviously not one whose lengths and diameters are amenable to rapid, if any, experimentation on the test-bed to determine their effect on the performance characteristics of the engine. This is the perfect situation for the application of an engine simulation which will inexpensively predict the performance characteristics while the designer changes the lengths and diameters of any pipe, or pipes, in the entire exhaust system. As usual, it is very inefficient to make such input data changes to a simulation without a basic understanding of the tuning effects that are taking place within the system. With this basic understanding, empiricisms can be derived that will guide you faster toward a set of input data for a collector exhaust system, allowing you to home the simulation in on the optimum design with the minimum number of data iterations.

Plate 6.2 The collector exhaust systems on a V8 offshore racing engine.
(Courtesy of Douglas Hahn)

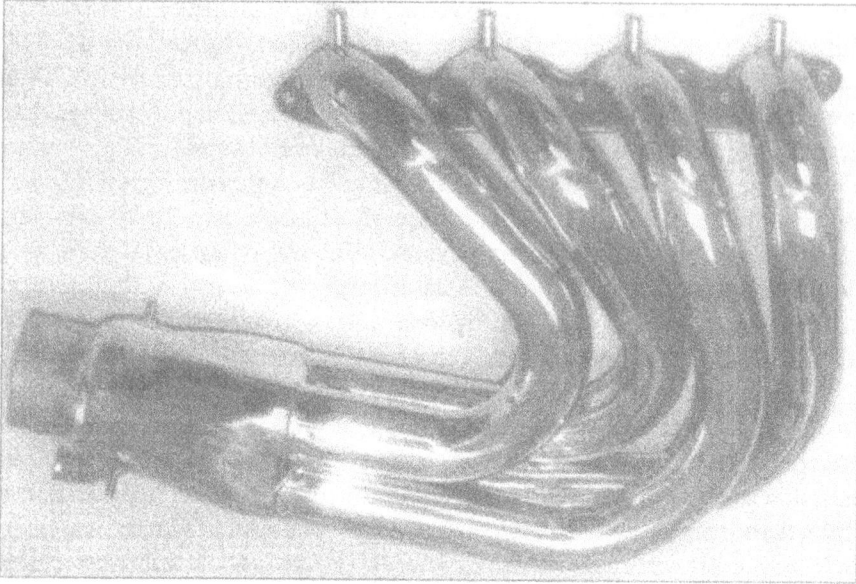

Plate 6.3 The exhaust primary pipes and collector for marine applications.
(Courtesy of Innovation Marine)

As usual, predictions of pressure wave motion within such a collector pipe system are the first steps toward both understanding the system and developing an empirical design procedure. Within the computation for the "standard" ITC four-cylinder engine and its collector exhaust system, as in Fig. 6.21, two pressure transducers are placed in the exhaust system. One is in the exhaust pipe connected to the cylinder numbered 1, before the collector, at some 30 mm before the exit to the final tailpipe, and the other is in the tailpipe, after the collector, at 30 mm from the tailpipe entrance from the collector. Under these circumstances and with transducers at these locations, the superposition pressure diagrams are recorded by the simulation at 8000 rpm, and are shown in Fig. 6.69. These correspond to those which could be measured by a transducer at the same physical locations.

Before the collector, and much as similarly shown in Fig. 6.68 at 6500 rpm for a straight pipe, the arrival of the now-steepened exhaust pulse is seen at 180° atdc. After the collector, the transmission of this pulse is observed, together with the transmitted pulses from the other three cylinders, each of which has arrived at the collector at 180° crank-angle intervals. Suction wave reflections of each transmitted exhaust pressure wave into the tailpipe, at its open end to the atmosphere, can also be observed in Fig. 6.69. The picture in Fig. 6.69 is a classic example of the misleading nature of measured, i.e., superposition, pressure diagrams.

To illuminate this remark, the leftward and rightward running waves are extracted from the simulation output at the same transducer positions in the exhaust pipe and the tailpipe, where the atmosphere is at its right-hand end, precisely as sketched in Fig. 6.21. They are plotted separately in Figs. 6.70 and 6.71, respectively.

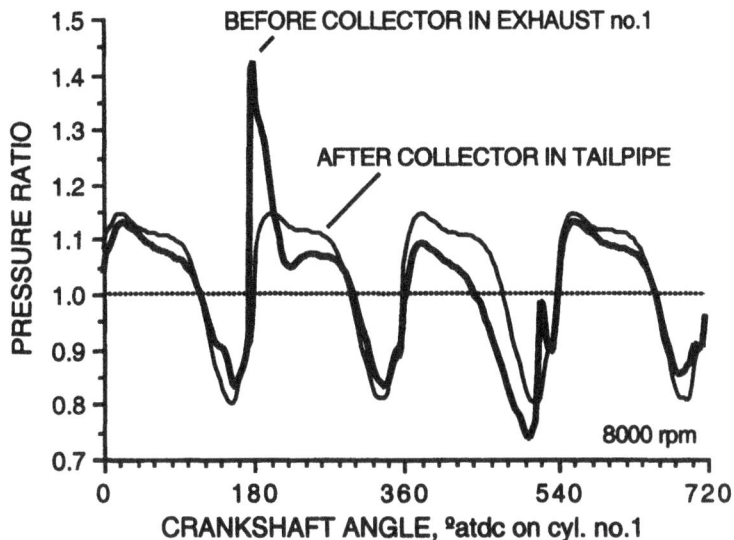

Fig. 6.69 Superposition pressures before and after the collector at 8000 rpm.

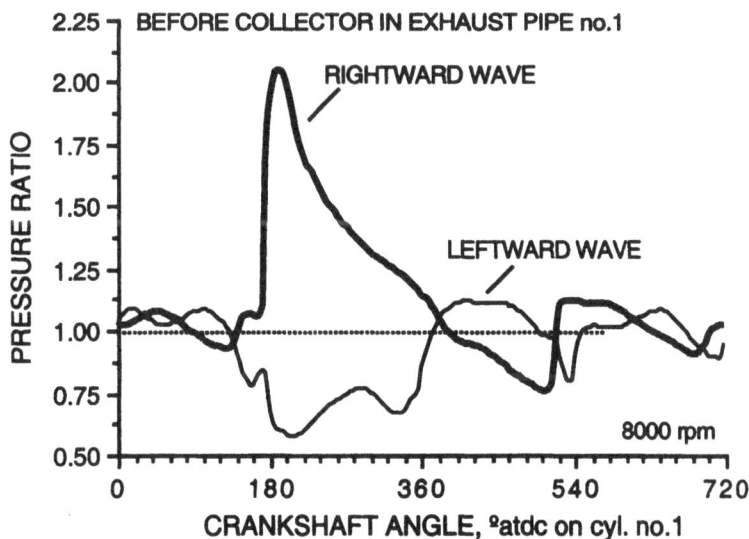

Fig. 6.70 Leftward and rightward waves before the collector at 8000 rpm.

In Fig. 6.70, in the primary exhaust pipe, the (rightward) exhaust wave and its (leftward) reflection at the collector are now seen clearly to be as predicted in Sec. 6.4.5 and as shown in Fig. 6.68 as a superposition diagram. This latter pulse is the suction wave that eventually arrives at the exhaust valve, steepened at its tail, and that enhances cylinder charging during the valve overlap period, as shown in Figs. 6.25 and 6.26. You may be surprised to see that there is very little evidence in Fig. 6.70 that this pipe is even connected to the other three cylinders of the engine at the collector.

Fig. 6.71 shows the leftward and rightward wave trains at the transducer located at the beginning of the tailpipe, i.e., just after the collector. These are remarkable profiles, of which the superposition diagram in Fig. 6.69 does not even give a hint. Individual exhaust (rightward) waves of some 1.35 atm, which never dip below atmospheric pressure, are meeting (leftward) suction reflections, which drop down to 0.78 atm and barely rise above the atmospheric pressure level. The result is considerable symmetry of tuning for all four cylinders. Such is the strength of suction from these well-timed wave trains in the tailpipe that there is almost no backflow at the collector into the other three primary exhaust pipes upon the arrival of each high-amplitude exhaust pulse in its primary exhaust pipe at the collector. This is why, in Fig. 6.70, there is no sign of activity from the other three cylinders. The amplitude of each primary exhaust pulse is seen in Fig. 6.70 to be 2.0 atm—the nominal limit that I postulated in Sec. 6.1.9, which is basically controlled by the manifold-to-port area ratio, C_{em}.

That backflow into each primary pipe is prevented in a well-designed collector can be observed by asking the simulation to provide what is still unmeasurable today, i.e., the dynamic mass flow time-history at the two transducer positions. This is shown in Fig. 6.72. As each exhaust pulse arrives in turn at the collector, the exiting mass flow in the tailpipe is always positive, i.e., rightward toward the atmosphere. Only minor quantities of backflow, into the primary exhaust pipe for cylinder number 1, occur at several periods during the cycle. For example, in Fig. 6.72, the small amount of backflow observed around the 470° period coincides with the low-amplitude, leftward compression wave transmitted into the primary exhaust pipe in Fig. 6.70.

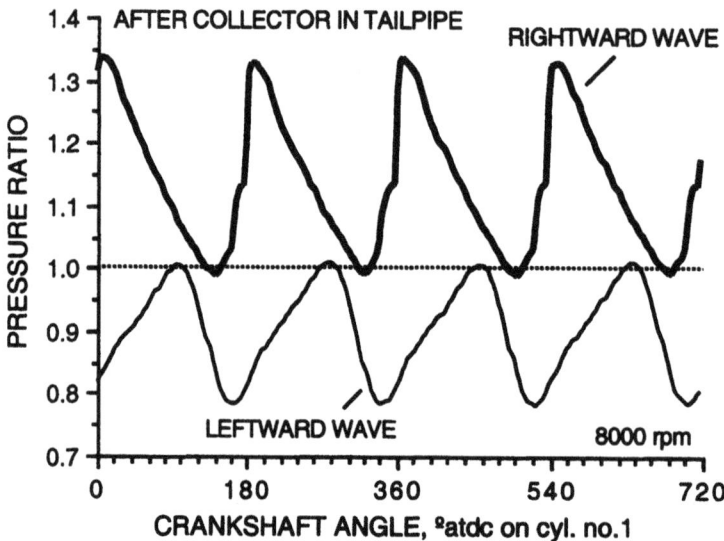

Fig. 6.71 Leftward and rightward waves after the collector at 8000 rpm.

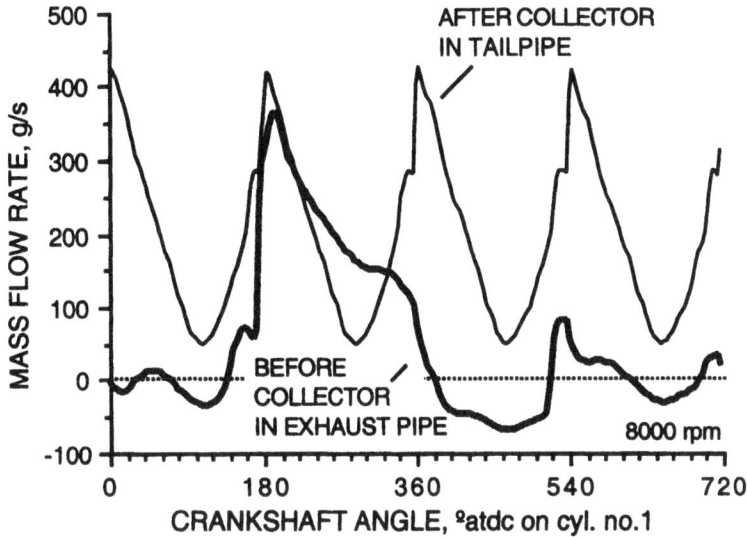

Fig. 6.72 Mass flow rate before and after the collector at 8000 rpm.

In short, the well-designed collector system permits each primary exhaust pipe to behave as if it were a straight pipe entering the atmosphere, with the tailpipe wave trains acting to prevent inter-cylinder cross-talk. This explains why the straight pipe and the collector system are seen in Fig. 6.60 to produce almost identical torque characteristics.

The basic tuning action of the tailpipe between the collector exit and the next major expansion in area, be it the atmosphere as in this case, or another branch, plenum, etc., is now manifest. Each exhaust compression pulse that arrives at the collector is transmitted into the tailpipe and reflects as an expansion wave at the next "open end." Optimal tailpipe tuning results when that suction reflection arrives back at the collector junction in time to meet the next exhaust pulse from another cylinder arriving down its primary exhaust pipe. For a four-cylinder engine, this means that the optimum return period for the double traverse of the tailpipe of the transmitted compression pulse and its suction reflection is 180° crank angle. By definition, for a three-cylinder engine with an equivalent 3-1 collector design, this return period is 240° crank angle; for an equivalent five-cylinder collector system, it is 144° crank angle; and for a six-cylinder collector design, it is 120° crank-angle. Before indulging in empiricism to represent this information, consider some further information from the simulation of the ITC engine.

The Effect of the Tailpipe Length on the Tuning of a Collector Exhaust System

This length phases the exhaust tailpipe tuning at the collector. To understand more about this behavior, three further computations are performed for the ITC engine, in which this length is changed from the "standard" value of 900 mm shown in Fig. 6.21 to 850, 950, and 1000 mm, i.e., -50 mm, +50 mm, and +100 mm, respectively. The calculations are conducted and the results are plotted in Fig. 6.73 as power gain or loss over the speed range. In these results, we see the usual mixture of "good" news and "bad" news. The standard length of 900 mm proves to be the best overall length, but a length of 850 mm does marginally enhance the performance at 7500 rpm, operating in conjunction with a primary exhaust pipe length that is known to be optimized at 7220 rpm. The results for the tailpipe length of 850 mm at an engine speed of 7500 rpm will be used as empirical evidence that the fundamental tuning criterion is satisfied by providing 180° phased suction reflections at the collector. Considering the previous bad experience of attempting a fundamental analysis, and in the presence of the same "wave-interference during superposition" problems that confused the issue in Sec. 6.4.5, the simplistic empirical approach finally, and successfully, followed there is repeated. To assist in handling designs where the temperature in the tailpipe may be different from that in the ITC engine, note that a scan of the temperature records in the tailpipe of that engine shows a mean temperature of 530°C.

All of the fundamental constants seen in the basic tuning equation, Eq. 6.4.3, are lumped to become a single exhaust tailpipe tuning factor, C_{tp}, giving an empirical relationship for the collector exhaust tailpipe length L_{tp}, optimized at an engine speed N, for an exhaust collector system where a number of (preferably evenly firing) cylinders, n_{cy}, feed their exhaust primary pipes into that collector:

$$L_{tp} = \frac{C_{tp}\sqrt{(T_{ex} + 273)}}{n_{cy} \times N} \quad mm \qquad (6.4.11)$$

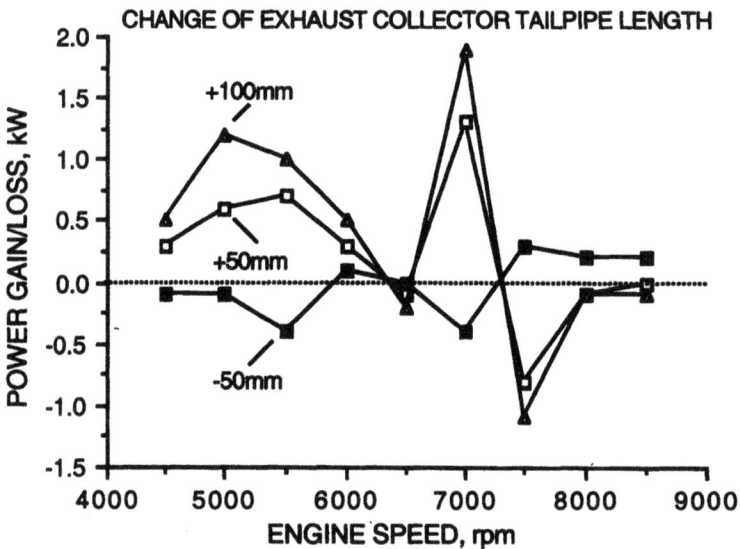

Fig. 6.73 Effect of collector tailpipe length on the performance of the ITC engine.

From the known facts, the value of the exhaust tuning factor for the collector tailpipe length, C_{tp}, is given by:

$$C_{tp} = \frac{L_{tp} \times n_{cy} \times N}{\sqrt{\left(T_{ex} + 273\right)}} = \frac{850 \times 4 \times 7500}{\sqrt{803}} = 9 \times 10^5 \qquad (6.4.12)$$

In this text, the only other engine with this type of collector system is the Nissan Infiniti engine, which has two collector exhaust systems, each of which is fed by the four cylinders on each bank of the V8 engine, as sketched in Fig. 5.50. The measured exhaust tailpipe length is 660 mm, which, from the application of Eq. 6.4.11, using the same temperature data, corresponds to an optimized tuned speed of 9660 rpm. There is some evidence from the measured and calculated torque curves, in Fig. 5.52, that this empirical estimate for the collector tailpipe length is quite logical in terms of the known performance outcome, particularly in terms of the flattening of the shallow intake ramming trough that is seen in Fig. 5.59 to extend from 9000 to 9500 rpm. In the early stages of the design of this Nissan engine, in which I was involved, the ability to empirically predict numbers for dimensions such as this tailpipe length considerably reduced the number of design iterations, and associated computer time, when running the GPB engine simulation.

The Effect of the Tailpipe Diameter on the Tuning of a Collector Exhaust System
The expansion area ratio, from all of the primary pipes joining the collector to the tailpipe, has control over the amplitude of the reflection of each exhaust pulse arriving at the collector from each cylinder in sequence. This effect is discussed more fundamentally in Secs. 2.13 and 2.14. An exhaust pulse arriving down any one pipe sees a sudden expansion of area, A_{out}, facing it, which is comprised of all other exhaust primary pipes plus the tailpipe. The area ratio of expansion required to produce effective reflection of a given exhaust pulse at any sudden increase of area has an approximate value of 6. A more general expression of this expansion area ratio criterion for a collector, C_{coll}, for an exhaust collector system that is fed by a number of exhaust primary pipes coming from a number of cylinders n_{cy}, with exhaust pipes of diameter d_{ep}, a tailpipe of diameter d_{tp}, through a common collector can be expressed as follows.

Collector area ratio: $\qquad C_{coll} = \dfrac{A_{out}}{A_{in}} = \dfrac{\left(n_{cy} - 1\right)d_{ep}^2 + d_{tp}^2}{d_{ep}^2} \qquad (6.4.13)$

Inserting the known geometry from the standard ITC data, as seen in Fig. 6.21, gives:

$$C_{coll} = \frac{3 \times 44.2^2 + 79.6^2}{44.2^2} = 6.24 \tag{6.4.14}$$

Inserting the equivalent numbers from the Nissan Infiniti data, as seen in Fig. 5.50, gives,

$$C_{coll} = \frac{3 \times 51^2 + 88.9^2}{51^2} = 6.04 \tag{6.4.15}$$

In short, the Nissan Infiniti and ITC engines data are in fairly close agreement on the optimum value of the empirical coefficient, C_{coll}, for the design of the diameter of the tailpipe after the collector.

A few simulations are conducted on the ITC engine to determine if there is more, or better, information available on this topic. If the "standard" data value for the tailpipe diameter, d_{tp}, of 79.6 mm is changed to 75, 85, and 90 mm, i.e., approximately -5 mm, +5 mm, and +10 mm, respectively, this gives C_{coll} values of 5.9, 6.7, and 7.1, respectively. The data values are so changed within the computation, and the results are plotted in Fig. 6.74 as a gain or loss of power output over the engine speed range. It can be seen that the engine is relatively insensitive to these changes, as there is very little variation of power over the "standard" data values, where C_{coll} is 6.24. However, the minor gain for the smaller tailpipe diameter in the peak

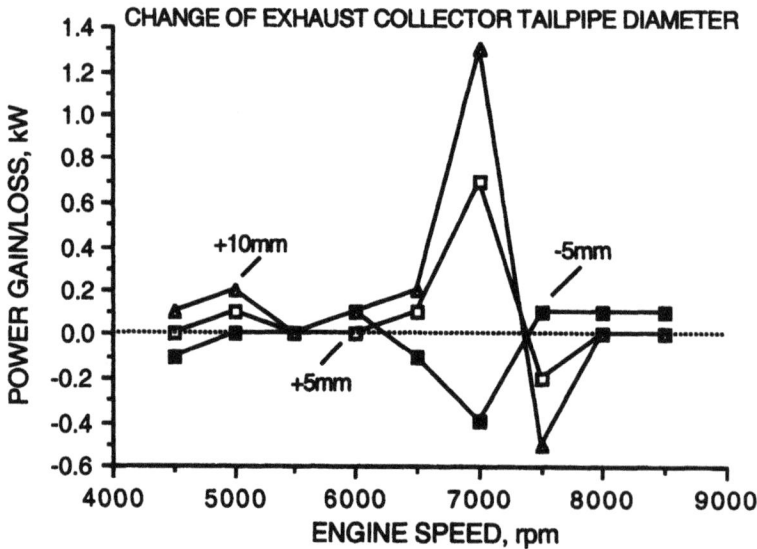

Fig. 6.74 Effect of collector tailpipe diameter on the performance of the ITC engine.

power zone from 7500 to 8500 rpm, where C_{coll} is 5.9, and its correspondence with the 6.04 value optimized for the Nissan engine, tips the balance in favor of offering a C_{coll} value of 6.0 as the best overall estimate for the expansion area ratio at an exhaust collector.

The Effect of the Branch Angle on the Tuning of a Collector Exhaust System

After the above exposition, it will come as no surprise that, because of the strength of wave-train suction at the collector, the inter-pipe, or branch, angle at the collector is not a hugely significant criterion in the highly-tuned engine. Within the computation using the "standard" data for the ITC engine, each pipe is mutually at a narrow 15° angle from each other, with the tailpipe symmetrically facing the exhaust primary pipes. In short, the sketch in Fig. 6.21 is visually accurate. This data value is changed to 10°, 20°, 25°, and 30°, the computations are repeated, and the output is analyzed. The results are presented in Fig. 6.75. The "standard" branch angle of 15° is the best, i.e., with a zero power gain or loss, with all other branch angles losing at some engine speeds as much as they gain at others. Nevertheless, it is clear that the narrower branch angles of 10° and 15° are the best, and above these values losses, however small, are occurring. It is interesting that the most visible power losses are at engine speeds where the exhaust tuning is least effective, i.e., below 6000 rpm.

6.5 Concluding Remarks on Empiricism for Engine Optimization

It is possible to specify empirical factors for high-performance engines which are effective for basic design and development purposes. These empirical design factors are known as:

(i) Time-areas per unit swept volume

These can be used to specify the cylinder head geometry so as to attain the requisite bmep (or torque and power) characteristics for a high-performance engine at a given

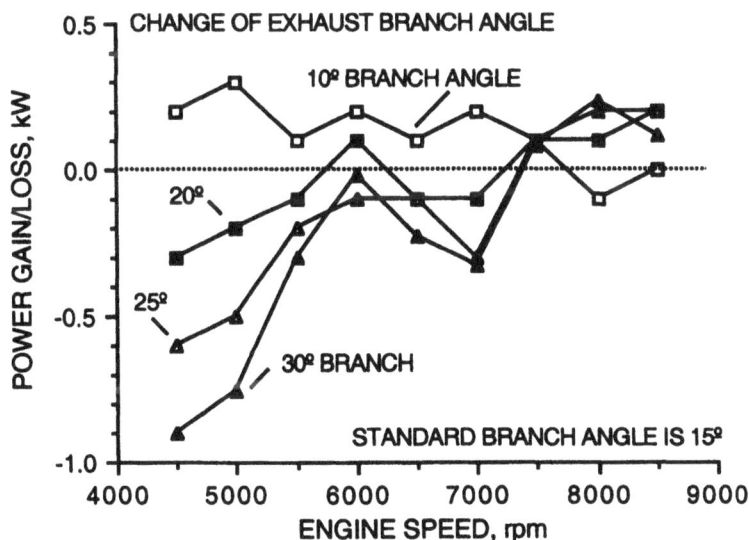

Fig. 6.75 Effect of collector tailpipe length on the performance of the ITC engine.

engine speed. The time-areas can be split into individual segments representing the exhaust blowdown, exhaust pumping, intake pumping, intake ramming, exhaust overlap, and intake overlap valve periods. Each individual segment criterion should be met or the overall design may be poorly matched and the target performance levels not realized as a consequence.

(ii) Manifold-to-port area ratios
These can be estimated to attain either a narrow band or broad band performance from a high-performance engine.

(iii) Discharge coefficients
These can be used to estimate the maximum lift of the valves either at the initial design or at the development phase.

(iv) Intake ramming and exhaust primary pipe tuning factors
It is possible to specify empirical relationships for the lengths of the intake tract and the exhaust primary pipes.

(vi) Multi-branch exhaust collector system design
It is possible to specify empirical relationships for the lengths, diameters, and branch angles of the ducting in collector pipe exhaust systems for multi-cylinder engines.

As far as I am aware, the above empiricism for the basic design of a high-performance engine and its ducting is a unique approach. Be that as it may, the optimization of an engine, by using a combination of the empiricism described here and an accurate engine simulation model, is an approach that can reduce the number of design iterations or experimental prototypes. The use of empiricism on its own, or engine simulation on its own, as a design process can lead to elements of that design being sufficiently poorly matched that significant performance loss may well be the outcome.

References for Chapter 6

6.1 W.J.D. Annand and G.E. Roe, "Gas Flow in the Internal Combustion Engine," Foulis, England, June 1974.

6.2 S. Yagi, K. Fujiwara, and N. Kuroki, "Total Engine Loss and Engine Output Characteristics in Four-Stroke Spark-Ignition Engines," Fourth Graz Two-Wheeler Symposium, Technische Universität, Graz, Austria, 8-9 April 1991, pp. 169-184.

6.3 S. Yagi, A. Ishizuya, and I. Fuji, "Research and Development of High Speed, High-Performance, Small Displacement Honda Engines," SAE International Congress, 1970, SAE Paper No. 700122.

6.4 H.A. Newlyn, "An Analysis of the Probable Effects of Engine Geometry on the Inlet Pipe Pressure, Flow and Volumetric Efficiency," SAE Motorsports Engineering Conference and Exposition, Dearborn, Mich., December 1994, SAE Paper No. 942531, pp. 221-232.

6.5 I. Fukutani and E. Watanabe, "An Analysis of the Volumetric Efficiency Characteristics of 4-Stroke Cycle Engines Using the Mean Mach Index," 1979, SAE Paper No. 790480.

6.6 D. Broome, "Induction Ram," *Automobile Engineer*, Parts 1, 2, and 3, published in April, May, and June, respectively, 1969.

6.7   C. Taylor, J. Livengood, and D. Tsai, "Dynamics in the Inlet System of a Four-Stroke Single-Cylinder Engine," Trans. ASME, 1955, pp. 1133-1145.

6.8   G.P. Blair and J.H. McConnell, "Unsteady Gas Flow Through High-Specific-Output 4-Stroke Cycle Engines," SAE Farm, Construction & Industrial Machinery and Powerplant Meetings, Milwaukee, Wisc. September 1974, SAE Paper No. 740736.

6.9   G.P. Blair, "Correlation of Measured and Calculated Performance Characteristics of Motorcycle Engines," Funfe Zweiradtagung, Technische Universität, Graz, Austria, 22-23 April 1993, pp. 5-16

6.10   G.P. Blair, *Design and Simulation of Two-Stroke Engines*, R-161, Society of Automotive Engineers, Warrendale, Pa., March 1996.

6.11   T. Morel and L.A. LaPointe, "Concurrent Simulation and Testing Concept in Engine Development," SAE International Congress, Detroit, Mich., 1994, SAE Paper No. 940207.

6.12   A.A. Boretti, G. Cantore, E. Mattarelli, and F. Preziosi, "Experimental and Computational Analysis of a High-Performance Motorcycle Engine," SAE Motorsports Engineering Conference and Exposition, Dearborn, Mich., December 8-10, 1996, SAE Paper No. 962526.

6.13   H. Seifert, "Erfahrungen mit einem Mathematischen Modell zur Simulation von Arbeitsfahren in Verbrennungsmotoren," MTZ, 39, (1978), Nr.7/8, pp. 321-325.

6.14   J.M. Novak and R.A. Kach, "Computer Optimisation of Camshaft Lift Profiles for a NASCAR V-8 Engine with Restrictor Plate," SAE Motorsports Engineering Conference and Exposition, Dearborn Mich., December 8-10, 1996, SAE Paper No. 962514.

6.15   D.E. Winterbone, R.J. Pearson, and Y. Zhao, "Numerical Simulation of Intake and Exhaust Flows in a High-Speed Multi-Cylinder Petrol Engine using the Lax-Wendroff Method," IMechE Conference C430/038, 1991.

6.16   G.P. Blair and F.M. Drouin, "The Relationship between Discharge Coefficients and the Accuracy of Engine Simulation," SAE Motorsports Engineering Conference and Exposition, Dearborn Mich., December 8-10, 1996, SAE Paper No. 962527.

6.17   V. Willoughby, *Classic Motorcycles*, Hamlyn, London, 1975, p. 176.

6.18   S.J. Kirkpatrick, G.P. Blair, R. Fleck, and R.K. McMullan, "Experimental Evaluation of 1D Computer Codes for the Simulation of Unsteady Gas Flow Through Engines—a First Phase," SAE International Off-Highway Meeting, Milwaukee, Wisc. September 1994, SAE Paper No. 941685, p. 18.

6.19   G.P. Blair, S.J. Kirkpatrick, and R. Fleck, "Experimental Validation of 1-D Modelling Codes for a Pipe Containing Gas of Varying Properties," Society of Automotive Engineers, International Congress, Detroit, Mich., March 1995, SAE Paper No. 950275, pp. 93-106.

6.20   G.P. Blair, S.J. Kirkpatrick, D.O. Mackey, and R. Fleck, "Experimental Validation of 1-D Modelling Codes for a Pipe System Containing Area Discontinuities," Society of Automotive Engineers, International Congress, Detroit, Mich., March 1995, SAE Paper No. 950276, pp. 107-120.

Chapter 7

Reduction of Noise Emission from Four-Stroke Engines

7.0 Introduction

The subject of noise emission from internal combustion engines, and its reduction, is a specialized topic. Many textbooks and technical papers have been written on this topic, so it is not possible to completely cover it in a single chapter in this book. Instead, I intend to orient you to the noise emission associated with four-stroke engines and the problems of silencer design for this particular type of power unit. (The international reader should note that the American word for "silencer" is "muffler.") Many texts discuss silencer design as if the only engine in existence is the spark-ignited, multi-cylinder, four-stroke automobile engine, and they merely describe the techniques applicable to this type of engine—a practice that can create considerable misconceptions for the designers of silencers for other four-stroke engines used, say, in lawnmowers, motorcycles, or boats where the space allocations for the silencers assume different priorities than they do in the automobile. Thus, one of the goals of this chapter is to dispel such misconceptions.

The opening section of this chapter gives some general background on the subject of noise, indeed it repeats in brief what may be found more completely in other books and papers, which are suitably referenced for your wider education. This opening section is included so that the remainder of the chapter will be more immediately meaningful to the tyro.

Later sections deal with the fundamental nature of pressure-wave-created noise emission and the theoretical methods available for its prediction. There is also a discussion on the future of the technology of silencer design. The succeeding sections cover the more empirical approaches to silencer design, the acoustic predictions of which will be checked by an engine simulation. The outcome is pragmatic advice on the design of intake and exhaust silencers for single and multi-cylinder four-stroke engines, and on the extent to which acoustic empiricism is of assistance in this matter. In the examples presented in this chapter, the design approach is to examine the trade-off in noise reduction with the loss of power normally caused by muffling the engine, and the extent to which system tuning can alleviate the reduction in engine performance.

7.1 Noise

In many texts, you will find "noise" described as "unwanted sound." This is a rather loose description, because noise that is "wanted" by some may be "unwanted" by others. Thirty years may have passed, but I can still hear Mike Hailwood power his inimitable way around the 1967 Ulster Grand Prix on the 250 cm^3 Honda "straight" six-cylinder racer. The engine produced about 65 totally unsilenced horsepower at 18,000 rpm on open megaphone exhaust pipes. It emitted a unique noise that was music to the ears of a hundred thousand racing motorcycle fans and, I feel, was of a quality that even Mozart might have relished. The Ulster Grand Prix is held on public roads through rolling countryside behind Belfast. A nearby farmer, the owner of a thousand chickens vainly trying to lay eggs, doubtless viewed the very same sound from an alternate standpoint. This particular example illustrates the quite subjective nature of noise assessment; another might be my attitude toward loud "pop" music. Nevertheless, between the limits of the threshold of human hearing and the threshold of physical damage to the human ear, it is possible to measure the pressure level caused by sound and to assign an experimental number to that value. This number does not detail whether the sound is "wanted" or "unwanted." As already pointed out, to some it will always be described as "noise."

7.1.1 Transmission of Sound

As discussed in Sec. 2.1.2, sound propagates in three dimensions from a source through the air, or a gas, as the medium of its transmission. The fundamental theory for this propagation is presented in Sec. 2.1.2. The speed of the propagation of a wave of acoustic amplitude is given by a_0, where:

$$a_0 = \sqrt{\gamma R T_0} = \sqrt{\frac{\gamma p_0}{\rho_0}} \qquad (7.1.1)$$

As shown in Sec. 2.1.6, at 25°C, the value of the ratio of specific heats, γ, is 1.4 for air and can be taken as 1.375 for exhaust gas. At such room-temperature conditions, the value of the gas constant, R, is 287 J/kgK for air and 291 J/kgK for the exhaust gas. In acoustic calculations for sound wave attenuation in silencers, treating exhaust gas as air produces errors of no real significance. For example, if the temperature is raised to 500 K, where γ and R for air and exhaust gas are taken from Sec. 2.1.6 and Table 2.1.3,

$$a_{air} = \sqrt{1.373 \times 287 \times 500} = 444 \text{ m/s}$$

$$a_{exhaust} = \sqrt{1.35 \times 290.8 \times 500} = 443 \text{ m/s}$$

Because exhaust gas in a four-stroke engine often contains a small amount of air which is short-circuited during the charging process during the overlap valve period, this reduces the already negligible error even further.

7.1.2 Intensity and Loudness of Sound

The propagation of pressure waves is already covered thoroughly in Sec. 2.1, so it is not necessary to repeat it here. Sound waves are but very small pressure waves. However, the propagation of these small pressure pulses in air, following one after the other, varying in both spacing and amplitude, gives rise to the human perception of the pitch and the amplitude of the sound. The frequency of the pressure pulsations produces the pitch and the amplitude produces the loudness. The human ear can detect frequencies ranging from 20 Hz to 20 kHz, although as one becomes older, the upper limit of that spectrum shortens to about 12 kHz. For an alternative introductory view of this topic, consult the books by Annand and Roe [7.33] and by Taylor [7.11].

More particularly, the intensity, I, is used to denote the physical energy of a sound, and loudness, β, *as defined in this book* is the human perception of that intensity in terms of sound pressure level. I am well aware that the term "loudness" is often defined differently in other texts [7.33], but because this word is most meaningful to you, and to any listener, as a way to express perceived noise level as measured by pressure, I feel justified in using it in that context within this text.

The relationship between intensity and loudness is fixed for sounds that have a pure tone, or pitch, i.e., sounds that are composed of sinusoidal pressure waves of a given frequency. For real sounds, that relationship is more complex. The intensity of the sound, being an energy value, is denoted in units of W/m^2. Noise meters, being basically pressure transducers, record the "effective sound pressure level," which is the root-mean-square (rms) of the pressure fluctuation about the mean pressure caused by the sound pressure waves. This rms pressure fluctuation is denoted by dp, and in a gas medium with a density ρ, and a reference speed of sound a_0, the intensity is related to the square of the rms sound pressure level by:

$$I = \frac{dp^2}{\rho a_0} \qquad (7.1.2)$$

The pressure rise, dp, can be visually observed in Plates 2.1 to 2.4 as it propagates away from the end of an exhaust pipe.

The level of intensity that can be recorded by the human ear is considerable, ranging from 1 pW/m^2 to 1 W/m^2. The human eardrum, our personal pressure transducer, will oscillate from an imperceptible level at the minimum intensity level up to about 0.01 mm at the highest level when a sensation of pain is produced by the nervous system as a warning of impending damage. To simplify this wide variation in physical sensation, a logarithmic scale is used to denote loudness, the units of which are the bel, symbolized B. But even this unit is too large for general use, so it is divided into ten sub-divisions called decibels, symbolized dB. The loudness of a sound is denoted by comparing its intensity level on this logarithmic scale to the "threshold

Design and Simulation of Four-Stroke Engines

of hearing," which is at an intensity, I_0, of 1.0 pW/m^2 or a rms pressure fluctuation, dp_0, of 0.00002 Pa or 0.0002 µbar. Thus, the intensity level of a sound, I_{L1}, where the actual intensity is I_1, is given by:

$$I_{L1} = \log_{10}\left(\frac{I_1}{I_0}\right) \text{ B}$$ (7.1.3)

$$= 10\log_{10}\left(\frac{I_1}{I_0}\right) \text{ dB}$$

In a corresponding fashion, a sound pressure level, β_1, where the actual rms pressure fluctuation is dp_1, is given by:

$$\beta_1 = \log_{10}\left(\frac{I_1}{I_0}\right) \text{ B}$$

or, $$\beta_1 = 10\log_{10}\left(\frac{I_1}{I_0}\right) = 10\log_{10}\left(\frac{dp_1}{dp_0}\right)^2 = 20\log_{10}\left(\frac{dp_1}{dp_0}\right) \text{ dB}$$ (7.1.4)

7.1.3 Loudness When There Are Several Sources of Sound

Imagine that you are exposed to two sources of sound of intensities I_1 and I_2. These two sources would separately produce sound pressure levels of β_1 and β_2, in dB units. Consequently, from Eq. 7.1.4:

$$I_1 = I_0 \text{ antilog}_{10}\left(\frac{\beta_1}{10}\right) \quad I_2 = I_0 \text{ antilog}_{10}\left(\frac{\beta_2}{10}\right)$$

The absolute intensity that you experience from both sources simultaneously is the superposition value, I_s, where:

$$I_s = I_1 + I_2$$

Hence, the total sound pressure level experienced from both sources is β_s, where:

$$\beta_s = 10\log_{10}\left(\frac{I_s}{I_0}\right) = 10\log_{10}\left\{\text{antilog}\left(\frac{\beta_1}{10}\right) + \text{antilog}\left(\frac{\beta_2}{10}\right)\right\}$$ (7.1.5)

Consider two simple cases:

   (i) You are exposed to two equal sources of sound that are at 100 dB.

   (ii) You are exposed to two sources of sound, one at 90 dB and the other at 100 dB.

First, consider case (i), using Eq. 7.1.5:

$$
\begin{aligned}
\beta_s &= 10\log_{10}\left\{\mathbf{antilog}\left(\frac{\beta_1}{10}\right) + \mathbf{antilog}\left(\frac{\beta_2}{10}\right)\right\} \\[2mm]
&= 10\log_{10}\left\{\mathbf{antilog}\left(\frac{100}{10}\right) + \mathbf{antilog}\left(\frac{100}{10}\right)\right\} \\[2mm]
&= 103.01 \quad \mathrm{dB}
\end{aligned}
\tag{7.1.6}
$$

Second, consider case (ii):

$$
\begin{aligned}
\beta_s &= 10\log_{10}\left\{\mathbf{antilog}\left(\frac{\beta_1}{10}\right) + \mathbf{antilog}\left(\frac{\beta_2}{10}\right)\right\} \\[2mm]
&= 10\log_{10}\left\{\mathbf{antilog}\left(\frac{100}{10}\right) + \mathbf{antilog}\left(\frac{90}{10}\right)\right\} \\[2mm]
&= 100.41 \quad \mathrm{dB}
\end{aligned}
\tag{7.1.7}
$$

    In case (i), it is clear that the addition of two sound sources, each equal to 100 dB, produces an overall sound pressure level of 103.01 dB, a rise of just 3.01 dB due to a logarithmic scale being used to attempt to simulate the response characteristics of the human ear. To physically support this mathematical contention, remember that the noise of an entire brass band does not appear to be so much in excess of one trumpet player at full throttle.

    In case (ii), the addition of the second weaker source at 90 dB to the noisier one at 100 dB produces a negligible increase in loudness level, just 0.41 dB above the noisier source. The physical analogy is that the addition of one more trumpet player to the aforementioned brass band does not raise significantly the total noise level as perceived by the listener.

    Within these simple examples is a fundamental message to the designer of engine silencers: If an engine has several different sources of noise, the loudest will swamp all others in the overall sound pressure level. The identification and muffling of that major noise source becomes the first priority for the designer of that silencer. Expanded discussion on these topics is presented in the books by Harris [7.12] and Beranek [7.13] and further on in this chapter.

7.1.4 Measurement of Noise and the Noise-Frequency Spectrum

The *noisemeter*, an instrument used to measure noise, is basically a microphone connected to an amplifier, with the system calibrated to read in the units of dB. Usually the device is internally programmed to read either the total sound pressure level, which is known as the linear value, i.e., dBlin, or the level on an A-weighted or B-weighted scale to represent the response of the human ear to loudness as a function of frequency. The A-weighted scale is the more common of the two scales, and the units are recorded appropriately as dBA. To put some numbers to this weighting effect, consider that the A-weighted scale reduces the recorded sound below the dBlin level by 30 dB at 50 Hz, 19 dB at 100 Hz, 3dB at 500 Hz, zero at 1000 Hz, then increases it by about 1 dB between 2 and 4kHz, before tailing off to reduce it by 10 dB at 20 kHz. The implications behind this weighting effect are that high frequencies between 1000 and 4000 Hz are very irritating to the human ear, to such an extent that a noise recording 100 dBlin at around 100 Hz only sounds as loud as 81dB at 1000 Hz, hence it is recorded as 81dBA. To quote another example, the same overall sound pressure level at 3000 Hz appears to be as loud as 101 dB at 1000 Hz, and is noted as 101 dBA. An example of the effect of A-weighting on noise over a range of frequencies is shown in Fig. 7.44, which depicts the exhaust noise from a motorcycle engine.

Equally common is for the noisemeter to be capable of a frequency analysis, i.e., to record the noise spectra over discrete bands of frequency. Usually these are carried out over one-octave bands or, more finely, over one-third octave bands. A typical one-octave filter set on a noisemeter would have switchable filters to record the noise about 31.25 Hz, 62.5 Hz, 125 Hz, 250 Hz, 500 Hz, 1000 Hz, 2000 Hz, 4000 Hz, 8000 Hz and 16,000 Hz. A one-third octave filter set carries out this function in even narrower steps of frequency change. The latest advances in microelectronics and cnjfuter-assisted data capture allow this process of frequency analysis to be carried out in even finer detail. Many portable noisemeters today will record and display the noise spectra at 6.25 Hz intervals.

The ability of a measurement system to record the noise-frequency spectrum is very important to the researcher who is attempting to silence, say, the exhaust system of a particular engine. As described in Sec. 7.1.3 regarding the addition of noise levels from several sources, the noisiest frequency band in the measured spectrum is the band that must be silenced first and foremost as it makes the largest contribution to the overall sound pressure level. The identification of the frequency band of that major noise component is the first step toward it eradication as a noise source.

The measurement of noise is a tedious experimental technique in that a set procedure is not only desirable, but essential. Seemingly innocent parameters, such as the height of the microphone from the ground during a test, or the reflectivity of the surface of the ground in the vicinity of the testing, e.g., grass, gravel, or tarmac, can have a major influence on the numerical value of the decibels recorded from the identical engine or machine. This has given rise to a plethora of apparently unrelated test procedures, such as those in the SAE standards [7.16]. In actual fact, the logic behind the formulation of these test procedures is quite impeccable, and any reader embarking on a silencer design and development exercise will be wise to study them thoroughly and implement them to the letter during experimentation.

7.2 Noise Sources in a Simple Four-Stroke Engine

The sources of noise emanating from a four-stroke engine are illustrated in Fig. 7.1. The obvious ones are the intake and the exhaust system, in which gas pressure waves are present, as discussed at length in Chapter 2. As these pressure waves propagate into the atmosphere they produce noise. The series of photographs in Chapter 2, Plates 2.1 to 2.4, illustrate the rapid nature of the pressure rise propagating into the atmosphere and toward the ear of the listener. The common belief is that the exhaust is the noisier of the two sources, and in general this is true. However, the most rudimentary of exhaust silencers will almost inevitably leave an unsilenced intake system as the noisier of these two sources, so the intake system also requires silencing to the same level and extent as the exhaust system.

You will realize that any vibrating solid surface can act as a noise source, in the same manner as does the vibrating diaphragm of a loudspeaker, hence the pressure signal emanating

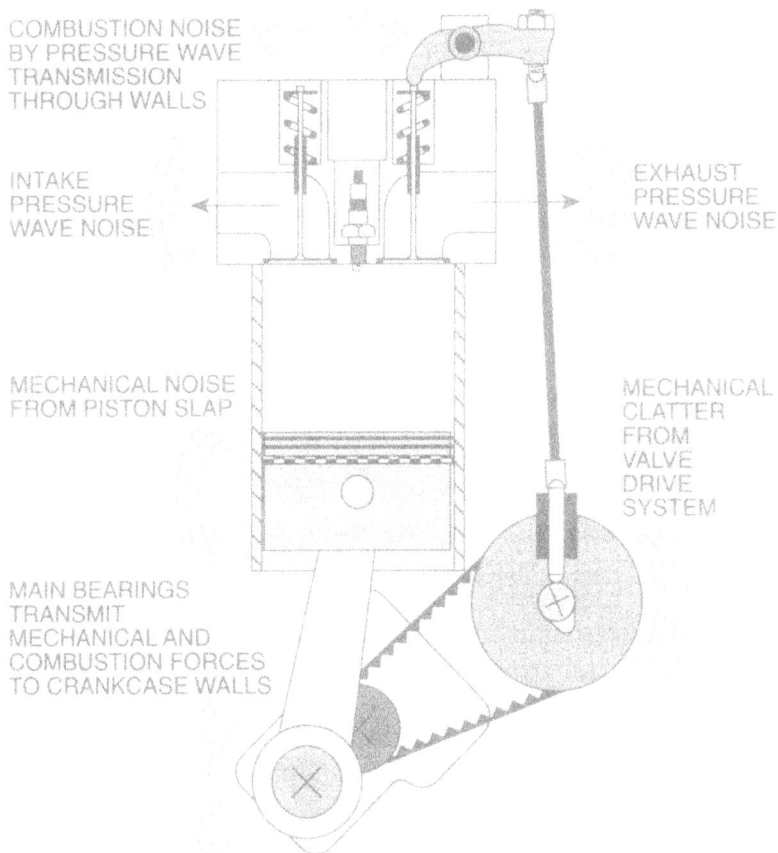

Fig. 7.1 Various sources of noise from a four-stroke engine.

703

from the combustion pressure rise is transmitted through the cylinder walls and can be propagated away from the outer surface of the cylinder and the cylinder head. If the engine is air-cooled, the cylinder and head finning is ideally suited to serve as metal diaphragms for this very purpose. The need to control this form of noise transmission is obvious, and you may have observed that air-cooled motorcycle engines have rubber damping plugs inserted between the cooling fins for this very purpose. The problem is greatly eased by the use of liquid (water) cooling of a cylinder and cylinder head, as the intervening liquid layer acts as a damper to noise transmission. But this does not totally solve the problem, for the combustion noise can be transmitted through the piston to the connecting rod, via the bearings to the crankcase walls, and ultimately to the atmosphere. The use of pressure-fed plain bearings on an engine crankshaft is superior to the use of ball, roller, or needle roller bearings as a means of suppressing this type of noise transmission, because the hydrodynamic oil film in the plain bearings does not so readily transmit the combustion vibrations, because it, too, acts as a damper.

The rotating bearings, particularly rolling element bearings, of an engine produce noise characteristics of their own, emanating from the vibrations of the mechanical components. Another mechanical noise source is that from piston slap, as the piston rocks on the gudgeon pin within the cylinder walls around the tdc and bdc positions. The valve train may be driven by gears or a chain, and the tappet clearances provide the mechanical clatter that can significantly add to overall engine noise. Such are the developments over this century in the mechanical design of the four-stroke engine that the modern spark-ignition automobile engine, running at idle with the bonnet (hood) up, is about as quiet as a well-oiled sewing machine.

Because all vibrating metal and plastic surfaces will produce noise in the manner of the loudspeaker, the outer surfaces of the exhaust and intake silencers may also need to be damped if the noises they emit are louder than those from the residual pressure waves at the silencer outlets. However, the discussion in this chapter regarding noise suppression concentrates mainly on the pressure-wave-generated noise from the exhaust and the inlet systems.

7.3 The Different Silencing Problems of Two-Stroke and Four-Stroke Engines

In the matter of silencing, the four-stroke engine has some advantages and disadvantages compared to an equivalent two-stroke engine [7.32]. These are cataloged below and discussed at greater length in succeeding sections as the need arises.

The Advantages of the Four-Stroke Engine
  (i)   The frequency of creation of gas pressure waves is half as great as in the two-stroke engine; this is an advantage because humans dislike exposure to higher-frequency noise.
 (ii)   The poppet valves of the four-stroke engine open more slowly than the ports in a two-stroke engine, thus the pressure wave fronts are more gradual, thereby creating fewer high-frequency noise components within the sound spectrum.
(iii)   Two-stroke engines conventionally use ball, roller, and needle roller element bearings, and these tend to be noisy compared to the pressure-fed hydrodynamic bearings used almost universally in four-stroke engines.

(iv) The intake system of the four-stroke engine does not contain the noise-creating devices often found in two-stroke engines, such as reed valves. Reed valves, much as in a clarinet or an oboe, can flutter or "honk" with an annoying frequency at certain engine speeds, creating a noise that is very difficult to eradicate.

The Disadvantages of the Four-Stroke Engine
(i) The two-stroke engine requires a tuned exhaust pipe to produce a high specific output, and as this is achieved by choking the final outlet diameter, greatly simplifying the design of an effective exhaust silencer. By contrast, muffling the exhaust of a four-stroke engine by resistive elements almost inevitably raises the "back pressure" on the engine cylinder(s) and reduces the air mass that can be breathed. Hence, a high specific power output is difficult, but not impossible, to maintain in a well-silenced four-stroke engine.
(ii) In the two-stroke engine, the conventional crankcase pump induces air by pumping with a low compression ratio and, because this reduces the maximum values of air intake particle velocity encountered in its time-history, the high-frequency content of the sound produced is reduced. The four-stroke engine gulps air forcefully from the atmosphere with a high compression ratio on each intake stroke. This raises its intake noise level and any simplistic restrictive approach to suppress that noise reduces the delivery ratio, and inevitably the power output, very considerably.
(iii) The peak combustion pressures are much higher in the equivalent four-stroke cycle engine, so the noise induced by those higher pressures is increased via all of the transmission components of the cylinder, cylinder head, piston, and crankshaft.

7.4 Some Fundamentals of Silencer Design
If you study the textbooks or technical papers on acoustics and on silencer design, such as those referenced below, you will find them to be full of empirical design equations for the many basic types of silencers found in internal combustion engines. But no matter how useful and informative these equations may be, you will get the feeling that you are not acquiring a fundamental understanding of the subject, particularly because the empiricism being applied is based on the propagation of acoustic waves, i.e., waves of infinitesimal amplitude, rather than finite-amplitude waves, i.e., waves of the very considerable amplitude to be found in the inlet and exhaust systems of the internal combustion engine. The acoustic approach to this theory is somewhat reminiscent of that for the heat transfer, one that produces an almost infinite plethora of empirical equations, about whose practical implementation the authors cheerfully admit rather large error bands. This has always seemed to me to be a most unsatisfactory state of affairs. Consequently, I instigated a research program at QUB many years ago to determine if it was possible to predict the noise spectrum emanating from the exhaust systems of internal combustion engines using an approach based on the calculation of the propagation of finite-amplitude waves. In the early years, the computations were conducted by the older method of characteristics [7.21], and more recently, by the more accurate methods described in Chapter 2. This resulted in several technical publications by Spechko, Coates, Mackey and Blair [7.1-7.3, 7.28-7.29]. A much more complete exposition of the work of Coates and Mackey is presented in their doctoral theses [7.17, 7.30].

7.4.1 The Theoretical Work of Coates

In these publications [7.3, 7.17], a theoretical solution is produced which shows that the sound pressure level at any point in space beyond the termination of an exhaust system into the atmosphere is—not empirically, but directly—capable of being calculated. The amplitude of the nth frequency component of the sound pressure, p_n, is shown to be primarily a complex function of: (a) the instantaneous mass flow rate leaving the end of the pipe system, $\dot{m}$, and (b) the location of the measuring microphone in both distance and directivity from the pipe end, together with other parameters of some lesser significance.

For any sinusoidal variation, the mean square sound pressure level, p_{rms}, of that nth frequency component is given by:

$$p^2_{rms\,n} = \frac{p^2_n}{2} \qquad (7.4.1)$$

Coates [7.3] shows that the mean square sound pressure emanating from a sinusoidal efflux from a pipe of diameter d is given by:

$$\left(p_{rms}\right)^2 = \frac{1}{2}\left(\frac{\pi d^2}{4}\right)^2\left(\frac{f_n}{r_m}\right)^2|\rho c|^2\left\{\frac{2J_1\left(\frac{\pi f_n d \sin\theta}{a_0}\right)}{\frac{\pi f_n d \sin\theta}{a_0}}\right\}^2 \qquad (7.4.2)$$

where J_1 is the notation for the first order Bessel function of the terms within. It can be seen from Eq. 7.4.2 that the combination of the second and fourth terms is the instantaneous mass flow rate at the aperture of an exhaust, or intake, system. The final term, in the curly brackets, indicates directivity of the sound, where θ is the angle between the receiving microphone from the center-line particle flow direction of the aperture. The variable r_m is the distance from the microphone to the pipe aperture leading to the atmosphere. From the theory in Sec. 2.2.3, Eq. 2.2.11 in particular, the instantaneous mass flow rate, $\dot{m}$, at the aperture to the atmosphere is given by:

$$\dot{m} = G_5 a_0 \rho_0\left(C_d \frac{\pi}{4}d^2\right)\left(X_i + X_r - 1\right)^{G5}\left(X_i - X_r\right) \qquad (7.4.3)$$

where the X_i and X_r values are the time-related incident and reflected pressure amplitude ratios of the pressure pulsations at the aperture to the atmosphere. An engine simulation of the type demonstrated in Chapter 5, using the theory of Chapter 2, inherently computes the instantaneous mass flow rate at every section of the engine and its ducting, including the inlet and exhaust apertures to the atmosphere. For a noise analysis within the simulation, the instantaneous mass flow rate at the aperture to the atmosphere is collected at each time step within the computation and a Fourier analysis of the resulting periodic function is performed numerically to include all harmonics within the audible range, giving a series of the form:

$$\dot{m}_t = \phi_0 + \left(\phi_{a1} \sin \omega t + \phi_{b1} \cos \omega t\right)......\left(\phi_{an} \sin n\omega t + \phi_{bn} \cos n\omega t\right) \qquad (7.4.4)$$

The amplitude of each harmonic is then used in Eq. 7.4.2 to calculate the mean square sound pressure due to that particular nth component:

$$\beta_n = 10 \log_{10} \left(\frac{p_{rms\,n}}{dp_0} \right)^2 \quad dB \qquad (7.4.5)$$

Hence, the overall mean square sound pressure, β_s, can be obtained by simple addition of each of the harmonics, using an extension of the procedure for the addition of sound energy, as seen in Eq. 7.1.5.

How successful that can be may be judged from some of the results presented by Coates [7.3]. Coates's work was carried out using the "homentropic method of characteristics" for the unsteady gas flow along the duct, and employed isentropic pipe end boundary conditions. Both of those approaches are shown by Kirkpatrick et al. [2.42, 5.20, 5.21] to be of a lesser accuracy than the methods presented in Chapter 2. The implication of this statement is that Coates's ability to accurately calculate mass flow rates at the pipe exit, upon which accurate noise computation is predicated, is impaired. However, because Coates's experiments on exhaust silencing were conducted in cold air with an "engine simulator," these theoretical errors are greatly reduced. Consequently, this research work is ongoing at QUB, using the theory shown in Chapter 2, to investigate the order of improvement in the accuracy of prediction of noise transmitted into space from the ducting of engines through silencers. The recent work of Mackey et al. [7.28-7.30] is solid evidence of this commitment and of the improvement in accuracy between measurement and simulation. The most recent work in this area, by Mawhinney [7.31], indicates that further sophistication and accuracy has been attained by employing the latest GPB simulation code allied to Coates's original theoretical approach for the noise propagation. There are other engine simulation codes that calculate both performance and noise characteristics, such as that presented by Silvestri [7.41], where the prediction of noise in the atmosphere beyond the intake system is estimated as a relative value based on the in-pipe pressure regime and not by the absolute methods of Coates [7.3].

7.4.2 The Experimental Work of Coates

The experimental rig used by Coates is described clearly in his technical paper [7.3], but a summary here will aid the discussion of the experimental results and their correlation with the theoretical calculations. The exhaust system is simulated by a rotary valve which allows realistic exhaust pressure pulses of cold air to be blown down into a pipe system at any desired cyclic speed for those exhaust pressure pulsations. The various pipe systems attached to the exhaust simulator are shown in Fig. 7.2, and are defined as SYSTEMS 1 to 4. Briefly, these systems are as follows:

SYSTEM 1 is a plain, straight 1.83 m pipe of 28.6 mm diameter, and is completely unsilenced.

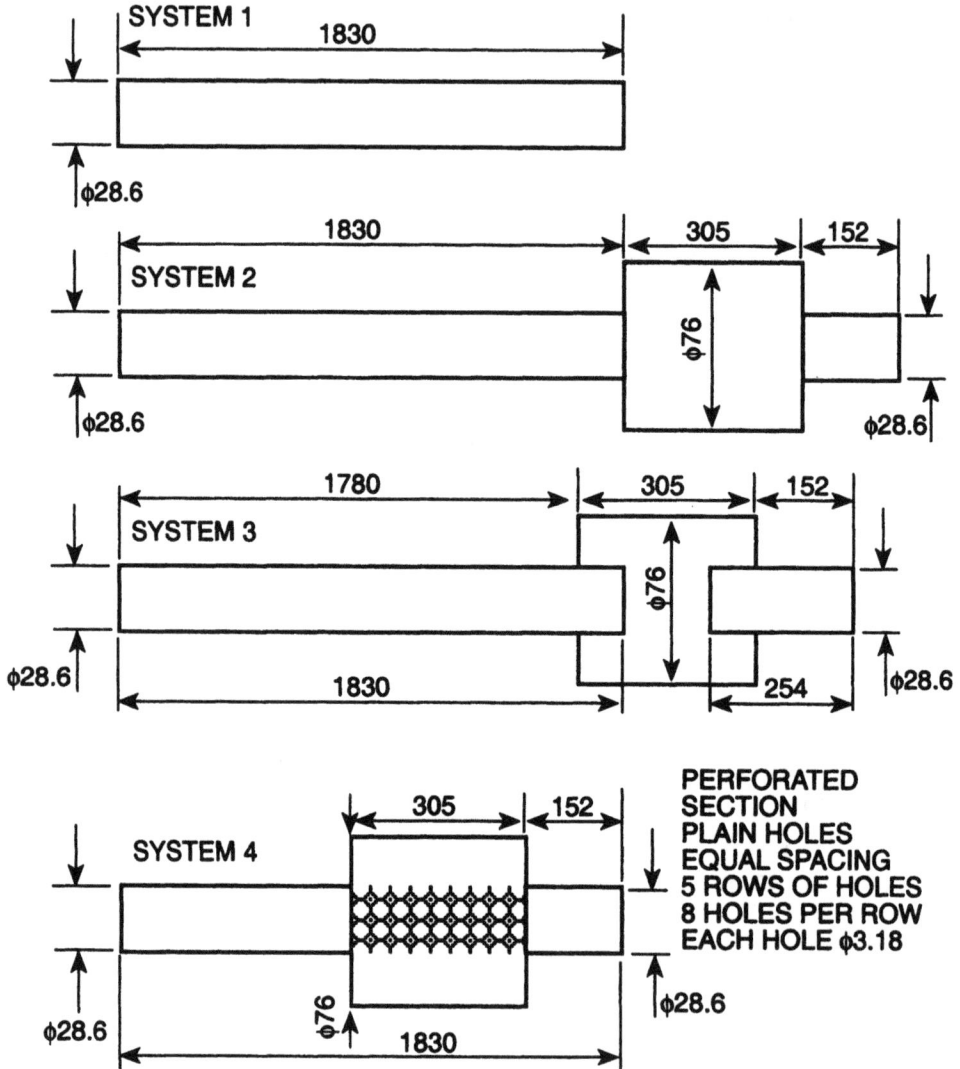

Fig. 7.2 Various exhaust systems and silencers used by Blair and Coates [7.3].

SYSTEM 2 has a 1.83 m plain pipe of 28.6 mm diameter culminating in what is termed a diffusing silencer, which is 305 mm long and 76 mm diameter. The tail pipe, of equal size to the entering pipe, is 152 mm long.

SYSTEM 3 is almost identical to SYSTEM 2, but the entry and exit pipes are reentrant into the diffusing silencer so that they are 102 mm apart within the chamber.

SYSTEM 4 has what is defined as a side-resonant silencer placed before the end of the 1.83 m pipe, and the 28.6 mm diameter through-pipe has 40 holes, of 3.18 mm diameter, drilled into it.

More formalized sketches of diffusing, side resonant, and absorption silencers are presented in Figs. 7.7 to 7.9. Further discussion of their silencing effects, based on an acoustic analysis, is presented in Sec. 7.5. It is sufficient to remark at this juncture that:

(i)   The intent of a diffusing silencer is to absorb all noise at frequencies other than those at which the box will resonate. Those frequencies that are not absorbed are called the pass-bands.

(ii) The intent of a side-resonant silencer is to completely absorb noise of a specific frequency, such as the fundamental exhaust pulse frequency of an engine.

(iii) The intent of an absorption silencer is to behave as a diffusing silencer, in which the packing absorbs the resonating noise at the pass-band frequencies.

The pressure-time histories within these various systems, and the one-third octave noise spectrograms emanating from them, were recorded. Of interest are the noise spectra, and these are shown for SYSTEMS 1 to 4 in Figs. 7.3 to 7.6, respectively, at 2000 exhaust pulses/min, i.e., the fundamental frequency is 33.3 Hz. The noise spectra are presented in the units of overall sound pressure level, dBlin, as a function of frequency. There are four spectrograms on any given figure: the one at the top was taken with the measuring microphone placed directly in line with the pipe end, and the directivity angle in this case is declared as zero; the others correspond to directivity angles of 30, 60, and 90°, respectively. On any given figure, the solid line represents the measured noise spectra and the dashed line that resulting from the theoretical solution outlined in Sec. 7.4.1, as calculated by Coates at that time. It can be seen that there is a very high degree of correlation between the calculated and measured noise spectra, particularly at frequency levels below 2000 Hz.

Perhaps the most important conclusion to be drawn from these results is that the designers of silencing systems, having a first priority of muffling the noisiest frequency, can use an unsteady gas-dynamic simulation program to predict the mass flow rate at the pipe termination to atmosphere, be it the inlet or the exhaust system, and from that cyclic mass flow rate calculation can determine the noisiest frequency to be tackled by the silencer.

Of direct interest is the silencing effect of the various silencer elements that Coates attached to the exhaust pipe, SYSTEMS 2 to 4, the noise spectra for which are shown in Figs. 7.4 to 7.6. SYSTEM 2, a simple diffusing silencer without reentrant pipes, can be seen to reduce the noise level of the fundamental frequency at 133 Hz from 116 dB to 104 dB, an attenuation of 12 dB. From the discussion in Sec. 7.1.2, an attenuation of 12 dB is a considerable level of noise reduction. It will also be observed that a large "hole," or strong attenuation, has been created at 400 Hz in the noise spectra of SYSTEM 2. In Sec. 7.5.1, this will be a source of further comment when an empirical frequency analysis is attempted for this particular design.

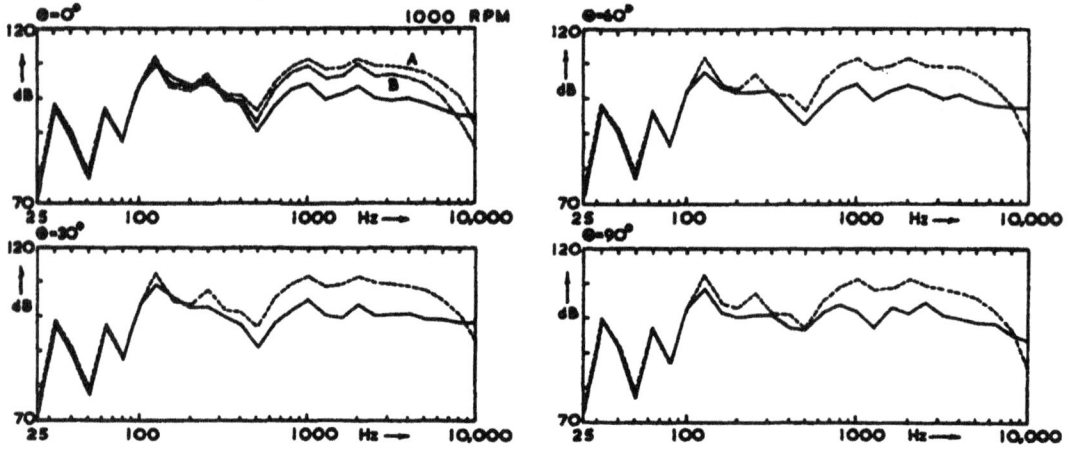

Fig. 7.3 One-third octave noise spectra measured from SYSTEM 1.

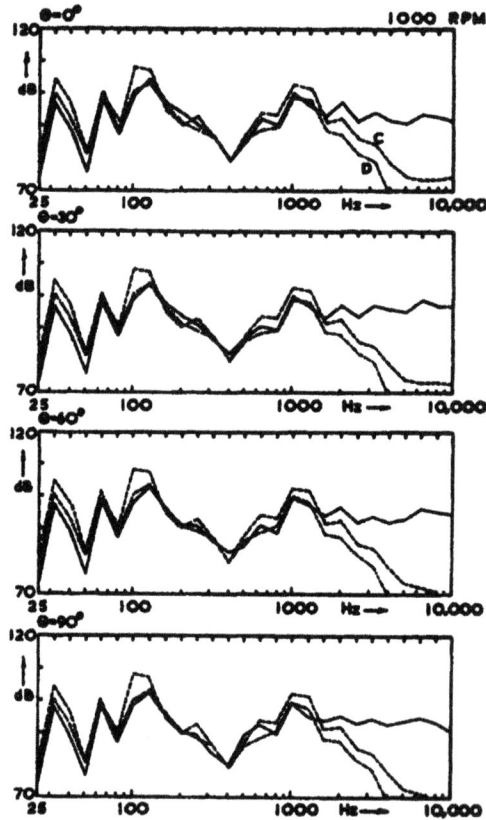

Fig. 7.4 One-third octave noise spectra measured from SYSTEM 2.

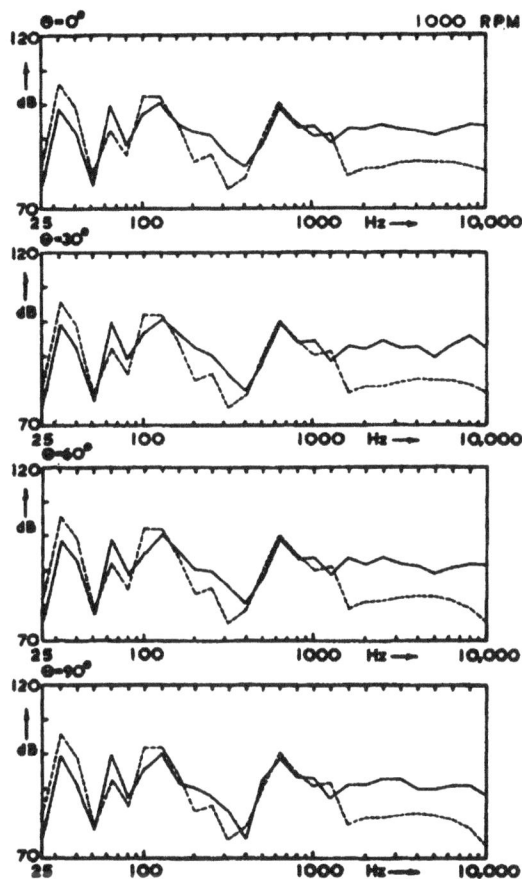

Fig. 7.5 One-third octave noise spectra measured from SYSTEM 3.

The noise spectra for SYSTEM 3, a diffusing silencer with reentrant pipes, are shown in Fig. 7.5. The noise level of the fundamental frequency of 133 Hz has been reduced further to 98 dB. The "hole" of high attenuation is now at 300 Hz and is deeper than that recorded by SYSTEM 2.

The noise spectra for SYSTEM 4, a side-resonant silencer, are shown in Fig. 7.6. The noise level of the fundamental frequency of 133 Hz is nearly as quiet as in SYSTEM 2, but a new attenuation hole has appeared at a higher frequency, about 400-500 Hz. This, too, will be commented on in Sec. 7.5.2 when an empirical acoustic analysis is presented for this type of silencer. It can also be seen that the noise level at higher frequencies, i.e., above 1000 Hz, is reduced considerably from the unsilenced SYSTEM 1.

The most important conclusion from this work by Coates is that noise propagation into space from a pipe system, with or without silencing elements, can be predicted by a theoretical calculation based on the motion of finite amplitude waves propagating within the pipe system

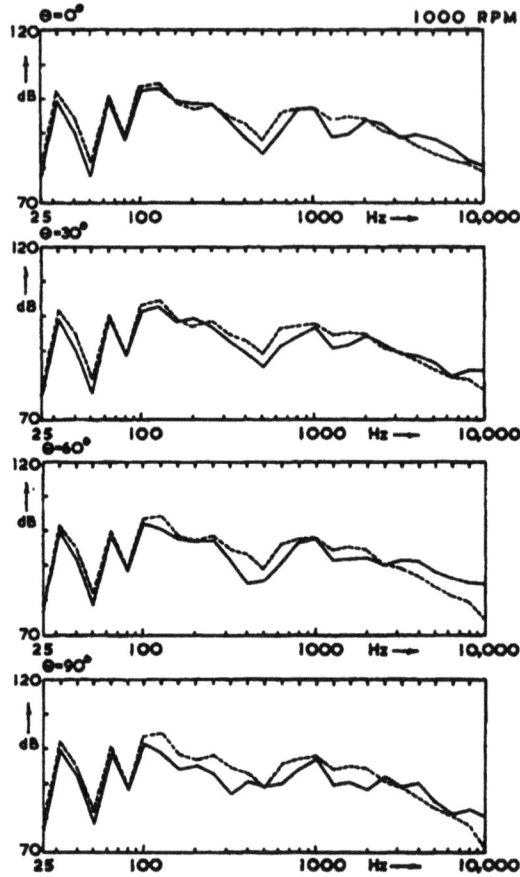

Fig. 7.6 One-third octave noise spectra measured from SYSTEM 4.

to the pipe termination at the atmosphere. In other words, designers of silencers for internal combustion engines do not have to rely on empiricism based on acoustics for their designs, be they for silencers for the intake or the exhaust system.

7.4.3 Future Work in the Prediction of Silencer Behavior

It has always seemed to me that this pioneering work of Coates [7.3, 7.17] has never received the recognition it deserves. Worse, it has tended to be ignored [7.25, 7.35, 7.36], in part because those involved in silencer design do not tackle their problems using theory based on unsteady gas dynamics, but persist with theory based on acoustics, i.e., pressure waves with an infinitesimal amplitude [7.24]. No matter how mathematically sophisticated that acoustic theory may be, it is fundamentally the wrong approach to calculate the noise emission created by the propagation of finite-amplitude pressure waves through silencers into the atmosphere. A subsidiary reason—and Coates comments on this—is that silencer elements tend to have somewhat complicated geometries and have gas particle flow characteristics that are often three-dimensional in nature. Because unsteady gas-dynamic calculations are nominally one-dimensional, their potential for accurately predicting the pressure wave reflection and transmission characteristics of real silencer elements is definitely compromised. Nevertheless, the fundamental accuracy of the Coates's theoretical premise is clearly demonstrated and should be the preferred route through which all researchers in this subject handle the real problems posed by the complex geometry of engine silencers. Mawhinney [7.31] demonstrated excellent correlation between measured and calculated noise spectra in many complex silencer geometries. However, the theoretical means for better accuracy are now available.

Secs. 3.1 and 3.2 show how Computational Fluid Dynamics (CFD) is employed to solve the three-dimensional flow behavior at pipe ends or at the cylinders. There is no logical reason why this same theoretical approach cannot be employed for the design of the most complex of silencers. The engine, together with its inlet and exhaust systems, is modeled by the unsteady gas dynamics method given in Chapter 2, and described further in Chapters 5 and 6, up to the boundaries of the various silencers, and its temporal output is fed to a CFD program for three-dimensional pursuit through the entire silencer geometry. The result of the CFD calculation is a more accurate prediction of the mass flow rate spectrum entering the atmosphere, from which the ensuing noise spectra can still be correctly assessed by the methods of Coates [7.3]. This becomes a realistic design process for the engine, together with its ducting and silencers, rather than the empirically based acoustics methods still in vogue today [7.18, 7.26]. Acoustics theory can be relegated to the useful role that befits all empiricism, namely providing more appropriate initial estimates of the geometry of silencer systems as input data, so that the combined UGD/CFD simulation is not so wasteful of expensive human and computer time. That this is no professorial pipe dream is confirmed in the work by Mackey et al. [7.29-7.30].

7.5 Acoustic Theory for Silencer Attenuation Characteristics

The behavior of silencers in terms of acoustic theory is described in many texts and papers, of which the book by Annand and Roe [7.33] deserves the most attention. Further study material may be found in the references at the end of this chapter. This section will concentrate on the characteristics of the most common types of exhaust silencers, the diffusing, side-resonant, and absorption silencers shown in Figs. 7.7 to 7.10, and the low-pass intake silencer sketched in Fig. 7.11.

7.5.1 The Diffusing Type of Exhaust Silencer

A sketch of a diffusing silencer element is shown in Fig. 7.7. This type of silencer element has entry and exit pipes of areas A_1 and A_2, respectively, which, if of circular cross section, have diameters of d_1 and d_2, respectively. The pipes can be reentrant into the box with lengths L_1 and L_2, respectively. The final tail pipe leaving the box has a length L_t. The box has a length L_b and a cross-sectional area A_b, or a diameter d_b, should the box be of a circular cross section. It is quite common, for reasons of manufacturing simplicity, for both the pipes and the box of a silencer to be circular in cross section. If the pipes are of the reentrant type, then any theory calling for a computation of the volume of the box, V_b, should take into account the box volume occupied by those pipes, thus:

$$V_b = A_b L_b - (A_1 L_1 + A_2 L_2) \qquad (7.5.1)$$

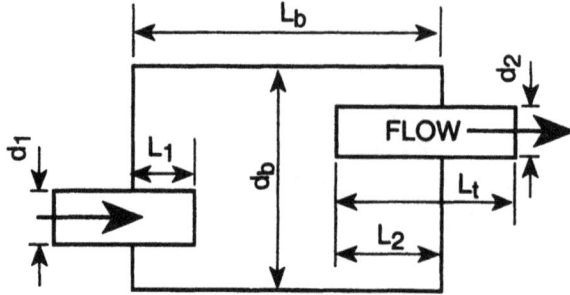

Fig. 7.7 Significant dimensions of a diffusing silencer element.

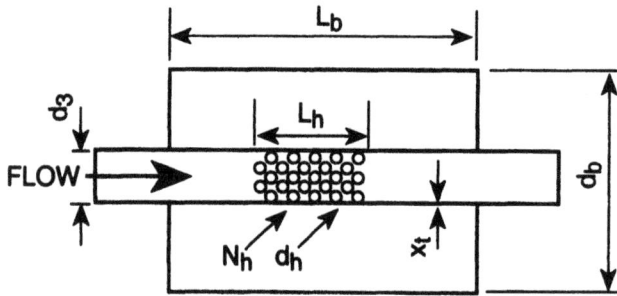

Fig. 7.8 Significant dimensions of a side-resonant silencer element.

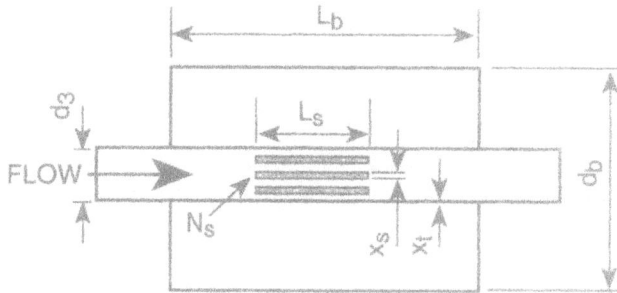

Fig. 7.9 Significant dimensions of a side-resonant silencer element with slits.

The transmission loss, or attenuation, β_{tr}, of a diffusing silencer is basically a function of two parameters: the expansion ratio, A_r, i.e.,

$$A_{r1} = \frac{A_b}{A_1} \qquad A_{r2} = \frac{A_b}{A_2}$$

and the relationship between the wavelength of the sound, Λ, and the length of the box, L_b.

714

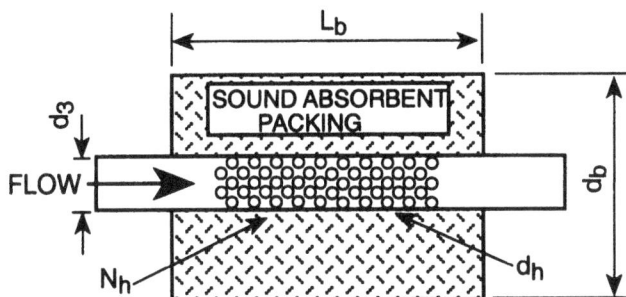

Fig. 7.10 Significant dimensions of an absorption silencer.

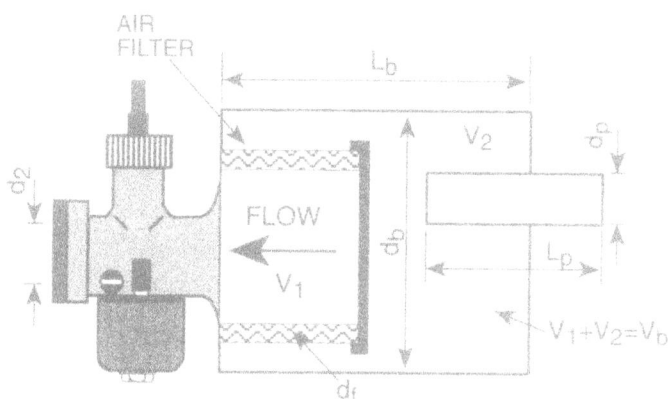

Fig. 7.11 Significant dimensions of a low-pass intake silencer.

The wavelength of the sound is connected to the frequency of that sound, f, and to the acoustic velocity, a_0. The acoustic velocity is calculated by Eq. 7.1.1. The wavelength, Λ, is found from:

$$\Lambda = \frac{a_0}{f} \qquad (7.5.2)$$

Because the frequency of the gas-borne noise arriving into the diffusing silencer varies, the box will resonate, much like an organ pipe, at various integer amounts of half the wavelength [7.33], in other words at,

$$\Lambda = 2L_b, \ \frac{2L_b}{2}, \ \frac{2L_b}{3}, \ \frac{2L_b}{4}, \ \frac{2L_b}{5}, \text{ etc.}$$

From Eq. 7.5.2, this means at frequencies of,

$$f = \frac{a_0}{2L_b}, \frac{2a_0}{2L_b}, \frac{3a_0}{2L_b}, \frac{4a_0}{2L_b}, \frac{5a_0}{2L_b}, \text{ etc.}$$

At these frequencies, the silencer will provide a transmission loss of zero, i.e., no silencing effect at all, and such frequencies are known as the "pass bands." It is also possible for the silencer to resonate in the transverse direction through the diametral dimension, d_b, and provide further pass-band frequencies at what would normally be a rather high frequency level.

Many empirical equations exist for the transmission loss of such a silencer, but Kato and Ishikawa [7.18] state that the theoretical solution of Fukuda and Izumi [7.9] is found to be useful. The relationship of Fukuda and Izumi [7.9] is as follows for the transmission loss of a diffusing silencer, β_{tr}, in dB units:

$$\beta_{tr} = 10\log_{10}\left(A_{r2}F(\kappa,L)\right)^2 \quad \text{dB} \tag{7.5.3}$$

where
$$F(\kappa,L) = \frac{\sin(\kappa L_b) \times \sin(\kappa L_t)}{\cos(\kappa L_1) \times \cos(\kappa L_2)} \tag{7.5.4}$$

and
$$\kappa = \frac{2\pi f}{a_0} \tag{7.5.5}$$

Not surprisingly, in such empirical relationships there are many correcting factors offered to modify the basic relationship to cope with the effects of gas particle velocity, end correction effects for pipes that are either reentrant or flush with the box walls, boxes that are lined with absorbent material, but not sufficiently dense as to be called an absorption silencer, etc. It will be left to you to pursue these myriad formulae, found in the References, should you be so inclined.

To assist you in the use of this theory for design purposes, a simple computer program listing, in BASIC, can be found in an appendix to one of my previous books [7.34], as Prog. 8.1. The attenuation equations programmed are those given by Fukuda, Eqs. 7.5.3 to 7.5.5.

To determine if such a design program is useful in a practical sense, an analysis of SYSTEM 2 and SYSTEM 3, as presented Coates [7.3], is attempted, and the results are shown in Figs. 7.12 and 7.13. These figures show the computer screen output from Prog. 8.1, which are plots of the attenuation in dB as a function of noise frequency up to a maximum of 4 kHz, beyond which frequency most experts agree that the diffusing silencer will have little silencing effect. The theory would continue to predict some attenuation to the highest frequency levels, indeed beyond the upper threshold of hearing. Also displayed on the computer screen are the input data for the geometry of that diffusing silencer according to the symbolism presented in Fig. 7.7 and in this section. These input data are in the more conventional length dimensions of mm units, with temperatures as Celsius values. The data values are converted to strict SI units before applying the theory in Eqs. 7.5.3 to 7.5.5.

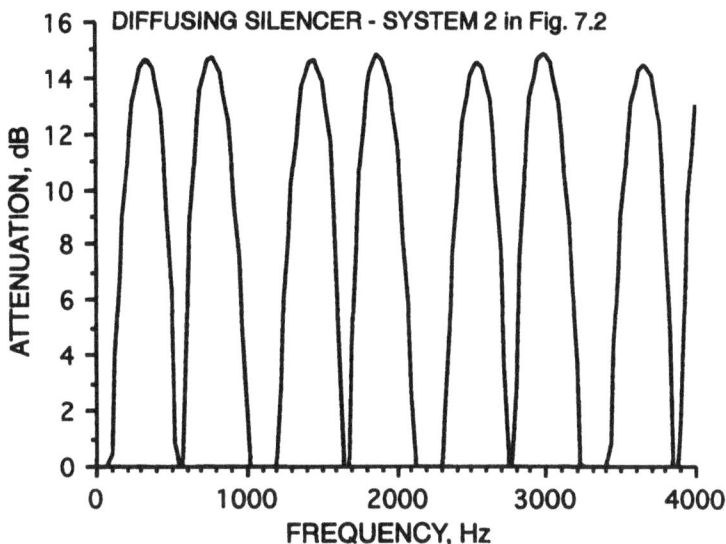

Fig. 7.12 Calculation of the attenuation characteristics of Coates's SYSTEM 2.

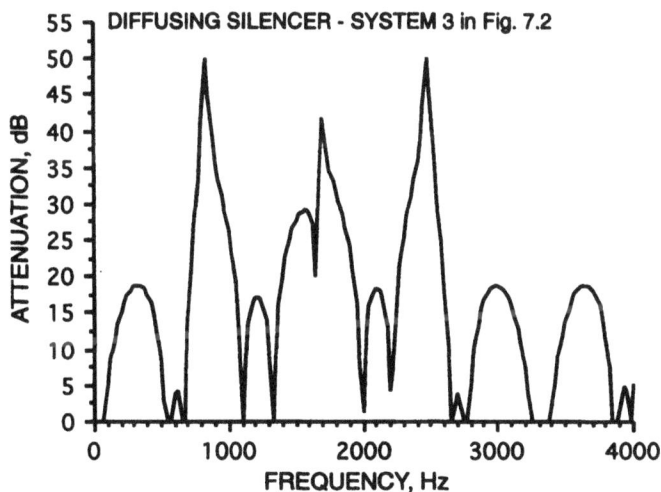

Fig. 7.13 Calculation of the attenuation characteristics of Coates's SYSTEM 3.

The measured noise characteristics of SYTEM 2 are shown in Fig. 7.4. The attenuation of SYSTEM 2, as predicted by the theory of Fukuda and shown in Fig. 7.12, has a first major attenuation of 14 dB at 320 Hz, and the first two pass-band frequencies are at 550 and 1100 Hz. If you examine the measured noise frequency spectrum in Fig. 7.4, you will see that an attenuation hole of 12 dB is created at a frequency of 400 Hz. Thus, the correspondence with the theory of Fukuda, with regard to this primary criterion, is quite good and gives some confidence in its application for this particular function. There is also some evidence of the narrow pass band at 550 Hz and there is no doubt about the considerable pass-band frequency at 1100 Hz in the measured spectrum. There is no sign of the predicted attenuation, nor the pass-band holes, at frequencies above 1.5 kHz in the measured spectrum. There is also no evidence from the empirical solution of the reason for the attenuation in the measured spectrum of the fundamental pulsation frequency of 133 Hz, as commented on in Sec. 7.4.2. The general conclusion, as far as SYSTEM 2 is concerned, is that the Fukuda acoustic solution is a useful empirical calculation for a diffusing silencer up to a frequency of 1.5 kHz.

The measured noise characteristics of SYTEM 3 are shown in Fig. 7.5. The attenuation of SYSTEM 3, as predicted by the theory of Fukuda and shown in Fig. 7.13, has a first major attenuation of 19 dB at 320 Hz, i.e., greater than that of SYSTEM 2. It is true that the measured silencing effect of SYSTEM 3 is greater than that of SYSTEM 2, but this is not the case at a frequency of 400 Hz, where the measured attenuations are both at a maximum and are virtually identical. The theory would predict that the pass-band hole of SYSTEM 2 at 1100 Hz would not be so marked for SYSTEM 3, and this can be observed in the measured noise spectrum. The prediction by Fukuda theory of a pass-band at 600 Hz is clearly seen in the measured noise spectrum. Again, there is little useful correlation between theory and experiment at higher frequencies above 1500 Hz.

The general conclusion to be drawn, admittedly from limited evidence, is that Prog. 8.1, codifying the theory of Fukuda, is a useful empirical approach for the design of diffusing silencers up to a frequency of about 1500 Hz.

7.5.2 The Side-Resonant Type of Exhaust Silencer

The fundamental behavior of this type of silencer is to absorb a relatively narrow band of sound frequency through the resonance of the side cavity at its natural frequency [7.36].

The Side-Resonant Silencer with Round Holes in a Central Duct

The sketch in Fig. 7.8 shows a silencing chamber of length L_b and area A_b, or diameter d_b, if the cross-section is circular. The exhaust pipe is usually located centrally within the silencer body, with an area A_3, or diameter d_3, if it is a round pipe, and has a pipe wall thickness x_t. The connection to the cavity, whose volume is V_b, is via holes, or a slit, or via a short pipe in some designs. The usual practice is to employ a number of holes, N_h, each of area A_h, or diameter d_h, if they are round holes for reasons of manufacturing simplicity. The volume of the resonant cavity is V_b, where, if all cross sections are circular:

$$V_b = A_b L_b - \frac{\pi L_b (d_3 + 2x_t)^2}{4} \qquad (7.5.6)$$

According to Kato and Ishikawa [7.18], the length L_h occupied by the holes should not exceed the pipe diameter d_3, otherwise the system should be treated theoretically as a diffusing silencer (however, you will see later that I find this advice to be of limited value when applied in the context of an actual engine). The natural frequency of the side-resonant system, f_{sr}, is given by Davis et al. [7.4] as:

$$f_{sr} = \frac{a_0}{2\pi}\sqrt{\frac{K_h}{V_b}} \qquad (7.5.7)$$

where K_h, the conductivity of the holes that make up the opening, is calculated from:

$$K_h = \frac{N_h A_h}{x_t + 0.8 A_h} \qquad (7.5.8)$$

The attenuation or transmission loss in dB of this type of silencer, β_{tr}, is given by Davis et al. [7.4] as:

$$\beta_{tr} = 10\log_{10}\left(1 + Z^2\right) \qquad (7.5.9)$$

where the term Z is found from:

$$Z = \frac{\dfrac{\sqrt{K_h V_b}}{2A_3}}{\dfrac{f}{f_{sr}} - \dfrac{f_{sr}}{f}} \qquad (7.5.10)$$

When the applied noise frequency, f, is equal to the resonant frequency, f_{sr}, the value of the term Z becomes infinite, as does the resultant noise attenuation in Eq. 7.5.9. Clearly, this is an impractical result, but it does give credence to the view that such a silencer should have a considerable attenuation level in the region of the natural frequency of the side-resonant cavity and connecting passage.

To assist you with the use of this theory for design purposes, a simple computer program listing, in BASIC, can be found in an appendix to one of my previous books [7.34], as Prog. 8.2. The attenuation equations programmed are those given by Davis, Eqs. 7.5.6 to 7.5.10.

To demonstrate the use of this acoustic design approach, and to determine if the predictions emanating from it are of practical use to the designer of side-resonant elements within an exhaust muffler for a four-stroke engine, the geometrical and experimental data pertaining to SYSTEM 4 from Coates [7.3] are inserted as data, and the results are illustrated in Fig. 7.14.

This figure represents the computer screen picture as seen by a user of Prog. 8.2. The information displayed is the input data for, in this instance, SYSTEM 4. You may well wonder about 10°C being the declared data value for exhaust temperature. This is because the simulation work of Coates was conducted using a rotor valve delivering compression waves in cold air to represent "exhaust" pulses.

The calculated sound attenuation of SYSTEM 4, as seen in Fig. 7.14, shows a large transmission loss of 50 dB at 729 Hz, with further attenuation stretching to 4.0 kHz. If you examine the measured noise spectrum of SYSTEM 4 in Fig. 7.6, and compare it to the spectrum for the unsilenced SYSTEM 1 in Fig. 7.3, you will see that there is attenuation produced by this side-resonant silencer at 400-500 Hz. The attenuation peak of the measured noise spectra is not at 729 Hz. However, the actual noise level at 400 Hz is only slightly lower than that of the unsilenced SYSTEM 1. Of much greater interest to the designer is the visibly strong attenuation given by SYSTEM 4 stretching up to 10 kHz, which is not present in either of the diffusing silencer designs of SYSTEMS 2 and 3. It will also be noted that the calculated attenuation stretches down to 100 Hz, i.e., below the fundamental frequency of 133 Hz seen in Coates's experiments depicted in Fig. 7.3, and that the measured diagram in Fig. 7.6 does show attenuation of that frequency at least equal to the level observed for the diffusing silencers.

This limited evidence shows that the theoretical work of Davis et al. [7.4], is reasonably useful in determining the fundamental muffling frequency of a side-resonant silencer, and that such a device is also effective, in practical terms, for attenuating noise at frequencies above 1 kHz.

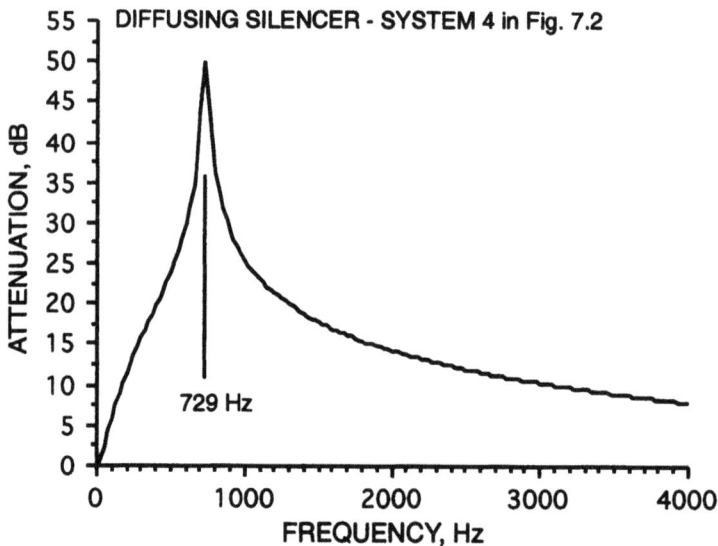

Fig. 7.14 Calculation of the attenuation characteristics of Coates's SYSTEM 4.

The Side-Resonant Silencer with Slits in a Central Duct

The sketch in Fig. 7.9, which is very similar to Fig. 7.8, shows a silencing chamber of length L_b and area A_b. The exhaust pipe is normally located centrally in the silencer, and is of area A_3, or diameter d_3, if it is a round pipe, and has a pipe wall thickness of x_t. The connection to the cavity, whose volume is V_b, is via a slit, or a number of slits, N_s. Each slit is of length L_s and of width x_s.

The theoretical solution for the behavior of this type of side-resonant silencer is virtually identical to that given above and expressed as Eqs. 7.5.6 to 7.5.10. Only the relationship for the conductivity of the opening, K_s, is different, and is given by Annand and Roe [7.33] as:

$$K_s = \frac{N_s x_s L_s}{x_t + 0.92\kappa_s\sqrt{x_s L_s}} \tag{7.5.11}$$

where the parameter κ_s is directly related to the length-to-width ratio of the slot by:

$$\kappa_s = 1.0287 - \sqrt{1.0579 - 0.0088763\times\left(120.09 - \frac{L_s}{x_s}\right)} \tag{7.5.12}$$

The solution for the transmission loss, β_{tr}, of this variation of the side-resonant silencer is obtained by solving the same equations as before, Eqs. 7.5.6 to 7.5.10, replacing the conductivity for the round holes, K_h, with that for the slits, K_s.

7.5.3 The Absorption Type of Exhaust Silencer

It is generally held that an absorption silencer, as illustrated in Fig. 7.10, acts as a diffusing silencer [7.18], and that the effect of the packing is to absorb noise in the pass-bands appropriate to a diffusing silencer. The experimental evidence, when large-amplitude waves are employed [7.3, 7.30], shows considerable, persistent attenuation at frequencies beyond 1 kHz in a fashion not provided by a diffusing silencer. The fundamental geometry of the system is for a pipe of area A_3, passing through a box of length L_b and area A_b. If the cross sections are circular, the diameters are d_3 and d_b, respectively. The holes through the pipe are normally round and are N_h in number, each of area A_h, or diameter d_h. The physical geometry is very similar to that of the side-resonant silencer, except the total cross-sectional area of all of the holes is probably in excess of five times the pipe area, A_3. This is the reason that the theoretical acoustic analysis is said to be more akin to a diffusing silencer than a side-resonant silencer, particularly because the area ratio of holes to pipe in a side-resonant silencer is normally less than unity. Expressed numerically, the following are conventional empiricism:

$$\text{Absorption silencer} \qquad \frac{N_h A_h}{A_3} > 5$$

$$\text{Side - resonant silencer} \qquad \frac{N_h A_h}{A_3} < 1$$

Experimental evidence shows that the absorption silencer is very effective at attenuating high frequencies, i.e., above 2 kHz, and is relatively ineffective below about 400 Hz. Most authorities agree that the theoretical design of an absorption silencer, either by the methodology of Coates [7.3] or by an acoustic procedure, is somewhat difficult because it depends significantly on the absorption capability of the packing material. The packing material is usually a glass-reinforced fiber material or mineral wool.

Thus, you should attempt the design using acoustic side-resonator theory, but arrange for as many holes of an appropriate diameter in the central pipe as is pragmatic. If the hole size is too large, say in excess of 3.5 mm diameter, the packing material within the cavity will be blown or shaken into the exhaust stream. If the hole size is too small, say less than 2.0 mm diameter, the particulates in the exhaust gas of a four-stroke engine may ultimately seal them over, thereby rendering the silencing system ineffective. The normal hole size used in such silencers is between 2.0 and 3.5 mm diameter.

One detailed aspect of design that warrants special comment is the configuration of the holes in the perforated section of the silencer. The conventional manufacturing method is to roll, then seam weld, the central pipe from a flat sheet of perforated mild steel. Although this produces an acceptable design for the perforated pipe, a superior methodology is to manufacture the pipe by the same method, but from a mild steel sheet in which the holes have been somewhat coarsely "stabbed," rather than cleanly excised. Indeed, the stabbed holes can be readily manufactured in pre-formed round pipe by an internal expanding tool. The result is as sketched in Fig. 7.15 for the finished pipe section. Needless to add, this method raises the discharge coefficient, C_d, for flow into the side-chamber from the central pipe [7.37]. This has the added effect of reducing the turbulent eddies produced by the gas flowing past the sharp edges of the clean-cut holes of conventional perforations. Such turbulence has an irritating high-frequency content; recall that a whistle produces noise by these same edge effects.

Positioning an Absorption Silencer Segment

As commented on above, the published literature on this topic [7.33] agrees on several facets of the design, but the principal function is to remove the high-frequency end of the noise spectra. The high-frequency portion of any noise spectrum emanates from two significant sources: the first is the sharp pressure rise at the front of the exhaust pulse, always steepening as it travels along the exhaust duct (see Fig. 6.68), and the second is the turbulence generated by the gas particle flow passing sharp edges or corners, i.e., any protuberance into the gas stream. The considerable noise content inherent in turbulence is quite visible in Plate 2.4, where the whirling smoke-ring of particles is seen to follow the spherical pressure wave front into the atmosphere. The implication is that an absorption silencer element should always be the final segment of a series of elements making up a complete muffler. The logic behind this

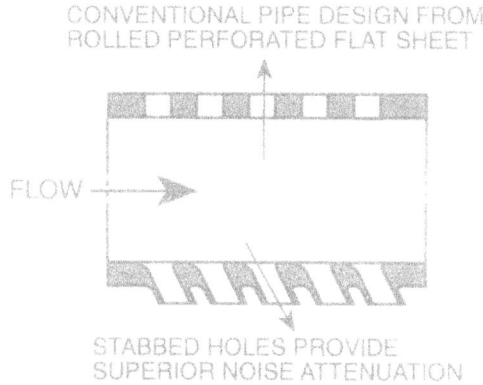

Fig. 7.15 Reduced turbulence noise in absorption silencers by using stabbed perforations.

statement is that there is little point in muffling the high-frequency portion of the noise spectrum with an absorption silencer element if this is followed by a diffusing silencer, because this would produce further turbulence-generated noise from the eddies in the wake off any sharp, protruding pipe edges. This may be impeccable logic for two-stroke engines, or low-performance four-stroke engines, but for a high-specific-output engine, a reversed layout can be applicable (see Sec. 7.6.4 and Fig. 7.31).

7.5.4 The Laminar Flow Exhaust Silencer

One of the most interesting and effective silencers, and one that is very simple to design and manufacture, is that suggested by Roe and others [7.27, 7.30, 7.32]. A variation on that theme, developed at QUB, is sketched in Fig. 7.16. This has been shown to be applicable to a four-stroke motorcycle [7.23].

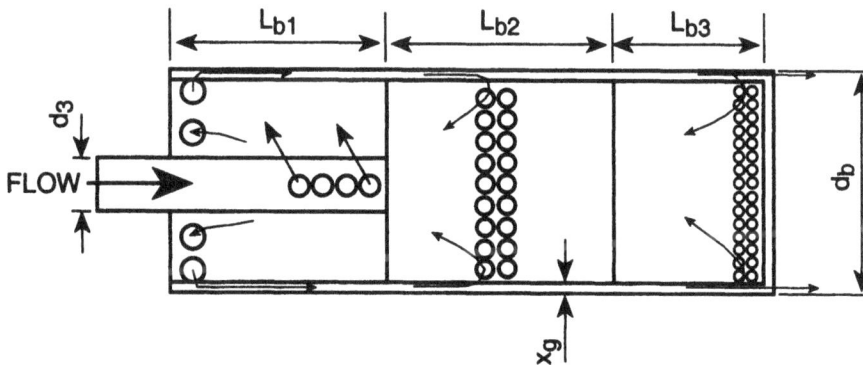

Fig. 7.16 A QUB design for a laminar flow silencer.

The basic principle is that the exhaust gas flows through the final exit section of the silencer, which is a narrow annular gap of radial dimension x_g. The word "laminar" has been used to describe this final phase of the outflow. For that to be an accurate description, the Reynolds number in the annular section would have to be less than 2000. The Reynolds number in the gap is given by:

$$\mathbf{Re}_g = \frac{\rho_g c_g d_g}{\mu_g} \tag{7.5.13}$$

where d_g is the hydraulic diameter at this location and the subscript "g" denotes the local gas properties.

Let it be assumed that the area of flow, A_g, in the final annulus phase is equated to the entering duct flow area, A_3. This is a somewhat generous proportionality if very effective silencing is required, but can be used as a design criteria if the device is to muffle a high-performance engine.

$$A_3 = \frac{\pi}{4} d_3^2 = A_g = \frac{\pi}{4}\left(d_b^2 - (d_b - 2x_g)^2 \right) \tag{7.5.14}$$

In this case, the radial gap dimension, x_g, is approximately given by:

$$x_g = \frac{d_b - \sqrt{d_b^2 - d_3^2}}{2} \tag{7.5.15}$$

The hydraulic diameter, d_g, is evaluated as:

$$d_g = \frac{4 \times \text{Area of gap}}{\text{wetted perimeter}} = \frac{d_b^2 - (d_b - 2x_g)^2}{2d_b - 2x_g} = 2x_g \tag{7.5.16}$$

To put some numbers to this contention, consider that if the pipe diameter, d_3, is 36 mm and the inside silencer body diameter, d_b, is 90 mm, then the gap dimension, x_g, becomes 3.76 mm. Taking the argument further, using the theory of Sec. 2.3.1, assume that an exhaust pressure pulse of 1.2 atm appears in the entry duct at a reference temperature of 500°C. The theory in Sec. 2.3.1 shows that the Reynolds number in the entry duct is 37,900, the local particle velocity is 73.5 m/s, and the friction factor for the flow is 0.0057. To maintain the mass flow rate without undue restriction, if it is assumed that equality of particle velocity is maintained in the annular gap, to complement the assumption of equality of flow area, then the Reynolds number in the final annulus is 7900 and the friction factor is 0.008. The convection heat transfer coefficients, C_h, in the entry duct and the final annulus are predicted as 173 and 244 W/m²K, respectively. If the mean velocity of the flow in the final annulus section is reduced below that of the pressure pulsations in the entering duct, then it can be observed that the possibility of laminar flow occurring is quite real.

Irrespective of the occurrence of laminar flow, the viscous effects and the area over which they apply, is significantly increased. This decreases turbulence levels and dampens other acoustic oscillations in a manner that is significantly greater than for conventional silencers. Heat transfer rate is increased, as is the area over which it is applied in the annulus, so that, too, will provide a sharper temperature profile along the annulus, thus giving continuous wave reflections to further dampen the pressure oscillations (see Sec. 2.5).

There are several other factors that contribute toward the overall effectiveness of this silencer:

(i) The efflux of hot exhaust gas occurs through an annular ring into air that is at a much lower temperature and higher density. There is a large contact area between the two surfaces. This produces considerable damping of the turbulence in the exiting exhaust gas plume, thereby reducing the high-frequency noise inherent in that source. This approach to silencing turbulence noise is conventional practice for aircraft gas turbines.

(ii) The central body houses side-resonant cavities that can be used to tune out particular frequencies with a high noise content. The flow passing the entrances to these cavities is moving at right angles to those apertures, and at particle velocities that are closer to acoustic levels than those in the conventional silencers shown in Figs. 7.7 to 7.9. Thus, the assumptions inherent in acoustic theory for Helmholz resonators [7.36] are approached more closely, and the application of that theory can be applied with a little more confidence in the quality of the outcome.

(iii) The exhaust pressure pulsations entering the first chamber, a diffusing silencer section, have the normal and considerable amplitude associated with such waves. The first box reduces the magnitude of the pressure oscillation before it enters the annulus en route to the atmosphere, passing the several resonant cavities as it goes. However, the outside skin of the silencer does not experience the pressure forces due to the full magnitude of the primary oscillation in the first box, as is the case with the other silencers shown in Figs. 7.7 to 7.9, and so the outside skin vibrates less than it does in those other silencers. Later on, Sec. 7.7 presents a discussion whereby double-skinning of the outside of a conventional silencer may be necessary to prevent that vibration from being a significant source of noise. That necessity may be eliminated by using a laminar flow silencer.

7.5.5 *The Manufacture of Exhaust Silencers*

Most automotive exhaust silencers are manufactured by rolling and stamping various segments from sheet steel to form the body of the silencer and making the internal, inlet, and outlet connections from round tubing. The steel is normally mild steel, and is either externally plated or painted to provide a more presentable finished article. Where corrosion protection is vital, or where durability or appearance are also considerations, stainless steel is often used instead. Such an exhaust silencer is shown attached to the Ryobi engine in Fig. 5.2. Typical automotive silencers manufactured by these construction techniques are shown in Plates 7.1 and 7.2. The silencers have been cut open to reveal the internal silencing elements used and the manufacturing techniques employed.

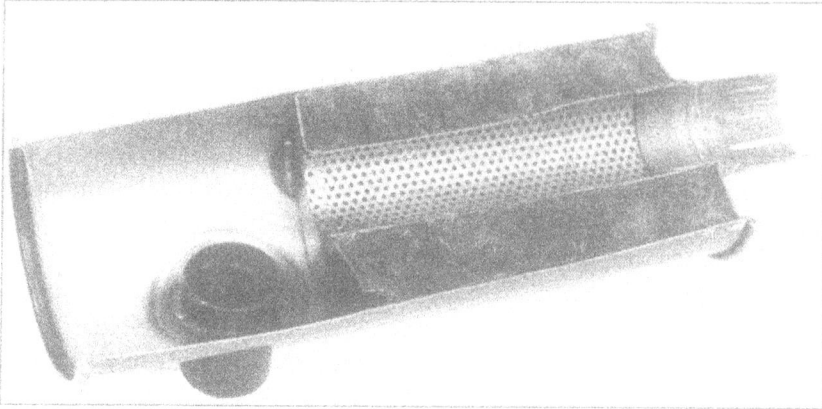

Plate 7.1 An exhaust silencer with diffusing and absorption elements.

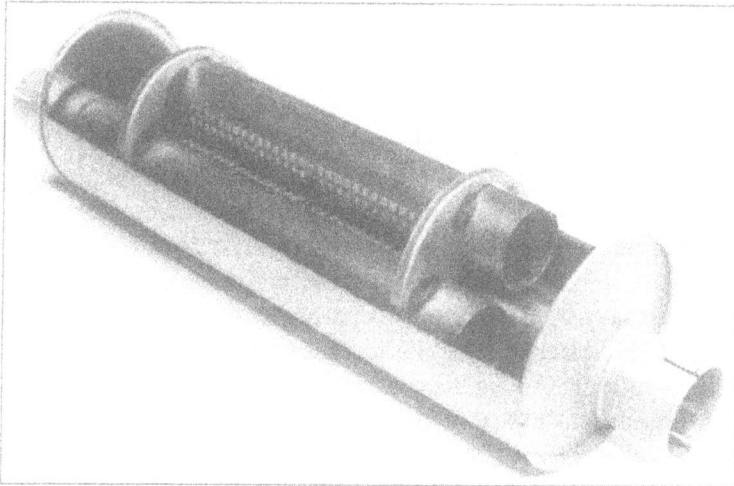

Plate 7.2 An exhaust silencer with diffusing and side-resonator elements.

In Plate 7.1, the entrance to the silencer is at the right and the exit to the silencer is at the left. This is a two-box design. The first silencer element is an absorption silencer. There are round holes in the central perforated pipe, which has been made by rolling perforated flat sheet into a round pipe and seam-welding it. The packing filling the absorption silencer element is clearly visible. The body of the silencer is rolled from flat sheet, both ends are stampings, and the end connections are made by crimping the end stampings to the central body. The central baffle is another stamping, and is spot-welded to the central body. This baffle separates the absorption element from the diffusing element to the left, where the short reentrant exit tube can be seen welded at its aperture through the central body. Because the central baffle is not sealed by welding around the body, but is merely spot-welded in several locations, gas leakage will occur from the space containing the packing into the plenum of the diffusing silencing element. Consequently, some of the packing may well get blown into the atmosphere with the passage of time. Such construction is typical of the automotive industry where silencers are built with equal concern for manufacturing cost and silencing effectiveness. The layout of this silencer, in conceptual terms, is similar to that shown in Fig. 7.31.

Plate 7.2 shows what is termed in the automotive industry as a "reactive" silencer. This has three boxes and is bidirectional, i.e., either end can be the entrance or the exit. It also has diffusing elements with reentrant pipes at both ends. In the center box, a side-resonator element, a perforated pipe connects the two diffusing silencer elements. The perforations are made by internal stabbing, much as recommended in Fig. 7.15, although the quality of the stabbed aperture shown in Plate 7.2 is not so effective from a silencing standpoint. The entry and exit pipes also connect directly to the two diffusing elements and, in passing through the central box, communicate to it as a side-resonator through similar stabbed holes in each pipe. The two center baffles are stampings and the three tubes are spot-welded to them. By definition, there is gas leakage from box to box as each tube passes through a baffle, which further complicates the simulation of such a silencer. The end connections to the central body are effected by crimping, as with the silencer in Plate 7.1, and this does provide good gas sealing. The entry and exit pipes are seam-welded to their respective end stampings, so that a very good gas seal can be obtained at those locations.

7.5.6 *Silencing the Intake System*

Probably the simplest and most effective form of intake silencer is the type sketched in Fig. 7.11. The geometry illustrated in this figure is for a single-cylinder engine, but the same basic arrangement applies to a multi-cylinder design [7.33, 7.38, 7.39]. The typical multi-cylinder arrangement for a current automobile is shown in Fig. 5.62, but you can also visualize an intake silencer box being placed and sealed on the flat plate atop the intake stacks of the V8 engine shown in Plate 6.2.

The Acoustic Design of the Low-Pass Intake Silencer

The intake silencer sketched in Fig. 7.11 provides the basic mechanism of induction silencing, being a volume connected to the induction system, however many air intakes to the several cylinders there may be. On the atmospheric side of this box is a pipe of length L_b and area A_b, or diameter d_b, if it has a circular cross section. The air cleaner can be placed within the box, helping to absorb the various high-frequency components emanating from the edges of throttles or carburetor slides. Assuming a conventional filter design, i.e., a design that is reasonably transparent to the air flow, this has no real effect on the silencing behavior of the box volume, V_b.

This type of silencer is known in acoustic theory as a low-pass device and, by that reckoning, has no silencing capability below its lowest resonating frequency.

The natural frequency of this type of silencer [7.33] is given by f_i:

$$f_i = \frac{a_0}{2\pi} \sqrt{\frac{A_p}{L_{eff} V_b}} \qquad (7.5.17)$$

where the effective length of the intake pipe, L_{eff}, is related to the actual length, L_p, and the diameter, d_p, by:

$$L_{eff} = L_p + \frac{\pi d_p}{4} \qquad (7.5.18)$$

In designing this type of silencer, the natural frequency of the silencer, f_i, is set to correspond to the natural frequency of the induction pulses from the engine, f_e. By this design criterion, the engine forcing frequency of direct design interest, f_e, is that corresponding to the engine speed of rotation, N_e, and the number of cylinders, n_{cy},

$$f_e = \frac{n_{cy} \times N_e}{120} \qquad (7.5.19)$$

and, because the intake silencer is of the low-pass type, the engine speed in question is at the lower end of the usable speed band. Annand and Roe [7.33] give some further advice on this particular matter. However, as can be seen in Sec. 6.3 where intake tuning is discussed, there may be three, four, or five pressure oscillations on an any intake pressure trace, so this minimum forcing frequency can be up to five times that calculated by Eq. 7.5.19. This point is emphasized in Sec. 7.6.5.

There are several other criteria to be satisfied as well in this design process, such as ensuring that the total box volume and the intake pipe area are sufficiently large so as not to choke the engine induction process and reduce the delivery ratio. Such a calculation requires the engine simulation model, extended to include the intake and exhaust silencers [7.21, 7.32, 7.41]. However, an approximate guide to such parameters is given by the following relationship, where n_{cy} is the number of cylinders, V_{sv} is the swept volume of any one cylinder, and d_2 is the flow diameter of the intake duct or the carburetor, as shown in Fig. 7.11:

$$8 \times V_{sv} \times n_{cy}^{0.5} < V_b < 12 \times V_{sv} \times n_{cy}^{0.5} \qquad (7.5.20)$$

$$0.85 \times d_2 \times n_{cy}^{0.4} < d_p < 1.15 \times d_2 \times n_{cy}^{0.4} \qquad (7.5.21)$$

7.6 Engine Simulation to Include the Noise Characteristics

Noise characteristics can be obtained by simulation, using the same software code employed earlier in Chapters 5 to 7, but incorporating the theory of Sec. 7.4.1. By definition, noise reduction for any engine implies providing space to accommodate the volumes of the several silencing elements displayed in Figs. 7.7 to 7.11. Noise reduction is easily accomplished if large volumes can be made available for the silencers. There is no need for an engine simulation code to design silencers if such is the case. However, in practice, unlimited space is rarely available for this purpose. Each engine application brings its own peculiar problems in the form of trade-off between noise reduction and the attainment of the requisite performance characteristics.

7.6.1 The Space Available for Silencer Elements
Hand-Held Power Tools and Other Industrial Engines

For a hand-held power tool, such as the Ryobi engine discussed in Sec. 5.4, the small bulk required of the entire powerplant precludes the availability of adequate space for effective silencers, be they intake or exhaust silencers. The problems in this regard are sketched in Fig. 5.6. There are other types of industrial four-stroke engines with untuned, or relatively untuned, exhaust and intake systems, and these are to be found on agricultural and electricity-generating equipment, where space for the entire powerplant may not be at the same premium as it is on a hand-held power tool such as a chainsaw or a brushcutter.

The most difficult engine for which to design an adequate silencing characteristic is the hand-held power tool, i.e., the chainsaw, or a similar device. Because noise is a function of the square of the dp, or pressure, fluctuation as seen in Eq. 7.1.2, it is self-evident that the larger is the silencer volume into which the exhaust pulses are "dumped," the lesser will be the dp value transmitted into the atmosphere. The designer of the hand-held power tool is continually looking for every available cubic centimeter of space to use for part of the exhaust and intake silencers. The space may be so minimal that the two-box design shown in Fig. 5.6 becomes an impossible luxury. The designer is left with the basic option of trimming out the fundamental firing frequency using a diffusing or side-resonant type of silencer as described in Sec. 7.5. As a rule of thumb, the designer of a silencer for the hand-held power tool will know that there are going to be real silencing difficulties if the total volume available for a silencer is not at least ten times greater than the cylinder swept volume. If this is not the case, then the only design methodology remaining is to choke the exhaust system by a restrictive silencer, thereby reducing the delivery ratio, and accept the consequential loss of power. Further design complications arise in many applications because of the legal necessity of incorporating a spark arrestor in the final tail pipe. By definition, a spark arrestor is a form of area restriction, as it has to inhibit, and more importantly extinguish, any small glowing particles of carbon leaving the silencer. The need for a spark arrestor is obvious for engines such as chainsaws and brushcutters, where the device is used in an environment with a known fire hazard.

Outboard Motors

Another example where space for the power unit is at a premium is in the outboard motor, where the virtually unsilenced exhaust gas is directed underwater into the propeller wash. This provides very effective exhaust silencing without any particular acoustic skills being required of the designer. Whether this underwater racket is offensive to fish is not known. The intake silencing can be conducted very effectively using the voluminous engine cowl as an intake silencer box. Hence, although the space for mufflers appears limited, compared to the space available in the hand-held power tool it is actually relatively generous from an acoustic design standpoint.

Inboard Marine Engines

By law, all marine engines must use water-cooled exhaust systems (see Plates 6.2 and 6.3). This means that all exhaust pipes are double-skinned and water-cooled, and water is dumped into the exhaust gas in the final exit pipe to the atmosphere. Even if the exhaust is not pumped underwater, this water spray is of a high density and acts most effectively as a silencing medium. Within the engine compartment of most boat designs, adequate space can normally be found to provide very effective intake silencing.

Automobiles

It is rarely difficult to find the necessary space to locate an effective intake silencer under the bonnet (hood) of a car, or an exhaust system and mufflers underneath the vehicle body.

Motorcycles

Space on a motorcycle, or a moped or scooter, is clearly limited. Motorcycles, particularly high-performance machines aimed at the on-road market are normally equipped with fairings, i.e., so-called streamlining, to reduce drag and increase rider comfort at high speed. Underneath the front and rear segments of that fairing, or even under segments such as the seat area, it is often possible to find a reasonable amount of space for adequate intake silencing. The use of shiny, large, tubular exhaust silencers has become something of a styling feature in recent times; perhaps an imitation of the silencers seen on Superbike racing machines has had some fashion influence on this matter. However, some of these silencers appear to me to be large enough to quiet a Volvo truck. So, adequate space for good exhaust silencing on on-road machines is apparently not the problem it was earlier in my career [7.33].

This is not the case, however, for off-road machines, which are stripped of all non-essential components to reduce bulk and weight and improve maneuverability. Here, the four-stroke engine designer has a real problem because engine performance is at a premium, so restrictive silencers are not a design option, with the less-bulky and lighter high-specific-output two-stroke engine giving the four-stroke engine such stiff competition in this market area.

7.6.2 Design Examples to Illustrate Silencing Design by Simulation

I will now take you on a "design journey," by means of the engine simulation, through the intake and exhaust silencing of a high-performance engine. It is easy to design silencers that muffle a power unit, but lose power en route. The design task, therefore, is to meet the noise legislation that every product today must satisfy, but lose the minimum possible power in the

length and 120 mm diameter, with butted joints at either end to the entry and exit pipes. The theory for this is given in Sec. 2.12, but without any restriction at each butted joint location.

A Diffusing Silencer with Reentrant Pipes

Within the simulation, and as sketched in Fig. 7.17, the reentrant pipes are treated as branches and the box center section becomes a reduced-length pipe of 120 mm diameter. The reentrant end segments of the silencer box are treated as pipes of an appropriately reduced area, with a three-way branch at one end and a throttled (closed) segment at the other end. The theory for three-way branches used within the simulation is found in Sec. 2.14, and that for closed pipe ends in Sec. 2.7. The branch angles for the pipes at the three-way branch are inserted into the simulation data structure just as they are sketched, 0°, 0° and 180°, and 0°, 180° and 180°, respectively. During the course of the simulation of the "ITC" single motorcycle engine, the reentrant lengths, L_1 and L_2, are both set equal, to 60, 100, and 140 mm, respectively, in three separate modeling tests, with the central pipe length appropriately readjusted in each case.

A Side-Resonant Silencer

This is sketched in Fig. 7.18, together with a layout of the approach to its simulation. It is shown treated as a three-way branch, where the side-holes form a short pipe to a plenum that is 3842 cm^3. The reduction in box volume from 4523 cm^3 is due to the volume occupied by the (1.2 mm wall thickness) pipe running through the silencer. You will observe that the four holes in the pipe, each of 22.1 mm diameter, in total constitute equality of length and area with the main exhaust pipe. According to Kato and Ishikawa [7.18], this is the upper limit of acceptability for such a design. It is possible to treat this junction in an even more complex fashion as a series of branched pipes, and eliminate the plenum, but the outcome in simulation terms is little different from that illustrated unless the holes are not centrally located within the side-resonator element. The theory for three-way branches and that for plenums, used within the simulation, are found in Secs. 2.14 and 2.16, respectively.

An Absorption Silencer

This is sketched in Fig. 7.19, where the perforations along the pipe are 1580 in number, each with a diameter of 3.0 mm. The totality of area of these holes is 11,168 mm^2. The pipe flow area is 1534 mm^2, and this represents an area ratio of 7.28 compared to the ratio of unity for the side-resonant silencer. As shown in Fig. 7.19, any four of the 3.0 mm perforations are laid out on a 5.5 mm grid, i.e., the edge-to-edge clearance of each perforation is 2.5 mm. This gives rise to the concept of opacity, O_p, for the perforations of an absorption silencer, defined as follows:

$$O_p = \frac{\text{perforation area}}{\text{pre - perforation area}} = \frac{\frac{\pi}{4}d_h^2}{\text{hole spacing}^2} \frac{\frac{\pi}{4}3.0^2}{5.5^2} = 0.23 \qquad (7.6.1)$$

where the "pre-perforation area" is the pipe area before perforations are added to any segment of that pipe.

A simple calculation reveals that 1580 holes will occupy a pre-perforation area of 47,800 mm^2 of the center pipe and, because the pipe has an inside diameter of 44.2 mm and an inside circumference of 139 mm, the perforations take up some 345 mm of the pipe length. In short, they fit comfortably within the maximum length of 400 mm of this absorption silencer. Such is the necessary, however mundane, arithmetic to which a designer must stoop. Incidentally, a perforation opacity of 20% is quite conventional in the design of an absorption silencer, and opacities as high as 30% are not uncommon.

The method adopted within the simulation for an absorption silencer is not dissimilar to that for the side-resonant silencer, shown in Fig. 7.18, but the damping effect of the packing on the gas flow within the silencer box must be taken into account. The analytic technique used to do this is sketched in Fig. 7.20.

The silencer box is split into a series of branches and interconnected plenums. The number of plenums required depends on the actual length of the absorption silencer and the mesh length selected for that particular pipe. For most engine simulations, a 20 mm mesh length is used for an exhaust pipe, so plenums spaced at between 60 and 100 mm will represent an absorption silencer quite accurately. If the number of plenums selected is n, this means that the absorption silencer is represented by n branches and n-1 interconnecting pipes between each plenum. Each pipe to each plenum has an area appropriate to the number of holes it represents:

$$A_1, A_2, \ldots A_n = \frac{A_t}{n} = \frac{N_h \times \frac{\pi}{4} d_h^2}{n} \qquad (7.6.2)$$

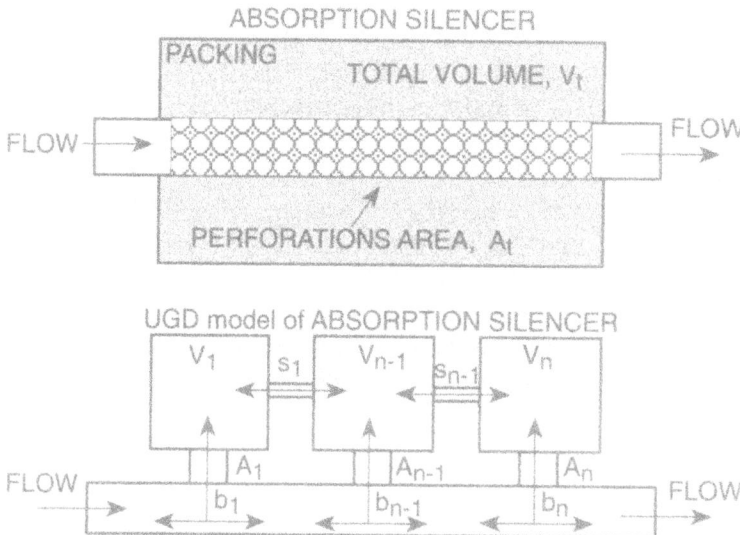

Fig. 7.20 Unsteady gas dynamic analytic technique to simulate an absorption silencer.

Each plenum has a volume equivalent to that obtained by splitting the total silencer box volume, V_t, into n plenums, using the symbolism of Figs. 7.10 and 7.20:

$$V_1, V_2, ... V_n = \frac{V_t}{n} = \frac{L_b \times \frac{\pi}{4}\left(d_b^2 - \left(d_3 + 2x_t\right)^2\right)}{n} \qquad (7.6.3)$$

The pipes, A_1 to A_n, are made as short as the simulation permits to simulate the fact that each perforation is only as long as the pipe wall thickness, x_t. The spacing of the plenum interconnecting pipes, s_1 to s_{n-1}, is normally made equal to that for the branches. The diameter, d_s, of each of these interconnecting pipes is meant to reflect the damping behavior of the packing, or the losses and reflections of the three-dimensional flow within the box volume if there is no packing at all. Naturally, the tighter the packing, the smaller is this diameter, and the largest diameter to be used represents an unpacked silencer. The following rule of thumb for this pipe diameter, d_s, should prove useful:

$$d_s = \frac{\sqrt{d_b^2 - \left(d_3 + 2x_t\right)^2}}{\rho_p} \qquad (7.6.4)$$

where the value of the packing density coefficient, ρ_p, has a value of 6 for a tightly-packed silencer, 4 for a loosely-packed silencer, and 3 for an unpacked silencer box.

To simulate the absorption silencer shown in Fig. 7.19, the plenum integer, n, is selected as 4 and the interconnecting pipe diameter, d_s, as 20.0 mm, i.e., by Eq. 7.6.4 the value of the packing density coefficient, ρ_p, is indexed as 5.5, indicating a fairly tightly packed silencer. By Eq. 7.6.3, each plenum volume is 950 cm^3 and, by Eq. 7.6.2, each pipe to each plenum has a diameter of 59.6 mm. The branch at each plenum entrance is spaced at 120 mm intervals.

7.6.3 The Noise and Performance Characteristics of the Exhaust Silencing Elements
In this first series of results from the simulation, the intake system is as yet unsilenced with the bellmouth end of the intake tract entering the atmosphere directly.

The Plenum, Diffusing, Side-Resonant, and Absorption Exhaust Silencers
Figs. 7.21 and 7.22 show the torque, as bmep, and exhaust noise produced by the engine over the 4000-9000 rpm speed range with a straight 900 mm unsilenced pipe, and also with the diffusing silencer, the side-resonant silencer, and the absorption silencer attached to it. The exhaust noise is recorded in dBlin units by the simulation at 1.0 m from the final pipe outlet in each case.

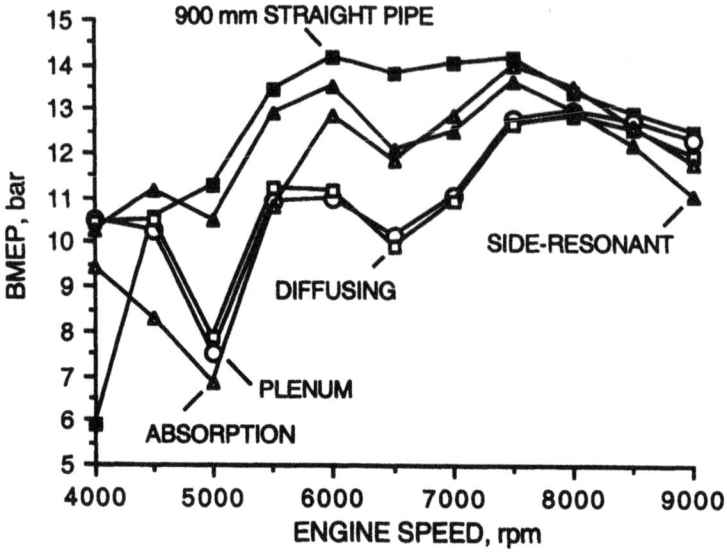

Fig. 7.21 Torque characteristics of plenum, diffusing, side-resonant, and absorption silencers.

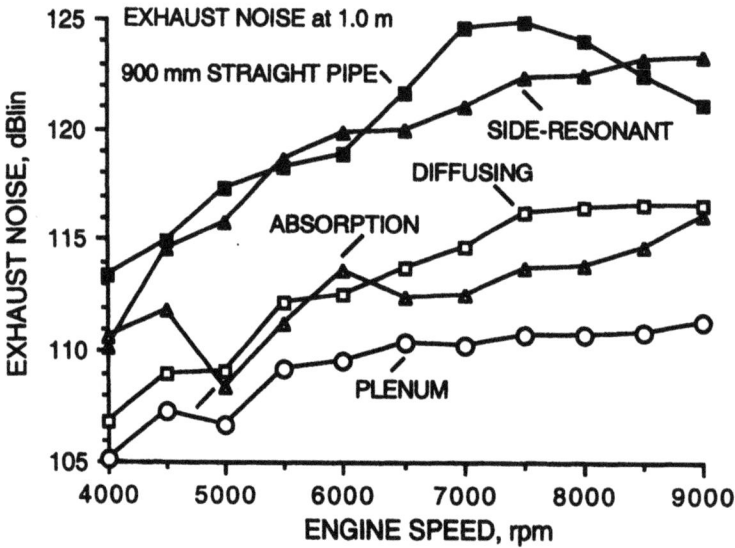

Fig. 7.22 Noise characteristics of plenum, diffusing, side-resonant, and absorption silencers.

Fig. 7.22 shows that, unsurprisingly, the unsilenced pipe creates the most noise, but a side-resonant silencer performs equally ineffectively in its attenuation of the exhaust noise. Only at a few engine speed locations does the unsilenced pipe show any signs of giving some noise reduction.

The plenum silencer muffles the best of this group, but the calculation is presented here only to show that it can be employed to predict equality of power performance with a diffusing silencer; this is evident from Fig. 7.21. Because there is no theoretical treatment of internal pressure waves within a plenum, the prediction of exhaust noise can be treated with the veritable "pinch of salt." Nevertheless, because the preparation of input data for the engine simulation for the other silencers is more complex than for a plenum, its ability to accurately predict the same performance characteristics as most diffusing silencers can be a useful time-saver in engine performance development. You will note that this simplification is employed in Sec. 5.9.1 (see Fig. 5.63 in particular) when simulating the sports car engine.

Ignoring the noisy side-resonant silencer, the silencer that gives torque performance nearest to that of the unsilenced pipe is the absorption silencer. The absorption silencer outperforms the diffusing silencer in terms of both power and noise reduction, with some 3 dBlin advantage, particularly in the regions of high engine speed and torque. Both silencers reduce the exhaust noise of the unsilenced pipe by some 10 dB. It is also clear, in Fig. 7.21, that both of these silencers give a serious loss of torque at 5000 rpm, which in terms of a design for a motorcycle would make the machine somewhat difficult to ride.

The Effect on Noise and Performance of the Reentrant Pipes in the Diffusing Silencer

The results of the simulation are shown in Figs. 7.23 and 7.24. Included for comparison are the results for the 900 mm straight pipe and the diffusing silencer. The 60 mm reentrant pipes provide less power than the diffusing silencer, but almost identical noise levels. The 100 mm reentrant pipes provide torque levels somewhat similar to those of the diffusing silencer, with somewhat similar noise levels, quieter at some engine speeds but noisier at others. The 140 mm reentrant pipes provide more torque than the diffusing silencer, but still less than the absorption silencer, and give noise levels almost identical to those of the unsilenced 900 mm straight pipe. While the range of reentrant lengths investigated is limited, the evidence shows that there is little acoustic or power benefit from the use of reentrant pipes in a diffusing silencer, and the further they are inserted, the lesser is the silencing effect attained.

The Relevance of Acoustic Analysis in the Design of Silencers

Figs. 7.25 to 7.29 show the noise attenuation characteristics of the several silencers tested in the engine simulation compared to those predicted by the acoustic theory given in Sec. 7.5. The geometries of the several silencers given in Figs. 7.17 to 7.19 are inserted into the relevant acoustic theory in Eqs. 7.5.1 to 7.5.10, and are plotted in Figs. 7.25 to 7.29. A temperature of 500°C is used within the acoustic theory for the exhaust gas in the silencer boxes and their

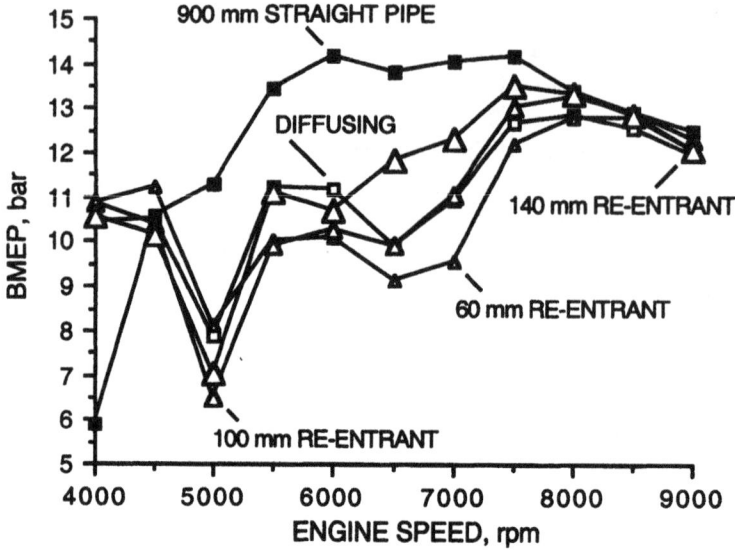

Fig. 7.23 Torque characteristics of reentrant pipes in a diffusing silencer.

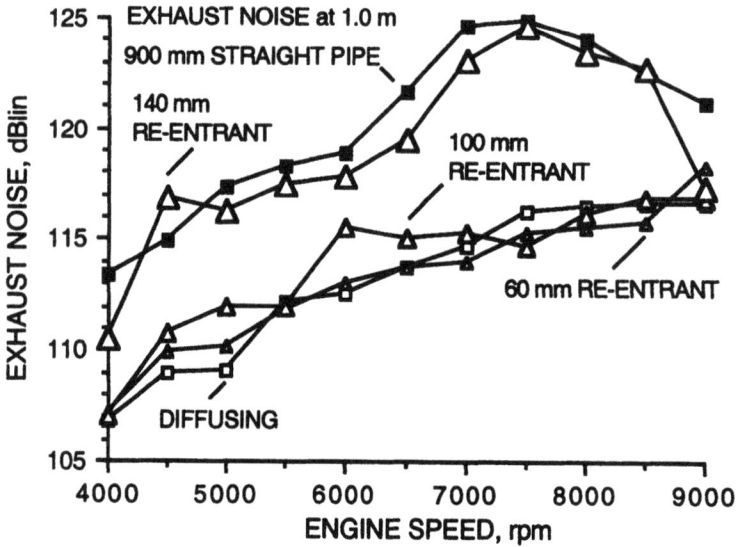

Fig. 7.24 Noise characteristics of reentrant pipes in a diffusing silencer.

pipes. The output of the engine simulations with the several silencers, giving both the overall noise levels and the noise spectra at 1.0 m from the exhaust outlet, are analyzed to find the noise attenuation characteristics of each of the several silencers compared to those of the unsilenced 900 mm straight pipe. The engine speed of 7500 rpm is selected for this purpose because it is the location of peak torque in most instances. At this engine speed, each of the noise spectra, in dBlin units, is subtracted from the noise spectrum of the unsilenced 900 mm straight pipe in order to determine the noise attenuation at each harmonic of the fundamental exhaust frequency, up to 2 kHz. At 7500 rpm in this single-cylinder engine, that fundamental exhaust frequency is 62.5 Hz. These are the noise attenuation spectra of the several silencers that are plotted in Figs. 7.25 to 7.29.

The predictions of the acoustic theory fare badly in Figs. 7.25 to 7.29. There is no perceived correlation with that predicted by the unsteady gas dynamic simulation. Because Mackey [7.30], in recent times, and Coates [7.3], in earlier days, have compared the noise characteristics calculated by the simulation code for a wide range of silencers with measured data, and have shown not only good correlation, but the same trends in noise characteristics as noted above, you may take these conclusions seriously.

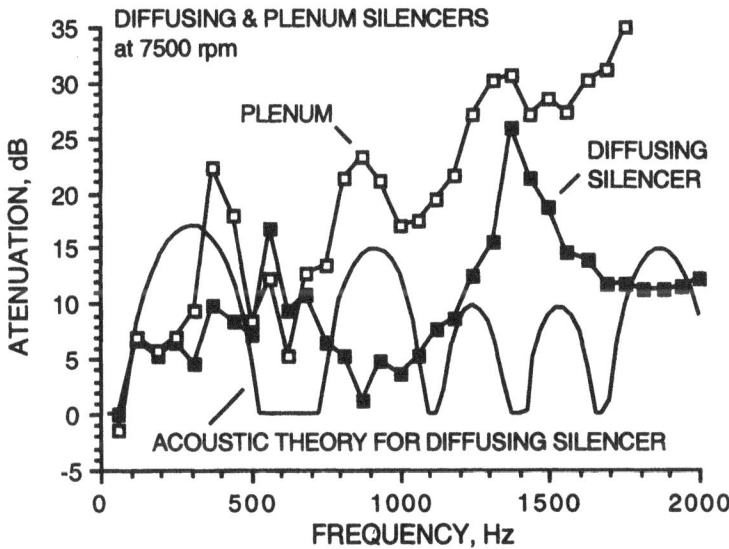

Fig. 7.25 Noise attenuation of plenum and diffusing silencers.

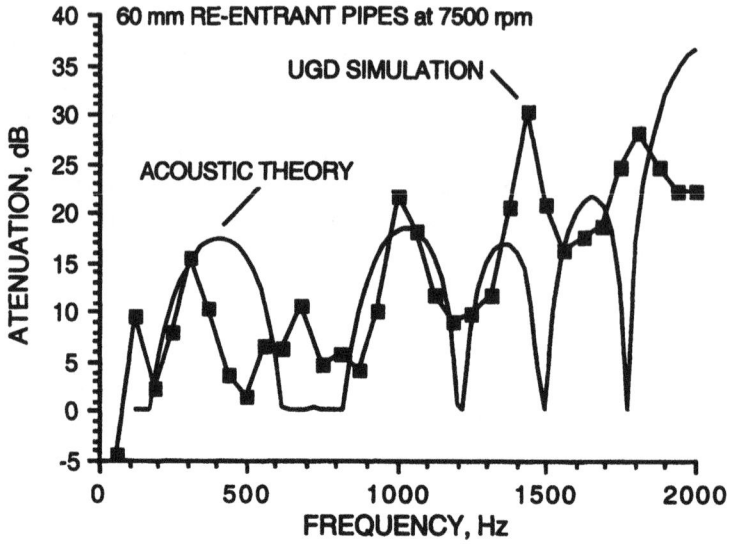

Fig. 7.26 Noise attenuation of a diffusing silencer with 60 mm reentrant pipes.

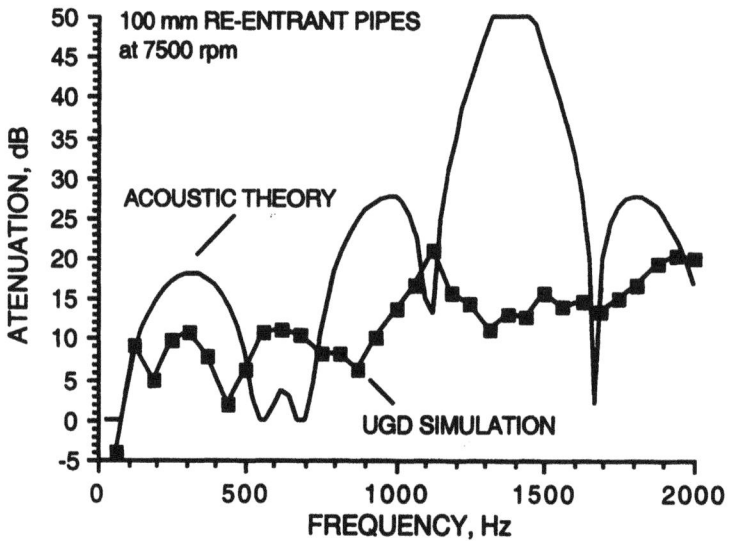

Fig. 7.27 Noise attenuation of a diffusing silencer with 100 mm reentrant pipes.

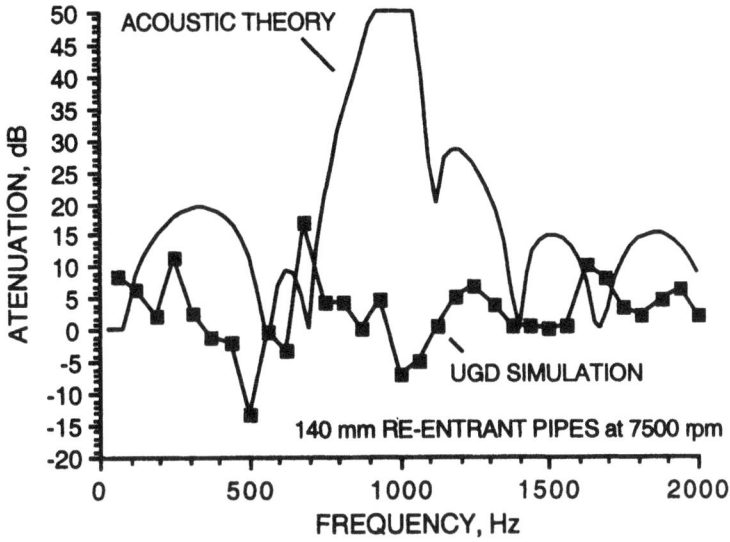

Fig. 7.28 Noise attenuation of a diffusing silencer with 140 mm reentrant pipes.

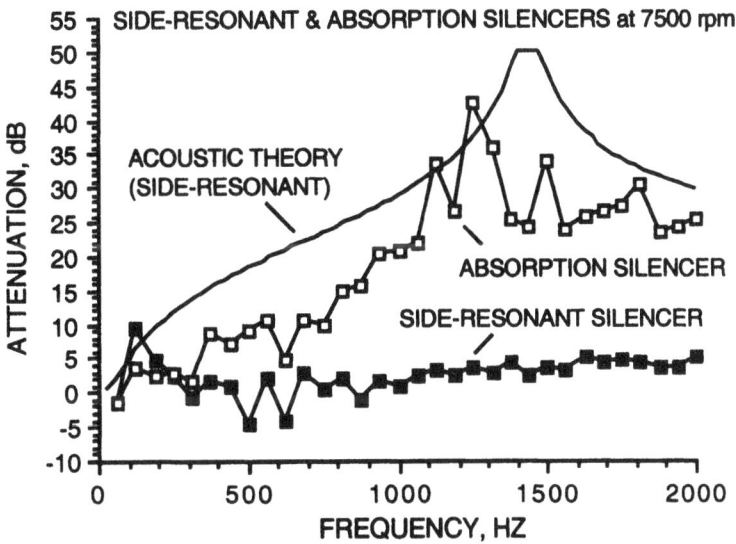

Fig. 7.29 Noise attenuation of a side-resonant silencer.

There are some results of some interest, one being that of the simulated attenuation of the absorption silencer which has been added to the data for the side-resonant silencer in Fig. 7.29. You will note that the attenuation of the absorption silencer, as predicted by the simulation, conforms quite closely to that predicted by the acoustic theory for the side-resonant silencer. On the other hand, there is no indication that the side-resonant silencer is doing any muffling at all. Although this result may baffle the acousticians, it will surprise nobody with experience of unsteady gas dynamics because, with an equal-area, three-way branch at the side-resonant aperture, as shown in Fig. 7.18, some 50% of the original exhaust pulse must proceed directly into the atmosphere. The resulting noise scenario becomes somewhat analogous to that of a two-trumpet duet when one of them stops playing.

Another interesting point is the attenuation of the plenum and diffusing silencers, in Fig. 7.25, where the engine simulation of the plenum silencer has more in common with the acoustic prediction for the diffusing silencer than the engine simulation of the diffusing silencer. In particular, the acoustic prediction of attenuation around 900 Hz for the diffusing silencer, which is noticeably absent in its simulated spectrum, can be seen clearly in the simulated spectrum for the plenum silencer.

The unsteady gas dynamic simulation, together with its associated sound transmission theory, produces noise spectra that are quite regular in profile. To illustrate this point, Fig. 7.30 shows the exhaust noise spectra at 7500 rpm for the straight pipe and two of the diffusing silencers, i.e., one with, and one without, a reentrant tube.

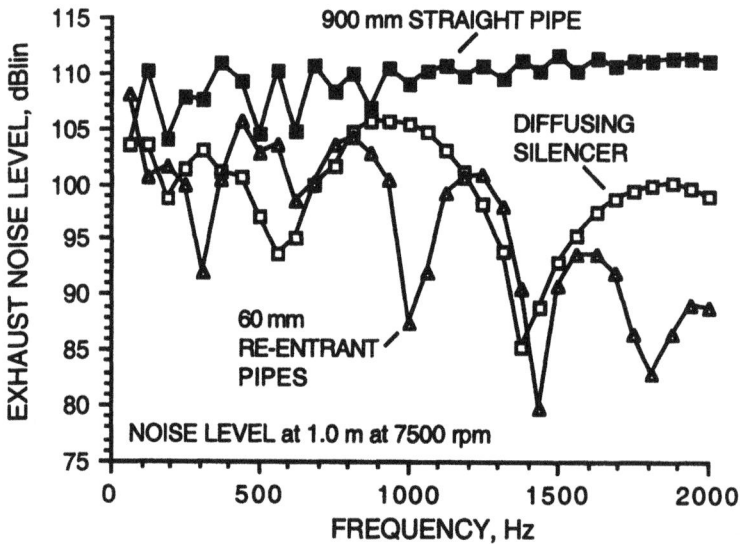

Fig. 7.30 Noise spectrum at 7500 rpm of a straight pipe and diffusing silencers.

In Fig. 7.30, the straight pipe is shown to produce a noise spectrum that rises with frequency, i.e., the classic, sharp "bark" of the unsilenced racing motorcycle on an open pipe at high speed and full throttle (see also Fig. 5.34 for the G50, which is just as loud with its exhaust megaphone). In Fig. 7.30, the noise level of the unsilenced pipe rises fairly continuously with increasing frequency. The two diffusing silencers exhibit the type of regular sinusoidal profile that the acoustic theory also demonstrates; however, as already noted in Figs. 7.25 to 7.29, there is no correspondence in the locations of the frequencies of the maximum and minimum noise attenuation. The effect of the reentrant pipe on the diffusing silencer, in this case 60 mm in length, is fairly clear in that it increases the number of peaks and troughs, and lowers the noise levels somewhat—but at the expense of the lower torque levels attained (see Fig. 7.23).

I have included this amount of detailed discussion, including the geometry of the actual silencers involved, in the hope that some acoustician will demonstrate to me that acoustic theory is available which will better simulate the effect of these silencers on this simple, high-performance single-cylinder engine. If this should actually happen, I shall be most pleased to be so educated, for effective empiricism is always useful to the engine modeler in reducing the numbers of data iterations when using an engine simulation during the design or development process.

7.6.4 A Two-Box Exhaust Silencer Design for the Single-Cylinder Motorcycle Engine

It is quite clear that a single-box silencer, i.e., any of those shown in Figs. 7.17 to 7.19, is not going to provide enough silencing to satisfy current motorcycle noise legislation and to maintain a high bmep and power level. If any of these silencers were sufficiently choked as to muffle adequately the exhaust noise, then the already-reduced power output would fall unacceptably. The addition of a second silencing box is the only design option. The design is shown in Fig. 7.31 as attached to the end of the exhaust down-pipe at 900 mm from the exhaust valve. It can be seen that this is an addition of the absorption silencer in Fig. 7.19 with a simple diffusing silencer, as in Fig. 7.17, but of half the length and with a slightly choked outlet pipe.

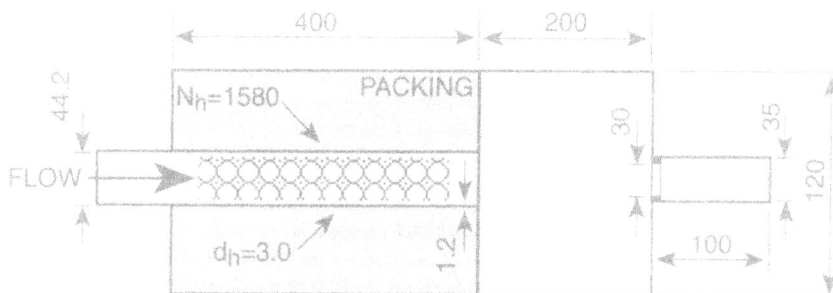

Fig. 7.31 A two-box silencer design for the single-cylinder motorcycle engine.

The total muffler length of 600 mm could be accommodated on a motorcycle. In terms of layout, it is quite similar to that shown in Plate 7.1. The theoretical treatment within the simulation of the two silencer elements is described above; the intake system is as yet unsilenced.

The effects predicted by the simulation on the torque produced by, and the exhaust noise emitted by, the engine are shown in Figs. 7.32 and 7.33, respectively. In Fig. 7.32, it can be seen that the engine torque levels are virtually restored to those of the unsilenced 900 mm straight pipe, and are better than with either the diffusing silencer, or the absorption silencer, acting alone. The bmep output for the diffusing and absorption silencers is also shown in Fig. 7.33 to make clear the torque advantage gained by the two-box exhaust silencer. In Fig. 7.33, the noise attenuation around the peak torque region for the two-box silencer is increased to some 20 dBlin, compared to 10 dBlin for the absorption or diffusing silencers acting alone. The noise data for the latter two silencers are also shown in Fig. 7.33 to assist with the comparisons.

The 7500 rpm region at full load gives the highest exhaust noise, so the noise spectrum for the two-box silencer, together with that of the unsilenced pipe and the original diffusing and absorption silencers, is shown in Fig. 7.34. The two-box silencer exhibits a falling noise spectrum, with a significant drop in the noise output for the first eight harmonics up to 600 Hz. By 2000 Hz, the attenuation has increased to 30 dBlin over the unsilenced pipe.

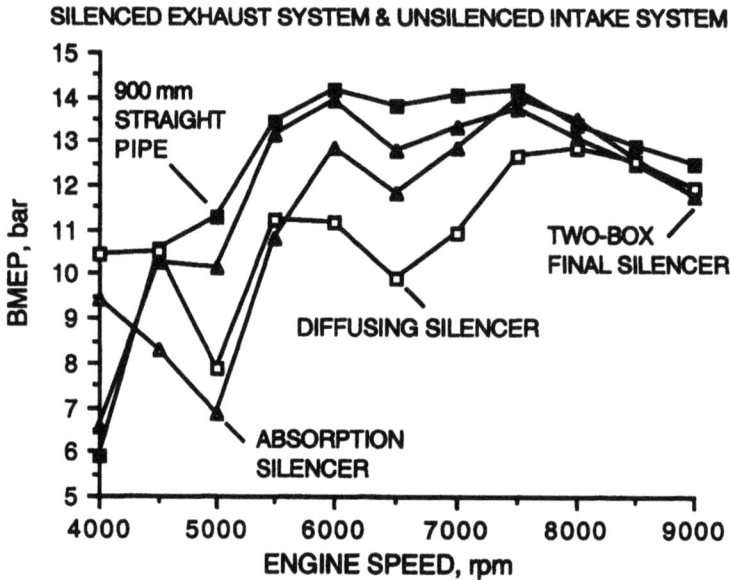

Fig. 7.32 Torque characteristics of the engine with the two-box exhaust silencer.

EXHAUST NOISE LEVEL at 1.0 m

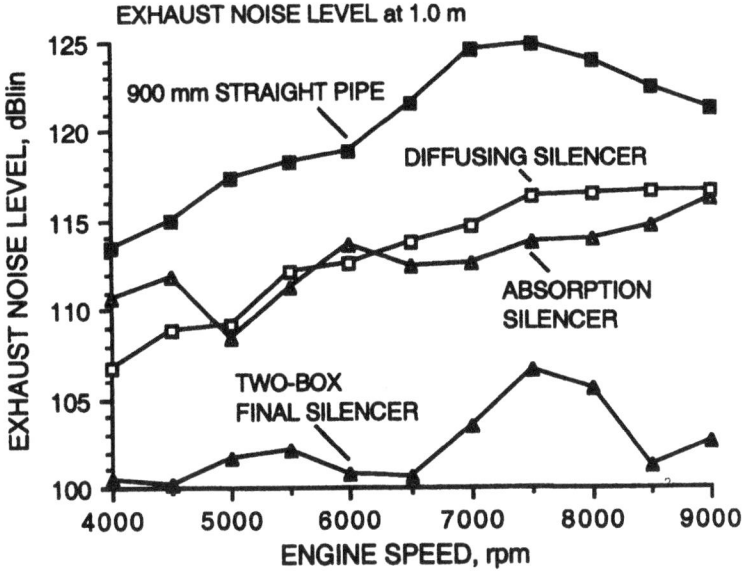

Fig. 7.33 Noise characteristics of the engine with the two-box exhaust silencer.

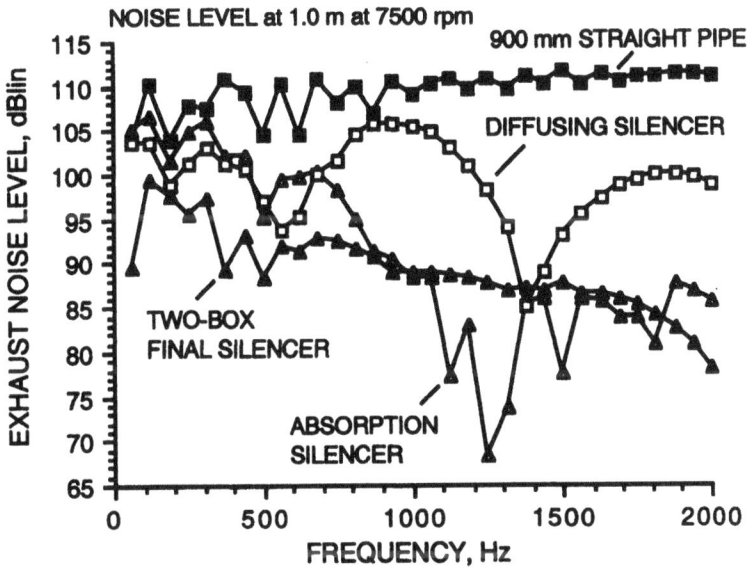

NOISE LEVEL at 1.0 m at 7500 rpm

Fig. 7.34 Noise spectrum at 7500 rpm with the two-box exhaust silencer.

745

Exhaust Mass Flow Rates at the Tailpipe Exit

Sec. 7.4.1 describes the theoretical basis for the analytic technique for noise prediction within the engine simulation. In Eq. 7.4.2, the sound pressure level and spectrum are shown to be functions of the instantaneous mass flow rate, not the gas particle velocity [7.35], at the atmospheric end of the system being analyzed. The time-varying mass flow rate is inherently calculated by the simulation, so it is readily obtained as output data for the several silencer configurations investigated. The instantaneous mass flow rates at the exhaust outlet are presented in Fig. 7.35 for the unsilenced straight pipe, the simple diffusing silencer of Fig. 7.17, the absorption silencer of Fig. 7.19, and the two-box silencer shown in Fig. 7.31. All of the mass flow rate diagrams in Fig. 7.35 are recorded by the simulation at 7500 rpm.

The mass flow rate for the unsilenced pipe shows the typical sharp spike of an exhaust pulse that has steepened. You have already seen the pressure diagrams at this very pipe end, albeit at 6500 rpm, in Fig. 6.68. Compared to this "steep-fronted" wave, the diffusing and absorption silencers exhibit lesser mass flow oscillations at the pipe end, and of significance in terms of high-frequency noise reduction, the slope of the initial exhaust outflow is reduced for both, but the absorption silencer is superior to the diffusing silencer in this regard. The two-box silencer shows the lowest mass flow oscillations at the pipe end and the shallowest slope on the outflow of the initial exhaust pulse. Because the unsilenced pipe, and all of the silencers, produce nearly the same power output at this speed, their net exiting mass flows will be very similar, i.e., their delivery ratios will be somewhat similar. Clearly, the ideal, quietest, silencer would be the device that provided a constant mass flow of about 45 g/s over the complete, 720° cycle. It is interesting to note that this would be only 6% of the maximum value exhibited by the unsilenced pipe. However difficult to meet, the ultimate design objective for ideal silencing is quite clear: reduce the exit mass flow rate to a constant velocity stream.

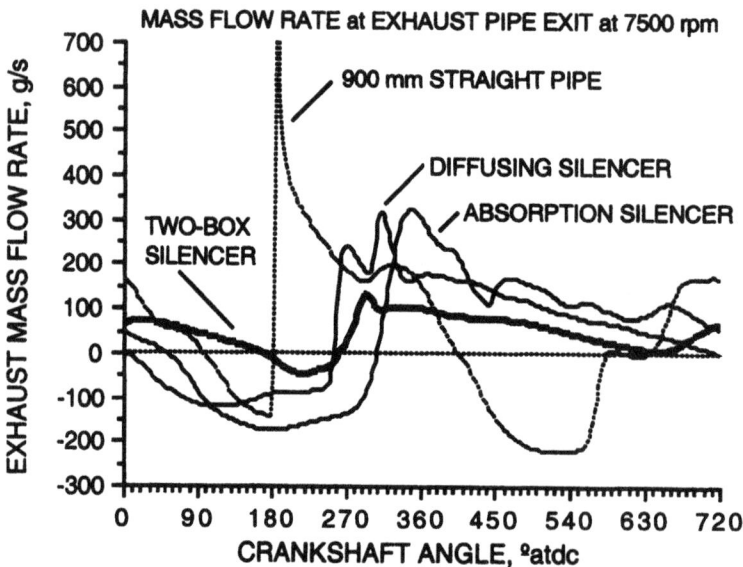

Fig. 7.35 Mass flow rates at 7500 rpm at the end of the exhaust pipe.

It is now obvious why the side-resonator silencer, of Fig. 7.18, is so ineffective. The mass flow rate of the unsilenced pipe arrives at the side-resonator branch. Only 50% of the steep-fronted mass flow rate flows sideways into its plenum at the equal-area three-way branch, leaving the other 50%, still steep-fronted, to proceed into the atmosphere.

Having reduced the exhaust noise of the motorcycle engine sufficiently to potentially meet an 80 dBA target in a full load, bypass acceleration test at 15 m, it is time to turn our attention to quieting the presently unsilenced intake system.

7.6.5 Silencing the Intake System of the Single-Cylinder Motorcycle Engine

In Sec. 7.5.6, the fundamental acoustic basis is established for the design of a low-pass intake silencer of the type illustrated in Fig. 7.11. The translation into a simple intake silencer box is straightforward, by ignoring the action of the air filter on the assumption that it is a low-pressure-loss, foam-element, or paper-based device. The current technology in air filters has become quite sophisticated in that 1 μm dust/dirt particles can now be successfully removed by such filter elements, without eventually clogging the filter element, and while providing a remarkably low pressure drop across the filter element. The translation of the basic design shown in Fig. 7.11 into one installed on the "ITC" single motorcycle engine, is shown in Fig. 7.36. The face of the intake silencer box mates with the bellmouth at the other end of the 400 mm long intake tract from the intake valve.

The dimensions of the box volume and the intake pipe are recommended by Eqs. 7.5.20 and 7.5.21, as long as the low-pass characteristics determined by Eqs. 7.5.17 and 7.5.19 are satisfied. A numerical interpretation of Eqs. 7.5.20 and 7.5.21 recommends the use of an intake silencer box volume with a minimum value, V_b, of 5000 cm^3 and a pipe diameter, d_p, of 45.4 mm.

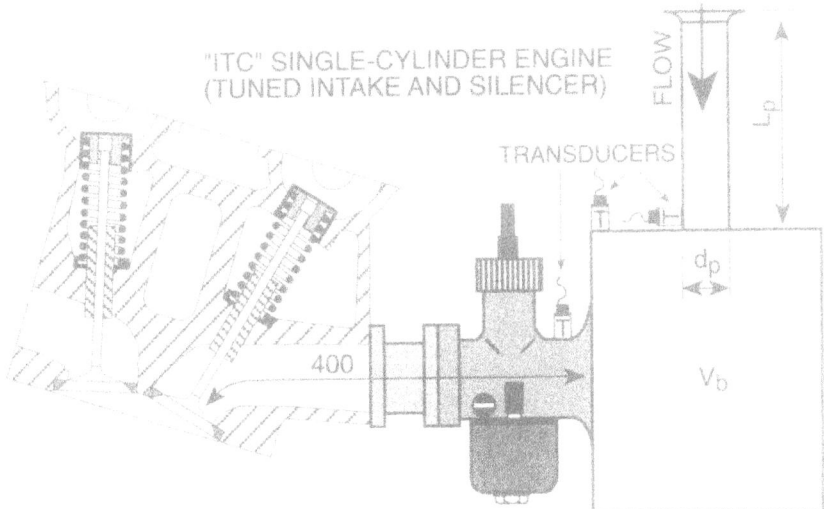

Fig. 7.36 The intake silencer dimensions for the "ITC" single motorcycle engine.

Considering the torque curve shown in Fig. 7.32, the engine is unlikely to be used at full throttle much below 4000 rpm. This is confirmed in Fig. 6.61, where the engine airflow curves at full load commence to climb rapidly at 4000 rpm. Using this data value as the low-pass forcing frequency, f_e, at 4000 rpm this is translated by Eq. 7.5.19 into 33.3 Hz. Assuming an air temperature of 20°C, the acoustic velocity, a_0, from Eq. 7.1.1, is 343 m/s. From Eq. 7.5.15, the intake silencing length, L_{eff}, is given by:

$$L_{eff} = \left(\frac{a_0}{2\pi f_i}\right)^2 \frac{A_p}{V_b} = \left(\frac{343}{66.6\pi}\right)^2 \frac{\frac{\pi}{4} \times \left(\frac{45.4}{1000}\right)^2}{5000 \times 10^{-6}} = 0.87 \text{ m} \qquad (7.6.5)$$

In this case, from Eq. 7.5.18, the intake silencer tube length, L_p, is found as:

$$L_p = L_{eff} - \frac{\pi d_p}{4} = 870 - \frac{\pi \times 45.4}{4} = 834 \text{ mm} \qquad (7.6.6)$$

The results of this combination of acoustic analysis and empirical advice, giving an intake box volume of some 5 liters capacity and an intake silencer tube that is 834 mm in length with a diameter of 45.4 mm, might suggest a device that is somewhat large to fit on a motorcycle. The fuel tank of a motorcyle normally holds about 15 or 20 liters, so you can get a mental picture of the size of the proposed intake box. An intake silencer tube with a length of some 800 mm may sound like an unlikely prospect to fit on a motorcycle, but it does not have to be straight and can be packaged as a curved, folded, and molded plastic device. The simulation is used to evaluate the effect of this device on the motorcycle engine.

The Effect of Intake Silencing on the Noise and Performance Characteristics
The single-cylinder motorcycle engine is simulated as fitted with the two-box-exhaust silencer and the intake silencer, the various geometrical parameters of which are varied to investigate their effect on both the intake noise and the engine performance characteristics.
In the first instance, the intake box volume, V_b, is fixed at 5000 cm^3, the intake silencing tube diameter, d_p, is set at 45 mm, and the length of the intake silencing tube, L_p, is varied from 100 to 850 mm. The simulation is run at 7500 rpm, i.e., in the region of peak torque, to determine the gains or losses on the maximum engine performance. The results from the simulation are plotted in Fig. 7.37. It is immediately apparent, by comparison with Fig. 7.32 which is for an unsilenced intake system but the same exhaust silencer, that the bmep has dropped from 14 bar to 12.8 bar, even with an optimum silencer tube length, L_p, of 800 mm. This is a considerable torque loss of some 9%, which is not good news. But in a sense, there is some good news in that the acoustic analysis has predicted a reasonably accurate value for the optimum length of the intake silencer tube. However, from Fig. 7.37, it can be seen that a 600 mm

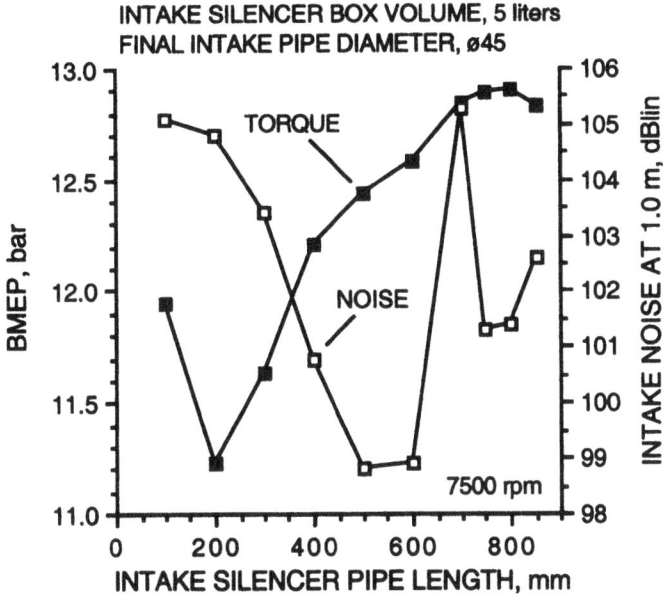

INTAKE SILENCER BOX VOLUME, 5 liters
FINAL INTAKE PIPE DIAMETER, ø45

Fig. 7.37 Effect of intake silencer tube length on performance at 7500 rpm.

length is quieter by some 4 dBlin than the 800 mm length, although a further fall in bmep, of 0.3 bar, is experienced. It is clear that there are engine tuning implications in the design of an intake silencer system for a single-cylinder, high-performance engine.

With the intake silencer tube length, L_p, fixed at 800 mm, and diameter, d_p, fixed at 45 mm, the box volume, V_b, is changed over a wide range, and the simulation results are shown in Fig. 7.38. It can be seen that there is no perceived benefit on engine performance with a box volume greater than 5 liters, although the noise would drop further by the considerable amount of some 3 dBlin. In case there is some trade-off, in terms of the difficulty of fitting either a long intake silencer tube or a large box, the simulation is repeated with a length, L_p, of 100 mm and various silencer box volumes, V_b. The simulation output, drawn to the same scale as Fig. 7.38, is shown in Fig. 7.39.

Although the system is much quieter at 7500 rpm, even the largest intake silencer box volume of 7 liters is down on power by some 10% with this shorter tube, compared to that exhibited by the 5 liter volume and 800 mm length combination given in Fig. 7.38. In short, the 5 liter box volume, as recommended by Eq. 7.5.20, is a good compromise between power attainment and noise reduction.

The last trade-off to be examined is that for the diameter of the silencer tube, d_p. In a series of tests it is varied within the simulation from 39 to 51 mm, and the noise and performance characteristics are plotted in Fig. 7.40. It can be seen that the optimum diameter, from a performance standpoint, is given by a silencer tube diameter of 48 mm, but this is at the expense of a considerable increase of the intake noise level. The power change from varying the intake

INTAKE SILENCER PIPE LENGTH, 800 mm
FINAL INTAKE PIPE DIAMETER, ø45

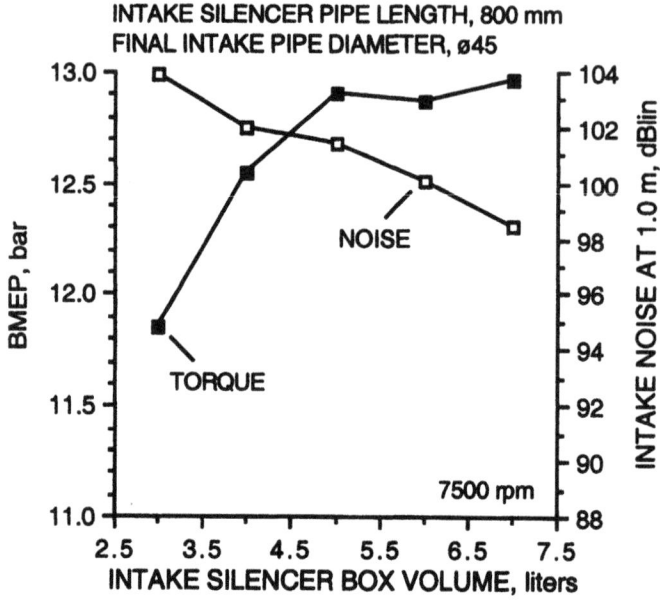

Fig. 7.38 Effect of intake silencer box volume on performance at 7500 rpm.

INTAKE SILENCER PIPE LENGTH, 100 mm
FINAL INTAKE PIPE DIAMETER, ø45

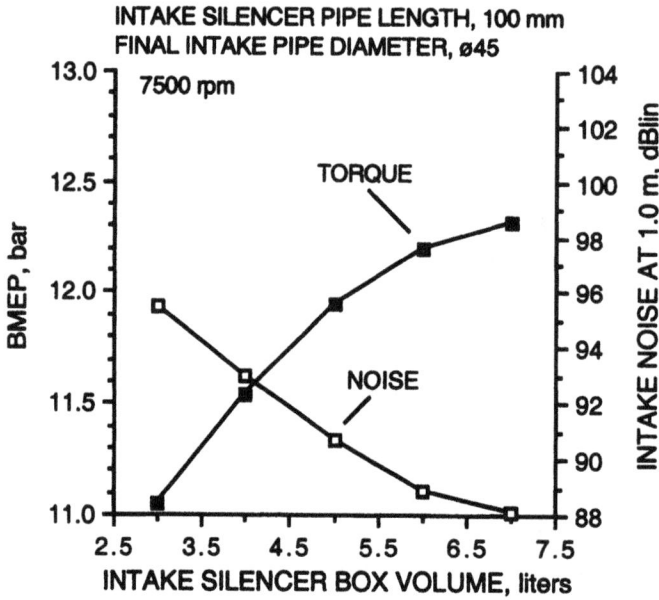

Fig. 7.39 Effect of intake volume, with a short tube, on performance at 7500 rpm.

750

silencer pipe diameter from 39 mm to 48 mm is 3%, whereas the intake noise increases by some 4 dBlin. Recall, from Eq. 7.1.6, that this increase is equivalent to doubling the intensity, i.e., one trumpet player is joined by another. The silencer tube diameter of 45 mm, as selected by Eq. 7.5.21, is seen to be a reasonable compromise between power attainment and noise reduction.

Insight into the Effect of Intake Silencer Tube Length on Noise and Performance

It is reasonably obvious that the larger the intake silencer box volume and the intake silencer tube diameter, the better will be the engine performance and the lower will be the intake noise. The effect of the length of the intake silencer tube is less obvious, and is not explained by the acoustic empiricism found in Eq. 7.5.17. The noise fundamentals can only be found by applying the theory of Coates [7.2, 7.3], and its theoretical expression in Eq. 7.4.2, within an engine simulation. The explanation for the effect of differing silencer tube lengths, observed in Fig. 7.37, is only possible by asking the simulation to provide data on the pressure-time, and mass flow rate-time, histories at the intake silencer box.

We have already seen, in Fig. 7.32, that a well-optimized exhaust silencer barely reduces the power and torque characteristics from an unsilenced, open pipe. The design of the intake silencer box for this motorcycle engine, of which the data in Fig. 7.37 represent my best design efforts, has reduced the torque at 7500 rpm by some 9%. However, the intake noise reduction is very considerable, for it drops from 116.5 dBlin (107.8 dBA) in the unsilenced configura-

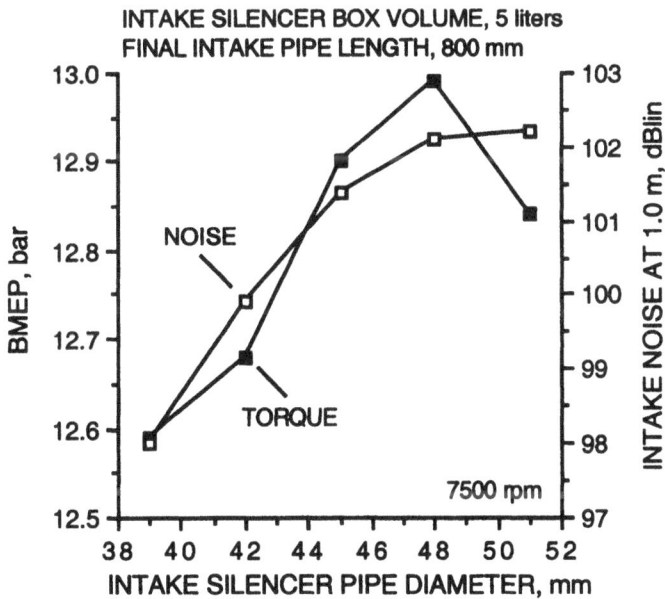

Fig. 7.40 Effect of intake silencer pipe diameter on performance at 7500 rpm.

tion to 101.3 dBlin (91.3 dBA) with the intake silencer added. The noise data are recorded by the simulation at 1.0 m from the inlet to the intake silencer system. The reason for the torque reduction is reasonably obvious: the mean intake plenum pressure must have fallen by some 9% below the atmospheric pressure, as the engine will gulp the same volume of air with, or without, an intake silencer box attached. It is the mass of ingested air, not the volume, that has been reduced because the mean air density in the plenum drops by some 9% as the intake process proceeds. The reason for the noise reduction is also reasonably obvious: the plenum volume damps the pressure oscillations, i.e., the mass flow rate oscillations, which travel up the silencer tube to arrive at the final inlet position to the atmosphere. However, nothing has been written above to explain why one length of silencer tube, compared to another, gives both a torque gain and a noise reduction. As has been the case many times throughout this text, the answers to all such conundrums are found in the temporal variations of the gas properties within the engine simulation.

Within the input data for the simulation, transducers are placed at the bellmouth end of the intake tract, within the intake plenum, and at the plenum end of the intake silencer duct, as sketched in Fig. 7.36. Fig. 7.37 shows that an intake silencer tube length, L_p, of 200 mm gives poor noise and performance characteristics and a length of 800 mm gives optimum values. The simulation is conducted at 7500 rpm for those two lengths, and the outcome at the three transducer locations is plotted in Figs. 7.41 and 7.42.

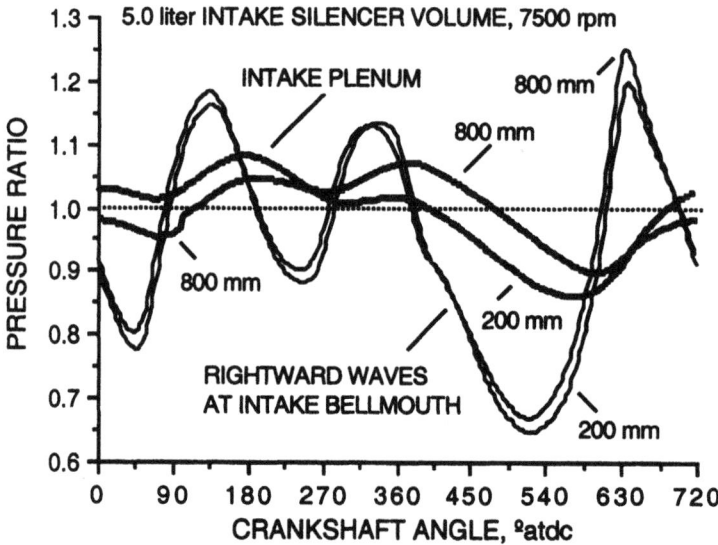

Fig. 7.41 Pressures in the intake plenum and intake pipe at 7500 rpm.

752

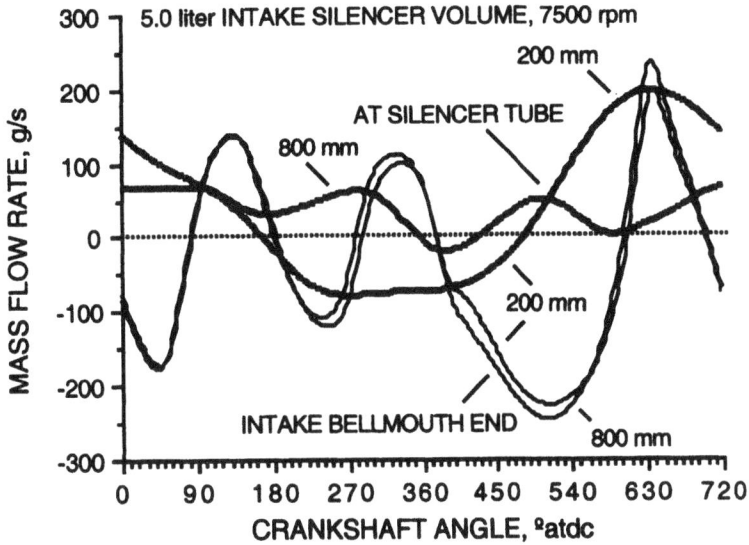

Fig. 7.42 Air mass flow rates in the intake pipe and silencer tube at 7500 rpm.

Fig. 7.41 shows the rightward pressure waves as they arrive at the bellmouth from the engine at 7500 rpm. It can be seen that there are three oscillations on these pressure waves, so the forcing function on the intake silencer box is three times that of the engine fundamental frequency, i.e., at 187.5 Hz, which implies that the third harmonic should exhibit the highest noise content in the intake noise spectrum. In Fig. 7.41, the rightward wave pressure signals, with either of the two silencer pipe lengths of 200 or 800 mm, are virtually identical, although that for the 200 mm silencer tube inhales with a slightly deeper breath during the main suction process from 540 to 600 °atdc. This is the period where attention must be focused. The optimum plenum pressure in this pseudo-atmosphere during this juncture, as it would be in the unsilenced scenario, is 1.0 atm. The intake plenum pressures for these two cases, plotted as heavier lines, reveal that the silencer box pressure dips more sharply into the sub-atmospheric region when the intake silencer tube length is 200 mm. That for the 800 mm length also falls to a sub-atmospheric pressure, but the first half of the main suction process is conducted with the plenum pressure above 1.0 atm. This is due to ramming by the intake silencer tube to coincide with the onset of the main engine suction. The 200 mm silencer tube also provides ramming, but it occurs too early at 180 °atdc, and the plenum pressure falls back to 1.0 atm by the time the engine suction commences at 370 °atdc. The entire process is, no more and no less, a replay of the intake ramming technique debated in Sec. 6.3, but with the ramming location changed from the intake valve to the plenum end of the intake silencer tube, and in a plenum where there are no valves to open or close either of the pipes.

This interpretation is confirmed by the mass flow rate-time histories shown in Fig. 7.42. These are shown for the transducer positions at the bellmouth end of the intake tract and the plenum end of the silencer tube. The profile of the main suction gulp at the bellmouth mimics that for the rightward pressure waves in Fig. 7.41. Negative values on the graph indicate outflow from the plenum. Positive values on the graph indicate inflow to the plenum.

Note that the 800 mm long silencer tube, experiencing the higher plenum pressure throughout, permits the intake process to ingest a greater mass flow rate, i.e., giving an increased delivery ratio and more torque. At the same time, the oscillations in the mass flow rate trace for the 800 mm long silencer tube, although they are greater in number, are much lower in amplitude. In addition, because they propagate to the atmospheric end of this pipe, they create much less noise in space.

Also note that the 800 mm long silencer tube fills the plenum up to the point where the main intake suction process starts at 370 °atdc and even provides a subsidiary ram right in the middle of it. Conversely, the 200 mm long silencer tube empties the plenum prior to the main suction process, tending to lower its pressure up to, and beyond, the point where it starts at 370 °atdc, and then rams in air at 630 °atdc after the main event is over. Without an engine simulation, these tuning effects would forever remain an unsolved mystery, as pressure transducers at these same transducer locations cannot provide the information seen in Figs. 7.42 and 7.43. Holmes would have liked it and, doubtless, declared it to Watson as "elementary."

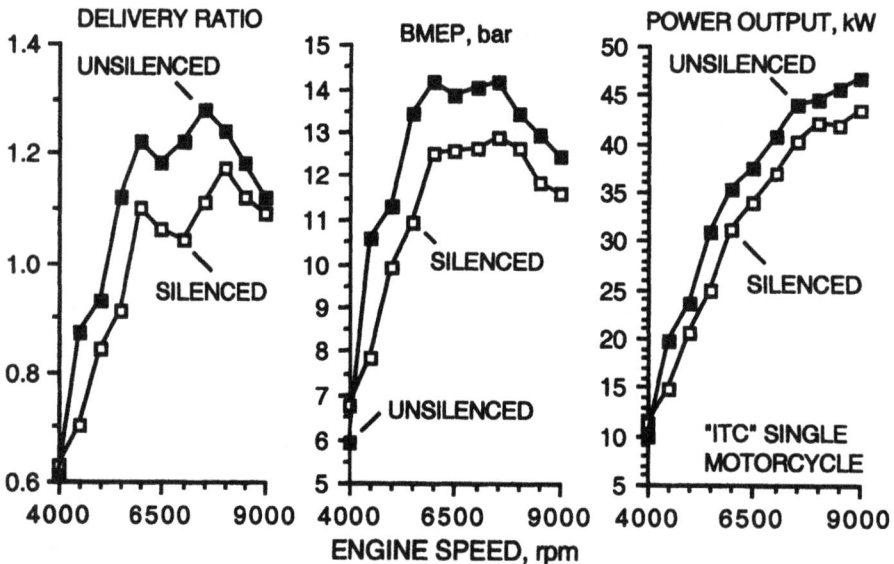

Fig. 7.43 Performance characteristics of the motorcycle engine with silencers.

7.6.6 The Single-Cylinder Motorcycle Engine with Intake and Exhaust Silencing

In the final design for the intake silencer sketched in Fig. 7.36, the intake box volume, V_b, is 5000 cm^3, the intake silencing tube diameter, d_p, is 45 mm, and the length of the intake silencing tube, L_p, is 800 mm. The exhaust silencer is the two-box design shown in Fig. 7.31. The performance characteristics for power, torque, and airflow are shown in Fig. 7.43, and are compared to those for the unsilenced engine with open pipes, where the total intake length, L_{it}, is 400 mm and the straight exhaust pipe length, L_{et}, is 900 mm. You will recall that the intake and exhaust silencers are attached to these two pipes. The final outcome is that the muffled engine drops some 3.2 kW, yet produces a high specific power output of 43.5 kW (58.3 bhp) at 9000 rpm, i.e., 87 kW/liter or 116.6 bhp/liter. The loss of peak power by silencing is just 7%.

The airflow curve, as delivery ratio, shows a fall in its maximum value from 1.28 to 1.17 through silencing, i.e., some 9%. It is interesting to note that the location of the intake ramming peaks and troughs, although unchanged in number, have been shifted upward in engine speed by 500 rpm due to the installation of the mufflers.

The torque curve of the silenced engine no longer shows a torque dip at 6500 rpm, which improves the driveability of the motorcycle. From this same standpoint, the silenced motorcycle engine must be revved within an engine speed band from 5000 to 9000 rpm, in order to attain good acceleration performance from it. The almost linear power curve under full-load conditions, even from 4000 rpm, will give the rider quite a "rush" during acceleration on such a machine, as the power output quadruples over that speed range.

In the unsilenced configuration, the minimum brake specific fuel consumption is at 6000 rpm, and is 276.4 g/kWh, whereas it is 302.8 g/kWh at maximum power at 9000 rpm. In the final, silenced design, the minimum brake specific fuel consumption is also at 6000 rpm, and is 286.1 g/kWh, whereas it is 314.1 g/kWh at maximum power at 9000 rpm. In short, the thermal efficiency of the engine is reduced by silencing, although not as severely if that silencing had not been optimized. In this context, you will recall that all simulation of this motorcycle engine, based on the "ITC" single, is conducted assuming unleaded fuel at an equivalence ratio, λ, of 0.88, or an air-fuel ratio, AFR, of 12.5. This is a mixture strength close to that which provides maximum power, and is very typical of that currently used for high performance motorcycles at full load. Leaning the mixture, i.e., at AFR values approaching the stoichiometric, would improve the fuel economy, but that carries the potential of incurring detonation under full-load conditions in this high-performance engine.

The noise characteristics of the motorcycle engine with silencers are, in the first instance, predicted at a distance of 1.0 m from the outlets of the intake and exhaust silencers, and they are plotted in Fig. 7.44 over the operating engine speed range at full load. Also shown are the noise characteristics of the unsilenced engine. These are plotted in both noise units, i.e., as dBlin, and also on an A-weighted scale as dBA values. The A-weighted scale, as discussed in Sec. 7.1.4, more closely represents the noise level as perceived by *Homo sapiens*, and dBA units are the units used to state limits in most noise legislation. The opinions of bats, birds, and beagles have, apparently, not been taken into consideration in the environmentally-conscious world of noise legislation.

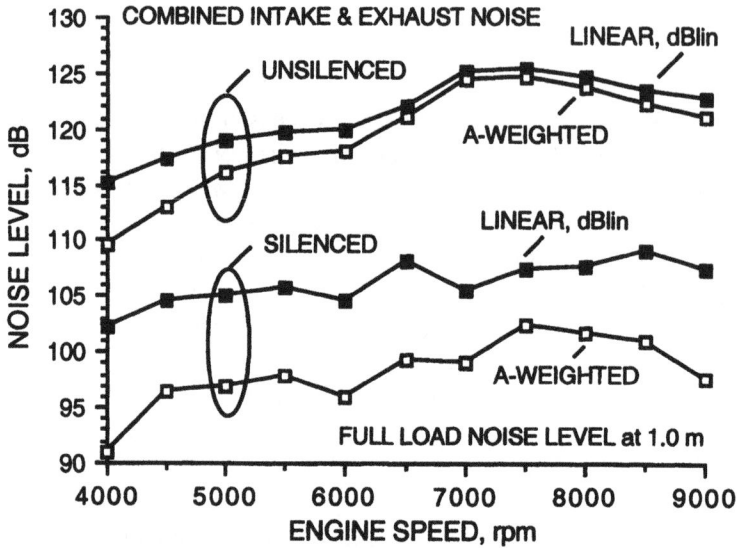

Fig. 7.44 Noise characteristics of the motorcycle engine with silencers.

In Fig. 7.44, at a range of 1.0 m, it can be seen that the unsilenced engine has a maximum noise output of some 125 dB, either on the linear or A-weighted scale. The explanation for this can be found in Fig. 7.34, where the predominant exhaust noise in the spectrum occurs at a high frequency. In Fig. 7.44, the silenced motorcycle is seen to have a maximum noise level of 109.1 dBlin or 101.7 dBA. There is further insight into the separation between linear and A-weighted noise for the silenced engine to be found in Fig. 7.34, which shows that the predominant noise in the exhaust spectrum is at low frequencies.

For the silenced engine, these noise levels represent maximum noise reductions of 16 dBlin and 23 dBA, which are very considerable, but absolutely necessary to meet current noise legislation for motorcycles. The noise reduction is accomplished with only a 7% drop in maximum power output from the unsilenced state, so you can see that there is really no excuse for the streets to be populated by noisy four-stroke motorcycles [7.23]—or noisy two-stroke motorcycles, for that matter [7.32].

Meeting Motorcycle Noise Legislation

The current scene in motorcycle noise legislation is that a motorcycle is required to meet an 80 dBA limit on a full-load acceleration test, such as that specified by SAE J47 or SAE J331 [7.16]. The procedure in SAE J47 is intended to record the maximum possible noise that a motorcycle can produce, and that in SAE J331 is intended to record the noise produced by the machine in conventional usage. As a motorcyclist, I shall forbear to comment upon which of the two test procedures I consider to be the more relevant in terms of the phrase "conventional usage." The test point is located at 15.0 m from a motorcycle as it is accelerated past under full load. To determine how the silenced single-cylinder motorcycle—the test vehicle in this chapter—would fare under such a full-load acceleration test, the simulation is asked to provide the noise characteristics at full load over the speed range, just as in the data run that gave us Fig. 7.44, but with the microphone placed at 15.0 m from the pipe outlets rather than 1.0 m. The result is plotted in Fig. 7.45 in the legislated units of decibels on an A-weighted scale.

The maximum noise level emitted by the intake and exhaust sources is seen to be 78.9 dBA at 7500 rpm. This number conforms to the highest value that could be recorded under the more stringent SAE standard, J47. Under the procedure in SAE J331, it would most probably be about 75 dBA. Of course, this decibel value does not include windage, tire, or mechanical noise contributions. However, on the assumption that these contributions are no more than equal to that of the combined intake and exhaust noise, that would raise the maximum noise output recorded by the motorcycle by about 3 dBA, to some 78 dBA under SAE J331 at best, or 81.9 dBA under SAE J47 in a worst case scenario. In short, the motorcycle engine example is one of high performance which also exhibits the potential to meet the most stringent, current noise legislation.

Fig. 7.45 also shows the separate contributions made to the total noise by the intake and exhaust systems. The maximum noise is made by the exhaust system at 7500 rpm. It is noisier than the intake system at all other speeds throughout the engine speed range except at 6500 and 8500 rpm. When you examine this graph, think in terms of Eq. 7.1.5, but also bear in mind my frequently used analogy of the effect of two trumpet players, either both playing loudly or

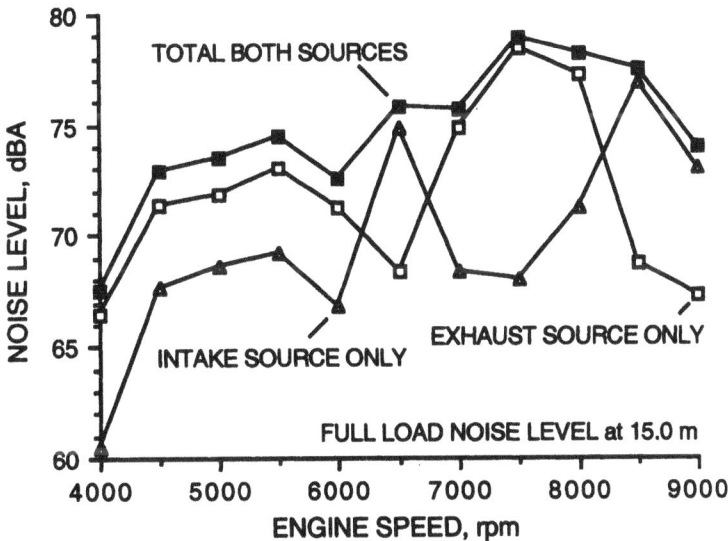

Fig. 7.45 Motorcycle engine noise characteristics recorded by SAE J47.

one of them playing more softly than the other! You can see that it is the intake system that keeps the noise high at the peak power point of 9000 rpm. Hence, the opening gambits in any further noise reduction strategy for this particular motorcycle engine would be to lower the exhaust noise at 7500 and 8000 rpm and the intake noise at 6500 and 8500 rpm.

The exhaust noise spectrum of the two-box exhaust silencer at 7500 rpm is shown in Fig. 7.34, albeit in dBlin units. From the simulation output that gave us Fig. 7.34, I can report that the noisiest frequency on the dBA scale is 750 Hz, so this is the frequency that must be primarily suppressed during any development process.

The intake noise spectra, in dBA units, at the problem engine speeds of 6500 and 8500 rpm, are shown in Fig. 7.46. These are recorded by the simulation at 15.0 m from the intake silencer outlet to the atmosphere. Observe, as forecast by Figs. 7.41 and 7.42, that the noisiest frequency at 8500 rpm is the third harmonic of the engine fundamental frequency, which is at 212.5 Hz. The noisiest frequency at 6500 rpm is the fourth harmonic, which is at 216.7 Hz; that there are four intake pressure waves per cycle at this speed is confirmed by Fig. 6.55. In short, there are three pressure pulsations on the intake pressure trace at 8500 rpm and four at 6500 rpm. At 8500 rpm, the sixth harmonic is almost equally as loud as the third harmonic, and there is also a suggestion that the high-frequency noise, above 1500 Hz, is rising. At 6500 rpm, the eighth harmonic is slightly quieter than the fourth. In any further development exercise to reduce the intake noise levels, there are two frequency bands that must be suppressed as a major priority, i.e., around 210 Hz and around 420 Hz.

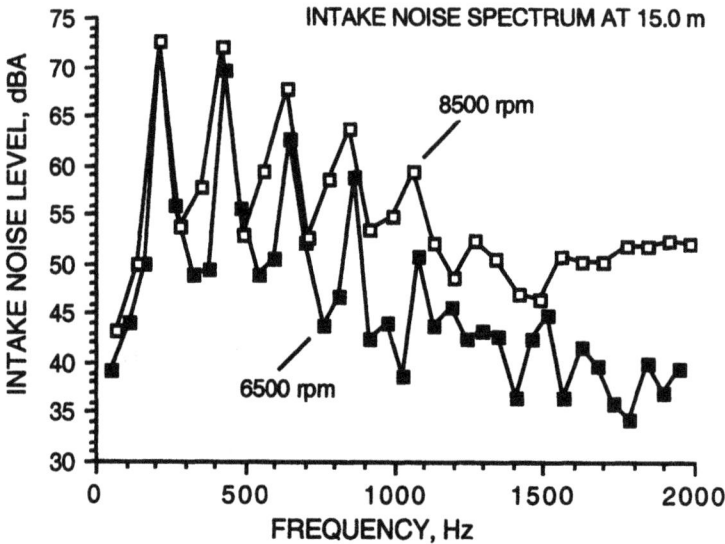

Fig. 7.46 Intake noise spectra of the silenced motorcycle at 6500 and 8500 rpm.

7.6.7 Intake Silencing in Multi-Cylinder Engines

In Sec. 5.9, a sports car engine is described and its design modeled extensively, with the exception of that for the intake and exhaust silencers. Because the above discussion on exhaust silencers is quite complete, further discussion on the detailed design of an exhaust silencer for the sports car engine is not necessary. However, the discussion on intake silencing, thus far, has focused on a single-cylinder engine, and the lessons learned there may not be so readily applicable to a multi-cylinder engine. In particular, it would be useful to have confirmed the extent to which the empirical advice on intake silencing, proffered in Sec. 7.5.6, is, or is not, applicable to a multi-cylinder engine.

In Sec. 5.9.1, where the geometry of the sports car engine is presented, Fig. 5.62 shows the intake ducting of the engine, including the intake silencer. It contains a dual-length intake duct. During an acceleration test to meet noise legislation, such as in SAE J986, the engine is unlikely to reach the 5250 rpm switching point from the longer to the shorter intake tract. It is highly probable that the entirety of any noise test would be conducted with the engine breathing through the longer intake tract. Thus, I propose, for the sake of brevity, to discuss only the design of the intake silencer when the longer intake tract is indexed by the engine management system, under full load between 1750 rpm and 5250 rpm.

The intake system of this sports car engine has much in common with the current designs for most automobiles. Such engines typically contain tuned intake runners to each cylinder from a plenum, which is connected through a throttle-body to another silencing box and air filter. This is inherently capable of being well-silenced as it is basically a two-box arrangement, unlike the simpler single-box design for the motorcycle engine as discussed above. Because it is more capable of being well-silenced, and because the first plenum box controls the intake ramming tuning, such a system is sometimes not as thoroughly optimized as it might be by the design engineer. Perhaps this is less liable to happen for a sports car engine, as the trade-off in noise and performance tends to be more at the forefront of the designer's mind.

The sports car engine has four cylinders, each with a swept volume, V_{sv}, of 500 cm^3. Consider what the empiricism, in the form of Eqs. 7.5.17 to 7.5.21, has to say about this intake silencer. Using the nomenclature of Fig. 7.11, Eq. 7.5.20 recommends an intake silencer box volume, V_b, of 10 liters. Eq. 7.5.21 recommends an intake silencer tube diameter, d_p, of 64.4 mm, based on an intake duct diameter, d_2, from Fig. 5.61, of 37 mm. Note, from Fig. 5.62, that all of the performance modeling in Chapter 5 is conducted with an intake silencer box volume of 10 liters and an intake silencer tube diameter of 65 mm. But what about the intake silencer tube length, L_p, which is found to be a critical item for the single-cylinder motorcycle engine in terms of both noise and performance?

As seen in Fig. 5.64, at full load the lowest engine speed to be used is 1500 rpm. Let this be considered as the speed that gives forcing frequencies of any interest or strength for the low-pass filter. The forcing frequency, f_e, at 1500 rpm is given by Eq. 7.5.19 as 50 Hz. Assuming an air temperature of 20°C, the acoustic velocity, a_0, from Eq. 7.1.1, is 343 m/s. The intake silencing length, L_{eff}, from Eq. 7.5.15, is given by:

$$L_{eff} = \left(\frac{a_0}{2\pi f_i} \right)^2 \frac{A_p}{V_b} = \left(\frac{343}{100\pi} \right)^2 \frac{\frac{\pi}{4} \times \left(\frac{65}{1000} \right)^2}{10000 \times 10^{-6}} = 0.396 \text{ m}$$

In this case, from Eq. 7.5.18, the intake silencer tube length, L_p, is found as:

$$L_p = L_{eff} - \frac{\pi d_p}{4} = 396 - \frac{\pi \times 65}{4} = 345 \text{ mm}$$

The result of this combination of acoustic analysis and empirical advice, i.e., an intake box volume of some 10 liters capacity, with an intake silencer tube that is 345 mm in length and 65 mm in diameter, is a device that is located under the bonnet (hood) of the automobile. Fitting it there does not pose the same packaging problem as designing a similar device for the more-exposed motorcycle. Note that the initial performance development, as shown in Fig. 5.62, is conducted in Sec. 5.9 with an intake silencer tube that is 300 mm in length. The simulation is used to investigate the optimized nature of this initial design decision. Because the peak torque for this engine is located at 4500 rpm, that speed is selected to examine the effect of a wide variation of intake silencer box geometries on the noise and performance characteristics. Put another way, changes are made to the box volumes and the silencer tube lengths and diameters and inserted into the engine simulation along with the remainder of the input data bank for the sports car engine.

Fig. 7.47 shows the results of the simulation. Within the simulation:
(a) When the intake silencer tube lengths are varied from 250 to 800 mm, the box volume is retained at 10 liters and the silencer tube diameter is set to 65 mm.
(b) When the intake silencer volume is varied from 5 to 10 liters, the silencer tube diameter is kept at 65 mm, and the silencer tube length is fixed at 300 mm.
(c) Because, in (a) 800 mm is found to be one of the best lengths for the intake silencer tube, when the intake silencer tube diameter is varied from 55 to 80 mm, the box volume is set to 10 liters, and the silencer tube length is fixed at 800 mm.

In Fig. 7.47, you will not be surprised to see that the larger is the volume of the box, the lower is the noise. There is a considerable noise increase over the range of box volumes of nearly 8 dBA. Although it is not plotted, there is no perceptible change of engine torque over this range of box volumes, as there is almost no primary intake ram tuning coming from the silencer box.

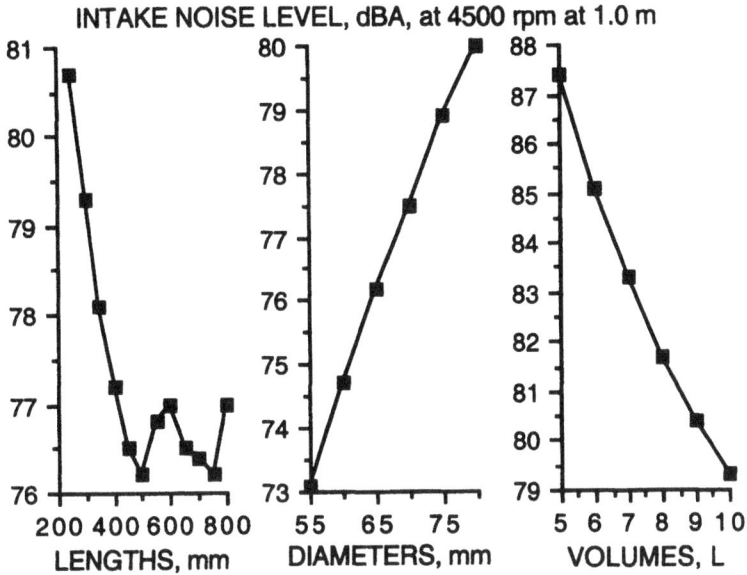

Fig. 7.47 Effect of intake silencer geometry on sports car noise characteristics.

Similarly, i.e., as anticipated from the results for the single-cylinder motorcycle engine, the larger is the diameter of the intake silencer tube, the higher is the noise level. It rises by some 7 dBA over the range of pipe diameters investigated, a range that corresponds closely to that recommended by Eq. 7.5.21. There is an increase of power with increasing silencer tube diameter, but it is miniscule, increasing almost linearly by only 1.1% over the diametral range examined.

As with the motorcycle engine, with this sports car engine the length of the intake silencer tube is critical in reducing the intake noise. Tube lengths between 500 and 800 mm provide the maximum noise reduction and the length of 345 mm, recommended by empiricism through Eqs. 7.5.17 to 7.5.19, is shown to be indifferent advice. There is a 4 dBA noise penalty for selecting the incorrect length of intake silencer tube. However, a tube length of 345 mm is a better starting point for optimization than the value of 300 mm that I started with in Fig. 5.62! There is no perceived change of engine torque over this range of intake silencer tube lengths. This is because, unlike with the single-cylinder motorcycle engine, the plenum pressure that controls primary intake ramming is not that of the intake silencer box.

To ensure that these recommendations are applicable over the engine speed range from 1750 to 5250 rpm, and still with the longer primary intake duct indexed, the simulation is repeated with a fixed value of intake silencer box volume of 10 liters, and an intake silencer tube diameter of 65 mm, as in Fig. 5.62, but with intake silencer tube lengths of 300, 500, and 700 mm. The results are shown in Fig. 7.47.

Fig. 7.47 shows that the engine speed of 2750 rpm is a region of high intake noise. My original selection of 300 mm, and one presumes that of 345 mm as well, gives intake noise that is an unacceptable 4 or 5 dBA higher than at other engine speeds. Although the longer tract of 700 mm gives the most noise reduction at 2750 rpm, it becomes equally loud at 5250 rpm.

However, because the maximum intake noise given by the 500 and 700 mm tube lengths is 80 dBA, that is still 5 dBA less than that given by the 300 mm tube length at 2750 rpm. An examination, not plotted, of the noise spectra at 2750 rpm for the three intake silencer tube lengths of 300, 500, and 700 mm shows that the fourth harmonic, at 91.7 Hz, is the loudest. The maximum noise at this frequency, for the three intake silencer tube lengths, is 84.6, 80.0, and 77.1 dBA, respectively. Clearly, in any future development of this intake silencer this frequency should be suppressed as the first major priority. However, even a noise level of 85 dBA at a microphone range of 1.0 m means a relatively "silent" intake system—considering the point made earlier regarding the relative difficulty of the design problem. The double-plenum intake arrangement of the current multi-cylinder automobile engine is an inherently quiet design, and this sometimes causes the designers of such devices to assume that intake silencing is not a difficult design problem. This is an attitude not found among those who have to design and develop hand-held power tools or single-cylinder mopeds and motorcycles.

7.7 Concluding Remarks on Noise Reduction
The Relevance of Silencer Volume for a Hand-Held Power Tool

The space to accommodate an adequate volume of exhaust or intake silencer for a simple four-stroke engine is often a critical design element. Here, the messages to the designer are obvious: (1) to silence such an engine, the frequency at which maximum noise is being created is the one that must be tackled as a first priority, and (2) for the silencing the exhaust or intake of the small industrial engine, every extra cubic centimeter of silencer box volume is well worth the ingenuity spent in its acquisition.

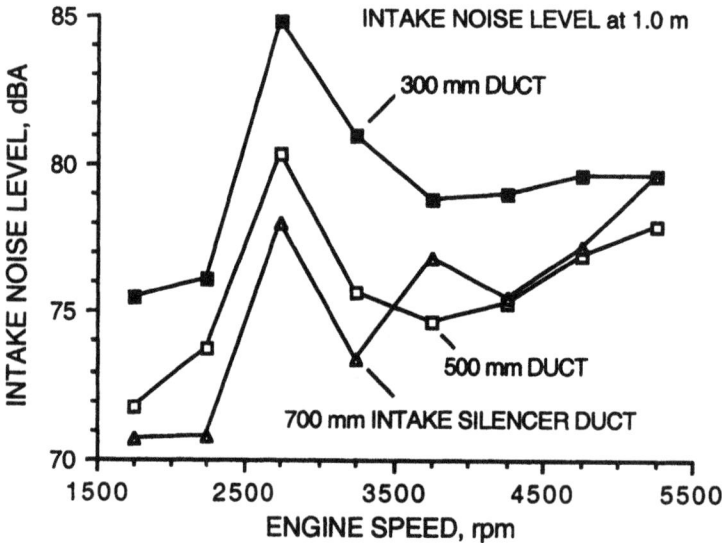

Fig. 7.48 Effect of intake silencer tube length on sports car noise characteristics.

Materials for the Manufacture of Silencers

The discussions concerning the design of silencers do not dwell on the materials to be used in their mechanical construction, so some information on this topic is offered here. Custom and practice show that silencers are typically fabricated from mild steel sheet metal of some 1.1-1.3 mm thickness. This may be acceptable for a racing engine, where the mass of the exhaust system, as a ratio of the total machine mass, must be as low as possible. However, for a production engine, particularly when the exhaust pulses become steep-fronted, a thin metal outer skin of a silencer plenum will act as the diaphragm of a metal loudspeaker, causing considerable noise transmission to the atmosphere [7.26]. Therefore, the designer must inevitably introduce double-skinning of the more sensitive parts of the system, and often a layer of damping material is inserted between that double-skinned surface. Occasionally, it is found to be adequate to line the internal surface of the silencer with a high-temperature plastic damping compound, and this is clearly an economic alternative in many cases.

It is common today to manufacture the intake silencer—indeed, even the entire intake system—from a "plastic" material rather than metal. Because such materials are not as stiff as metal, this increases the potential for the skin to drum and possibly provide more noise transmission to the atmosphere than that emitted by the pressure wave system at its outlets. Because an engine simulation provides the time-varying pressures at every point throughout the entire geometry of the engine, i.e., both intake and exhaust systems and within the cylinder, it is an ideal tool for the mechanical designer of the engine hardware. The information from the simulation can be included in an FEM (finite element model) of the entire structure so as to more accurately predict the "loudspeaker" effects on its surfaces. Look, for example, at Fig. 7.42, and note that the pressure forcing function on the skin of an intake system can be predicted and therefore can be included by the designer within the FEM analysis of the system structure in order to improve the accuracy of its design [7.38, 7.39, 7.42]. Many of the research workers in this area persist with acoustic modeling of such silencers, then ally it to a sophisticated FEM analysis, apparently unaware that 1D modeling exists.

The Primary Objectives

When silencing the four-stroke engine, or indeed any reciprocating engine, the principal concepts that the designer or developer must retain at the forefront of his or her thought processes are:

(i)   Identify the principal source of noise from the engine, be it mechanical, induction, or exhaust noise, and suppress that as a first priority. Should the single loudest source remain unmuffled, the other sources of noise may be completely eliminated, but the total noise from the engine will be virtually unaffected.

(ii) The technique for suppressing the noise from any given source is to determine the frequency spectrum of that noise and to tackle, as a first priority, the frequency band with the highest noise content. Should that remain unmuffled, the other frequency bands may be completely silenced, but the total noise content will virtually remain at the original level.

(iii) In the identification of noise sources, remember that noise level, or loudness, is a function of the square of the distance from any given source [7.3] to the microphone recording the noise level of that source. Thus, placing the microphone close to the inlet tract end, or the exhaust pipe end, or a gearbox, or a cylinder head, in a manner that differentially distances the other sources, permits an initial and approximate identification of the relative contributions of the several sources of noise to the total noise output.

References for Chapter 7

7.1 G.P. Blair and J.A. Spechko, "Sound Pressure Levels Generated by Internal Combustion Engine Exhaust Systems," SAE Automotive Congress, Detroit, Mich., January, 1972, SAE Paper No. 720155.

7.2 G.P. Blair and S.W. Coates, "Noise Produced by Unsteady Exhaust Efflux from an Internal Combustion Engine," SAE Automotive Congress, Detroit, Mich., January, 1973, SAE Paper No. 730160.

7.3 S.W. Coates and G.P. Blair, "Further Studies of Noise Characteristics of Internal Combustion Engines," SAE Farm, Construction and Industrial Machinery Meeting, Milwaukee, Wisc., September, 1974, SAE Paper No. 740713.

7.4 D.D. Davis, G.M. Stokes, D. Moore, and G.L. Stevens, "Theoretical and Experimental Investigation of Mufflers," NACA Report No. 1192, 1954.

7.5 P.O.A.L. Davies and R.J. Alfredson, "The Radiation of Sound from an Engine Exhaust," *J. Sound Vib.*, Vol. 13, p. 389, 1970.

7.6 P.O.A.L. Davies, "The Design of Silencers for Internal Combustion Engines," *J. Sound Vib.*, Vol. 1, p. 185, 1964.

7.7 P.O.A.L. Davies, "The Design of Silencers for Internal Combustion Engine Exhaust Systems," Conference on Vibration and Noise in Motor Vehicles, *Proc.I.Mech.E.*, 1972.

7.8 G.H. Trengrouse and F.K. Bannister, "The Reduction of Noise Generated by Pressures Waves of Finite Amplitude–Attenuation Due to Rows of Holes in Pipes," University of Birmingham, Research Report No. 40, 1964.

7.9 M. Fukuda and H. Izumi, "Kuudougata Shouonki no Tokusei ni Kansuru Kenkyuu," *Trans. Japan Soc. Mech. E,* Vol. 34, No. 263, p. 1294, 1968.

7.10 P.M. Nelson, "Some Aspects of Motorcycle Noise and Annoyance," SAE Paper No. 850982.

7.11 R. Taylor, *Noise*, Penguin, London, 1975.

7.12 C.M. Harris, *Handbook of Noise Control*, McGraw-Hill, New York, 1957.

7.13 L.L. Beranek, *Noise Reduction*, McGraw-Hill, New York, 1960.

7.14 M.B. Johnston, "Exhaust Port Shapes for Sound and Power," SAE Farm, Construction and Industrial Machinery Meeting, Milwaukee, Wisc., September, 1973, SAE Paper No. 730815.

7.15 K. Groth and N. Kania, "Modifications on the Intake Ports with the Aim to Reduce the Noise of a Two-Stroke Crankcase-Scavenged Engine," SAE International Off-Highway Meeting and Exposition, Milwaukee, Wisc., September, 1982, SAE Paper No. 821069.

7.16 SAE Noise Standards, SAE Handbook, Volume 1, pp. 14.01-14.120:

(a) SAE J47, Maximum Sound Level Potential for Motorcycles, October 1993.

(b) SAE J331, Sound Levels for Motorcycles, October 1992.

(c) SAE J1287, Measurement of Exhaust Sound Levels of Stationary Motorcycles, June 1993.

(d) SAE J192, Exterior Sound Level for Snowmobiles, March 1985.

(e) SAE J986, Sound Level for Passenger Cars and Light Trucks, August 1994.

(f) SAE J1030, Maximum Sound Level for Passenger Cars and Light Trucks, February 1987.

(g) SAE J34, Exterior Sound Level Procedure for Pleasure Motorboats, December 1991.

(h) SAE J1174, Operator Ear Sound Level Measurement Procedure for Small Engine Powered Equipment, March 1985.

(i) SAE J1175, Bystander Sound Level Measurement Procedure for Small Engine Powered Equipment, March 1985.

(j) SAE J1074, Engine Sound Level Measurement Procedure, February 1987.

(k) SAE J1207, Measurement Procedure for Determination of Silencer Effectiveness in Reducing Engine Intake or Exhaust Sound, February 1987.

7.17 S.W. Coates, "The Prediction of Exhaust Noise Characteristics of Internal Combustion Engines," Doctoral Thesis, The Queen's University of Belfast, April, 1974.

7.18 E. Kato and R. Ishikawa, "Motorcycle Noise," Dritte Grazer Zweiradtagung, Technische Universitat, Graz, Austria, 13-14 April, 1989.

7.19 N.A. Hall, *Thermodynamics of Fluid Flow*, Longmans, London, 1957.

7.20 G.M. Jenkins and D.G. Watts, *Spectral Analysis and Its Applications*, Holden-Day, San Francisco, 1968.

7.21 G.P. Blair, "Computer-Aided Design of Small Two-Stroke Engines for Both Performance Characteristics and Noise Levels," I.Mech.E Automobile Division Conference on Small Internal Combustion Engines, Isle of Man, May 31-June 2, 1978, Paper No. C120/78.

7.22 SAE J335b, Multiposition Small Engine Exhaust System Fire Ignition Suppression.

7.23 G.P. Blair, W.L. Cahoon, and C.T. Yohpe, "Design of Exhaust Systems for V-Twin Motorcycle Engines to Meet Silencing and Performance Criteria," SAE Motorsports Engineering Conference and Exposition, Dearborn Mich., December 5-8, 1994, SAE Paper No. 942514, p. 10.

7.24 C.V. Beidl, K. Salzberger, A. Strebenz, and M. Buchberger, "Theoretical and Empirical Approaches to Reduce the Noise Emission of Intake Silencers and Exhaust Systems with Catalysts," Funfe Grazer Zweiradtagung, Technische Universitat, Graz, Austria, 22-23 April, 1993.

7.25 R. Kamiya, "Prediction of Intake and Exhaust System Noise," Fourth Grazer Zweiradtagung, Technische Universitat, Graz, Austria, 8-9 April, 1991.

7.26 C. Poli and A. Pucci, "Vibro-Acoustical Analysis of a New Two-Stroke Engine Exhaust Muffler for Increasing its Sound Deadening," ATA/SAE Small Engine Technology Conference, Pisa, 1-3 December, 1993, ATA/SAE 931572, pp. 855-863.

7.27 G.E. Roe, "The Silencing of a High Performance Motorcycle," *J. Sound Vib.*, 33, 1974.

7.28 D.O. Mackey, G.P. Blair, and R.Fleck, "Correlation of Simulated and Measured Noise Emissions and Unsteady Gas Dynamic Flow from Engine Ducting," SAE International Off-Highway Meeting, Indianapolis, Ind., August 1996, SAE Paper No. 961806, pp. 137-161.

7.29 D.O. Mackey, G.P. Blair, and R.Fleck, "Correlation of Simulated and Measured Noise Using a Combined 1D/3D Technique," SAE International Congress, Detroit, Mich., February 26-28, 1997, SAE Paper No. 970801, pp. 105-121.

7.30 D.O. Mackey, "A Further Study of Noise Emitted by Unsteady Gas Flow Through Silencer Elements," PhD Thesis, The Queen's University of Belfast, September 1986.

7.31 G.H. Mawhinney, "An Evaluation of Unsteady Gas Flow through Silencer Elements," PhD Thesis, The Queen's University of Belfast, May 1999.

7.32 G.P. Blair, *Design and Simulation of Two-Stroke Engines*, R-161, SAE, Warrendale, Pa., 1996.

7.33 W.J.D. Annand and G.E. Roe, *Gas Flow in the Internal Combustion Engine*, Foulis, Yeovil, Somerset, 1974.

7.34 G.P. Blair, *The Basic Design of Two-Stroke Engines*, R-104, Society of Automotive Engineers, Warrendale, Pa., 1990.

7.35 A. Onorati, "Numerical Simulation of Exhaust Flows and Tailpipe Noise of a Small Single-Cylinder Diesel Engine," Small Engine Technology Conference, Milwaukee, September 1995, SAE Paper No. 951755, pp. 1-11.

7.36 P.C. Lai and W. Soedel, "Free Gas Pulsation of Helmholtz Resonator Attached to a Thin Muffler Element," SAE International Congress, Detroit, Mich., February 1998, SAE Paper No. 980281.

7.37 A. Broatch, J.P. Brunel, F. Payri, and A.J. Torregrosa, "Pressure Loss Characterization of Perforated Ducts," SAE International Congress, Detroit, Mich., February 1998, SAE Paper No. 980282.

7.38 J.S. Lim and J.D. Kostum, "Global Acoustic Sensitivity Analysis Applied to the Reduction of Shell Noise Radiation of a Simulated Engine Air Induction System Component," SAE International Congress, Detroit, Mich., February 1998, SAE Paper No. 980280.

7.39 G.P. Sievewright, "Using Acoustic Optimisation to Improve Sound Quality of a High Performance Engine Air Intake Manifold", SAE International Congress, Detroit, Mich., February 1998, SAE Paper No. 980273.

7.40 D.J. Scholl, S.G. Russ, C. Davis, and T. Barash, "Volume Acoustic Modes of Spark-Ignited Internal Combustion Chambers," SAE International Congress, Detroit, Mich., February 1998, SAE Paper No. 980893.

7.41 J.J. Silvestri, T. Morel, and M. Costello, "Study of Intake System Wave Dynamics and Acoustics by Simulation and Experiment," SAE International Congress, Detroit, Mich., February 1994, SAE Paper No. 940206.

7.42 G.P. Sievewright, "Noise Optimisation of Air Intake Manifolds," *Automotive Engineering*, October 1998, pp. 85-87.

p. 25   Eq. 1.4.5 should read:

$$F_{tdc} + G_{tdc} = \sqrt{\left(L_{cr} + L_{ct}\right)^2 - D^2}$$

Eq 1.4.6 should read:

$$\theta_{tdc} = \tan^{-1}\left(\frac{D}{F_{tdc} + G_{tdc}}\right)$$

Eq. 1.4.8 should read:

$$F_{tdc} = -L_{ct}\cos\theta_{tdc} + \sqrt{\left(L_{cr} + L_{ct}\right)^2 - D^2}$$

Eq. 1.4.9 should read:

$$F_{bdc} = \sqrt{\left(L_{cr} - L_{ct}\right)^2 - D^2}$$

p. 26   Eq. 1.4.18 should read:

$$H_t = \sqrt{\left(L_{cr} + L_{ct}\right)^2 - D^2} - \sqrt{L_{cr}^2 - \left(L_{ct}\sin\theta - D\right)^2} - L_{ct}\cos\theta$$

p. 412   Paragraph 4, last line, "enging" should read "engine."

pp. 568 and 569   In the captions of Figs. 5.36(b) through 5.40, "Ducati 995" should read "Ducati 955."

pp. 294 and 296   Figures 2.58(a) and 2.58(b), referred to on these pages, are missing from the text. These two figures are reproduced below:

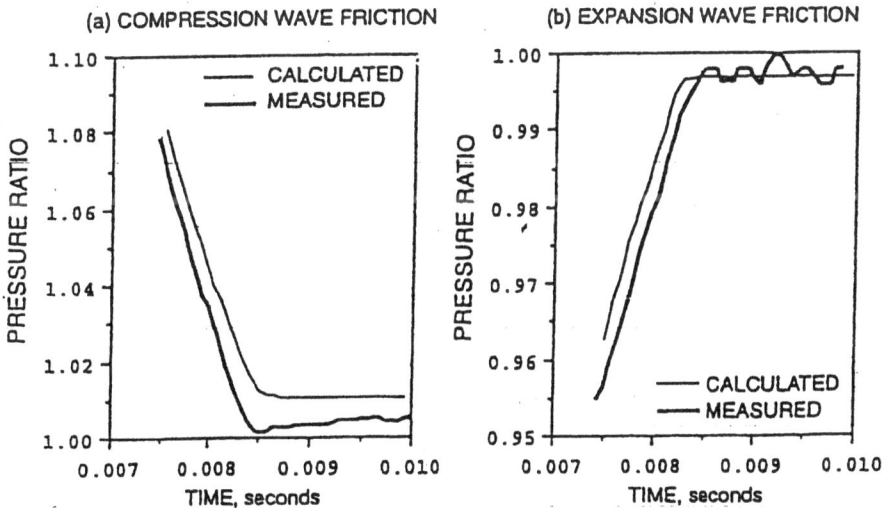

Fig.2.58 Friction causing wave reflections from compression and expansion waves.

Postscript

I can think of no better postscript for this book than to continue the unimaginative pattern of applying the doggerel verse that originally appeared on the flyleaf of the previous book. The fact is that the sentiments expressed then are even more apt in this book than they were in that dedicated to the design of two-stroke engines, particularly as the four-stroke engine is highlighted in a winning context, Jack Williams and Joe Craig and their 7R AJS and Manx Norton racers are rightly lauded in the text, and the last line became prophetic.

THE SECOND MULLED TOAST

When as a student a long time ago
my books gave no theory glimmers,
why two-strokes ended in second place slow,
and four-strokes were always the winners.

Williams and Craig were heroes enough
whose singles thumped to Tornagrough,
such as black 7R or silver Manx,
on open megas they enthused the cranks.

Wallace and Bannister gave me the start
into an unsteady gas dynamic art,
where lambdas and betas meshed in toil
for thirty years consumed midnight oil.

With the parrot on Bush a mental penny
into slot in brain fell quite uncanny.
Lubrication of grey cells finally gave
an alternative way to follow a wave.

That student curiosity is sated today
and many would describe that as winning.
Is this then the end of the way?
No, learning is aye a beginning.

Gordon Blair, July 1994

...and finally, to answer those many queries...

Tornagrough (pronounced as "tawernagruff") is a bend on the Dundrod circuit where the Ulster Grand Prix is held and at which spot I watched Juan Manuel Fangio flip a BRM backward and was a flag marshal forty years ago when John Surtees won on the MV Agusta...and Bush is the local colloquialism for the amber nectar produced at the world's oldest licensed whiskey distillery at Bushmills, Co. Antrim, a modicum of which has been known to ease the pain caused by reading a book containing excessive quantities of thermodynamics and unsteady gas dynamics...

Appendix

Computer Software and Engine Simulation Model

1. Computer Software for Education and Design

The following software is available from SAE as a single executable file for installation on either IBM® PC (or compatible) or Macintosh® computers. This software has been written by the author of this book and contains a series of programs based on the theory within the various chapters of the book. These can be easily accessed in a user-friendly format.

Each program contains an information page detailing precisely the theory being employed and its location within the book. Included are tips and hints on how to use the program most effectively.

Each program also contains a set of input data straight from the relevant section of the book. By clicking the 'calculate' button, the program produces the same numbers and graphics that appear on the printed page. The ensuing variation of, and computation of, input data greatly enhances the discussion presented in the book.

Many of the programs allow input and output data to be inserted into other graphics packages, retrieved for further design activity, or transmitted to others for, say, manufacturing purposes.

The following is a complete listing of the programs in this software package:

1. Engine and Crank Geometry (based on theory in Chapter 1)
2. Basic Engine Performance (based on theory in Chapter 1)
3. Four-Valve Head Layout (based on theory in Chapter 1)
4. Two-Valve Head Layout (based on theory in Chapter 1)
5. Exhaust Valve Design (based on theory in Chapters 1 and 6)
6. Intake Valve Design (based on theory in Chapters 1 and 6)
7. Air Standard Otto Cycle (based on theory in Chapter 1)
8. Air Standard Diesel Cycle (based on theory in Chapter 1)
9. Otto Cycle (Phased Burn) (based on theory in Chapters 1 and 4)
10. Diesel Cycle (Phased Burn) (based on theory in Chapters 1 and 4)
11. Simple Wave Flow (based on theory in Chapter 2)
12. Wave Superposition (based on theory in Chapter 2)
13. Friction and Heat Transfer (based on theory in Chapter 2)

14.	Temperature Discontinuity (based on theory in Chapter 2)
15.	Restricted Pipe (based on theory in Chapter 2)
16.	Three-Way Branched Duct (based on theory in Chapter 2)
17.	Cylinder-Pipe Boundary (based on theory in Chapter 2)
18.	Valve Discharge Coefficient (based on theory in Chapter 3)
19.	Intake Ramming (based on theory in Chapter 6)
20.	Exhaust Tuning (based on theory in Chapter 6)
21.	Diffusing Silencer (based on theory in Chapter 7)
22.	Side-Resonant Silencer (based on theory in Chapter 7)

For more information on this or other SAE products, contact SAE Customer Sales and Support at 724-776-4970; fax 724-776-0790; e-mail: publications@sae.org; web: www.sae.org/BOOKSTORE.

2. Engine Simulation Model

An engine simulation model, similar to the GPB simulation model described in this book, is available for installation on IBM® PC (or compatible) computers. This model is based on the theory within the various chapters of the book. The application of the model is discussed in Chapters 5–7.

The GPB simulation model described within this book was written by the author exclusively for the Macintosh® Power PC computer, whereas that written for the IBM® PC platform has been produced through the joint efforts of the author and the personnel of OPTIMUM Power Technology of Bridgeville, Pa.

The engine simulation model for the IBM® PC platform will simulate a single-cylinder naturally-aspirated four-stroke engine with the following attributes:

- Almost any two-valve or four-valve cylinder head
- Almost any intake ducting geometry and simple silencer
- Almost any exhaust ducting geometry and simple silencers

In addition, the model allows the following:

- Selection from a menu of Diesel and Otto cycle combustion processes
- Varying of Otto and Diesel fueling over a sufficiently wide range
- Selection from a limited menu of discharge coefficient maps for valves, etc.
- Selection from a limited menu of diesel and gasoline fuels

The input data for the simulation have been prepared in a user-friendly format, and are supplied in a separate program with the simulation. This program contains information and help pages to assist the user.

The simulation model and the input preparer package permit the user to insert input and output data into other graphics packages and retrieve the input data for further design activity. The output data appears as a prediction of the engine performance characteristics, including power, torque, fuel consumption, airflow, emissions and noise characteristics, etc., and the associated time-varying data of pressures, temperatures, etc., much as presented within

Chapters 5–7 of the book. Users must have Microsoft® Excel available on the computer to observe the graphic output data from the simulation.

For more information on this or other SAE products, contact SAE Customer Sales and Support at 724-776-4970; fax 724-776-0790; e-mail: publications@sae.org; web: www.sae.org/BOOKSTORE.

Index

About the Author

Dr. Gordon P. Blair is a world-renowned expert on the design and development of two-stroke engines. Since 1997, he is Professor Emeritus at The Queen's University of Belfast (QUB), where he has taught a couple of generations of students and has single-mindedly carried out research into the internal combustion engine. This book reveals that a much greater proportion of that research than is generally appreciated was into four-stroke engines.

Professor Blair began his academic career by holding the post of Assistant Professor at New Mexico State University from 1962–1964 and then returned to Belfast. Apart from holding the Chair of Mechanical Engineering at QUB for some twenty years, Professor Blair has served the University as Dean of the Faculty of Engineering and as Pro-Vice Chancellor. In addition, Professor Blair has been a consultant for many well-known companies, such as Ford, General Motors, Mercury Marine, Volvo, and Yamaha, and international sporting bodies such as the Union Internationale Motonautique and the Federation Internationale Motocycliste. He has been named a Fellow of the Society of Automotive Engineers, the Institution of Mechanical Engineers, and the Royal Academy of Engineering. Professor Blair has been recognized by Her Majesty The Queen by the award of a CBE (Commander of the Order of the British Empire).

www.ingramcontent.com/pod-product-compliance
Lightning Source LLC
Chambersburg PA
CBHW062009190326
41458CB00009B/3023